Precambrian Crustal Evolution in the North Atlantic Region

Geological Society Special Publications
Series Editor A. J. FLEET

GEOLOGICAL SOCIETY SPECIAL PUBLICATION NO. 112

Precambrian Crustal Evolution in the North Atlantic Region

EDITED BY

T. S. BREWER

Department of Geology
University of Leicester
Leicester, UK

1996

Published by
The Geological Society
London

THE GEOLOGICAL SOCIETY

The Society was founded in 1807 as the Geological Society of London and is the oldest geological society in the world. It received its Royal Charter in 1825 for the purpose of 'investigating the mineral structure of the Earth'. The Society is Britain's national society for geology with a membership of 8000. It has countrywide coverage and approximately 1000 members reside overseas. The Society is responsible for all aspects of the geological sciences including professional matters. The Society has its own publishing house, which produces the Society's international journals, books and maps, and which acts as the European distributor for publications of the American Association of Petroleum Geologists, SEPM and the Geological Society of America.

Fellowship is open to those holding a recognized honours degree in geology or cognate subject and who have at least two years' relevant postgraduate experience, or who have not less than six years' experience in geology or a cognate subject. A Fellow who has not less than five years' relevant postgraduate experience in the practice of geology may apply for validation and, subject to approval, may be able to use the designatory letters C Geol (Chartered Geologist).

Further information about the Society is available from the Membership Manager, The Geological Society, Burlington House, Piccadilly, London W1V 0JU, UK. The Society is a Registered Charity, No. 210161.

Published by the Geological Society from:
The Geological Society Publishing House
Unit 7
Brassmill Enterprise Centre
Brassmill Lane
Bath BA1 3JN
UK
(*Orders*: Tel 01225 445046
Fax 01225 442836)

First published 1996

The publisher makes no representation, express or implied, with regard to the accuracy of the information contained in this book and cannot accept any legal responsibility for any errors or omissions that may be made.

© The Geological Society 1996. All rights reserved. No reproduction, copy or transmission of this publication may be made without written permission. No paragraph of this publication may be reproduced, copied or transmitted save with the provisions of the Copyright Licensing Agency, 90 Tottenham Court Road, London W1P 9HE. Users registered with the Copyright Clearance Center, 27 Congress Street, Salem, MA 01970, USA: the item-fee code for this publication is 0305-8719/96/$07.00.

British Library Cataloguing in Publication Data

A catalogue record for this book is available from the British Library

ISBN 1-897799-62-4

Typeset by EJS Chemical Composition,
Midsomer Norton, Bath BA3 4BQ, UK

Printed by The Alden Press, Osney Mead, Oxford, UK

Distributors
USA
AAPG Bookstore
PO Box 979
Tulsa
OK 74101-0979
USA
(*Orders*: Tel (918) 584-2555
Fax (918) 548-0469)

Australia
Australian Mineral Foundation
63 Conyngham Street
Glenside
South Australia 5065
Australia
(*Orders*: Tel (08) 379-0444
Fax (08) 379-4634)

India
Affiliated East-West Press PVT Ltd
G-1/16 Ansari Road
New Delhi 110 002
India
(*Orders*: Tel (11) 327-9113
Fax (11) 326-0538)

Japan
Kanda Book Trading Co.
Tanikawa Building
3–2 Kanda Surugadai
Chiyoda-Ku
Tokyo 101
Japan
(*Orders*: Tel (03) 3255-3497
Fax (03) 3255-3495)

Contents

PREFACE	vii
SNYDER, D. B., LUCAS, S. B. & MCBRIDE, J. H. Crustal and mantle reflectors from Palaeoproterozoic orogens and their relation to arc–continent collisions	1
ROLLINSON, H. R. Tonalite–trondhjemite–granodiorite magmatism and the genesis of Lewisian crust during the Archaean	25
VREVSKY, A., KRIMSKY, R. & SVETOV, S. Rare earth and isotopic (Nd, O) heterogeneity of the Archaean mantle, Baltic Shield	43
BIBIKOVA, E. V., SKIÖLD, T. & BOGDANOVA, S. V. Age and geodynamic aspects of the oldest rocks in the Precambrian Belomorian Belt of the Baltic (Fennoscandian) Shield	55
BOGDANOVA, S. V. High-grade metamorphism of 2.45–2.4 Ga age in mafic intrusions of the Belomorian Belt in the northeastern Baltic Shield	69
VAN KRANENDONK, M. J. & WARDLE, R. J. Burwell domain of the Palaeoproterozoic Torngat Orogen, northeastern Canada: tilted cross-section of a magmatic arc caught between a rock and a hard place	91
RIVERS, T., MENGEL, F., SCOTT, D. J., CAMPBELL, L. M. & GOULET, N. Torngat Orogen – a Palaeoproterozoic example of a narrow doubly vergent collisional orogen	117
WARDLE, R. J. & VAN KRANENDONK, M. J. The Palaeoproterozoic Southeastern Churchill Province of Labrador–Quebec, Canada: orogenic development as a consequence of oblique collision and indentation	137
KERR, A., RYAN, B., GOWER, C. F. & WARDLE, R. J. The Makkovik Province: extension of the Ketilidian Mobile Belt in mainland North America	155
CHADWICK, B. & GARDE, A. A. Palaeoproterozoic oblique plate convergence in South Greenland: a reappraisal of the Ketilidian Orogen	179
GOWER, C. F. The evolution of the Grenville Province in eastern Labrador, Canada	197
STARMER, I. C. Accretion, rifting, rotation and collision in the North Atlantic supercontinent, 1700–950 Ma	219
WIKSTRÖM, A., SKIÖLD, T. & ÖHLANDER, B. The relationship between 1.88 Ga old magmatism and the Baltic–Bothnian shear zone in northern Sweden	249
CONNELLY, J. N. & ÅHÄLL, K.-I. The Mesoproterozoic cratonization of Baltica – new age constraints from SW Sweden	261
MENUGE, J. F. & BREWER, T. S. Mesoproterozoic anorogenic magmatism in southern Norway	275
CONNELLY, J. N., BERGLUND, J. & LARSON, S. Å. Thermotectonic evolution of the Eastern Segment of southwestern Sweden: tectonic constraints from U–Pb geochronology	297
PAGE, L. M., STEPHENS, M. B. & WAHLGREN, C.-H. $^{40}Ar/^{39}Ar$ geochronological constraints on the tectonothermal evolution of the Eastern Segment of the Sveconorwegian Orogen, south-central Sweden	315
MERTANEN, S., PESONEN, L. J. & HUHMA, H. Palaeomagnetism and Sm–Nd ages of the Neoproterozoic diabase dykes in Laanila and Kautokeino, northern Fennoscandia	331
EMMETT, T. F. The provenance of pre-Scandian continental flakes within the Caledonide Orogen of south-central Norway	359
FITCHES, W. R., PEARCE, N. J. G., EVANS, J. A. & MUIR, R. J. Provenance of late Proterozoic Dalradian tillite clasts, Inner Hebrides, Scotland	367
Index	379

Preface

Recent interest in Precambrian crustal evolution in the North Atlantic Region has, in part, been driven by IGCP project 275 'The Deep Geology of the Baltic Shield' and project 371 'Structure and Correlation of the Precambrian in North East Europe and the North Atlantic Realm (COPENA)'. Project 275 acted as a catalyst for the integration of geological, geophysical, geochemical and geochronological data across national boundaries, particularly with the former Soviet Union. During the life of project 275 the BABEL geophysical project demonstrated the advantage of large multinational projects and focused international attention on aspects of Precambrian crustal evolution. At the same time LITHOPROBE in Canada and North America was demonstrating the advantage of large scale multidisciplinary projects to the study of Precambrian crustal evolution on the opposite side of the Atlantic.

In September 1994 IGCP project 371 'Structure and Correlation of the Precambrian in North East Europe and the North Atlantic Realm (COPENA)' held its inaugural meeting in Nottingham. This project aims to bring together Earth Scientists from each side of the Atlantic to focus attention on Precambrian crustal evolution. From this approach an up-scaling from 'local' tectonic modelling to large-scale investigation of major Precambrian orogens will be possible. This should ultimately allow for the identification of how the Precambrian crust in the North Atlantic realm was created, amalgamated and reworked during the early history of the Earth.

The results in this volume provide a synthesis of papers presented during the Nottingham meeting, which spans the geographical realm of COPENA. Financial support for the Nottingham meeting was provided by the Geological Society of London, the Royal Society and the IGCP. The success of the meeting was in part due to support from Janet Baker and Ian Starmer, while numerous reviewers provided support during the development of this volume.

Crustal and mantle reflectors from Palaeoproterozoic orogens and their relation to arc–continent collisions

D. B. SNYDER[1], S. B. LUCAS[2] & J. H. McBRIDE[1]

[1] *BIRPS, Bullard Laboratories, Madingley Road, Cambridge CB3 0EZ, UK*
[2] *Geological Survey Canada, 601 Booth Street, Ottawa, Ontario K1A 0E8, Canada*

Abstract: Two prominent geological features characterize the Palaeoproterozoic orogenic belts of Laurentia and Baltica: (1) Archaean cratons form the stable footwall during early stages of convergence and throughout crustal collision; and (2) juvenile, predominantly arc-derived crust was accreted to Archaean cratons through arc–continent collision. Seismic reflection profiling over the Svecofennian (Baltic Shield), Lewisian (British & Irish Isles) and Trans-Hudson (Canadian Shield) segments of an arguably once continuous orogenic belt has provided geometries of reflectors throughout both Palaeoproterozoic and Archaean crustal blocks as well as laterally coherent mantle reflectors. Two BABEL deep seismic reflection profiles within the Baltic Shield revealed structures along an irregularly shaped boundary between the juvenile 2.0–1.8 Ga Svecofennian domain and an Archaean craton (Karelia Province). An important result of the survey is the approximately 100 km horizontal offset between the inferred mantle suture, or palaeo-subduction boundary, and the geochemically-mapped crustal suture between juvenile Proterozoic crust and the Archaean craton. On the other side of the Atlantic, LITHOPROBE reflection profiles across the Trans-Hudson Orogen in central Canada reveal that 1.92–1.83 Ga juvenile arc and oceanic terranes of the Reindeer Zone form an allochthonous carapace about an Archaean basement block. Unexpectedly, the juvenile allochthons dip beneath the bounding Superior and Hearne cratons, defining a crustal-scale culmination in the core of the Orogen. The British and Irish Isles form an important bridge between the Palaeoproterozoic orogens of Laurentia and those of Baltica. Twenty deep reflection profiles on the continental shelf north and west of Scotland have traced a mantle reflector for over 800 km along its reconstructed strike, which is at a high angle to younger structures. This feature dips away from the reconstructed cratonic nucleus of Laurentia and may trace the Palaeoproterozoic (1.7–1.9 Ga) suture between the Archaean Lewisian block and accreted juvenile crust represented by the *c.* 1.8 Ga Rhinns complex. A common theme that emerges from the integration of all these geological and seismic results is the role played by the Archaean lithosphere during arc–continent collision. In all cases studied, juvenile 1.9–1.8 Ga lithosphere was delaminated and its crustal flakes overrode the Archaean margins. This consistency reflects the relative strength and durability of Archaean crust/lithosphere, and suggests that large parts of the lithosphere underlying detached and flaked Palaeoproterozoic juvenile terranes, such as the Svecofennian or Reindeer Zone terranes, may be Archaean in age but modified during Proterozoic tectonism.

Over the past decade, deep seismic reflection profiles of Precambrian orogenic belts (e.g. Green *et al.* 1988; BABEL Working Group 1993; Nelson *et al.* 1993; Lewry *et al.* 1994) have revealed crust and mantle reflectors that are geometrically similar to structures at modern plate convergence zones (e.g. Choukroune & ECORS Team 1989; Pfiffner *et al.* 1990). Whereas the pattern of crustal reflectivity generally complements collisional architecture inferred from geologic studies (e.g. Gaál & Gorbatschev 1987; Lucas 1989; Lewry *et al.* 1990), the mantle reflectors provide particularly important clues to the processes of Precambrian plate convergence and collision.

For over a decade unequivocal reflections from the uppermost mantle have been observed on deep seismic reflection profiles, most commonly on marine profiles (McGeary & Warner 1985; Flack *et al.* 1990; Lie & Husebye 1994; Calvert *et al.* 1995). Many of these observations consist of single features recorded on a lone profile (e.g. Flack *et al.* 1990; Best 1991). However, a majority of the mantle features were substantiated by their consistent appearance on local grids or neighbouring parallel profiles (Flack & Warner 1990; BABEL Working Group 1993; Baird *et al.* 1995). These deep mantle reflectors occur predominantly beneath outcrops of Proterozoic or Archaean continental crust (e.g. Calvert *et al.* 1995), excluding those reflectors clearly associated with Wadati–Benioff zones and active subduction (e.g. Davey & Stern 1990). Several of the best studied mantle reflectors lie along the *c.* 1.7–2.0 Ga boundaries between Archaean cratons and

Palaeoproterozic terrains that occur around the present North Atlantic (Fig. 1). Palaeoproterozoic ages for both crust and mantle events in these areas are inferred from the continuity of reflectors with dated crustal structures (Lucas *et al.* 1994), the lack of younger major thermo-tectonic events (BABEL Working Group 1993; Lewry *et al.* 1994), or superposition relationships in which reflectors associated with dated crustal deformations cut other, presumably older, reflectors (Snyder & Flack 1990).

Largely independent geological, geochronological and geochemical studies in the Baltic Shield (Gaál & Gorbatschev 1987; Öhlander *et al.* 1993), the British Isles (Muir *et al.* 1992), Greenland (Kalsbeek *et al.* 1987, Bridgewater *et al.* 1990), and the Canadian Shield (Hoffman 1988; Bickford *et al.* 1990; Lewry *et al.* 1990; St-Onge *et al.* 1992; Van Kranendonk *et al.* 1993; Lucas *et al.* 1996; Stern *et al.* 1995*a*, *b*) have established that the interval 1.9–1.8 Ga was characterized by the generation of juvenile crust and the accretion of juvenile terranes onto Archaean cratonic margins throughout Laurentia and Baltica (Fig. 2). These Palaeoproterozoic belts formed an orogenic system on the scale of the Tethyan Alpine–Himalayan Orogen, and preserve remnants of the destroyed oceanic realm in the form of ophiolites (Kontinen 1987; Scott *et al.* 1991; St-Onge *et al.* 1992; Stern *et al.* 1995*b*), oceanic island arcs (Syme 1990; Stern *et al.* 1995*a*) and accretionary collages (Park 1991; Lucas *et al.* 1996). As with the modern Alpine–Himalayan Orogen, the individual components of the Precambrian Orogen are better understood than their linkage and interrelationships along-strike, although recent syntheses made significant advances (e.g. Patchett & Arndt 1986; Hoffman 1988; Park 1991; Muir *et al.* 1992; Van Kranendonk *et al.* 1993).

Seismic reflector geometries along segments of this proposed Proterozoic Orogen provide important constraints on former convergence and collision processes. To date, no attempt has been made to compare the structure of Palaeoproterozoic orogenic segments as interpreted from deep seismic reflection data, or to examine similarities and differences in accretion and collision events between segments. In this paper, the principal results of seismic reflection profiling of the Svecofennian Orogen, the Laxfordian deformation region within the Lewisian terrane, and the Trans-Hudson Orogen in its Ungava, Hudson Bay and Western segments (Fig. 1) are reviewed. Description of the currently available deep seismic data that is relevant to understanding this Palaeoproterozoic orogen follows a similar synthesis of data along the trans-Atlantic Caledonian/Appalachian Orogen (Hall & Quinlan 1994). A unifying theme from interpretation of the seismic data is that the crustal portions of Palaeoproterzoic juvenile terranes appear to have been largely detached from their lithospheric mantle during the process of accretion to, or juxtaposition between, relatively rigid Archaean cratons. This result underlines the importance of delamination and flake tectonics in crustal growth and recycling, via the mechanisms of accretion of juvenile crust and the subduction of mantle and possibly lower crust.

Terrain links and correlations

Recent compilations of age and structural data have led to a reconstruction of Proterozoic Laurentia, reuniting Archaean blocks and Proterozoic belts observed in Labrador and Greenland (Van Kranendonk *et al.* 1993). This reconstruction requires only small translations to produce reasonable terrane fits. Attempts to make reconstructions that include the Precambrian blocks of the British Isles and the Baltic are substantially more difficult and involve much larger translations that developed over longer periods of time. At present, no general agreement exists as to the fit between Laurentia and Baltica following the 1.8–1.7 Ga accretion/collision events (compare Wardle *et al.* 1986; Gower 1990; Kalsbeek *et al.* 1993; Park 1994).

Recent advances in quantifying the palaeolocation of Baltica up to 600 Ma ago using palaeomagnetic data have already produced some surprising implications concerning Vendian continental assemblages (Torsvik *et al.* 1992; Soper 1994). The implied drift history for Baltica indicates that an anticlockwise rotation of nearly 180° occurred between 525–425 Ma ago, coeval with a more general northward drift throughout the Palaeozoic (Torsvik *et al.* 1992; Personen *et al.* 1991). This orientation enables the juxtaposition of a combined northern Scotland–eastern Greenland block and the Tornquist Zone region of central Europe across a major rift zone in order to explain extensive rift sequence sediments in these areas (Soper 1994).

Reconstructions showing Laurentia and Baltica at 1.8–1.7 Ga must be extrapolated from the Vendian (600 Ma) configuration using much inferior palaeomagnetic constraints. Most reconstructions have therefore attempted to link regions of Grenvillian and Sveco-norwegian deformation (c. 1000 Ma) and older trends (Patchett & Arndt 1986; Hoffman 1988; Gower 1990). Park (1994) has recently reviewed these reconstructions, introducing structural trend data and drawing upon the palaeomagnetic data of Patchett & Bylund (1977) for the period 1.0–0.9 Ga, Patchett *et al.* (1978) for 1.26–1.19 Ga, and Piper (1976) for 1.90–1.25 Ga. However, lack of well dated palaeopoles, inadequate distribution of precise U–Pb ages and

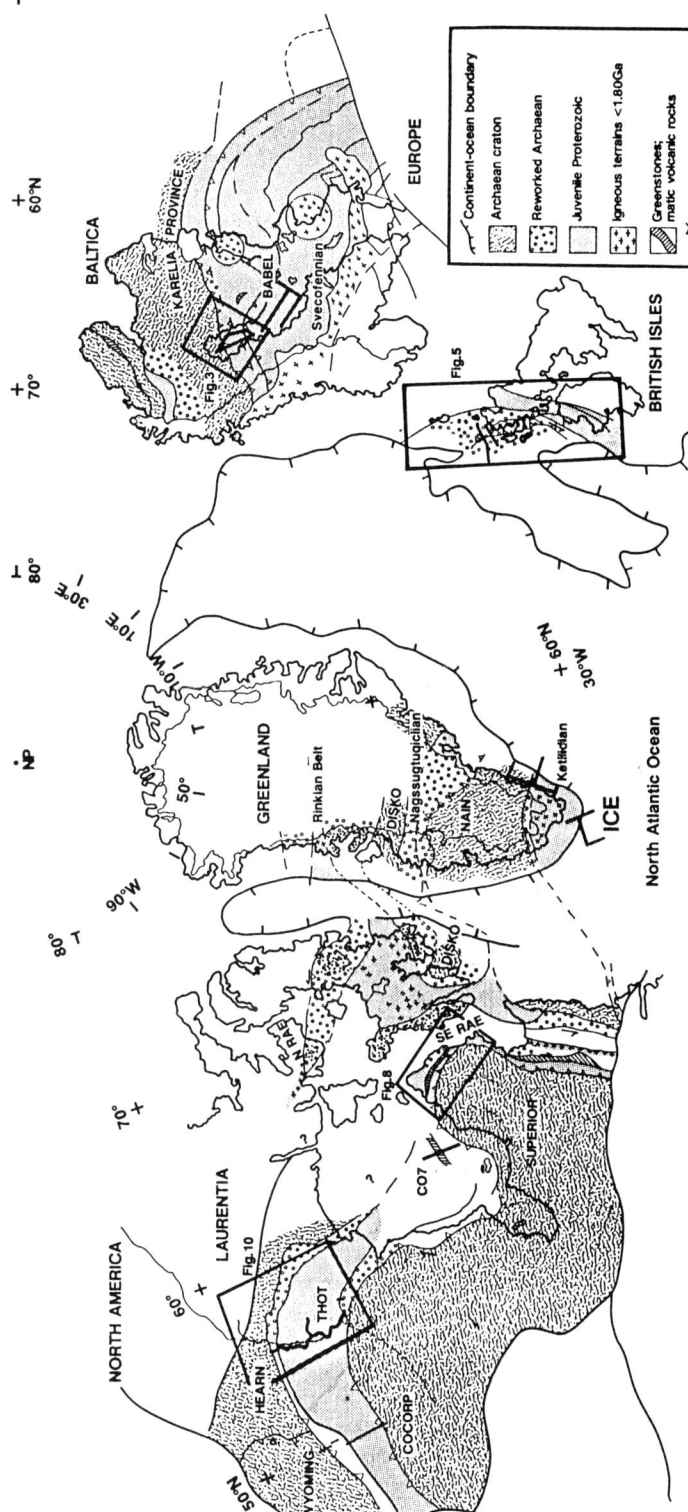

Fig. 1. Present-day distribution of some of the elements of the Laurentian and Baltic palaeocontinents that are discussed in the paper. Solid lines show locations of relevant deep seismic reflection profiles: COCORP & THOT are Trans-Hudson orogen (western segment) surveys, CO7 is the Hudson Bay segment section, ICE are recent lines acquired by the Danish Lithosphere Centre not yet processed or interpreted but included for completeness, BABEL are the Baltic Shield profiles. Shading indicates areas of Archaean craton and reworked Archaean crust that bound juvenile Palaeoproterozoic terrains.

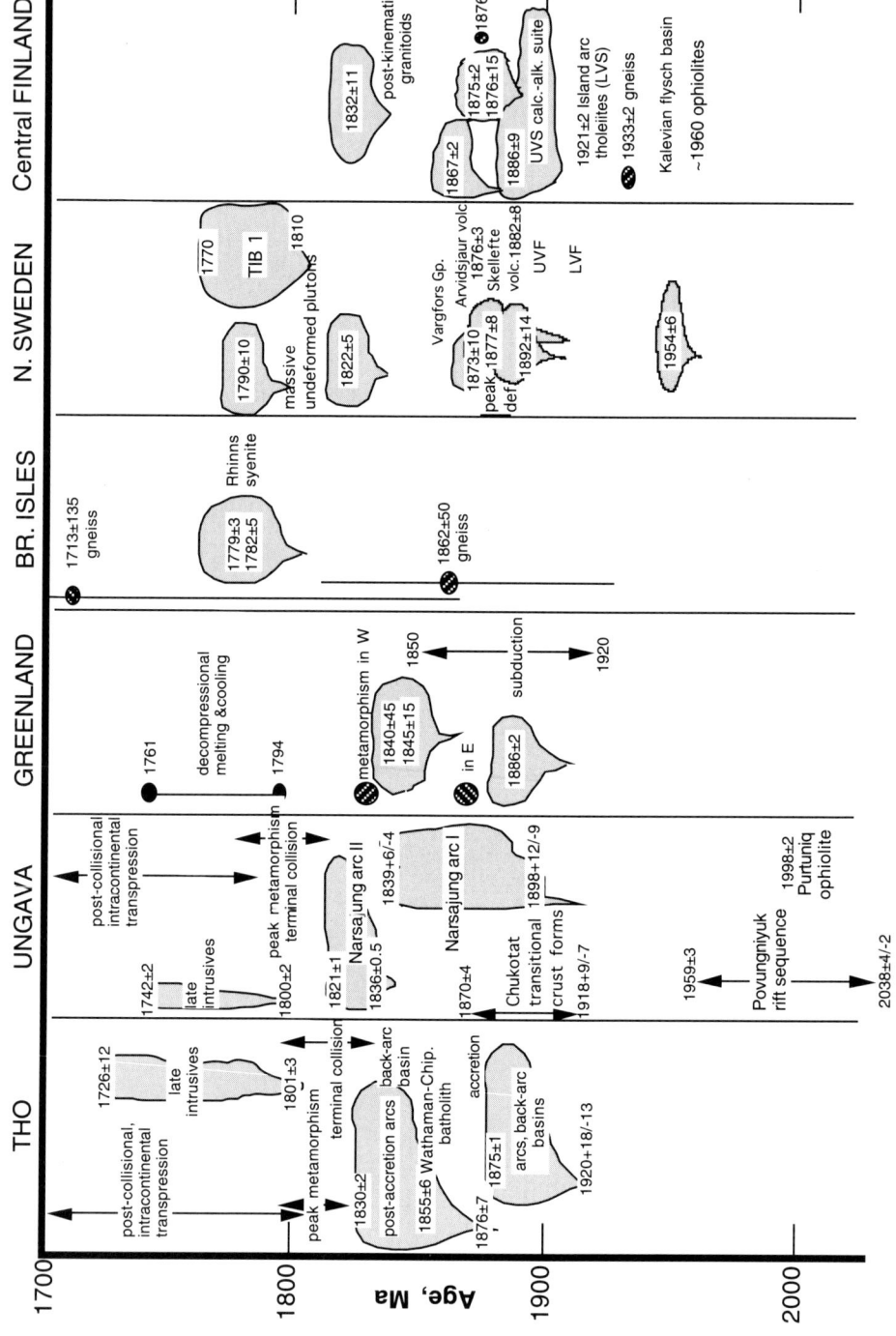

significant gaps in the geology and tectonic history of Palaeoproterozoic belts on either side of the Atlantic render 1.8–1.7 Ga reconstructions uncertain at best.

Phanerozoic orogenic belts typically have distinct segments along their strike, each of which evolve diachronously over tens of millions of years. Although the limited geological and geochronological database reduces our confidence in three-dimensional correlation of Proterozoic terranes, an abundance of clear seismic reflectors observed throughout the crust and in the uppermost mantle of many key Precambrian terrains partly compensates. The clear reflections are in part due to good transmission of seismic waves within the crystalline rocks at or near the surface. These reflectors can aid in defining sometimes complex structural geometries between terranes and will be the primary observations used in the following analysis of individual orogenic segments.

The Svecofennian Orogen of the Baltic Shield

Rocks associated with the Svecofennian Orogen comprise three principal tectonic domains (Figs 1 & 3): (1) the Karelia Province of the Archaean domain, composed of 3.1–2.6 Ga tonalitic–trondhjemitic crust that is covered by the 2.3–2.1 Ga Jatulian platform sequence (Gaál & Gorbatschev 1987); (2) a Palaeoproterozoic Svecofennian–Archaean boundary zone, composed of mobilized and reworked Archaean basement rocks (Tuisku & Laajoki 1990 and references therein), 1.96 Ga ophiolites (cf. Kontinen 1987), and the 2.0–1.9 Ga Kalevian Group of turbiditic, continental margin rocks (Gaál & Gorbatschev 1987 and references therein); and (3) the Svecofennian domain of 1.93–1.87 granitoid plutons, calc-alkaline volcanic suite rocks and thick turbidite sequences. The Svecofennian domain is further subdivided into a northern and southern volcanic belt and the intervening metasediment-

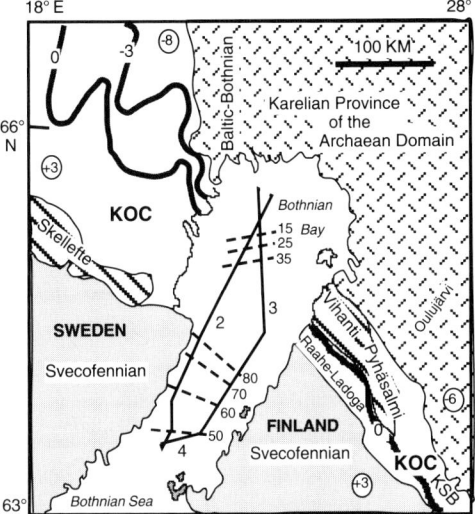

Fig. 3. Location map showing the northern Gulf of Bothnia, also called Bothnian Bay, between Sweden and Finland and BABEL deep seismic reflection profiles 2–4. The dashed contours are depth in kilometres to prominent reflection boundaries (see Fig. 4), from north to south: 15–35 km, the southern edge of bright reflections associated with the Archaean margin; 50–80 km, mantle reflector. Encircled numbers and thick contour line show initial ε_{Nd} values for that area (Huhma 1986; Öhlander et al. 1993); negative values are interpreted to indicate interaction with Archaean crust. The Skellefte and Vihanti-Pyhäsalmi districts are here considered to represent the crustal suture and contain mixed continental, oceanic and arc rocks. KOC, Kalevian–Outokumpu Collage; KSB, Kainu Schist Belt.

dominated Bothnian Basin (Gaál & Gorbatschev 1987; Park 1991, and references therein). This domain may include a younger collisional belt extending from SE Finland to central Sweden that is associated with 1.84–1.81 Ga potassic granites (Ehlers et al. 1993). Although the terms Karelian

---

Fig. 2. Comparative time stratigraphic columns for the early Proterozoic terranes discussed: the Trans-Hudson orogen of western Canada, the Ungava orogen of eastern Canada, Greenland, the northern British Isles, and the Skellefte and Pyhalsi mining districts of the Baltic. Geochronological data, in Ma, as compiled from Bickford et al. (1990), Parrish (1989, pers. comm. 1994), Kalsbeek et al. (1987), Skiöld (1988), Gordon et al. (1990), Machado (1990), Ansdell et al. (1992), Meyers et al. (1992), Muir et al. (1992), St-Onge et al. (1992), Weihed et al. (1992), Kousa et al. (1996), Machado et al. (1993), Van Kranendonk et al. (1993), Wasström (1993), Ansdell & Norman (1995), Ansdell et al. (1996), David et al. (1996), Fedorowich et al. (1995), Lucas et al. (1995) and Stern et al. (1995b). All dates for magmatic rocks are U–Pb determinations on zircons or baddeleyite. Dating methods on metamorphism vary but range from Rb–Sr whole rock ages for Scandinavian studies, to U–Pb geochronology with metamorphic zircon, monazite and titanite and Ar–Ar analysis of biotite, hornblende and K-feldspar for Trans-Hudson orogen studies. This table is not intended to be exhaustive, only representative of the hundreds of age determinations available and their relative precisions.

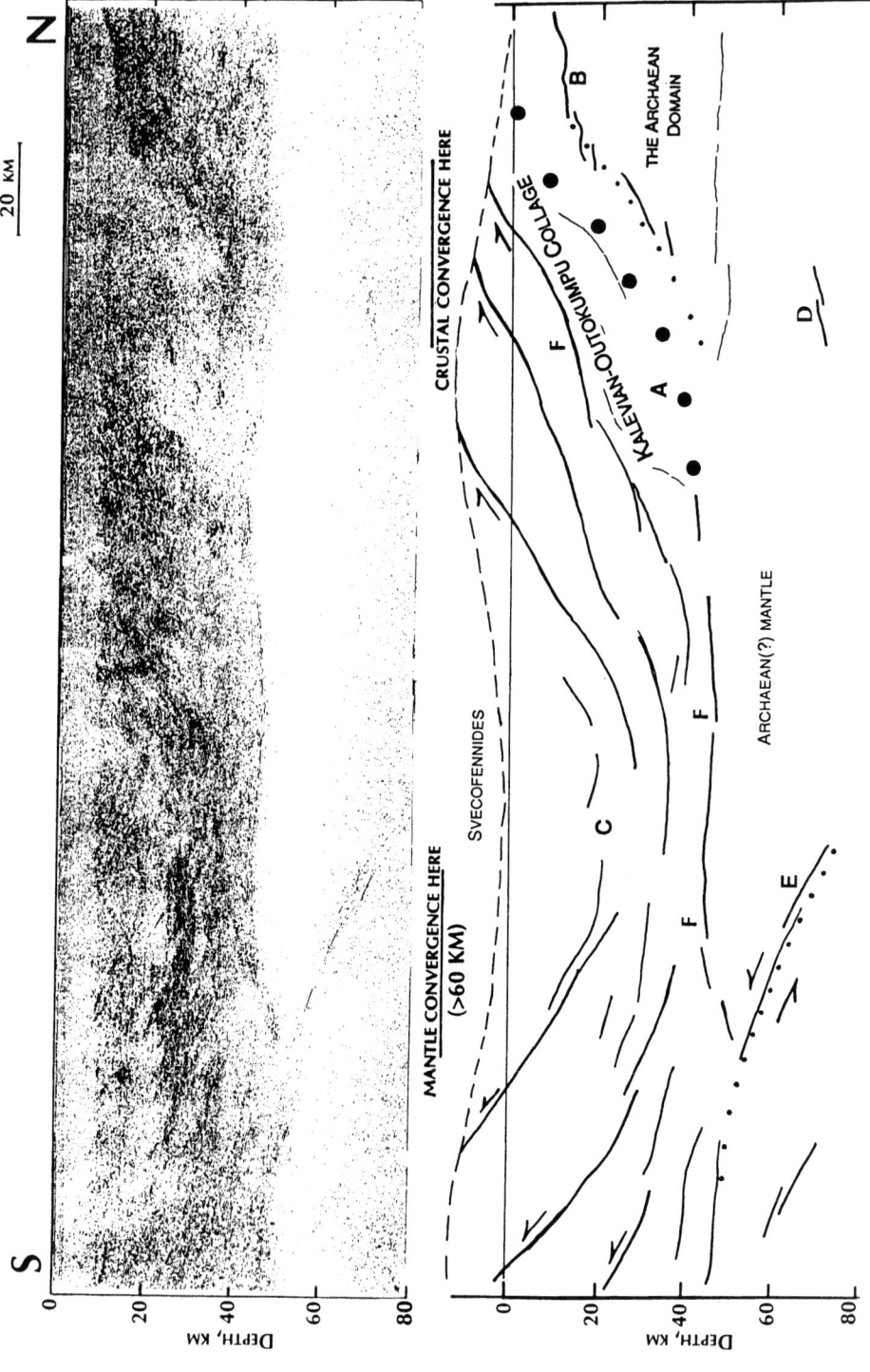

Province and Svecofennian now have widely accepted usage, the Archaean–Proterozoic boundary zone has been associated with diverse nomenclature over the years. In this paper it is called the 'Kalevian–Outokumpu Collage' (KOC in Fig. 3) because it contains reworked Archaean crust and Proterozoic turbiditic and ophiolitic rocks that were interleaved prior to collision with the Svecofennian terrane.

Within the Baltic Shield, two sub-parallel deep seismic reflection lines from the BABEL survey in the Bothnian Bay (Fig. 3) show similar reflection geometries (BABEL Working Group 1993). These lines provided the first observations that constrain three-dimensional geometry of a crustal suture zone between the combined Archaean Karelia Province and juvenile Palaeoproterozoic crust of the Svecofennides (Fig. 4). This suture was first inferred from geological surface mapping (e.g. Gaál & Gorbatschev 1987 and references therein) and later correlated with a prominent, discontinuous conductive zone (Korja & Hjelt 1993). Here we will consider this crustal suture to lie within the Kalevian–Outokumpu Collage domain and to be most easily recognized as coinciding with the mineralized zones of Skellefte, Vihanti-Pyhäsalmi (Figs 3 & 4) and the 'Outokumpu association' (Gaál & Gorbatschev 1987).

Geochemical and isotopic studies of Svecofennian plutons in the Baltic Shield indicate two suites of island arc rocks: an early Svecofennian (1.93–1.91 Ga) arc of primitive island arc tholeiites built above a subduction zone presumed to dip away from the Archaean craton (southward), and a late Svecofennian (1.89–1.875 Ga) calc-alkaline arc suite overlapping the final suture between Svecofennian and Archaean terranes (Weihed et al. 1992; Kousa et al. 1996). The assembly of crustal elements within the Kalevian–Outokumpu Collage may have occurred above a southward-dipping subduction zone, as suggested by the absence of an Andean-type arc of the appropriate age (c. 1.93–1.91 Ga) on the Karelian margin. However, the observation that the prominent mantle reflector (E, Fig. 4) dips to the north implies a subduction polarity reversal. Evidence for an Andean-type arc associated with late north-dipping subduction is provided by the 1.89–1.875 Ga calc-alkaline granitoid plutons, the calc-alkaline Arvidsjaur volcanic rocks and the Kiruna porphyries. These units, and more certainly the overlying Vargfors Group of northern Sweden (Skiöld 1988), have been interpreted to 'stich' the interleaved Archaean and Proterozoic (2.0–1.9 Ga) units within the Kalevian–Outokumpu Collage in northern Finland and Sweden (Park 1991). Thus, a reversal in subduction polarity (from south to north dipping) at c. 1.9 Ga is suggested by these studies.

Archaean–Proterozoic crustal boundary

The BABEL survey covered the full width of the Svecofennides but only the two northernmost lines explored Archaean crust of the Baltic Shield (BABEL Working Group 1993). BABEL line 3 shows a south-dipping zone within the crust (A in Fig. 4), about 50 km wide, that contains few reflections and an indistinct base to reflectivity. This contrasts strongly with the pronounced reflectivity to c. 45 km depths observed to both the north and south of this zone. The less reflective zone correlates along the strike of linear positive gravity anomalies with the mineralized Vihanti-Pyhäsalmi district in Finland (Fig. 3) (BABEL Working Group 1993), which is similar to the Skellefte district in Sweden. Both of these areas are marked by c. 1.9 Ga volcanic arc and oceanic assemblages (Weihed et al. 1992). The highly reflective part of the crust to the north between 10–30 km depths (B in Fig. 4) may represent banded ironstones and black schists of high seismic impedances interlayered with shales and sandstones of low impedance within the Jatulian passive margin sequence as observed overlying basement onshore in Finland.

The Kalevian–Outokumpu Collage is a seismically less reflective zone (A, Fig. 4) coinciding with the SW to NE transition from positive to negative values of initial ε_{Nd} isotope ratios (Fig. 3) in post-1.87 Ga plutons (Sweden: Öhlander et al. 1993; Finland: Huhma 1986 and Huhma et al. 1991). The negative initial ε_{Nd} values are inferred to indicate that the sampled plutons interacted with crust of Archaean age. The change in ε_{Nd} values is therefore consistent with the interpretation that the south-dipping zone of low reflectivity is a crustal suture between the juvenile Svecofennian terrane (Huhma et al. 1991; Weihed

Fig. 4. Deep seismic reflection profile of BABEL lines 3 & 4 (Fig. 1). The section has first been migrated using a FK transform (Stolt) algorithm, then depth converted, both steps using a two-dimensional velocity model derived from the far-offset land station records (BABEL Working Group 1993). Southward-dipping reflections predominate in the northern half of the section, northward-dipping ones in the south. The drawing immediately below indicates the primary features on the section that are discussed in the text (A–F). Small dots indicate reflectors or features contoured in Fig. 4. Large dots deliniate the inferred crustal suture zone. The mantle convergence estimate is simply the line length of reflector E.

et al. 1992) and the Archaean Karelian craton. As defined by the ϵ_{Nd} values, this suture lies within the complex Kalevian–Outokumpu Collage (Kontinen 1987; Huhma et al. 1991; Öhlander et al. 1993). The suture's trend, as determined from contouring the northern edge of the less reflective zone at 15–35 km depths using BABEL lines 3 and 2, appears to diverge from the more regional northwest–southeast trend of the buried footwall (Archaean) margin as indicated by the ϵ_{Nd} values reported by Öhlander et al. (1993) and Huhma (1986) (Fig. 3). However, lack of systematic regional coverage with either the reflection seismic or Nd-isotopic data sets makes it difficult to attach too much significance to this observation.

Reflective crust continues southward for over 200 km, although packets of reflections switch from southward to northward dips at about 120 km (south) (C in Fig. 4) into the Svecofennian terrane, defining a crustal-scale synform. This dip reversal has not been definitively documented by onshore geological studies, although metamorphic grade generally decreases to the south near the onshore projection of the dip reversal zone (Weihed et al. 1992). If both sets of dipping structures are broadly coeval, this style of crustal deformation matches that predicted for doubly vergent orogenic zones by recent numerical modeling (Beaumont & Quinlan 1994). Alternatively, the bivergence may be due to a complex orogenic evolution in which earlier northward-directed thrusting evolved into southward-directed thrusting concurrent with the 1.88–1.87 Ga magmatism associated with a south-facing Andean-type arc built on the Kalevian–Outokumpu Collage. Beneath the doubly vergent crustal reflectors lies a particularly distinct basal layer of horizontal reflectors at Moho depths (Fig. 4) (BABEL Working Group 1993).

Wide-angle (far offset) reflections and refractions recorded at land stations provide velocities at Moho depths. The base of reflectivity coincides with the transition to velocities greater than 7.5 km s^{-1}, and as such is interpreted as the Moho (BABEL Working Group 1993). Only a few strong reflectors lie in the mantle here. One 20-km long segment (D in Fig. 4) occurs at 70 km depth directly beneath the less reflective zone in the crust (A in Fig. 4). A second, more prominent mantle reflection zone near the southern end of BABEL line 4 (E in Fig. 4) dips northward from the Moho at 50 km to 80 km depths. This reflector has been interpreted as the trace of the top of a c. 1.88 Ga subduction zone (BABEL Working Group 1993). Another, less prominent set of reflections, parallel to the E reflections (Fig. 4), occurs about 50 km to the south and may represent a layer within the subducted slab or possibly a second, younger subduction zone. Modelling of reflection data just southwest of this mantle reflection indicates thin (c. 100 m) layers of alternating high and low velocity rocks in the lower crust (Lindsey & Snyder 1994). These layers were interpreted as either underthrust oceanic crust and sediments within a subduction zone or as cumulate layering at the base of an arc.

The prominent reflections observed throughout the crust on northern BABEL seismic sections are interpreted to outline structural geometries resulting from final convergence and collision between the reworked Karelian cratonic margin and Svecofennian juvenile terrane at c. 1.9–1.8 Ga. The architecture of the Svecofennian Orogen as outlined by the crust and mantle reflectors is inferred to have survived largely intact since its formation. The only significant post-orogenic sedimentary units in the region are a few hundred metres of Cambro-Ordivician limestones beneath the Gulf of Bothnia and c. 1.35 Ga Jotnian sandstones, which are locally up to several kilometres thick and occur in graben structures (BABEL Working Group 1993, and references therein).

Estimating Svecofennian collisional geometries using BABEL sections

The highest grade of metamorphism in the orogen is exposed in parts of the Svecofennian terrane near the Kalevian–Outokumpu Collage (e.g. Korsman et al. 1984; Weihed et al. 1992), which corresponds to the part of BABEL line 3/4 with dense dipping reflections (to either side of C in Fig. 4). Mineral assemblages in the Mn-bearing metapelites of the Kainu Schist Belt (KSB) include sillimanite, chlorite, cordierite, biotite, garnet and staurolite. Thermobarometric and geochronological studies indicate that these rocks were uplifted from ≤ 20 km depths at c. 1.87 Ga during late-Svecofennian deformation associated with thrust/nappe stacking (Fig. 3) (Tuisku & Laajoki 1990). In Sweden, uplift and crustal thickening occurred toward the end of the Svecofennian orogeny, from 1.87–1.82 Ga, and exposed greenschist to lower amphibolite facies rocks (Weihed et al. 1992). The BABEL lines show that reflectors can be traced from near surface to 40 km depths immediately to the south of the inferred crustal suture zone in the Kalevian–Outokumpu Collage. If rocks presently at the surface originally came from c. 20 km depths along shear zones dipping c. 30° through the crust (F in Fig. 4), then simple geometric considerations or balancing of cross-sectional areas imply >35 km of differential horizontal displacement (Fig. 4). The increased metamorphic grades and greater density of dipping reflectors beneath the region immediately south of the Skellefte and Vihanti-Pyhäsalmi districts (Fig. 3) suggests that crustal shortening during the final amalgamation of Svecofennian domain with the Archaean margin was greatest in the part of the Svecofennian crust corresponding to the 'horizontal tectonic regime' described by Gaál & Gorbatschev (1987).

Interpretation of the only available evidence of mantle structures, the 60 km long, downdip trace of a prominent northward dipping mantle reflector, indicates that >60 km of shortening within the uppermost mantle occurred along a thrust zone located 100 km to the south of the crustal suture. This geometry requires that the crust of the northern Svecofennian domain delaminated from its juvenile mantle, and that horizontal shortening at the location of the crustal suture was transferred along a décollement at Moho depth to the location of the northward-dipping mantle subduction zone. Subhorizontal reflections are bright and dense over a broad range of depths between 25–45 km (C–F, Fig. 4), possibly indicating that the lower crust acted as a detachment zone (e.g. Hurich et al. 1985; Reston 1988).

Seismic reflectors discussed here provide structural geometries presumed to be indicative of the last stages of convergence between the juvenile Svecofennian terrains and Archaean blocks of the Baltic craton. Reflectors associated with possible early convergence structures and subduction zones dipping away from the Archaean craton were probably overprinted or destroyed. The observed mantle reflectors indicate late-stage subduction beneath the cratonic margin, with the mantle shear zone offset by about 100 km from the older crustal suture zone located in the Kalevian–Outokumpu Collage. Some of the mantle lithosphere of this cratonic margin may include accreted juvenile mantle from the Svecofennian terranes, but high-strength Archaean cratonic lithosphere probably controlled the position of the final mantle subduction zone observed as mantle reflectors. These reflector geometries are consistent with previous tectonic models based on field mapping and geochemical/isotopic studies (e.g. Gaál 1990). The overall geometry of the orogen suggests that the Svecofennian arc(s) were accreted ('obducted') as crustal flakes and were subsequently translated northward above the footwall Archaean lithosphere.

The Rhinns complex and Lewisian terrain of the British Isles

Precambrian studies within the British Isles were long dominated by analysis of the Archaean Lewisian basement complex (Park & Tarney 1987) and, to a lesser extent, of the Neoproterozoic Dalradian–Moinian–Torridonian supracrustal rocks (e.g. Rogers et al. 1989; Dickin & Bowes 1991; Soper 1994). The Lewisian is a relatively small block of Archaean (3.3 Ga: Burton et al. 1994) crust, intruded by the 1.7–2.1 Ga Scourie dykes and reworked during the Palaeoproterozoic Laxfordian thermo-tectonic event (Park & Tarney 1987). The age of Laxfordian tectonism is poorly constrained at 1.9–1.7 Ga. Granite sheets along the Laxford Front yielded an U–Pb zircon age of 1678^{+20}_{-10} Ma (van Breeman et al. 1971), whereas Rb–Sr whole rock metamorphism ages of 1862 ± 50 and 1713 ± 135 Ma were obtained for quartzofeldspathic gneisses (Lambert & Holland 1972). Laxfordian deformation has been related to thrusting of high-level crustal blocks to the NW on a combination of low-angle shear zones and steep lateral ramps (Coward & Park 1987; Park 1994).

Recent geochemical and isotopic analysis showed that at least one terrane of exposed Precambrian rocks, the Rhinns Complex (Malin block of Park 1994), represents new crust that formed during the 1.9–1.7 Ga Laxfordian event (Fig. 5). This block contains a syenite with U–Pb zircon crystallization ages of 1782 ± 5 and 1779 ± 3 Ma (Marcontonio et al. 1988; Daly et al. 1991). Concurrence of crystallization and Nd model ages of both the syenite and a related gabbro indicate derivation from a depleted mantle source at c. 1.8 Ga (Muir et al. 1992).

The Rhinns Complex is limited in outcrop and cannot be linked with any part of the Lewisian complex. However, it is coeval with Laxfordian deformation and with rocks that intrude the Lewisian basement. Based on studies of inherited Proterozoic zircons and Proterozoic Nd model ages in Caledonian granites of mainland Scotland, the Rhinns Complex may also occur beneath much of the northern British Isles covered by Moine and Dalradian metasedimentary rocks (Fig. 5) (Frost & O'Nions 1985; Dickin & Bowes 1991; Muir et al. 1992). If the Rhinns Complex outcrops represent only a very small part of an extensive juvenile Proterozoic terrane, then its juxtaposition with the reworked Laxfordian parts of the Lewisian complex may be analogous to the accretion of the juvenile Svecofennian terrane to the reworked margin of the Archaean Karelian craton in the Baltic Shield.

The Flannan and W mantle reflectors

The Palaeoproterozoic age of the Rhinns Complex is of particular interest to the present discussion because the British Isles contain the densest coverage of publicly available deep seismic reflection profiles in the world and the largest number of mantle reflections (Flack et al. 1990). Two of these

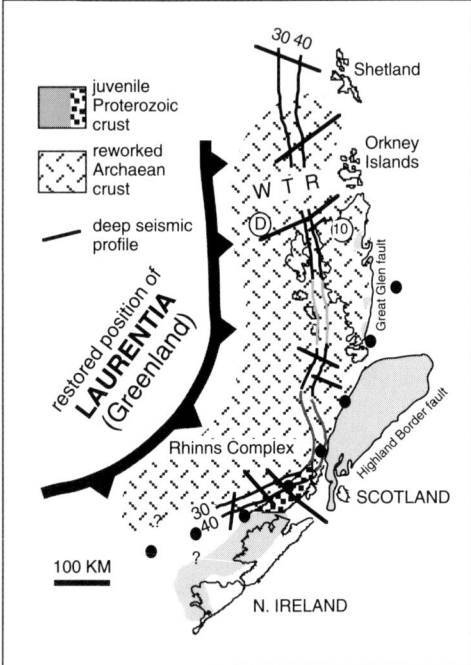

Fig. 5. Map of reworked Archaean and Palaeoproterozoic Laurentian terrains of the British Isles, as inferred from outcrop and isotopic studies of xenoliths in intrusions (stipple pattern) (Muir et al. 1992). The large dots deliniate the inferred crustal suture between reworked Archaean crust to the upper left and juvenile Palaeoproterozoic crust to the lower right. The tectonic blocks are reconstructed at c. 400 Ma, post-Caledonian thrusting, but prior to Caledonian strike-slip and Cenozoic extension (see Snyder & Flack 1990). Labelled contours indicate 30 and 40 km depths of the mantle W-reflector. WTR is the Wyville–Thompson Ridge. Encircled letter D and 10 indicate the locations of the deep seismic profiles of Fig. 6.

Fig. 5). The mantle reflector dips away from the reconstructed cratonic nucleus of Laurentia in a recent model of Vendian (0.55–0.62 Ga) tectonics (Soper 1994), but lies beneath Lewisian crust that was part of Laurentia at that time. If Dalradian rocks cover a juvenile Proterozic terrane that includes the Rhinns Complex, then the correlation of Dalradian rocks with the W-reflector implies that they occur near the boundary between reworked Archaean Lewisian crust and crust newly accreted 1.9–1.7 Ga ago (Fig. 5). In this scenario, the mantle reflector dips away from the reworked Archaean crust and beneath the juvenile Proterozoic crust at 25–45 km depths. The palinspastic restorations imply a pre-Caledonian age for the W-reflector (Snyder & Flack 1990), but cannot uniquely designate its origin as due to either 1.9–1.7 Ga convergence or Vendian rifting.

The existence of deep seismic reflection profile grids showing mantle reflections beneath the Rhinns Complex provides evidence of a potential relationship between dipping mantle reflectors and Precambrian tectonics in at least one part of the British Isles. Recent studies of deep reflection profiles west of Shetland (Fig. 5) add several more mantle reflectors along-strike to the NNE from those already interpreted as the W-reflector (McBride et al. 1995). The unusual combination of eastward-dipping mantle reflectors along 800 km of the former margin of Laurentia and their rarity elsewhere within the lithosphere of the British Isles makes these reflectors an important feature both to seismologists trying to understand the cause of the reflection and to geologists trying to determine its age and origin. Even if the reflectors result from Vendian (c. 0.6 Ga) opening of the Iapetus Ocean, their spatial correlation with major Palaeoproterozoic tectonic boundaries suggests that they may have reutilised structures related to Laxfordian (1.9–1.7 Ga) arc-continent collision.

mantle features, the Flannan and W-reflectors, were identified on more than 20 reflection profiles over a strike length of greater than 800 km (Figs 5 & 6; Snyder & Flack 1990). The two mantle reflectors are coincident on some sections (Fig 6b), but it is the subhorizontal part of the W-reflector that underlies much of northern Scotland and is traced for 800 km (Fig. 6a).

Palinspastic restorations of late Caledonian deformation (Snyder & Flack 1990) produced a reconstructed W-reflector that followed the previously inferred east-facing pre-Caledonian margin of Laurentia (Soper 1994), and that tracks outcrops of Dalradian metasedimentary rocks originally deposited on the margin of Laurentia during the opening of Iapetus (Harris et al. 1995;

The Trans-Hudson Orogen of Laurentia

Understanding of the crustal structure and tectonic history of the 2.1 to 1.7 Ga Trans-Hudson Orogen in northern Canada has increased substantially over the past ten years, due to a combination of geological mapping, multidisciplinary studies and deep seismic reflection profiling programs (summarized in Hoffman 1988, 1989; Lewry & Collerson 1990; St. Onge et al. 1992; Van Kranendonk et al. 1993; Lucas et al. 1993; Lewry et al. 1994). The name 'Trans-Hudson' Orogen follows from its continuity around the northwestern margin of Superior Province, extending from the Ungava Peninsula (Cape Smith Belt, Ungava

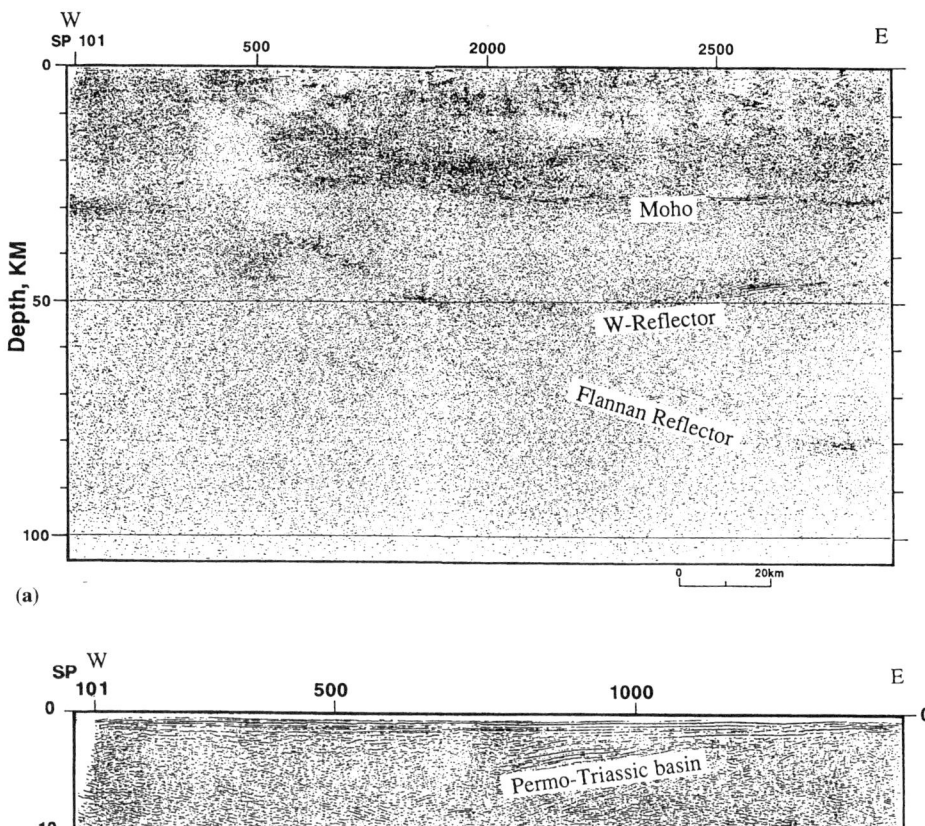

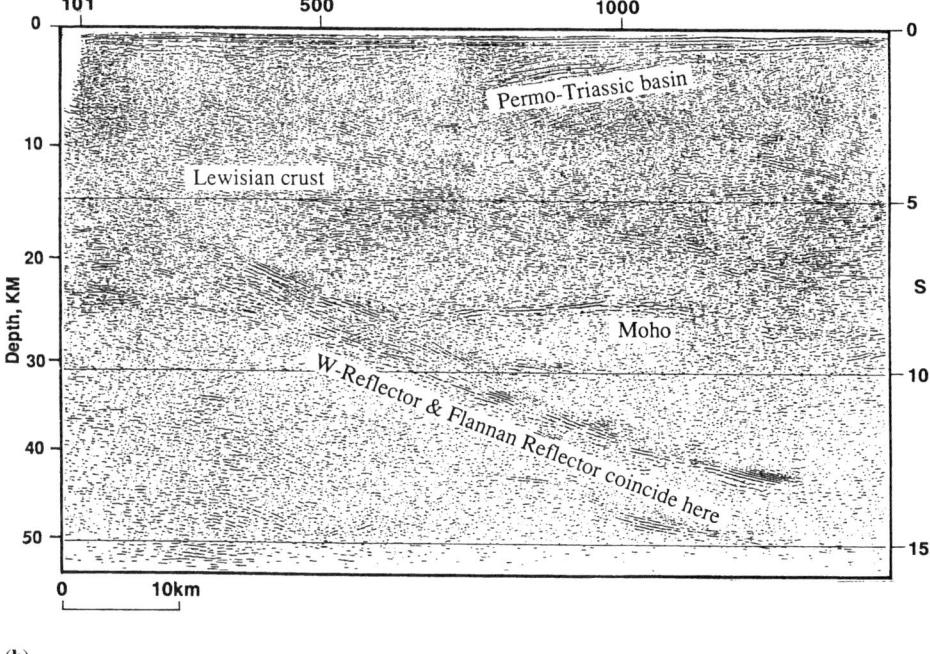

Fig. 6. Deep seismic reflection profiles, (**a**) the DRUM and (**b**) GRID 10 lines from north of Scotland (lines labelled in Fig. 5). The sections were migrated using the Stolt algorithm and a velocity of 6 km s^{-1}. The DRUM depth-converted section shows reflections between 30 and 80 km depths from an inferred mantle shear zone possibly reactivated as a Permo-Triassic normal fault zone, as a Caledonian thrust zone, or as a Vendian extensional structure and c.1.8 Ga subduction or convergence zone. The Moho is at 25–28 km depths.

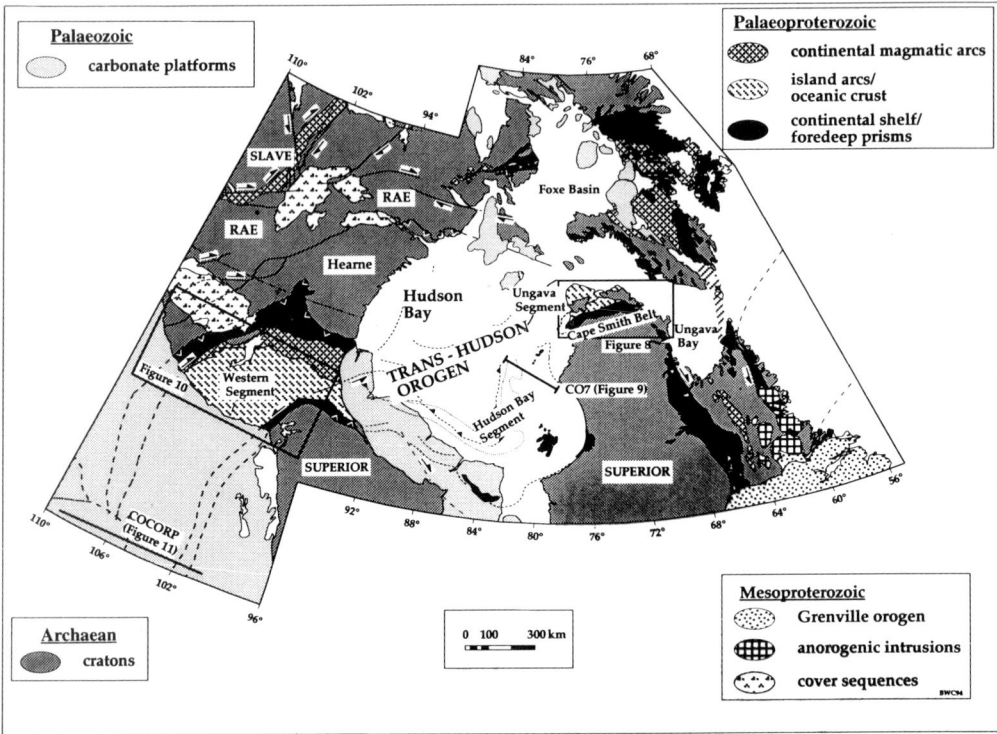

Fig. 7. Map of the Palaeoproterozoic (2.0–1.7 Ga) orogenic belts of the eastern Canadian shield and Hudson Bay region, after Hoffman (1988). The Ungava, Hudson Bay and Western segments of the Trans-Hudson Orogen are discussed here. FRB is the Fox River Belt. Locations of Figs 8, 10 and seismic reflection lines shown in Fig 9 (CO7) and 11 (COCORP) are shown.

segment: Fig. 7) across Hudson Bay to the extensive shield exposure of the Western segment in Manitoba and Saskatchewan, west of 96° longitude (Fig. 7; Hoffman 1988; Lewry & Collerson 1990). The foreland to the Trans-Hudson Orogen is the Archean Superior craton, stable since about 2.6 Ga and characterized by relatively thick lithosphere (Grand 1987) interpreted to have formed during Neoarchaean accretion tectonics (Silver & Chan 1991; Hoffman 1990). An autochthonous to parautochthonous rift sequence occurs along much of Superior craton's northern and western margins ('Circum-Superior Belt' of Baragar & Scoates 1981), and is marked by mafic–ultramafic intrusive and extrusive rocks that range in age from 2.04 to 1.87 Ga (Fig. 2; Parrish 1989; St-Onge et al. 1992; Machado et al. 1993). A reworked crustal suture separates the Superior craton and its rift margin sequences from the allochthonous terranes that characterize the interior of the orogen. These terranes have been shown to be dominantly juvenile and include a 2.00 Ga ophiolite (Scott et al. 1991, 1992) as well as 1.92–1.82 Ga arc and back-arc assemblages (e.g. Parrish 1989; Bickford et al. 1990; Syme 1990; Thom et al. 1990; St-Onge et al. 1992; Lucas et al. 1996; Stern et al. 1995a, b; Dunphy 1995). In the following sections we examine the crustal structure of the suture zone between Superior craton and the allochthonous terranes in three locations (Fig. 7): (1) the Ungava segment, using the geological control offered by substantial structural relief (cf. Lucas 1989); (2) the Hudson Bay segment, using an interpretation of a commercial marine reflection profile acquired in the southeast part of the bay (Roksandic 1987); and (3) the Western segment, employing LITHO-PROBE and COCORP deep seismic reflection data (Nelson et al. 1993; Lewry et al. 1994; Lucas et al. 1994; White et al. 1994; Baird et al. 1995).

Ungava segment

Three principal tectonostratigraphic domains are represented on the northern Ungava peninsula of Quebec (Fig. 8): (1) Archaean Superior craton basement, (2) Palaeoproterozoic rift margin/

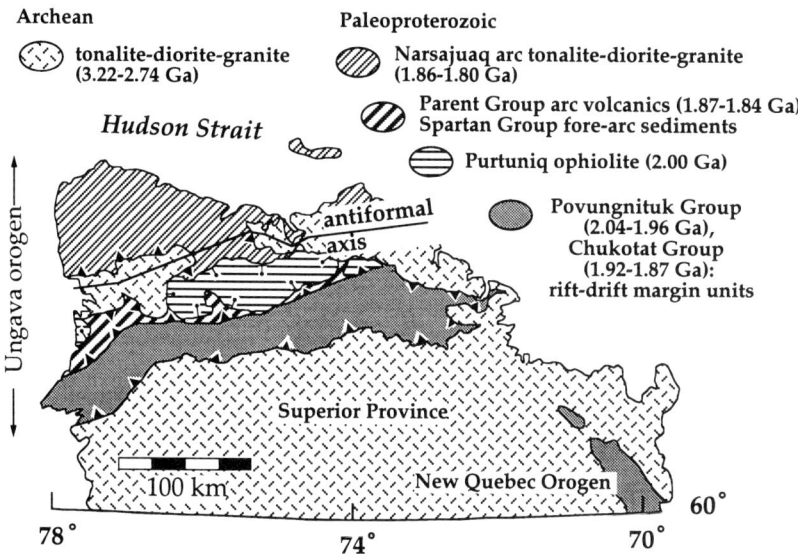

Fig. 8. Map of the Ungava Peninsula in northern Quebec (St. Onge *et al.* 1992, 1996; Lucas & St-Onge 1995). See Figs 1 and 7 for location.

oceanic units, and (3) Palaeoproterozoic convergent margin units (St-Onge *et al.* 1992). Superior craton represents the stratigraphic basement to domain (2) as well as the structural basement to both domains (2) and (3). The Palaeoproterozoic rift/oceanic units include: (1) a 2.00 Ga ophiolite (Parrish 1989; Scott *et al.* 1991, 1992); (2) the Povungnituk Group, a 2.04–1.96 Ga continental rift sequence of sedimentary and volcanic rocks, (Parrish 1989; St-Onge *et al.* 1992; Machado *et al.* 1993); and (3) the Chukotat Group, a sequence of basalts with geochemical signatures of transitional crust and bracketed in age at 1.92–1.87 Ga (Francis *et al.* 1983; St-Onge *et al.* 1992). The convergent margin units have U–Pb zircon ages ranging from 1.90 to 1.82 Ga (Parrish 1989; St-Onge *et al.* 1992; Machado *et al.* 1993). These units comprise supracrustal sequences inferred to represent arc (Parent Group) and forearc (Spartan Group) deposits, and plutonic and metasedimentary rocks thought to represent a mid-crustal segment of the Narsajuaq magmatic arc (Lucas *et al.* 1992; St-Onge *et al.* 1992). This arc was built in part on oceanic crust and in part on older continental crust (Dunphy 1995), in response to initial northward subduction of an oceanic basin now represented by the Chukotat Group (Fig. 9) (Lucas *et al.* 1992; St-Onge *et al.* 1992).

The tectonic history of the Ungava orogen is characterized by episodes of increased structural–metamorphic activity that are inferred to both pre-date and postdate a collision between the Narsajuaq arc terrane and the northern continental margin of the Superior Province (cf. Lucas & St-Onge 1992). Distinct pre-collisional tectonic histories are documented for Palaeoproterozoic rocks forming the lower-plate (*c.* 2.04–1.92 Ga rift-to-drift margin sequence) and the upper-plate (*c.* 2.00 Ga ophiolitic and *c.* 1.90–1.82 Ga magmatic arc units) of the Ungava Orogen. The lower plate units preserved in the external part of the orogen (Cape Smith Belt) record the development of a thrust belt characterized by south-verging faults ramping up from a basal décollement located at the basement-cover contact (Lucas 1989). Lucas & St-Onge (1992) have suggested that initial contraction of the north-facing Superior continental margin occurred at approximately 1.87 Ga. The ophiolitic and arc units were accreted to the older thrust belt along south-verging faults which re-imbricated the thrust belt by *c.* 1.80 Ga, the age of collisional granites (Lucas & St-Onge 1992; R. Parrish, per. comm. 1994). Thickening and consequent exhumation resulted in greenschist- to amphibolite-facies metamorphism of lower-plate cover units (St-Onge & Lucas 1991; Bégin 1992), retrogression of lower-plate basement gneisses (St-Onge & Lucas 1995), and retrogression of syn-magmatic high-T assemblages in Narsajuaq terrane rocks (Lucas & St-Onge 1992).

Shortening continued with the thrusting replaced by folding of both the upper-plate units and lower-plate cover and Archaean basement (Lucas & Byrne 1992). Geological mapping indicates that the Superior Province rocks exposed north of the Cape

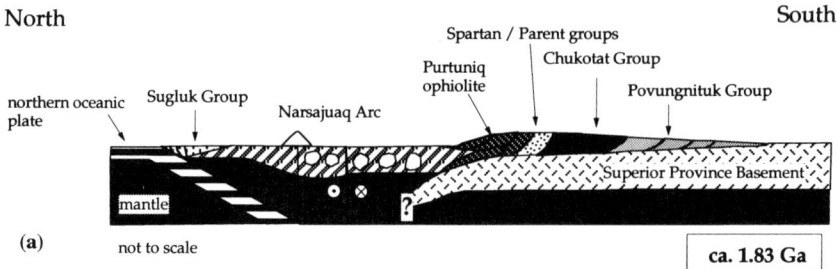

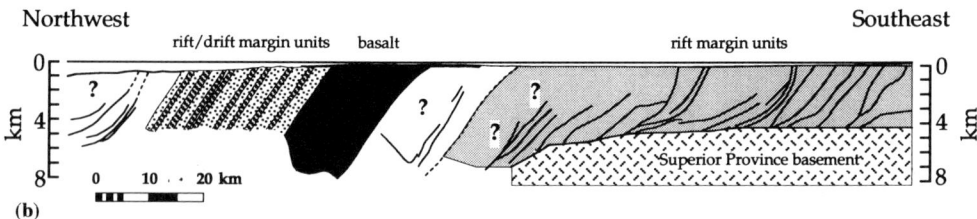

Fig. 9. (a) Schematic block diagram showing the possible arrangement of tectonic elements in the Ungava segment of the Trans-Hudson Orogen at c. 1.83 Ga (St-Onge et al. 1992); (b) interpretative section of a shallow seismic reflection–aeromagnetic–gravity profile (Roksandic 1987) from Hudson Bay 500 km to the west of the Ungava orogen exposures (see CO7 in Figs 1 and 7 for location).

Smith Belt lie in the core of a major, regional-scale antiform (St-Onge et al. 1996). Cross-folding of this structure has generated a regional fold-interference geometry, with a structural depression in the central part of the orogen passing up-plunge into the two basement-cored culminations to the east and west (Lucas & St-Onge 1995; Fig. 8). The overall result of the post-collisional folding episodes is that Archaean basement, Palaeoproterozoic allochthons and the suture zone that separates them are all exposed in oblique section (Lucas 1989).

Upper-plate arc and ophiolitic units can be shown to have ≥100 km of displacement with respect to autochthonous basement (Fig. 8; Lucas & St-Onge 1992). One implication is that both the ophiolite and the arc massif must have been delaminated as they overrode the continental margin, as there is no evidence of mantle material at the sole of either the ophiolite or the accreted arc. A subduction polarity flip between early imbrication of the continental margin (c. 1.87 Ga) and late arc magmatism (1.84–1.82 Ga) has been inferred by St-Onge et al. (1992; Fig. 9a).

Hudson Bay segment

Relatively little is known about the tectonic history and crustal structure of the Trans-Hudson orogen in Hudson Bay and its borderlands due to difficulty of access and limited exposure of Palaeoproterozoic units. Extrapolations across the bay have relied on aeromagnetic and gravity data (e.g. Hoffman 1989), geological and geochemical data from the few islands and shore exposures (Baragar & Scoates 1981; Chandler 1988), and a lone published seismic reflection profile (Roksandic 1987). Superior craton lithosphere is inferred to underlie the craton-verging thrust belts that have been traced from the Belcher Islands to the Fox River Belt (Fig. 7; Hoffman 1988, 1989; Hynes 1991). Roksandic's (1987) highly speculative interpretation of marine shallow reflection data and potential field data (Fig. 9b) is consistent with the geological (Hoffman 1988) and flexural (Hynes 1991) models that predict that Archaean basement extends into Hudson Bay beneath the craton-verging thrust belt. No evidence exists of an arc built on autochthonous Superior crust anywhere in the Hudson Bay borderland, from the Ungava Orogen to the Fox River Belt (Fig. 7), consistent with the interpretation that Superior craton formed the lower (underthrust) plate in the orogen. Hoffman (1988) proposed that much of Hudson Bay is underlain by juvenile terranes, thus implying that Superior craton and its Palaeoproterozoic continental margin sequence formed the footwall to crustal thrusts carrying the juvenile terranes, similar to the Ungava Orogen (Fig. 8).

Western segment

The western segment of the Trans-Hudson Orogen comprises four principal tectonic domains (Lewry & Collerson 1990; Fig. 10): (1) a narrow eastern foreland, the Thompson Belt (Bleeker 1990); (2) a broad collage of dominantly juvenile arc and oceanic terranes, the Reindeer Zone, that structurally overlies an 'exotic' Archaean block exposed in three small basement windows (Stauffer 1984; Bickford et al. 1990' Lewry et al. 1990); (3) an Andean-type continental margin batholith, the Wathaman-Chipewyan Batholith (Meyers et al. 1992 and references therein); and (4) a broad, reworked northwestern hinterland, the Cree Lake Zone of the Hearne craton (Bickford et al. 1994). Plate convergence and collision in the Trans-Hudson Orogen spans the period from 1.9 to 1.7 Ga. Generation of arc and back-arc crust within the Reindeer Zone occurred between 1.92 and 1.87 Ga (Bickford et al. 1990; Syme 1990; Thom et al. 1990; Stern et al. 1995a, b) and was followed by intra-oceanic accretion at c. 1.88–1.87 Ga (Lucas et al. 1996). The interval from 1.87–1.83 Ga was marked by development of post-accretion arcs (Bickford et al. 1990; Lucas et al. 1996), including the Wathaman Batholith at 1.86–1.85 Ga following accretion of the La Ronge–Lynn Lake arcs to the Hearne craton (Fig. 10; Bickford et al. 1990; Meyers et al. 1992). The absence of arc magmatism on the Hearne margin prior to 1.86 Ga suggests that this margin formed the lower plate during collision with the La Ronge and Lynn Lake arcs. Cessation of Wathaman Batholith magmatism by 1.85 Ga implies termination of northwest-dipping subduction along the Hearne margin due to collision with other juvenile terranes (Ansdell et al. 1996). The period 1.85–1.83 Ga was characterized by widespread non-marine sedimentation and volcanism and back-arc basin extension and sedimentation (Ansdell et al. 1992, 1996; David et al. 1996). Terminal collision occurred between 1.83 and 1.80 Ga (Bickford et al. 1990; Gordon et al. 1990; Ansdell & Norman 1995), and involved the Reindeer Zone and trailing Hearne craton, the 'exotic' Archean block and Superior craton. Post-collisional deformation continued until c. 1.69 Ga (Machado 1990; Fedorowich et al. 1995).

In 1991, LITHOPROBE acquired more than 1100 km of seismic reflection data across the Trans-Hudson Orogen (Fig. 10), extending from Superior craton to the western hinterland of the Hearne craton (Lewry et al. 1994; White et al. 1994) and including cross-lines for three-dimensional control (Lucas et al. 1994). Seismic sections reveal a broadly symmetric crustal

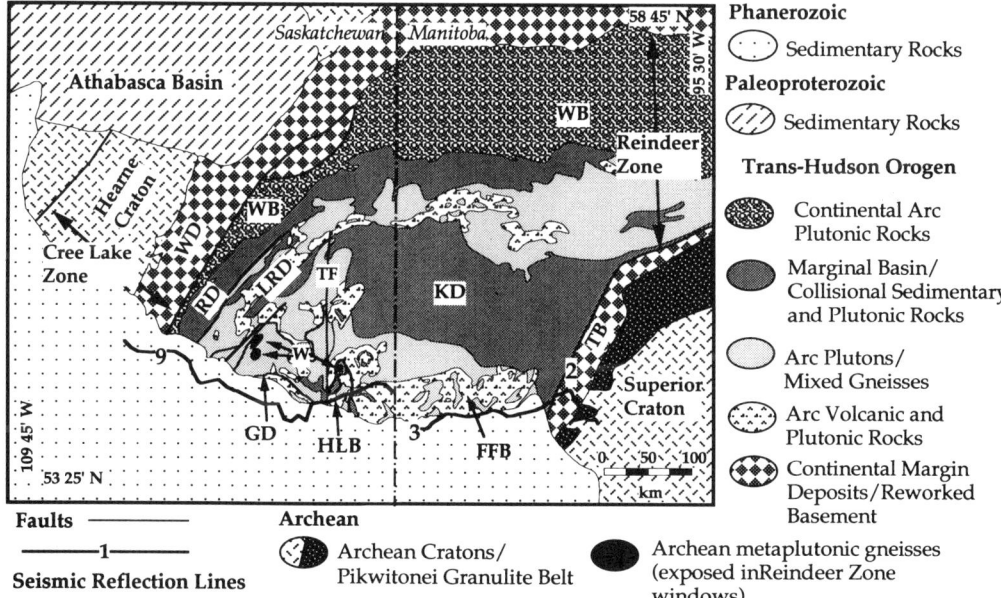

Fig. 10. Geological map of the Tran-Hudson Orogen (after Hoffman 1988), showing the location of the LITHO-PROBE seismic reflection profiles. W indicates the location of Archaean basement windows within the Reindeer Zone. FFB = Flin Flon Belt; GD = Glennie Domain; HLB = Hanson Lake Belt; KD = Kisseynew Domain; LRD = La Ronge Domain; RD = Rottenstone Domain; TB = Thompson Belt; TF = Tabbernor Fault Zone; WB = Wathaman–Chipewyan Batholith & WD = Wollaston Domain.

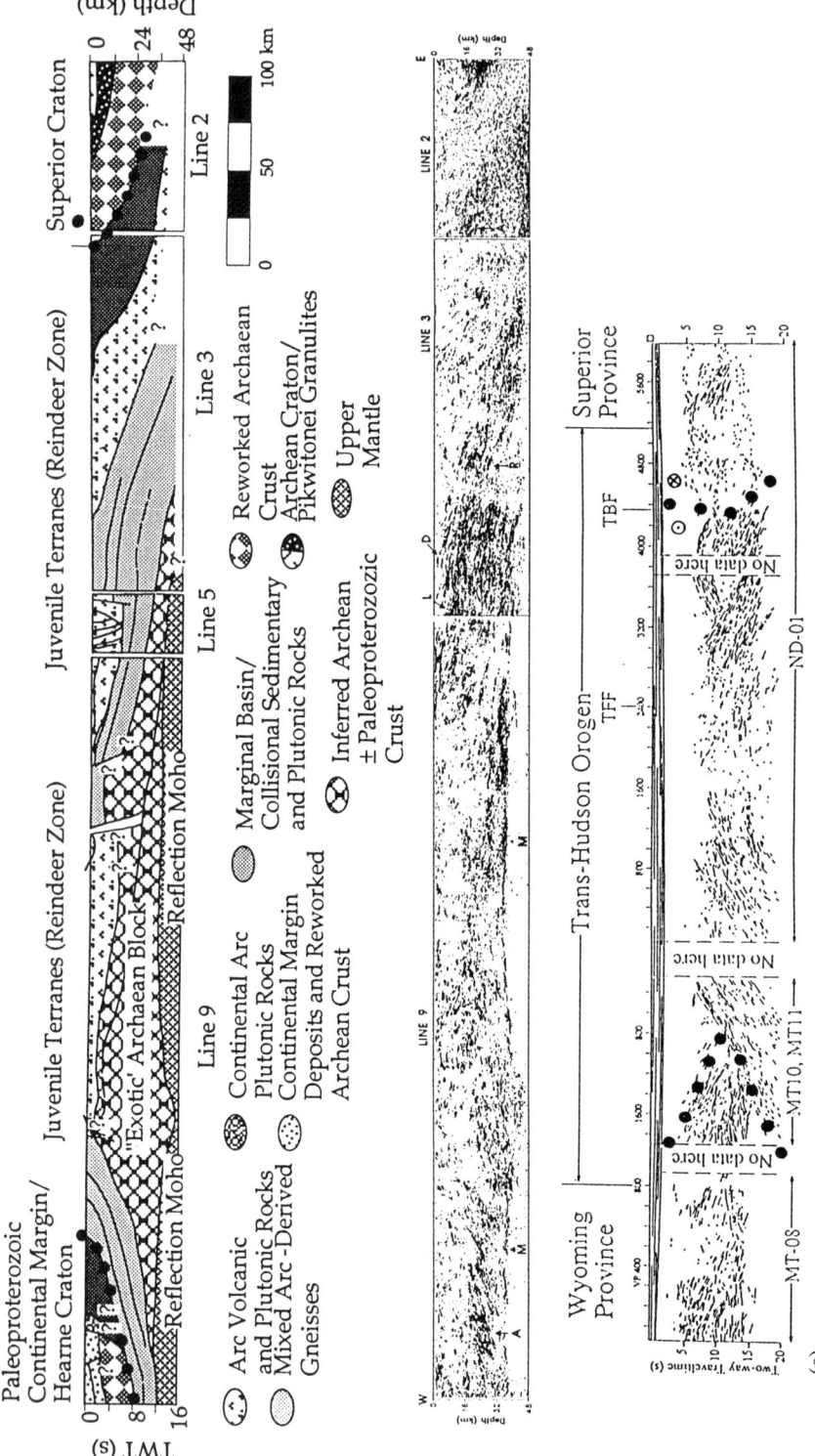

Fig. 11. (a) Interpretative geotectonic section for lines 2, 3 & 9 of the LITHOPROBE Tran-Hudson Orogen Transect (see Fig. 10 for location: after Lucas et al. 1993; Lewry et al. 1994). Large dots deliniate the inferred crustal suture zones. (b) Automated line drawing of coherency filtered and migrated data on which the above interpretation was made (also see Lewry et al. 1994; White et al. 1994). Depth conversion was made assuming a constant velocity of 6 km s⁻¹. (c) Unmigrated, interpretative line drawing of the COCORP deep seismic reflection sections across the Trans-Hudson Orogen and Williston Basin of the northernmost USA (Nelson et al. 1993; Baird et al. 1995), see Figs 1 and 7 for location. The Proterozoic crust of the Trans-Hudson Orogen underthrusts the Archaean blocks along both of its margins in both areas surveyed by these seismic profiles. Large dots deliniate the inferred crustal suture zone. All the sections are at the same scale with no vertical exaggeration.

reflection pattern about a culmination in west-central Trans-Hudson Orogen, in which reflective zones dip east and west from the near surface to the lower crust (Fig. 11). Reflections dip beneath the margins of the bounding Archaean Hearne and Superior cratons (Fig. 11). A remarkably similar crustal structure was imaged in a COCORP reflection survey across the buried Trans-Hudson Orogen approximately 500 km to the south (Nelson *et al.* 1993; Baird *et al.* 1995). Reflections from the Moho were observed for over 500 km across the western Trans-Hudson Orogen, with a local crustal root occurring beneath the culmination where crustal thickness increases from 36 to 45 km (Fig. 11). Interpretation of wide-angle reflection data from a major refraction experiment across THO corroborates this result (Nemeth & Hajnal 1996). The culmination is interpreted as being cored by a block of Archean crust that is exposed in the Reindeer Zone basement windows (Fig. 10; Lewry *et al.* 1990, 1994).

Interpretation of the Trans-Hudson Orogen reflection data is challenging because of the complex collisional history involving the Hearne and Superior cratons, the largely buried 'exotic' Archaean block and the juvenile 1.9–1.8 Ga terranes. Although most pre-collisional structural geometries have probably been obliterated by collisional and post-collisional deformation (Lucas *et al.* 1993), the crustal structure of the western part of the orogen is consistent with inferred subduction beneath the Hearne margin at 1.86–1.85 Ga (e.g. Meyers *et al.* 1992; Bickford *et al.* 1990). However, geological and geochronological studies indicate that the actual reflectors are probably associated with collisional thrusting and the development of tectonic layering during regional metamorphism at 1.83–1.80 Ga (Bickford *et al.* 1994; Lewry *et al.* 1994). In fact, much of the ductile structure within the interior of the orogen (Reindeer Zone) is associated with southwestward thrusting of the juvenile allochthons over the 'exotic' Archaean block at 1.83–1.80 Ga (Lewry *et al.* 1990, 1994; Lucas *et al.* 1994). Oblique collision of the orogen with Superior craton occurred relatively late during this event (c. 1.82–1.80 Ga; Bleeker 1990; Machado 1990; Ansdell *et al.* 1996). The Superior (Thompson Belt)–Reindeer Zone suture is represented at surface by a strike-slip fault zone (Bleeker 1990), and at depth by steeply east-dipping reflections (White *et al.* 1994; White & Lucas 1994). The prominent structural geometries imaged across the orogen (e.g. crustal-scale culmination, reflections dipping beneath the Superior craton margin, Moho reflections east of the crustal root) are attributed to post-collisional deformation involving transpression and longitudinal extension of the orogen (Lucas *et al.* 1994; White *et al.* 1994; Hajnal *et al.* 1995).

Despite the obliquity of Trans-Hudson Orogen tectonics and the intensity of post-collisional deformation, some important observations can be made concerning arc–continent collision processes. First, the stack of allochthonous terranes in the Reindeer Zone appears to only involve slices of juvenile crust. This implies that the juvenile terranes were delaminated during accretion–collision events, with their lithospheric mantle being underthrust (subducted?) and the crustal flakes being imbricated. It seems likely that the colliding arcs/accretionary collages were delaminated against the three Archaean elements in the orogen: Superior craton, Hearne craton, and the 'exotic' block. An implication is that much of the crust and lithospheric mantle underlying the Reindeer Zone is either Archaean in age or was derived from Archaean-aged lithosphere (Lucas *et al.* 1993; Lewry *et al.* 1994).

The Ketilidian and Nagssugtoqidian terrains of Greenland

One key piece of Laurentian Precambrian terrain was only recently explored by very limited reflection seismic profiling (ICE in Fig. 1). The Ketilidian and Nagsugtoqidian terrains of southern Greenland are separated by the broad expanse of the Nain craton (Fig. 1). The Ketilidian terrain and correlative Makkovik Orogen in Labrador have felsic magmatic rocks dated at c. 1.86 Ga and syntectonic granites in the folded migmatite zone dated at 1845 ± 15 and 1840 ± 45 Ma (Fig. 2) (Kalsbeek *et al.* 1987, and references therein). Monzonite and titanite dates of 1.794–1.761 Ga indicate a time of decompressional melting and cooling (e.g. Schärer *et al.* 1988).

The c. 300 km wide Nagsugtoqidian belt consists mainly of reworked Archaean gneisses bordering a central zone of Proterozoic metasediments and calc-alkaline volcanic and plutonic rocks. Isotopic evidence indicates that the igneous rocks contain no significant contribution from Archaean crust; isotope and geochemistry studies on these rocks suggest that they formed over a subduction zone between 1.920–1.850 Ga ago (Kalsbeek *et al.* 1987). Age data from the northwestern part of the orogen range from 1.845–1.822 Ga and are interpreted to date high-grade metamorphism that was followed by south-directed thrusting (Bridgewater *et al.* 1990). The Proterozoic and Archaean gneisses are distinguishable only using isotopic criteria (Kalsbeek *et al.* 1987). Karlsbeek *et al.* (1993) and Park (1994) discuss possible correlations between these terrains and the Lewisian, but subduction polarity associated with the igneous rocks is unknown. No fully processed deep seismic reflection data are available at present.

Discussion: a general model for arc–continent collisions

The interval 2.0–1.8 Ga has been recognized for some time as an important period of crustal growth and continental assembly (Patchett & Arndt 1986; Gaál & Gorbatschev 1987; Hoffman 1988, 1989; Bickford et al. 1990; Lewry & Collerson 1990; Van Kranendonk et al. 1993). The global amount of juvenile Palaeoproterozoic lithosphere created, accreted to existing continents and ultimately preserved has yet to be fully assessed, owing in large part to a lack of geometrical, geochemical and isotopic constraints on the allochthonous juvenile terranes within Palaeoproterozoic orogens (e.g. Reindeer Zone, Svecofennides). The tectonic histories and crustal processes associated with the the Palaeoproterozoic orogenic belts of Laurentia and Baltica are increasingly better understood from field and laboratory studies. However, virtually no consideration has been given to their deep crustal and mantle structure even though it is implied in many tectonic models. The seismic images of mantle reflectors in both the BABEL and BIRPS profiles discussed in this paper underscore the importance of considering the mantle lithosphere in tectonic models for the evolution of these orogens.

With the availability of deep seismic reflection data from a number of individual orogenic segments, as well as geological constraints derived from the study of well exposed oblique crustal sections, it is now possible to make some general comments concerning crustal structure and accretion mechanisms associated with Palaeoproterozoic orogens in Laurentia and Baltica. In the three examples studied here, juvenile Palaeoproterozoic arc, oceanic and composite terranes appear to have been detached from their lower crust/mantle during accretion to Archaean margins (Fig. 12). In the Reindeer Zone (Western segment) and Cape Smith Belt/Narsajuaq arc (Ungava segment), a mid- to lower crustal depth of detachment for the terranes can be directly observed where they overlie Archaean basement (Lewry et al. 1990, 1994; Lucas & St-Onge 1992; St-Onge et al. 1992). Juvenile terranes caught in the convergence zones are thus only crustal flakes, imbricated and internally deformed during final collision and subsequent tectonic events. BABEL reflections imply at least partial subduction of the Svecofennide mantle lithosphere. Subduction of juvenile mantle during the accretion of La Ronge arc crust to the Hearne craton is likewise consistent with both geological and seismic data for the western margin of the Trans-Hudson Orogen (Lewry et al. 1994), although there are no prominent mantle reflections to provide direct evidence of lithospheric subduction.

Archaean lithosphere appears to have played a similar role as the stable footwall to thrusting in both the Baltic and Laurentian orogenic segments, but important differences exist. In the case of the Svecofennides and the western margin of the Trans-Hudson Orogen, latest subduction occurred beneath the margin of an Archaean continent and resulted in Andean-type magmatic arcs being built across older crustal sutures related to arc-continent collision (Fig. 12). In the Trans-Hudson Orogen, the Andean arc is represented by the 1.86–1.85 Ga

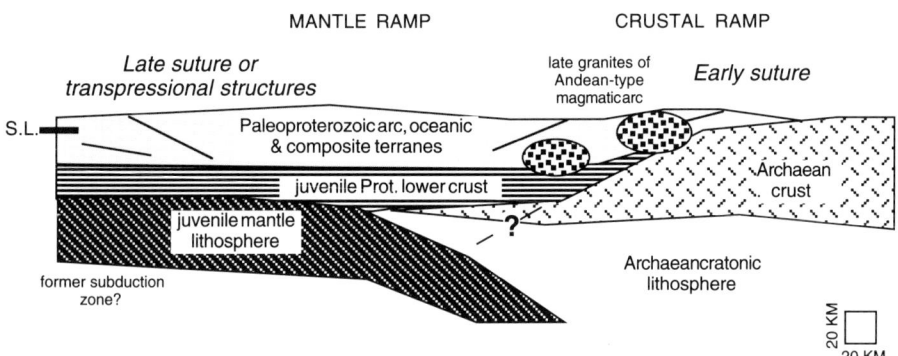

Fig. 12. Cartoon showing a generalized tectonic model for the convergence of a juvenile Proterozoic arc with Archaean cratonic lithosphere. The model is based on deep seismic reflector geometries and geological observations. Mantle subduction polarity reversal late in the convergence cycle results in symmetrical, doubly vergent thrust structures in the crust (for an alternative model, see Beaumont & Quinlan 1994). Late orogenic plutons of an Andean-type arc stitch the older crustal suture. The Svecofennian Orogen and western (Hearne craton) margin of the Trans-Hudson Orogen are examples.

Wathaman Batholith, whereas in the Baltic Shield it is manifest in the 1.89–1.87 Ga plutonic (Kiruna porphyries) and volcanic (Arvidsjaur) rocks in Sweden. However, the early histories of both of these convergent margin segments suggest that Archaean craton was attached to the subducting plate, implying that subduction polarity reversed during the last stages of convergence and consolidation (Gaál & Gorbatschev 1987; Bickford *et al.* 1990; Park 1991; Lewry *et al.* 1994).

In contrast, the Superior margin along the Trans-Hudson Orogen is not characterized by Andean-type arcs (cf. Baragar & Scoates 1981; Hoffman 1988), and is interpreted to have acted as the autochthonous footwall to crustal thrusts throughout 1.9–1.8 Ga collisional tectonism (e.g. Fig. 9a). Geological evidence from the Ungava segment (Cape Smith Belt) indicates that the Superior Province lithosphere was overridden by allochthonous arc and oceanic terranes (Lucas 1989; Lucas & St-Onge 1992), which were detached within the middle to lower crust (Scott *et al.* 1991, 1992; St-Onge *et al.* 1992). The tectonic history of Narsajuaq arc in the Ungava segment (Fig. 8) appears to require a subduction polarity reversal (from north- to south-dipping) outboard of the Superior margin at c. 1.84 Ga (Fig. 9a; St-Onge *et al.* 1992). However, an Andean-type arc was not built on the Superior basement or Cape Smith Thrust Belt during post-polarity reversal magmatism (1.84–1.82 Ga; Parrish 1989; St-Onge *et al.* 1992) because it was in a back-arc setting.

A particularly significant conclusion of this study is that the crust of juvenile arc, oceanic and composite terranes was detached from its lower crust and lithospheric mantle during the process of accretion to Archaean cratonic margins. Remnant juvenile lithospheric crust/mantle was either underplated and stacked against the Archaean cratonic mantle or recycled into the asthenosphere. This form of flake tectonics reflects the relative strength and durability of Archaean lithosphere, and implies large parts of the lower crust and lithospheric mantle underlying Palaeoproterozoic juvenile terranes may be Archaean in age. Archaean domains should therefore be assumed to be significantly larger than their surface exposures unless specifically delimited by geochemical, isotopic and geophysical surveys.

Post-collisional collapse of the orogens in response to delamination of the lithospheric mantle (cf. Nelson 1991) is not indicated by either the seismically-imaged structure or the geology of the Trans-Hudson or Svecofennian orogens. Potential tests of this proposed model include: (1) geological studies of deep crustal sections of these origins to search for accreted Palaeoproterozoic mantle slices; and (2) tracer isotope studies on lower crustal/ mantle xenoliths entrained in kimberlites or other younger magmatic rocks that may indicate the age and origin of lithosphere beneath Proterozoic juvenile terranes such as the Svecofennides, the Rhinns Complex, and the Reindeer Zone.

BIRPS is funded by the British Natural Environment Research Council and BIRPS' Industrial Associates Program (Amerada Hess Ltd, Amoco, BP Exploration Operating Company Ltd, Chevron (UK) Ltd, Conoco (UK) Ltd, Mobil North Sea Ltd and Shell (UK) Ltd. BIRPS seismic sections are available at the cost of reproduction from the British Geological Survey (Marine Geophysics Programme Manager), West Mains Road, Edinburgh EH9 3LA, UK. LITHOPROBE is funded by a Collaborative Special Project and Program grant from the Natural Sciences and Engineering Research Council of Canada and by the Geological Survey of Canada (GSC). We gratefully acknowledge the contributions of Marc St-Onge, Dave Scott, Jan Dunphy and Randy Parrish to the Ungava Orogen study; and Zoli Hajnal, Don White, John Lewry and Ron Clowes to the Trans-Hudson Orogen LITHOPROBE Transect. Roland Gorbatschev, Dick Wardle and Don White are thanked for their thorough, constructive reviews of the manuscript. Cambridge Earth Science contribution 4395. Geological Survey of Canada Contribution Number 26895. LITHOPROBE publication number 700.

References

ANSDELL, K. M. & NORMAN, A. R. 1995. U-Pb geochronology and tectonic development of the southern flank of the Kisseynew Domain, Trans-Hudson Orogen, Canada. *Precambrian Research*, **72**, 147–167.

——, KYSER, K., STAUFFER, M. & EDWARDS, G. 1992. Age and source of detrital zircons from the Missi Group: A Proterozoic molasse deposit, Trans-Hudson Orogen, Canada. *Canadian Journal of Earth Sciences*, **29**, 2583–2594.

——, LUCAS, S. B., CONNORS, K. A. & STERN, R. A. 1996. Kisseynew metasedimentary gneiss belt, Trans-Hudson Orogen (Canada): Back-arc origin and collisional inversion. *Geology* (in press).

BABEL WORKING GROUP 1993. Integrated seismic studies of the Baltic Shield using data in the Gulf of Bothnia region. *Geophysical Journal International*, **112**, 305–324.

BAIRD, D. J., KNAPP, J. H., STEER, D. N., BROWN, L. D. & NELSON, K. D. 1995. Upper-mantle reflectivity beneath the Williston basin, phase-change Moho, and the origin of intracratonic basins. *Geology*, **23**, 431–434.

BARAGAR, W. R. A. & SCOATES, R. F. J. 1981. The Circum-Superior Belt: a Proterozoic plate margin? *In:* KRONER, A. (ed.) *Precambrian Plate Tectonics*. Elsevier, Amsterdam, 297–330.

BÉGIN, N. J. 1992. Contrasting mineral isograd sequences

in metabasites of the Cape Smith Belt, northern Québec, Canada: three new bathograds for mafic rocks. *Journal of Metamorphic Geology*, **10**, 685–704.

BEAUMONT, C. & QUINLAN, G. 1994. A geodynamic framework for interpreting crustal scale seismic reflectivity patterns in compressional orogens. *Geophysical Journal International*, **116**, 754–783.

BEST, J. M. 1991. Mantle reflections beneath the Montana Great Plains on COCORP seismic reflection data. *Journal Geophysical Research*, **96**, 4279–4288.

BICKFORD, M.E., COLLERSON, K.D. & LEWRY, J.F. 1994. Crustal history of the Rae and Hearne provinces, southwestern Canadian Shield, Saskatchewan: constraints from geochronologic and isotopic data. *Precambrian Research*, **68**, 1–21.

——, ST-COLLERSON, K. D., LEWRY, J. F., VAN SCHMUS, W. R. & CHIARENZELLI, J. R. 1990. Proterozoic collisional tectonism in the Trans-Hudson orogen, Saskatchawan. *Geology*, **18**, 14–18.

BLEEKER, W. 1990. New structural-metamorphic constraints on Early Proterozoic oblique collision along the Thompson nickel belt, northern Manitoba. *In:* LEWRY, J. F. & STAUFFER, M. R. (eds) *The Early Proterozoic Trans-Hudson Orogen of North America*. Geological Association of Canada, Special Paper, **37**, 57–74.

BRIDGEWATER, D., AUSTRHEIM, H., HANSEN, B. T., MENGEL, F., PEDERSEN, S. & WINTER, J. 1990. The Proterozoic Nagssugtoqidian mobile belt of southeast Greenland: a link between the eastern Canadian and Baltic Shields. *Geocience Canada*, **17**, 305–310.

BURTON, K. W., COHEN, A. S., O'NIONS, R. K. & O'HARA, M. J. 1994. Archaean crustal development in the Lewisian complex of northwest Scotland. *Nature*, **370**, 552–555.

CALVERT, A. J., SAWYER, E. W., DAVIS, W. J. & LUDDEN, J. N. 1995. Archaean subduction inferred from seismic images of a mantle suture in the Superior Province. *Nature*, **375**, 670-674.

CHANDLER, F. W. 1988. The Early Proterozoic Richmond Gulf graben, east coast of Hudson Bay, Quebec. *Geological Survey of Canada, Bulletin*, **362**.

CHOUKROUNE, P. & ECORS TEAM 1989. The ECORS Pyrenean deep seismic profile reflection data and the overall structure of an orogenic belt. *Tectonics*, **8**, 23–39.

COWARD, M. P. & PARK, R. G. 1987. The role of midcrustal shear zones in the Early Proterozoic evolution of the Lewisian. *In:* PARK, F. G. & TARNEY, J. (eds) *Evolution of the Lewisian and Comparable Precambrian High Grade Terrains*. Geological Society, London, Special Publications, **27**, 127–138.

DALY, S., DALY, J. S., MUIR, R. J. & CLIFF, R. A. 1991. A precise U-Pb zircon age for the Inishtrahull syenitic gneiss, County Donegal, Ireland. *Journal of the Geological Society, London*, **148**, 639–642.

DAVEY, F. J. & STERN, T. A. 1990. Crustal seismic observations across the convergent plate boundary, North Island, New Zealand. *Tectonophysics*, **173**, 283–296.

DAVID, J., BAILES, A. & MACHADO, N. 1996. Evolution of the Snow Lake portion of the Paleoproterozoic Flin Flon and Kisseynew belts, Trans-Hudson Orogen, Manitoba, Canada. *Precambrian Research* (in press).

DICKIN, A. P. & BOWES, D. R. 1991. Isotopic evidence for the extent of early Proterozoic basement in Scotland and northwest Ireland. *Geological Magazine*, **128**, 385–388.

DUNPHY, J. M. 1995. *Magmatic evolution and crustal accretion in the Early Proterozoic: The geology and geochemistry of the Narsajuaq terrane, Ungava Orogen, northern Québec*. PhD thesis, Université de Montréal, Montréal, Canada.

EHLERS, C., LINDROOS, A. & SELONEN, O. 1993. The late Svecofennian granite-migmatite zone of southern Finland – a belt of transgressive deformation and granite emplacement. *Precambrian Research*, **64**, 295–309.

FEDOROWICH, J. S., KERRICH, R. & STAUFFER, M. R. 1995. Geodynamic evolution and thermal history of the central Flin Flon Domain Trans-Hudson Orogen: Constraints from structural development, $^{40}Ar/^{39}Ar$, and stable isotope geothermometry. *Tectonics*, **14**, 472–503.

FLACK, C. A. & WARNER, M. R. 1990. Three-dimensional mapping of seismic reflections form the crust and upper mantle, northwest of Scotland. *Tectonophysics*, **173**, 469–481.

——, KLEMPERER, S. L., MCGEARY, S. E., SNYDER, D. B. & WARNER, M. R. 1990. The reflective upper mantle of the British Isles. *Geology*, **18**, 528–532.

FRANCIS, D. M., LUDDEN, J. N. & HYNES, A. J. 1983. Magma evolution in a Proterozoic rifting environment. *Journal of Petrology*, **24**, 556-582.

FROST, C. D. & O'NIONS, R. K. 1985. Caledonian magma genesis and crustal recycling. *Journal of Petrology*, **26**, 515–544.

GAÁL, G. 1990. Tectonic styles of early Proterozoic ore deposition in the Fennoscandian shield. *Precambrian Research*, **46**, 83–114.

—— & GORBATSCHEV, R. 1987. An outline of the Precambrian evolution of the Baltic Shield. *Precambrian Research*, **35**, 15–52.

GORDON, T. M., HUNT, P. A., BAILES, A. H. & SYME, E. C. 1990. U-Pb ages from the Flin Flon and Kisseynew belts, Manitoba: Chronology of crust formation at an Early Proterozoic accretionary margin. *In:* LEWRY, J. F. & STAUFFER, M. R. (eds) *The Early Proterozoic Trans-Hudson Orogen of North America*. Geologic Association, Canada, Special Papers, **37**, 177–199.

GOWER, C. A. 1990. Mid-Proterozoic evolution of the eastern Grenville Province, Canada. *Geologiska Fören. Stockholm Förh.*, **112**, 127–139.

GRAND, S. P. 1994. Mantle shear structure beneath the Americas and surrounding oceans. *Journal of Geophysical Research*, **99**, 11 591–11 621.

GREEN, A. G., MILKEREIT, B., DAVIDSON, A., SPENCER, C., HUTCHINSON, D. R. *ET AL.* 1988. Crustal structure of the Grenville front and adjacent terranes. *Geology*, **16**, 788–792.

HAJNAL, Z., LUCAS, S., WHITE, D., LEWRY, J., BEZDAN, S., STAUFFER, M. & THOMAS, M. 1996. Seismic

Reflection Images of Strike-Slip Faults and Linked Detachments in the Trans-Hudson Orogen. *Tectonics* (in press).

HALL, J. & QUINLAN, G. 1994 A collisional crustal fabric pattern recognised from seismic reflection profiles of the Appalachian/Caledonide orogen. *Tectonophysics*, **232**, 31–43.

HARRIS, A. L., HASELOCK, P. J., KENNEDY, M. J. & MENDUM, J. R. 1995. The Dalradian Supergroup in Scotland, Shetland and Ireland. *In*: GIBBONS, W. & HARRIS, A. L. (eds) *A Revised Correlation of Precambrian Rocks in the British Isles*. Geological Society of London, Special Reports, **22**, 33–53.

HOFFMAN, P. F. 1988. United plates of America, the birth of a craton: early Proterozoic assembly and growth of Laurentia. *Annual Review Earth Planetary Sciences*, **16**, 543–603.

—— 1989. Precambrian geology and tectonic history of North America. *In*: BALLY, A. W. & PALMER, A. R. (eds) *The Geology of North America – An Overview*. Geological Society of America, **A**, 447–512.

—— 1990. Geological constraints on the origin of the mantle root beneath the Canadian shield. *Philosophical Transactions Royal Society London*, **A331**, 523–532.

HUHMA, H. 1986. Sm-Nd, U-Pb and Pb-Pb isotopic evidence for the origin of the Early Proterozoic Svecokarelian crust in Finland. *Geological Survey of Finland Bulletin*, **337**.

——, CLAESSON, S., KINNY, P. D. & WILLIAMS, I. S. 1991. The growth of Early Proterozoic crust: new evidence from Svecofennian detrital zircons. *Terra Nova*, **3**, 175–179.

HURICH, C., SMITHSON, S., FOUNTAIN, D. & HUMPREYS, M. 1985. Seismic evidence of mylonite reflectivity and deep structure in the Kettle Dome metamorphic complex, Washington. *Geology*, **13**, 577–580.

HYNES, A. J. 1991. The gravity field of eastern Hudson Bay: Evidence for a flexural origin for the Hudson Bay (Nastapoka) arc? *Tectonics*, **10**, 722–728.

KALSBEEK, F., AUSTRHEIM, H., BRIDGEWATER, D., HANSEN, B. T., PEDERSEN, S. & TAYLOR, R. N. 1993. Geochronology of the Ammassalik area, South-East Greenland, and comparisons with the Lewisian of Scotland and the Nagssugtoqidian of West Greenland. *Precambrian Research*, **62**, 239–270.

——, PIDGEON, R. T. & TAYLOR, R. N. 1987. Nagssugtoqidian mobile of West Greenland: a cryptic 1850 Ma suture between two Archaean continents – chemical and isotopic evidence. *Earth & Panetary Science Letters*, **85**, 365–385.

KONTINEN, A. 1987. An Early Proterozoic ophiolite – The Jorma mafic-ultramafic complex, northeastern Finland. *Precambrian Research*, **35**, 313–342.

KORJA, T. & HJELT, S-E. 1993. Electromagnetic studies in the Fennoscandian shield – electrical conductivity of the Precambrian crust. *Physics of the Earth & Planetary Interiors*, **81**, 107–138.

KORSMAN, K., HÖLTTA, P., HAUTALA, T. & WASENIUS, P. 1984. Metamorphism as an indicator of the evolution and structure of the crust in eastern Finland. *Geological Survey of Finland Bulletin*, **328**.

KOUSA, J., MARTTILA, E. & VAASJOKI, M. 1996. Petrology, geochemistry, and timing of Paleoproterozoic metavolcanic rocks in the Pyhäjärvi region, central Finland. *Geological Survey of Finland, Special Paper*, in press.

LAMBERT, R. ST. J. & HOLLAND, J. G. 1972. A geochronological study of the Lewisian from Loch Laxford to Durness, Sutherland, N. W. Scotland. *Journal of the Geological Society, London*, **128**, 3–19.

LEWRY, J. F. & COLLERSON, K. D. 1990. The Trans-Hudson orogen: extent, subdivision, and problems. *In*: LEWRY, J. F. & STAUFFER, M. R. (eds) *The Early Proterozoic Trans-Hudson Orogen of North America*. Geological Association Canada, Special Paper, **37**, 1–14.

——, HAJNAL, Z., GREEN, A., LUCAS, S. B., WHITE, D. ET AL. 1994. Structure of a Paleoproterozoic continent-continent collision zone: a LITHO-PROBE seismic reflection profile across the Trans-Hudson Orogen, Canada. *Tectonophysics*, **232**, 143–160.

——, THOMAS, D. J., MACDONALD, R. & CHIARENZELLI, J. 1990. Structural relations in accreted terranes of the Trans-Hudson Orogen, Saskatchewan: Telescoping in a collisional regime? *In*: LEWRY, J. F. & STAUFFER, M. R. (eds) *The Early Proterozoic Trans-Hudson Orogen of North America*. Geological Association Canada, Special Paper, **37**, 75–94.

LIE, J. E. & HUSEBYE, E. S. 1994. Simple-shear deformation of the Skaggerrak lithosphere during the formation of the Oslo Rift. *Tectonophysics*, **232**, 133–141.

LINDSEY, G. & SNYDER, D. 1994. Pre-critical wide-angle reflections from the Baltic Shield: evidence for a 1.8 Ga subduction complex. *Tectonophysics*, **232**, 179–194.

LUCAS, S. B. 1989. Structural evolution of the Cape Smith Thrust Belt and the role of out-of-sequence faulting in the thickening of mountain belt. *Tectonics*, **8**, 655–676.

—— & BYRNE, T. 1992. Footwall involvement during arc-continent collision, Ungava orogen, northern Canada. *Journal of the Geological Society, London*, **149**, 237-248.

—— & ST-ONGE, M. R. 1992. Terrane accretion in the internal zone of the Ungava orogen, northern Quebec. Part 2: structural and metamorphic history. *Canadian Journal of Earth Science*, **29**, 765–782.

—— & —— 1995. Syn-tectonic magmatism and the development of compositional layering, Ungava Orogen (northern Quebec, Canada). *Journal of Structural Geology*, **17**, 475-491.

——, PARRISH, R. R. & DUNPHY, J. M. 1992. Long-lived continent-ocean interaction in the Early Proterozoic Ungava orogen, northern Quebec, Canada. *Geology*, **20**, 113–116.

——, GREEN, A., HAJNAL, Z., WHITE, D., LEWRY, J. ET AL. 1993. Deep seismic profile across a Proterozoic collision zone: Surprises at depth. *Nature*, **363**, 339-342.

——, STERN, R. A. & SYME, E. C. 1996. Intra-oceanic tectonics and the development of continental crust: 1.92–1.84 Ga evolution of the Flin Flon Belt

(Canada). *Geological Society of America Bulletin*, (in press).

——, WHITE, D., HAJNAL, Z., LEWRY, J., GREEN, A. ET AL. 1994. Three-dimensional collisional structure of the Trans-Hudson Orogen, Canada. *Tectonophysics*, **232**, 161–178.

MACHADO, N. 1990. Timing of collisional events in the Trans-Hudson Orogen: Evidence from U-Pb geochronology for the New Quebec Orogen, the Thompson Belt and the Reindeer Zone (Manitoba and Saskatchewan). *In*: LEWRY, J. F. & STAUFFER, M. R. (eds) *The Early Proterozoic Trans-Hudson Orogen of North America*. Geological Association Canada, Special Paper, **37**, 433–441.

——, DAVID, J., SCOTT, D. J., LAMOTHE, D., PHILIPPE, S. & GARIEPY, C. 1993. U-Pb geochronology of the western Cape Smith Belt, Canada: new insights on the age of initial rifting and arc magmatism. *Precambrian Research*, **63**, 211–223.

MARCONTONIO, F., DICKIN, A. P., MCNUTT, R. H. & HEAMAN, L. M. 1988. A 1,800 million-year-old Proterozoic gneiss terrane in Islay with implications for the crustal structure and evolution of Britain. *Nature*, **335**, 62–64.

MCBRIDE, J. H., SNYDER, D. B., TATE, M. P., ENGLAND, R. W. & HOBBS, R. W. 1995. Upper-mantle reflector structure and origin beneath the Scottish Caledonides, *Tectonics*, **14**, 1351–1367.

MCGEARY, S. & WARNER, M. 1985. Seismic profiling the continental lithosphere. *Nature*, **17**, 795–797.

MEYERS, M. T., BICKFORD, M. E. & LEWRY, J. F. 1992. The Wathaman batholith: an Early Proterozoic continental arc in the Trans-Hudson orogenic belt, Canada. *Geological Society of America Bulletin*, **104**, 1073–1085.

MUIR, R. J., FITCHES, W. R. & MALTMAN, A. J. 1992. Rhinns complex: a missing link in the Proterozoic basement of the North Atlantic region. *Geology*, **20**, 1043–1046.

NELSON, K. D. 1991. A unified view of craton evolution motivated by recent deep seismic reflection and refraction results. *Geophysical Journal International*, **105**, 25–35.

——, BAIRD, D. J., WALTERS, J. J., HAUCK, M., BROWN, L. D. ET AL. 1993. Trans-Hudson orogen and Williston basin in Montana and North Dakota: New COCORP deep-profiling results. *Geology*, **21**, 447–450.

NEMETH, B. & HAJNAL, Z. 1996. Moho signature from wide angle reflections: Preliminary results of the 1993 Trans-Hudson Orogen refraction experiment. *Tectonophysics* (in press).

ÖHLANDER, B., SKIÖLD, T., ELMING, S-Å, CLAESSON, S., NISCA, D.H. & BABEL WORKING GROUP 1993. Delineation and character of the Archæan–Proterozoic boundary in northern Sweden. *Precambrian Res*earch, **64**, 67–84.

PARK, A.F. 1991. Continental growth by accretion: a tectonostratigraphic terrane analysis of the evolution of the western and central Baltic Shield, 2.50 to 1.75 Ga. *Geological Society of America Bulletin*, **103**, 522–537.

PARK, R. G. 1994. Early Proterozoic tectonic overview of the northrn British Isles and neighbouring terrains in Laurentia and Baltica. *Precambrian Research*, **68**, 65–79.

—— & TARNEY, J. (eds) 1987. *Evolution of the Lewisian and Comparable Precambrian High Grade Terrains*. Geological Society, London, Special Publications, **27**.

PARRISH, R. R. 1989. U-Pb geochronology of the Cape Smith Belt and Sugluk block, northern Quebec. *Geoscience Canada*, **16**, 126–130.

PATCHETT, P. J. & ARNDT, T. 1986. Nd isotopes of 1.9–1.7 Ga crustal genesis. *Earth & Planetary Science Letters*, **78**, 329–338.

—— & BYLUND, G. 1977. Age of Grenville Belt magnetization: Rb–Sr and paleomagnetic evidence from Swedish dolerites. *Earth & Planetary Science Letters*, **35**, 92–104.

——, —— & UPTON, B. G. J. 1978. Paleomagnetism and the Grenville Orogeny: New Rb–Sr ages from dolerites in Canada and Greenland, *Earth & Planetary Science Letters*, **40**, 349–364.

PERSONEN, L. J., BYLUND, G., TORSVIK, T. H., ELMING, S-Å & MERTANEN, S. 1991. Catalogue of paleomagnetic directions and poles from Fennoscandia: Archean to Tertiary. *Tectonophysics*, **195**, 151–207.

PFIFFNER, O. A., FREI, W., VALASEK, P., STÄUBLE, P., M., LEVATO, L. ET AL. 1990. Crustal shortening in the Alpine orogen: results from deep sesimic reflection profiling in the eastern Swiss Alps, Line NFP 20-east. *Tectonics*, **9**, 1327–1355.

PIPER, J. D. A. 1976. Paleomagnetic evidence for a Proterozoic supercontinent. *Philosophical Transactions of the Royal Astronomical Society*, **A288**, 469–490.

RESTON, T. J. 1988. Evidence for shear zones in the lower crust offshore Britain. *Tectonics*, **7**, 929–945.

ROGERS, G., DEMPSTER, T. J., BLUCK, B. J. & TANNER, P. W. G. 1989, A high-precision U-Pb age for the Ben Vuirich granite: Implications for the evolution of the Scottish Dalradian Supergroup. *Journal of the Geological Society London*, **146**, 789–798.

ROKSANDIC, M. M. 1987. The tectonics and evolution of the Hudson Bay region, *In*: BEAUMONT, C. & TANKARD, A. J. (eds) *Sedimentary Basins and Basin-forming Mechanisms*. Canadian Society of Petroleum Geologists, Memoir, **12**, 507–518.

SCHÄRER, U., KROUGH, T. E., WARDLE, R. J., RYAN, B. & GANDHI, S. S. 1988. U-Pb ages of early and middle Proterozoic volcanism and metamorphism in the Makkovik Orogen, Labrador. *Canadian Journal of Earth Sciences*, **25**, 1098–1107.

SCOTT, D. J., HELMSTAEDT, H. & BICKLE, M. J. 1992. Purtuniq ophiolite, Cape Smith Belt, northern Quebec, Canada: A reconstructed section of Early Proterozoic oceanic crust. *Geology*, **20**, 173–176.

——, ST-ONGE, M. R., LUCAS, S. B. & HELMSTAEDT, H. 1991. Geology and chemistry of the early Proterozoic Purtuniq ophiolite, Cape Smith Belt, northern Quebec, Canada. *In*: PETERS, TJ. (ed.) *Ophiolite Genesis and Evolution of the Oceanic Lithosphere*. Kluwer, Dordrecht, 825–857.

SILVER, P. G. & CHAN, W. W. 1991. Shear-wave splitting and subcontinental mantle deformation. *Journal of Geophysical Research*, **96**, 16429–16454.

SKIÖLD, T. 1988. Implications of new U-Pb zircon chronology to early Proterozoic crustal accretion in northern Sweden. *Precambrian Research*, **38**, 147–164.

SNYDER, D. B. & FLACK, C. A. 1990. A Caledonian age for reflections within the mantle lithosphere north and west of Scotland. *Tectonics*, **9**, 903–922.

SOPER, N. J. 1994. Neoproterozoic sedimentation on the northeastern margin of Laurentia and the opening of Iapetus. *Geological Magazine*, **131**, 291–299.

STAUFFER, M.R. 1984. Manikewan: an Early Proterozoic ocean in central Canada, its igneous history and orogenic closure. *Precambrian Research*, **25**, 257-281.

ST-ONGE, M. R. & LUCAS, S. B. 1991. Evolution of regional metamorphism in the Cape Smith Thrust Belt (northern Québec, Canada): interaction of tectonic and thermal processes. *Journal of Metamorphic Geology*, **9**, 515–534.

—— & —— 1995. Large-scale fluid infiltration, metasomatism and re-equilibration of Archean basement granulites during Paleoproterozoic thrust belt construction, Ungava Orogen, Canada. *Journal of Metamorphic Geology*, **13**, 509–535.

——, & PARRISH, R. R. 1992. Terrane accretion in the internal zone of the Ungava orogen, northern Quebec. Part 1: tectonostratigraphic assemblages and their tectonic implications. *Canadian Journal of Earth Sciences*, **29**, 746–764.

——, & SCOTT, D. J. 1996. The Ungava Orogen and the Cape Smith Thrust Belt. *In*: WIT, M. J. & ASHWAL, L. D. (eds) *Tectonic Evolution of Greenstone Belts*. Oxford Monographs on Geology and Geophysics Series. Oxford University Press. (in press).

STERN, R. A., SYME, E. C., BAILES, A. H. & LUCAS, S. B. 1995a. Paleoproterozoic (1.90–1.86 Ga) arc volcanism in the Flin Flon Belt, Trans-Hudson Orogen, Canada. *Contributions to Mineralogy and Petrology*, **119**, 117–141.

——, —— & LUCAS, S. B. 1995b. Geochemistry of 1.9 Ga MORB- and OIB-like basalts from the Amisk collage, Flin Flon Belt, Canada: Evidence for an intra-oceanic origin. *Geochimica et Cosmochimica Acta*, **59**, 3131–3154.

SYME, E. C. 1990. Stratigraphy and geochemistry of the Lynn Lake and Flin Flon metavolcanic belts, Manitoba. *In*: LEWRY, J. F. & STAUFFER, M. R. (eds) *The Early Proterozoic Trans-Hudson Orogen of North America*. Geological Association Canada, Special Paper, **37**, 143–162.

THOM, A., ARNDT, N. T., CHAUVEL, C. & STAUFFER, M. R. 1990. Flin Flon and western La Ronge belts, Saskatchewan: Products of Proterozoic subduction-related volcanism. *In*: LEWRY, J. F. & STAUFFER, M. R. (eds) *The Early Proterozoic Trans-Hudson Orogen of North America*. Geological Association Canada, Special Paper, **37**, 163–176.

TORSVIK, T. H., SMETHURST, M. A., VAN DER VOO, R., TRENCH, A., ABRAHAMSEN, N. & HALVORSEN, E. 1992. Baltica. A synopsis of Vendian–Permian paleomagnetic data and their paleotectonic implications. *Earth Science Reviews*, **33**, 133–152.

TUISKU, P. & LAAJOKI, K. 1990. Metamorphic and structural evolutional of the Early Proterozoic Puolankajärvi Formation, Finland – II. The pressure–temperature–deformation–composition path. *Journal of Metamorphic Geology*, **8**, 375–391.

VAN BREEMEN, O., AFTALION, M. & PIDGEON, R T. 1971. The age of the granitic injection complex of Harris, Outer Hebrides. *Scottish Journal of Geology*, **7**, 139–152.

VAN KRANENDONK, M. J., ST-ONGE, M. R. & HENDERSON, J. R. 1993. Early Proterozoic tectonic assembly of northeast Laurentia through multiple indentations. *Precambrian Research*, **63**, 325–342.

WARDLE, R. J., RIVERS, T., GOWER, C. F., NUNN, G. A. G. & THOMAS, A. 1986. The northeastern Grenville province: new insights. *In*: MOORE, J. M., DAVIDSON, A. & BAER, A. J. (eds) *The Grenville Province*. Geological Association Canada, Special Paper, **31**, 13–29.

WASSTRÖM, A. 1993. The Knaften granitoids of Västerbotten County, northern Sweden. *In*: LUNDQVIST, T. (ed) *Radiometric Dating results*. Geological Survey of Sweden Research Papers, **C 823**, 60–64.

WEIHED, P., BERGMAN, J. & BERGSTRÖM, U. 1992. Metallogeny and tectonic evolution of the Early Proterozoic Skellefte district, northern Sweden. *Precambrian Research*, **58**, 143–167.

WHITE, D. J. & LUCAS, S. B. 1994. A closer look at the Superior Boundary Zone. *Trans-Hudson Orogen Transect, LITHOPROBE Report*, **38**, 35–41.

——, ——, HAJNAL, Z., GREEN, A. G., LEWRY, J. F. ET AL. 1994. Paleoproterozoic thick-skinned tectonics: LITHOPROBE seismic reflection results from the eastern Trans-Hudson Orogen. *Canadian Journal of Earth Sciences*, **31**, 458–469.

ZWANZIG, H. V. 1990. Kisseynew gneiss belt in Manitoba: stratigraphy, structure, and tectonic evolution. *In*: LEWRY, J. F. & STAUFFER, M. R. (eds) *The Early Proterozoic Trans-Hudson Orogen of North America*. Geological Association Canada, Special Paper, **37**, 95–120.

Tonalite–trondhjemite–granodiorite magmatism and the genesis of Lewisian crust during the Archaean

H. R. ROLLINSON

Department of Geography and Geology, Cheltenham and Gloucester College of Higher Education, Francis Close Hall, Swindon Road, Cheltenham GL50 4AZ, UK

Abstract: This paper reviews the major and trace element and isotopic geochemistry of the Lewisian tonalite–trondhjemite–granodiorite/granite (TTG) gneisses from the mainland Lewisian. The main crust-forming event in the Lewisian was at *c.* 3.0 Ga, although some supracrustal rocks preserve evidence for an earlier history. The granulites of the Central Region of the Lewisian preserve evidence of a younger episode of crust generation and a model is proposed for these rocks in which new crust generated at *c.* 2.6 Ga is contaminated with *c.* 3.0 Ga crust. It is proposed that the 'depletion' in the elements U, Th, K & Rb in the Central Region TTG granulites is original, i.e. related to the process of crust generation rather than of granulite facies metamorphism as previously supposed.

TTG genesis in the Lewisian is the product of the dehydration melting of amphibolite at >18 kb depth and > 800°C. Differences in the levels of U, Th, K & Rb (LILE) between the TTG gneisses of the different regions of the mainland Lewisian is attributed to differences in the partial melting process. LILE-depleted rocks experienced dehydration followed by dehydration melting, whereas LILE-rich rocks are the product of dehydration melting alone.

The Lewisian gneisses of northwest Scotland form the eastern part of the Archaean North Atlantic craton. While they represent only a fraction of the areal extent of the craton and preserve only a limited part of its geological history, they have been subjected to a disproportionate amount of scrutiny – largely as a result of their accessibility and good exposure – and offer important insights into the processes of crust genesis.

The purpose of this paper is to review the geochemistry of the Lewisian gneisses with a view to investigating the more general problem of Archaean crust generation. Samples have been selected in order to avoid the effects of late events such as Proterozoic crustal reworking, in order to elucidate the primary (igneous) crust-forming processes. Over the past three decades a large number of major element, trace element and isotopic studies have been made on the Lewisian gneisses. The results reported here are drawn from a geochemical database, compiled from the literature, for the mainland Lewisian.

Sample-suites have been investigated from the three main areas of the mainland Lewisian (Fig. 1). For the purposes of this study they are defined as:

- the **Northern Region**, north of Loch Laxford, and metamorphosed to amphibolite grade;
- the **Central Region**, the areas around Scourie and Assynt, which preserve a granulite facies mineralogy;
- the **Southern Region**, the area south of Gruinard Bay where the gneisses show evidence of hornblende granulite facies metamorphism (Field 1978) to Torridon where they are metamorphosed to amphibolite grade.

The Lewisian gneisses are particularly instructive in developing models for the generation of Archaean continental crust for four reasons: (i) their geochemistry is well characterized; (ii) the gneiss complex has been mapped in some detail and so the geological relationships between analysed units are known; (iii) the timing of the crust forming events is moderately well known; and (iv) variations in metamorphic grade within the gneiss complex suggest that more than one crustal level is preserved, thus a comparison can be made between the processes of lower crust (granulite facies) and middle crust (amphibolite facies) generation.

Comparative major element geochemistry

The dominant rock-types of Archaean age in the Lewisian are igneous in origin and belong to the tonalite–trondhjemite–granodiorite suite (TTG) of magmas, typical of many Archaean Cratons. Older supracrustal rocks (metasediments amphibolites and mafic granulites) are ubiquitous but form a small proportion of the complex as a whole. Layered intrusions comprising ultramafic rocks

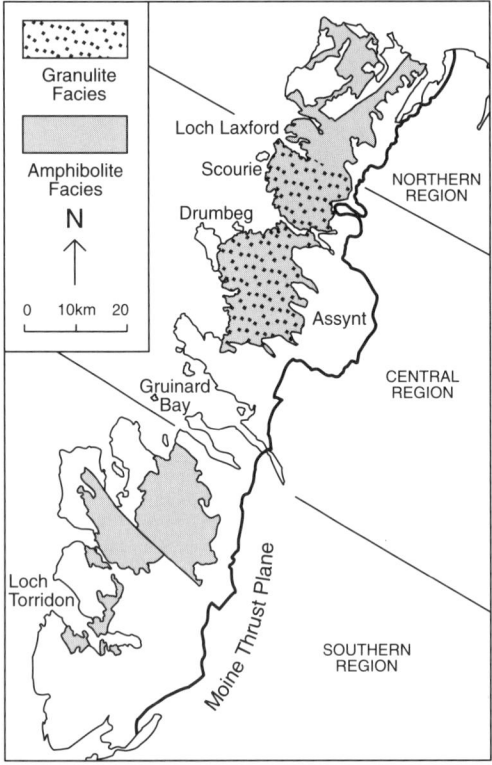

Fig. 1. Geological map of the Lewisian gneisses of northwest Scotland showing the regional subdivisions used in this paper.

Gneisses from the Northern Region have silica contents in the range 60–72 wt% and feldspar normative compositions which cluster around the tonalite–trondhjemite–granodiorite 'triple point' (Fig. 2a), typical of TTG suites from many Archaean terrains (Martin 1993). TTG gneisses dominate the area although amphibolite bodies and rare metagreywacke sediments are also present in the Laxford Front area (Davies 1976).

Rocks from the Central Region can be subdivided into two groups on the basis of their field relationships. 'Older tonalitic gneisses' form more than 90% of the outcrop of felsic rocks in this area and these are intruded by younger 'felsic sheets'. Older tonalitic gneisses vary in composition from diorite to trondhjemite (55–77 wt% SiO_2). Older gneisses with granodioritic and granitic compositions have not been found. On a normative Ab–An–Or plot samples from an area of about 8 km^2 around Scourie plot exclusively in the tonalite field, whereas gneisses from a larger area (about 80 km^2) around Drumbeg range from tonalitic to trondhjemitic in composition (Fig. 2b). This may reflect compositional differences between individual igneous intrusions in the protolith to the gneiss complex. Younger felsic sheets plotted in Fig. 2b vary in composition from tonalite to granite (SiO_2 65–77 wt%) and are thought to be related to one another by fractional crystallization (Rollinson 1994). Tarney & Weaver (1987) point out that this region contains a high proportion of mafic and ultramafic masses, associated with which are rare metasediments, and that the tonalitic gneisses contain vast numbers of smaller mafic/ultramafic enclaves.

TTGs from the Southern Region can also be subdivided into two suites on the basis of their age. A group of older gneisses plot principally as tonalites (SiO_2 = 61–68 wt%), whereas younger gneisses plot on a trondhjemite–granodiorite (SiO_2 = 69–75 wt%) trend (Fig. 2c). In contrast to the Central Region, however, in the Gruinard Bay and Torridon areas the dominant rock-types are the trondhjemite–granodiorite–granite suite. The older tonalitic gneisses contain amphibolite inclusions (Rollinson 1987) and form only a small fraction of the total gneiss complex. Gneisses of dioritic composition are also reported by Fowler (1986).

Thus, the principal difference in bulk composition between the regions is reflected in the proportion of tonalitic (and dioritic) rocks present. Tonalites and diorites are most common in the Central Region and yet form only a minor component in the gneisses of the Southern Region. In the Northern Region diorites have not been recorded and tonalites form a small part of the gneiss complex. Sheraton et al. (1973) argued that these differences were primary.

and gabbros intrude and are intruded by the TTG suite but form only a minor part of the Gneiss Complex.

Rock compositions calculated in terms of their normative feldspar content are plotted in Fig. 2, on the Ab–An–Or classification diagram of O'Connor (1965) as modified by Barker (1979). The majority of samples plot in the tonalite and trondhjemite fields although rocks with granodiorite and granitic compositions are also present. Strictly, tonalites with high normative An-contents are more correctly classified as diorites and quartz-diorites on the normative Q'-ANOR classification diagram of Streckeisen & Le Maitre (1979).

Previous workers (notably Holland & Lambert 1973; Sheraton et al. 1973) have demonstrated that there are compositional differences between the three regions of the mainland Lewsian investigated in this study. These differences are chiefly in the levels of Na, Ca and K and are illustrated here by means of the sample CIPW-normative feldspar composition in Ab–An–Or plots (Fig. 2) and are briefly described below.

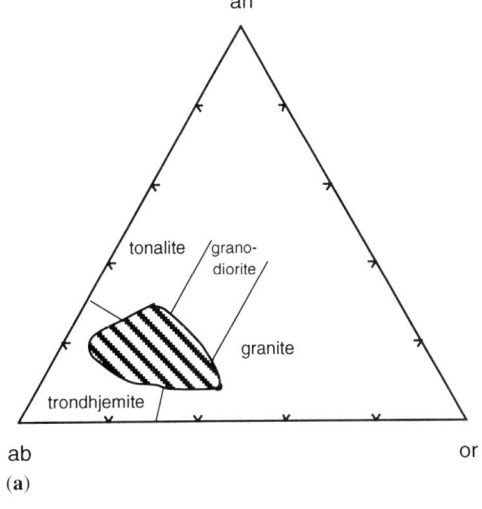
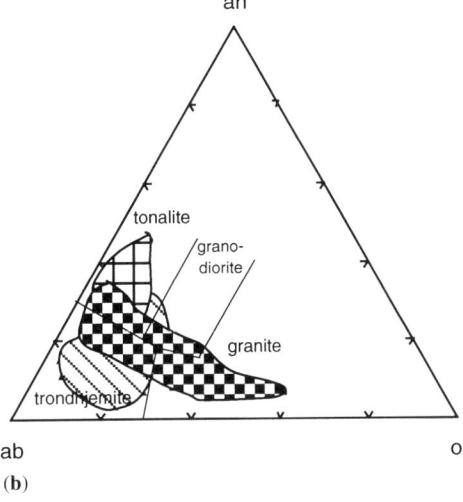
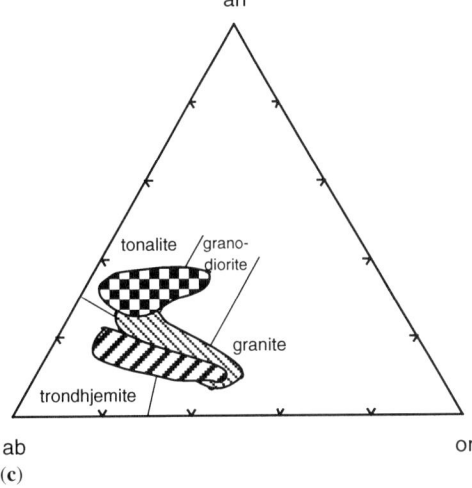

Fig. 2. Normative feldspar compositions for the Lewisian gneisses plotted on the O'Connor (1965) classification diagram using the field boundaries of Barker (1979). (a) Northern Region gneisses (data from Weaver & Tarney 1981; Rollinson unpubl.); (b) Central Region gneisses – cross-grid shading, gneisses from the Scourie area (data from Rollinson 1978), diagonal shading, gneisses from the Assynt area (data from Weaver & Tarney 1980), check shading, felsic sheets at Scourie (data from Rollinson 1978); (c) Southern Region gneisses – check shading, older gneisses at Gruinard Bay, light diagonal shading, main gneiss suite at Gruinard Bay, dark diagonal shading, main gneiss suite at Torridon (data from Rollinson 1978).

Comparative trace element chemistry

Rare earth element geochemistry

A compilation of rare earth element (REE) patterns for gneisses from the Northern, Central and Southern regions of the Lewisian is presented in Fig. 3(a–n). The REE patterns are subdivided by region (north to south) and also into groups according to rock type (this broadly correlates with silica content) and steepness of the REE pattern, measured here by the normalized La/Yb ratio.

Two general observations may be made of the data presented in Fig. 3. Firstly, there is a simple pattern of increasing steepness of REE pattern with increasing silica content. Secondly, REE patterns for a given rock composition are very similar throughout the three regions of the gneiss complex. Differences between the regions are more a function of the differing proportions of rock-types found than in the chemistry of the gneisses themselves (cf. Weaver & Tarney 1981).

The largest and most complete REE dataset is for the Central Region and these are discussed first. Diorites and quartz-diorites (Fig. 3d–e) have relatively flat REE patterns $(La/Yb)_N = 4.4$–7.2 in the lower range of silica contents (55–58 wt% SiO_2) increasing to $(La/Yb)_N = 12.5$ 34.5 at $SiO_2 = 57$–62 wt%. Tonalites ($SiO_2 = 62$–68 wt%) have REE patterns similar to the more felsic diorites, although with slightly lower total REE abundances (Fig. 3f).

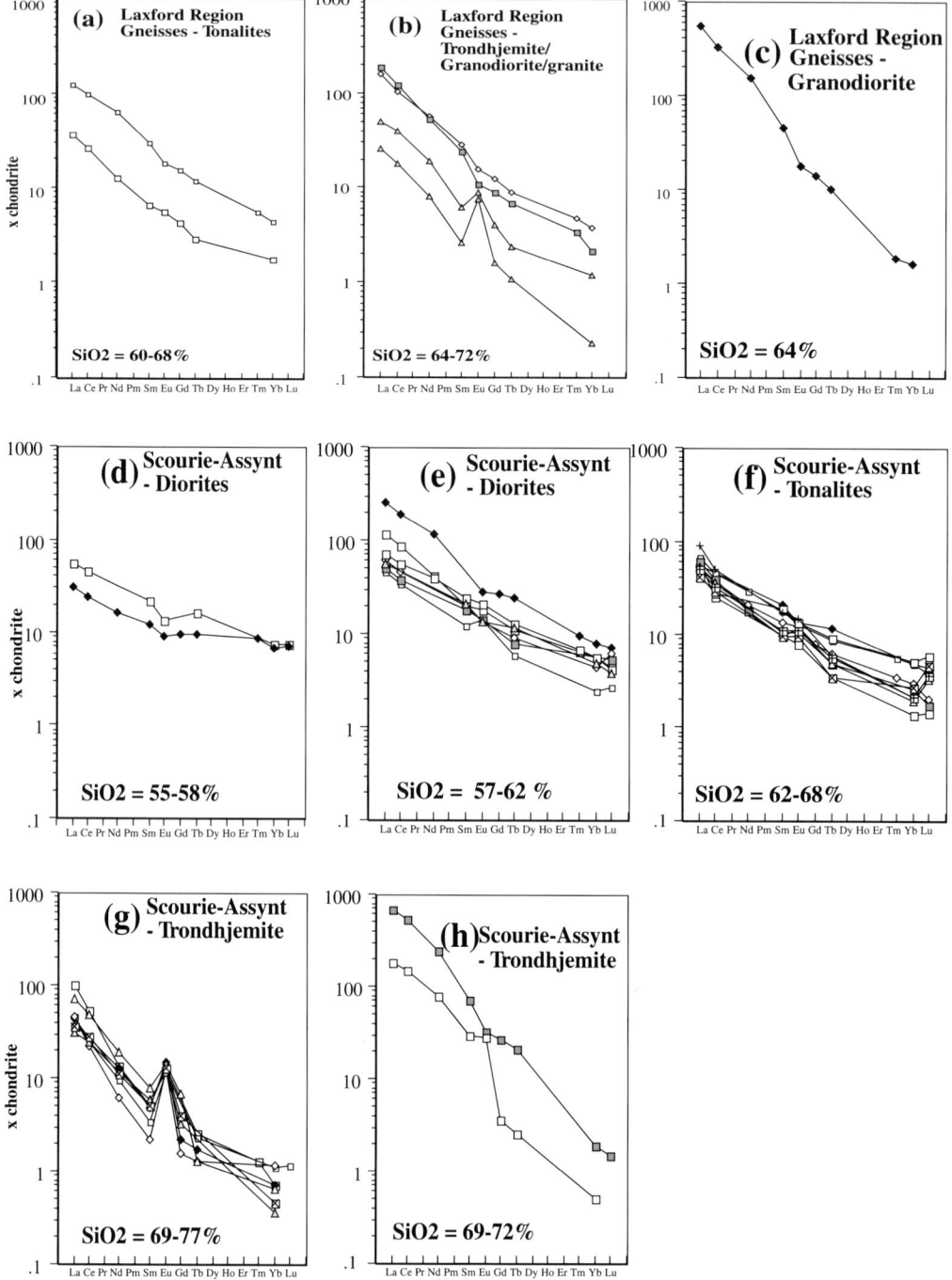

Trondjemites (SiO$_2$ = 69–77 wt%) are subdivided into two groups. Most samples have steep REE patterns (Fig. 3g, (La/Yb)$_N$ = 39–140) and marked positive Eu anomalies; a few samples, however, have very steep patterns (Fig. 3h (La/Yb)$_N$ = 350–370), higher total REE concentrations and a less marked Eu anomaly. Felsic sheets of TTG composition intrusive into the main TTG gneisses are

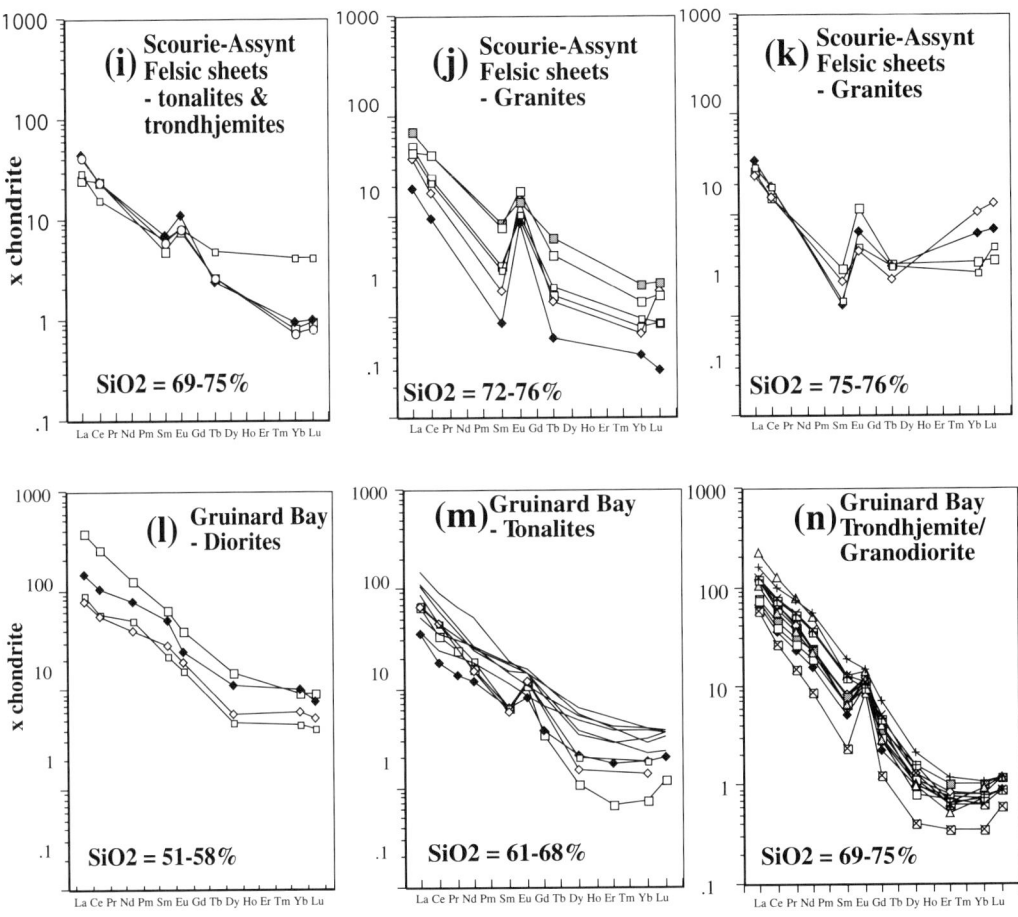

Fig. 3. Chondrite normalized REE plots from the Lewisian gneisses classified according to rock-type and region. (**a–c**) Northern Region – data from Weaver & Tarney (1981); (**d–k**) Central Region – data from Rollinson (1994), Pride & Muecke (1980, 1982), Weaver & Tarney (1980); (**l–n**) Southern Region – data from Rollinson & Fowler (1987), Fowler & Plant (1987). (Normalizing values from Nakamura 1974.)

subdivided here into tonalites and trondhjemites (SiO_2 = 69–75 wt%) and granites (SiO_2 = 72–76 wt%). REE patterns for the tonalite–trondhjemite group have $(La/Yb)_N$ ratios in the range 7.3–56.8 and show positive Eu anomalies (Fig. 3i). REE patterns for the granites are subdivided into those with steep patterns ($(La/Yb)_N$ = 15.3–31.8) and large positive Eu anomalies (Fig. 3j), and those enriched in heavy REE (Fig. 3k) as a consequence of garnet and zircon accumulation (Rollinson 1994).

Gneisses from the Northern Region conform to the general pattern outlined above for the Central Region (Fig. 3a–c). Rocks of dioritic composition have not been described from this area but tonalites (SiO_2 = 60–68 wt%) have moderately steep REE patterns $(La/Yb)_N$ = 20–26, whereas trondhjemites and granodiorites (SiO_2 = 64–72 wt%) fall into two groups. Most rocks have steep REE patterns ($(La/Yb)_N$ = 41–112) and with decreasing total REE content develop a positive Eu anomaly. One sample, however, has a much higher total REE content and a very steep REE pattern ($(La/Yb)_N$ = 346).

In the Southern Region, gneisses of dioritic composition (SiO_2 = 51–58 wt%) have high total REE concentrations and moderately steep patterns ($(La/Yb)_N$ = 13–42) (Fig. 3l). Tonalites (SiO_2 = 61–68 wt%) have REE patterns with a similar steepness, but with lower total REE concentrations and which develop a positive Eu anomaly at lower REE concentrations (Fig. 3m). A similar feature is noted for the trondhjemites and granodiorites (SiO_2 = 69–75 wt%), although in these rocks the REE patterns are steeper (Fig. 3n, $(La/Yb)_N$ = 78–313).

Weaver & Tarney (1981) argued that the REE geochemistry in the Northern and Central regions of the Lewisian are very similar. Here, this observation is extended to include the Southern Region and it is concluded that the mainland Lewisian is uniform with respect to its REE chemistry. The small differences between the regions are a function of the different proportions of rock types present.

Primitive mantle normalized multi-element diagrams

Primitive mantle normalized multi-element diagrams (spiderdiagrams) have been plotted for selected trace elements from TTGs from the Northern, Central and Southern regions of the Lewisian gneisses using the normalizing values of Sun & McDonough (1989). A plot of normalized arithmetic mean values is given in Fig. 4 for TTGs from each region. These are terrain averages and, in part, reflect a bias towards the dominant rock-type in each region. Nevertheless, a similar result is obtained when the mean values for the tonalites from each region are compared, indicating that the apparent differences between the regions are real. Two important observations can be made from these plots, as follows.

1. The depletion of 'mobile' elements. TTGs from the Central Region are depleted in the elements K, Rb, Th, and U relative to TTGs from the Northern and Southern regions. Whitehouse (1989b) has shown that the Central Region gneisses are also depleted in Pb. Element depletion in the Central Region granulites has been discussed extensively by Sheraton *et al.* (1973), Rollinson & Windley (1980b), Weaver & Tarney (1980, 1981). Several authors (see, for example, Moorbath *et al.* 1969 and Whitehouse 1989b) have linked the depletion of K, Rb, U and Th to the granulite facies metamorphism of the Central Region. There are two principal lines of evidence. Firstly, the study of Pb isotopes in the Central Region indicates that the granulite facies TTGs crystallized at 2680 Ma from a strongly U-depleted source (Moorbath *et al.* 1969; Chapman & Moorbath 1977). This granulite facies crystallization is 250 Ma younger than recorded crust-formation ages for other parts of the gneiss complex. It is commonly assumed, therefore, that TTG gneisses in the Central Region formed at the same time as the gneisses in the Northern and Southern regions (at about 2900–2950 Ma) but were metamorphosed to granulite grade at 2680 Ma during which event there was substantial U-loss (together with the loss of K, Rb and Th). Secondly, Fowler (1986) showed that mean values for TTGs from hornblende granulites at Gruinard Bay show depletion levels intermediate between the extremes of the highly depleted Central Region granulites and the undepleted amphibolite facies gneisses of the Northern Region. He argued that element depletion is correlated with metamorphic grade.

An alternative model was proposed by Tarney & Weaver (1987) who showed that sediments within the Central Region had 'normal' K/Rb ratios and yet zircons in the tonalitic gneisses had crystallized

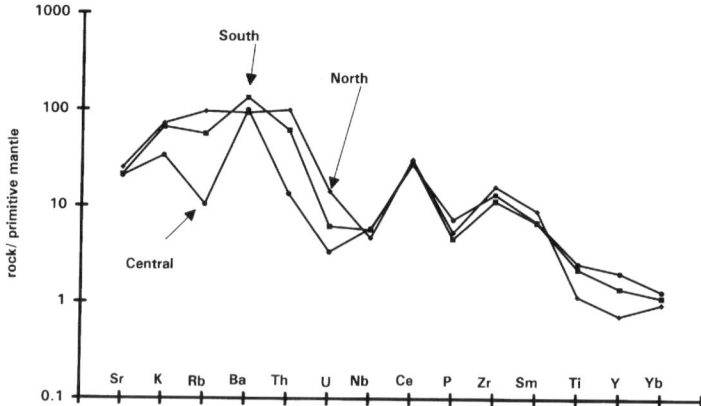

Fig. 4. Primitive mantle normalized multi-element diagrams for Lewisian TTG gneisses. Mean values are plotted for amphibolite facies Northern Region (North), granulite facies Central Region older gneisses (Central), amphibolite facies Southern Region (South). Normalizing values are from Sun & McDonough (1989). Data for U are taken from Fowler (1986).

in a very U-poor environment. They suggested that element depletion in the Central Region granulites was original and part of the process of crust formation. Evidence presented elsewhere in this paper supports the view of Tarney & Weaver (1987) and their argument is developed more fully below.

2. *The similarity of 'immobile' elements.* The immobile high field strength elements (Nb–Yb on Fig. 4) show similar values for TTGs from each of the three regions. Fowler (1986) made a similar observation using Harker diagrams. This is important since it has been suggested that the REE, in particular the elements Nd and Sm, were fractionated during granulite facies metamorphism (Whitehouse 1988). Figure 4 shows that Ce and Sm values are very similar for all regions and does not support the idea of the mobility of the light and middle REE during granulite facies metamorphism – a view which is also supported by the Nd isotopic studies of Cohen *et al.* (1991). These authors argued that the wide range of ε_{Nd} values in the gneisses of the Scourie area at 2.7 Ga indicate minimal exchange of Nd between the different closely associated rock-types during metamorphism. Concentrations of Ti and Yb are lower for the Northern Region TTGs and this may reflect the more granodioritic and less ferromagnesian-rich bulk composition of the terrain.

Geochronology and isotope geochemistry

It is now possible, using the recent Pb–Pb, U–Pb and Nd isotope studies of Whitehouse (1989*a*, *b*), Cohen *et al.* (1991), Burton *et al.* (1994), Corfu *et al.* (1994) and Kinny & Friend (1994) to make comparisons between the three regions of the Lewisian on the Scottish Mainland.

Nd isotopic studies

All the available Nd isotope data for the Lewisian are plotted on an ε_{Nd} versus time diagram in Fig. 5. This diagram compares the Nd-isotopic composition of a sample with that of possible mantle sources at the time of formation and shows the change in Nd-isotopic composition from formation to the present. Two models for the isotopic evolution of the mantle are shown – (i) the CHUR model in which the primitive mantle is assumed to be undifferentiated and chondritic in character (CHUR has a constant zero value on an ε_{Nd} versus time diagram); (ii) the depleted mantle (DM) model in which the composition of the mantle is assumed to be light REE depleted and increases in ε_{Nd} towards the present. In this study, the depleted mantle curve of Goldstein *et al.* (1984) is preferred to that of DePaolo (1981) used in earlier studies (cf. Nagler & Stille 1993).

The following inferences are made about the geological and Nd isotopic history of the Lewisian from the ε_{Nd} versus time diagram:

- Sm–Nd isochron ages for the TTG gneisses of the Northern Region and the Southern Region (Whitehouse 1989*a*) indicate that the main crust-forming event in the Lewisian, was at about 2.96 Ga. Independent evidence to support a major mantle differentiation event at *c.* 3.0 Ga comes from the work of Waters *et al.* (1990) who showed that the 2.0 Ga Scourie dykes from the Central Region sampled ancient lithosphere which was isolated from the convecting mantle at 3.08 Ga.

- ε_{Nd} values calculated from isochrons for the source region of the Northern Region and the Southern Region TTG gneisses (Whitehouse 1989*a*) plot close to the depleted mantle curve and imply a depleted mantle source. A similar result was obtained by Waters *et al.* (1990) who proposed a 3.08 Ga mantle source for the Scourie dykes with an $\varepsilon_{Nd} = 4.18 \pm 1.57$ (see Fig. 5).

- Layered gabbro-peridotite intrusions from the Central Region formed between 2.91 and 2.71 Ga (Whitehouse 1989*a*). Similar intrusions from South Uist also formed in this time interval (Whitehouse 1993). The ε_{Nd} values calculated from the isochrons for these intrusions are lower than those expected from the Goldstein *et al.* (1984) depleted mantle curve. It is possible that they were derived from a mantle source region which was different from the precursor to the TTGs. Alternatively, the gabbro-peridotite intrusions are contaminated with Nd derived from older continental crust (Cohen *et al.* 1991).

- If a depleted mantle source is assumed for metasediments from Scourie in the Central Region, then model ages calculated using the data of Cohen *et al.* (1991) are between 3.1 to 3.23 Ga. This suggests that there was old continental crust in this region which predates the main crust-forming event at *c.* 3.0 Ga. Support for the existence of older crustal material also comes from Nd mineral isochron ages of *c.* 3.3 Ga (Burton *et al.* 1994) for amphibolites from Gruinard Bay.

- The young age (*c.* 2.6 ± 0.155 Ga) and the negative ε_{Nd} value (= –2.4 ± 1.9) for the gneisses of the Central Region make them anomalous amongst the TTG gneisses (Whitehouse 1988, 1989*a*) and suggest that the crust-forming process in the Central Region was different from that in the Northern and Central regions. Two explanations of the very low ε_{Nd} have been proposed: (1) Whitehouse (1988) explained the

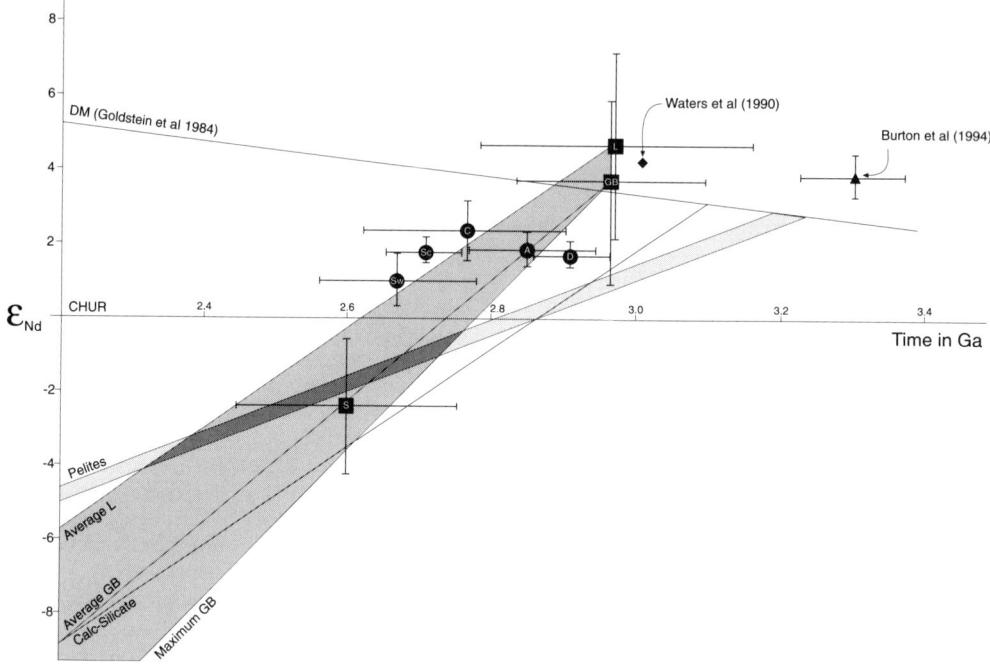

Fig. 5. ε_{Nd} versus time (in Ga) diagram for the Lewisian gneisses. CHUR – mantle ε_{Nd} evolution based upon a chondritic model for the earth; DM (solid line) – depleted mantle ε_{Nd} evolution after Goldstein *et al.* (1984). Filled squares – whole-rock isochron ages for TTG gneisses from Laxford (L) in the Northern Region, Gruinard Bay (GB) in the Southern Region and Scourie (S) in the Central Region. Filled circles – whole-rock isochron ages for mafic – ultramafic intrusions at Scourie (Central Region) (Sw, data from Whitehouse 1989*a*); Sc, data from Cohen *et al.* 1991), Corodale, South Uist (C), Achiltibuie (Central Region) (A) and Drumbeg (Central Region) (D). Filled triangle – mineral isochron for amphibolite from Gruinard Bay (Southern Region) from Burton *et al.* (1994). Filled diamond – mantle source region sampled by Scourie dykes (from Waters *et al.* 1990). The shaded area (light stipple) shows the evolution of ε_{Nd} in the TTGs from the Northern and Southern regions with time. The shaded area (dark stipple) and the calc-silicate line show the ε_{Nd} evolution of Lewisian sediments with time. Data from Hamilton *et al.* (1979), Whitehouse (1989*a*, 1993), Cohen *et al.* (1991), Burton *et al.* (1994) and Waters *et al.* (1990).

anomaly in terms of Sm/Nd fractionation during granulite facies metamorphism; (2) Cohen *et al.* (1991) suggested that it represents evidence for an earlier crustal history. This could imply the melting of older crust, mixing with new crust either in the source or during emplacement (cf. Arndt & Goldstein 1987).

Pb isotopic studies

TTGs from the Northern Region of the Lewisian have a Pb isochron age of 2950 ± 70 Ma and were derived from a low U/Pb source ($\mu_1 = 7.82$) (Whitehouse & Moorbath 1986; Whitehouse 1989*b*). This age agrees with Sm–Nd isochron and Nd–DM model ages calculated for these gneisses and is thought to represent the time of crust formation. This interpretation is consistent with that of Whitehouse & Moorbath (1986), but not with that of Whitehouse (1989*b*) who regarded the 2950 Ma Pb–Pb isochron for the Northern Region TTG gneisses as spuriously old and an artefact of regional proportional Pb loss from the gneisses.

A Pb isotope errorchron (MSWD = 4.3) of 2800 ± 200 Ma ($\mu_1 = 7.71$) for TTGs from the Southern Region (Whitehouse 1989*b*) is lower than but, within error of, the Sm–Nd isochron age and T_{DM} Nd model ages for this area. This result does not negate the crust formation age of about 2950 Ma indicated by the Nd isotope data above.

Lead isotope studies on TTGs from the Central Region yield isochron ages of 2680 ± 60 Ma ($\mu_1 = 7.7$) (Chapman & Moorbath 1977) and 2790 ± 180 Ma ($\mu_1 = 7.78$) (Whitehouse 1989*b*). These isochron ages are anomalous inasmuch as they indicate a younger age than that for the Northern and Southern regions, although U/Pb ratios are the same in all regions. The data of Chapman & Moorbath (1977) and Whitehouse (1989) are plotted together with data points for

TTGs from Chapman (1978) and Cohen et al. (1991) on a $^{207}Pb/^{204}Pb$ versus $^{206}Pb/^{204}Pb$ isochron diagram (Fig. 6) and shown relative to a single-stage lead isotope evolution curve with a μ_1 value ($^{238}U/^{204}Pb$) of 7.75. When examined in detail the results show a scatter of values between reference lines for 2.65 Ma and 2.95 Ma. There are two possible interpretations of these results. Previous workers (see Whitehouse 1989b) have argued that gneisses which formed at c. 2.95 were metamorphosed to granulite grade at c. 2.65 Ga. Pb-isotope homogenisation (together with U and Pb loss) during granulite facies metamorphism can account for the present day distribution of lead isotope ratios. Alternatively, after the method of Taylor et al. (1980), the $^{207}Pb/^{204}Pb$ versus $^{206}Pb/^{204}Pb$ plot (Fig. 6) can be interpreted as the contamination of 2.65 Ga crust with older 2.95 Ga crust, of the type found in the Northern and Southern regions of the Lewisian.

U–Pb isotopic studies of zircon

U–Pb single zircon studies from the Northern Region, near Laxford Bridge and the Central Region, between the Laxford Front and Badcall are reported by Corfu et al. (1994). These authors show that:

(1) zircon from granulite facies gneisses in the Central Region has an age of 2711 Ma. This result is regarded as more precise than the zircon size-fraction age of 2660 ± 20 Ma from the same region (Pidgeon & Bowes 1972). There is also an indication of the presence of older material (2790 +50/–40 and 2790 +220/–170 Ma);
(2) zircon from the 'later' felsic sheets in the Central Region has an age of 2720 +40/–30 Ma;
(3) zircons from the Northern Region indicate an apparent age of 2882 Ma;
(4) there was new zircon growth in the Central Region between 2480 and 2490 Ma. This is interpreted as a high-grade metamorphic event.

Preliminary results from ion-probe studies of U–Pb in zircons from the Scourie area of the Central Region by Kinny & Friend (1994) show a scatter of ages along the concordia curve between 2500 Ma and 2950 Ga. The 2950 Ma event is interpreted as that of tonalitic magmatism and crust formation whereas the 2500 Ma event is interpreted as the timing of granulite facies metamorphism. Zircons from the Northern Region show maximum ages of 2850 Ma although a single core has an age of 3400 Ma. The 2500 Ma event is less obvious in this region.

These results are consistent with the existence of 'old' (> 3.0 Ga) crust in the Northern Region, crust formation in the Northern and Central regions between c. 2.95 Ga and c. 2.7 Ga, and highlight the importance of an event at about 2.5 Ga previously recognised by Humphries & Cliff (1982).

Crust-forming processes in the Lewisian

The timing of crust formation

The main crust-forming event in the Lewisian was at 3.0 Ga. This event is recognized in all regions of

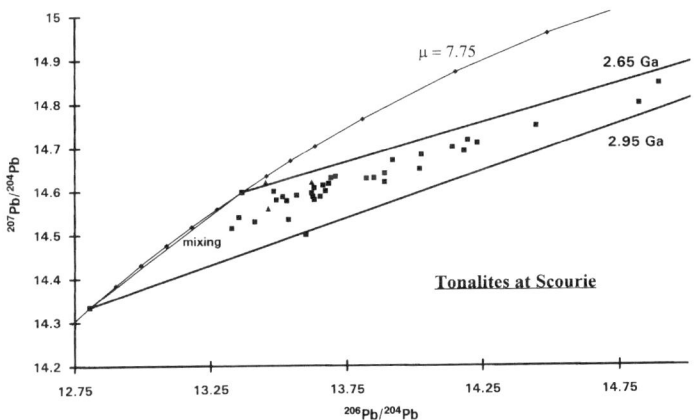

Fig. 6. $^{207}Pb/^{204}Pb$ versus $^{206}Pb/^{204}Pb$ for TTG gneisses from the Scourie area of the Central Region, showing a single-stage Pb isotope evolution curve with a μ-value of 7.75, isochrons at 2.95 and 2.65 Ga and a mixing line for 2.65 magmas contaminated with 2.95 Ma crust. Most data points define a line with a slope parallel to the 2.65 Ga isochron but plot below the isochron. This suggests contamination with older crust. Data from Chapman & Moorbath (1977), Chapman (1978), Whitehouse (1989b) and Cohen et al. (1991).

the Lewisian but is particularly well developed in the Northern and Southern regions. TTG crust was derived from a mantle source with $\varepsilon_{Nd} = +4$ (Fig. 5) and $\mu_1 = 7.75$. Indirect evidence for this crust-forming event comes from the work of Waters *et al.* (1990) who showed that the 2.0 Ga Scourie dykes sampled ancient lithosphere ($\varepsilon_{Nd} = 4.18 \pm 1.57$) which was isolated from the convecting mantle at 3.08 Ga. There is evidence from Nd, Pb and U–Pb isotopic systems to suggest that older crust, pre-dating the 3.0 Ga event was present in all regions of the Lewisian. This older crust has not been well characterized.

The timing of crust formation in the Central Region is more complex. It is argued here on the basis of Nd and Pb isotopes that the crust of the Central Region is younger and crystallized between 2.6 and 2.7 Ga. In addition it is geochemically anomalous and is characterized by: (i) a low $\varepsilon_{Nd} = -2.4 \pm 1.9$; (ii) a scattered distribution of data points on a $^{207}Pb/^{204}Pb$ versus $^{206}Pb/^{204}Pb$ plot between 2.65 and 2.95 Ga reference lines; (iii) a range of U–Pb zircon crystallization ages between 2.95 and 2.71 Ga; (iv) low concentrations of U, Th, K, Rb and Pb.

Previous workers have interpreted the younger age and anomalous geochemistry as indicative of the granulite facies metamorphism at between 2.6 and 2.7 Ga, of *c.* 3.0 Ga crust. It has been argued that during this metamorphism the elements Sm and Nd were fractionated, U and other LIL elements were removed and that the Pb and Nd isotope systems reset. It has been shown above that the redistribution of Sm and Nd is considered unlikely, a view also supported by Cohen *et al.* (1991) and this has led to an alternative explanation of the low ε_{Nd} value. However, an alternative explanation of the Nd isotopes also provides a new way to interpret other features of the Central Region TTGs.

Low ε_{Nd} values in TTG gneisses can be taken to indicate the involvement of older continental crust in the process of TTG genesis. Taken alone, the low ε_{Nd} for the Central Region TTGs could indicate remelting of older (3.0 Ga) felsic crust (Fig. 5). However, this explanation is not favoured because the REE would show prominent negative Eu anomalies, U and Rb levels would be higher and the μ_1 value would also be higher. None of these features are observed and a preferred model is the contamination of new crust by old, after the manner described by Arndt & Goldstein (1987). Crustal contamination of newly generated 2.65 Ga crust by older 3.0 Ga crust can explain the low ε_{Nd} value, the scatter of Pb isotopic compositions (see Taylor *et al.* 1980) and the range of U–Pb zircon ages.

Field evidence for crustal components of different ages in the Central Region of the Lewisian was reported by Davies (1975). This work was discounted after Chapman & Moorbath (1977) showed from a Pb-isotopic study of supposed 'older gneisses' that there was no evidence in the Lewisian for the existence of 'very old' 3.7 Ga crust of the type found in west Greenland. What was not considered in the Chapman & Moorbath (1977) study was evidence for the more subtle contamination of *c.* 2.65 Ga crust with older 2.95 Ga crust. The interpretation of the isotopic evidence presented here suggests that the 'older gneisses' described by Davies from the Central Region are significant.

If the interpretation presented here is correct and the crust in the Central Region is younger than other parts of the gneiss complex then the crust-formation age of *c.* 2.6–2.7 Ga is the same as that previously attributed to the Badcallian granulite facies metamorphism. This raises raises the possibility that the processes of crust-formation and granulite facies metamorphism are interrelated in this area (cf. Tarney & Weaver 1987). It should be noted that the interpretation of the Lewisian proposed here follows that of Sheraton *et al.* (1973) but is the converse of that suggested by Sutton & Watson (1951) who regarded the Northern Region as younger than the Central Region.

Thus it is concluded that whilst the entire crust-forming processes in the Lewisian spanned 600 Ma, between *c.* 3.4 Ga and *c.* 2.65 Ga, that there were two principal crust-forming events one at *c.* 2.95 Ga and one at *c.* 2.65 Ga. The isotopic data used here do not support crust formation at 2.4 Ga (Burton *et al.* 1994), a point also made by Fowler *et al.* (1995). The status of the granulite facies event(s) in the Central Region is uncertain. It is proposed here that it was part of the process of crust formation at *c.* 2.65 Ga, although the data of Corfu *et al.* (1994) and Kinny & Friend (1994) suggest that it may have been later at 2.48–2.49 Ga.

Processes of crust formation

The results of recent experimental studies (Rapp *et al.* 1991; Rushmer 1991) show that TTG liquids are generated by the partial melting of amphibolite in equilibrium with a garnet amphibolite or eclogite residue. According to this model TTG melts are produced from the mantle in two stages. Firstly, basalt is produced by the partial melting of depleted mantle. This is followed by the hydration and partial melting of the basalt to form TTG liquids.

In this section, Nd isotopic data and the REE chemistry are used to constrain the processes by which the Lewisian TTG's were generated at 3.0 Ga. Insights from other trace elements are discussed in the following section. Following

Rollinson & Fowler (1987), the emphasis is upon samples from the Southern Region because this is where potential protolith material is best preserved and where crust generation is not made complex by the possibility of contamination with pre-existing crust. The aim is to identify the residual mineralogy in equilibrium with the TTG melts and to estimate the likely P–T conditions at which melt and residue co-existed.

The REE data presented in Fig. 3 are summarized in a normalized (La/Yb) versus Yb plot (Fig. 7), in which total REE concentration (represented by Yb_N) is plotted against the steepness of the REE pattern $(La/Yb)_N$ This diagram shows the asymptotic trend, typical of Archaean TTGs, first described by Martin (1986).

An amphibolitic source for the TTG suite. It was shown above from Nd isotope studies that the Lewisian TTG gneisses in the Northern and Southern regions were derived from a depleted mantle source. The melting of a depleted mantle source is modelled on a $(La/Yb)_N$ versus Yb_N plot in Fig. 8. A comparison of the mantle melting curves in Fig. 8 and the data for the TTG gneisses in Fig. 7 shows that the $(La/Yb)_N$ ratios are too high and the Yb_N values too low for the TTG suite to be the product of direct partial melting of depleted mantle. However, three amphibolites described by Rollinson (1987) from Gruinard Bay do plot close to the calculated mantle melting curves in Fig. 8, suggesting that they are partial melts of depleted mantle. Sample 104 approximates to a small degree of partial melting of garnet lherzolite, whereas sample 115 has a composition which approximates to a higher percentage of melting (c. 30%). Sample 146 has a composition consistent with about 20% partial melting of spinel lherzolite with depleted LREE chemistry.

Calculations for the REE based upon the two-stage model outlined above are presented in Figs 9 & 10. The results of batch melting calculations for the REE in Gruinard Bay amphibolites 115, 146 and 104 show that only amphibolite 104 (slightly enriched in LREE) has the appropriate $(La/Yb)_N$ ratio and Yb_N value to be a suitable source for the TTG suite of the Southern Region. The results of calculations based upon the partial melting of amphibolite 104 in equilibrium with garnet amphibolite and eclogitic residues are shown in Fig. 9. Similar amphibolites from Gruinard Bay have been dated at 3.3 Ga (Burton et al. 1994).

TTGs from the Southern Region. **Tonalites** from the Southern Region show a pattern of increasing La/Yb ratios and the development of a positive Eu anomaly with decreasing total REE (Fig. 3m) resulting principally from hornblende fractionation (Rollinson & Fowler 1987). Thus the tonalites with higher Yb_N values and lower $(La/Yb)_N$ ratios are thought to be most representative of parental liquids. These may be modelled by the partial melting of garnet amphibolite. A low percentage of garnet in the residue (<10%) means that the measured REE ratios can be duplicated with melt fractions of less than 50% (Fig. 9). The full REE pattern was modelled using the partition coefficients for andesitic liquids tabulated in Rollinson (1993), 30–40% melting in equilibrium with a residue with the composition 6% garnet 59% clinopyroxene 20% hornblende 15% plagioclase. Plagioclase is required in the residue

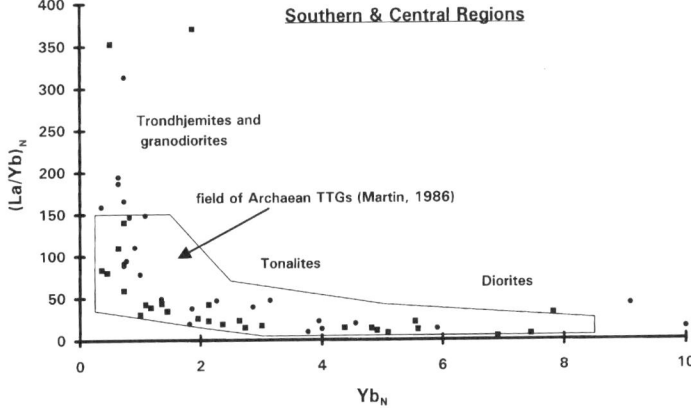

Fig. 7. $(La/Yb)_N$ versus Yb_N plot for dioritic, tonalitic, trondhjemitic and granodioritic gneisses from the Central and Southern regions of the Lewisian. Symbols: black squares, Central Region gneisses; circles, Southern Region gneisses. Field of Archaean TTGs from Martin (1986).

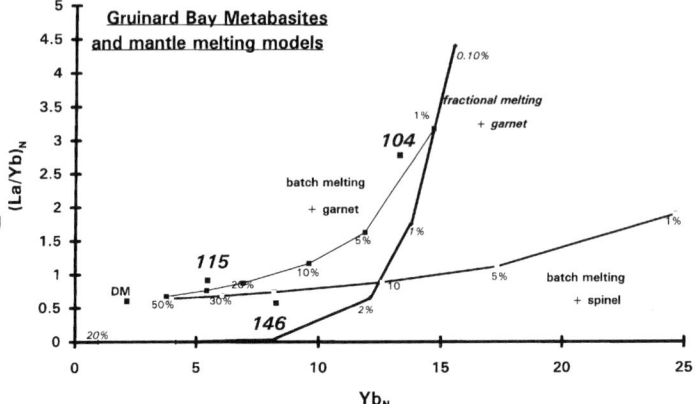

Fig. 8. $(La/Yb)_N$ versus Yb_N diagram showing partial melting trends for depleted mantle with the mineralogy of garnet and spinel lherzolite. Batch melting of garnet lherzolite ($ol_{0.59}$ $op_{0.15}$ $cp_{0.25}$ $grt_{0.005}$) and spinel lherzolite ($ol_{0.55}$ $op_{0.15}$ $cp_{0.25}$ $spin_{0.05}$) plotted for 1%, 5%, 10%, 20%, 30% and 50% melting. Fractional melting of garnet lherzolite ($ol_{0.59}$ $op_{0.15}$ $cp_{0.25}$ $grt_{0.005}$) plotted for 0.1%, 1%, 2% 5%, 10% and 20% melting. DM = depleted mantle (Ce = 1.3, Yb = 2.15 times chondrite). Samples 104, 115, and 146 are amphibolites from Gruinard Bay. Partition coefficients from Luais & Hawkesworth (1994).

in order to satisfy the measured Sr and Eu levels in the tonalites.

Trondhjemites and granodiorites from the Southern Region show similar REE patterns to those of the tonalites – increasing positive Eu anomaly with decreasing total REE concentrations (Fig. 3n) – although $(La/Yb)_N$ ratios are higher. The variability in the suite can be explained by hornblende and plagioclase fractionation (Rollinson & Fowler, 1987). The majority of trondhjemite samples have REE contents which may be modelled by 10–30% partial melting in equilibrium with an eclogitic source Fig. 9). Rocks with more extreme $(La/Yb)_N$ ratios (>150) are enriched in light REE and are thought to contain accumulations of light REE-enriched minor phases.

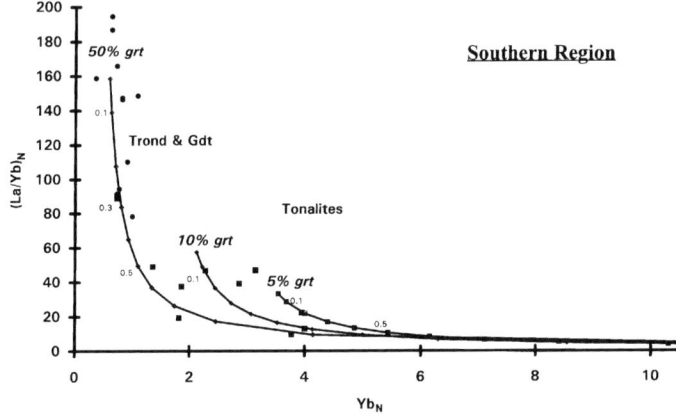

Fig. 9. $(La/Yb)_N$ versus Yb_N diagram showing the composition of TTGs from the Southern Region compared with batch partial melting trends for amphibolite 104 in equilibrium with an eclogitic residue (40% cpx, 50% grt, 10% hbl). Melt compositions calculated using rhyolite partition coefficients from Rollinson (1993). Melt compositions in equilibrium with garnet amphibolite/basic granulite residues (10% plag, 20% hbl, 60% cpx 10% grt) and (15% plag, 20% hbl, 60% cpx 5% grt) calculated using andesite partition coefficients from Rollinson (1993). Melting intervals shown on partial melting curves at 5% and then every 10%. Tonalites, black squares; Trondhjemites and granodiorites, circles.

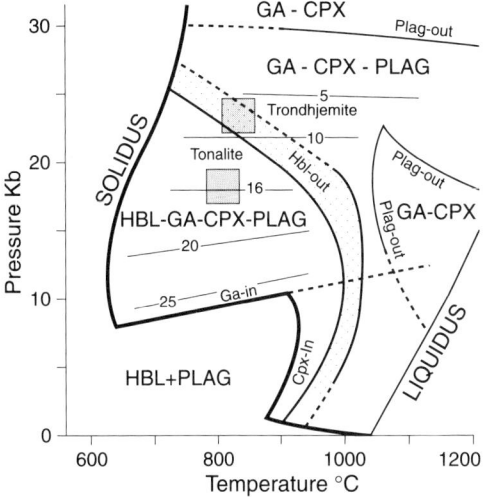

Fig. 10. Phase diagram showing the stability fields of the minerals hornblende (hbl) garnet (ga), clinopyroxene (cpx), plagioclase (plag) and TTG melt for the dehydration melting of amphibolite. The faint dashed lines show the percentage of plagioclase present in the solid residue. Based on Wyllie & Wolf (1993) with additional data taken from Rapp (1994), Rapp et al. (1991) and Rushmer (1991, 1993). Estimated compositions of residues of Lewisian tonalites and trondhjemites are shown as shaded boxes.

The full REE pattern was modelled using the partition coefficients for rhyolitic liquids tabulated in Rollinson (1993), 30% melting and a residue with the composition 40% garnet 50% clinopyroxene 10% hornblende. The presence of plagioclase cannot be ruled out but it must form less than 10% of the residue. Similarly, the presence of hornblende is not essential but it cannot be eliminated.

Interpretation of results in the light of the phase diagram of Wyllie & Wolf (1993)

The mineral residue in equilibrium with the TTG melts calculated for the Lewisian Gneisses can be used to estimate the P–T conditions of partial melting of a mafic source. Wyllie & Wolf (1993) present a phase diagram for the co-existence of TTG melts with garnet, amphibole clinopyroxene and plagioclase during the dehydration melting of amphibolite over a range of pressures and temperatures. This diagram (Fig. 10) has been modfied here using additional experimental data from Rapp (1994), Rapp et al. (1991) and Rushmer (1991, 1993) to show the fields in which TTG melts co-exist with differing mineralogical residues. It should be noted, however that the position of the field boundaries is the subject of some debate (see Wyllie & Wolf 1993) and their positions are dependent upon the bulk composition of the amphibolite used in the melting experiment.

The calculated residues for the tonalites of the Southern Region (6% garnet, 59% clinopyroxene, 20% hornblende, 15% plagioclase) indicate that the tonalitic melts could be generated over a wide temperature and pressure range within the Hbl–Ga–Cpx–Plag–melt stability field. However, the percentage of plagioclase in the residue places the assemblage in the higher pressure part of the stability field, and the moderate amount of hornblende places the assemblage away from the hornblende-out curve. Probable conditions for the co-existence of tonalite and residue are about 18 kb, 800°C (Fig. 10).

The calculated residue for trondhjemites and granodiorites from the Southern Region (40% garnet, 50% clinopyroxene, 10% hornblende, ± plagioclase) is eclogitic with minimal hornblende and plagioclase. Compared with the tonalites, the lower plagioclase and higher garnet contents of the residue suggest melting at higher pressures, whilst the lower hornblende content suggests higher temperatures or higher pressures. Probable conditions for the existence of trondhjemites and melt are > 22 kb, > 800°C (Fig. 10).

The results of these calculations suggest slight differences in the probable P–T conditions of melting of a mafic source for the tonalites and trondhjemites, but more importantly indicate that the likely conditions of partial melting were at depths greater than those of 'normal' 30 km thick continental crust. Thus, unless the amphibolite source was already part of over-thickened continental crust, these results are at variance with the proposal of Burton et al. (1994) that the Southern Region trondhjemites formed by intracrustal melting from an amphibolitic source. These results also disagree with the proposal by Ireland et al. (1994) who suggested that melting takes place in the diamond stability field.

Depletion in granulites – a product of crustal reworking or crust formation?

It has long been recognized that the granulite facies TTG gneisses of the Central Region have extremely low concentrations of U. This has been attributed to depletion during granulite facies metamorphism a view which was challenged by Tarney & Weaver (1987). Here, the evidence for granulite facies element depletion is reviewed with the purpose of showing that it is not related to the metamorphism. Firstly, it should be noted that U/Pb

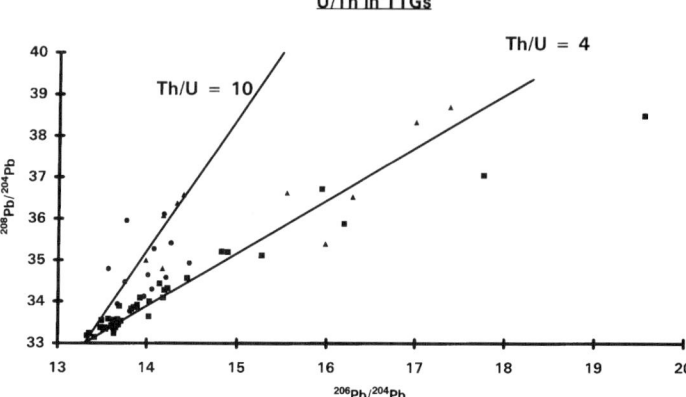

Fig. 11. $^{208}Pb/^{204}Pb$ versus $^{206}Pb/^{204}Pb$ plot to illustrate Th–U ratios in Lewisian TTGs. Samples from the Central Region (squares) define a trend which has time integrated Th–U ratio close to 4.0. Sample from the Northern Region (triangles) and the Southern Region (circles) show more scatter, indicating, for the most part, post-magmatic U loss. Data from Whitehouse (1989b), Chapman & Moorbath (1977).

ratios are low throughout the Lewisian and are not restricted to the granulite facies region. Most TTG gneisses in the mainland Lewisian have evolved with U/Pb ratios less than that of the mantle as is evidenced by their position close to the mantle evolution curve on a $^{207-206}Pb$ evolution diagram (Fig. 6). This suggests that all the TTGs have experienced some U loss. Secondly, zircons from the granulite facies Central Region of the Lewisian have much lower U levels than those from the Northern Region and appear to have crystallized from a U-poor melt (Pidgeon & Bowes 1972; Corfu et al. 1994). This suggests that the low U levels are magmatic.

U loss may be further constrained by U/Th ratios. The average terrestrial ratio of U/Th is between 3.5 and 4.0 (Rogers & Adams 1978). This value is not normally altered during partial melting but may be changed by fluid-driven U mobility or by the fractionation of a U-rich, Th-poor phase such as zircon. TTGs from the granulite facies Central Region show a well-correlated trend on a $^{208}Pb/^{204}Pb$ vs. $^{206}Pb/^{204}Pb$ diagram (Fig. 11) which corresponds to a time-integrated Th–U ratio of about 4, a value close to the average terrestrial ratio. In the Northern Region, U/Th ratios vary between 3.2 and 9.5, whereas in the Southern Region they are between 1.1 and 22 (Whitehouse 1989b), indicating that in the amphibolite facies regions U and Th are behaving independently and for most samples reflect the preferential loss of U, probably in a fluid phase. The close agreement between Pb–Pb and Sm–Nd ages in the Northern Region suggests that U loss took place during or immediately after rock formation, whereas the lack of such agreement in the Southern Region suggests that the U loss postdates the time of formation and the Pb-errorchron age of 2.8 ± 0.2 Ga gives the approximate time of U loss (Taylor et al. 1991). The close correspondence of the Th/U ratio in the dry granulite facies rocks of the Central region to the average terrestrial ratio strongly suggests that the observed Th/U ratios in the Central Region are magmatic. These data do not support the simple concept of the granulite facies depletion of U and its retention at lower metamorphic grades and further point towards low initial U concentrations in the igneous protolith of the Central Region magmas.

Less well constrained is the depletion in K and Rb in the Central Region granulites. Rudnick et al. (1985), in a study of element depletion in granulites worldwide, made two important observation about the Lewisian granulites. Firstly, that K/Rb ratios in the Central Region granulites are extremely high and atypical of granulites in general, and secondly that it is difficult to document K depletion in granulites for there is no firm criterion by which it may be measured.

These arguments support the view of Tarney & Weaver (1987) that the low concentrations of K, Rb, U and Th in the Central Region granulites are not the product of granulite facies metamorphism, but rather are a feature of the igneous protolith of the granulites. If this is so, then: (i) the case for a granulite facies metamorphism as a distinct event, separate from that of crust formation at c. 2.65 Ga is less necessary and the isotopic data for the Central Region can plausibly be interpreted as the time of crust formation; (ii) the low K–TTG magmas (exclusively tonalites and trondhjemites) of the Central Region are original igneous compositions.

It is important to establish a mechanism by which the low K, Rb, U and Th contents of the TTG magmas in the Central Region were produced. Whilst it is not difficult to produce low-K magmas through crystal-melt equilibria, it is more difficult to produce high K/Rb ratios by this process unless biotite or phlogopite is present as a residual phase. Thus, it is unlikely that the element depletion was achieved through partial melting or crystal fractionation processes alone. More plausible is the removal of elements from the source of the TTG suite via a fluid phase – a model proposed by Tarney & Weaver (1987). The composition of the fluid can be constrained from the elements mobilized, for whilst U is soluble in both H_2O- and CO_2-rich fluids, Rb is only soluble in an H_2O-rich fluid. It is likely, therefore, that element depletion took place via an H_2O-rich fluid removed from the source of the TTG magmas prior to partial melting.

The line of reasoning presented here is consistent with one of the models for TTG genesis. High pressure and temperature dehydration melting experiments on amphibolite produce melts of TTG compositions (Rapp et al. 1991; Rushmer 1991). Under lower temperature conditions hydrous minerals in a metabasalt will break down releasing volatiles which can transport the more water-soluble elements. This mechanism is thought to be responsible for some of the element fluxing adjacent to modern subduction zones (Saunders et al. 1991) and is seen here as a viable mechanism for the depletion of a basaltic source prior to partial melting and the generation of TTG magmas. Partial melts of a basaltic slab are documented from some modern subduction zones. Such rocks are known as adakites (Defant and Drummond 1990) and have a number of key geochemical similarities to Archaean TTGs. These melts typically have K/Rb ratios of 500–600, but some have ratios in excess of 1000 (Saunders et al. 1987). The different levels of depletion in the three regions of the Lewisian described here reflect the contrast between the processes of dehydration melting and dehydration followed by melting.

Discussion

The working hypothesis adopted here is that the Lewisian TTG magmas are the product of the partial melting of a mafic source at about 3.0 Ga. This model is consistent with the results of experimental studies on TTG suites in general. Further, the REE chemistry of the Lewisian TTGs is consistent with this hypothesis and old amphibolite fragments – some of which may be remnants of parental material – are present in the gneiss complex. The results of trace element modelling in this study suggest that the TTGs were produced by the partial melting at >18 kb and 800°C.

The nature of the parent amphibolite is the subject of conjecture. Burton et al. (1994) have shown that the 3.3 Ga amphibolites at Gruinard Bay are a possible source for some TTGs, although it is apparent from Fig. 5 that it is not possible to derive the TTGs of the Northern and Southern regions from this source. Nevertheless, other amphibolites from the Gruinard Bay area are plausible protolith material (see Fig. 8). A recent theoretical study by Peacock et al. (1994) shows that a subducting slab will only melt when the ocean floor is very young and warm. Thus, if Archaean ocean floor is parental to TTGs, then the amphibolite protolith will have an age indistinguishable from that of the resultant melts. In this context it is interesting to note that there may be an overlap between the age of the Scourie More layered body and that proposed for crust formation in the Central Region. The Scourie More layered intrusion has a Sm–Nd isochron age of 2.707 ± 0.052 Ga interpreted as the time of differentiation of the parent body from the mantle (Cohen et al. 1991). This age is within error of the U–Pb zircon ages for this area of Corfu et al. (1994) and the Nd and Pb isotope data reviewed earlier in this paper.

There are important geochemical differences between the granulite facies Region Central of the Lewisian and the amphibolite facies regions of the north and south, namely:

- the bulk composition of the Central Region is more tonalitic;
- there are differences in the isotopic composition of the Central Region suggesting that it formed in a later crust-forming event at about 2.65 Ga;
- the Central Region is depleted in LILE, here interpreted as an indication that the TTGs of the Central Region were derived from an element-depleted dehydrated source.

These differences in geochemistry may be attributed to differences in the process of crust generation which, in turn, reflects differences in the thermal conditions of partial melting. It is suggested that in the amphibolite facies North and South regions dehydration melting of amphibolite-produced hydrous TTG magmas. In contrast, the TTG magmas of the Central Region are the product of a two-stage process in which a mafic source which experienced (partial) dehydration (during LILE were lost) and then partial melting. For example, if the mafic protolith for the Central Region followed a steeper P–T trajectory than that for the Northern and Southern regions, it could cross the hornblende-out curve before crossing the (nearly dry) solidus and would dehydrate before it melted. It is possible that the partial melting of a dry protolith produced melts with a very low water content which crystallized with an anhydrous mineralogy and that the Central Region TTG granulites were never hydrous. There is a very large time interval between crust generation in the North and South regions and that in the Central Region (c. 300 Ma) and it is unlikely that the two events can be integrated into a single model for crust generation.

Conclusions

(1) TTG gneisses in the Lewisian formed from a depleted mantle source at c. 3.0 Ga. Their genesis is consistent with the dehydration melting of amphibolite at pressures of 18–22 kb and c. 800°C. Amphibolites at Gruinard Bay have trace element compositions which are consistent with this model.
(2) Element depletion in the Central Region, previously attributed to the effects of granulite facies metamorphism, is now thought to be a feature of the basaltic parent to the TTG magmas. It is thought that the elements U, Th, K and Rb were lost from the parent basalt in a hydrous fluid prior to partial melting.
(3) The TTG gneisses of the Central Region of the Lewisian are anomalous when compared with those of the Southern and Northern regions with respect to

- their age – they are in part younger than similar gneisses elsewhere in the mainland Lewisian and indicate a later crust-generation event;
- their isotopic composition – they are contaminated with an older crustal component;
- the composition of the original magmas – they are less potassic;
- their Rb, U, Th contents – they are depleted relative to concentrations in amphibolite facies regions;
- metamorphic grade

The compositional differences are attributed to the process of crust-formation, rather than the effects of later metamorphism and imply that there were differences in the process of crust formation between the Central Region and Northern and Southern regions of the Lewisian.

I thank Mike Fowler and John Tarney for their thoughtful comments on this manuscript. Research into Lewisian geochemistry is supported by Cheltenham and Gloucester College of Higher Education.

References

ARNDT, N. T. & GOLDSTEIN, S. L. 1987. Use and abuse of crust-formation ages. *Geology*, **15**, 893–895.

BARKER, F. 1979. Trondhjemite: Definition, environment and hypotheses of origin. *In*: BARKER, F. (ed.) *Trondhjemites, Dacites and Related Rocks*. Elsevier, Amsterdam, 1–12.

BURTON, K. W., COHEN, A. S., O'NIONS R. K. & O'HARA, M. J. 1994. Archaean crustal development in the Lewisian complex of northwest Scotland. *Nature*, **370**. 552–555.

CHAPMAN, H. J. 1978. *Geochronology and Isotope Geochemistry of Precambrian Rocks from NW Scotland*. D.Phil thesis, University of Oxford.

—— & MOORBATH, S. 1977. Lead isotope measurements from the oldest recognised Lewisian gneisses of northwest Scotland. *Nature*, **268**, 41–42.

COHEN, A. S., O'NIONS, R. K. & O'HARA, M. J. 1991. Chronology and mechanisms of depletion in Lewisian granulites. *Contributions to Mineralogy and Petrology*, **106**, 142–153.

CORFU, F., HEAMAN, L. M. & ROGERS, G. 1994. Polymetamorphic evolution of the Lewisian complex, NW Scotland, as recorded by U–Pb isotopic compositions of zircon, titanite and rutile. *Contributions to Mineralogy and Petrology*, **117**, 215–228.

DAVIES, F. B. 1975. Origin and ancient history of gneisses older than 2,800 Myr in Lewisian Complex. *Nature*, **258**, 589–591.

—— 1976. Early Scourian structures in the Scourie–Laxford region and their bearing on the evolution of the Laxford Front. *Journal of the Geological Society*, **132**, 543–554.

DEFANT, M. J. & DRUMMOND M. S. 1990, Derivation of some modern arc magmas by melting of young subducted lithosphere. *Nature*, **347**, 662–665.

DEPAOLO, D. J. 1981, Neodymium isotopes in the Colorado Front Range and crust-mantle evolution in the Proterozoic. *Nature*, **291**, 193–196.

FIELD, D. 1978. Granulites at Gruinard Bay. *Scottish Journal of Geology*, **14**, 359–361.

FOWLER, M. B. 1986. Large ion lithophile element characteristics of an amphibolite facies to granulite facies transition at Gruinard bay, North-west Scotland. *Journal of Metamorphic Geology*, **4**, 345–59.

—— & PLANT, J. A. 1987. Rare earth element geochemistry of Lewisian grey gneisses from Gruinard Bay. *Scottish Journal of Geology*, **23**, 193–202.

——, FRIEND, C. R. L. & WHITEHOUSE, M. J. 1995. Crust formation in the Lewisian. *Nature*, **375**, 366–67.

GOLDSTEIN, S. L., O'NIONS, R. K. & HAMILTON, P. J. 1984. A Sm–Nd study of atmospheric dusts and particulates from major river systems. *Earth and Planetary Science Letters*, **70**, 221–236.

HAMILTON, P. J., EVENSEN, N. M., O'NIONS, R. K. & TARNEY, J. 1979. Sm-Nd systematics of Lewisian gneisses: implications for the origin of granulites. *Nature*, **277**, 25–28.

HOLLAND, J. G. & LAMBERT, R. St. J. 1973. Comparative major element geochemistry of the Lewisian of the mainland of Scotland. *In*: PARK, R. G. & TARNEY, J. (eds) *The Early Precambrian of Scotland and related rocks of Greenland*. University of Keele, 51–62.

HUMPHRIES, F. J. & CLIFF, R. A. 1982. Sm–Nd dating and cooling history of Scourian granulites, Sutherland. *Nature*, **295**, 515–517.

IRELAND, T. R., RUDNICK, R. L. & SPETIUS, Z. 1994. Trace elements in diamond inclusions from eclogites reveal link to Archaean granites. *Earth and Planetary Science Letters*, **128**, 199–213.

KINNY, P. & FRIEND, C. 1994. A new insight into Lewisian chronology (Abstract). *Mineralogical Magazine*, **58A**, 481–482.

LUAIS, B. & HAWKESWORTH, C. J. 1994. The generation of continental crust: an integrated study of crust-forming processes in the Archaean of Zimbabwe. *Journal of Petrology*, **35**, 43–93.

MARTIN, H. 1986. Effect of steeper Archaean geothermal gradient on geochemistry of subduction-zone magmas. *Geology*, **14**, 753–756.

—— 1993. The mechanisms of petrogenesis of the Archaean continental crust – Comparison with modern processes. *Lithos*, **30**, 373–388.

MOORBATH, S., WELKE, H. & GALE, N. H. 1969. The significance of lead isotope studies in ancient, high-grade metamorphic basement complexes, as exemplified by the Lewisian rocks of northwest Scotland. *Earth and Planetary Science Letters*, **6**, 245–256.

NAGLER, T. F & STILLE, P. 1993. Remarks on depleted mantle evolution models used for Nd model age calculation. *Schweitzerische Mineralogische und Petrographische Mitteilungen*, **73**, 375–381.

NAKAMURA, N. 1974. Determination of REE, Ba, Fe, Mg and K in carbonaceous and ordinary chondrites. *Geochimica et Cosmochimica Acta*, **38**, 757–775.

O'CONNOR, J. T. 1965. *A classification for quartz-rich igneous rocks based upon feldspar ratios*. United States Geological Survey Professional Paper 525-B, 79–84.

PEACOCK, S. M., RUSHMER, T. & THOMPSON, A. B. 1994. Partial melting of subducting oceanic crust. *Earth and Planetary Science Letters*, **121**, 227–244.

PIDGEON, R. T. & BOWES, D. R. 1972. Zircon U–Pb ages of granulites from the Central Region of the Lewisian, northwestern Scotland. *Geological Magazine*, **109**, 247–58.

PRIDE, C. & MUECKE, G. K. 1980. Rare earth element geochemistry of the Scourian Complex, NW Scotland – evidence for the granite–granulite link. *Contributions to Mineralogy and Petrology*, **73**, 403–412.

—— & —— 1982. Geochemistry and origin of granitic rocks, Scourian Complex, N. W. Scotland. *Contributions to Mineralogy and Petrology*, **80**, 379–385.

RAPP, R. P. 1994. The amphibole-out phase boundary in melted metabasalt and its control over melt fraction and composition. (Abstract) *Eos*, **75**, Spring meeting supplement, 359.

——, WATSON, E. B. & MILLER, C. F. 1991. Partial melting of amphibolite/eclogite and the origin of Archaean trondhjemites and tonalites. *Precambrian Research*, **51**, 1–25.

ROGERS, J. J. W. & ADAMS, J. A. S. 1978. Th: Abundances in common igneous rocks. *In*: WEDEPOHL, K. H. (ed.) *Handbook of Geochemistry*, Springer-Verlag, Berlin, 90-E-1 – 90-E-12.

ROLLINSON, H. R. 1978. *Geochemical studies on the Scourian Complex, NW Scotland*. Thesis, University of Leicester.

—— 1987. Early basic magmatism in the evolution of Archaean high-grade gneiss terrains: an example from the Lewisian of NW Scotland. *Mineralogical Magazine*, **51**, 345–55.

—— 1993. *Using Geochemical Data: evaluation, presentation, interpretation*. Longmans, Harlow.

—— 1994. Origin of felsic sheets in the Scourian granulites: new evidence from rare earth elements. *Scottish Journal of Geology*, **30**, 121–129

—— & FOWLER, M. B. 1987. The magmatic evolution of the Scourian Complex at Gruinard Bay. *In*: PARK, R. G. & TARNEY, J. (eds) *The Evolution of the Lewisian and Comparable Precambrian High-Grade Terrains*. Geological Society, London, Special Publication, **27**, 57–71.

—— & WINDLEY, B. F. 1980a. Geochemistry and origin of an Archaean granulite grade tonalite-rondhjemite–granite suite from Scourie, NW Scotland. *Contributions to Mineralogy and Petrology*, **72**, 265–81.

—— & —— 1980b. Selective elemental depletion during metamorphism of Archaean granulites, Scourie NW Scotland. *Contributions to Mineralogy and Petrology*, **72**, 257–64.

RUDNICK, R. L., MCLENNAN, S. M. & TAYLOR, S. R. 1985. Large ion lithophile elements in rocks from high pressure granulite facies terrains. *Geochimica et Cosmochimica Acta*, **49**, 1645–55.

RUSHMER, T. 1991. Partial melting of two amphibolites: contrasting experimental results under fluid absent conditions. *Contributions to Mineralogy and Petrology*, **107**, 41–59.

—— 1993. Experimental high-pressure granulites: some applications to natural mafic xenolith suites and Archean granulite terranes. *Geology*, **21**, 411–414.

SAUNDERS, A. D., NORRY, M. J. & TARNEY, J. 1991. Fluid influence on the trace element compositions of subduction zone magmas. *Philosophical Transactions of the Royal Society of London*, **A335**, 377–392.

——, ROGERS, G., MARRINER, G. F., TERRELL, D. J. & VERMA, S. P. 1987. Geochemistry of Cenozoic volcanic rocks, Baja California, Mexico: implications for the petrogenesis of post-subduction magmas. *Journal of Volcanology and Geothermal Research*, **32**, 223–245.

SHERATON, J. W., SKINNER, A. C. & TARNEY, J., 1973. The geochemistry of the Scourian gneisses of the Assynt District. *In*: PARK, R. G. & TARNEY, J. (eds) *The Early Precambrian of Scotland and related rocks of Greenland.* University of Keele, 13–30.

STRECKEISEN, A. & LEMAITRE, R. W. 1979. A chemical approximation to the modal QAPF classification of igneous rocks. *Nues Yahrb. Mineral. Abh.*, **136**, 169–206.

SUN, S. S. & MCDONOUGH W. F. 1989. Chemical and isotopic systematics of oceanic basalts: implications for mantle composition and processes. *In*: SAUNDERS, A. D. & NORRY, M. J. (eds) *Magmatism in the Ocean Basins.* Geological Society, London, Special Publication, **42**, 313–345.

SUTTON, J. & WATSON, J. V. 1951. The pre-Torridonian metamorphic history of the Loch Torridon and Scourie areas in the north-west Highlands, and its bearing on the chronological classification of the Lewisian. *Quarterly Journal of the Geological Society of London*, **106**, 241–307.

TARNEY, J. & WEAVER, B. L. 1987. Geochemistry of the Scourian Complex: petrogenesis and tectonic models. *In*: PARK, R. G. & TARNEY, J. (eds) *The evolution of the Lewisian and comparable Precambrian high-grade terrains.* Geological Society Special Publication **27**, 45–56.

TAYLOR, P. N., KRAMERS, J. D., MOORBATH, S., WILSON, J. F., ORPEN, J. L. & MARTIN, A. 1991. Pb/Pb, Sm–Nd and Rb–Sr geochronology in the Archaean Craton in Zimbabwe. *Precambrian Research*, **87**, 175–196.

——, MOORBATH, S., GOODWIN, R. & PETRYKOWSKI, A. C. 1980. Crustal contamination as an indicator of the existence of early Archaean continental crust: Pb-isotopic evidence from the late Archaean gneisses of west Greenland. *Geochimica et Cosmochimica Acta*, **44**, 1437–1453.

WATERS, F. G., COHEN, A. S., O'NIONS, R. K. & O'HARA, M. J. 1990. Development of Archaean lithosphere deduced from chronology and isotope chemistry of Scourie dykes. *Earth and Planetary Science Letters*, **97**, 241–55.

WEAVER, B. L. & TARNEY, J. 1980. Rare earth geochemistry of Lewisian granulite facies gneisses, northwest Scotland: implications for the petrogenesis of the Archaean lower continental crust. *Earth and Planetary Science Letters*, **51**, 279–96.

—— & —— 1981. Lewisian gneiss geochemistry and Archaean crustal development models. *Earth and Planetary Science Letters*, **55**, 171–80.

WHITEHOUSE, M. J. 1988. Granulite facies Nd-isotopic homogenisation in the Lewisian complex of northwest Scotland. *Nature*, **331**, 705–707.

—— 1989a. Sm–Nd evidence for diachronous crustal accretion in the Lewisian complex of northwest Scotland. *Tectonophysics*, **161**, 245–56.

—— 1989b. Pb-isotopic evidence for U–Th–Pb behaviour in a prograde amphibolite to granulite transition from the Lewisian complex of northwest Scotland: implications for Pb-Pb dating. *Geochimica et Cosmochimica Acta*, **53**, 717–724.

—— 1993. Age of the Corodale Gneisses, South Uist. *Scottish Journal of Geology*, **29**, 1–7.

—— & MOORBATH, S. 1986. Pb–Pb systematics of Lewisian gneisses – implications for crustal differentiation. *Nature*, **319**, 488–489.

WYLLIE, P. J. & WOLF, M. B. 1993. Amphibolite dehydration-melting: sorting out the solidus. *In*: PRITCHARD, H. M., ALABASTER, T., HARRIS, N. B. W. & NEARY, C. R. (eds) *Magmatic Processes and Plate Tectonics.* Geological Society, London, Special Publication, **76**, 405–416

Rare earth and isotopic (Nd, O) heterogeneity of the Archaean mantle, Baltic Shield

A. VREVSKY, R. KRIMSKY & S. SVETOV

Institute of Precambrian Geology and Geochronology, Makarova emb.2, St. Petersburg, 199034, Russia

Abstract: Investigations of major and trace element geochemistry, whole-rock Sm–Nd and O isotopic composition of Archaean komatiites from the Baltic Shield show clear evidence of mantle source heterogeneity.
    The komatiites were generated by 50–60% partial melting of the mantle at P c. 35 kb, without garnet in the residue. Three different geochemical types of mantle source can be defined on the basis of HREE concentration: Type I, II and III sources yield komatiites with HREE abundances between 0.5–2, 2–3 and 3.5–5 Chondritic levels respectively. These types of mantle source can be correlated with three U–Pb zircon age groups of greenstone belts and their Nd and O isotopic composition.

Archaean greenstone belts constitute prominent features of Precambrian Shield areas. The komatiites are common and important stratigraphic members of the volcanic sequences of the Archaean greenstone belts throughout the world. They constitute one of the best lithologies for investigation of the chemical and isotopic evolution of the Archaean mantle because they are produced by large degrees of mantle partial melting. Consequently, abundance ratios of such incompatible elements as Ti, Zr, Y, and REE may be similar to those in mantle source of the komatiite melts. If garnet or clinopyroxene did not remain in the residue at the high degrees of melting, the relative concentrations of Al, Ca and HREE would also be similar to those of the mantle source. However, it must be remembered that while komatiites were erupted with a temperature of 1560–1650°C (Green 1979; Nisbet *et al.* 1993) they can contaminate crustal material during their upwards migration or extrusion (Arndt & Jochum 1990). Komatiites are also prone to alteration, resulting in the destruction of their primary geochemical composition (see Arndt *et al.* 1989). The rocks analysed in this study are metamorphosed in different degrees from greenschist (Hautavaara belt) to amphibolite facies (greenstone belts of the Kola peninsula and North Karelia), and therefore care must be taken in interpreting their geochemical signatures with a view to understanding the nature of their mantle source region.

Aims and objectives

The major aim of investigation has been the documentation of differences and similarities of the chemical (especially REE and trace elements) and isotopic (Nd, O) composition and the petrogenesis of komatiites from the Late Archaean greenstone belts of the Baltic Shield in order to explore the nature of the source of the komatiites as a tool in reconstructing the evolutionary history of the Archaean mantle–crust system of the Eastern Baltic Shield. Sample collection was concentrated especially on spinifex-textured komatiites from the greenstone belts of Kola peninsula, East, Central and West Karelia. (Spinifex-textured flows were chosen because they are more closely representative of primary liquid compositions and, in comparison to cumulate zones, appear to preserve most of their original chemical signatures during metamorphism.) Published data (Jahn *et al.* 1980) were used for the komatiites of Kuhmo-Suommusalmi belt in East Finland (Fig. 1). In the greenstone belts of the Baltic Shield, komatiites typically occur within the lower horizons of the volcanic stratigraphy. However, in E. Finland (Suommusalmi belt), W. Karelia (Kostomuksha belt) and N. Karelia (Hizovaara belt) they also occur at higher levels, intercalated with calc-alkaline volcanics and sediments (Fig. 2). Based on U–Pb and Sm–Nd whole-rock data, the greenstone belts could be subdivided into three groups of distinctly different ages (see Fig. 3).The preservation of the original spinifex-textured parts of the lava flows sampled in this study is variable according to location: good preservation occurs in the Kamennoozero and Palaj-Lamba belts but textures are only weakly preserved in the Hizovaara, Korva and Hautavaara belts. However, in all instances some semblance of spinifex texture could be determined either in outcrops or in thin sections.

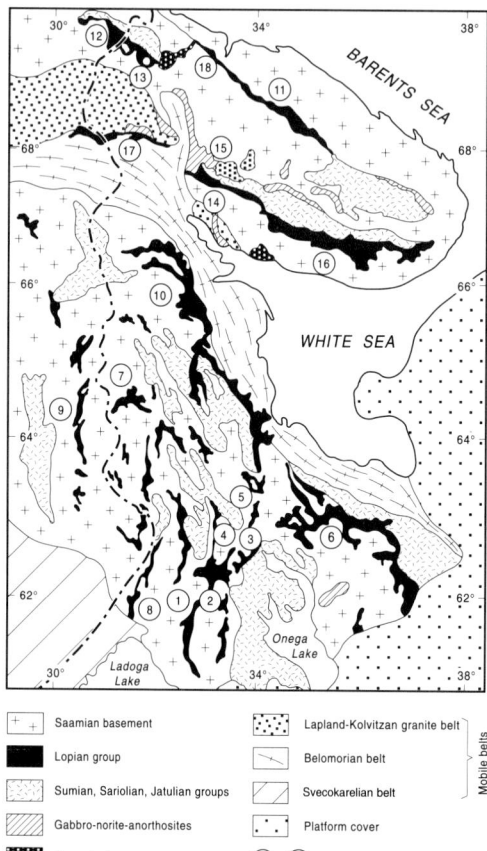

Fig. 1. Geological sketch map of the Eastern Baltic Shield. Numbers refer to specific greenstone belts: 1, Hautavaara; 2, Kojkary; 3, Palaj-Lamba; 4, Oster; 5, Paraandovo; 6, Kamennoozero; 7, Kostamuksha; 8, Jalonvaara; 9, Kumho; 10, Hizovaara; 11, Polmos-Porso; 12, Kaskamsky; 13, Allarehenskaya; 14, Priimandrovskya; 15, Olenegorskaya; 16, Terskaya; 17, Korva; 18, Ura-Guba.

Mineralogically all samples consist entirely of secondary assemblages with few relicts of previous magmatic minerals (olivine, pyroxene, chromite).

Analytical techniques

Concentration of major elements, Rb, Sr, Ti, Y, Zr, Ni, Cr and Co were determined by the XRF method. For the trace elements uncertainties are of the order of 10% for concentrations lower than 30 ppm, while for concentrations greater than 30 ppm, uncertainties are reduced to 3%. REEs were measured in the National Geophysics Research Institute by the isotope dilution mass spectrometry (MS VG-354) method (Jahn et al. 1980). Analytical errors, including chemical preparation, blank effects, uncertainties in spike calibration and mass spectrometry runs, are estimated at about 5% for LREE and 3% for HREE. Oxygen was extracted using ClF3 (Borthwick & Harmon 1982) and analysed as CO_2 gas on $\frac{1}{2}$ triple-collector MI-1201 (Russian) mass spectrometer. 13 analyses of the NBS-28 standard were measured relative to SMOW giving a δO^{18} value of +9.57+/–0.12% (1σ). 11 analyses of the NBS-30 standard give a δO^{18} value of +4.84+/–0.12% (1σ); the estimated absolute accuracy for the total method is +/–0.22%. $^{143}Nd/^{144}Nd$ and $^{147}Sm/^{144}Nd$ ratios are determined following the procedure described by Richard et al. (1976). $^{143}Nd/^{144}Nd$ ratio were measured using an 8-collector Finnigan MAT-261 mass spectrometer in static mode. All the ratios are normalized against $^{143}Nd/^{144}Nd$ = 0.7212 for isotopic fractionation effect. The reproducibility of $^{143}Nd/^{144}Nd$ ratio in La Jolla Nd standard measures is 0.005% (2σ). Nd and Sm blanks are lower than 0.2 ng and 0.1 ng, respectively. Average results for 10 separate runs of BCR-1 standard are: Sm = 6.65 Mg g^{-1}, Nd = 28.6 Mg g^{-1}, $^{147}Sm/^{144}Nd$ = 0.13802 +/–3, $^{143}Nd/^{144}Nd$ = 0.12628 +/–5.

Major and trace element geochemistry

Once the analyses have been corrected to anhydrous totals, komatiites vary in MgO content between 24–34% and those with MgO greater than 34% represented the cumulate parts of the flows. Komatiites of Kola peninsula are relatively rich in total Fe in comparison with Karelian komatiites (ΣFeO = 13–16%) (Fig. 4a). They contain 1500–2000 ppm Ni (Fig. 4b) and 30–90 ppm Sr (Fig. 4c), 1–3% CaO (Fig. 4c), 2–3% Al_2O_3 (Fig. 4a) and have near chondritic Y/Zr and TiO_2/Zr ratios (Figs 4d, e). The komatiites of Karelia and E. Finland are the most Al-rich (Al_2O_3 = 6–8%; Fig. 4a) of those analysed in this study. Finally, the komatiites of N. Karelia are more comparable with the Kola komatiites in Al and Sr contents and Y/Zr ratio (Figs 4a, c, e). In general, the komatiites of the Baltic Shield have wide variations of CaO/Al_2O_3 and Al_2O_3/TiO_2 ratios. Such variations could be explained by alteration of the komatiites or by specific magma generation processes. In magmatic processes only garnet fractionation during partial melting can produce komatiite melt enriched in Ca and Ti and depleted in Al and HREE. On the plots of $(Gd/Yb)_N$ vs. Al_2O_3/TiO_2 and CaO/Al_2O_3 (Fig. 5) the komatiites are located in the fields of Al-undepleted komatiites (Jahn et al. 1982).

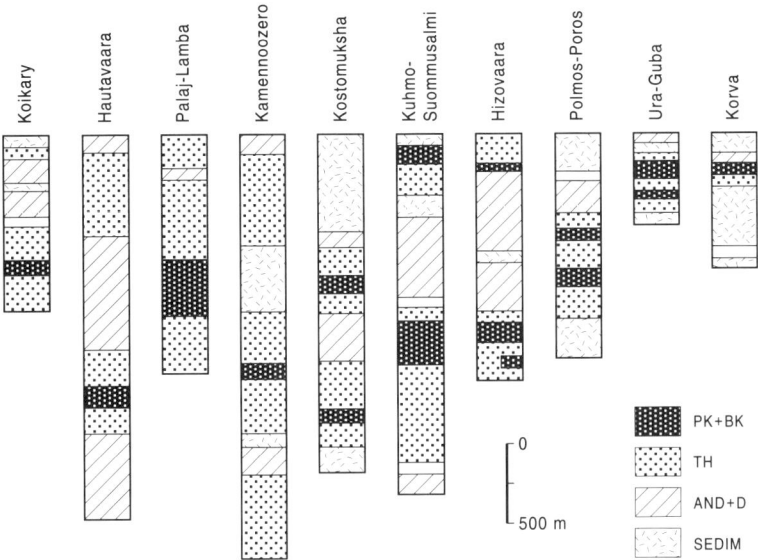

Fig. 2. Stratigraphic sections of the main Archaean greenstone belts of the Baltic Shield.

Generally, this means that garnet has not played an important role during the generation of the komatiitic melts, although a minor possible exception is found in the komatiites from Suommusalmi belt (E. Finland). Several authors have suggested that variations of CaO·Al$_2$O$_3$ ratio in komatiites are the result of migration of Ca in metamorphic processes (Arndt & Nesbitt 1982; Beswick 1982; Ludden & Gelinas 1982). That this may be the cause of the observed variation is illustrated by unsystematic changing of Sr against CaO (Fig. 4c).

REE distribution patterns

Komatiites of the Baltic Shield have also very different REE distribution patterns, which can be subdivided into four distinct classes. All these classes lie within the first typological group of

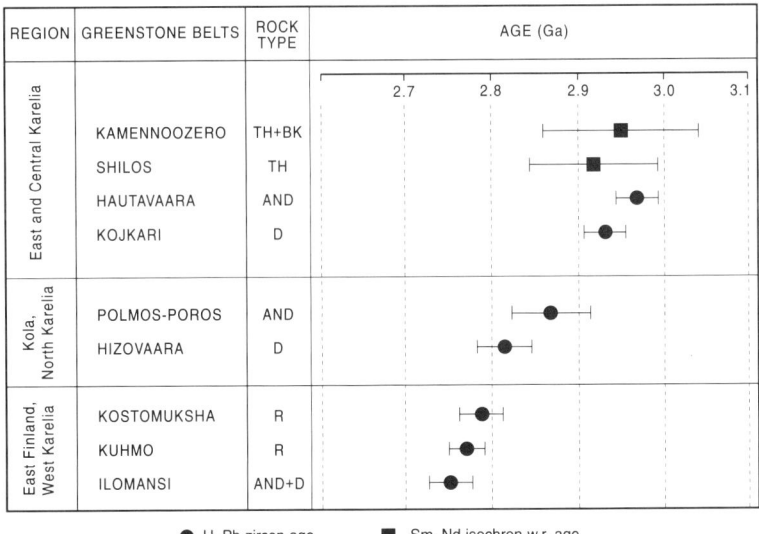

Fig. 3. Isotopic age of the volcanic rocks from the Archaean greenstone belts.

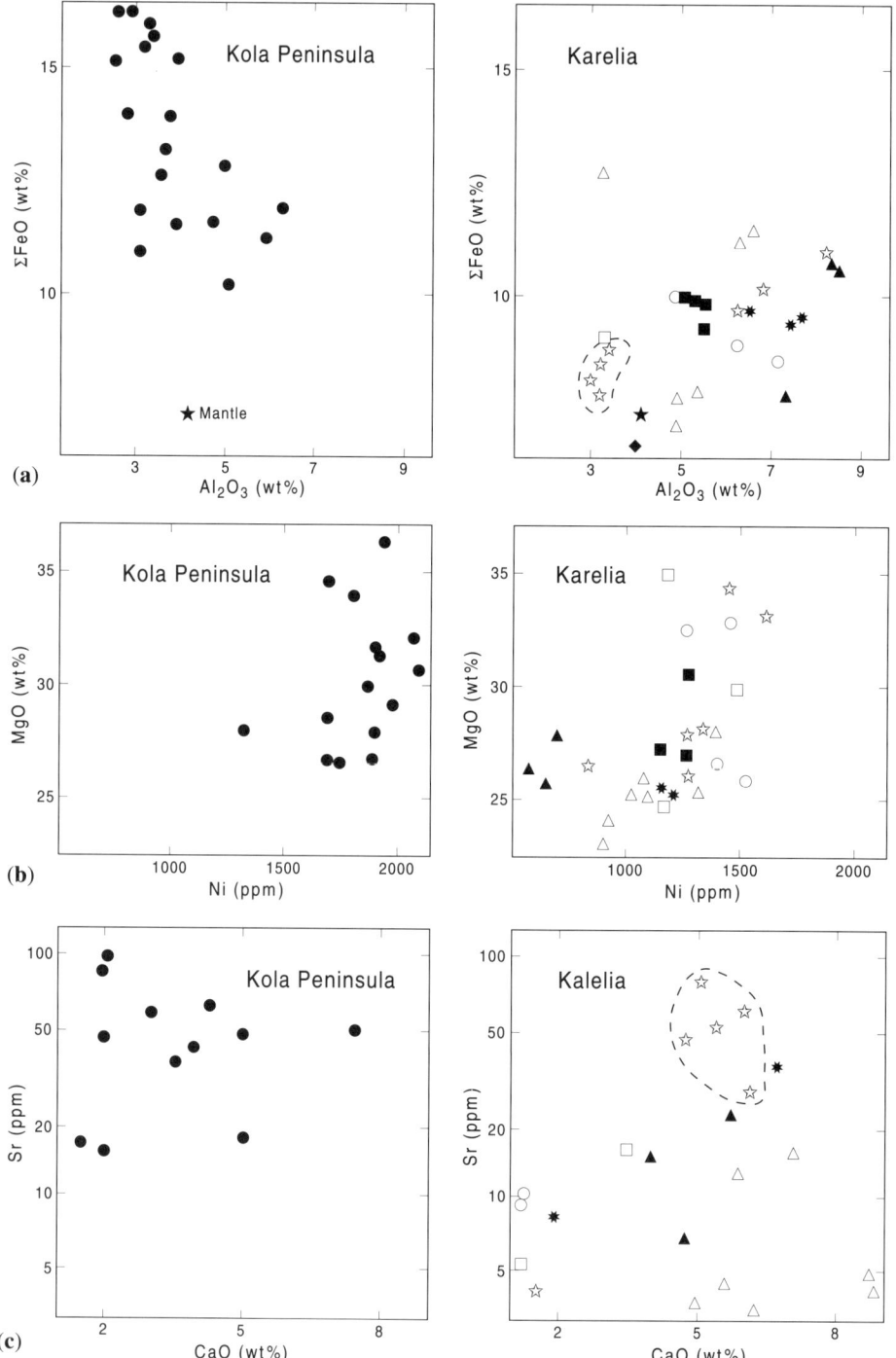

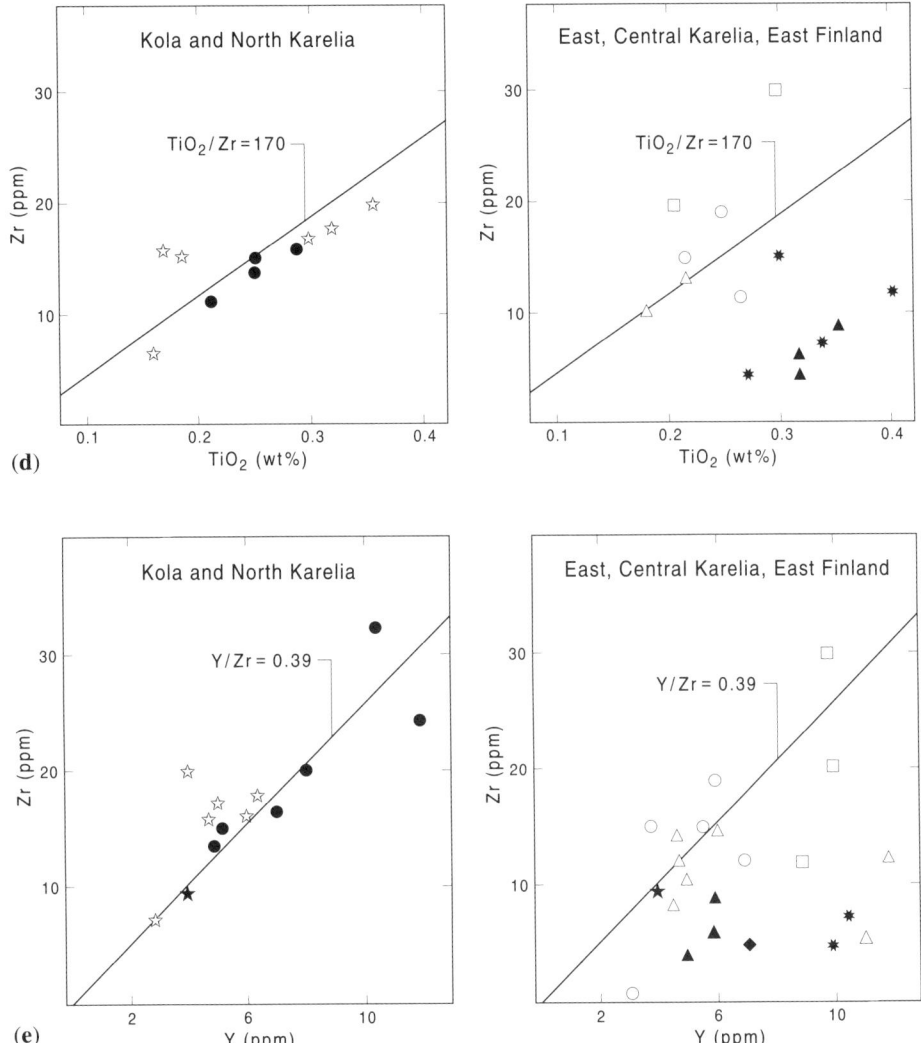

Fig. 4. Geochemical variations within the komatiites from selected greenstones in the Baltic Shield.

komatiites based on the HREE classification of to Jahn *et al.* (1982):

- Class I: with flat chondrite REE distribution [$(Ce/Yb)_N = 0.8–1.2$] and different overall REE content, includes komatiites from Kola peninsula, Central and East Karelia (Figs 6–8);
- Class II: komatiites with different degrees of LREE depletion [$(La/Sm)_N = 0.31–0.70$] and flat HREE [$(Gd/Yb)_N = 1.1–1.2$], occuring in almost all greenstone belts of the Baltic Shield (Central and East Karelia, Eastern Finland, Kola: Figs 6–8);
- Class III: with slight LREE enrichment [$(La/Sm)_N = .5–1.6$] and relatively flat HREE [$(Gd/Yb)_N = 1.1–1.3$] patterns – characteristic of komatiites from Central Karelia (Hautavaara belt) and E. Finland (Suommusalmi belt) (Figs 6–8);
- Class IV: includes komatiites with I–III classes REE distribution patterns, but with different degrees of negative Eu-anomaly ($Eu/Eu^* = 0.4–0.8$: Figs 6–8). In particular the komatiites of W. Karelia (Kostomuksha belt), and one sample of serpentinite from the Suommusalmi belt (E. Finland) with a uniquely large negative Eu anomaly up to 70% ($Eu/Eu^* = 0.04–0.1$) fall within this last group.

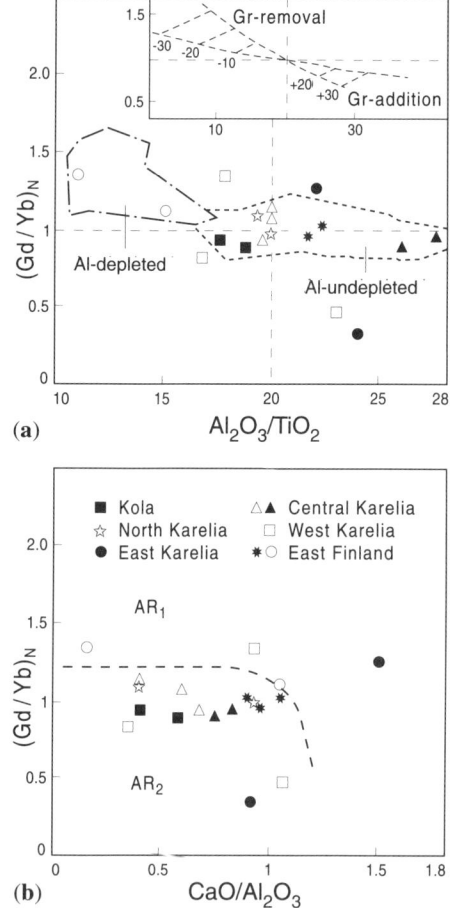

(a)

(b)

Fig. 5. Variation of Al_2O_3/TiO_2 and CaO/Al_2O_3 versus $(Gd/Yb)_N$ in the peridotitic komatiites from the Baltic Shield. Al-depleted and undepleted komatiites from Jahn *et al.* (1982) and Arndt (1986). Models of garnet addition and removal after Ohtani *et al.* (1986).

Petrogenesis

Modelling of the 4 geochemical classes of komatiites was attempted using petrological model calculations based on the equation of Shaw (1970) for partial melting and Rayleigh fractionation law. The results of modelling are summarized below:

- Class I: flat chondritic REE distributions for the high magnesian liquids (MgO > 25–30%) might be produced at 50–60% single-stage partial melting of the mantle source at P *c.* 35 kb. Compositionally similar (especially in MgO content) komatiites which differ in overall REE abundances can be explained by differences in the total overall REE content in their source.
- Class II: komatiites depleted in LREE and with flat HREE can be produced by secondary 25–30% partial melting at P = 25 kb of the mantle source, after 10–15% initial melt extraction. The composition of the first melt extracted could be basalt-komatiite, tholeiitic or dioritic, depending on the degree of LREE depletion.
- Class III: komatiites with slight LREE enrichment and slight HREE fractionation could be produced by partial melting of the mantle source at high pressure (P > 40 kb) with separation of garnet in residue (Ohtani *et al.* 1989). In this case komatiites must have CaO/Al_2O_3 and $(Gd/Yb)_N$ ratios more than 1.0, but as it has been shown earlier (Fig. 5) the komatiites of the Baltic Shield are all Al-undepleted. Moreover, the experiments suggest that at such pressures partial melts will be more alkaline in composition. Therefore, a more reasonable explanation of LREE enrichment of the komatiites is contamination by crustal material (e.g. Arndt & Jenner 1986). Komatiites in the Suommusalmi and Hizavaara belts with these LREE enriched patterns are

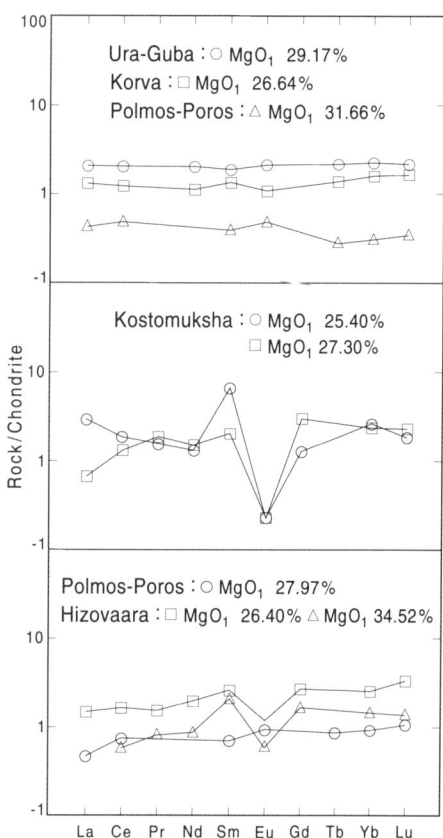

Fig. 6. REE distributions in the Type I komatiites.

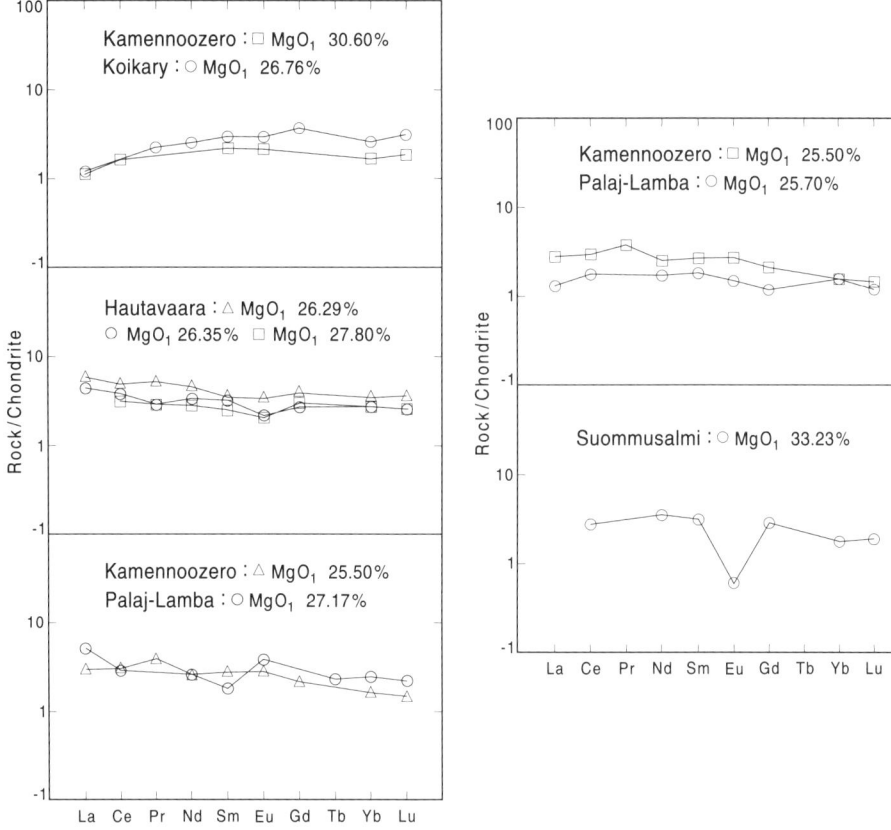

Fig. 7. REE distributions in the Type II komatiites.

underlined by andesite and dacite volcanics which may prove suitable contaminants in modifying REE patterns.

- Class IV: REE distributions with a negative Eu anomaly could be explained in different ways: by plagioclase fractionation, oxygen fugacity or by post-magmatic alteration. Plagioclase extraction from komatiite generation is unlikely, because this mineral does not appear on the liquids until low, less than 10% MgO values (Sun & Nesbitt 1978). So the most reasonable explanation of the negative Eu anomaly of up to 40% is the migration of Eu in hydrothermal post-volcanic fluid flow (as occurs in present day black smokers) with low oxygen fugacity (fO_2), as was experimentally demonstrated by Jahn et al. (1980) for MORB.

Sm–Nd isotopic systematics

The discussion of rare earth patterns given above suggests that either the geochemical komatiite

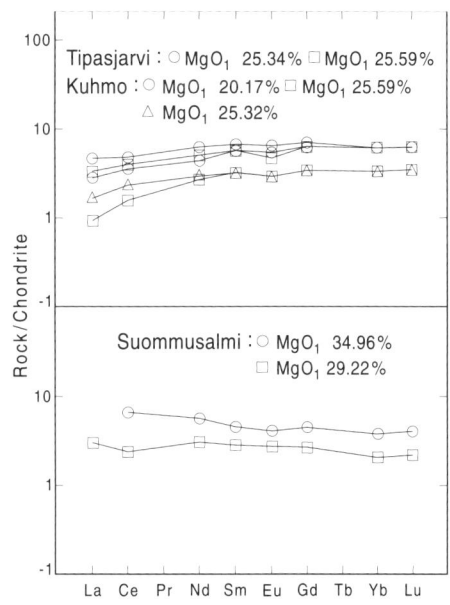

Fig. 8. REE distribution in the Type III komatiites.

mantle source was spatially heterogeneous or evolved with time. Thus further investigations were attempted using Sm–Nd and O isotope systematics in order to enhance increase our understanding of the Archaean mantle.

The komatiites of the Baltic Shield have initial ε_{Nd} values, calculated on U–Pb zircon age of overlying volcanics, ranging from +1.7 to +4.2 (Fig. 9). The isotopic data can be correlated to the three supposed geochemical types of komatiite mantle sources given above, and also with the 3 groups of greenstone belts associated with different U–Pb zircon ages of volcanism.

1. ε_{Nd} of +3.7 to +2.6. The komatiites with ε_{Nd} (T) from +3.7 to +2.6 (Table. 1) represent the greenstone belts of the 2.8–2.9 Ga age group. Their mantle source has overall REE content (particularly HREE) of 0.4–2.0 × chondrite. The fact that 7 data points form a good linear array with MSWD = 3.6 (Fig. 10) favours the hypothesis that komatiites from Kola peninsula and North and West Karelia (Kostomuksha belt) were erupted during a time span of (2.78+/–0.14 Ga) and that the isotopic characteristics of their respective mantle source were similar (ε_{Nd} = +3.4). This TDM model age of 2.78 +/–0.14 Ga is equal within errors to the U–Pb isochron zircon age (Table 1) which is thought to represent the time of volcanism in the greenstone belts. It should be noted that these samples are depleted in LREE to different degrees but have similar (flat) HREE distribution patterns (Figs 6–8). If such LREE heterogeneity is not reflected by ε_{Nd} values registered in the rocks, it could mean that LREE depletion occurred not long before the magma generation.

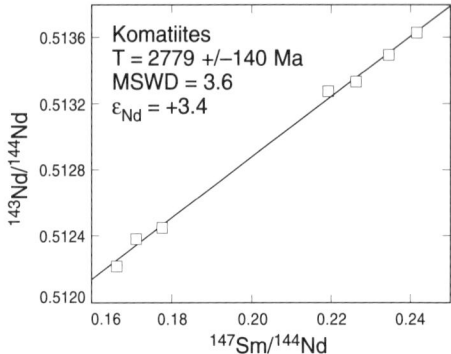

Fig. 10. Sm–Nd isochron plot for the komatiites of the Baltic Shield. Sample numbers from Table 1.

2. ε_{Nd} of +1.7 and +1.3. Two samples from Hautavaara and Palaj-Lamba belts has considerably lower ε_{Nd} values (+1.7 and +1.3, respectively; Table 1). This can be attributed to one of three processes.

(i) Contamination: mixing x parts of crust rock and one part of mantle-derived material (PK) to yield the isotopic composition of Hautavaara and Palaj-Lamba komatiites is described by the two-component mixing equation (Faure 1986):

$$x = \frac{C_{Nd}(DM[\varepsilon_{Nd}(PK) - \varepsilon_{Nd}(DM)]}{C_{Nd}(C)[\varepsilon_{Nd}(C) - \varepsilon_{Nd}(PkK)]},$$

where DM is the komatiites with $\varepsilon_{Nd(2.9)}$ = 4.0, PK is the komatiites of the Hautavaara and Palaj-Lamba belts and C is crustal rock. This was calculated in order to diminish $\varepsilon_{Nd(2.9)}$ = +4.0 from DM to the values of +1.7 and +1.3 observed in Hautavaara and Palaj-Lamba komatiites, 10–12% of underlining dacites (ε_{Nd} +/–1.35), or 30–25% of tholeiites ε_{Nd} = +0.5, or 8–5% of tonalities (ε_{Nd} = –0.3) must have been assimilated. These estimates seem unreasonable in the light of Zr, Ti and Y concentration (Figs 4d, e) in komatiites.

(ii) Secondary REE mobility: migration of Nd could have occurred during the Proterozoic (Svecokarelian) tectono-metamorphic event, as has been assumed for Tipasjarvi and Suommusalmi (Finland) komatiites (Tourpin et al. 1991; Gruau et al. 1992). This is unlikely because these rocks also yield Pb–Pb whole-rock and K–Ar amphibole Archaean ages (2940+/–62 and 2.540+/–40 Ma respectively). As these systems are much more sensitive than the Sm–Nd isotopic system to alteration processes it would appear unlikely that they would remain undisturbed and retain

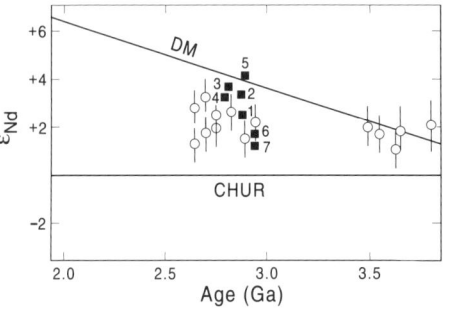

Fig. 9. Variation of ε_{Nd} versus time for the Archaean komatiites.

■ Komatiites of the Baltic Shield
1 Ura-Guba, 2 Polmos-Poros, 3 Hizovaara, 4 Kostomuksha, 5 Kamennoozero, 6 Hautavaara, 7 Palaj-Lamba

◊ Komatiites from different Archaean regions

Table 1. Sm–Nd isotopic whole-rock data for the peridotitic komatiites of the Baltic Shield

N	Sample	Location	Nd (ppm)	Sm (ppm)	$^{147}Sm/^{144}Nd$	$^{143}Nd/^{144}Nd$ +/–2s*	Age (Ga)	ε_{Nd} (T)
1	9016	Ura-Guba	1.674	0.626	0.2267	0.513339+/–14	2.87	+2.6
2	200	Polmos-Poros	2.108	0.576	0.1658	0.512214+/–10	2.87	+3.2
3	200-a	Polmos-Poros	1.739	0.509	0.1778	0.512445+/–12	2.87	+3.3
4	753	Korva	1.482	0.419	0.1713	0.512378+/–13	2.75	+3.7
5	576-4	Hizovaara	1.989	0.419	0.2199	0.513273+/–15	2.80	+3.8
6	746	Kostomuksha	2.790	1.110	0.2419	0.513630+/–8	2.79	+3.3
7	737-2	Kostomuksha	3.240	1.250	0.2349	0.513498+/–9	2.79	+3.2
8	224	Kamennoozero	2.835	0.765	0.1637	0.512214+/–11	2.92	+4.2
9	427-2	Hautavaara	1.989	0.613	0.1869	0.512530+/–8	2.94	+1.7
10	275-10	Palaj-Lamba	0.849	0.304	0.2171	0.513102+/–13	2.93	+1.3

$^{143}Nd/^{144}Nd$ ratios are normalized to La Jolla $^{143}Nd/^{144}Nd$ = 0.511860. Chondritic values used to calculate ε_{Nd} (T) were $^{143}Nd/^{144}Nd$ = 0.512636 and $^{147}Sm/^{144}Nd$ = 0.1967 (Wasserburg et al. 1981).
* Uncertainties in the last decimal place.

Archaean ages throughout a metamorphic episode which affected Sm–Nd systematics.
(iii) If processes (i) and (ii) do not yield satisfactory explanations for the low ε_{Nd} values (+1.3 to +1.7) in komatiites from Hautavaara and Palaj-Lamba belts, it can be assumed that the mantle reservoir with which these komatiites re-equilibrated, was relatively 'enriched'.

3. ε_{Nd} of +4.2. The most depleted ε_{Nd} value (+4.2) was observed in komatiites from the Kamennoozero belt (E. Karelia) (Table 1, N 8). Such a high ε_{Nd} value might be explained by depletion of the mantle source by removal of a significant mass of the continental crust, represented by adjacent gneisses of Vodlozero region, with a U–Pb zircon age of 3.14+/–0.03 Ga and Sm–Nd model age of >3.3 Ga (ε_{Nd} = –2 to +1) (Lobach-Zhuchenko et al. 1993). In this case, it might be expected that in the other parts of the Baltic Shield less volume of continental material would have been extracted from the mantle before 2.9 Ga.

Discussion

In the present day, temperature variations within the convecting mantle, away from subduction zones, are thought to be small and the potential temperature of the upper mantle is estimated at c. 1280°C (McKenzie & Bickle 1988; Campbell & Griffiths 1993). The occurrence of high Mg komatiites in Archaean greenstone belts has convinced most geologists that the Archaean mantle was appreciably hotter than the mantle today. If 1580–1650°C is taken as the eruption temperature of peridotitic komatiite (MgO > 20%, Green 1979; Nisbet et al. 1993), then the mantle source may have been at temperatures of c. 2000°C at pressures of 18 Gpa, which would correspound with a mantle potential temperature of 1900°C. These estimated temperatures suggest that Archaean komatiites could have formed in response to hot matle plumes (Campbell et al. 1989; Campbell & Griffiths 1992; Nisbet et al. 1993; Lasher & Arndt 1995).

Laboratory and theoretical studies of plume dynamics suggest that a plume cannot displace the overlying mantle and ascend rapidly until a large amount of bouyancy has collected in the spherical head (Campbell et al. 1989; Campbell & Griffiths 1992). This dynamic model predicts that the head of an Archaean plume would likley have had a diameter between 600 to 800 km when it impacted with the base of the lithosphere. If however, the head of the plume moved away from its source it is likely that the plume conduit became inclined. In such a senario the material in the conduit will rise as an inclined plume and may entrain surrounding mantle. The net result of such a model is that the head could contain a mixture of mantle material derived from different melting regimes.

From the whole geochemical and isotopic data a model is proposed for the Baltic Shield komatiites which involves an inclined mushroom-shaped mantle plume. In such a model the following stages are indentified:

(1) Early stage (3.05–2.9 Ga): mixing of material derived from moderately depleted and undepleted mantle to produce the komatiites from East and Central Karelia. These komatiites are characterized by high Al_2O_3 (6–10 %), moderately low FeO (6–13%), low Ti/Zr and

Y/Zr ratios, slightly depleted LREE and positive ε_{Nd} (+4.2).

(2) Middle Stage (2.9–2.8 Ga): lateral spreading of plume producing komatiites in the Kola Pennisula, North Karelia and Kostamuksha. These komatiites are characterised by high MgO (28–36%) and FeO (10–20%), chondritic Y/Zr ratios, flat LREE profiles and ε_{Nd} ranging from +2.6 to +3.7, all of which probably reflects melts derived from the axial zone of the plume.

(3) Late Stage (2.8–2.7 Ga): melting of strongly depleted mantle which is marginal to the plume head. This produces the komatiites of Eastern Finland which are characterized by low $(Ce/Sm)_N$, and low Ti/Zr and Y/Zr ratios.

Conclusions

The investigation of the major and trace element geochemistry, whole-rock Nd and O isotopic composition and petrogenesis of komatiites of the Eastern Baltic Shield shows a good correlation of all these features, which could be more explained by lateral long-term heterogeneity or evolution of their mantle source in the Late Archaean, that lay beneath the Eastern Baltic Shield.

(1) The komatiites were generated by 50–60% partial melting of the mantle source (P c. 35 kb) without garnet in the residue.

(2) There were several geochemical mantle source types for komatiites with different overall HREE content:

Type I: 0.4–2.0 × chondrite (depleted)
Type II: 2.0–3.0 × chondrite (normal)
Type III: 3.0–5.0 × chondrite (enriched)

(3) The most depleted in REE mantle source, assigned to Kola peninsula and N. Karelia, were enriched in total Fe and Sr and depleted in Al_2O_3 and CaO compared with the other parts of the shield.

(4) The three geochemical types of mantle source are conforms to three assumed age groups of greenstone belts volcanism:

Group 1: 3.05–2.9 Ga = type II (East, Central Karelia)
Group 2: 2.9–2.8 Ga = type I (Kola peninsula, N. Karelia)
Group 3: 2.8–2.7 Ga = type III (W. Karelia, E. Finland)

(5) Komatiites of Type I and Group 2 have $\varepsilon_{Nd}(T)$ values ranging from +3.7 to +2.6 and form a Sm–Nd whole-rock isochron of 2.78+/–0.14 Ga (SNd =+3.4); this TDM model age is equal within error to the U–Pb zircon ages, which are thought to represent the time of volcanism in the greenstone belts.

(6) The mantle source of Eastern and Central Karelia komatiites, with normal REE distribution patterns (Type II), was heterogeneous in initial Nd isotopic composition: enriched (ε_{Nd} = +4.2) in East Karelia and depleted (ε_{Nd} = +1.3 to +1.7) in Central. Karelia.

(7) There is good negative correlation of $\varepsilon_{Nd}(T)$ and $\delta^{18}O$ whole-rock values, which could be explained by assumption that Sm–Nd and O isotopic systems of komatiites mantle source evolve simultaneously in Late Archaean.

This research was held in co-operation with the Indian National Geophysical Research Institute and sponsored by the International Science Foundation, Grant R1J000. The author would like to thank Drs K. Gopalan, R. Srinivasan and B. Rao for their help in ICP REE and trace element analysis.

References

ARNDT, N. T. 1986. Spinifex and swirling komatiite in a lava lake, Munro township, Canada. 1986. *Precambrian Research*, **34**, 139–155.

—— & JENNER, G. A. 1986. Crustally contaminated komatiites and basalts from Kambalda, Western Australia. *Chemical Geology*, **56**, 229–255.

—— & JOCHUM, K. P. 1990. Komatiites: unreliable witness of the Archaean mantle. *Third International Archaean Symposium, Abstract volume*, 147–148.

—— & NESBITT, R. W. 1982. Geochemistry of the Munro Township basalts. *In*: ARNDT, N. T. & NISBET, E. G. (eds) *Komatiites*. George Allen and Unwin, London, 309–330.

——, TEIXEIRA, N. A. & WHITE, W. M. 1989. Bizarre geochemistry of komatiites from the Crixás greenstone belt, Brazil. *Contributions to Mineralogy and Petrology*, **101**, 187–197.

BESWICK, J. 1982. Some geochemical aspects of alteration and genetic relations in komatiite suites. *In*: ARNDT, N. T. & NISBET, E. G. (eds) *Komatiites*. George Allen and Unwin, London, 281–308.

BORTHWICK, I. & HARMON, R. S. 1982. A note regarding as an alternative to BrF5 for oxygen isotope analysis. *Geochimica et Cosmochimica Acta*, **46**, 1665–1668.

CAMPBELL, I. H. & GRIFFITHS, R. W. 1992, The changing nature of mantle hotspots through time: implications for the chemical evolution of the mantle. *Journal of Petrology*, **92**, 497–523.

—— & —— 1993. The evolution of the mantle's chemical structure. *Lithos*, **30**, 389–399.

——, —— & HILL, R. I. 1989. Melting in an Archaean mantle plume: heads it,s basals, tails it's komatiites. *Nature*, **339**, 697–699.

FAURE, G. 1986. *Principles of Isotopic Geology* (2nd edn). Wiley, New York.

GREEN, D. H. 1979. Genesis of Archaean peridotite magmas and constraints on Archaean geothermal gradient and tectonics. *Geology*, **3**, 15–18.

GRUAU, G., TOURPIN, S., FOURCADE, S & BLIAS, S. 1992. Loss of isotopic (Nd,O) and chemical (REE) memory during metamorphism of komatiites: new evidence from eastern Finland. *Contributions to Mineralogy and Petrology*, **112**, 66–82.

JAHN, B.-M., AUVRAY, B., BLIAS, CAPDEVILA, R., CORNICHET, J., VIDAL, P. & HAMEURT, J. 1980. Trace elements geochemistry and petrogenesis of Finnish greenstone belts. *Journal of Petrology*, **21**, 201–244.

——, GRUAU, G. & GLIKSON, A. Y. 1982. Komatiites of the Onverwacht Group, S. Africa: REE geochemistry, Sm/Nd age and mantle evolution. *Contributions to Mineralogy and Petrology*, **80**, 25–40.

LASHER, C. M & ARNDT, N. T. 1995. REE and Nd isotope geochemistry, petrogenesis and volcanic evolution of contaminated komatiites at Kambalda, Western Australia. *Lithos*, **34**, 127–157.

LOBACH-ZHUCHENCO, S. B., CHEKULAYEV, V. P., LEVEHENKOV, O. A., CHEKULEAV, V. P & KRYLOV, I. N. 1993. Archaean rocks from south-eastern Karelia (Karelian granite greenstone terrain). *Precambrian Research*, **62**, 375–397.

LUDDEN, J. N. & GELINAS, L. 1982. Trace element characteristics of komatiites and komatiitic basalts from the Abitibi metavolcanic belt of Quebec. In: ARNDT, N. T. & NISBET, E. G. (eds) *Komatiites*. George Allen and Unwin, London, 331–346.

MCKENZIE, D. P. & BICKLE, M. J. 1988. The volume and composition of melt generated by extension of the lithosphere. *Journal of Petrology*, **29**, 625–679.

NISBET, E. G., CHEADLE, M. J., ARNDT, N. T. & BICKLE, M. J. 1993. Constraining the potential temperature of the Archaean mantle: a review of the evidence from komatiites. *Lithos*, **30**, 291–307.

OHTANI, E., KATO, T. & SAWAMOTO, H. 1986. Melting of a model chondritic mantle to 20 GPa. *Nature*, **322**, 352–353.

——, KAWABE, I., MORIYAMA, J. & NAGATA, Y. 1989. Partitioning of elements between majorite garnet and melt and implication for petrogenesis of komatiite. *Contributions to Mineralogy and Petrology*, **103**, 263–269.

RICHARD, P., SHIMIZI, N., & ALLEGRE, C. J. 1976. ^{143}Nd/^{144}Nd, a nature tracer: an application to oceanic basalts. *Earth and Planetary Science Letters*, **31**, 269–273.

SHAW, D. M. 1970. Trace element fractionation during anatexis. *Geochimica et Cosmochimica Acta*, **34**, 237–243.

SUN, S.-S. & NESBITT, R. W. 1978. Petrogenesis of Archaean ultrabasic and basic volcanics: evidence from rare earth elements. *Contributions to Mineralogy and Petrology*, **65**, 301–325.

TOURPIN, S., GRUAU, G., BLIAS, S. & FORCADE, S. 1991. Resetting of REE, and Nd and Sr isotopes during carbonitization of komatiite flow from Finland. *Chemical Geology*, **90**, 15–29.

Age and geodynamic aspects of the oldest rocks in the Precambrian Belomorian Belt of the Baltic (Fennoscandian) Shield

E. V. BIBIKOVA[1], T. SKIÖLD[2] & S. V. BOGDANOVA[3]

[1] *Vernadsky Institute of Geochemistry and Analytical Chemistry of the Russian Academy of Sciences, Ul. Kosygina 19, 117 334 Moscow, Russia*
[2] *Laboratory for Isotope Geology, Swedish Museum of Natural History, PO Box 50007, S-10405 Stockholm, Sweden*
[3] *Department of Mineralogy and Petrology, Institute of Geology, Lund University, Sölvegatan 13, S-223 62 Lund, Sweden*

Abstract: The Belomorian Belt in the White Sea region was until recently considered the oldest core of the Baltic Shield but is now recognized to be essentially Late Archaean. This study reports U-Pb zircon ages and whole-rock Nd-isotopic results for aluminous metasedimentary gneisses and tonalitic–trondhjemitic–granodioritic intrusions. Metamorphic zircons from the aluminous gneisses date the earliest high-grade metamorphism to about 2.86 Ga. This was a low- to moderate-pressure granulite-facies event. Multigrain fractions of detrital zircons from these gneisses preserve little memory of protolith ages, but the whole-rock Sm-Nd DM-model ages are below 3.0 Ga which excludes dominantly Early to Middle Archaean detritus sources and indicates a short crustal prehistory. For the tonalitic–trondhjemitic intrusive rocks, U-Pb zircon and Sm-Nd model ages define two generations formed *c.* 2.8 and 2.74–2.72 Ga ago. The oldest events in the Belomorian Belt can thus be correlated with Late Archaean developments in the adjoining Karelian Province further southwest in the Baltic Shield, but occurred at an active continental margin. This supports a Late Archaean accretional/collisional relationship between these two crustal segments.

In the Archaean domain of the Baltic Shield, the Belomorian Mobile Belt extends along the shores of the White Sea (Fig. 1) and separates the crustal province in Karelia from Archaean terrains in the Kola Peninsula furthest in the northeast. Although the Belomorian Belt was until recently considered to be the oldest core of the Baltic Shield (e.g. Stenar' 1988), most of its rocks and events of metamorphism have yielded Late Archaean U-Pb zircon ages (Tugarinov & Bibikova 1980; Bibikova *et al.* 1993b; Bogdanova & Bibikova 1993). However, until now, the oldest history of the belt was poorly constrained which prevented detailed correlation of its Archaean with that of the granite–greenstone terrain in Karelia and with other crustal provinces of the Baltic Shield and other cratons.

According to recent structural studies, the Belomorian Belt is a Late Archaean collisional pile of metasedimentary, metavolcanic and meta-plutonic rocks which has been folded several times in the Archaean and differently in the Palaeoproterozoic (Miller & Milkevich 1995).

The main lithological components of the Belomorian tectonostratigraphy are garnet- and kyanite-bearing aluminous gneisses ('the Chupa strata'), metavolcanic amphibolites and amphibole-bearing gneisses ('the Khetolambina strata'), and tonalitic–trondhjemitic-granodioritic (TTG) gneisses (the Keret' and other 'strata'). These can be traced along the whole *c.* 600 km exposed length of the Belomorian Belt (Fig. 1). All the rocks have been multiply metamorphosed and migmatized already during the Archaean (e.g. Glebovitsky *et al.* 1978; Volodichev 1990).

Previous age determinations on unfractionated zircon concentrates with large average discordancies from various samples of the Chupa aluminous gneisses have produced an upper intercept age of *c.* 2740 Ma (Tugarinov & Bibikova 1980). This age was explained as being due to the presence of detrital as well as metamorphic zircons of different ages.

The zircon ages of TTG-rocks from the central and northeastern parts of the Belomorian Belt range between 2.8 and 2.7 Ga (Tugarinov & Bibikova 1980; Bogdanova & Bibikova 1993; Daly *et al.* 1993). A similar age range has been obtained for metaandesites in the western Belomorian Belt (U-Pb zircon, Bibikova *et al.* 1995b) and for the zircons from the metavolcanics at the northeastern margin of the Belomorian Belt (Mitrofanov & Pozhilenko 1991; Daly *et al.* 1993).

The present study was carried out to shed additional light on the earliest development of the Belomorian Belt. It reports new U-Pb zircon ages

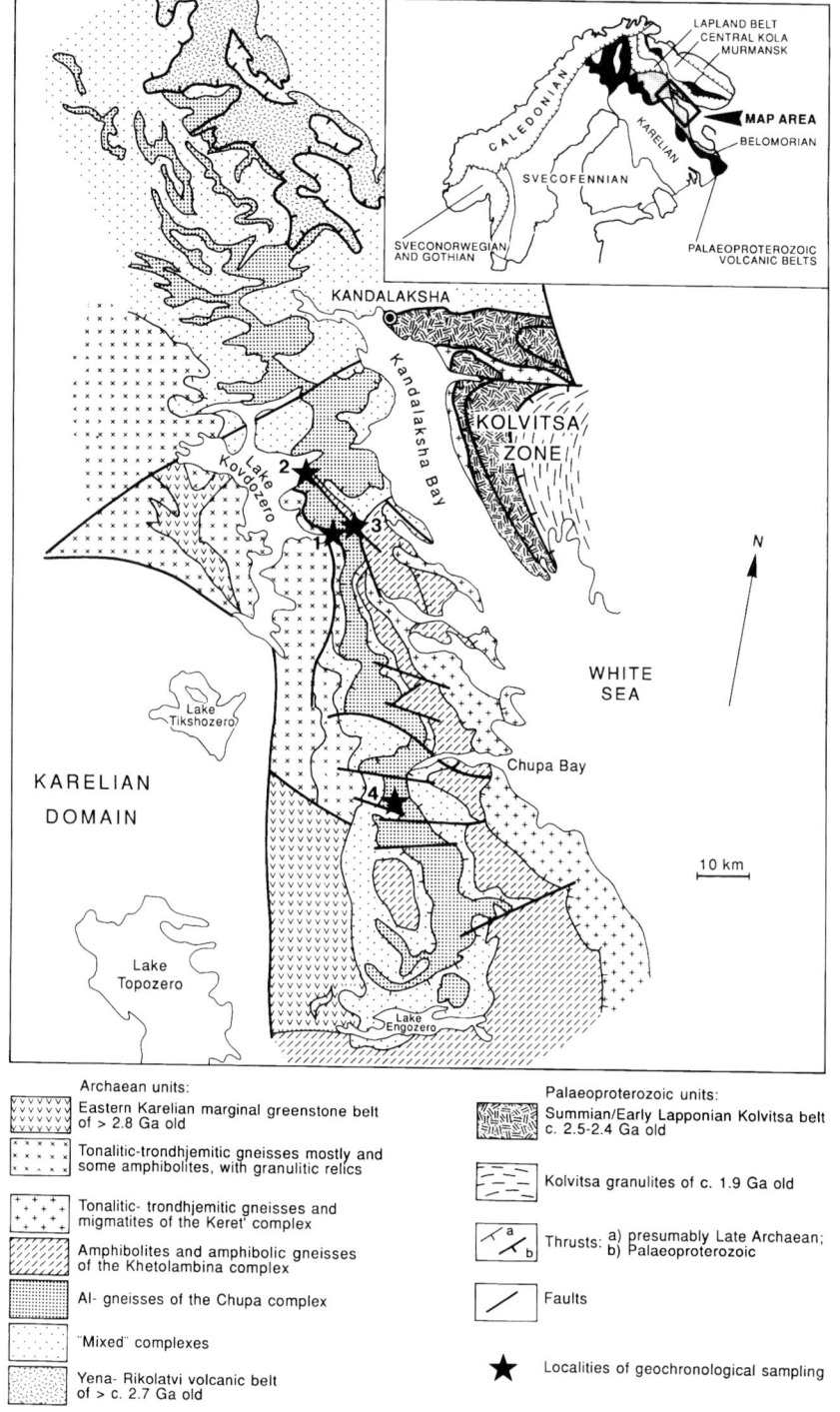

Fig. 1. Sketch map of the Belomorian Belt (modified after Bogdanova & Bibikova 1993 and Miller & Milkevich 1994, with the latter authors' permission). The sites of sampling for the geochronological study are: 1- Tupaya Bay, two different localities, 2 - Seryak, 3 - Lyagkomina, 4 - Chupa.

and Sm-Nd isotopic results for the aluminous gneisses and the TTG rocks of the Belomorian Belt that are considered to be its oldest components.

The aluminous gneisses of the Chupa unit

The widespread occurrence of aluminous gneisses is a characteristic feature of the Belomorian Belt (Fig. 1) that distinguishes it from the adjacent Karelian granite–greenstone terrain.

The ages of these aluminous gneisses have been poorly constrained and their origins may vary. While most are metamorphosed equivalents of essentially metasedimentary lithologies such as greywackes, tuffites and pelites (e.g. Shurkin et al. 1962), others appear to be products of metamorphic alteration of igneous salient rocks (Volodichev 1990). It has also been observed that aluminous gneisses mark thrusts and zones of migmatization and retrograde metamorphism formed during the different stages of the tectonic development of the Belomorian Belt (e.g. Salie et al. 1985). In some places, garnet–kyanite–mica gneisses are distinctly metasomatic products formed from high-Al tonalites or intermediate metavolcanics occurring within the development area of the aluminous gneisses. Because the aluminous gneisses of the Chupa 'strata' experienced a more complex metamorphic development than other Belomorian rock units, equilibration of the metamorphic assemblages has not always been attained. This is particularly true of areas where the rocks have been reworked in the Palaeoproterozoic.

The most frequent mineral assemblages in the aluminous gneisses are: garnet ± sillimanite + biotite + plagioclase + quartz, and garnet + kyanite + biotite ± muscovite + plagioclase ± K-feldspar + quartz. However, the parageneses range from the granulite- and upper amphibolite facies to those of the lower amphibolite facies. Elevated pressures reaching 10 to 12 kbar have been very character-istic of the metamorphic development of the aluminous gneisses both in the Late Archaean and in the Palaeoproterozoic (Volodichev 1990).

The samples used for age determination have metasedimentary protoliths. They were chosen from localities at various levels of the Belomorian tectonostratigraphy and feature different structural and metamorphic histories.

Samples 203, 208 and 602 represent fine- to medium-grained garnet + biotite ± sillimanite + plagioclase + quartz gneisses from the rock complex around the Tupaya Bay of Lake Kovdozero, northeastern Karelia (site 1 in Fig.1). In this locality, Palaeoproterozoic reworking is weak and several Archaean igneous and metamorphic events have been dated previously (Bibikova et al. 1993a, b; Bogdanova & Bibikova 1993; Bibikova et al. 1995a). The earliest mineral association is made up of high-Mg garnet, biotite, plagioclase (oligoclase to andesine) and rare K-feldspar. The cores of the garnets contain inclusions of sillimanite and Ti-rich biotite. A subsequently formed mineral assemblage is very similar to the first one, but its high-Mg garnet forms overgrowths on the crystals of early garnet as well as separate grains that occur together with flakes of Ti-biotite arranged along a foliation. This latter paragenesis was formed at 750–800° C and c. 7 to 8.5 kbar, while the early metamorphism reached c. 750° C at 6 to 7 kbar (e.g. Volodichev 1990; Drygova's data in Bibikova et al. 1993b; Glebovitsky et al. 1993). Finally, a superimposed stage of amphibolite-facies metamorphism resulted in the development of yet another generation of garnet and biotite associated with kyanite. The PT-conditions during that stage were 650–700° C at c. 9 kbar (Bogdanova, unpublished data).

Samples 91011, TS1 and TS2 from Lyagkomina (site 3 in Fig. 1) were taken from a zone of marked Palaeoproterozoic structural and metamorphic reworking of the aluminous gneisses situated c. 5 km to the northeast of the Tupaya Bay locality (for details cf. Bibikova et al. 1993b). The grain sizes of the kyanite- and mica-bearing gneisses in this zone vary from very fine to coarse. The fine-grained gneisses, which preserve signs of the earliest deformation, are often found in relic lenses and in the cores of rootless isoclinal folds surrounded by coarser kyanite- and mica-bearing gneisses that were developed mostly during the Palaeoproterozoic folding and metamorphism. Sample Lg92, used for Sm-Nd model age determination, and sample T-44 Seryak (site 2 in Fig. 1) derive from two such relic bodies of fine-grained, little reworked gneisses that are less reworked.

Estimates of the PT conditions of metamorphism for the earliest, Archaean stages of deformation in the Lyagkomina locality are 750–800° C at 10–12 kbar (peak metamorphism). The retrograde stage was marked by temperatures dropping to 570° C at 7–8 kbar. The PT-conditions in the coarse-grained gneisses which outline the Palaeoproterozoic folds and carry large kyanites that define a lineation ranged between 650° C at 8 kbar and 530° C at 5.5 kbar (Bogdanova, unpublished data briefly referred to in Bibikova et al. 1993b).

Sample 556 Chupa from site 4 in Fig. 1 represents zircon fractions from a rock previously studied by Tugarinov & Bibikova (1980). This is a coarse-grained garnet–mica gneiss typical of the rocks in the central part of the Belomorian Belt which have been affected markedly by

Palaeoproterozoic anatectic processes and are cut by 1.8–1.75 Ga old pegmatites (e.g. Tugarinov & Bibikova 1980). Here, the PT- conditions of metamorphism were between 650 and 500° C at 8 to 5 kbar (cf. Salie 1985).

Because of the complex metamorphic history of the aluminous gneisses and their mostly metasedimentary origin, this study paid particular attention to assessing the morphology of the zircons before using them for the U-Pb isotopic work.

The zircons in the studied samples are highly heterogeneous. At least two generations and several morphological types can be recognized.

Nearly half of the zircon population in sample 208 from the Tupaya Bay locality is represented by transparent, homogeneous grains with brilliant luster and light colour. They vary from short-prismatic to isometric in shape and are always multifaceted (Fig. 2A & C). Such a morphology is often explained by the crystallization of the zircons under granulite-facies conditions (e. g. Masamichi & Hi 1983). The rest of the zircons consist of brownish, semi-transparent or opaque, highly fractured grains. Oval as well as more elongated shapes can be observed. The surfaces of many of the grains have been corroded but are often healed by transparent, multifaceted overgrowths with a bright luster. A range of internal structures indicates variable origins of the zircons. Some grains carry cores suggesting a largely detrital provenance (Fig. 2B & D). Many are zoned and possibly grew during retrograde migmatization.

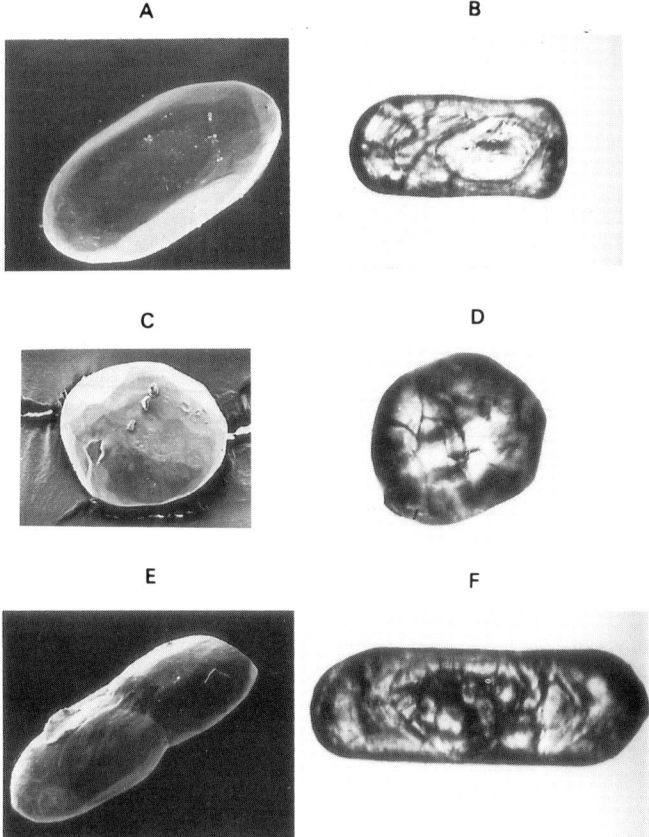

Fig. 2. Photomicrographs showing examples of the dated zircons from the aluminous and TTG gneisses of the Belomorian Belt. The crystals are between 100 and 125 μm wide. (**A**) Scanning electron microscope (SEM) photograph of short prismatic, well-formed transparent zircon crystals from the aluminous gneiss of the Tupaya Bay locality (sample 602); (**B**) zircon grain with a core of igneous, finely zoned zircon from the same sample (photograph in transmitted light); (**C**) SEM photograph of isometric, multifaceted, transparent ('granulitic') zircon from the Tupaya Bay locality (sample 208); (**D**) isometric dark zircon from the same sample (photograph in transmitted light); (**E**) SEM photograph of igneous 'dumb-bells' zircon crystals from a meta-tonalite at Tupaya Bay (sample 195); (**F**) typical crystals of the prismatic zircons with fine igneous zonation (sample 195; photograph in transmitted light).

The same kinds of zircon as in sample 208 are present also in sample 602, but in that case 70% belong to the short-prismatic to isometric type.

Short-prismatic, transparent zircons predominate in sample T-44 from the Seryak locality. In sample 91011 from the Lyagkomina locality there are long-prismatic transparent zircons in addition to the types described above.

In sample 556 from Chupa, the zircons are mainly oval, brownish and semi-transparent with perceptible detrital cores. However, isometric, transparent crystals form about 20% of the zircon population.

Tonalitic–trondhjemitic–granodioritic (TTG) gneisses

TTG-gneisses form another widespread group of rocks in the Belomorian Belt. Similarly to the aluminous gneisses and the amphibolites, they have undergone complex deformation (Kotov & Zinger 1985). The predominant mode of occurrence is in sheet-like bodies which are mostly concordant with their host rocks but have strongly tectonized contacts. Together, the different lithologies form banded units. Due to this, many of the TTG gneisses in the Belomorian Belt have previously been regarded as metasedimentary in origin (e.g. Shurkin et al. 1962). However, in a number of localities the igneous nature of the TTG rocks is evident. In the Tupaya Bay locality (Fig.1, site 1), for instance, the tonalites represented by sample 195 have clearly intruded country-rock amphibolites and contain breccia fragments of these rocks. Distinct, small relics of fine-grained aluminous gneisses also occur in the tonalitic body of the Lyagkomina locality (Fig. 1, site 3). Igneous textures are common and the chemical compositions similarly indicate igneous origins (cf. Kotov & Zinger 1985).

The meta-tonalites of samples 195, 91012 and 91013 are composed of medium-grained subhedral andesine (An 40-45) with numerous inclusions of epidote, long-prismatic apatite, allanite, well formed amphibole and biotite, and quartz. These minerals represent an igneous assemblage, whereas a later generation of biotite, bluish-green amphibole and garnet outlines a foliation in the fine-grained, partly recrystallized matrix and is metamorphic. Sample 91013 is a strongly migmatized tonalitic gneiss with new microcline overgrowing the older textures.

The zircons from these tonalites are prismatic with smooth terminations, grey in colour, transparent and semi-transparent, and show fine internal zoning (Figs 2E & F). There are characteristic 'dumb-bell' intergrowths of the zircon crystals which are typical of igneous zircons in tonalitic rocks (Bibikova 1989).

The zircons from the TTG-gneisses of the Lyagkomina locality (samples 91012 and 91013) are similarly prismatic, semi-transparent and highly fractured. In addition, however, there also occur metamorphic, very long-prismatic crystals as well as isometric and transparent ones.

Analytical procedure

The U-Pb and Sm-Nd isotope analyses were carried out in the Laboratory for Isotope Geology of the Swedish Museum of Natural History in Stockholm.

The U-Pb method. For isotopic dating, the zircons were separated into size- and morphological fractions which were upgraded further under the microscope until limited numbers of crystals remained. Some fractions were abraded. The analytical procedure followed that of Krogh (1973, 1982). One-inch minibombs, cation chromatography in 50 µl resin and sample loading in a clean-air environment gave reproducible total lead blanks of about 30 pg. A ^{205}Pb spike was used for small samples, while some of the larger ones were aliquoted, and a ^{208}Pb spike was used. Both spikes were mixed with uranium spike containing approximately equal amounts of ^{233}U and ^{235}U. The mass-spectrometric measurements were carried out with a Finnigan MAT 261 apparatus using static multicollector detection. Lead was loaded as nitrate on a single rhenium filament with a weak phosphoric acid and silica gel. Uranium was run in double filament configuration. Regression analysis was performed according to Ludwig (1991) and the ages are given at the 2σ level. The error in the Pb/U isotopic ratios is 0.3%.

For sample 556, the isotope analyses were performed at the Vernadsky Institute of Geochemistry and Analytical Chemistry in Moscow. The chemistry of large zircon from this sample was determined after Krogh (1973), the blank being 300 pg. A Cameca TSN 206A mass-spectrometer was used for the isotopic measurements. The precision of the Pb and U isotopic ratios was 1%.

The Sm-Nd method. All the data reported are from whole-rock samples. The rocks were dissolved in Krogh-type teflon bombs at 200°C for 3 days. They were spiked with a mixed ^{147}Sm-^{150}Nd tracer. REE were separated as a group using HCl in standard cation exchange columns for chromatography. Sm and Nd were subsequently separated from each other and from the rest of the REE in a second cation exchange column with alpha-hydroxyisobutyric acid that was pH-adjusted

Table 1. U-Pb zircon data for Al-gneisses and TTG rocks from the Belomorian Belt

Fraction (μm)	Weight (mg)	U (ppm)	Pb rad (ppm)	$^{206}Pb/^{204}Pb$ meas.	Radiogenic (at. %)* ^{206}Pb	^{208}Pb	Atomic ratios $^{206}Pb/^{238}U$	$^{207}Pb/^{235}U$	$^{207}Pb/^{206}Pb$ (Ma)
1	2	3	4	5	6	7	8	9	10
Al-gneisses, the Tupaya Bay locality									
Sample 208									
>125, ast, 6xx	0.107	175	90.0	17 500	78.07	6.29	0.4664	12.9450	2837 ± 1
125–100, ast	0.090	213	104.9	14 000	76.89	7.77	0.4403	12.1690	2830 ± 1
>125, ast	0.119	224	88.3	8 200	78.40	6.21	0.3593	9.7710	2803 ± 3
100–75, art	0.090	222	125.9	515	77.96	6.96	0.5125	13.7330	2779 ± 10
100–75, rt	0.206	246	133.1	4 000	78.16	6.78	0.4905	13.1040	2774 ± 2
<125, ard	0.104	827	436.1	14 000	80.39	4.65	0.4923	12.6830	2715 ± 2
125–100, adr	0.106	808	404.1	8 600	81.91	3.05	0.4759	12.1010	2693 ± 2
100–75, adr	0.147	737	375.1	3 000	80.17	5.02	0.4686	11.9900	2703 ± 18
Ru	1.760	60	18.1	11 000	89.66	0.74	0.3149	4.6740	1760 ± 3
Sample 203									
Ru, red	1.842	38.5	11.4	2 920	90.27	0.12	0.3124	4.6060	1748 ± 2
Ru, green	1.524	14.5	12.5	6 500	89.57	0.90	0.3131	4.6180	1749 ± 7
Sample 602									
>100, ast	0.124	255	145.4	17 200	77.59	7.52	0.5135	13.6490	2766 ± 1
100–75, ast	0.149	290	161.4	20 000	77.77	7.48	0.5031	13.2270	2748 ± 1
<75, st	0.079	366	189.5	19 000	79.00	6.44	0.4749	12.1250	2700 ± 2
Al-gneisses, the Seryak locality									
Sample T-44									
>100, ast	0.035	243	134.0	55	79.35	5.71	0.5090	13.2730	2735 ± 63
>100, ast	0.104	366	199.7	4 915	77.91	7.68	0.4934	12.6460	2706 ± 1
<75, sd	0.088	717	286.5	972	83.23	3.71	0.3862	8.3980	2431 ± 1
Al-gneisses, the Lyagkomina locality									
Sample 91011									
104–90, avt	2.700	481	209.2	8 400	82.10	4.25	0.4155	9.5280	2521 ± 2
75–60, avt	1.500	509	223.2	6 000	82.53	3.75	0.4210	9.6540	2521 ± 4
60–45, lt	1.700	538	185.6	2 700	79.52	8.86	0.3200	6.4200	2302 ± 5
75–60, lt	1.200	694	246.7	2 700	78.76	9.19	0.3256	6.8670	2379 ± 3

THE BELOMORIAN BELT OF THE BALTIC SHIELD

>104, avt, 29xx	0.046	556	244.2	6 000	83.59	2.87	0.4268	9.5330	2477 ± 5
>104, avt, 10xx	0.028	881	375.3	5 650	83.79	2.91	0.4152	9.0890	2443 ± 1
90–75, ast	0.020	274	147.5	222	79.99	5.36	0.5001	12.6280	2681 ± 4
Ru, yellow	0.278	37	11.4	1 700	87.25	3.36	0.3136	4.5610	1758 ± 7
Al-gneisses, the Chupa locality									
Sample 556									
125–100, rb	6.200	370	166.9	4 170	82.93	3.22	0.4333	9.7990	2497 ± 2
100–75, rb	2.800	356	154.5	17 240	82.46	4.20	0.4169	9.2460	2464 ± 1
<75, rb	8.600	192	87.1	7 400	76.83	10.76	0.4025	8.8700	2454 ± 1
TTG rocks, the Tupaya Bay locality									
Sample 195									
>100, apt	0.077	181	110.2	80	75.93	9.38	0.5359	14.3680	2780 ± 24
100–75, apt	0.053	394	211.2	3 180	79.98	5.27	0.4982	12.7290	2701 ± 1
<75, apt	0.084	426	222.4	975	80.91	4.32	0.4904	12.4030	2684 ± 1
>100, apb	0.077	563	292.6	3 720	81.00	4.25	0.4890	12.3340	2680 ± 1
TTG rocks, the Lyagkomina locality									
Sample 91012									
>45, av	0.021	804	198.5	394	83.45	2.45	0.2395	5.5790	2547 ± 5
>45, al	0.043	924	236.8	515	82.78	3.24	0.2466	5.7420	2547 ± 4
75–60, as	0.072	971	284.7	860	83.09	2.51	0.2832	6.7430	2584 ± 3
75–60, as, 12xx	0.012	820	282.7	134	81.20	4.72	0.3253	7.7760	2591 ± 3
Ru, yellow	1.634	15	4.6	530	88.83	1.59	0.3173	4.7180	1763 ± 9
Ru, green	1.056	18	5.3	1 150	87.50	2.98	0.3158	4.7370	1779 ± 7
Sample 91013									
>75, lt	0.056	529	172.3	711	82.85	4.76	0.3137	6.4720	234 ± 5
<75, lt	0.079	552	182.0	560	80.93	6.16	0.3103	6.8250	2451 ± 5
90–75, av	0.064	468	191.1	780	80.37	5.86	0.3816	9.0130	2570 ± 2
Fragments, rs	0.128	304	123.2	1 850	82.77	3.97	0.3896	8.6070	2458 ± 2
rs	0.123	343	142.8	6 250	83.36	2.64	0.4062	9.0130	2465 ± 2
<75, av	0.052	470	161.1	583	82.91	3.83	0.3303	7.2860	2456 ± 3
<75, as	0.015	555	193.4	165	81.68	4.99	0.3310	7.4480	2489 ± 4
Titanite	0.519	82	31.2	1 120	76.50	14.30	0.3364	5.5780	1960 ± 8

* corrected for lead blank and common lead (Stacey & Kramers 1975)
Errors in U/Pb ratios are 0.3%
Letters are: a, abraded; b, brownish; d, dark; l, long prismatic; p, prismatic; r, isometric; rs, rose fragments; Ru, rutile; s, short prismatic; t, transparent; v, oval; xx, number of crystals

to 4.7 under slightly elevated pressure. The total blanks for Nd and Sm were about 300 and 100 pg, respectively. The samples were loaded as chlorides on Re filaments and run as metal ions in a double filament configuration on a Finnigan MAT 261 mass-spectrometer. All calculations were performed using the constants recommended by Wasserburg et al. (1981), but with the fractionation of Nd isotope ratios being corrected relative to ^{146}Nd/^{144}Nd = 0.7219.

Analytical results and discussion

The analytical data for the U-Pb zircon work are given in Table 1, the zircon data also being shown graphically in the conventional Concordia diagrams of Figs 3–5. The Sm-Nd data are presented in Table 2 and plotted as initial epsilon-Nd values versus age in Fig. 6.

The aluminous gneisses

The two zircon types in these rocks differ considerably in their contents of uranium. These are approximately four times higher in the dark grains than in the light ones. The uranium concentrations in the light, transparent zircons vary between the different localities. The zircons from Tupaya Bay have the lowest uranium contents, while those from the Lyagkomina locality are high in uranium (up to 900 ppm).

The complex history of the zircons is seen in the Concordia diagrams (Figs 3A–E) constructed for the different sites. It is evident that no good isochron would be obtained by regressing all these analyses together.

However, three thoroughly abraded fractions of the short-prismatic, light, transparent zircons from sample 208 of the Tupaya Bay site have the highest ^{207}Pb/^{206}Pb ages in excess of 2800 Ma (Table 1). These well-shaped zircons low in uranium are also the ones that best represent the zircon material interpreted as having been crystallized during granulite facies metamorphism. Together the three fractions define a discordia line with an upper intercept age of 2855 ± 3 Ma (Fig. 3A). The lower intercept is not far away from origo, indicating that nearly recent lead loss predominated strongly over loss of lead caused by late metamorphism or other diffusion. This makes it possible to consider the 2855 Ma age as an age reflecting zircon crystallization during a distinct metamorphic, geological event. The other zircon fractions in Fig. 3A plot inside a triangle delimited by lines joining the 2855 Ma point on the Concordia, defined by the age of the three fractions of abraded light zircons, with a point close to zero

and another at approximately 1750 Ma. The latter, incidentally, is close to the end of the Palaeoproterozoic ('Svecofennian') orogeny in the Lapland-Kola region. The geological significance of the 1750-Ma date is also brought out isotopically by the nearly concordant, c. 1750 Ma U-Pb ages of rutile from samples 203, 208, 11011 and 11012 (Table 1) which mark the time when the U-Pb isotopic system of that mineral was closed at c. 400° C (cf. Corfu 1988). These are shown on Fig. 3A as well as Figs 4 & 5). The peak of Palaeoproterozoic metamorphism in the studied area must consequently have occurred somewhat earlier. Thus the combined evidence accounts for at least two disturbances of the U-Pb systems in the zircons from the Tupaya Bay samples 203 and 208. One occurred in the Palaeoproterozoic (Svecofennian) while the other was nearly recent. From Figs 3B–E, the Palaeoproterozoic event appears to have affected almost all of the Archaean zircons except those used to define the 2855 Ma age. Also the zircons from sample 602 plot close to a discordia reference line (marked 'Ref. line' in Figs 3 & 4) with Concordia intercepts at c. 1.75 and 2.86 Ga. An upper-intercept age of c. 2.82 Ga has been calculated for that sample by additional isotopic work that is being published in a different context (Bibikova et al. 1995a).

The long-prismatic zircons from Lyagkomina (Fig. 3D) either lost nearly all radiogenic lead during the Palaeoproterozoic or crystallized during the Palaeoproterozoic, incorporating small amounts of older radiogenic lead.

In Fig. 4, the analytical points of nearly all zircon fractions from the aluminous gneisses (except the ones representing the Palaeoproterozoic long-prismatic zircons from the Lyagkomina locality) are plotted together in a single diagram which also shows the reference line connecting the 2.86 and 1.75 Ga points on the Concordia. The purpose of including Fig. 4 in the paper is to demonstrate the diversity of metamorphic imprints on the zircons from the different localities in the Belomorian Belt rather than to calculate a multi-locality regressional age. From Fig. 4 it is seen that Palaeoproterozoic alteration was most pronounced in the Lyagkomina and Chupa areas as the zircons from these localities have the most disturbed U-Pb systems. This is in agreement with field observations.

Previous work has demonstrated major metamorphic events in the Belomorian Belt at about 2700–2600 Ma and 1900–1800 Ma (Bibikova et al. 1993b; Bogdanova & Bibikova 1993; Bibikova et al. 1995a). From Fig. 4 it appears evident that these events have to varying degrees affected the U-Pb isotopic systems of the Archaean zircons of the aluminous gneisses which, as stated above,

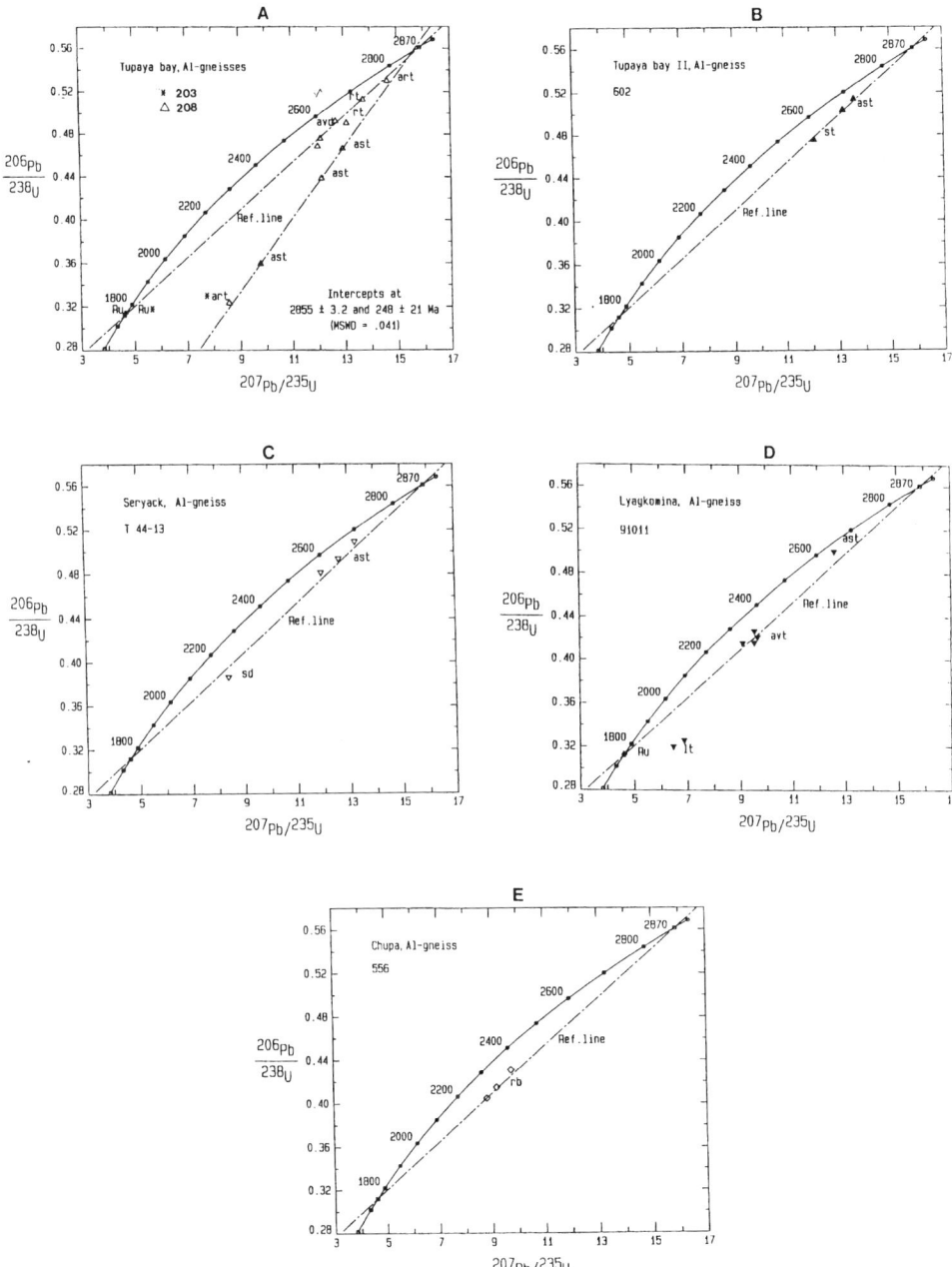

Fig. 3. Concordia-diagrams A–E for the zircons from aluminous gneisses in different sites of the Belomorian Belt. In each of these diagrams, a reference line ('Ref. line') joining the 2860 and 1750 Ma Concordia intercepts has been drawn. This line is based on the relatively precise dating of the 'ast'-zircons in (A) and the *c.* 1750 Ma concordant ages of rutiles. See text and list of the abbreviation symbols in Table 1 for further explanation.

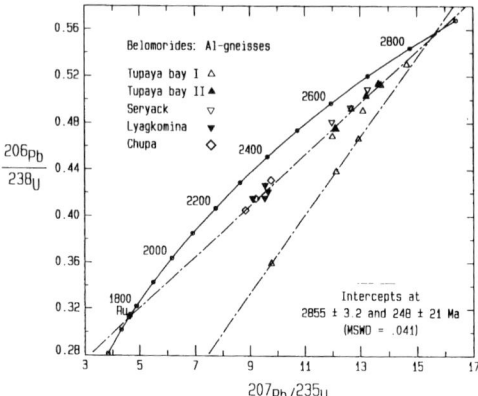

Fig. 4. Compound Concordia diagram for all the dated zircons from the aluminous gneisses of the Belomorian Belt, except the light, transparent ones in sample 208.

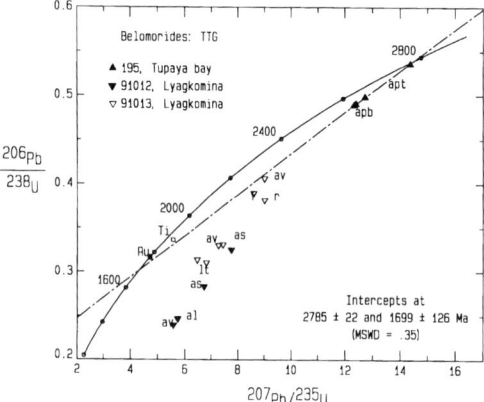

Fig. 5. Concordia diagram for the analysed zircons from the oldest TTG-gneisses of the Belomorian Belt.

predominantly reflect the time of the earliest high-grade metamorphic event rather than the primary age of the source detrital rocks. The multiphase metamorphic history that caused metamictization and substantial loss of lead in the high-U detrital zircons prevents determination of their original ages.

In order to constrain the ages of the protoliths of the aluminous gneisses the study therefore took in the Sm-Nd isotopic systems in the whole-rock samples from the Tupaya Bay and Lyagkomina localities in the Belomorian Belt (Fig.1 and Table 2). The DM-model ages range between 3000 and 2860 Ma. When related to the c. 2860 Ma ages of the zircons formed during the earliest high-grade metamorphism, the epsilon-Nd values close to or slightly above zero indicate a short prehistory of the protoliths.

The TTG-gneisses

Three zircon fractions from sample 195 of the Tupaya Bay locality plot on a discordia line with an upper-intercept age of 2785 ± 22 Ma (Fig. 5). The lower intercept age around 1700 Ma once more indicates Svecofennian disturbance of the isotopic systems. A very similar age of 2787 ± 21 Ma has previously been obtained for zircon xenocrysts in a 2580 Ma old anatectic trondhjemite from the same locality (Bogdanova & Bibikova 1993).

The zircons of the TTG-rocks from Lyagkomina are characterized by extremely high contents of uranium which reach 1000 ppm and appear to be the reason for the highly discordant plots of the analysed fractions. No meaningful discordias can be calculated. Nevertheless all the data points are situated within a triangle connecting the c. 2800 Ma, 1800 Ma and 0 Ma points on the Concordia line (Fig. 5). These zircons could therefore be similar in age to the TTG zircons from Tupaya Bay and also strongly disturbed at least twice, once in the Palaeoproterozoic and another time nearly recently. However, they could also be

Table 2. *Sm-Nd results of whole-rock aluminous gneisses and TTG of the Belomorian Belt*

Sample	Age	^{147}Sm/^{144}Nd	^{143}Nd/^{144}ND	ε_{Nd} T	CHUR	T DM
91011 Al-gneiss	2.85	0.1068	0.510958 ± 9	0.24	2.83	3.01
91012 Tonalite	2.70	0.0947	0.510872 ± 5	0.99	2.63	2.81
91013 Tonalite	2.70	0.1043	0.510976 ± 8	−0.30	2.72	2.91
92077 Al-gn 208	2.85	0.1108	0.511048 ± 9	0.51	2.80	2.99
92080 Al-gn Lg92	2.85	0.1007	0.510940 ± 17	2.12	2.68	2.86
92082 Al-gn TS1	2.85	0.1121	0.511156 ± 6	2.16	2.65	2.86
92083 Al-gn TS2	2.85	0.1098	0.511104 ± 9	1.98	2.68	2.87
92084 Tonalite TS3	2.70	0.1035	0.511016 ± 12	0.76	2.64	2.83

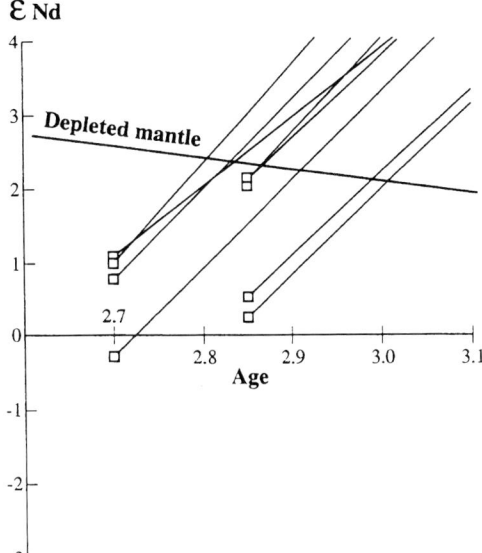

Fig. 6. Initial whole-rock ε_{Nd} versus age diagram for the analysed of aluminous- and TTG-gneisses in the Belomorian Belt.

much younger because ages of 2740–2720 Ma have been obtained for TTG-gneisses from other localities in the Belomorian Belt (Bogdanova & Bibikova 1993; Bibikova *et al.* 1995*b*). Thus there occurred at least two stages of emplacement of TTG plutonic rocks in the Late Archaean around 2.8 Ga and 2.74–2.72 Ga ago, respectively.

The Sm-Nd DM-model ages for the Lyagkomina TTG samples (Table 2) are between 2900 and 2800 Ma with ε Nd-values close to zero which indicates short crustal residence also of these rocks.

Geodynamic consequences

Although our study does not date the deposition of the meta-sedimentary aluminous gneisses which are regarded as the oldest rocks of the Belomorian Belt ('the Chupa strata'), the age of these rocks is constrained by the Sm-Nd isotopic depleted-mantle model ages which are less than 3 Ga if the De Paolo simplification (1981) of the depleted-mantle curve is accepted for the Belomorian Belt in the Late Archaean (Fig. 6). The epsilon-Nd values related to the 2.86 Ga age of metamorphism are around or somewhat above zero which implies that the metasedimentary rocks do not contain substantial amounts of much older detrital material.

The oldest recognized plutonic rocks have ages of *c.* 2.8 Ga. Similar results have also been reported by Daly *et al.* (1993) and Timmerman & Daly (1995) for TTG-gneisses and metavolcanics in the northeastern part of the Belomorian Belt and by Bibikova *et al.* (1995*b*) for metavolcanics in its western part.

In consequence, the Belomorian Belt was formed in the Late Archaean and no Early and not even Middle Archaean rocks are or were present in quantity. This once more confirms that the Belomorian Belt is not the ancient core of the Baltic Shield and does not contain equivalents of the oldest rocks present in the Archaean province of Karelia where at least the 3.2 Ga old TTG plutonics separate *c.* 3.0 Ga old and later riftogenic greenstone belts (cf. e.g. Kröner & Compston 1990; Sergeyev *et al.* 1990; Lobach-Zhuchenko *et al.* 1993; Vaasjoki *et al.* 1993).

The oldest known metamorphism in the Belomorian Belt, now dated at *c.* 2.86 Ga, and subsequent episodes of TTG-rock formation and Late Archaean metamorphism correlate in time (Bibikova 1989; Levchenkov *et al.* 1989) with the later stages of greenstone-belt development in the Karelian Archaean granite–greenstone terrain and with the major, 'Rebolian' event of deformation and metamorphism.

Later developments differ to the extent that although crust-forming processes occurred simultaneously in the Belomorian Belt and along the eastern margin of the Karelian Province, the interior parts of that province had been consolidated somewhat earlier. Postmetamorphic potassic granites thus appeared in the central parts of the Karelian Province between 2.8 and 2.7 Ga ago (e.g. Levchenkov *et al.* 1989), while petrologically equivalent rocks in the Belomorian Belt were emplaced significally later, between 2.5 and 2.45 Ga ago although the 2.44 granites also in Karelia (Bibikova *et al.* 1993*a*, *b*). In view of these circumstances and taking into account also the presence of repeatedly activated thrust planes along the Karelian–Belomorian boundary and inside the Belomorian Belt (e.g. Miller & Milkevich 1994), a model involving Late Archaean accretional–collisional relationships between a continental domain in Karelia and a mobile belt in the White Sea region appears most adequate.

Within the Belomorian Belt itself, this study suggests that the earliest metamorphism and deformation occurred *c.* 2.86 Ga ago as recorded by the zircons of the aluminous gneisses which form the oldest rock unit in the belt. A somewhat younger igneous crystallization of andesitic metavolcanics of calc-alkaline affinity in the western part of the Belomorian Belt has been dated by the U-Pb zircon method at *c.* 2.82 Ga (Bibikova *et al.* 1995*b*). The close association in time of low- to moderate-pressure amphibolite to granulite facies metamorphism in the meta-sediments and calc-alkaline volcanism suggest a continental-marginal volcanic-arc environment during the early stages of Belomorian Belt development. In this context, it is notable that a number of Late Archaean greenstone belts outline the entire length of the northeastern edge of the Karelian Province (e.g. Kozhevnikov 1992) also overlapping into the Belomorian Belt (Slabunov 1992). These greenstone belts are similarly characterized by calc-alkaline magmatism and may have been formed in an island-arc environment subsequently accreted to the Karelian Province.

The following increase of the P/T ratios of granulitic metamorphism (Bibikova *et al.* 1995a) was related in time to the emplacement of additional, 2.74–2.72 Ga old TTG-plutonics (Bogdanova & Bibikova 1993) and may well mark a collisional stage of the orogenic process that took place between *c.* 2.8 and 2.7 Ga ago. About 2.45 Ga ago, the occurrence of bimodal gabbro-potassic granite igneous associations in the Belomorian Belt signals that the formation of Late Archaean crust in the Baltic Shield had ended some time before.

Conclusions

- Our Sm-Nd isotopic whole-rock data from the oldest aluminous meta-sedimentary and TTG meta-plutonic gneisses in the Belomorian Belt indicate no Early or Middle Archaean ages either for these rocks or the mass of the protoliths of the meta-sediments. The Belomorian Belt is therefore a Late Archaean crustal segment and cannot be the ancient core of the Baltic Shield.
- An early, high-grade metamorphic event in the Belomorian Belt appears to have taken place *c.* 2.86 Ga ago. During this event, the P/T ratios were low to moderate.
- The oldest recognized group of TTG plutonic rocks was formed *c.* 2.8 Ga ago, but there also exists a later generation of *c.* 2.74–2.72 Ga old plutonics which intruded concomitantly with an increase of the P/T ratios of metamorphism.
- The formation of continental crust in the Belomorian Belt can be correlated with post-3 Ga developments in the Archaean Province in Karelia, but many rocks differ lithologically and older, Middle Archaean rocks are present in the Karelian Province. The overall evidence is compatible with the one-time existence of a continental nucleus in Karelia that bordered a mobile belt in the present Belomorian region.
- The combined results of geochronological and PT work suggest a passage from volcanic-arc to collisional environments in the Belomorian Belt between *c.* 2.8 and 2.7 Ga ago.

This study was carried out within the framework of the Archaean Working Group of IGCP project 275 'Deep Geology of the Baltic/Fennoscandian Shield'. Financial support from the Swedish Royal Academy of Sciences allowed Elena Bibikova to work at the Laboratory for Isotope Geology of the Swedish Museum of Natural History in Stockholm. Grants from the Swedish Natural Sciences Research Council (NFR) supported the study. For the underlying field work in the Belomorian Belt by S. Bogdanova, funding by the Geological Institute of the then USSR Academy of Sciences (Moscow) is acknowledged.

We thank Prof. R. Gorbatschev (Lund), Dr Yu. Miller (St Petersburg) and Dr A. Slabunov (Petrozavodsk) for interesting discussions. Prof. Gorbatschev also critically revised and substantially improved the manuscript. Dr H. Huhma proposed several useful revisions.

References

BIBIKOVA, E. V. 1989. *Uranium-Lead Geochronology for the Early Development Stages of Ancient Shields*. Nauka, Moscow, 1–180 (in Russian).

——, BOGDANOVA, M. N. & SKIÖLD, T. 1995a. New U-Pb isotope data for the Archaean of the North-West Belomorian area. *Doklady (Transactions) of the Russian Academy of Sciences*, in press (in Russian).

——, GRACHEVA, T. V. & MAKAROV, V. A. 1993a. U-Pb isotopic dating of the Early Precambrian K-granites in the northwest Belomorian (White Sea) area. *Abstracts of 'The Svecofennian Domain' Sympoium*, IGCP Project 275, Turku, Finland, 13.

——, SKIÖLD, T., BOGDANOVA, S. V., DRUGOVA, G. M. & LOBACH-ZHUCHENKO, S. B. 1993b. Geochronology of the Belomorides: an interpretation of the multistage geological history. *Geokhimiya*, **10**, 1393–1411 (in Russian).

——, SLABUNOV, A. I., KIRNOZOVA, T. I. & MAKAROV, V. A. 1995b. U-Pb zircon ages of the rocks of the Keret granite-greenstone system in the junction zone between the Karelian and Belomorian structures of the Baltic Shield. *Doklady (Transactions) of the Russian Academy of Sciences*, in press (in Russian).

BOGDANOVA, S. V. & BIBIKOVA, E. V. 1993. The 'Saamian' of the Belomorian Mobile Belt: new geochronological constraints. *Precambrian Research*, **64**, 131–152.

CORFU, F. 1988. Differential response of the U-Pb system in coexisting accessory minerals. *Contributions to Mineralogy and Petrology*, **98**, 312–325.

DALY, S. J., MITROFANOV, F. P. & MOROZOVA, L. N. 1993. Late Archaean Sm-Nd model ages from the Voche-Lambina area: implications for the age distribution of Archaean crust in the Kola Peninsula, Russia. *Precambrian Research*, **64**, 189–195.

DE PAOLO, D. J. 1981. A neodymium and strontium isotopic study of the Mesozoic calc-alkalic granitic batholiths of the Sierra Nevada and Peninsular Ranges, California. *Journal of Geophysical Research*, **86**, 10470–10488.

GAÁL, G. & GORBATSCHEV, R. 1987. An outline of the Precambrian evolution of the Baltic Shield. *Precambrian Research*, **35**, 15–52.

GLEBOVITSKY, V. A., DOOK, V. L. & SHARKOV, E. V. 1978. Endogenic processes. *In*: KRATTS, K. O. (ed.) *The Crust of the Eastern Baltic Shield*, Nauka, Leningrad, 112–171 (in Russian).

——, VAPNIK, Ye. A., SEDOVA, I. S. & SEMENOV, A. P. 1993. Significance of fluid inclusions and rock fluid compositions in interpreting polymetamorphic complexes in the Tupaya Guba region, Lake Kovdozero. *Geochemisrty International*, **30**, 58–72.

KOTOV, A. B. & ZINGER, T. F. 1985. Structural-compositional characteristics of migmatites and granitoids of the Belomorian megacomplex. *In*: MITROFANOV, F. P. (ed.) *Migmatization and Granitoid Formation in Different Thermodynamic Regimes*, Nauka, Leningrad, 88–93 (in Russian).

KOZHEVNIKOV, V. N. 1992. Geological evolution of North Karelian Archaean structures. *Abstracts of the International IGCP Symposium on Projects 275 and 257*, Petrozavodsk, September, 7–17, 1992, 43–44.

KROGH, T. E. 1973. A low contamination method for hydrothermal decomposition of zircons and extraction of U and Pb for isotopic age determinations, *Geochimica et Cosmochimica Acta*, **3**, 485–495.

—— 1982. Improved accuracy of U-Pb zircon ages by the creation of more concordant systems using an air abrassion technique. *Geochimica et Cosmochimica Acta*, **4**, 637–649.

KRÖNER, A. & COMPSTON, W. 1990. Archaean tonalitic gneiss of Finnish Lapland revisited: zircon ion-microprobe ages. *Contributions to Mineralogy and Petrology*, **104**, 348–352.

LEVCHENKOV, O. A., LOBACHZHUCHENKO, S. B. & SERGEEV, S. A. 1989. The geochronology of the Karelian granite–greenstone terrain. *In*: LEVSKY, L. K. & LEVCHENKOV, O. A. (eds) *Isotope Geochronology of the Precambrian*, Nauka, Leningrad, 63–72 (in Russian).

LOBACH-ZHUCHENKO, S. B., CHEKULAEV, V. P., SERGEEV, S. A., LEVCHENKOV, O. A. & KRYLOV, I. N. 1993. Archaean rocks from southeastern Karelia (Karelian granite-greenstone terrain). *Precambrian Research*, **62**, 375–397.

LUDWIG, K. R. 1991. ISOPLOT program. *USA Geological Survey Open-file*, report 91.

MASAMICHI, M. & HI, T. 1983. Prediction of crystal structures of minerals under extremal conditions by method of energy minimization. *Journal of Mineralogical Society of Japan*, **16**, 217–221.

MILLER, Yu. V. & MILKEVICH, R. I. 1995. Folded-nappe structure of the Belomorian tectonic zone and its relationships with the Karelian granite–greenstone terrain. *Geotektonika*, in press.

MITROFANOV, F. P. & POZHILENKO, V. I. (eds) 1991. *The Archaean Voche-Lambina Geodynamic Testing Ground of the Kola Peninsula*. Kola Science Centre of the Russian Academy of Sciences, Apatity (in Russian).

SALIE, M. E. 1985. Thermodynamic parameters of metamorphism of the belomorian rocks (the Chupa segment). *In*: KRATTS, K. O. (ed.) *Geology and Pegmatites of the Belomorides*, Nauka, Leningrad, 134–145 (in Russian).

SERGEYEV, S. A., BIBIKOVA, E. V., LEVCHENKOV, O. A., LOBACH-ZHUCHENKO, S. B., YAKOVLEVA, S. Z. ET AL. 1990. Isotopic geochronology of the Vodlozero gneiss complex. *Geochemistry International*, **27**, 65–74.

SHURKIN, K. A., GORLOV, N. V., SALIE, M. E., DOOK, V. L. & NIKITIN, YU. V. 1962. *The Belomorian complex of Northern Karelia and the southwestern Kola Peninsula*, The USSR Academy Science Publishing House, Moscow–Leningrad, 1–306 (in Russian).

SLABUNOV, A. I. 1992. The Lake Keret' area. *In*: KOZHEVNIKOV, V. N., SLABUNOV, A. I. & SYSTRA, Y. Y. (compilers) *Guidebook of the Geological Excursion on the Archaean of northern Karelia*. Karelian Research Centre, Petrozavodsk, 28–35.

STACEY, J. S. & KRAMERS, J. D. 1975. Approximation of terrestrial lead isotope evolution by a two-stage model. *Earth and Planetary Science Letters*, **2**, 207–221.

STENAR', M. M. 1988. Stratigraphy of Archaean deposits in Soviet Karelia. In: Archaean Geology of the Fennoscandian Shield. *Geological Survey of Finland Special Paper*, **4**, 7–14.

TIMMERMAN, M. J. & DALY, S. J. 1995. Sm-Nd evidence for late Archaean crust formation in the Lapland-Kola Mobile Belt, Kola Penenisula, Russia and Norway. *Precambrian Research*, **72**, 97–107.

TUGARINOV, A. I. & BIBIKOVA, E. V. 1980. *Geochronology of the Baltic Shield by Zircon Age Determinations*. Nauka, Moscow, 1–130 (in Russian).

VAASJOKI, M., SORJONEN-WARD, P. & LAVIKAINEN, S. 1993. U-Pb age determinations and sulfide Pb-Pb characteristics from the late Archaean Hattu schist belt, eastern Finland. *In*: NURMI, P. A. & SORJONEN-WARD, P. (eds) *Geological development, gold mineralization and exploration methods in the late Archaean Hattu schist belt, Ilomantsi, eastern Finland*, Geological Survey of Finland, Special Paper **17**, 103–131.

VOLODICHEV, O. I. 1990. *The Belomorian Complex of Karelia (Geology and Petrology)*. Nauka, Leningrad (in Russian).

WASSERBURG, G. J., JAKOBSEN, S. B., DEPAOLO, D. J., MCGULLOCH, M. T. & WEN, T. 1981. Precise determination of Sm/Nd ratios, Sm and Nd isotope abundances in standard solutions. *Geochimica et Cosmochimica Acta*, **45**, 2311–2323.

High-grade metamorphism of 2.45–2.4 Ga age in mafic intrusions of the Belomorian Belt in the northeastern Baltic Shield

S. V. BOGDANOVA

Department of Mineralogy and Petrology, Institute of Geology, Lund University, Sölvegatan 13, S-223 62 LUND, Sweden

Abstract: In the Late Archaean Belomorian Belt of the Baltic Shield, shearing and high-grade metamorphism occurred in association with bimodal gabbroic–granitic igneous activity between 2.45 and 2.40 Ga. That period is generally regarded as one of anorogenic rifting and extension. In the studied case of the Tolstik intrusion, the time of metamorphism is constrained by the 2434 ± 7 Ma age of a metamorphosed mafic intrusion and the 2405 ± 20 Ma age of a subsequently intruded potassic granite. Several generations of mafic dykes allow a detailed study of the metamorphic evolution during that time span.

The conditions of post-intrusion peak metamorphism in the mafic intrusion were 700–800° C and 11–12 kb which was followed first by near-isothermal decompression and, after the loss of c. 5 km crustal thickness, by near-isobaric cooling to c. 600° C at 8–9 kb.

These results suggest metamorphism possibly triggered by igneous intrusions into previously thickened crust, but the P/T-ratios of metamorphism were moderately high. Later followed crustal extension concomitantly with continuous igneous input and the P/T- ratios decreased. Finally, a post-extensional period of cooling ensued as igneous activity ceased.

Employing recent geochronological work as a salient point (Bibikova *et al.* 1993*a*; Bogdanova & Bibikova 1993), the present study addresses problems of metamorphism and deformation in the northeastern part of the Baltic Shield during an apparently 'anorogenic' period marked by rifting and crustal extension in the earliest Palaeoproterozoic between *c.* 2.5. and 2.0 Ga.

At the current level of knowledge, (e.g. Gorbatschev & Bogdanova 1993), two major periods of orogenic geodynamic activity are recognized to account for the formation of most of the presently known continental crust in the northwestern part of the East European Craton.

One of these is Late Archaean, the other Palaeoproterozoic. The Late Archaean period featured several orogenic events between *c.* 2.9 and 2.7 Ga which can be interpreted in terms of first accretional and later collisional orogeny (e.g. Gaál & Gorbatschev 1987; Miller & Milkevich 1995; Bibikova *et al.* this volume). The Palaeoproterozoic culmination between *c.* 2.0 and 1.6 Ga , comprised events previously grouped into the Svecofennian and Gothian–Kongsbergian orogenies. During the earlier part of this Palaeoproterozoic development (*c.* 2.0–1.8 Ga), accretional formation of large amounts of new continental crust took place in the central Baltic Shield (Patchett & Kouvo 1986; Patchett *et al.* 1987) and in its continuation in the basement of the East European Craton beneath platformal sediments (Puura & Huhma 1993;

Bogdanova *et al.* 1994). Semi-simultaneously, however, collisional orogeny that reworked large swaths of Archaean crust occurred in the Lapland–Kola Mobile Belt in the extreme north of Fennoscandia (Marker *et al.* 1993). A continuation of the Lapland–Kola Belt can be traced tentatively across the Atlantic to Greenland and Canada (Bridgwater *et al.* 1990).

The period between the Late Archaean and Palaeoproterozoic culminations of orogenic activity was marked by quite different geodynamic regimes. From *c.* 2.45 Ga onwards, the Archaean nucleus of the Baltic Shield was extensively rifted and partly dispersed (e.g. Glebovitsky *et al.* 1978; Gaál & Gorbatschev 1987; Bibikova *et al.* 1990; Gaál 1990; Mints 1993). Attendant igneous activity led to the formation of extensive rift-related volcanic belts, mafic dyke swarms, layered mafic intrusions, anorthositic gabbro and some charnockites and granites (Huhma *et al.* 1990; Puchtel *et al.* 1991; Balashov *et al.* 1993; Mitrofanov *et al.* 1993). Similar Palaeoproterozoic effects have also been reported from the Archaean crust in Scotland (Weaver & Tarney 1981; Park & Tarney 1987; Barooach & Bowes 1990; Cadman *et al.* 1990), Greenland (Kalsbeek *et al.* 1993), and Canada (Mengel & Rivers 1991).

While the earliest Palaeoproterozoic in the present North Atlantic realm was thus marked by tension in the crust, there also occurred substantial strike-slip faulting and attendant folding and metamorphism associated with igneous activity during the period 2.45–2.30 Ga. This deformation is known in Russia under various names such as the Seletsk (Glebovitsky et al. 1978) and the Loukhi folding (Stenar' 1988), or simply F_2-folding (e.g. Stepanov 1981; Volodichev 1990). Balagansky et al. (1986) used the term 'Kolvitsa folding' to describe a similar structural development. However, they considered it to be Late Archaean.

In these contexts, the crustal segment named the Belomorian Belt stands apart in the northeastern Baltic Shield by having been the preferred site of strike-slip shearing, folding, and variegated ultramafic and gabbroic to granitic plutonic igneous activity. In contrast, it strikingly lacks the extensive early Proterozoic platform cover that characterizes substantial parts of the Archaean Karelian Province adjoining the Belomorian Belt in the southwest (Fig. 1). Thus the Belomorian Belt apparently represents a deeper earliest Proterozoic crustal section than its neighbours and is particularly favourable to study the relationships between earliest Proterozoic plutonism and metamorphism.

The Belomorian Belt is exposed for about 600 km between the edge of the sedimentary cover

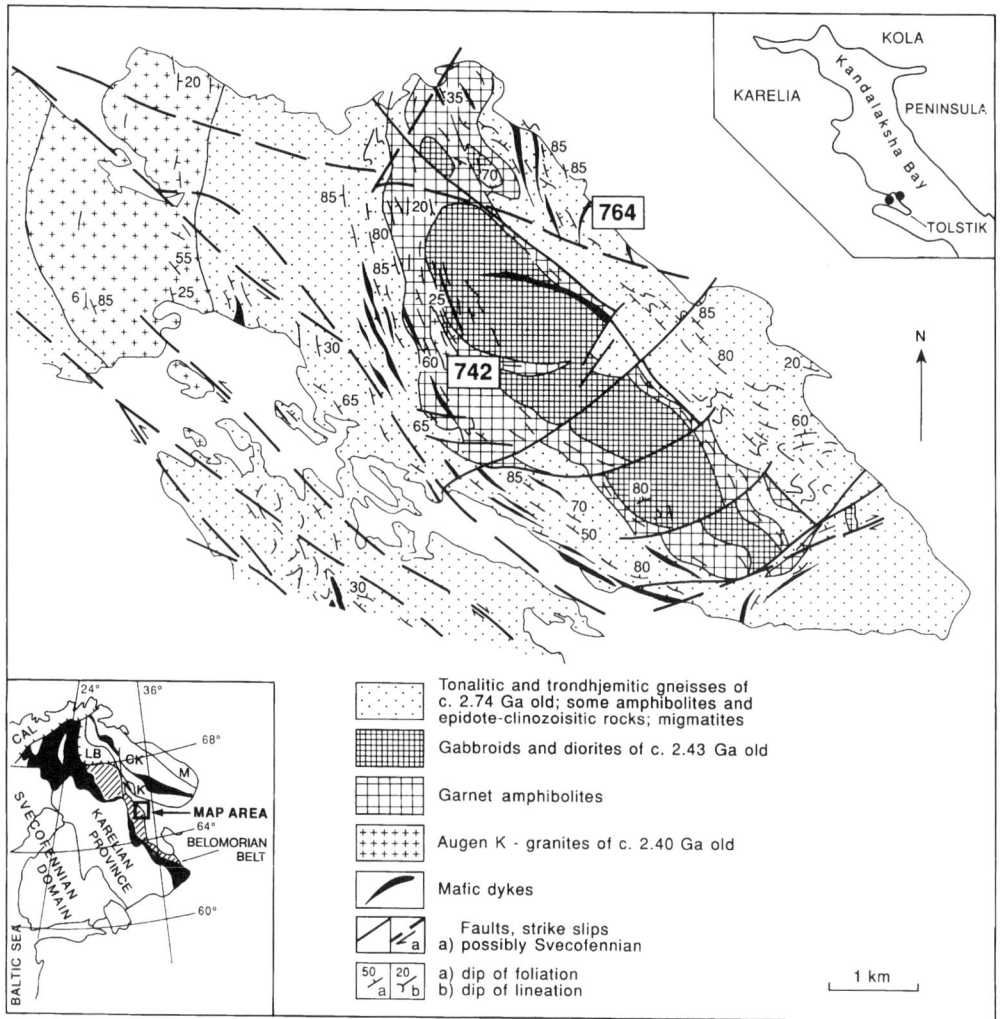

Fig. 1. Sketch map of the Tolstik mafic intrusion (modified after Bogdanova & Bibikova 1993, and an unpublished map by Yu. Miller, used with his permission). The localities sampled for the P–T work are 742 and 764. The letter symbols in the inset showing the eastern Baltic Shield are: M, Murmansk terrane; CK, Central Kola terrane; LB, Lapland Granulite Belt; K, Kolvitsa zone; CAL, Caledonides. The most important areas of early Palaeoproterozoic rift-related volcanics are shown in black.

of the East European Platform in the southeast and the Lapland Granulite Belt in the northwest (Fig. 1, inset). It is richer in Late Archaean basaltic metavolcanics and aluminous metasediments, and features stronger penetrative deformation than the adjoining Archaean crustal segments. Throughout the Late Archaean, the Belomorian Belt differed from the Karelian Archaean Province as the mobile belt in a mobile belt–foreland relationship (Gáal & Gorbatschev 1987). During the Palaeoproterozoic collisional orogeny, the Belomorian Belt was at the presently south-southwestern margin of the Lapland–Kola Mobile Belt and underwent refolding, metamorphism and granite emplacement at c. 1.95–1.8 Ga. Rutile cooling ages are around 1.75 Ga (Bibikova et al. this volume). At least the northern margin of the belt was overridden at that time by the Lapland Granulite Belt, and the continuation of that belt exposed in the Kolvitsa area further southeast, in the southwesternmost Kola Peninsula.

Igneous activity in the Belomorian Belt during the earliest Palaeoproterozoic is mostly represented by ultramafic, mafic, gabbroic to anorthositic, dioritic, charnockitic and granitic, mostly minor dismembered intrusions (e.g. Stepanov 1981; Stepanov & Slabunov 1989). These are age equivalents of c. 2.45–2.40 Ga old mafic intrusions and the earliest among the Palaeoproterozoic ultramafic, mafic and acid hypabyssals and volcanics elsewhere in the Archaean Domain of the Baltic Shield.

According to Glebovitsky et al. (1978) and Volodichev (1990), the pressures of earliest Palaeoproterozoic metamorphism in the Belomorian Belt exceeded 12 kb. In consequence, eclogite-like mineral assemblages in mafic rocks and metapelitic kyanite + orthoclase/microcline gneisses were formed. However, the precise age of this metamorphism is uncertain, and similar pressures were reached also between 1.9 and 1.8 Ga at the sole of the Lapland Granulite Belt (e.g. Glebovitsky et al. 1978; Krill 1985) due to its overthrusting onto the Belomorian Belt . Thus the investigation of metamorphic conditions in the earliest Palaeoproterozoic before c. 2.0 Ga, and even the correlation of igneous rocks of that age in the Belomorian Belt, are hampered by the metamorphic and deformational effects of later reworking during collisional orogeny in the Lapland–Kola Mobile Belt. Fortunately, however, neither deformation nor metamorphic re-equilibration were wholly comprehensive during the collisional development at 1.9–1.8 Ga. The effects of collision decrease rapidly away from the axis of the Lapland–Kola Mobile Belt and from the thrust front of the Lapland Granulite Belt. Easily recognizable recrystallization is largely limited to the vicinity of major shear zones.

As far as earliest Palaeoproterozoic deformation is concerned, Volodichev in a thorough study of the Belomorian Belt (1990) concluded that the tectonic patterns of the 2.4–2.0 Ga period were governed by deformation along a system of steep, anastomozing zones of ductile shearing which separate lenses of less deformed rock (cf. also Bogdanova & Yefimov 1993). As described by Volodichev, the axial planes of the 'Seletsk' (F_2 in his book) folds associated with the ductile shearing trend NW, NS, and NE, many of them featuring vertical or subvertical hinge lines. The folding pattern thus appears to be due to the effects of strike-slip movements rather than to variations of the regional stress field. Applying current terminology, this amounts to transpressive and drag folding.

For the present detailed study, the Tolstik Peninsula on the southwestern shore of the northwesternmost arm of the White Sea (the Kandalaksha Bay) was selected. The area features a gabbroic-dioritic intrusion of 2434 ± 7 Ma age, succeeded by a 2405 ± 20 Ma intrusion of potassic granites (U–Pb zircon ages, Bogdanova & Bibikova 1993). In addition, there are three generations of mafic dykes. The earliest of these is coeval with the gabbro and diorite intrusion, the intermediate one is later than the gabbro and the diorite but earlier than the potassic granite, while the youngest one intruded also the granite. Shearing occurred repeatedly, concomitantly with and later than this igneous activity. During that time, the whole area was also subjected to metamorphism that formed high-grade metamorphic parageneses in the mafic intrusion and the two earliest generations of mafic dykes, but did not affect the granite. No major metamorphism or deformation of 1.9–1.8 Ga age is recognized except locally, and the geochronological study showed that there are no indications of major disturbances of this age in the zircon U–Pb isotopic systems. For instance, no later rims are seen around the zircons of the granite. These relationships make the Tolstik area suitable for the study of the metamorphism that occurred between 2.43 and 2.40 Ga. Because the structural, textural and intrusive relationships of the studied rocks are of utmost importance to assess the age of the metamorphism, they are considered in particular detail in the following sections.

Structure and composition of the Tolstik intrusion

In the Tolstik Peninsula, a large body of mafic igneous rocks, comprising gabbros, leucogabbros, anorthositic gabbros and quartz diorites, intruded migmatized tonalitic-trondhjemitic gneisses which have a 2741 ± 43 Ma U–Pb zircon age. The U–Pb zircon crystallization age of the mafic rocks is

Table 1. *Representative analyses of rocks from the Tolstik intrusion and mafic dykes*

	Gabbros and diorites of the main intrusion								Outlier of the main gabbro			Dyke I	Dyke II		Dyke III
Sample	742-1	B2-6	B6-2	742-25	742-1b	SB742	B1-2	742-1c	SB764-5	SB764-6	SB764-9a	SB764-13	S-1786-2	S-1822-2	SB742-4	742-3
SiO_2	48.32	48.56	50	51.89	52.15	52.62	54.28	58.91	51.37	50.46	50.16	50.06	54.8	55.56	56.57	48.91
TiO_2	1.82	2.24	1.09	1.82	1.58	0.82	0.88	1.36	1.02	1.06	1.05	1.12	0.64	0.49	1.41	1.02
Al_2O_3	10.57	15.04	15.01	13.12	13.54	16.89	15.14	12.53	14.3	10.7	15.4	14.13	13.23	13.3	12.32	14.71
Fe_2O_3	4.98	3.29	1.27	3.86	3.24	1.63	1.31	3.4	0.52	3.45	13.31*	13.86*	3.36	1.62	3.08	2.97
FeO	15.52	14.48	11.36	11.58	11.52	10.02	10.36	8.31	11.53	9.07	nd	nd	6.31	7.9	10.98	9.66
MnO	0.22	0.2	0.2	0.15	0.17	0.11	0.15	0.12	0.18	0.24	0.19	0.19	0.17	0.18	0.15	0.19
MgO	5.54	4.13	7.07	4.47	4.16	2.93	5.72	2.85	6.75	7.77	5.96	6.73	7.04	7.46	3.1	5.68
CaO	8.25	9.04	11.1	8.52	8.12	8.24	7.79	5.65	10.8	12.3	10.21	10.53	9.8	8.99	6.41	9.93
Na_2O	2.28	1.79	2.03	2.91	2.74	3.92	2.34	2.83	2.19	2.08	2.48	2.1	2.49	2.26	3.08	3.83
K_2O	0.67	0.41	0.31	0.79	0.83	0.8	0.86	1.51	0.26	1.06	0.13	0.23	1.33	1.16	1.55	1.16
P_2O_5	0.1	nd	nd	0.12	0.17	0.14	nd	0.36	0.07	0.11	0.12	0.1	0.1	0.04	0.24	0.12
Total	99.90	100.60	100.48	99.35	99.95	99.71	99.92	99.41	99.87	99.50	99.50	99.60	100.32	100.07	100.03	99.62

Analyses of samples prefixed 'B' are from Balagansky *et al.* 1986, table 2; 'SB' by the present author; 'S' from Stepanov & Slabunov 1989, table 47; the others by M. Bogdanova (with the author's permission). *, Fe oxides determined as total Fe_2O_3; nd, not determined.

2434 ± 7 Ma. These rocks were, in turn, intruded by a 2405 ± 20 Ma potassic granite (Fig.1; ages cf. Bogdanova & Bibikova 1993). A large, c. 1 km thick sheet-like intrusion of similar granites is found in the vicinity of the mafic body to the west (Fig. 1).

The strongly deformed mafic intrusion forms a lopolith-like body that is c. 6 km long and up to 2 km thick (Fig.1). Its contacts with the surrounding tonalitic–trondhjemitic–granodioritic (TTG) gneisses are mostly tectonized and semi-concordant. A set of sub-parallel NE-trending faults subdivides the Tolstik body into a number of segments. Some of these appear to have been somewhat rotated in relationship to each other. Several NW-trending strike-slip faults of presumably 1.9–1.8 Ga age have been mapped in the studied area (Yu. Miller, pers. comm., cf. Fig.1). However, these are characterized by lower-grade metamorphism and cut across the earlier structural pattern of the Tolstik Peninsula. Pre-1.9 Ga recumbent multiple folding with sub-horizontal hinge lines and mineral stretching lineation have been described in the Tolstik Peninsula and its vicinity (e.g. Balagansky et al. 1986). These structures are well-preserved in and around the Tolstik mafic intrusion, where several stages of magmatic emplacement, deformation and metamorphism have been demonstrated (Yefimov et al. 1987). Indications of at least four surges of basic and intermediate melts can be observed in the body. The order of intrusion is gabbro, leucogabbro and anorthositic gabbro, quartz diorite, and granodiorite.

All these rocks belong to a single igneous suite intermediate in composition between calc-alkaline and tholeiitic (Table 1, Fig. 2a). The 2405 Ma old potassic granite and somewhat later aplitic granites also join the same chemical trend. According to the classification of Debon & Le Fort (1983), the overall bimodal magmatic association belongs to a partly metaluminous, but mostly peraluminous type with a negative A–B tendency (Fig. 2b). The latter is most common in magmatic associations arising from the hybridization of mantle melts and crustal materials (Debon & Le Fort 1983). This is in good agreement with the rift-related tectonic setting prevailing at that time.

In the central, least altered part of the Tolstik intrusion, the medium-grained meta-gabbros and meta-diorites have igneous textures. At the margins of the mafic intrusion, however, and along strike-slip shears and some transverse faults inside it, the rocks have sheared fabrics and are commonly metamorphosed to markedly foliated and folded garnet amphibolites and garnet–amphibole gneisses. Where shearing is very strong, particularly in gabbro apophyses in the country-rock gneisses, fine banding is developed. In these rocks, the metamorphic minerals form elongated aggregates defining planar, and mostly very gently plunging linear fabrics.

The principal igneous minerals are hypersthene and clinopyroxene (exsolved from original high-T pigeonitic pyroxene), plagioclase (An 40 to 55), ilmenite, magnetite and titanomagnetite. The metamorphic mineral assemblage comprises garnet, amphibole, diopside, sphene and epidote. These minerals either occur as individual crystals or form fine-grained coronas mantling the igneous mafic minerals. Igneous greenish-brown high-Ti hornblende and reddish brown high-Ti biotite are common in the meta-diorites and granodiorites. They are also partly replaced by metamorphic garnet + amphibole ± biotite + plagioclase + quartz intergrowths, commonly containing clinopyroxene. The metamorphic amphibole and biotite mark a gently plunging (20–25°) mineral lineation striking subparallel to the margins of the mafic body.

Notable features of the development of the Tolstik igneous body are repeated, episodic strike-slip movements and metamorphic recrystallization that accompanied the intrusion of the melts. Each magma surge thus intruded and brecciated rocks that had previously been sheared and metamorphosed to some extent (Fig. 3a & b).

Numerous metamorphosed mafic dykes cut the mafic intrusion and the country-rock gneisses, some of them being, in turn, cut by the potassic granite. This makes it possible to constrain the timing of metamorphism between 2434 and 2405 Ma.

The mafic dykes associated with the Tolstik intrusion can be subdivided into three categories:

(1) mafic dykes similar in composition to the Tolstik gabbroic rocks. These can be interpreted as feeder-dykes or apophyses. They only cut the late Archaean TTG-gneisses;
(2) mafic dykes that intruded the mafic body and its surroundings before the emplacement of the potassic granites, and
(3) mafic dykes that also intruded the potassic granites, but appear to be older than the 1.9–1.8 Ga deformation in the region.

In addition, some minor, chunky bodies of gabbro which compositionally and texturally resemble the Tolstik gabbro occur outside the main Tolstik intrusion.

Mafic dykes related to the Tolstik gabbro

The dykes which may be feeder dykes or apophyses of the Tolstik gabbro are found on the northeastern shore of the Tolstik Peninsula where they form a 340–355° striking dyke swarm

74 S. V. BOGDANOVA

(a)

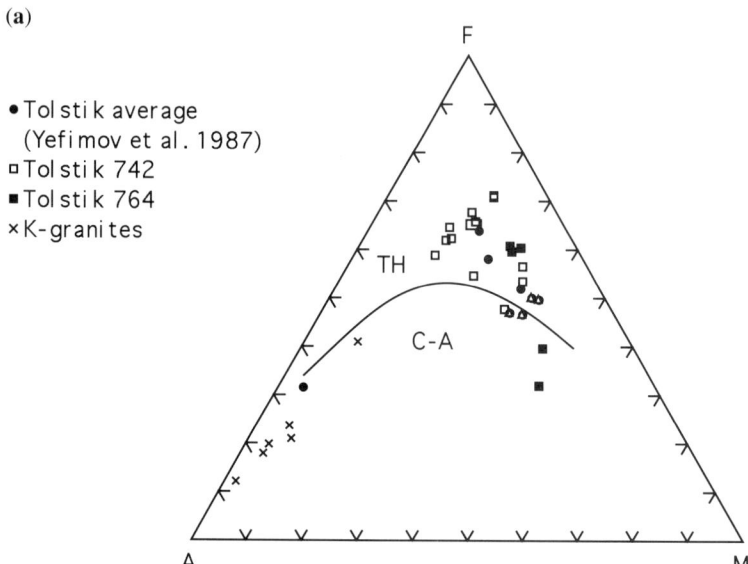

- Tolstik average (Yefimov et al. 1987)
- Tolstik 742
- Tolstik 764
× K-granites

(b)

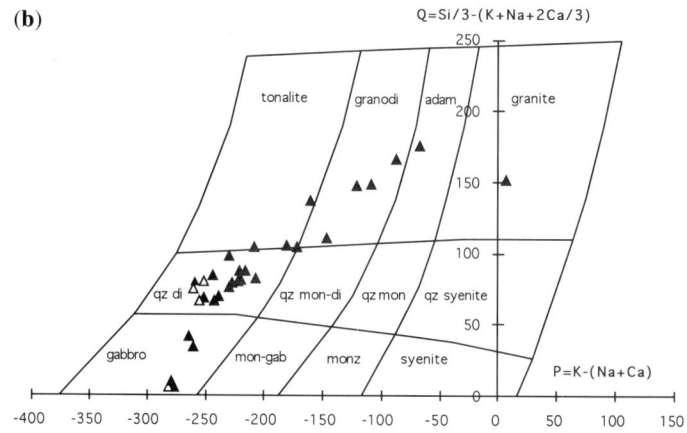

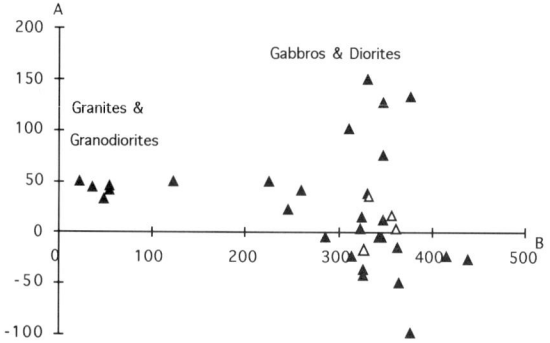

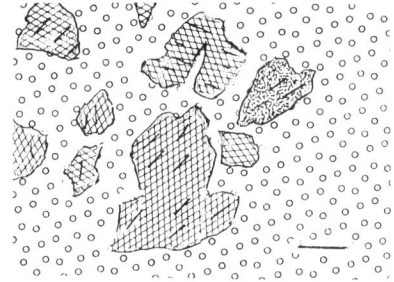

(a)

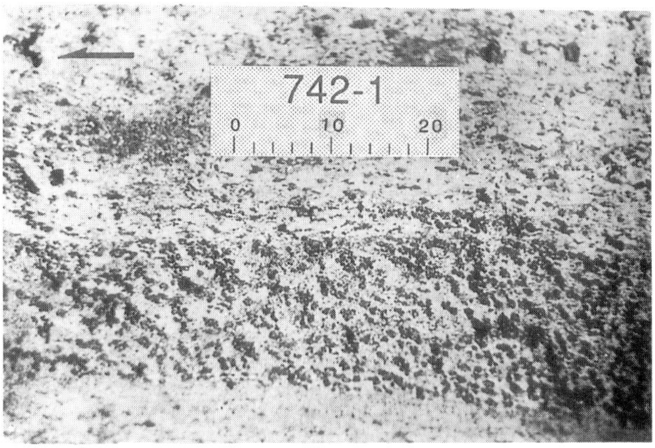

(b)

Fig. 3. Examples of relationships between different melts in the Tolstik intrusion. (**a**) Igneous breccia of previously sheared meta-gabbro and meta-diorite in the potassic 'augen' granites, locality 742. Note the angular shapes of the rotated mafic inclusions and their misaligned foliation. Sketch after a photograph. The scale bar is 50 mm. (**b**) A fragment of Tolstik diorite with its foliation truncated by the enclosing potassic 'augen' granite, locality 742. The scale bar is 20 cm.

(Yefimov *et al.* 1987). This swarm is made up of individual dykes spaced at intervals of 5–20 m. The dykes are usually less than 1 m wide and the longest can be followed for distances of more than 100 m. They clearly cut the multiply folded tonalitic–trondhjemitic gneisses (Fig. 4a). Close to the dyke swarm, approximately 20 m from the dyke that was sampled in locality 764 for the study of its metamorphic parageneses (Figs 1, 4a & 5), there is a body of meta-gabbro that is very similar to the Tolstik gabbro (Fig. 5). This is probably an outlier of the main gabbro body or possibly even an apophyse connected with that body. The gabbro outlier is *c.* 20 m across. Sharply outlined, small inclusions of the host rocks occur in its marginal part.

The mafic dykes appear fairly homogeneous but, under the microscope, are observed to have metamorphic fabrics with granoblastic, fine-grained aggregates of newly formed diopside, garnet,

Fig. 2. Chemical characteristics of the Tolstik intrusion and its mafic dykes. (**a**) AFM, diagram showing the transitional position of the Tolstik intrusions between the fields of tholeiitic and calc-alkaline rocks, and their overall bimodal character. TH and C–A mark the tholeiitic and calc-alkaline fields, respectively. (**b**) Q–P and A–B diagrams of Debon & Le Fort (1983) for the chemical–mineralogical classification of the Tolstik igneous rocks. The abbreviations are: adam, adamellite; granodi, granodiorite; qz di, quartz diorite; qz mon-di, quartz monzodiorite; qz mon, quartz monzonite; qz syenite, quartz syenite; mon-gab, monzogabbro; monz, monzonite. Open triangles indicate the compositions mafic-dyke rocks.

Fig. 4. Photographs showing some of the field relationships of mafic dykes of the first generation. Northeastern shore of the Tolstik Peninsula, locality 764. The strike of the dykes is NW 330–355°. (**a**) A 1 m thick mafic dyke of the Tolstik-type rocks that cross-cuts the $c.$ 2740 Ma banded, multiply folded TTG gneisses. The margins of the dyke are dark and rich in hornblende. The lighter inner part is mainly composed of garnet, clinopyroxene and plagioclase. (**b**) A different thinner, Tolstik-type dyke affected by dextral strike-slip deformation together with the country gneisses and completely transformed into garnet amphibolite.

plagioclase, quartz, and some brownish hornblende. Within these aggregates there are remains of igneous clinopyroxene with exsolution patches. These relics form a kind of microscopic 'augen' (cf. Figs 9a & b). Remaining outlines of ophitic, elongated plagioclase laths and subhedral grains of igneous clinopyroxene are also observed.

Similar metamorphic, newly crystallized mineral aggregates occur in the gabbro where they can be observed even by the naked eye and define a well-developed lineation. The entire gabbro body is strongly sheared and foliated. Small folds with axial planes sub-parallel to the foliation have flat-lying hinge lines that plunge 25–30° N.

A later strike-slip deformation which affected dykes, meta-gabbro and country rocks alike, caused the formation of a marked foliation. This foliation trends sub-parallel to the dyke contacts and is marked by veins of coarse microcline granites and pegmatitic granites which also fill tension fractures in the dykes (Figs. 4b & 5).

The gabbro body similarly contains veins of these granites which fill tension cracks and outline boudins formed by shearing. Along the contacts of both the meta-gabbro and the mafic dykes there are rims of dark, fine- to medium-grained amphibolite (Fig. 4a). The thinner dykes are wholly amphibolized (Fig. 4b).

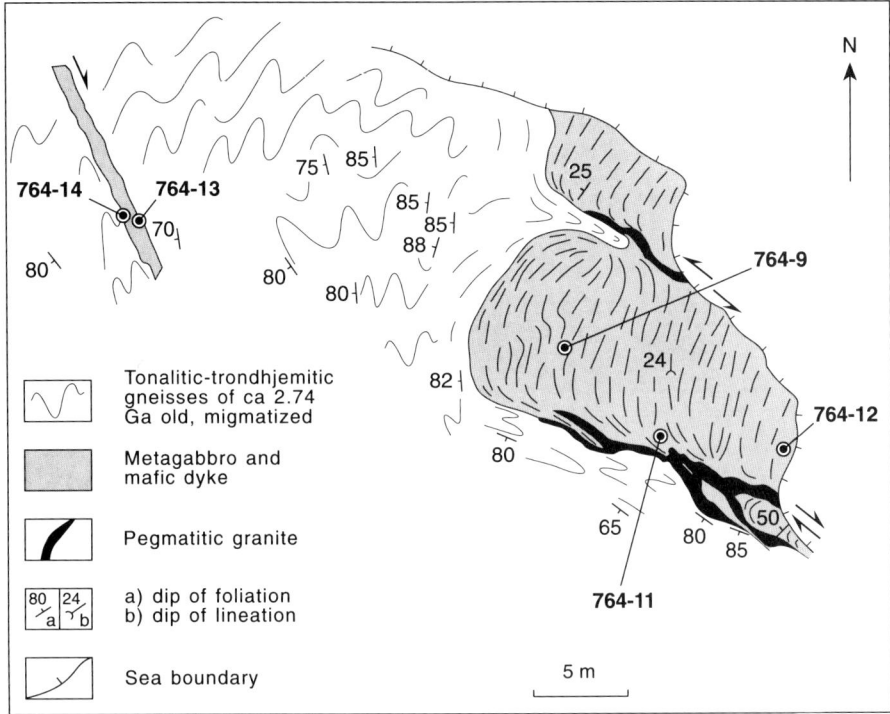

Fig. 5. Sketch map of the outlier of the Tolstik-type gabbro and a dyke of the first generation (cf. Fig. 4a) in locality 764 (Fig. 1). There are signs of multiple strike-slip movements marked by an early, gentle stretching lineation within the outlier body and later, sinistral strike-slip structures outlined by pegmatitic granites (modified after an unpublished map by M. N. Bogdanova, S. V. Bogdanova and L. P. Popova, made in 1986).

Mafic dykes that intruded the gabbros and diorites before the emplacement of the potassic granites

The mafic dykes of this group cut the 2434 Ma meta-gabbros and meta-diorites in locality 742 and elsewhere (Fig. 1) but are themselves found as large xenoliths, sharply delimited minor, angular fragments and boudinaged relics within the potassic augen granites of 2405 Ma age (Figs 6a & b). They differ from the dykes of the oldest group both in chemistry (Table 1) and in mineral composition. The dykes of the second generation are mostly dioritic and contain primary igneous biotite and andesine as well as metamorphic diopside, garnet , hornblende, magnetite, oligoclase, and quartz. There are also patches of pyrite in ilmenite which probably mark sub-solidus decomposition. The biotites and the relics of primary opaque minerals are overgrown by amphibole.

The boundaries of the dyke fragments in the granites and the dykes themselves are outlined by thin amphibole-rich rims. Where the dykes of this group cut the meta-gabbros and meta-diorites, they exhibit typical extension (dilation) patterns with a dextral component of deformation (Fig. 7; cf. Cadman et al. 1990). Similar strike-slip patterns are also frequent where quartz dioritic inclusions occur in the augen granites. These inclusions occasionally preserve an earlier foliation (Fig. 3a & b).

All these features indicate that the emplacement of the dykes of the second generation took place during strike-slip movements. The dyke strikes are generally 330–340°, similar to those of the first group.

The margins of these dykes are perceptibly chilled against the meta-gabbro and meta-diorite Metamorphic garnet often occurs in these margins (cf. Fig. 11a).

Potassic granites

Potassic, megacryst-rich, 'augen' granites and associated aplitic and pegmatitic dykes and veins (Figs 1 and 8) form lensoid bodies inside the

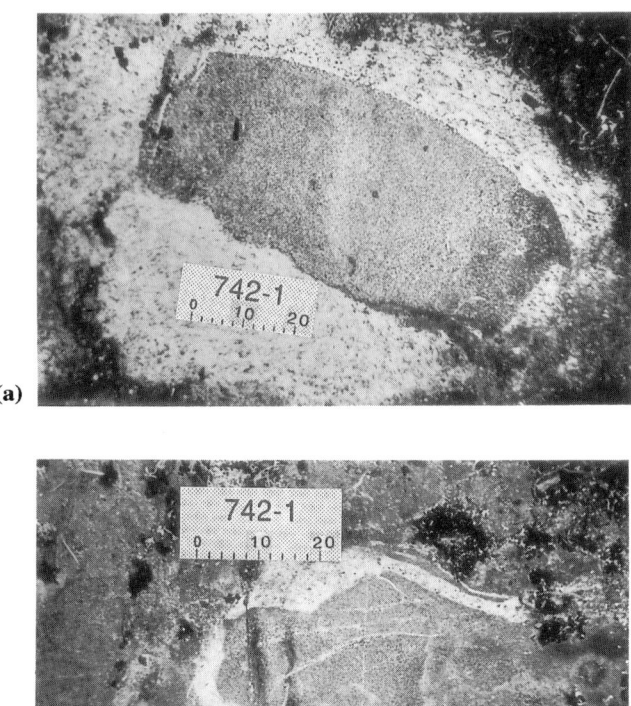

Fig. 6. Photographs showing some field relationships of the mafic dykes of the second generation in the southwestern part of the Tolstik intrusion. The scale bars are 20 cm. (**a**) Boudinaged fragment of a mafic dyke enclosed in the potassic 'augen' granite. (**b**) A larger fragment of a dyke with tension fractures filled by potassic 'augen' granite.

Tolstik mafic intrusion. These bodies are semi-concordant with the long axis of the intrusion and the foliation in the mafic rocks. The largest granite body is situated in the southwestern part of the intrusion. It has a length of 500 m and a maximum width of 80 m. The 'augen' granites have 'finger-like' contacts with the mafic rocks, which suggest dilation-controlled melt intrusion with a displacement component. The granites contain a lineation defined by amphibole and biotite which is orientated sub-parallel to the lineation of the mafic rocks. Nevertheless, as described above, the granite also contains rotated xenoliths of previously foliated meta-gabbro and meta-diorites (cf. Fig. 3a & b). Some of the foliation in the granite was formed earlier than the emplacement of the aplitic dykes and has been destroyed along the contacts with these dykes. A later foliation, which is present in all rock types on the Tolstik Peninsula, trends sub-parallel to the older foliation. It is distributed inhomogeneously and concentrated into distinct shear belts in which the granites are thoroughly foliated.

The potassic granites are coarse-grained, predominantly, inequigranular rocks. The megacrysts consist of microcline with grey orthoclase relics and subordinate oligoclase. They are set in a matrix of quartz, oligoclase (An 23 to 28), microcline, greenish-brown hornblende and biotite. Orthite, sphene and monazite are common accessories. Garnet and epidote occasionally appear in fine-grained zones of intense deformation.

Mafic dykes intruding the granites

The dykes of the youngest group intruded after the emplacement of the potassic 'augen' granites, and the aplitic and pegmatitic dykes succeeding these rocks (Fig. 8). These dykes also mainly strike NW ($c.\,330°$) like the older dykes and were

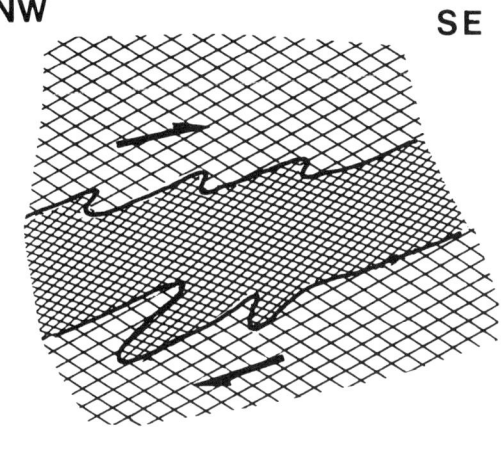

Fig. 7. Sketch of a dyke of the second generation that cross-cuts a leucocratic Tolstik meta-gabbro NE of locality 742 (Fig. 1). The contacts demonstrate a dextral extension–dilation pattern. The dyke strikes NW 330–340°.

similarly deformed by strike-slip movements after crystallization. Chilled margins and porphyritic textures are very common. This suggests shallower levels of magma crystallization and lower degrees of metamorphism than those prevailing when the dykes of the two earlier groups were formed.

The third-generation dyke rocks are greyish-green, fine-grained amphibolites made up of diopside, hornblende, biotite, andesine, magnetite, and quartz. Rare garnet is present. The textures are granoblastic, but the matrices are studded with accumulations of granoblastic plagioclase grains which possibly represent original feldspar megacrysts that reached 3 cm in length. These plagioclase patches only occur in the central parts of the dykes, while the dyke margins are more 'aphyric' and enriched in amphibole.

Thus, the formation and metamorphism of the Tolstik mafic intrusion, the dykes, and the potassic granites was accompanied by repeated intense shearing that followed similar directions. The shearing is manifested in the entire Tolstik Peninsula and adjacent areas but tends to be strongest in shear zones along the marginal parts of the intrusion, leaving its central part little reworked.

It is probable that this episodic shearing was the mechanism that caused the fragmentation of the main Tolstik intrusion and the relative rotation of its fragments. It also determined most of the present tectonic pattern in the studied area, which was thus apparently created during multistage magma emplacement between 2434 and 2405 Ma.

Metamorphism and P–T characteristics of the Tolstik mafic dykes

For P–T determinations of the 2.43–2.40 Ga metamorphism, the mafic dykes of the two first generations were chosen. Sampling was carried out at locality 764 (the first, gabbro-related generation of dykes) and locality 742 (the second dyke generation, cf. Fig. 1).

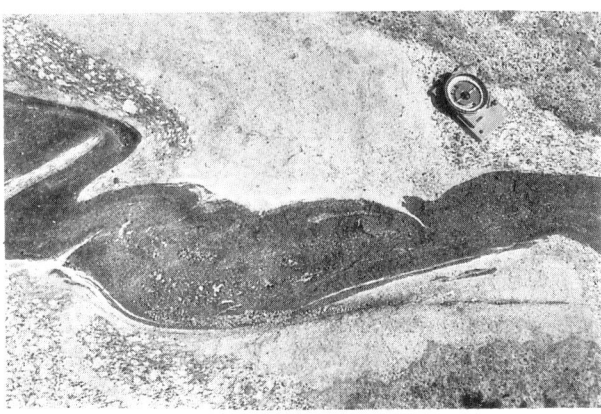

Fig. 8. Mafic dyke of the third generation cross-cuts the potassic 'augen' granite and an aplitic dyke. The mafic dyke was affected by nearly brittle, strike-slip deformation accompanied by blastomylonitization of the country rock granite. The inner parts of the dyke feature porphyritic texture.

Table 2. Representative microprobe analyses of minerals from the metagabbro and the mafic dyke, locality 764

Sample	Meta-gabbro 764														Dyke 764				
	764-9				764-11				764-12					764-13			764-14	
Mineral	GT core	PX core	AM rim	AM rim	FP	GT core	GT rim	PX	FP	GT core	PX core	AM core	FP	GT core	PX	FP	PX	AM
SiO_2	38.59	53.06	43.92	44.27	61.59	37.63	37.82	51.19	59.64	38.18	53.00	46.05	62.82	38.95	52.22	61.20	51.57	43.09
TiO_2	0.00	0.06	1.41	1.43	0.00	0.02	0.00	0.34	—	0.00	0.08	0.38	0.01	0.07	0.13	0.00	0.22	0.87
Al_2O_3	22.21	2.15	12.19	12.06	15.15	21.65	20.52	2.93	24.34	22.81	2.58	11.14	22.85	21.39	2.22	23.51	2.90	12.31
Fe_2O_3	—	—	—	—	—	0.07	2.72	0.97	—	—	—	—	—	—	—	—	—	—
FeO	26.96	8.03	13.53	14.12	0.00	25.41	23.64	8.08	0.05	26.72	9.55	13.42	0.08	24.38	8.20	0.31	9.84	17.05
MgO	5.40	13.10	11.73	11.75	0.00	4.31	4.16	12.33	—	4.79	12.85	12.78	0.02	4.85	12.88	0.01	11.16	9.24
MnO	0.62	0.05	0.00	0.03	0.00	0.49	0.59	0.03	0.01	0.68	0.05	0.12	0.03	0.73	0.13	0.02	0.37	0.23
CaO	7.46	22.46	11.15	11.31	6.30	10.28	10.60	22.65	7.18	8.21	20.78	11.11	4.53	8.06	23.11	6.23	22.77	11.83
Na_2O	0.06	0.65	1.57	1.80	7.66	—	—	0.52	7.21	0.00	0.83	1.11	8.49	0.02	0.67	7.68	0.90	1.57
K_2O	0.07	0.07	1.12	1.22	0.43	—	—	—	0.63	0.09	0.02	1.08	0.50	0.02	0.01	0.28	0.00	1.01
Cr_2O_3	0.04	0.00	0.00	0.00	0.00	0.00	0.02	0.06	—	0.00	0.02	0.00	0.02	0.00	0.15	0.02	0.21	0.03
H_2O	—	—	—	—	—	—	—	—	—	—	—	—	—	—	—	—	—	—
Total	101.40	99.64	96.63	97.98	101.13	99.86	100.07	99.10	99.05	101.48	99.76	97.19	99.35	98.48	99.71	99.26	99.95	97.23
Si	2.97	1.98	6.54	6.53	3.00	2.96	2.97	1.93	2.69	2.94	1.98	6.77	2.80	3.06	1.95	2.74	1.93	6.51
Al	2.01	0.10	2.14	2.10	0.87	2.00	1.90	0.13	1.29	2.07	0.11	1.93	1.20	1.98	0.10	1.24	0.13	2.19
Fe^3	0.07	—	0.00	0.00	0.00	0.00	0.16	0.03	0.00	0.07	—	0.00	0.00	—	0.05	0.00	0.05	0.00
Fe^2	1.66	0.25	1.69	1.74	0.00	1.67	1.55	0.26	0.00	1.65	0.30	1.65	0.00	1.60	0.21	0.01	0.26	2.15
Mg	0.62	0.73	2.60	2.58	0.00	0.51	0.49	0.69	0.00	0.55	0.71	2.80	0.00	0.57	0.72	0.00	0.62	2.08
Ca	0.61	0.90	1.78	1.79	0.33	0.87	0.89	0.92	0.35	0.68	0.83	1.75	0.22	0.68	0.92	0.30	0.92	1.91
Na	0.01	0.05	0.45	0.51	0.72	0.00	0.00	0.04	0.63	0.00	0.06	0.32	0.73	0.00	0.05	0.67	0.07	0.46
K	0.01	0.00	0.21	0.23	0.03	0.00	0.00	0.00	0.04	0.01	0.00	0.02	0.03	0.00	0.00	0.02	0.00	0.19
Ti	0.00	0.00	0.16	0.16	0.00	0.00	0.00	0.01	0.00	0.00	0.00	0.04	0.00	0.00	0.00	0.00	0.01	0.10
Mn	0.04	0.00	0.00	0.00	0.00	0.03	0.04	0.00	0.00	0.04	0.00	0.02	0.00	0.05	0.00	0.00	0.01	0.03
mg	27.15	74.42	60.72	59.72	—	23.23	23.88	73.13	—	24.99	70.57	62.94	27.98	26.19	77.37	5.00	70.71	149.14
Total Cat	8.00	4.00	15.57	15.64	4.94	8.04	8.00	4.00	5.00	8.00	4.00	15.48	4.98	7.95	4.00	4.98	4.00	15.63
Ox Equiv	12.00	6.00	23.00	23.00	8.00	12.00	12.00	6.00	8.00	12.00	6.00	23.00	8.00	12.00	6.00	8.00	6.00	23.00

GT, garnet; PX, pyroxene; AM, amphibole; FP, feldspar.

Table 3. *Representative microprobe analyses of metamorphic minerals from the mafic dyke of the second generation and from meta-diorite in contact*

Sample	Dyke 742							Meta-diorite 742							
	742d							S742a						742b	
Mineral	GT core	GT core	GT rim	PX incl	AM core	AM rim	AM rim	FP	GT1 core	GT1 core	AM1 core	AM1 core	GT2 rim	GT2 rim	AM2 core	AM2 rim	FP
SiO$_2$	38.06	37.46	37.01	51.94	41.34	41.11	41.57	60.02	38.22	37.93	40.37	40.47	37.13	37.19	41.15	41.14	61.51
TiO$_2$	—	—	—	0.21	1.63	1.48	1.58	—	—	—	1.41	1.30	—	—	1.85	1.91	—
Al$_2$O$_3$	21.80	21.59	21.25	2.36	12.28	12.46	12.77	24.60	21.74	21.76	12.95	12.73	21.11	21.42	13.07	13.00	23.01
Fe$_2$O$_3$	—	—	—	—	—	—	—	—	—	—	—	—	—	—	—	—	—
FeO	26.79	26.69	26.61	8.58	17.53	17.55	17.38	0.32	27.30	27.23	19.54	19.19	26.77	26.51	18.32	17.75	0.27
MgO	4.31	4.41	3.64	13.19	9.58	9.50	9.79	—	4.37	4.48	8.62	8.95	3.80	3.89	9.38	9.47	—
MnO	1.57	1.59	2.20	0.14	0.15	0.11	0.07	—	0.88	0.92	0.10	0.06	1.53	1.45	0.09	0.10	—
CaO	8.84	8.85	8.13	22.74	11.59	11.92	11.74	5.82	9.15	8.76	11.74	11.77	9.04	9.08	11.66	11.58	5.61
Na$_2$O	—	—	—	0.97	1.46	1.43	1.47	7.93	—	—	1.33	1.52	—	—	1.47	1.36	8.05
K$_2$O	—	—	—	—	1.73	1.74	1.70	0.29	—	—	1.76	1.74	—	—	1.60	1.61	0.24
Cr$_2$O$_3$	—	—	—	—	—	—	—	—	—	—	—	—	—	—	—	—	—
H$_2$O	—	—	—	—	—	—	—	—	—	—	—	—	—	—	1.99	1.98	—
Total	101.37	100.59	98.84	100.09	97.30	97.31	98.08	98.97	101.70	101.08	97.82	97.72	99.38	99.54	100.44	99.93	98.69
Si	2.95	2.92	2.95	1.92	6.30	6.28	6.28	2.70	2.95	2.94	6.19	6.21	2.94	2.94	6.21	6.23	2.77
Al	1.99	1.98	2.00	0.10	2.21	2.24	2.27	1.30	1.98	1.99	2.34	2.30	1.97	1.99	2.32	2.32	1.22
Fe^3	0.12	0.18	0.10	0.11	0.00	0.00	0.00	0.00	0.13	0.13	0.00	0.00	0.15	0.14	0.00	0.00	0.00
Fe^2	1.61	1.56	1.68	0.16	2.24	2.16	2.19	0.01	1.63	1.63	2.51	2.46	1.62	1.61	2.31	2.25	0.01
Mg	0.50	0.51	0.43	0.73	2.18	1.95	2.20	0.00	0.50	0.52	1.97	2.05	0.45	0.46	2.11	2.14	0.00
Ca	0.73	0.74	0.70	0.90	1.89	1.95	1.90	0.28	0.76	0.73	1.93	1.93	0.77	0.77	1.88	1.88	0.27
Na	0.00	0.00	0.00	0.07	0.43	0.42	0.43	0.69	0.00	0.00	0.40	0.45	0.00	0.00	0.43	0.40	0.70
K	0.00	0.00	0.00	0.00	0.34	0.34	0.33	0.02	0.00	0.00	0.34	0.34	0.00	0.00	0.31	0.31	0.01
Ti	0.00	0.00	0.00	0.01	0.19	0.17	0.18	0.00	0.00	0.00	0.16	0.15	0.00	0.00	0.21	0.22	0.01
Mn	0.10	0.11	0.15	0.00	0.02	0.02	0.01	0.00	0.06	0.06	0.01	0.01	0.10	0.10	0.01	0.01	0.00
mg	23.55	24.66	20.56	82.40	49.35	49.11	50.11	0.00	23.55	24.07	44.04	45.39	21.63	22.09	47.73	48.74	0.00
Total Cat	8.00	8.00	8.00	4.00	15.79	15.82	15.79	5.00	8.00	8.00	15.85	15.89	8.00	8.00	15.79	15.75	4.98
Ox Equiv	12.00	12.00	12.00	6.00	23.00	23.00	23.00	8.00	12.00	12.00	23.00	23.00	12.00	12.00	23.00	23.00	8.00

GT, garnet; AM, amphibole; PX, pyroxene; FP, feldspar.

To assess possible differences of metamorphism between dykes and the main Tolstik intrusion and to obtain P–T data as complete as possible for the studied period, the two sampled dykes were compared with gabbro and diorite from the same localities, i.e. with the gabbro in the apophyse or outlier of the main intrusion in locality 764, and with a diorite in locality 742. The latter belongs to the main Tolstik mafic intrusion and occurs in direct contact with the dykes of the second generation.

The representative mineral compositions given in Tables 2 and 3 were obtained using the JEOL 6400 equipment of the Institute of Geology, Lund University, and the CAMECA SX-50 microprobe at the National Microprobe Laboratory, Department of Earth Sciences, Uppsala University. No systematic differences were found when comparing analyses from the two instruments. All calculations of mineral compositions were carried out using the 'MINTAB' computer program (Rock & Carroll 1990).

Metamorphic textures, mineral assemblages and thermobarometry

Locality 764

In this locality, the meta-gabbro and the mafic dykes of the first generation display very similar mineral assemblages representing peak of metamorphism. These comprise garnet + clinopyroxene + ilmenite + rutile ± hornblende + plagioclase + quartz.

In the marginal parts of the mafic bodies, however, greenish hornblende, magnetite, sphene, and clinozoisite overgrow the earlier metamorphic assemblage and characterize a retrograde stage.

Fine-grained (mostly less than 0.5 mm) granoblastic textures, straight grain boundaries and the virtual absence of reaction rims between metamorphic minerals are evidence that the mineral assemblages attained equilibrium. Relics of igneous clinopyroxene with exsolution textures are, however, overgrown by metamorphic clinopyroxene and garnet. Some coarser garnet grains appear to have nucleated on and grown at the expense of primary ore minerals which are preserved in their cores (Figs 9a & b).

The metamorphic minerals in the dyke are mostly homogeneous. In particular, there is very slight variation of garnet composition (Table 2). The only exception is a more Fe-rich garnet that is in contact with opaque minerals. Clinopyroxenes have very similar compositions in the dyke and in the meta-gabbro. The K_D of the Fe–Mg distribution between garnet and clinopyroxene is mostly 0.33–0.34. The plagioclase is andesine (An 32–34) both in the cores and the rims. All these compositional features indicate good equilibration of the mineral assemblage and permit the use of core–core mineral compositions for P–T estimates for peak of metamorphism.

The internal parts of the gabbro and the sampled dyke, chemically distinct rims on the mineral grains are not developed well enough to determine the retrograde conditions. A better record is obtained from analyses of the minerals in the marginal zones of the mafic bodies. In these rocks, pargasitic–tschermakitic hornblende (cf. Fig. 13) is found in association with clinopyroxene and garnet. The latter are more Fe-rich than their equivalents in the peak -metamorphic assemblage (cf. sample 764-12, Table 2).

Locality 742

In this locality, samples from a mafic dyke of the second generation and the meta-diorite of the main Tolstik intrusion were used to estimate the P–T conditions.

Metamorphic mineral assemblages are present both in the wall rocks and in the dyke, and comprise garnet ± clinopyroxene ± ilmenite + magnetite + hornblende + plagioclase + quartz. K-feldspar and albite are found in some of the intergranular spaces.

Fine-grained, sometimes symplectite-like intergrowths of garnet, clinopyroxene, hornblende, plagioclase and quartz are very typical for this dyke and indicate simultaneous formation of the metamorphic minerals (Figs 10a & b). These intergrowths separate grains of annitic high-Ti biotite and ilmenite (Table 3), which are the only preserved igneous minerals.

Garnet occurs more often in the marginal zone of the dyke but is rare in its interior. It is essentially an almandine–pyrope–grossularite mixture generally similar to that in the rocks from locality 764, but the almandine and spessartite contents are higher. Compositional zoning is distinct in narrow diffusional rims along the contacts with plagioclase and hornblende. These rims have less Mg than the cores but higher contents of Fe, Mn and Ca (Fig. 11a).

Clinopyroxene inclusions within garnet (Fig. 10b) are very magnesian (Table 3) which is probably due to strong diffusional loss of Fe to the garnet. The plagioclase is unzoned oligoclase (An 27–28). All hornblendes are dominantly pargasitic with somewhat varying amounts of actinolite and tschermakite components. The hornblende from the dyke of the second generation is higher in Ti than the hornblende from the dyke of the first generation in locality 764 (Table 3). This is possibly due to differences in whole-rock composition (Table 1).

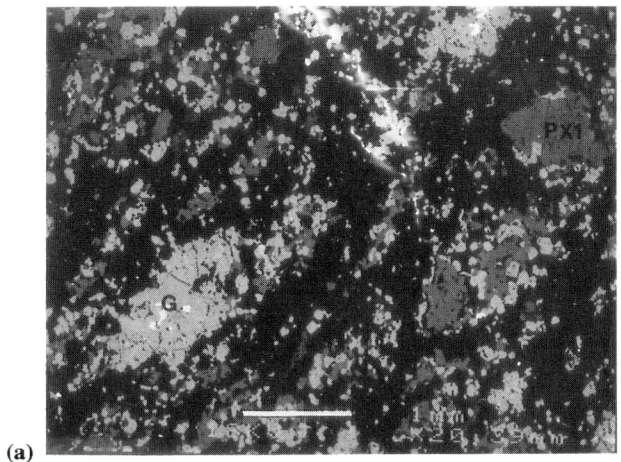

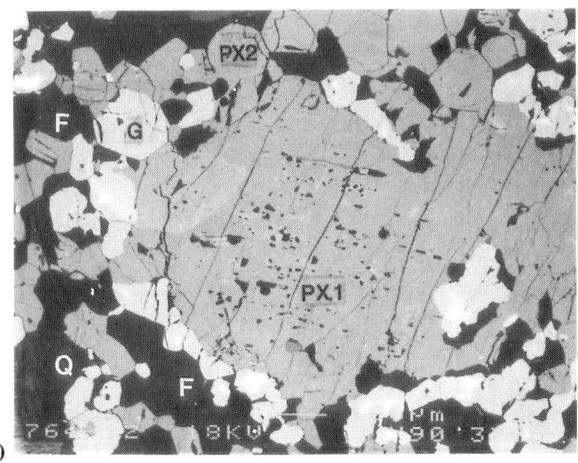

Fig. 9. Backscatter electron images of the textures of the mafic dykes of the first generation, locality 764. (**a**) Dyke-rock texture with lensoid aggregates of the metamorphic garnet–clinopyroxene–plagioclase–quartz assemblage including relics of igneous clinopyroxene. (**b**) Subhedral grain of igneous clinopyroxene (PX1) surrounded by metamorphic clinopyroxene (PX2), garnet (G), quartz (Q) and plagioclase (F).

The fabrics of the meta-diorite in this locality are very similar to those in the dyke, but the rock is generally more coarse-grained, with garnets reaching 1–1.5 mm in diameter (Figs. 10a & c). Two metamorphic mineral assemblages can be distinguished. The early one is represented by garnet and brownish-green hornblende with inclusions of igneous biotite and ore minerals. The plagioclase in contact with, or included in these minerals is oligoclase (An 27-28).

The apparently later metamorphic assemblage in the meta-diorite is similar to that of the dyke. It is made up of light-coloured hornblende, garnet, rare clinopyroxene, plagioclase and quartz. These minerals are often intergrown, and inclusions of the other minerals in garnet are very common.

Similar inclusion-rich garnet also forms overgrowths on garnets of the early assemblage (Fig. 10c). Microprobe profiles across such garnets show compositional discontinuity at the boundaries between the cores and the overgrowths (Fig. 11b). However, core and overgrowth garnet each have their own diffusional rims characterized by relative increases of Fe, Mn and Ca, and lower Mg contents. This records a regressive metamorphic stage in the diorite that preceded the formation of

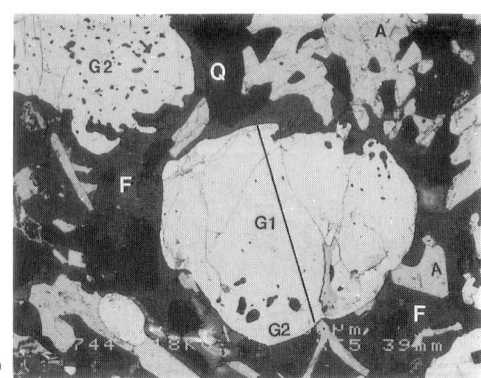

Fig. 10. (a) Backscatter electron image of the textures at the contact between a cutting mafic dyke of the second generation (bottom and left) and the wallrock Tolstik meta-diorite (top and right), locality 742. The white boxes mark sites of detail images (b) and (c); (b) is at the left, (c) in the centre of picture (a). The scale bar is 1 mm, the abbreviations used in (b) and (c) are G, garnet, A, hornblende, F, plagioclase, Q, quartz. Picture (b) shows the dyke with the dyke garnet G. In picture (c), the peak-metamorphic garnet of the meta-diorite (G1) is rimmed by inclusion-rich garnet (G2). The lines drawn across the garnets in (b) and (c) mark microprobe profiles in Figs 11a and b.

the second generation of garnet, and presumably also the intrusion and metamorphism of the second generation of dykes.

For the P–T estimates, the cores and rims of the two generations of garnet and hornblende were compared separately. The samples are described as 'core 1' and 'core 2' and 'rim 1' and 'rim 2', respectively, in Tables 3 and 4.

P–T estimates

The P–T estimates listed in Table 4 have been obtained by the use of various geothermometers and geobarometers. The textural and compositional features of the studied rocks indicate that the metamorphic assemblages attained equilibrium. This makes it possible to estimate the P–T conditions of both peak and retrograde metamorphism by the use of thermobarometry employing the core–core and rim–rim mineral relationships (Spear 1993). The computer program 'GEOTHERMOBAROMETRY' coded by M. Kohn, Rensselaer Polytechnic Institute, Troy, NY, was used for the calculations (Spear et al. 1991). That program contains many recent, experimentally obtained and consistent calibrations (cf. Spear, 1993).

The temperature estimates using the garnet–clinopyroxene thermometer of Powell (1985) and garnet–amphibole thermometer of Graham & Powell (1984) are similar. With one exception indicated in Table 4, the Pattison & Newton (1989) garnet–clinopyroxene thermometer, however, gives unrealistically low temperatures which are 50–150° C lower than the others.

The pressure estimates using the Newton & Perkins (1982) garnet–clinopyroxene–plagioclase–quartz barometer are always 2–3 kb lower than those obtained from the analogous barometers of Powell & Holland (1988) and Moecher et al. (1988). The Mg-end member calibration of the latter authors was used in this work because the Fe-end member calibration mostly indicated unrealistically low pressures with differences up to 3 kb. The garnet–hornblende thermometer of Graham & Powell (1984) in combination with the garnet–plagioclase–hornblende–quartz barometer of Kohn & Spear (1990) provide the most consistent estimates for the studied rocks. The coincidence of the estimates based on different calibrations supports the reliability of the obtained P–T parameters of metamorphism.

The P–T estimates of the peak metamorphism using the garnet–clinopyroxene–plagioclase–quartz paragenesis in the meta-gabbro and the mafic dyke of the first generation (locality 764) are at 700–710° C and 11 to 12 kb (meta-gabbro), and at

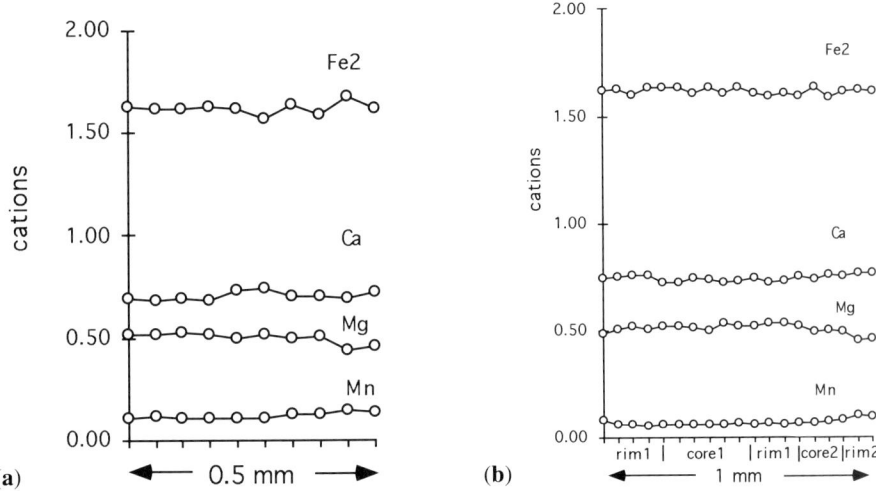

Fig. 11. Compositional-zoning profiles across garnets from the meta-diorite and the dyke, locality 742: (**a**) dyke (cf. Fig. 10b); (**b**) meta-diorite (cf. Fig. 10c).

Table 4. *P–T estimates of metamorphism for the meta-gabbro, meta-diorite and the mafic dykes of the Tolstik intrusion*

	T, (° C)		P (kb)			
Calibration	Grt–Cpx	Grt–Hbl	Grt–Pl–Cpx–Qtz			Grt–Pl–Hbl–Qtz
Author	P&N*; PL**	G & PL	N & P	PL & H	MH, Mg	K & S
Locality 764						
Gabbro:						
764-9 cores (4)	710**	680	9.5	11	11	9.5
764-9 rims (3)	700**	650	8	10.5	10	8.5
764-11 cores (3)	660*; 670**		8	11	10	
764-12 cores (1)	700**			12.5	12	
764-12 rims (1)		620				10
Dyke:						
764-13 cores (2)	660**	650	9	11.5	10.5	9.5
Locality 742						
Meta-diorite:						
742a core 1 (2)		800				12
742a rim1 (3)		780				12
742a core 2 (3)		740				11
742a rim 2 (3)		710				11
S742 core (1)		600				9–9.5
S742 rim (2)		550				8.5
Dyke:						
742 cores (4)		740				10
742 interiors (2)		725				10–9.5
742 rims-1 (2)		700				10
742 rims-2 (1)		680				9.5
742 rims (3)	480**		5.5	8.5–8	7.5–7	

P&N, Pattison & Newton (1989); PL, Powell (1985); G & PL, Graham & Powell (1984); N & P, Newton & Perkins (1982); PL & H, Powell & Holland (1988); MH, Mg, Moecher *et al.* (1988), Mg-end member; K & S, Mg, and Fe-end members, Kohn & Spear (1990), tschermakite model. Numbers in brackets indicate the number of mineral datasets used for the estimates. All calculated errors are less than ±25° C and ±1 kb.

665° C and 10–11 kb (dyke). The garnet–hornblende–plagioclase–quartz parageneses in these rocks indicate a lower temperature and pressure (635° C and c. 10 kb) which may correspond to a retrograde stage.

Similar pressures but higher temperatures were estimated for the peak metamorphism in the meta-diorite of the main Tolstik intrusion and the mafic dyke of the second generation (locality 742). Meta-diorite 742 records the highest P–T values of 800° C and 12 kb obtained from the garnet and hornblende cores of the early metamorphic assemblage. The cores of the same minerals of the later generation ('core 2' in Table 4) yield 740° and 11 kb, while the rims suggest 710° C and 11 kb. The lowest retrograde P–T values have been obtained from the garnet + hornblende + plagioclase + quartz assemblage in another sample of meta-diorite, labelled 'S742' in Tables 3 and 4. The determined P–T conditions are at 600 to 550° C, and 9–9.5 to 8.5 kb (Table 4).

The apparently highest metamorphic conditions in the cross-cutting dyke of the second generation are estimated at 740° and 10 kb. Note that this temperature is in good agreement with the T estimate for the second stage of metamorphism in the meta-diorite host. However, the obtained pressures differ by 1 kb (10%) which may or may not be due to minor compositional inhomogeneity or imperfect equilibration of the mineral phases.

A retrograde stage of metamorphism in the dyke is recorded by a range of P–T values obtained from diffusional rims at the contacts between garnet, hornblende and plagioclase, and garnet, clinopyroxene and plagioclase. The lowest temperature obtained is 480° C at c. 8 ± 1 kb (Table 4).

Discussion and conclusions

Despite the limited numbers of analysed mineral pairs, the P–T characteristics estimated by different calibrations indicate pressures of 11 ± 1 kb for the 2.43–2.40 Ga high-grade metamorphism in the Tolstik Peninsula. At some stage during that time, the studied crustal section was therefore at depths of 35–40 km, that is in the lower continental crust. The crustal thermal gradients were relatively low, ranging between 20 and 30° C km^{-1}.

These results may appear unexpected when seen against the background of presently dominant views that regimes of general tension, rifting and crust disruption prevailed between the end of the late Archaean orogeny (less than 2.7 Ga ago) and the beginning of the Palaeoproterozoic collisional orogeny in the Lapland–Kola Mobile Belt (at most c. 1.9 Ga ago). However, they are in line with the concept of a 'Seletsk folding' in the earliest Proterozoic (Glebovitsky et al. 1978), and also explain why Palaeoproterozoic platform-cover sediments are absent in the Belomorian Belt while they are common in the adjoining Archaean province of Karelia.

Although the Tolstik mafic intrusion is virtually identical in age with the layered mafic intrusions that were formed during the early stages of extension and rifting of the Archaean crust in Karelia and the Kola Peninsula (e.g. Alapieti et al. 1990), it had itself been affected by the 2.43–2.40 Ga high-grade, high-P metamorphism.

This study thus shows that at 2.45–2.40 Ga the central parts of the Belomorian Belt must have been at much greater depth than the adjoining Archaean provinces of Karelia and the Kola Peninsula. Even within the Belomorian Belt itself there is a marked contrast of depth level between its central and marginal parts. At the edge of the Karelian Province, 2.45 Ga old potassic granites were intruded into Archaean gneisses at a depth of only c. 7 km (Bibikova et al. 1993a, b). Substantial and selective uplift of the central Belomorian Belt, or its later upthrusting as an allochthonous terrane is therefore indicated.

Whether the thermal effects of the mafic igneous activity, of which the Tolstik intrusion is part, triggered the formation of new granulite facies mineral associations in a wider region is difficult to assess until the spatial distribution of the 2.43–2.40 Ga high-grade metamorphism and the occurrence of Tolstik-type mafic rocks are known better. Until very recently, all such rocks were considered Early Archaean, while at present the existence of 2.7–2.6 Ga, 2.45–2.4 Ga and 2.0–1.9 Ga generations of mafic intrusions is known (Alapieti et al. 1990; Balashov et al. 1993; Bogdanova & Bibikova 1993).

Despite the probability of thermal effects from the Tolstik mafic intrusions, the P/T ratios of metamorphism are moderately high and the crustal thermal gradients must have been low. This could indicate a thickened continental crust. However, it appears highly improbable that the P/T effects of a Late Archaean collisional orogeny that culminated at c. 2.7 Ga (e.g. Miller & Milkevich 1995) could have persisted until 2.45 Ga. Elevated P/T-ratios created by the stacking of cool crust should have decayed within a few tens of millions of years (e.g. Spear 1993). A more plausible explanation for the calculated pressure conditions is the general negative correlation between Moho heat flow and crustal thickness (e.g. Čermak 1982).

The present study, however, gives no indications of whether thick, cool crust was the ultimate result of Late Archaean crustal thickening, or whether it was due to crustal stacking somewhat before the intrusion of the Tolstik mafic rocks during, for instance, a 'Seletsk' tectonic event.

The alternative that the high-P granulite metamorphism is actually a manifestation of the 1.9–1.8 Ga collisional orogeny in the Lapland–Kola Mobile Belt must be considered. As mentioned above, however, very strong effects during this time in the studied rocks of the Tolstik Peninsula are contradicted by the absence of any detectable 1.9–1.8 Ga old influence on the U–Pb systems of the zircons, the lack of growth rims of that age around the igneous zircons, and the apparent absence of high-grade parageneses in the third and latest generation of mafic dykes. It is also contradicted by the field data which indicate that the high-grade metamorphism is older than the 2405 Ma old potassic granite, and the observation that high pressures prevailed both before and after the intrusion of the second generation of mafic dykes. In these dykes, only one episode of garnet growth occurred, whereas two generations of garnet are present in the meta-diorite intruded by the dykes. Collisional orogeny in the Lapland–Kola belt can, however, be one of the mechanisms that caused the exhumation of the rocks in the Tolstik Peninsula.

The nature and full extent of the 'Seletsk folding' in the Belomorian Belt are not yet known exhaustively. Shearing was prominent, but previous investigators (e.g. Volodichev 1990) also emphasize that high-P regimes occurred in at least part of the Belomorian Belt in the earliest Proterozoic even before the beginning of the 1.9–1.8 Ga 'Svecofennian' event in the Lapland–Kola region. However, the precise timing of that metamorphism was not known before the present study.

As far as the P–T paths are concerned, the lines connecting the P–T estimates based on core and rim mineral compositions tend to be complex (Fig. 12). The P–T paths for the minor meta-gabbro body and the mafic dykes sited outside the main Tolstik intrusion (in locality 764) indicate substantial decompression of up to 3 kb for some samples, but

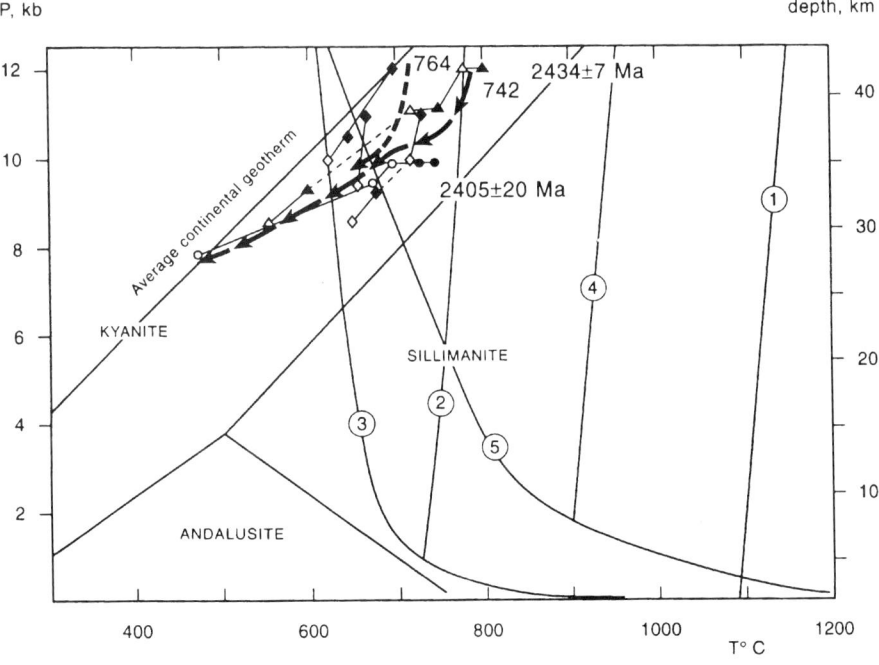

Fig. 12. Diagram showing P–T estimates for the studied rocks (cf. Table 4). The thick arrows mark generalized P–T paths for the metamorphic development of the Tolstik region between c. 2.43 and 2.4 Ga ago. The Al_2SiO_5 triple point is from Holdaway (1971). The curves are: 1, dry gabbro solidus (Wyllie et al. 1981); 2, biotite dehydration-melting (Le Breton & Thompson 1988); 3, H_2O-saturated granite solidus (Huang & Wyllie 1973); 4, amphibole dehydration-melting (Rushmer 1987); 5, H_2O-saturated olivine basalt solidus (Yoder & Tilley 1962). The symbols are: diamonds, metagabbro and mafic dyke of the first generation, locality 764; triangles, metadiorites of the main mafic intrusion, locality 742, circles, the mafic dyke of the second generation cross-cutting the meta-diorite, locality 742. Filled symbols show core–core and open symbols rim–rim P–T estimates.

only minor cooling. Metamorphism in these rocks deviated less from the model of a steady continental geotherm, while metamorphism in the main Tolstik intrusion began at higher temperatures. In the latter case, the P–T path gives evidence of a more complicated metamorphic history than for the minor intrusions. It records alternating isobaric cooling and near- isothermal decompression. Despite the limited number of analyses and analytical uncertainty, the results can be interpreted to indicate that after peak metamorphism at 700–800° C and 11–12 kb, decompression was near-isothermal, implying uplift. After the loss of about 5 km overlying crust, continued cooling was more isobaric. As a whole, this means that the emplacement of new melts in the main Tolstik intrusion continued throughout uplift and was accompanied by ductile shearing that affected the structural features and metamorphic textures of many of the rocks.

Altogether, this could be taken to indicate extension of thickened crust (cf. Sandiford & Powell 1986; Harley 1989; Thompson 1990; Spear 1993). However, all P–T paths except one are very short and terminate at temperatures above 600° C. The absence of a record of adaption of the mineral associations to still lower temperatures may be due to sluggish reaction kinetics in dry granulite systems (Bohlen 1987; Harley 1989) and does not necessarily indicate quenching by a subsequent rapid uplift of the Belomorian Belt. Indeed, during the $c.$ 25 Ma long period of high-grade metamorphism between 2430 and 2405 Ma, the rate of uplift due to extension and erosion was moderate, about 0.4 km Ma^{-1}, with a cooling rate of $c.$ 6° C Ma^{-1}. Modelling of the late uplift stage must therefore rely on geological arguments, until additional regional P–T data become available from other localities. In that context, the thermal role of the Tolstik-type magmatism must also be reconsidered.

In summary, the principal conclusions of the present study are:

- High-grade metamorphism at high pressures took place in the central Belomorian Belt between 2.43 and $c.$ 2.40 Ga, concomitantly with intrusion of mafic rocks on the Tolstik Peninsula.
- Elevated P/T ratios and the P–T paths indicate a preceding event of crustal thickening. However, it is still unclear whether that thickening occurred during collisional orogeny at $c.$ 2.7 Ga or more probably, was caused by crustal stacking during the previously described but not precisely dated 'Seletsk folding', shortly before the intrusion of the Tolstik mafic rocks.
- Preliminary P–T paths indicate an initial stage of rapid decompression which was accompanied by multiple magma emplacement, followed by a stage of near-isobaric cooling and continued, sub-horizontal shearing. This sequence of events fits deep-level extension of thickened continental crust along detachment zones (cf. Sandiford & Powell 1986) in the Belomorian Belt during the earliest Palaeoproterozoic.

The underlying field work for this study was done in 1986–1988 during a mapping campaign in the Tolstik Peninsula funded by the Geological Institute of the then USSR Academy of Science (Moscow). Collaboration with M. Bogdanova and M. Yefimov from the Geological Institute of the Kola Science Centre (Apatity) at that time is gratefully acknowledged. D. Ryabukhin, I. Tkachenko and E. Shirokova from Moscow University assisted well in the mapping. The study of the metamorphic assemblages was financed by a grant from the Swedish Natural Sciences Research Council and was carried out in the Microprobe Laboratory of the Geological Institute, Uppsala University, and at the Department of Mineralogy and Petrology, Institute of Geology, Lund University. Research engineers Hans Harrysson (Uppsala) and Takeshi Miyazu (Lund) kindly helped with microprobe work, providing guidance, training and equipment control. Britt Nyberg (Lund) as usual did an excellent work of drafting.

Fruitful discussions with Prof. R. Gorbatschev (Lund), Prof. V. Glebovitsky and Dr. Yu. Miller (St Petersburg) have greatly improved this paper. An anonymous referee provided excellent editing and many comments important for future work in the Belomorian Belt.

This paper is a contribution to IGCP Project 275 'Deep Geology of the Baltic/Fennoscandian Shield'.

References

ALAPIETI, T. T., FILEN, B. A., LAHTINEN, J. J., LAVROV, M. M., SMOLKIN, V. F. & VOITSEKHOVSKY, S. N. 1990. Early Proterozoic layered intrusions in the northeastern part of the Fennoscandian Shield. *Mineralogy and Petrology,* **42**, 1–22.

BALAGANSKY, V. V., BOGDANOVA, M. N. & KOZLOVA, N. E. 1986. *Structural–Metamorphic Evolution of the Northwestern Belomorian Belt.* Apatity, Kola Science Centre, 1–100 (in Russian).

BALASHOV, YU. A., BAYANOVA, T. B. & MITROFANOV, F. P. 1993. Isotope data on the age and genesis of layered basic-ultrabasic intrusions in the Kola Peninsula and northern Karelia, northeastern Baltic Shield. *Precambrian Research,* **64**, 197–206.

BAROOACH, B. P. & BOWES, D. R. 1990. Separation of early Proterozoic mafic dyke swarms by structural relationships in the Lewisian Complex, near Scourie, Scotland. *In:* PARKER, A. J., RICKWOOD,

P. C., & TUCKER, D. H. (eds) *Mafic Dykes and Emplacement Mechanisms*. A. A. Balkema, Rotterdam, Netherlands, 507–520.

BIBIKOVA, E. V., GRACHEVA, T. V. & MAKAROV, V. A. 1993a. U–Pb isotopic dating of the Early Precambrian K-granites in the northwest Belomorian (White Sea) area. *Abstracts of 'The Svecofennian Domain' Symposium*, IGCP-Project 275, Turku, Finland, p. 13.

——, MITROFANOV, F. P., BALASHOV, YU. A. & BOGDANOVA, S. V. 1990. The age and extent of the Archean in the Eastern Baltic Shield. *Abstracts of the Second Symposium on the Baltic Shield*, Lund, p. 16.

——, SKIÖLD, T., BOGDANOVA, S. V., DRUGOVA, G. M. & LOBACH-ZHUCHENKO, S. B. 1993b. Geochronology of the Belomorides: an interpretation of the multistage geological history. *Geokhimiya*, **10**, 1393–1411 (in Russian).

——, SKIÖLD, T., BOGDANOVA, S. V. 1996. Age and geodynamic aspects of the oldest rocks in the Precambrian Belmorian Belt of the Baltic (Fennoscandian) Sheild. *This volume*.

BOGDANOVA, M. N. & YEFIMOV, M. 1993. Origin of Parental Anorthosite Magmas, Tectonic and Metamorphic Processes in the Evolution of Anorthosites (Kolvitsa Anorthosite Associations). *Guidebook of Geological Excursion for IGCP Project 290*, Apatity, 1–61.

BOGDANOVA, S. V. & BIBIKOVA, E. V. 1993. The 'Saamian' of the Belomorian Mobile Belt: new geochronological constraints. *Precambrian Research*, **64**, 131–152.

——, BIBIKOVA, E. V. & GORBATSCHEV, R. 1994. Palaeoproterozoic U–Pb zircon ages from Belorussia: new geodynamic implications for the East European Craton. *Precambrian Research*, **68**, 231–240.

BOHLEN, S. R. 1987. Pressure–temperature–time paths and a tectonic model for the evolution of granulites. *Journal of Geology*, **95**, 617–632.

BRIDGWATER, D., AUSTRHEIM, H., HANSEN, B.T., MENGEL, F. PEDERSEN, S. & WINTER J., 1990. The Proterozoic Naggsugtoqidian mobile belt of southeast Greenland: a link between the eastern Canadian and Baltic Shields. *Geoscience Canada*, **17**, 305–310.

CADMAN, A., TARNEY, J. & PARK, R. G. 1990. Intrusion and crystallisation features in Proterozoic dyke swarms. *In:* PARKER, A. J., RICKWOOD, P. C. & TUCKER, D. H. (eds) *Mafic Dykes and Emplacement Mechanisms*, Balkema, Rotterdam, Netherlands, 13–24.

ČERMAK, V. 1982. Crustal temperature and mantle heat flow in Europe. *Tectonophysics*, **89**, 123–142.

DEBON, F. & LE FORT, P. 1983. A chemical-mineralogical classification of common plutonic rocks and associations. *Transactions of the Royal Society of Edinburgh: Earth Sciences*, **73**, 135–149.

GAÁL, G. 1990. Tectonic style of Early Proterozoic ore deposition in the Fennoscandian Shield. *Precambrian Research*, **46**, 83–114.

—— & GORBATSCHEV, R. 1987. An outline of the Precambrian evolution of the Baltic Shield. *Precambrian Research*, **35**, 15–52.

GLEBOVITSKY, V. A., DOOK, V. L. & SHARKOV, E. V. 1978. Endogenic processes. *In*: KRATZ, K. O. (ed.) *The Crust of the Eastern Baltic Shield*, Nauka, Leningrad, 112–171 (in Russian).

GORBATSCHEV, R. & BOGDANOVA, S. 1993. Frontiers in the Baltic Shield. *Precambrian Research*, **64**, 3–21.

GRAHAM, C. M. & POWELL, R. 1984. A garnet-hornblende geothermometer : calibration, testing, and application to the Pelona Schist, Southern California. *Journal of Metamorphic Geology*, **2**, 13–21.

HARLEY, S. L. 1989. The origins of granulites : a metamorphic perspective. *Geological Magazine*, **126**, 215–247.

HOLDAWAY, M. J. 1971. Stability of andalusite and the aluminum silicate phase diagram. *American Journal of Science*, **271**, 97–131.

HUANG, W. L. & WYLLIE P. J. 1973. Melting relation of muscovite-granite to a 35 kbar as a model for fusion of metamorphosed subducted oceanic sediments. *Contributions to Mineralogy and Petrology*, **42**, 1–14.

HUHMA, H., CLIFF, R. A., PERTTUNEN, V. & SAKKO, M. 1990. Sm–Nd and Pb isotopic study of mafic rocks associated with early Proterozoic continental rifting: the Peröpohja schist belt in northern Finland. *Contributions to Mineralogy and Petrology*, **104**, 369–379.

KALSBEEK, F., AUSTRHEIM, H., BRIDGWATER D., HANSEN, B. T., PEDERSEN, S. & TAYLOR, P. N. 1993. Geochronology of Archaean and Proterozoic events in the Ammassalik area, South-East Greenland, and comparisons with the Lewisian of Scotland and the Nagssugtoqidian of West Greenland. *Precambrian Research*, **62**, 239–270.

KOHN, M. J. & SEAR, F. S. 1990. Two new geobarometers for garnet amphibolites, with applications to southeastern Vermont. *American Mineralogist*, **75**, 89–96.

KRILL, A. G. 1985. Svecokarelian thrusting with thermal inversion in the Karasjok-Levajok area of the northern Baltic Shield. *In: Geology of Finnmark – a collection of papers. Norges geologiske undersøkelse*, Trondheim, 89–101.

LE BRETON, N. & THOMPSON, A. B. 1988. Fluid-absent (dehydration) melting of biotite in metapelites in the early stage of anatexis. *Contributions to Mineralogy and Petrology*, **99**, 226–237.

MARKER, M., BOGDANOVA, S., BRIDGWATER, D., GLEBOVITSKY, V., GORBATSCHEV, R., MENGEL, F. & MILLER, Yu. 1993. The Lapland–Kola Mobile Belt, Norhern Baltic Shield. *Terra nova, abstracts*, **5**, 319.

MENGEL, F. & RIVERS, T. 1991. Decompression reactions and P–T conditions in high-grade rocks, northern Labrador: P–T–t paths from individual samples and implications for Early Proterozoic tectonic evolution. *Journal of Petrology*, **1**, 139–167.

MILLER, Yu. V. & MILKEVICH, R. I. 1995. Folded-nappe structure of the Belomorian tectonic zone and its relationships with the Karelian granite-greenstone terrain. *Geotektonika*, **6**, 80–92 (in Russian).

MINTS, M. V. 1993. Palaeotectonic reconstructions of the Early Precambrian in the Eastern Baltic Shield.

I. Early Proterozoic. *Geotektonika*, **1**, 39–57 (in Russian).

MITROFANOV, F. P., BALAGANSKY, V. V., BALASHOV, YU. A., GANNIBAL, L. F., DOKUCHAEVA, V. S. ET AL. 1993. U–Pb age of gabbro-anorthosites in Kola, Russia. *Abstracts of 'The Svecofennian Domain' Symposium*, IGCP-Project 275, Turku, Finland, 42–43.

MOECHER, D. P., ESSENE, E. J. & ANOVITZ, L. M. 1988. Calculation and application of clino-pyroxene-garnet-plagioclase-quartz geobarometers. *Contributions to Mineralogy and Petrology*, **100**, 92–106.

NEWTON, R. C. & PERKINS, D. III. 1982. Thermodynamic calibration of geobarometers based on the assemblages garnet-plagioclase-orthopyroxene (clinopyroxene)-quartz. *American Mineralogist*, **67**, 203–222.

PARK, R. G. & TARNEY, J. 1987. The Lewisian complex: a typical Precambrian high-grade terrain. *In:* PARK, R. G. & TARNEY, J. (eds) *Evolution of the Lewisian Complex and Comparable Precambrian High Grade Terrains*. Geological Society, London Special Publication, **27**, 13–25.

PATCHETT, P. J. & KOUVO, O. 1986. Origin of continental crust of 1.9–1.7 Ga age: Nd isotopes and U–Pb zircon ages in the Svecokarelian terrain of South Finland.*Contributions to Mineralogy and Petrology*, **92**, 1–12.

——, GORBATSCHEV, R. & TODT, W. 1987. Origin of continental crust of 1.9–1.7 Ga age: Nd isotopes in the Svecokarelian terrains of Sweden. *Precambrian Research*, **35**, 145–160.

PATTISON, D. & NEWTON, R.S. 1989. Reversed experimental calibration of the garnet–clino-pyroxene Fe–Mg exchange thermometer. *Contributions to Mineralogy and Petrology*, **101**, 87–103.

POWELL, R. 1985. Regression diagnostics and robust regression in geothermometer / geobarometer calibration: the garnet-clinopyroxene geothermometer revised. *Journal of Metamorphic Geology*, **3**, 231–243.

—— & HOLLAND, T. J. B. 1988. An internally consistent dataset with uncertainties and correlations: 3. Applications to geobarometry, worked examples and a computer program. *Journal of Metamorphic Geology*, **6**, 173–204.

PUCHTEL, I. S., ZHURAVLEV, D. Z., KULIKOVA, V. S. & KULIKOVA, V. V. 1991. Petrography and Sm–Nd Age of a Differetiated Sheet of Komatiitic Basalt in the Vetra Belt, Baltic Shield. *Geochemistry International*, **12**, 14–22.

PUURA, V. & HUHMA, H. 1993. Palaeoproterozoic age of the East Baltic granulitic crust. *Precambrian Research*, **64**, 289–294.

ROCK, N. M. S. & CARROLL, G. W. 1990. MINTAB: A general-purpose mineral recalculation and tabulation program for Macintosh microcomputers. *American Mineralogist*, **75**, 424–430.

RUSHMER, T. 1987. Fluid-absent melting of amphibolite – experimental results at 8 kbar. *Terra Cognita*, **7**, 286.

SANDIFORD, K. & POWELL, R. 1986. Deep crustal metamorphism during continental extension: ancient and modern examples. *Earth and Planetary Science Letters*, **79**, 151–158.

SPEAR, F. S. 1993. *Metamorphic Phase Equilibria and Pressure–Temperature–Time Paths*. Mineralogical Society of America, Washington, D. C.

——, PEACOCK, S. M., KOHN, M. J., FLORENCE, F. P. & MENARD, T. 1991. Computer programs for petrologic P–T–t path calculations. *American Mineralogist*, **76**, 2009–2012.

STENAR', M. M. 1988. Stratigraphy of Archaean deposits in Soviet Karelia. *In:* Archaean Geology of the Fennoscandian Shield. *Geological Survey of Finland Special Paper*, **4**, 7–14.

STEPANOV, V. S. 1981. *The Precambrian Mafic Magmatism in the Western White Sea Region*. Nauka, Leningrad (in Russian).

—— & SLABUNOV, A. I. 1989. *Amphibolites and Early Mafic-Ultramafics of the Precambrian in Northern Karelia*. Nauka, Leningrad, 177 pp (in Russian).

THOMPSON, A. B. 1990. Heat, fluids, and melting in the granulite facies. *In:* VIELZEUF, D. & VIDAL, PH. (eds) *Granulites and Crustal Evolution*. Kluwer, Dordrecht, The Netherlands, 37–57.

VOLODICHEV, O. I. 1990. *The Belomorian Complex of Karelia (Geology and Petrology)*. Nauka, Leningrad (in Russian).

WEAVER, B. I. & TARNEY, J. 1981. The Scourie Dyke Suite: Petrogenesis and Geochemical Nature of the Proterozoic Sub-Continental Mantle. *Contributions to Mineralogy and Petrology*, **78**, 175–188.

WYLLIE, P. J., DONALDSON, C. H., IRVING A. J., KESSON, S. E., MERRILL, R. B. ET AL. 1981. Experimental petrology of basalts and their source rocks. *In:* Basaltic Volcanism Study Project. *Basaltic Volcanism on the Terrestrial Planets*, Pergamon, New York, 494–633.

YEFIMOV, M. M., BOGDANOVA, M. N. & BOGDANOVA, S. V. 1987. Periodicity and Archaean magmatism in the north-western White Sea region (the Tolstik key-area). *In: Mafic-ultramafic Magmatism in the Main Structures of the Kola Peninsula*. Kola Science Centre, Apatity, 15–29 (in Russian).

YODER, H. S. & TILLEY, C. E. 1962. Origin of basalt magmas: an experimental study of natural and synthetic rock systems. *Journal of Petrology*, **3**, 342–532.

Burwell domain of the Palaeoproterozoic Torngat Orogen, northeastern Canada: tilted cross-section of a magmatic arc caught between a rock and a hard place

MARTIN J. VAN KRANENDONK[1] & RICHARD J. WARDLE[2]

[1] *Continental Geoscience Division, Geological Survey of Canada, 601 Booth Street, Ottawa, Ontario, K1A 0E8, Canada*
(Present address: Department of Geology, University of Newcastle, Newcastle, NSW 2308, Australia)
[2] *Geological Survey Branch, Newfoundland Department of Mines and Energy, PO Box 8700, St John's Newfoundland, A1B 4J6, Canada*

Abstract: The northern segment of the Palaeoproterozoic Torngat Orogen is unique relative to the southern segment in that it is underlain by a 1910–1885 Ma suite of diorite–tonalite–granite rocks (DTG suite) emplaced into the margin of the Archaean Nain Province. Field and geo-chemical data indicate that it represents a subduction-generated Andean-type magmatic arc. Slivers of MORB-type, tholeiitic amphibolites and metasedimentary rocks within the DTG suite are interpreted as remnants of a backarc basin. Arc rocks are preserved in a tilted section from homogeneous, foliated tonalites at mid-amphibolite facies in the east, through gneissic tonalites, to massive orthopyroxene-bearing granitoid rocks of the Killinek charnockitic suite in the west. This assemblage is separated from the Archaean Rae Province, with which Nain Province collided, by a wide belt of turbiditic sedimentary rocks, now at granulite facies, known as the Tasiuyak gneiss. A younger suite (c. 1865 Ma) of mafic tonalites and megacrystic metagranites within the Nain arc suite represents either syn-collisional plutons or late additions to the continental magmatic arc.

The southern segment of the orogen differs from the northern segment because it preserves no evidence of the DTG suite on Nain Province margin, but instead contains a widespread suite of calc-alkaline plutons across the Tasiuyak gneiss/Rae Province margin, dated as 1880 Ma. We develop two testable models to account for these along-strike variations and the presence of arc magmas on both sides of the suture. In **Model A**, a flip in subduction polarity from east-dipping at 1910–1885 Ma to west-dipping at c. 1885 Ma is invoked, in which the c. 1865 Ma suite of plutons in the northern segment are interpreted to represent syn-collisional magmas generated during accretion of the Burwell arc onto Nain Province. In **Model B**, the younger plutons in the northern segment may be interpreted as a late phase of the arc built on Nain Province, which we suggest may have been accomplished during a phase of double subduction immediately prior to collisional orogeny.

The Nain–Rae collision at c. 1870–1860 Ma formed a doubly vergent thrust wedge, burying part of the arc in Burwell domain to depths of 35 km. During subsequent sinistral shear deformation at 1845–1822 Ma, Nain Province was buried obliquely beneath the still hot arc to depths of 40 km. At 1798–1780 Ma, crustal-scale disharmonic folding of the northern segment resulted in the simultaneous exhumation of the deeply-buried granulites of the arc and of the northwestern margin of the Nain Province across the Komaktorvik shear zone (KSZ). The KSZ was the site of significant translation between the Burwell domain and unreworked portions of the Nain Province, but it does not represent a fundamental plate boundary.

The Palaeoproterozoic Torngat Orogen of the northeastern Laurentian Shield is a north–south striking belt of strongly deformed Archaean and Palaeoproterozoic rocks that forms the deeply exhumed root of a collision zone between the Archaean blocks of the Nain and Rae provinces to the east and west, respectively (Fig. 1; Hoffman 1988, 1990; Wardle *et al.* 1990). The core of Torngat Orogen, along its entire strike length, is occupied by an 8–40 km wide belt of garnet paragneiss and diatexite known as the Tasiuyak gneiss (Wardle 1983). This unit, characterized by pronounced negative aeromagnetic anomalies, can be traced north onto southern Baffin Island over a distance of 1300 km (e.g. Hoffman 1990). In northern Labrador, the distribution of the Tasiuyak

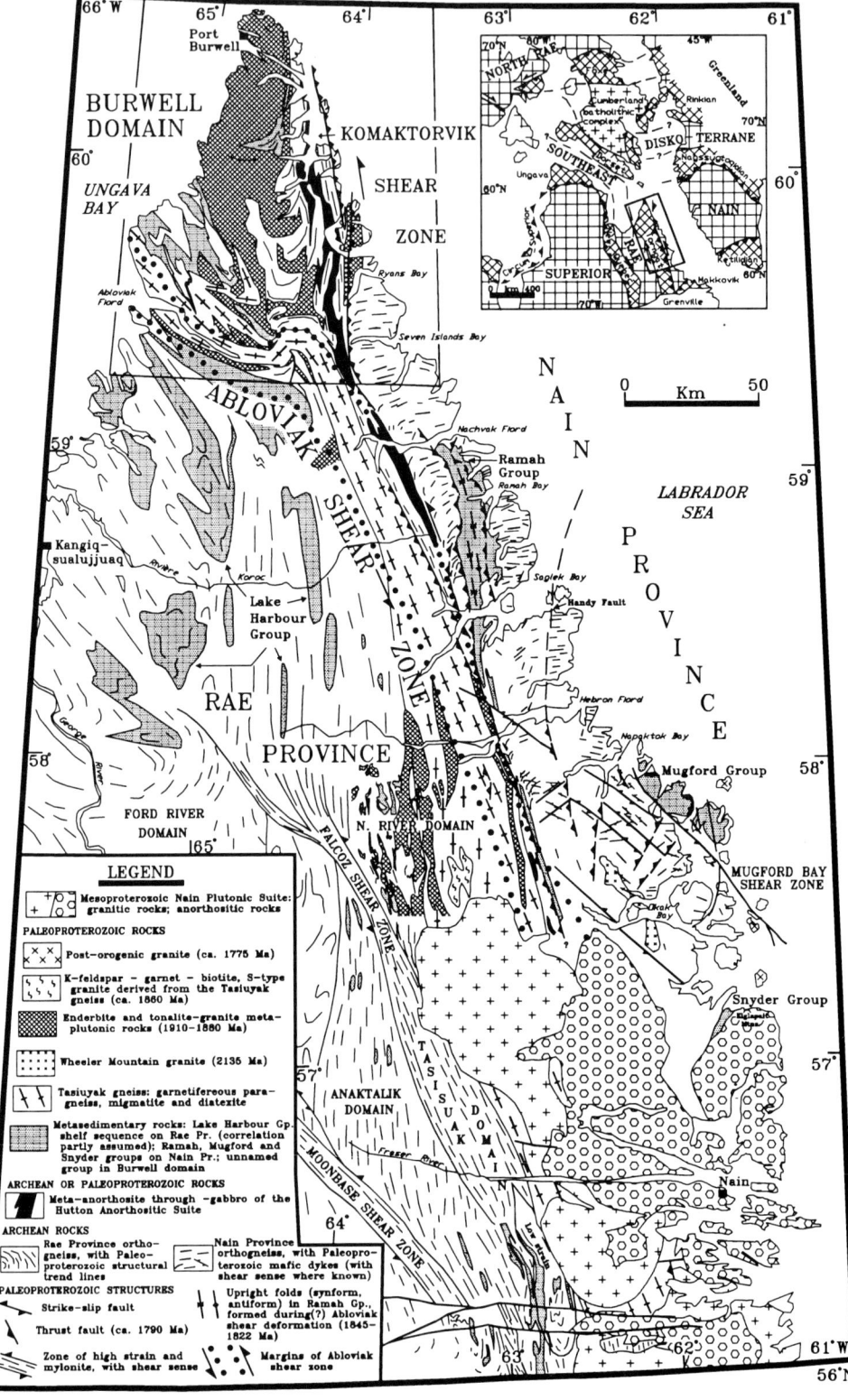

gneiss and the Abloviak shear zone in which it is largely deformed, is used to define two segments of the orogen. The southern segment strikes NNW–SSE and is straight over a distance of almost 600 km from Nachvak Fiord to the Grenville Orogen, although its southern extent is poorly known and largely obscured by glacial drift. In the north, however, lithologic and structural elements are deflected to the west around a triangular-shaped region of relatively low-strained crust known as the Burwell domain. This region is separated from the Nain Province to the east by the Komaktorvik shear zone, which is also marked by a pronounced aeromagnetic lineament (Fig. 1; Van Kranendonk et al. 1993a, after Korstgård et al. 1987).

Previous work has been concentrated in the southern segment of the Torngat Orogen between Nachvak Fiord and Nain, where it has been shown that following the development of arc magmatism on Rae Province margin at c. 1880 Ma, Torngat Orogen evolved episodically through three distinct tectono-metamorphic events: (1) oblique Nain–Rae collision at c. 1860 Ma and the development of a doubly-vergent thrust wedge under peak metamorphic conditions of 9.5 kb, 950°C in the core of the orogen; (2) formation of the sinistral Abloviak shear zone during a period of penetrative, widespread sinistral transpression at between 1845 and 1822 Ma under post-peak metamorphic conditions of 7.3 kb, 750°C; and (3) the formation of westside-up, reverse dsuctile mylonite zones along the eastern margin of the orogen, during orogenic reactivation between 1794–1786 Ma. This was followed by cooling from 1780 to 1740 Ma (structural data from Van Kranendonk & Ermanovics 1990; Ryan 1990; Van Kranendonk 1992, 1996: metamorphic data from Mengel & Rivers 1991; Van Kranendonk 1992, 1996; age data from Bertrand et al. 1993; Ryan et al. 1991; Mengel et al. 1991).

In the northern segment of the orogen, this scenario is complicated by the presence of the Burwell domain, which forms a wedge separating the Nain and Rae provinces. Burwell domain was originally described by Taylor (1979) as a mixture of Palaeoproterozoic granulite- and amphibolite-facies granitoid gneisses interspersed with bands of metasedimentary gneiss. On the basis of an irregular aeromagnetic signature, suggestions were later made that the Burwell domain represented either a block of Nain-type Archaean crust within an anastomosing shear zone (Korstgård et al.

1987), or an independent Archaean plate bounded by separate sutures along the southern and eastern margins now hidden within the Abloviak and Komaktorvik shear zones, respectively (Burwell terrane: Hoffman 1988; Burwell Province: Hoffman 1990). Further interpretation was hindered by a lack of detailed geological information until 1991, when a three-year mapping project of the Burwell domain was jointly initiated by the Geological Survey of Canada and the Newfoundland Department of Mines and Energy (Van Kranendonk & Scott 1992; Van Kranendonk et al. 1993b, 1994; Wardle et al. 1992, 1993, 1994).

Burwell domain is located at a critical triple junction between the southern segment of the Torngat Orogen and splays of Palaeoproterozoic deformation to the north in the Baffin Orogen (Dorset fold belt and an unnamed fold-thrust belt on southeastern Baffin Island) and the Nagsuggtoqidian Orogen of western Greenland (see inset of Fig. 1: Hoffman 1990; Van Kranendonk et al. 1993c). Research was carried out to determine: (1) the composition and age of the crust in Burwell domain, in order to identify whether or not it represents a separate Archaean plate; (2) the significance of the Komaktorvik shear zone as a potential suture zone; (3) the distribution of Palaeoproterozoic magmatism relative to Nain and Rae provinces and its relationships to that defined in the southern segment; and (4) the relationship of deformational episodes to those identified further south. What follows is a descriptive summary of the geology of the northern segment of the Torngat Orogen and our current ideas on possible models relating to the tectonic evolution of this region prior to deformation. A detailed description of the structural evolution is provided by Van Kranendonk & Wardle, in press). All U–Pb ages, unless otherwise referenced, are from Scott & Machado (1993, 1994, 1995).

Geology of the northern segment

Subdivisions and terminology

The geology of the northern segment of the Torngat Orogen is shown in Fig. 2. This region has been subdivided into a number of lithotectonic complexes and structural elements as shown in Fig. 3 (after Van Kranendonk et al. 1993a). **Lithotectonic complexes** comprise assemblages of individual rock units, suites and groups, and are

Fig. 1. Geology of the Palaeoproterozoic Torngat Orogen in northeastern Canada between Nain and Port Burwell. Inset figure shows pre-Mesozoic drift reconstruction of Archaean cratons and Palaeoproterozoic orogens in northeast Laurentia (after Van Kranendonk et al. 1993c and Hoffman 1989). Location of Fig. 2 indicated by square outline.

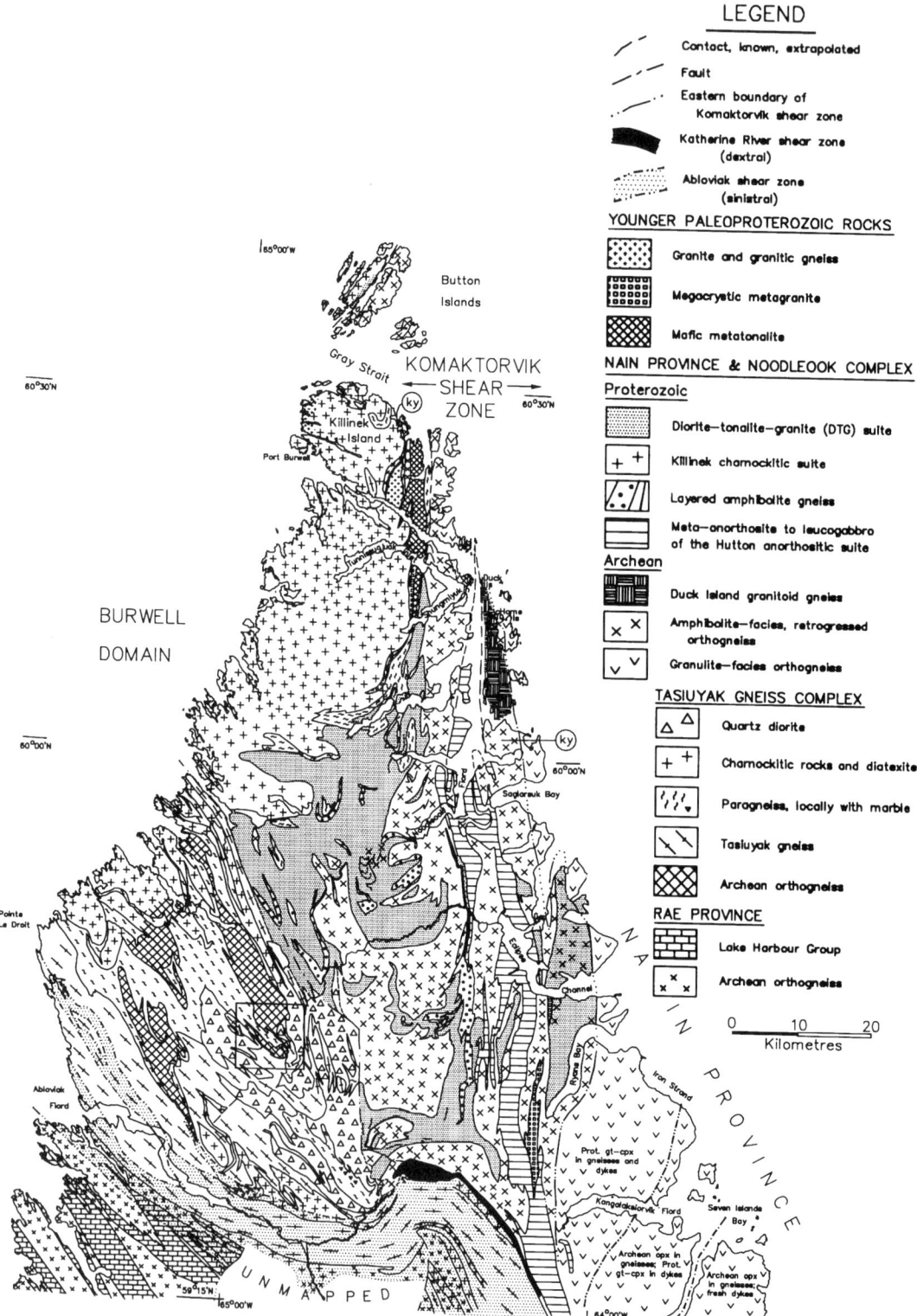

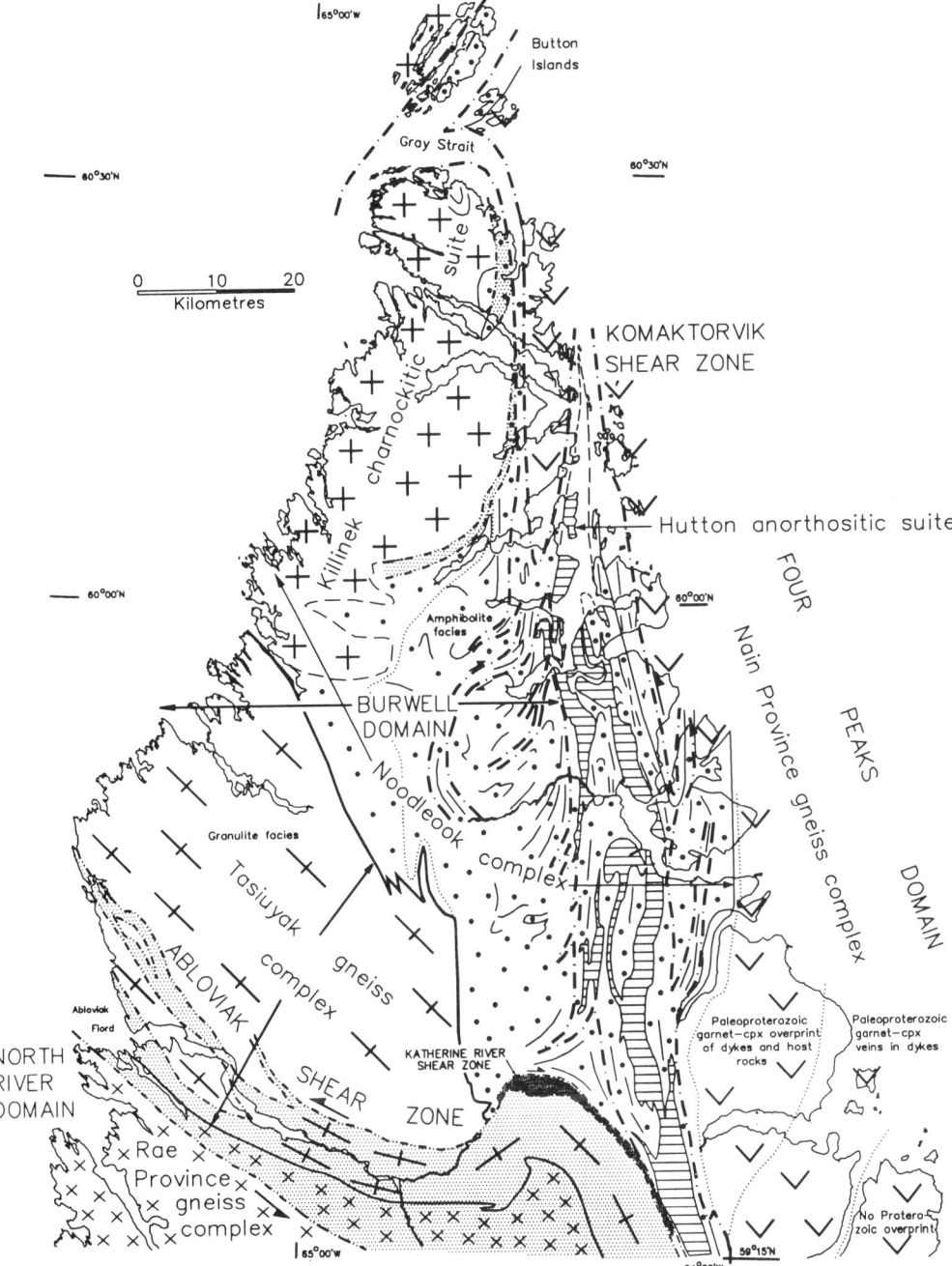

Fig. 3. Lithotectonic elements (complexes and suites) and structural divisions (DOMAINS, ZONES) of the northern segment of the Torngat Orogen. Four Peaks domain and Nain Province gneiss complex are coincident. Metamorphic zones are outlined by dotted lines. D_{n+2} shear zones outlined by light stipple, and the D_{n+3} Komaktorvik shear zone by heavy dash–dot line.

Fig. 2. Geology of the northern segment of the Torngat Orogen. Location of Fig. 7 indicated by small square outline. Dash–dot lines in southeast indicate the position of Palaeoproterozoic metamorphic isograds in Archaean Nain gneisses.

defined on the basis of restricted associations between some rock units, suites and groups, a restricted geographic distribution, and a continuity of assemblages along-strike. Individual rock units may transgress lithotectonic complex boundaries, but suites and groups do not.

The intensity and distribution of Palaeoproterozoic structures has been used to subdivide the northern segment of the orogen into six **structural elements**, including three domains of relatively low strain (Four Peaks, Burwell, and North River) separated by the Abloviak, Komaktorvik and Katherine River shear zones (Fig. 3). This structural zonation is superimposed on the lithotectonic complexes, many of which cross the structural domain boundaries. Structural elements are described in the section on structural geology.

Lithology: Nain Province gneiss complex

The granulite-facies Nain Province consists of ≤80% of leucocratic tonalitic orthogneiss and migmatite, which contains abundant leucosome veins and thin (30 cm to 2 m) layers and inclusions of mafic gneiss and rusty, biotite-rich paragneiss (see Wardle et al. 1994). Larger, more coherent rafts (500 m wide and up to 3 km long) within the gneisses contain ultramafic rocks (metaperidotite and metapyroxenite), compositionally-layered mafic gneiss (metagabbro) and homogeneous mafic rocks, anorthositic rocks, garnet–biotite ± sillimanite ± cordierite paragneiss, iron formation, and rate carbonate rocks.

Orthogneisses are cut by veins and small bodies of homogeneous, foliated to gneissic tonalite/ granodiorite, dated in one locality as 2834 ± 4 Ma (Fig. 4; Scott & Machado 1994). In the southeast, linear bodies of coarse-grained granodiorite/granite emplaced into the migmatic gneisses may represent larger bodies of this suite (Van Kranendonk 1994a). All rocks were then deformed under granulite-facies metamorphic conditions which is interpreted to have occurred at 2769 ± 5 Ma, the age of metamorphic zircon overgrowths from the previous sample. These rocks are in turn cut by white, biotite–allanite pegmatite sheets, up to 3 m wide, which are undated.

Avayalik dykes. Archaean gneisses and structures are cut by Palaeoproterozoic mafic dykes of the Avayalik swarm (Wardle et al. 1992). Dykes and their host rocks were affected by Palaeoproterozoic metamorphism which varies in grade across the northern segment of the orogen. In the southeast, orthopyroxene-bearing Archaean gneisses host E–W trending, sub-vertical Avayalik dykes with chilled margins and black feldspar megacrysts. Locally these undeformed and unmetamorphosed dykes have irregular, branching, or *en echelon* intrusion forms and shallow dips. Toward Kangalaksiorvik Fiord, these dykes are cut by garnet–clinopyroxene–quartz veins. Further west, dykes trend NE–SW and dip moderately to the SE, but are still undeformed, indicating that this change in orientation is primary. Both Archaean gneisses and Avayalik dykes in this area are recrystallized to a static metamorphic assemblage of garnet–clinopyroxene–quartz after hornblende–plagioclase, in the absence of associated deformation features (Van Kranendonk et al. 1994; Wardle et al. 1994).

A N–S striking, garnetiferous dyke in the Four Peaks domain, that is interpreted to belong to the Avayalik swarm, has yielded a U–Pb zircon upper intercept age of 1834 +7/–3 Ma, which Scott & Machado (1994) regard as the time of crystallization of the dyke (see Fig. 4 for location). Although it is possible that this date does reflect the intrusion age of the dyke, such an age is not in agreement with observations from regional mapping and geochronology which have shown that Avayalik dykes are not present in any Palaeoproterozoic supracrustal or metaplutonic units ≤ 1910 Ma (the age of the oldest dated Palaeoproterozoic metaplutonic rock: see below) within the Nain Province or Burwell domain. This therefore suggests that either two sets of dykes of similar appearance but different age have been grouped together in the Avayalik swarm, or that the age represents the time of metamorphic recrystallization of the dyke and crystallization of zircon, perhaps as a replacement of igneous baddeleyite. As the dyke contains metamorphic garnet, and the upper intercept age falls within the time of regional high-grade metamorphism within Torngat Orogen (D_{n+2}: see below), we provisionally interpret this age – subject to cnfirmation by work in progress – to represent the time of formation of the garnet–clinopyroxene assemblages within the Four Peaks domain (Van Kranendonk et al. 1994). The lower intercept age of c. 1740 Ma in the same dyke is clearly related to an amphibolite-facies overprint event (D_{n+3}: see below). The main set of dykes in the Avayalik swarm may be correlated with dykes identified further south, that have been dated as c. 2450 Ma (K–Ar and Rb–Sr whole-rock methods: Fahrig 1970; Taylor 1974) and/or c. 2234 Ma (U–Pb baddeleyite age on Kikkeravak dykes, in Ermanovics 1993).

Lithology: Rae Province gneiss complex

Archaean gneisses of suspected Archaean age. In the southwestern corner of the northern segment, Rae Province orthogneisses of suspected Archaean

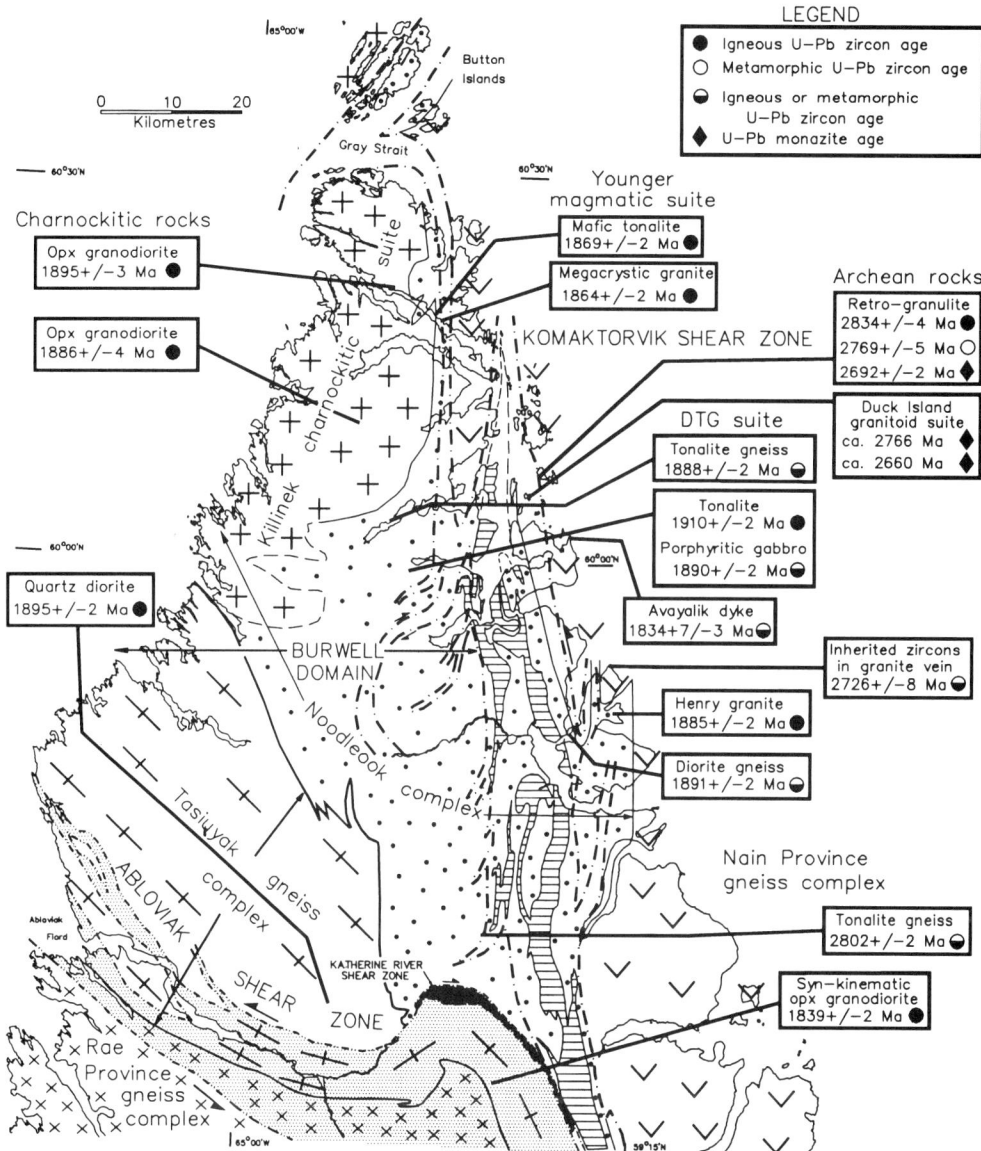

Fig. 4. Magmatic U–Pb zircon and monazite age data from the northern segment of the Torngat Orogen (data from Scott & Machado 1993, 1994).

age vary from white-weathering, homogeneous, tonalitic gneiss to agmatitic and schlieric textured migmatites in which 40–50% of the rock consists of inclusions of grey gneiss and mafic – or less commonly anorthositic – rocks floating in a granitic neosome. Rare mafic dykes have also been observed. Rae Province orthogneisses differ from their Nain Province counterparts and those in the Tasiuyak gneiss complex by having abundant granite and a paucity of ultramafic inclusions. No age data are available for these rocks in the northern segment of the orogen.

Lake Harbour Group. Orthogneisses of suspected Archaean age are infolded with metasedimentary rocks of the Palaeoproterozoic Lake Harbour Group (Jackson & Taylor 1972; Taylor 1979), including white-weathering marble, quartzite, garnet–sillimanite metapelite, rusty garnet–graphite–biotite ± sillimanite paragneiss, hornblende–biotite–plagioclase–quartz dioritic gneiss, and rare sheets of hornblende–plagioclase ± garnet ± orthopyroxene metagabbro. Contact relationships between the group and orthogneisses are sharp and strongly sheared near the Abloviak shear zone, but unsheared away from the zone where they may represent stratigraphic contacts or the traces of pre-metamorphic thrust faults.

Charnockitic rocks. Orthopyroxene-bearing granitoid rocks with a single set of planar and linear fabric elements are interlayered with Rae orthogneisses and Lake Harbour Group metasedimentary rocks. The charnockitic rocks are similar to those of the Killinek charnockite suite (see below) and are distinct from the migmatitic, heterogeneous gneisses of suspected Archaean age by virtue of their single fabric, the presence of coarse hypersthene crystals, and a buff weathering appearance.

Lithology: Noodleook complex

Noodleook complex is composed of four distinct rock suites, including: reworked Archaean gneisses with metamorphosed and deformed Avayalik dykes, the Hutton anorthositic suite, a Palaeoproterozoic diorite–tonalite–granite (DTG) suite of metaplutonic rocks, and orthopyroxene-bearing granitoid rocks of the Killinek charnockite suite (Figs 2, 3). A younger suite of Palaeoproterozoic magmatic rocks and a set of syn-tectonic mafic dykes comprise the remainder of the complex.

Reworked Archaean gneiss and deformed Avayalik dykes. Reworked Archaean gneisses of the Noodleook complex are identical in appearance to those of the Nain Province gneisss complex in terms of textural heterogeneity and structural complexity. One sample of migmatitic orthogneiss, without dykes, yielded a U–Pb zircon age of 2802 ± 2 Ma, confirming the Archaean age of the structurally most complex rocks (Fig. 4). In the eastern part of the complex, migmatitic tonalite/granodiorite orthogneisses at granulite facies are cut by schlieric pink and grey granite and granitoid gneiss of the Duck Island granitoid suite (Fig. 2; Van Kranendonk *et al.* 1994), which contains a penetrative, amphibolite-facies foliation that is cut by the Avayalik dykes. This observation suggests a period of late Archaean deformation at amphibolite facies, which is estimated to have occurred at *c.* 2660 Ma, the average U–Pb age of two monazite grains from a sample of the suite (Fig. 4). Migmatitic granitoid gneisses throughout the Noodleook complex contain deformed, metamorphosed and folded mafic dyke remnants that are interpreted to represent the western continuation of the Avayalik dyke swarm, as they cut Archaean structures but are affected by Palaeoproterozoic sets of structures. These rocks are partly to completely recrystallized as a result of Palaeoproterozoic amphibolite-facies metamorphism, dated as between *c.* 1780–1710 Ma (U–Pb ages of titanite: Scott & Machado 1994).

Hutton anorthositic suite. Reworked Archaean gneisses in the eastern part of the complex are host to a continuous body, up to 3 km wide and over 110 km long, of anorthositic and associated rocks of the Hutton anorthositic suite (Fig. 2; Taylor 1979; Van Kranendonk & Scott 1992; Wardle *et al.* 1992, 1993). Similar rocks occur as a disaggregated train of lenses and rafts for over 325 km along-strike, south to Okak Bay (see Fig. 1; Ermanovics *et al.* 1989). Textures vary from pristine, coarse-grained anorthosite with subophitic textures of blue igneous plagioclase feldspar (labradorite) and igneous orthopyroxene (variably recrystallized to hornblende–biotite–epidote), to medium-grained, granoblastic gneisses and mylonitic rocks. In low strained areas, igneous layering is locally preserved together with an extremely coarse texture in which individual hypersthene crystals range up to 60 cm in diameter. Gabbroic anorthosite compositions dominate the suite, but anorthosite, leucogabbro, rare ultramafic rocks, and rusty sulphide gossans also occur (Wardle *et al.* 1992, 1993).

Contact relationships between anorthositic rocks and Archaean gneisses are sharp and deformed, so that a relative age relationship between these units is indeterminate. However, the occurrence of a thin sheet of gabbroic anorthosite and leucogabbro within Archaean rocks just north of Saglarsuk Bay (Fig. 2) suggests the anorthositic suite may have been emplaced as a sill complex within the Archaean host rocks (Van Kranendonk & Wardle 1995*a*: cf. Van Kranendonk 1991). This interpretation is consistent with a much simpler set of structures within anorthositic rocks than in their migmatitic, polycyclic host rocks (see below). The Hutton anorthositic suite is cut by Avayalik dykes, recognized as such on the basis of containing characteristic black feldspar megacrysts and garnet–clinopyroxene metamorphic mineral assemblages.

Direct attempts at dating the Hutton anorthositic suite have so far yielded only metamorphic zircons (1795–1779 Ma: Scott & Machado 1993), but there is good field evidence for an older age of the suite because it is cut by: (1) a granite dyke, dated as 1804 +3/–2 Ma, which cuts migmatitic gneissosity in the suite; (2) megacrystic metagranite sheets, one of which has been dated outside of the Hutton anorthoisitic suite as ≥ 1864 Ma; (3) Palaeoproterozoic diorite–tonalite–granite sheets (DTG suite; see below), the equivalents of which have been dated outside of the Hutton anorthositic suite as 1910–1885 Ma; and (4) Avayalik dykes, which are considered to be > 1910 Ma (see discussion above, and Fig. 4 for age locations).

Diorite–tonalite–granite (DTG) suite. Archaean gneisses and Palaeoproterozoic supracrustal rocks of the Noodleook complex were intruded by a polyphase intrusive suite of Palaeoproterozoic metaplutonic rocks that varies in composition from gabbro and diorite, through tonalite, to granite (Fig. 5A, B: Van Kranendonk *et al.* 1993*b*). Tonalite is the most voluminous component west of the Hutton anorthositic suite, but diorite and granite are more common further east. In the eastern half of Burwell domain and across the Komaktorvik shear zone, DTG suite rocks are heterogeneously deformed at middle amphibolite facies and retain igneous textures in areas of low strain where the rocks contain only a small volume of leucosome veins (Fig. 5B). In the western half of the complex, however, rocks of the DTG suite are at transitional amphibolte- to granulite-facies and composed of gneissic tonalitic rocks with abundant inclusions of mafic amphibolite to mafic granulite.

The DTG suite displays a broad calc-alkaline trend on an A–F–M diagram (Van Kranendonk *et al.* 1994) and falls within the volcanic arc granitoid (VAG) field of trace element-based tectonic discrimination diagrams (e.g. Fig. 6a), leading to an interpretation that it represents a subduction-generated magmatic arc suite. Zircon dates from a variety of compositions in the suite fall in the range 1910–1885 Ma (Fig. 4). Sm–Nd results from two tonalite samples plot just below the depleted mantle curve at 1.91 Ga (ε_{Nd} of 2.685 and 1.448), suggesting the involvement of some Archaean crust (Archaean crust or Proterozoic sediments with an Archaean detrital component) in plutonic rocks with an otherwise relatively juvenile signature (Campbell 1994). Contamination by an Archaean source is supported by the presence of inherited Archaean zircons in these rocks (Scott & Machado 1995).

The DTG suite contains elongate inclusions and rafts of supracrustal rocks (Fig. 5E). Garnet–biotite ± sillimanite paragneiss occurs dominantly in the northern part of the Noodleook complex, particularly along the boundary between the Killinek charnockitic and DTG suites, but also as layers and rafts within the DTG suite to about 60°N (Fig. 2; Wardle *et al.* 1993). South of this latitude, paragneisses gives way to mafic gneisses characterized by a centimetre-scale compositional layering (Fig. 5C; Van Kranendonk *et al.* 1993*b*, 1994). Rare interlayers of paragneiss, calc-silicate and possible fragmental andesites in the mafic gneisses suggest a predominantly supracrustal origin for these rocks, so that the typical centimetre-scale layering of low-strained mafic gneisses is interpreted to be original compositional layering in flows and/or tuffs. Homogeneous, unlayered units in these sequences, ≤ 20 m thick, are interpreted to represent massive volvanic flows or sills, whereas the occurrence of layers and discordant sheets of hornblende diorite and rare units of quartz–feldspar porphyry suggest that subvolcanic intrusions are also present. Possible evidence for pillows was identified in one locality (Fig. 5D). Geochemical analyses of the layered mafic gneisses show them to be subalkaline, N-type MORB tholeiites with depleted LREE patterns (Fig. 6c).

An absolute age for the supracrustal rocks is unknown, but a Palaeoproterozoic age is inferred by the fact that they have a much simpler set of structures than those present in rocks of known Archaean age, lie in fault contact with rocks of known Archaean age, and do not contain Avayalik dykes (Van Kranendonk *et al.* 1994). A minimum age is provided by the cross-cutting, 1910 Ma DTG suite (Fig. 5E).

Killinek charnockitic suite. The northwestern part of the northern segment is dominated by homogeneous, medium- to coarse-grained, orthopyroxene-bearing granitoid rocks of the Killinek charnockitic suite (Figs 2, 3). The suite shows a range in composition from tonalite through granodiorite to granite, but is dominated by granodiorite (Wardle *et al.* 1993). Granoblastic textures are ubiquitous, such that orthopyroxene is interpreted to be of metamorphic, rather than igneous origin. These rocks contain rare xenoliths of ultramafic, anorthositic and mafic rocks, large rafts of Tasiuyak-like metasedimentary gneisses, and locally, orthogneiss inclusions with complex structures suggestive of an Archaean protolith. In the northern part of the northern segment, the Killinek charnockitic suite is separated from Archaean Nain gneisses and associated Palaeoproterozoic rocsks of the DTG suite by thin belts of metasediment. In the west-central part of the area, however, the relationship between the charnockitic and DTG suites is less clear and the transition

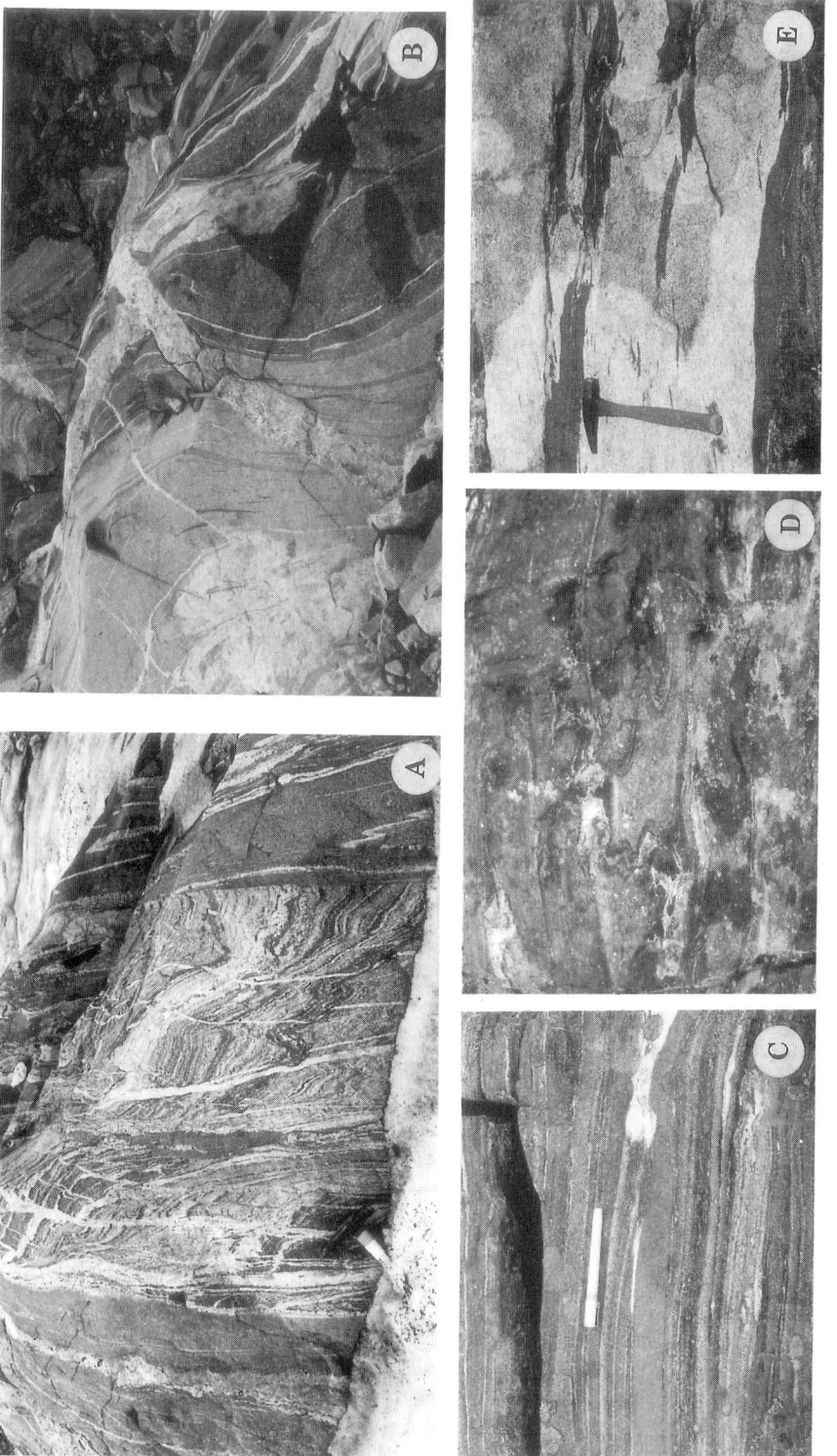

Fig. 5. (A) Foliated diorite–tonalite–granite veins and dykes of the DTG suite cutting Archaean gneissosity in Nain Province host rocks (GSC#1991-573h); (B) foliated, homogeneous tonalite of the DTG suite, dated at 1910 ± 2 Ma, with inclusions of plagioclase-phyric gabbro, cut by pegmatitic granite dykes (GSC#1994-776b); (C) layered mafic gneisses of suspected Palaeoproterozoic age (Pen scale = 13 cm) (GSC#1994-776a); (D) folded and flattened possible pillow structures within layered mafic gneisses (central 'pillow' = 15 × 25 cm) (GSC#1993-252t); (E) weakly foliated DTG tonalite with amphibolite xenoliths showing lobe-and-cusp fold structures due to flattening (GSC#1993-252j). Hammer head in (A), (B), and (E) is 15 cm.

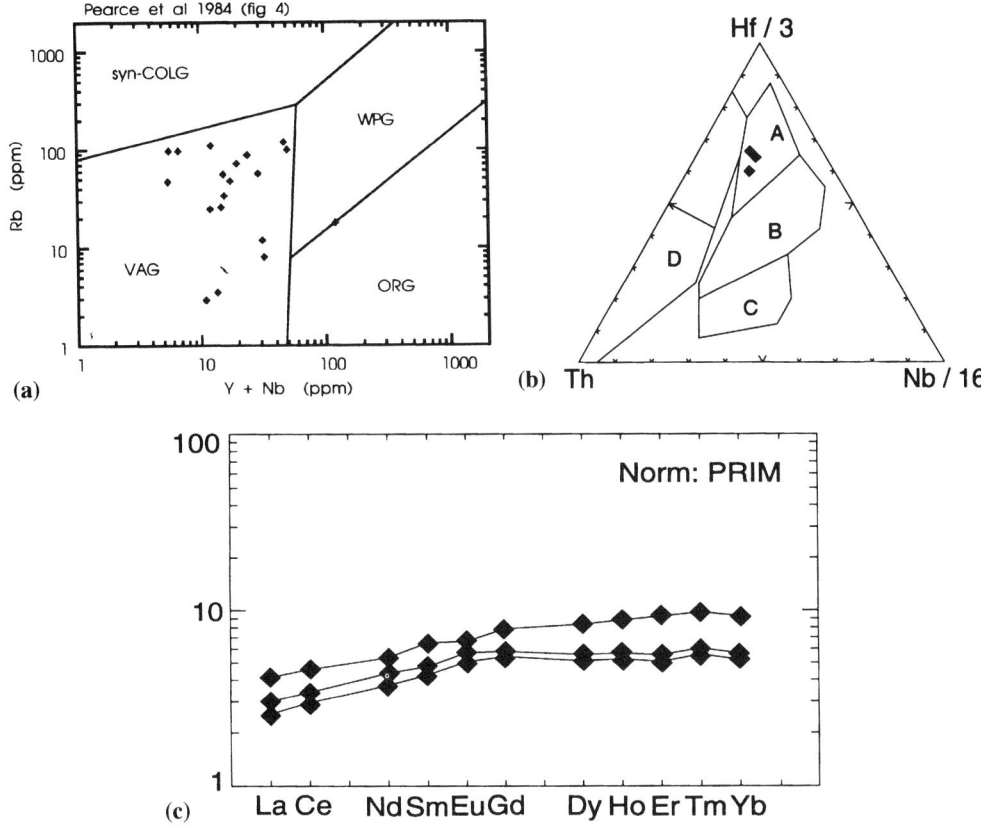

Fig. 6. (a) Rb vs. Y + Nb tectonomagmatic discrimination diagram for rocks of the calc-alkaline DTG suite, showing their volcanic-arc granite (VAG) affinity: syn-COLG, syn-collisional granites; WPG, within-plate granites; ORG, ocean-ridge granites (after Pearce *et al.* 1984); (b) Hf/3–Th–Nb/16 tectonomagmatic discrimination diagram for the layered mafic gneisses – field A, N-type MORB; B, P-type MORB; C, within-plate basalts; D, destructive plate margin basalts (after Wood 1980); (c) REE patterns of the layered mafic gneisses, normalized to primordial mantle.

occurs across a complex zone of heterogeneous metaplutonic gneisses with abundant mafic supracrustal enclaves (Van Kranendonk & Wardle 1995b). The transition in part coincides with the change from granulite facies in the west to retrogressed granulite facies in the east, and there is no evidence that the two suites are separate in age or origin (cf. Scott & Machado 1995).

Two samples of this suite were dated by the U–Pb method on zircon, and yielded ages of 1895 ± 3 Ma and 1886 ± 4 Ma. Some samples show highly radiogenic Pb isotopic signatures similar to that obtained from local metasedimentary rocks, suggestive of either derivation from partial melting, or assimilation of sedimentary protoliths. Other samples with less radiogenic Pb may be derived from melting of Archaean crust (D. Bridgwater, Geological Museum, Copenhagen, Denmark, pers. Comm. 1993).

Younger Palaeoproterozoic rocks. A younger suite of Palaeoproterozoic metaplutonic rocks that comprises plagioclase–porphyritic, mesocratic metatonalites (1869 +3/−2 Ma: colour index = 30–40) and megacrystic metagranites (minimum age of 1864 ± 2 Ma occurs within the northern part of the Komaktorvik shear zone along a structural boundary between the western limit of the Nain Province and the Killinek charnockitic suite that is marked by a thin sliver of metasedimentary rocks (Fig. 2). This 'Younger magmatic suite' is compositionally and geochronologically distinct from either the DTG or Killinek charnockitic suites and is geographically restricted to the DTG–Killinek charnockitic suites contact zone, although sheets of megacrystic granite also occur within the southern part of the Komaktorvik shear zone (Fig. 2; Van Kranendonk & Wardle 1994a, 1995a). Mesocratic metatonalites cut Nain gneisses and DTG suite rocks, but are in sharp (?)tectonic contact with the Killinek charnockitic suite. Megacrystic metagranites crosscut rocks of both the Killinek charnockitic suite and the Nain Province/DTG suite association, thus indicating a minimum age for tectonic stitching of these units, if they were, indeed, ever separate. Sm–Nd isotopic results from the mesocratic tonalites of −1.1 and −0.58, suggest a component of contamination by Archaean crust (Campbell 1994).

Syn-tectonic mafic dykes. Along the eastern margin of the Noodleook complex, restricted to within, or immediately adjacent to, the Komaktorvik shear zone, are one or more swarms of distinctive mafic dykes that cut the DTG suite and are therefore younger than the Avayalik swarm. One set of straight-margined dykes comprises equigranular amphibolites which have a maximum age of 1804 +3/−2 Ma as determined from a granite dyke which is cut by one of the dykes (age from D. Scott, GEOTOP, Université de Québec à Montréal, Canada, pers. comm. 1994). A second set of dykes that cuts Avayalik dykes (Reid 1994), is characterized by a distinctive olive-green weathering appearance and is composed of an equigranular to weakly diabasic matrix of horneblende-plagioclase within which is found occasional garnet and black plagioclase megacrysts. These appinite dykes (cf. Pitcher 1993) have highly irregular intrusion forms within the KSZ and are considered to be syn-tectonic with Komaktorvik shear deformation (Van Kranendonk *et al.* 1994).

Lithology: Tasiuyak gneiss complex

Granulite-facies rocks of the Tasiuyak gneiss complex (after Van Kranendonk & Ermanovics 1990) include extensive tracts of paragneiss and diatexite, rare units of ultramafic rocks, Palaeoproterozoic metaplutonic rocks, and tectonic slices of Archaean crust which contain deformed mafic dykes (Fig. 2; Van Kranendonk *et al.* 1994; Van Kranendonk & Wardle 1995*c*).

Paragneisses. The Tasiuyak gneiss complex is predominantly underlain by white and rusty weathering garnet ± sillimanite ± graphite paragneiss and associated diatexite collectively known as the Tasiuyak gneiss (Wardle 1983). Characterized by lilac-coloured garnets and sillimanite, the Tasiuyak gneiss varies from a compositionally layered paragneiss to a homogeneous unit of mylonitized diatexite within the Abloviak shear zone, where it contains a sub-vertical schistosity and sub-horizontal mineral elongation lineations. In low strain domains, the Tasiuyak gneiss is seen to contain a sub-horizontal compositional layering that varies between garnetiferous quartz arenites, through garnet–biotite–graphite quartzofeldspathic gneiss (semi-pelite and diatexitic granite), to garnet–sillimanite–biotite–graphite metapelite. This layering clearly contains a component of relict bedding in addition to a strong straightening/flattening component developed prior to Abloviak shear deformation (see following section on structures). The composition of the layers in the low strain domains is used to suggest that the Tasiuyak gneiss was originally derived from a thick, monotonous sequence of turbiditic metasediments.

In the northern part of the complex, the Tasiuyak gneiss grades into grey, migmatitic paragneisses characterized by red garnet and no sillimanite, that locally contains a ≤ 15 m thick unit of calc-silicate and marble (Van Kranendonk 1994*b*). This significance of this compositional change in the paragneisses is unclear, but unique to the northern segment of the orogen.

Archaean orthogneisses. Migmatitic tonalite/granodiorite orthogneisses with evidence of a complex structural and metamorphic history prior to Palaeoproterozoic deformation occur as fault-bounded panels withn the Tasiuyak gneiss complex (Figs 2 & 7: Van Kranendonk 1994*d*; Van Kranendonk *et al.* 1994). These polycyclic gneisses contain layers and rafts of mafic through anorthositic gneiss and paragneiss, and contain folded and disrupted remnants of discordant mafic dykes. The absence of such dykes in the adjacent paragneisses and the more complex structural history in the orthogneisses suggest an Archaean age for the latter, which has been confirmed by preliminary U–Pb age data (J. Connelly, Dept of Geology, University of Texas at Austin, pers. comm. 1995).

Panels of Archaean gneiss are separated from Palaeoproterozoic paragneisses by a 1–500 m thick contact zone characterized by discontinuous units of ultramafic rock (hornblendite), rusty, graphitic paragneisses, blue-grey graphitic and/or red Fe-rich quartzites, and rare layers of calc-silicate and/or impure marble (Fig. 7). These contact zones are interpreted to represent faults, as individual units within them are discontinuous and truncated against adjacent units. These fault zones formed early in the deformational history, as they are themselves folded and deformed by regionally developed structures assocated with formation of the Abloviak shear zone (see below).

Within the fault-bounded panels, migmatitic orthogneisses are locally to extensively homogenized to an equigranular, wispy-textured orthopyroxene-bearing tonalite/granodiorite with numerous mafic layers and lenses derived from

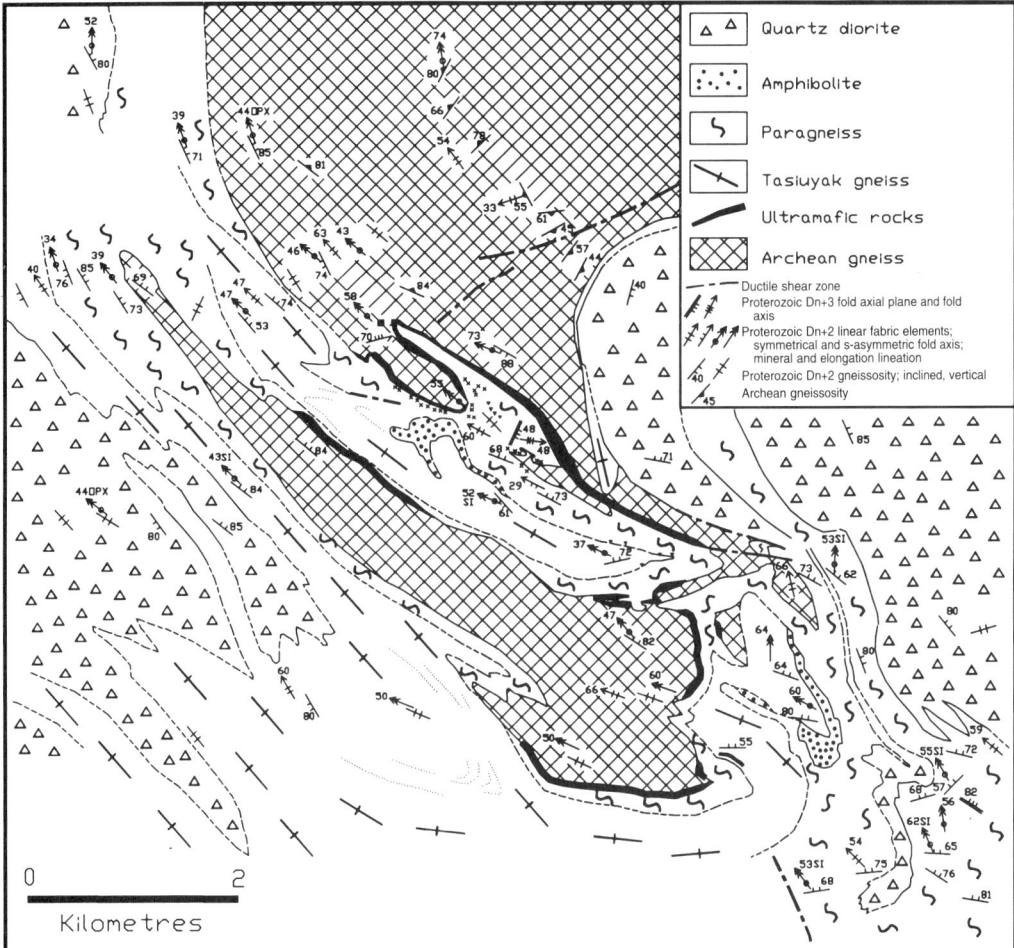

Fig. 7. Detailed map of fault-bounded slices of Archaean gneiss in the Tasiuyak gneiss complex (see Fig. 2 for location). Units other than Archaean gneiss are Palaeoproterozoic in age. Note the thick units of folded and boudinaged ultramafic rocks and of paragneiss along margins of the Archaean gneisses. In paragneiss: small x = occurrences of marble, calc-silicate and graphitic quartzite; small inverted triangles = rusty graphitic paragneiss.

remnant mafic dykes and mafic granulite gneisses. The homogeneous rocks have an identical appearance to those of the Killinek charnockitic suite further north, except that those in the Tasiuyak gneiss complex contain a greater proportion of mafic inclusions. Homogenization may have occurred either through assimilation of the older rocks by widespread intrusion of younger magmas, or through *in situ* partial melting of the Archaean gneisses.

Palaeoproterozoic metaplutonic rocks. Charnockitic rocks in the complex have a similar appearance to those of the Killinek charnockitic suite but commonly contain xenoliths of mafic granulite and layered orthogneiss and thus were extensively mapped as diatexite (Van Kranendonk 1994c; Van Kranendonk et al. 1994). These rocks, which do not contain mafic dykes, grade into the previously described areas of lithologically and structurally more complex orthogneisses with dykes, of suspected Archaean age. A Sm–Nd result from a sample of homogeneous, orthopyroxene tonalite from this area yielded an ε_{Nd} (@ 1.86 Ga, the time of interpreted high-grade metamorphism; see below) of –6.7, consistent with contamnation by Archaean crust and significantly more negative than for samples of the DTG suite (Campbell 1994). A lower-strained, homogeneous orthopyroxene tonalite from the bend of the Abloviak shear zone, lacking leucosome veins, is 1839 ± 3 Ma.

In the northern part of the complex, largely confined to within the grey, migmatitic paragneisses north of the Tasiuyak gneiss, are dark grey to black weathering, mesocratic quartz diorites with 20–30% of hornblende + orthopyroxene and a weak plagioclase–porphyritic texture (Fig. 2; see Van Kranendonk et al. 1993b). In contrast with the schlieric charnockitic rocks described above, which appear to be derived through melting of Archaean crust, quartz diorites outside the Abloviak shear zone are homogeneous, weakly strained bodies which show intrusive relationships against surrounding paragneisses and contain cognate xenoliths of finer-grained diorite. A sample of quartz diorite was dated as 1896 ± 2 Ma (Fig. 4).

Structural and metamorphic geology: structural elements

Three sets of Archaean structures in the Nain Province gneiss complex, and four regionally developed sets of Palaeoproterozoic structures across the rest of the northern segment have been identified (Van Kranendonk et al. 1993b). The Archaean history of the Nain Province gneiss complex is described by Wardle et al. (1994) and Van Kranendonk & Wardle (1995a) and will not be dealt with herein. For the purpose of this paper, the Archaean structures are collectively referred to as D_n. The progressive sets of Palaeoproterozoic structures are referrred to as D_{n+1}, etc., and each was accompanied by metamorphism (M_{n+1}, etc.). A more detailed description of Palaeoproterozoic deformational events is presented in Van Kranendonk & Wardle (in press).

The intensity and distribution of the D_{n+2} and D_{n+3} sets of Palaeoproterozoic structures have been used to subdivide the northern segment of the orogen into three domains of relatively low Palaeoproterozoic strain, described below, and three major shear zones (Fig. 3).

Four Peaks domain, coincident with the Nain Province gneiss complex, is unaffected by penetrative Palaeoproterozoic strain other than narrow shear zones (centimetre to decimetre thick) and discrete faults. Rocks within the domain were deformed and metamorphosed during Late Archaean orogeny at granulite facies and, subsequent to Avayalik dyke intrusion, affected by a static Palaeoproterozoic overprint of Mg-rich garnet–clinopyroxene assemblages (Van Kranendonk et al. 1994; Wardle et al. 1994).

Burwell domain is an area of tight to open folds and narrow high strain zones developed in rocks of the Noodleook and Tasiuyak gneiss complexes (Fig. 3). The domain is bounded to the south by the Abloviak and Katherine River shear zones, and to the east by the splayed western margin of the Komaktorvik shear zone. Metamorphic grade varies from granulite facies throughout the western half, to amphibolite facies in the eastern half, which is in part retrograde after the granulite facies, but in part also prograde in rocks that were never at granulite facies.

North River domain is named after the region west of the Abloviak shear zone in the Okak Bay area (Fig. 1; Van Kranendonk & Ermanovics 1990) and characterized by NW–SE trending, tight to open D_{n+2} folds of amphibolite- to granulite-facies rocks of the Rae Province gneiss complex. The dominant structures in this domain are related to the formation of the Abloviak shear zone (D_{n+2}), into which it grades. A southern limit for this domain was not mapped.

Structures and associated metamorphism

First phase structures. D_{n+1} structures include a migmatitic gneissosity developed in the granulite-facies rocks of the Tasiuyak gneiss complex and Killinek charnockitic suite, and a foliation or gneissic layering in some rocks of the DTG suite. High strain zones with local development of sheath folds occur along the contacts between layered mafic gneisses and reworked Archaean rocks of the Noodleook complex and are interpreted to reflect a phase of D_{n+1} thrusting that was responsible for the tectonic intercalation of rock units in this area (Van Kranendonk et al. 1993b). The sense of vergence of the thrusting is unknown due to the intensity of subsequent deformational events in this area. D_{n+1} structures die out to the east across the Komaktorvik shear zone, but in the North River domain, SW-verging 'D1' thrust faults were recognized by Goulet & Ciesielski (1990) between Lake Harbour Group and Archaean Rae basement.

D_{n+1} deformation in the western part of Burwell domain occurred under granulite-facies conditions, as indicated by a set of migmatitic structures and associated granulite-facies mineral assemblages that have been deformed by D_{n+1} structures (also at granulite facies). In the eastern part of Burwell domain, some amphibolite-facies rocks have never been to granulite facies, suggesting that M_{n+1} metamorphism decreased from west to east. The grade of M_{n+1} metamorphism in the North River domain is not known, due to subsequent granulite-facies metamorphism during D_{n+2}.

D_{n+1} structures are related in that they predate younger, better-defined sets of structures, but have not been directly dated in most parts of the northern segment, so that the possibility exists that they are of different age in different parts of the northern segment. Available U–Pb age constraints suggest that the D_{n+1} deformation occurred at between 1870 ± 3 Ma – the age of monazite in a sample of Tasiuyak gneiss – and c. 1860 Ma – the age of metamorphic zircon overgrowths in metasediment from the central part of the northern segment of the orogen (see Fig. 8).

Second phase structures. D_{n+1} planar fabric elements are deformed within NW-trending D_{n+2} folds and sinistral shear zones across the western half of the northern segment, including the 8–10 km wide **Abloviak shear zone (ASZ)**, which straddles the boundary between the Nain and Rae provinces along the entire strike length of the orogen and is developed primarily within the Tasiuyak gneiss (Figs 1, 3). The ASZ is characterized by sub-vertical mylonitic schistosities, sub-horizontal mineral elongation lineations defined by quartz, sillimanite and orthopyroxene, and isoclinal folds on sub-horizontal axes parallel to the lineations. Kinematic indicators within the zone are everywhere sinistral (Van Kranendonk et al. 1993b). West of the bend in the ASZ, the northern margin of the zone splays into several arms of high strain that wrap around low-strain augen in which D_{n+2} structures are restricted to open, upright folds and a weak axial-planar quartz foliation. The ASZ passes north into the Burwell domain. Throughout the Tasiuyak gneiss complex and Killinek charnockitic suites, upright, NW–SE striking D_{n+2} folds on moderately NW-plunging axes contain a penetrative, sub-vertical axial planar foliation and granulite-facies mineral elongation lineations. Two zones of high D_{n+2} strain along the eastern margin of the Killinek charnockitic suite are characterized by moderately-plunging L_{n+2} elongation lineations and are responsible for SE-directed uplift of the suite. These zones are rotated into parallelism with the western margin of the D_{n+3} Komaktorvik shear zone, which is characterized by south-plunging amphibolite-facies mineral elongation lineations.

Eastward within the southern part of the Burwell domain (in the Noodleook complex), D_{n+2} structures die out rapidly across a zone of interference between NW-trending D_{n+2} and SW-trending D_{n+3} folds, and were not recognized very far east of the granulite- to amphibolite-facies transition (Fig. 3). South of the ASZ, the North River domain is characterized by upright, tight, doubly-plunging folds which form large-scale dome and basin structures (cf. Taylor 1977).

M_{n+2} granulite-facies metamorphism accompanied D_{n+2} deformation across the Tasiuyak gneiss complex and northwestern part of the Noodleook complex underlain by the Killinek charnockitic suite. P–T estimates of ≤ 10 kb, 800°C have been obtained for this area by Mengel & Rivers (1994). As with the M_{n+1} metamorphism, M_{n+2} metamorphism dies out rapidly to the east in the southern part of Burwell domain, and was not recognized in the eastern half of the Noodleook complex. D_{n+2} deformation in this area is estimated to have occurred at 1843 ± 3 Ma, the age of a metamorphic zircon overgrowth in a sample of the Killinek charnockitic suite (Fig. 8). A post-D_{n+2}, pre-D_{n+3} pegmatite in the Tasiuyak gneiss, dated as 1824 ± 2 Ma, provides a minimum estimates for the age of the D_{n+2} deformation (Scott & Machado 1994).

In the Four Peaks domain, the prograding southeast to northwest development of static garnet–clinopyroxene assemblages is interpreted to have occurred during D_{n+2}, as suggested by the 1834 +7/–3 Ma age of zircon from a garnet-bearing Avayalik dyke in this region (see previous discussion): metamorphism occurred in the absence of a penetrative deformation during flexural burial of the Nain Province margin (see below). P–T estimates from this area range from 7.3 kb, 650°C in the north, to 10–13 kb, 700–800°C in the south (Mengel & Rivers 1994; Van Kranendonk & Wardle in press). Pressure estimates from garnet–clinopyroxene–plagioclase–quartz assemblages in rocks from the southern part of the domain **increase** from core to rim, suggesting that the M_{n+2} assemblages record progressive burial during D_{n+2} deformation (Van Kranendonk & Wardle in press).

Third phase structures. D_{n+3} structures at amphibolite facies include the Komaktorvik shear zone, the Katherine River shear zone, and large-scale, N–S trending folds of the Abloviak shear zone and rock units within the Burwell domain. The **Komaktorvik shear zone** (KSZ) is a broad (up to 12 km wide), NNW–SSE striking belt of D_{n+3} amphibolite-facies ductile shear concentrated across the Hutton anorthositic suite along the boundary between the Nain Province gneiss complex and the Noodleook complex (Fig. 3; Van Kranendonk et al. 1993a, b; Van Kranendonk & Wardle 1995c in press). The KSZ has splayed, saw-toothed margins with adjacent Burwell and Four Peaks domains and is characterized by steeply dipping protomylonitic to mylonitic schistosities, southerly-plunging mineral elongation lineations, and meso- to macro-scale, tight to open folds. Garnet–hornblende-bearing meta-dykes within the zone are tightly folded or transposed into parallelism with the gneissosity. High strain in the KSZ is concentrated within unattached, 1–5 km

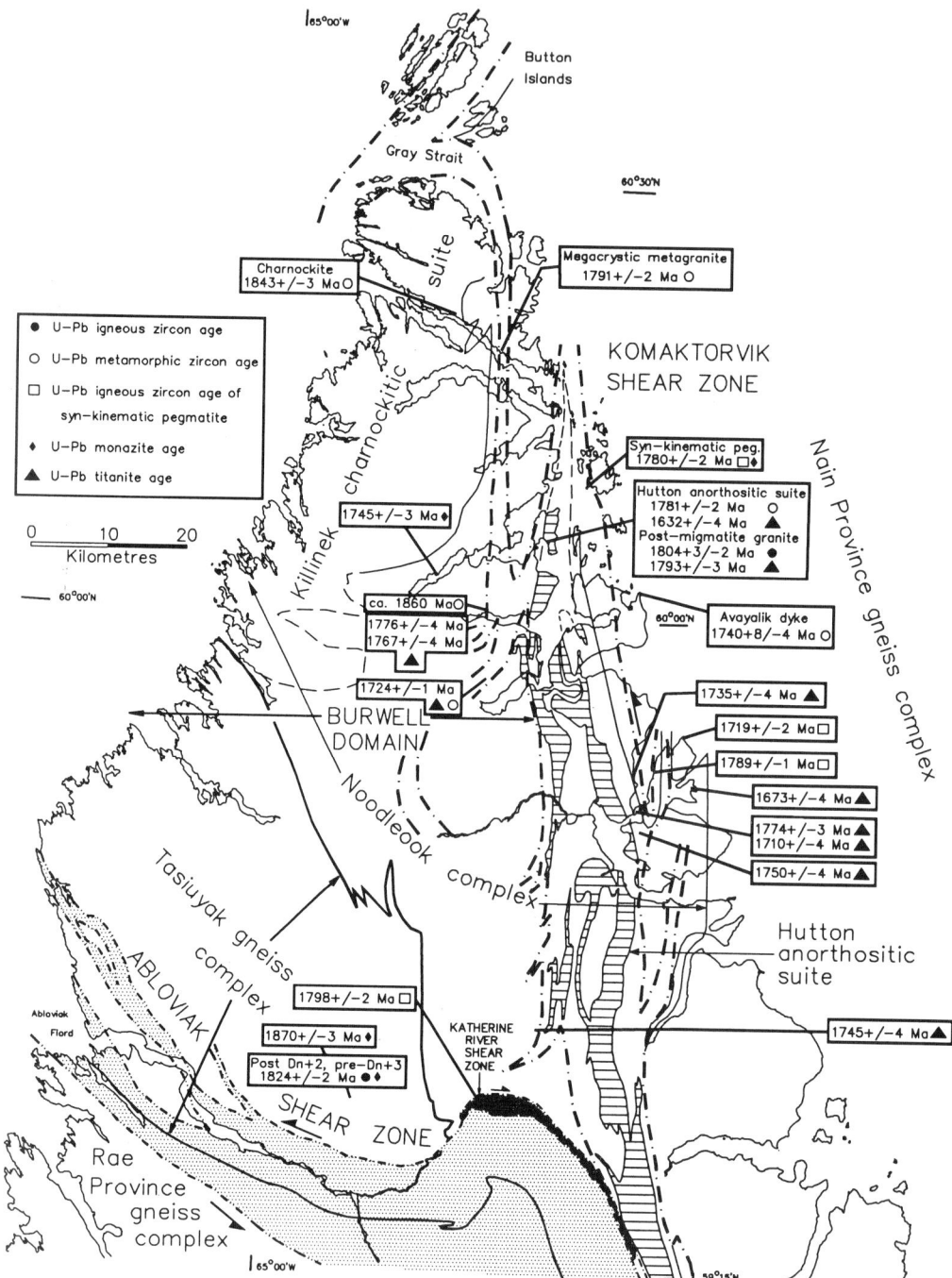

Fig. 8. Metamorphic U–Pb age data on zircon, monazite and titanite in the northern segment of the Torngat Orogen (data from Scott & Machado 1993, 1994).

wide, straight segments of porphyroclastic protomylonite, including a 30 km long segment from the southeastern corner of the northern segment to Ryans Bay, another from the bottom of Eclipse Harbour to 60°N, and a third, west of the Hutton anorthositic suite from 60°N to the top of Killinek Island. S_{n+3} foliations change across the KSZ from steeply east-dipping in the east, through the vertical, to steep westerly and northwesterly dips in the west. L_{n+3} linear fabric elements also vary in orientation across the northern segment, from steeply east-plunging in the east, through moderately south-plunging across the KSZ of the northern segment, to shallow southwest-plunging in the west. Kinematic indicators of east-side-up, sinistral oblique displacement characterize the eastern part of the zone (Van Kranendonk & Scott 1992), and rare kinematic indicators of oblique dextral, west-side-up displacement were observed in SW-striking splays of the KSZ into the Burwell domain (Van Kranendonk & Wardle 1995b).

The Katherine River shear zone (KRSZ) is an amphibolite-facies zone of dextral shear (D_{n+3}) located across the boundary between the Noodleook and Tasiuyak gneiss complexes along the northern margin of the ASZ (Fig. 3; Van Kranendonk 1994a). Deformation states within the zone vary from protomylonite and mylonite in the northwest, to mylonite and ultramylonite in the southeast (Van Kranendonk et al. 1993b). Mylonitic foliations are steeply dipping, whereas mineral elongation lineations vary from moderately W- to SW-plunging in the west, through sub-horizontal WNW–ESE plunging around the central-eastern portion, to moderately NW-plunging along the NW–SE striking segment of the shear zone (Van Kranendonk et al. 1993b; Van Kranendonk 1994a). At the northwestern corner of the zone, mylonite passes into brittle, NW–SE striking faults that link up with a major fault zone which separates the Noodleook and Tasiuyak gneiss complexes.

M_{n+3} metamorphism involved the widespread retrogression of granulite-facies mineral assemblages to garnet–horneblende assemblages. This occurred under P–T conditions of surprisingly high metamorphic grade, as determined from a number of samples within and immediately to the east of the KSZ, that record conditions of up to 12 kb 800°C (Mengel & Rivers 1994). Thermobarometric analysis of a garnet–horneblende mylonite in the southern part of the KSZ yielded P–T estimates indicating an **increase** in pressure from core to rim of syn-kinematic minerals, consistent with a model of reverse uplift of the Four Peaks domain over Burwell domain across the steeply east-dipping margin of the KSZ (see Van Kranendonk & Wardle in press). Amphibolite-facies recrystallization continued to lower P–T conditions during cooling and uplift (Mengel & Rivers 1994).

Dating of syn-tectonic pegmatites and of metamorphic zircons from rocks within the KSZ and KRSZ has shown that the main phase of amphibolite-facies deformation occurred at between c. 1798 and 1779 Ma, while titanite dates and a zircon age from a younger syn-tectonic pegmatite indicate prolonged cooling and localized deformation between 1776 and 1673 Ma (Fig. 8).

Fourth phase structures. Narrow zones of ultramylonite and fault breccia, commonly with pseudotachylite veins, constitute the youngest set of Palaeoproterozoic structures in the northern segment (D_{n+4}). In the KSZ, NNW–SSE striking faults, orientated sub-parallel to D_{n+3} shear fabrics with east-side-up kinematics, show west-side-up displacement. Further west, NW–SE striking faults contain a component of sinistral strike-slip displacement, and may be related to late-stage tightening of the bend in the Abloviak shear zone. This deformation occurred under amphibolite- to greenschist-facies conditions, and continued to shallow crustal levels where pseudotachylite formed.

Pre-deformational subduction/accretion models

Following late Archaean cratonization, the Nain Province was subjected to rifting in the earliest Proterozoic, which continued through to at least 2234 Ma (the age of Kikkertavak dykes). The age and origin of the Hutton anorthositic suite represents an enigma due to its size and restricted occurrence to the western margin of the Nain Province. Unfortunately, direct attempts at dating this suite have so far yielded only metamorphic zircons, so that either a Late Archaen or Palaeoproterozoic igneous age is possible. However, a Late Archaean age for the Hutton anorthositic suite is deemed unlikely, as no other unit within the Nain Province gneiss complex maintained such remarkable continuity through Late Archaean granulite-facies orogeny. Instead, the Hutton anorthositic suite is envisaged to have been emplaced during rifting of the Nain Province margin after the end of late Archaean tectonism (c. 2.55 Ga; Connelly & Ryan 1993), in an analogous setting to 2.49–2.48 Ga anorthositic rocks at the base of the Huron Supergroup (2.45–2.38 Ga) that mark the onset of rifting of the Superior Province margin and basin development (Krogh et al. 1984), and with the rift-related c. 2.45 Ga layered mafic intrusions in the Lapland–Kola Mobile Belt (Pechenga-Varzuga and Kolvitsa belts: Balashov et al. 1993;

Balagansky et al. 1994; R. Parrish & T. Frisch, Geological Survey of Canada, Ottawa, Canada, pers. comm. 1994). The absolute age of this rifting event is unknown.

Following rifting and mafic dyke emplacement, deposition of passive margin sequences occurred on both the Nain and Rae provinces margins in the Ramah and Lake Harbour groups in both the northern and southern segments of the Torngat Orogen. The origin of the extensive belt of Tasiuyak gneiss also remains enigmatic; proposals have been put forth that it was deposited in an intracratonic basin, as a continental rise sequence, or as an accretionary prism. In the former scenario, much of the detritus would be expected to be derived from the adjacent Archaean continent (Nain, or Rae provinces). However, the results from recent Sm–Nd and Pb–Pb studies on the Tasiuyak gneiss and Nain and Rae provinces show that the Tasiuyak gneiss contains detritus from either a Late Archaean source, or a mixture of Late Archaean and Palaeoproterozoic sources (Jackson & Hegner 1991; Thériault & Ermanovics 1994; M. Hamilton, Geological Survey of Canada, Ottawa, Canada, pers. comm. 1993), and that such a source could not be derived from the Nain Province, thereby ruling out the Tasiuyak gneiss as an intracratonic basin or continental rise association built on the Nain margin. As well, Sm–Nd results from the North River area show the Tasiuyak gneiss to have a significantly different source of detritus than nearby Lake Harbour Group remnants, thereby ruling out the Tasiuyak gneiss as a continental rise sequence built on the Rae margin (Thériault & Ermanovics 1994). These data suggest a source region distal to the present location of the sediment, implying significant along-strike transport, possibly in analogy with the far-travelled sediment in the Andean trench. This is supported by the preliminary results from detrital zircon work on a sample of Tasiuyak gneiss from the northern segment, which indicates a 2060–1940 Ma source for some of the detritus (Scott & Machado 1994). The discovery of fault-bounded panels of Archaean gneiss and of ultramafic rocks in the Tasiuyak gneiss complex may infer an accretionary wedge origin for the Tasiuyak gneiss.

A question remains as to which of the Nain or Rae provinces margins the Tasiuyak gneiss was built on. As preserved contacts between the Tasiuyak gneiss and adjacent units may everywhere represent pre-metamorphic thrust faults, it is impossible to determine on which margin it lay from field evidence. However, in the southern segment of the orogen, an 1880 Ma magmatic suite crosses the boundary between the Tasiuyak gneiss complex and the Rae Province, thereby pinning the Tasiuyak gneiss to the Rae side of the orogen (Van Kranendonk & Ermanovics 1990; Bertrand et al. 1993).

The distinct grey paragneiss unit in the northern part of the Tasiuyak gneiss complex is unique to this part of the orogen, and may indicate the presence of a separate sedimentary wedge that was juxtaposed with the Tasiuyak gneiss during Nain–Rae collision across a hidden suture. The observation that 1895 Ma quartz diorite intrusives are confined to within the grey migmatitic paragneisses and are of identical age to the DTG suite on Nain Province margin suggests that these rocks were located on Nain Province margin, although this requires further testing by isotopic and geochronological methods.

The northern segment of the Torngat Orogen is largely underlain by the DTG and Killinek charnockitic suites, that are in part intrusive into, and in part in fault contact with, the Nain Province margin. These rocks are interpreted on the basis of field and geochemical evidence cited above, to have formed as a continental magmatic arc during easterly subduction beneath the Nain Province at between 1910 and 1885 Ma. This large volume of plutons formed a continental promontory on Nain Province, which indented the Rae Province margin during collisional orogeny and was subsequently the locus of much structural complexity (Van Kranendonk & Wardle 1994c, in press). Slivers of mafic supracrustal rocks within the plutons are interpreted on the basis of their field characteristics and geochemistry to represent the remnants of a marginal (backarc?) basin which developed along the thinned western margin of the Nain Province during subduction.

This association of Palaeoproterozoic arc rocks is absent from the Nain Province margin in the southern segment of the Torngat Orogen where, instead, a suite of 1880 Ma calc-alkaline plutons in the Rae Province margin and Tasiuyak gneiss is interpreted to reflect a period of arc magmatism related to westerly subduction at this time (Van Kranendonk & Ermanovics 1990; Bertrand et al. 1993). The along-strike variations in the distribution of Palaeoproterozoic plutons from north to south in the Torngat Orogeny may be explained in a variety of ways and require critical evaluation through detailed Sm–Nd isotopic studies, as is currently underway (Campbell 1994). Two possible scenarios are outlined below, whose variation depends on the tectonic significance given to the younger magmatic suite within the northern segment (Van Kranendonk & Wardle 1994b).

In the first scenario (Model A), which we prefer, development of the magmatic arc on Nain Province is restricted to the period 1910–1895 Ma, after which subduction polarity flipped to westerly-dipping beneath the Rae Province margin and

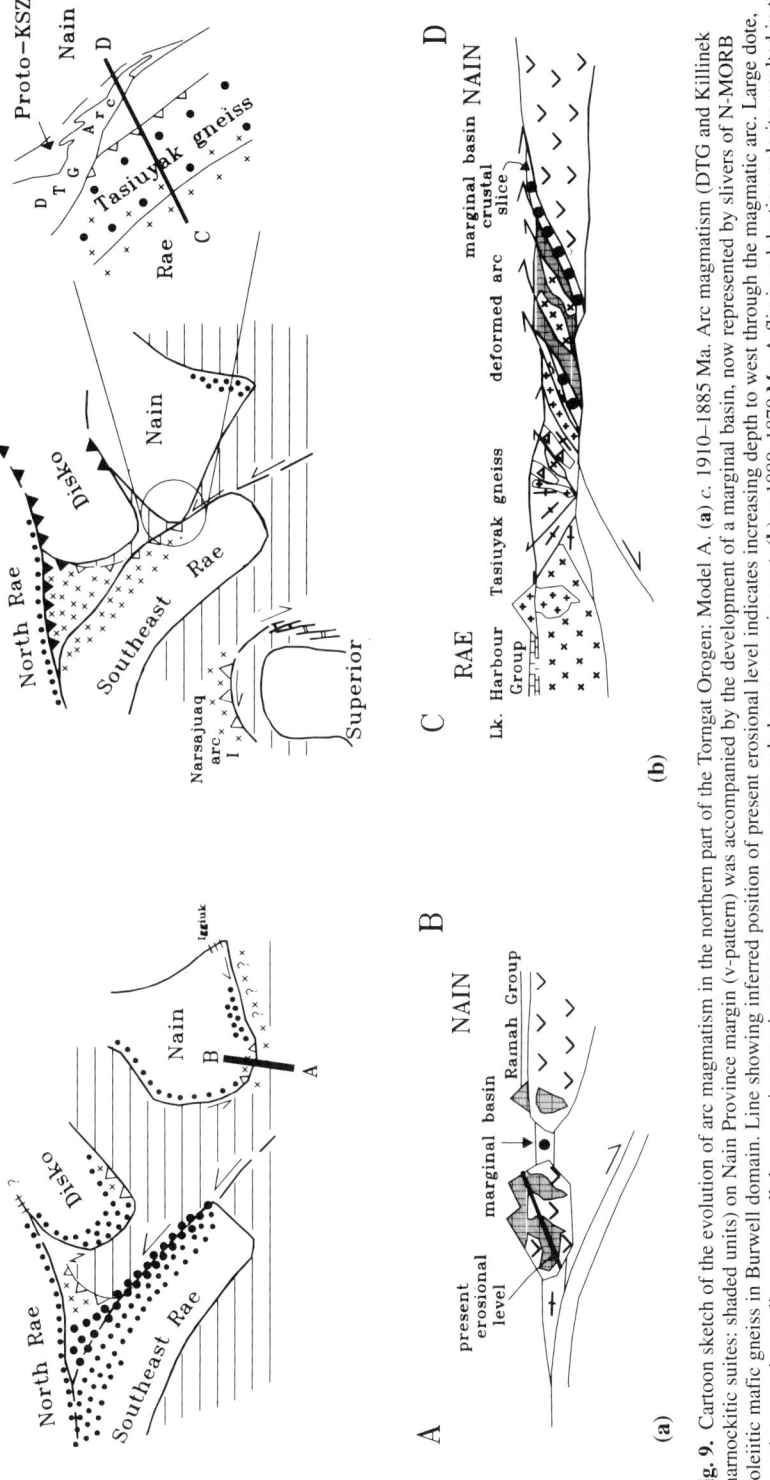

Fig. 9. Cartoon sketch of the evolution of arc magmatism in the northern part of the Torngat Orogen: Model A. (a) c. 1910–1885 Ma. Arc magmatism (DTG and Killinek charnockitic suites: shaded units) on Nain Province margin (v-pattern) was accompanied by the development of a marginal basin, now represented by slivers of N-MORB tholeiitic mafic gneiss in Burwell domain. Line showing inferred position of present erosional level indicates increasing depth to west through the magmatic arc. Large dote, accretionary prism sediments; small dots, passive margin sequences; x, arc magmas; ruled area, oceanic crust. (b) c. 1880–1870 Ma. A flip in subduction polarity resulted in the emplacement of metaplutonic rocks in Rae Province margin (+ pattern), followed closely by accretion and initial collision, when the syn-collisional 'Younger magmatic suite' was emplaced into the suture between the marginal arc and the remnant Nain craton. Plate reconstructions in plan views after Van Kranendonk et al. (1993c).

Tasiuyak gneiss (Fig. 9). In this model, Rae Province margin was predominantly a strike-slip boundary during formation of the early arc suite. The younger magmatic suite in this model is interpreted to represent syn-collisional magmas emplaced into the suture bvetween the marginal part of the arc and the Nain Province, based on its younger age than the DTG suite, distinct composition, and restricted geographical distribution.

In a second scenario (Model B), the c. 1865 Ma suite is interpreted to form an integral part of the northern magmatic arc (after Scott & Machado 1994). If so, this implies that arc magmatism occurred roughly contemporaneously on both sides of the orogen at between c. 1880 and 1864 Ma, following a model of double subduction based on analogy with the Solomon Sea plate (Fig. 10; cf. Cooper & Taylor 1987).

A critical test of these models is to determine the crystallization age of the monocyclic tonalitic plutons within Rae Province margin: if they are Palaeoproterozoic in age, do they fall within the range 1910–1885 Ma as determined for rocks pinned to the Nain Province, or do they fall within a possibly younger suite, ≤1880 Ma? In addition, further geochemical, isotopic and geochronological studies are required on the wide variety of Palaeoproterozoic metaplutonic rocks throughout the orogen in order to determine whether they are part of one, or several suites.

Structural evolution

The structural evolution of the northern segment of the Torngat Orogen is outlined in detail by Van Kranendonk & Wardle (in press) and summarized below (Fig. 11). The D_{n+1} phase of deformation was responsible for the formation of gneissic fabrics in Palaeoproterozoic units, thrust intercalation of some Palaeoproterozoic and Archaean units, and high-grade metamorphism in the wstern half of the Burwell domain. These structures formed during crustal thickening as a result of Nain–Rae continental collision at c. 1870–1860 Ma, probably in a doubly-vergent thrust wedge (see also Van Kranendonk 1996). Accretion of the magmatic arc onto the Nain Province margin was accompanied by emplacement of the younger magmatic suite and burial of the western half of the arc to depths of up to 35 km. Indentation by the continental promontory on Nain Province formed by the magmatic arc, caused a deflection in Rae Province margin during collision.

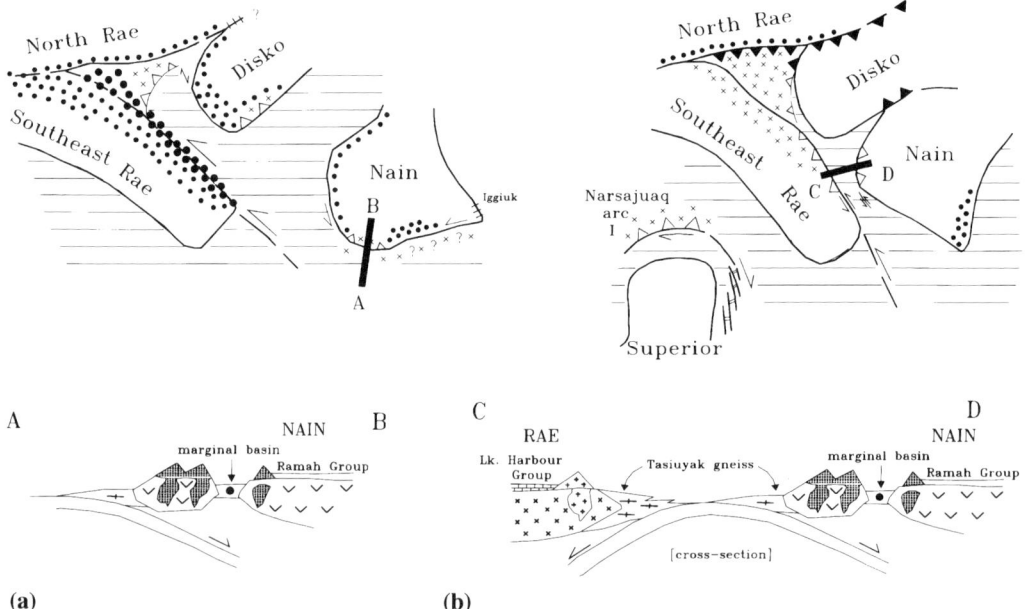

Fig. 10. Cartoon sketch of the evolution of arc magmatism in the northern part of the Torngat Orogen: Model B. (**a**) c. 1910–1885 Ma. Arc magmatism on Nain Province margin and marginal basin development, as in Model A above, followed in (**b**) c. 1880 Ma, by double subduction in the consumption of a triangular remnant of oceanic crust, and the resultant formation of contemporaneous magmatic suites on both sides of the orogen. Plate reconstructions in plan views after Van Kranendonk et al. (1993c). Same unit designators as in Fig. 9.

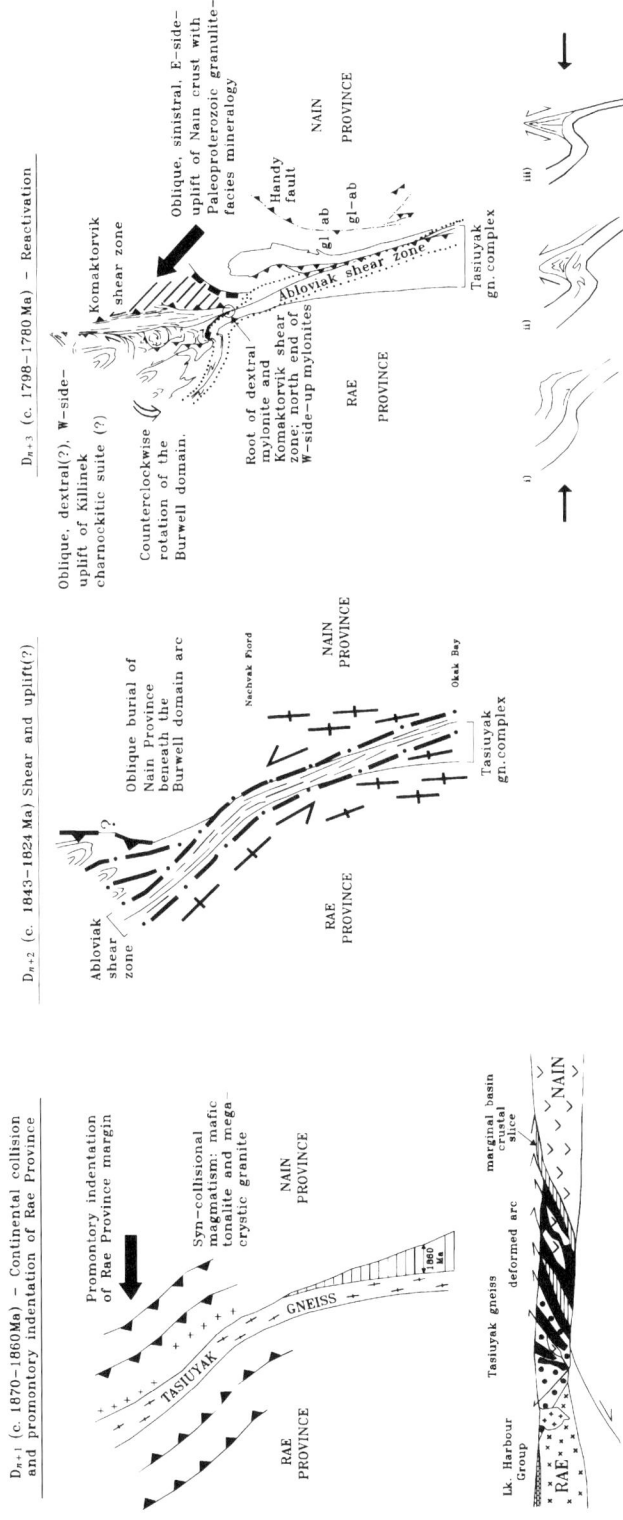

Fig. 11. Schematic structural evolution of the Torngat Orogen. During D_{n+1} continental collision, interaction of the continental promontory on Nain Province margin causes indentation of Rae Province margin. Northerly translation of Nain Province during D_{n+2} results in formation of the Abloviak shear zone and oblique burial of Nain Province beneath the Proterozoic arc. Reactivation of the orogen through progressive disharmonic folding of the Abloviak shear zone results in formation of the diffuse Komaktorvik shear zone and uplift of the granulite-facies Nain Province and deep-level parts of the arc during D_{n+3}.

D_{n+2} sinistral shear deformation, still at granulite facies, occurred at between c. 1843 and 1824 Ma. Shearing was concentrated within the Abloviak shear zone, but formed folds and smaller-scale shear zones throughout the western half of the Burwell domain. Shear deformation splayed to the northwest off of the deflection in the Rae Province margin (Fig. 11). Northward translation of the Nain Province resulted in oblique burial of its northwestern edge to depths of ≤ 40 km beneath the still hot and buoyant Palaeoproterozoic arc, and the formation of high-pressure garnet–clinopyroxene assemblages in the southern part of the Four Peaks domain (see also Wardle & Van Kranendonk this volume).

The broad KSZ, with its irregular geometry and splayed margins, does **not** mark the site of a fundamental plate boundary, as rocks of similar age and provenance occur on either side of it. Instead, the KSZ is interpreted to represent the site of concentrated deformation on the limbs of a crustal-scale fold, centred on the Hutton anorthositic complex, that formed during crustal-scale disharmonic folding of the Abloviak shear zone across the D_{n+1} flexure in Rae Province margin (Van Kranendonk et al. 1993b; Van Kranendonk & Wardle in press). The Katherine River shear zone is interpreted to represent the principal slip plane of the disharmonic fold. During folding and shear zone development, east-directed reverse shear along the western margin of the KSZ and west-directed reverse shear along the eastern margin of the KSZ resulted in simultaneous uplift of both the granulite-facies Four Peaks domain and the Killinek charnockitic suite, respectively. Uplift of these domains was not fast enough to cause recrystallization of the lower-grade rocks in the core of Burwell domain to higher P-T conditions (Van Kranendonk et al. 1993b; Van Kranendonk & Wardle 1995c, in press) D_{n+4} deformation represents in part a continuation of the D_{n+3} events, and in part the result of rebound of the internal parts of the orogen after burial during D_{n+3}.

Summary

The northern segment of the Torngat Orogen is largely underlain by a Palaeoproterozoic magmatic arc which was emplaced into the Nain Province margin at 1910–1895 Ma and formed a continental promontory. A variety of supracrustal rocks interlayered with the arc plutons and in thrust contact with the Archaean blocks on either side, are thought to represent the remnants of one or more accretionary sedimentary prisms and a marginal (backarc?) basin. The pre-deformational geometry of the major rock units in the northern segment is discussed in terms of the overall evolution of the orogen.

The arc is preserved in a tilted cross-section from east to west as a result of four distinct structural/metamorphic events that were in large part controlled by the presence of the arc trapped between the rigid Nain Province and the partly ductile Rae Province (rock and a hard place, respectively). Early deformation resulted from arc–continent and continent–continent collision at c. 1870–1860 Ma, which resulted in burial of the western half of the arc to depths of 35 km. Sinistral shear along the continental suture resulted in formation of the Abloviak shear zone and oblique burial of the Nain Province under the arc, to depths of ≤ 40 km. Subsequent reactivation of the orogen at c. 1798–1780 Ma resulted in crustal-scale disharmonic folding of the Abloviak shear zone and the formation of the Komaktorvik shear zone along the edge of the rifted Nain margin. Uplift of granulite-facies rocks in the western half of the Burwell domain and in the Nain Province occurred at this time. Rebound of the central part of the orogen across narrow zones of ultramylonite completes the Palaeoproterozoic tectonic evolution of this area.

We would like to thank F. Mengel, L. Campbell, D. Scott and other members of the Torngat project for their mapping efforts, excellent research, stimulating discussions and friendly times in the field. Fieldwork was supported by the Geological Survey of Canada and the Newfoundland Department of Natural Resources through a Canada–Newfoundland Agreement on Mineral Development. MVK would like to acknowledge the Natural Sciences and Engineering Research Council for providing a Government Visiting Fellowship at the Geological Survey of Canada. Production of Fig. 6 used the software program 'Newpet' developed at the Department of Earth Sciences at Memorial University of Newfoundland. Helpful reviews by I. Ermanovics, A. Hynes and T. Skulski helped clarify the views presented herein. Publication is with the permission of the Executive Director of the Newfoundland Geological Survey Branch. Lithoprobe contribution #659. Geological Survey of Canada contribution #14995.

References

BALAGANSKY V. V., TIMMERMAN, M. J. & KISLITSYN, R. V. 1994. 2.5–1.9 Ga magmatism, metamorphism and deformation in the southeastern branch of the Lapland Granulite Belt, Kola Peninsula, Russia. *Terra Nova*, **6** (Abst. Suppl. no. 2), 2.

BALASHOV, Y. A., BAYANOVA, T. B. & MITROFANOV, F. P. 1993. Isotope data on the age and genesis of layered basic–ultrabasic intrusions in the Kola Peninsula and northern Karelia, northeastern Baltic Shield. *Precambrian Research*, **64**, 197–205.

BERTRAND, J-M., RODDICK, J. C., VAN KRANENDONK, M. J. & ERMANOVICS, I. 1993. U–Pb geochronology of deformation and metamorphism across a central transect of the Early Proterozoic Torngat Orogen, North River map area, Labrador. *Canadian Journal of Earth Sciences*, **30**, 1470–1489.

CAMPBELL, L. M. 1994. Preliminary results from an Sm–Nd isotopic study of Proterozoic crustal formation and accretion in the Torngat Orogen, northern Labradoer. *In*: WARDLE, R. J. & HALL, J. (eds) *Eastern Canadian Shield Onshore–Offshore Transect (ECSOOT), Report of Transect Meeting*. The University of British Columbia, LITHOPROBE Secretariate, Report No. 36, 100–107.

CONNELLY, J. N. & RYAN, B. 1993. U–Pb constraints on the thermotectonic history of the Nain area. *In*: WARDLE, R. J. & HALL, J. (eds) *Eastern Canadian Shield Onshore–Offshore Transect (ECSOOT), Report of Transect Meeting*. The University of British Columbia, LITHOPROBE Secretatiate, Report No. 32, 137–144.

COOPER, P. & TAYLOR, B. 1987. Seismotectonics of New Guinea: A model for arc reversal following arc–continent collision. *Tectonics*, **6**, 53–67.

ERMANOVICS, I. 1993. *Geology of the Hopedale Block, southern Nain Province, and the adjacent Proterozoic terranes, Labrador, Newfoundland*. Geological Survey of Canada, Memoir **431**.

——, ——, CORRIVEAU, L. & MENGEL, F. 1989. The boundary zone of the Nain–Churchill Provinces in the North River–Nutak map areas, Labrador. *In*: *Current Research, Part C*. Geological Survey of Canada, Paper **89-1C**, 385–394.

FAHRIG, W. F. 1970. *In*: *Age determinations and geological studies, K–Ar isotopic ages, Report 9*. Geological Survey of Canada, Paper **70-2A**.

GOULET, N. & CIESIELSKI, A. 1990. The Abloviak shear zone and the NW Torngat Orogen, eastern Ungava Bay, Québec. *Geoscience Canada*, **17**, 269–273.

HOFFMAN, P. F. 1988. United Plates of America, the Birth of a Craton: Early Proterozoic Assembly and Growth of Laurentia. *Annual Review of Earth and Planetary Sciences*, **16**, 543–603.

—— 1989. Precambrian geology and tectonic history of North America. *In*: BALLY, A. W. & PALMER, A. R. (eds) *The Geology of North America – An overview*. Geological Society of America, The Geology of North America, v. **A**, 447–512.

—— 1990. Subdivision of Churchill Province and extent of the Trans-Hudson Orogen. *In*: LEWRY, J. F. & STAUFFER, M. R. (eds) *The Early Proterozoic Trans-Hudson Orogen of North America*. Geological Association of Canada, Special Paper, **37**, 15–40.

JACKSON, G. & HEGNER, E. 1991. Evolution of Late Archean to Early Proterozoic crust based on Nd isotopic data for Baffin Island and northern Quebec and Labrador. *Geological Association of Canada, program with Abstracts*, **16**, A59.

—— & TAYLOR, F. C. 1972. Correlation of major Aphebian rock units in the northeastern Canadian Shield. *Canadian Journal of Earth Sciences*, **9**, 1650–1669.

KORSTGÅRD, J., RYAN, A. B. & WARDLE, R. J. 1987. The boundary between Proterozoic and Archean blocks in central West Greenland and northern Labrador. *In*: PARK, R. G. & TARNEY, J. (eds) *Evolution of the Lewisian and Comparable Precambrian High Grade Terrains*. Geological Society, London, Special Publication, **27**, 247–259.

MENGEL, F. & RIVERS, T. 1991. Decompression reactions and P–T conditions in High-grade Rocks, Northern Labrador: P–T–t paths from individual samples and implications for Early Proterozoic tectonic evolution. *Journal of Petrology*, **32**, 139–167.

—— & —— 1994. 3-D architecture and thermal structure of a crustal-scale transcurrent shear zone – Report 4: P–T results from the northern Torngat transect. *In*: WARDLE, R. J. & HALL, J. (eds) *Eastern Canadian Shield Onshore–Offshore Transect (ECSOOT), Report of Transect Meeting*. The University of British Columbia, LITHOPROBE Secretariate, Report No. 36, 163–170.

——, —— & REYNOLDS, P. 1991. Lithotectonic elements and tectonic evolution of Torngat Orogen, Saglek Fiord, northern Labrador. *Canadian Journal of Earth Sciences*, **28**, 1407–1423.

PEARCE, J. A., HARRIS, N. B. W. & TINDLE, A. G. 1984. Trace element discrimination diagrams for the tectonic interpretation of granitic rocks. *Journal of Petrology*, **25**, 956–983.

PITCHER, W. S. 1993. *The Nature and Origin of Granites*. Blackie, Glasgow.

REID, L. 1994. *Structural and metamorphic interpretation of multiple sets of dykes on Home Island, northern Labrador*. BSc thesis, University of Ottawa, Ontario, Canada.

RYAN, A. B. 1990. Basement-cover relationships and metamorphic patterns in the foreland of the Torngat Orogen in the Saglek–Hebron area, Labrador. *Geoscience Canada*, **17**, 276–279.

—— (COMPILER), KROGH, T. E., HEAMAN, L., SCHÄRER, U., PHILIPPE, S. & OLIVER, G. 1991. On recent geochronological studies in the Nain Province, Churchill Province and Nain plutonic suite, north-central Labrador. *In*: *Current Research (1991)*. Newfoundland Department of Mines and Energy, Geological Survey Branch, Report **91-1**, 257–261.

SCOTT, D. J. & MACHADO, N. 1993. U–Pd geochronology of the northern Torngat Orogen: summary of progress to date. *In*: WARDLE, R. J. & HALL, J. (eds) *Eastern Canadian Shield Onshore–Offshore Transect (ECSOOT), Report of Transect Meeting*. The University of British Columbia, LITHOPROBE Secretariate, Report No. 32, 32–41.

—— & —— 1994. U–Pb geochronology of the northern Torngat Orogen: Results from work in 1993. *In*: WARDLE, R. J. & HALL, J. (eds) *Eastern Canadian Shield Onshore–Offshore Transect (ECSOOT), Report of Transect Meeting*. The University of British Columbia, LITHOPROBE Secretariate, Report No. 36, 141–155.

—— & —— 1995. U–Pb geochronology of the northern Torngat Orogen, Labrador, Canada: a record of Paleoproterozoic magmatism and deformatioin. *Precambrian Research*, **70**, 169–190.

TAYLOR, F. C. 1974. *In*: *Age determinations and geological studies, K–Ar isotopic ages, Report 12*. Geological Survey of Canada, Paper 74-2.

—— 1977. *Geology, Pointe Le Droit, Quebec–Newfoundland (Labrador)*. Geological Survey of Canada, Map 1429A, scale 1:250 000.
—— 1979. *Reconnaissance geology of a part of the Precambrian Shield, northeastern Quebec, northern Labrador and Northwest Territories*. Geological Survey of Canada, Memoir **393**.
THÉRIAULT, R. J. & ERMANOVICS, I. 1994. Nd-isotopic transect across the Torngat Orogen, North River–Nutak area. *In*: WARDLE, R. J. & HALL, J. (eds) *Eastern Canadian Shield Onshore–Offshore Transect (ECSOOT), Report of Transect Meeting*. The University of British Columbia, LITHOPROBE Secretiariate, Report No. 36, 119–133.
VAN KRANENDONK, M. J. 1991. A magmatic sheet origin for thin metagabbroic anorthosite units in the Fishog subdomain of the southern Central Gneiss Belt, Grenville Province, Ontario. *Canadian Journal of Earth Sciences*, **28**, 431–446.
—— 1992. *Archean and Early Proterozoic geology of a transect across the Nain Province and Torngat Orogen in the North River–Nutak map area, northern Labrador, Canada*. PhD thesis, Queen's University, Kingston, Ontario, Canada.
—— 1994a. *Geology, Tower Mountain, Newfoundland (Labrador)–Quebec*. Geological Survey of Canada, Open File 2828, 1:50 000 scale.
—— 1994b. *Geology, Mont Jacques-Rosseau, Quebec–Newfoundland (Labrador)*. Geological Survey of Canada, Open File 2738, 1:50 000 scale.
—— 1994c. *Geology, Riviere Lepers, Quebec–Newfoundland (Labrador)*. Geological Survey of Canada, Open File 2829, 1:50 000 scale.
—— 1994d. *Geology, Lac de Loriere, Newfoundland (Labrador)–Quebec*. Geological Survey of Canada, Open File 2925, 1:50 000 scale.
—— 1996. Multiple, discrete deformation and metamorphic events in the tectonic evolution of the Paleoproterozoic Torngat Orogen; Evidence from P–T–t–d paths in the North River map area, Labrador. *Tectonics*, **15**.
—— & ERMANOVICS, I. 1990. Structural evolution of the Hudsonian Torngat Orogen in the North River map area, Labrador: Evidence for east-west transpressive collision of Nain and Rae continental blocks. *Geoscience Canada*, **17**, 283–288.
—— & SCOTT, D. 1992. Preliminary report on the geology and structural evolution of the Komaktorvik Zone of the Early Proterozoic Torngat Orogen, Eclipse Harbour area, northern Labrador. *In: Current Research, Part C*. Geological Survey of Canada, Paper **92-1C**, 59–68.
—— & WARDLE, R. J. 1994a. *Geology, Ryans Bay, Newfoundland (Labrador)–Quebec*. Geological Survey of Canada, Open File 2926, 1:50 000 scale.
—— & —— 1994b. Geological synthesis and musings on possible subduction–accretion models in the formation of the northern Torngat Orogen. *In*: WARDLE, R. J. & HALL, J. (eds) *Eastern Canadian Shield Onshore–Offshore Transect (ECSOOT), Report of Transect Meeting*. The University of British Columbia, LITHOPROBE Secretiariate, Report No. 36, 62–80.

—— & —— 1994c. Promontory indentation, transpression and disharmonic folding in the formation of the Paleoproterozoic Torngat Orogen, northeastern Canada. *Terra Nova*, **6** (Abst. Suppl. no. 2), 20.
—— & —— 1995a. *Geology, Eclipse Harbour, Newfoundland (Labrador)–Quebec*. Geological Survey of Canada, Open File 2986, 1:50 000 scale.
—— & —— 1995b. *Geology, Le Baret (24P/15), Quebec–Newfoundland (Labrador)*. Geological Survey of Canada, Open File 2985, 1:50 000 scale.
—— & —— 1995c. *Geology of the Labrador Peninsula north of 59°59′N (25A, 24P and 14M), Newfoundland (Labrador), Québec and Northwest Territories*. Geological Survey of Canada, Open File 2927, 1:100 000 scale.
—— & —— in press. Crustal-scale flexural slip folding during late tectonic amplification of an orogenic boundary perturbation in the Paleoproterozoic Torngat Orogen, NE Canada. *Canadian Journal of Earth Sciences*.
——, ——, MENGEL, F. C., CAMPBELL, L. & REID, L. 1994. New results and summary of the Archean and Paleoproterozoic geology of the Burwell domain, northern Torngat Orogen, Labrador, Quebec and Northwest Territories. *In: Current Research, Part C*. Geological Survey of Canada, Paper **1994-C**, 321–332.
——, ——, ——, RYAN, B. & RIVERS, T. 1993a. Lithotectonic divisions of the northern part of the Torngat Orogen, Labrador, Quebec and N.W.T. *In*: WARDLE, R. J. & HALL, J. (eds) *Eastern Canadian Shield Onshore–Offshore Transect (ECSOOT), Report of Transect Meeting*. The University of British Columbia, LITHOPROBE Secretiariate, Report No. 32, 21–31.
——, GODIN, L., MENGEL, F., SCOTT, D., WARDLE, R., CAMPBELL, L. & BRIDGWATER, D. 1993b. Geology and structural develpment of the Archaean to Paleoproterozoic Burwell domain, northern Torngat Orogen, Labrador and Quebec. *In: Current Research, Part C*. Geological Survey of Canada, Paper **93-1C**, 329–340.
——, ST-ONGE, M. R. & HENDERSON, J. R. 1993c. Paleoproterozoic tectonic assembly of Northeast Laurentia through multiple indentations. *Precambrian Research*, **63**, 325–347.
WARDLE, R. J. 1983. Nain-Churchill Province cross-section, Nachvak Fiord, northern Labrador. *In: Current Research*. Newfoundland Department of Mines and Energy, Geological Survey branch, Report **83-1**, 68–89.
—— & VAN KRANENDONK, M. The Palaeoproterozoic Southeastern Churchill Province of Labrador–Quebec, Canada: orogenic development as a consequence of oblique collision and indentation. *This volume*.
——, BRIDGWATER, D., MENGEL, F., CAMPBELL, L., VAN KRANENDONK, M. J. *ET AL.* 1994. Mapping in the Torngat Orogen, northernmost Labrador: Report 3, The Nain Craton (including a note on ultramafic dyke occurrences in northernmost Labrador).

In: Current Research (1994). Newfoundland Department of Mines and Energy, Geological Survey Branch, Report **94-1**, 399–407.

——, RYAN, A. B., NUNN, G. A. C. & MENGEL, F. C. 1990. Labrador segment of the Trans-Hudson Orogen: Crustal development through oblique convergence and collision. *In*: LEWRY, J. F. & STAUFFER, M. R. (eds) *The Early Proterozoic Trans-Hudson Orogen of North America*. Geological Association of Canada, Special Paper, **37**, 353–369.

——, VAN KRANENDONK, M., MENGEL, F. & SCOTT, D. 1992. Geological mapping in the Torngat Orogen, northernmost Labrador: Preliminary results. *In: Current Research (1992)*. Newfoundland Department of Mines and Energy, Geological Survey Branch, Report **92-1**, 413–429.

——, ——, ——, ——, SCHWARZ, S. & RYAN, B. 1993. Geological mapping in the Torngat Orogen, northernmost Labrador: Report 2. *In: Current Research (1993)*. Newfoundland Department of Mines and Energy, Geological Survey Branch, Report **93-1**, 77–89.

WOOD, D. A. 1980. The application of a Th–Hf–Ta diagram to problems of tectonomagmatic classification and to establishing the nature of crustal contamination of basaltic lavas of the British Tertiary volcanic province. *Earth and Planetary Science Letters*, **50**, 11–30.

Torngat Orogen – a Palaeoproterozoic example of a narrow doubly vergent collisional orogen

TOBY RIVERS[1], FLEMMING MENGEL[2], DAVID J. SCOTT[3],
LISA M. CAMPBELL[4] & NORMAND GOULET[5]

[1] *Department of Earth Sciences, Memorial University of Newfoundland, St John's, Newfoundland, A1B 3X5, Canada*

[2] *Danish Lithosphere Centre, Østervoldgade 10, DK-1350 Copenhagen K, Denmark*

[3] *GEOTOP, Université du Québec à Montréal, PO Box 8888, Succursale A, Montréal, Québec, H3C 3P8, Canada*

[4] *CIRES and Department of Geological Sciences, Box 250, University of Colorado, Boulder, CO, 80309-0250, USA*

[5] *Département des sciences de la Terre, Université du Québec à Montréal, PO Box 8888, Succursale A, Montréal, Québec, H3C 3P8, Canada*

Abstract: Three widely spaced transects across northern Torngat Orogen demonstrate the broadly synclinorial structure and axial location of the monocyclic, supracrustal Tasiuyak gneiss complex in the centre of a doubly vergent orogen, with east-verging thrusting onto the Nain craton and west- or southwest-verging thrusting onto the Southeast Rae craton. On the basis of structural evidence, ages of detrital zircons and ε_{Nd} signature, the sedimentary protolith of the Tasiuyak gneiss is inferred to be unrelated to the autochthonous platformal sequences on the adjacent Nain and Southeast Rae cratons, and an origin as an allochthonous accretionary thrust wedge is proposed.

Structural studies of the Tasiuyak gneiss indicate the presence of early recumbent isoclinal folds (F_1) and associated stretching lineations (L_1), interpreted to have developed during thrust imbrication, that are refolded by later upright coaxial F_2 structures with opposing vergences towards the marginal cratons. The upright fabrics in the centre of the orogen can be traced without metamorphic break into mylonitic fabrics of the crustal-scale, sinistral transcurrent Abloviak shear zone, implying that the latter was initiated late in the structural history, but during peak metamorphic conditions.

The emplacement of the primary, mantle-derived, but variably crustally contaminated ($+3 > \varepsilon_{Nd} < -3$), calc-alkaline diorite–tonalite–granite intrusions between 1910–1870 Ma into the northwestern margin of the Nain craton suggests that subduction was easterly-directed, under the Nain craton. Association of these intrusions with crustally-derived (or severely crustally contaminated) charnockites (ε_{Nd} c. −6.7) of similar age (1895–1886 Ma) implies that substantial melting of Nain crust occurred at this time.

Crustal thickening during the collisional stage of Torngat Orogeny resulted in extensive reworking of the margins of the Southeast Rae and Nain cratons and widespread *in situ* generation of leucogranites at c. 1860 Ma in the Tasiuyak gneiss, and was coeval with emplacement of charnockite intrusions into the Tasiuyak gneiss between c. 1860–1840 Ma. Metamorphic grade varies from greenschist facies near Torngat Front in the east of the orogen, to uppermost amphibolite and granulite facies in Tasiuyak gneiss in Abloviak shear zone in the axis of the orogen. Within Komaktorvik shear zone, developed in reworked Nain gneisses along the east side of the orogen, metamorphic grade also attained granulite facies. Estimates of the peak P–T conditions along the present erosion surface are in the range 10–12 kb and 850°C in both Abloviak and Komaktorvik zones, implying that double crustal thicknesses were attained during Torngat Orogeny. Syntectonic uplift and retrogression in Abloviak shear zone down to 6–7 kb/750–650°C did not involve significant incursion of fluids, whereas in Komaktorvik shear zone syntectonic uplift and retrogression down to 4–6 kb/550–500°C involved significant hydration of the granulite facies assemblages. Available age determinations suggest that peak metamorphism was synchronous along the length of the orogen at c. 1870–1840 Ma, with transcurrent shearing occurring between c. 1840–1820 Ma in Abloviak shear zone. Syntectonic uplift, exhumation and retrogression, and movement on Komaktorvik shear zone continued until closure of geochronologic and geothermobarometric systems between c. 1790–1720 Ma.

A new tectonic model compatible with these constraints and inspired by the numerical modelling experiments of Willett *et al.* (1993, *Geology*, **21**, 371–374) is presented. The apparent symmetry of the doubly vergent orogen is shown to be a result of a fundamentally asymmetric process.

Torngat Orogen is a narrow, north–south trending orogenic belt situated between the Nain and Southeast Rae cratons in northern Labrador and Québec. On a regional scale, it displays a high degree of symmetry, with fold-thrust belts of opposing vergence on either side and a major transcurrent shear zone, the sinistral Abloviak shear zone, in the centre. This structural pattern has led several authors to propose that the orogen developed as a result of oblique collision between the Archaean Nain and Southeast Rae cratons (e.g. Mengel 1988; Hoffman 1990; Van Kranendonk & Ermanovics 1990; Mengel et al. 1991; Bertrand et al. 1993)

Our work is aimed at understanding the tectonic development of the major transcurrent shear zone and its relationships to the marginal fold-thrust belts. We focus in this paper on testing the oblique-collision model and in the process provide new information concerning the relationship between crustal thickening, as displayed by structures and mineral assemblages, and transcurrent displacement in the axial shear zone. Most rocks in Torngat Orogen display upper amphibolite to granulite facies metamorphic assemblages, thus in addition, the study presents an opportunity to examine the roots of an orogen and the transition towards the bounding forelands.

The principal lithotectonic elements in Torngat Orogen (Fig. 1) have been described by Van Kranendonk & Ermanovics (1990), Mengel et al. (1991), Van Kranendonk & Wardle (1994), Van Kranendonk et al. 1993a, b) and Wardle et al. (1992, 1993), and are described briefly below from east to west. **Nain craton** is an Archaean, amphibolite to granulite facies gneiss complex with a long and complex supracrustal, igneous and metamorphic history between 3.8 and 2.6 Ga (e.g. review in Bridgwater & Schiøtte 1991). It was intruded by several swarms of diabase dykes (e.g. **Napaktok** and **Avayalik** dykes) in the Palaeoproterozoic, and the western edge of the Nain craton, including the dykes, was strongly reworked under amphibolite and granulite facies conditions during Torngat Orogeny. Nain craton with Napaktok dykes is unconformably overlain by the Palaeoproterozoic **Ramah Group**, a predominantly sedimentary supracrustal succession disposed in a narrow east-verging, metamorphic fold-thrust belt in the southeastern part of Torngat Orogen.

In the northeastern part of Torngat Orogen, reworked Nain gneisses are intruded by large volumes of Palaeoproterozoic calc-alkaline plutonic rocks of the diorite–tonalite–granite (**DTG**) **suite**, emplaced between 1910–1870 Ma, and by large volumes of coeval (1895–1886 Ma) orthopyroxene–bearing granodiorite and quartz diorite (**Killinek charnockite suite**) (Scott & Machado 1994; Van Kranendonk & Wardle 1994). In the centre of the orogen is **Tasiuyak gneiss complex**, a Palaeoproterozoic, principally metasedimentary (semipelitic, pelitic, graphitic and sulphidic schists and quartzite) gneiss complex metamorphosed to upper amphibolite and granulite facies, with abundant intrusive sheets of leucogranite, minor charnockite and local slices of Archaean gneisses and slivers of mafic and ultramafic rocks (Bertrand et al. 1993; Van Kranendonk & Wardle 1994) that we argue in this paper developed as an accretionary thrust wedge. (This usage of Tasiuyak gneiss complex extends the term, originally defined by Van Kranendonk & Wardle (1994), to include the Tasiuyak gneiss and tectonically related units throughout the orogen). Further west, Tasiuyak gneiss complex is in contact with the **Southeast Rae craton**, a poorly understood Archaean block consisting of a variety of high grade (granulite and amphibolite facies) gneisses with emplacement ages between 2922 and 2597 Ma and evidence of a late Archaean metamorphism dated at 2572 Ma (see Van Kranendonk et al. 1993a for a summary of available data). It is unconformably overlain by the Palaeoproterozoic **Lake Harbour Group**, a platformal sequence including psammite, semi-pelite and marble that is folded into a southwest-verging metamorphic fold-thrust belt, and locally intruded by charnockitic (s.l.) plutons. Southeast Rae craton was pervasively reworked under ductile conditions during Torngat Orogeny.

Torngat Orogen is transected by a major system of anastomosing, ductile transcurrent shear zones with sinistral displacement. Principal components of the shear zone system are **Abloviak shear zone** (**ASZ**), developed in reworked gneisses of the Southeast Rae craton and in Tasiuyak gneiss, and **Komaktorvik shear zone** (**KSZ**) developed in reworked Nain gneisses and Palaeoproterozoic intrusions of the DTG suite (Fig. 1). ASZ and KSZ are juxtaposed throughout much of the orogen, but diverge south of Burwell domain (Fig. 1), where the shear zone fabric in the ASZ bends through almost 90°. (In contradistinction to Van Kranendonk et al. (1993a), but following the original proposal of Korstgård et al. (1987), we use the term KSZ to include **all** reworked Nain rocks with transcurrent shear fabrics east of the ASZ; thus KSZ in our terminology extends as far south as the Nain Plutonic Suite, Fig. 1).

Rocks of the **Burwell domain**, situated between the Abloviak and Komaktorvik shear zones in northernmost Labrador, have been subdivided into three lithotectonic units by Van Kranendonk & Wardle (1994). **Tasiuyak gneiss complex** occupies the south of the domain, and **Noodleook complex**

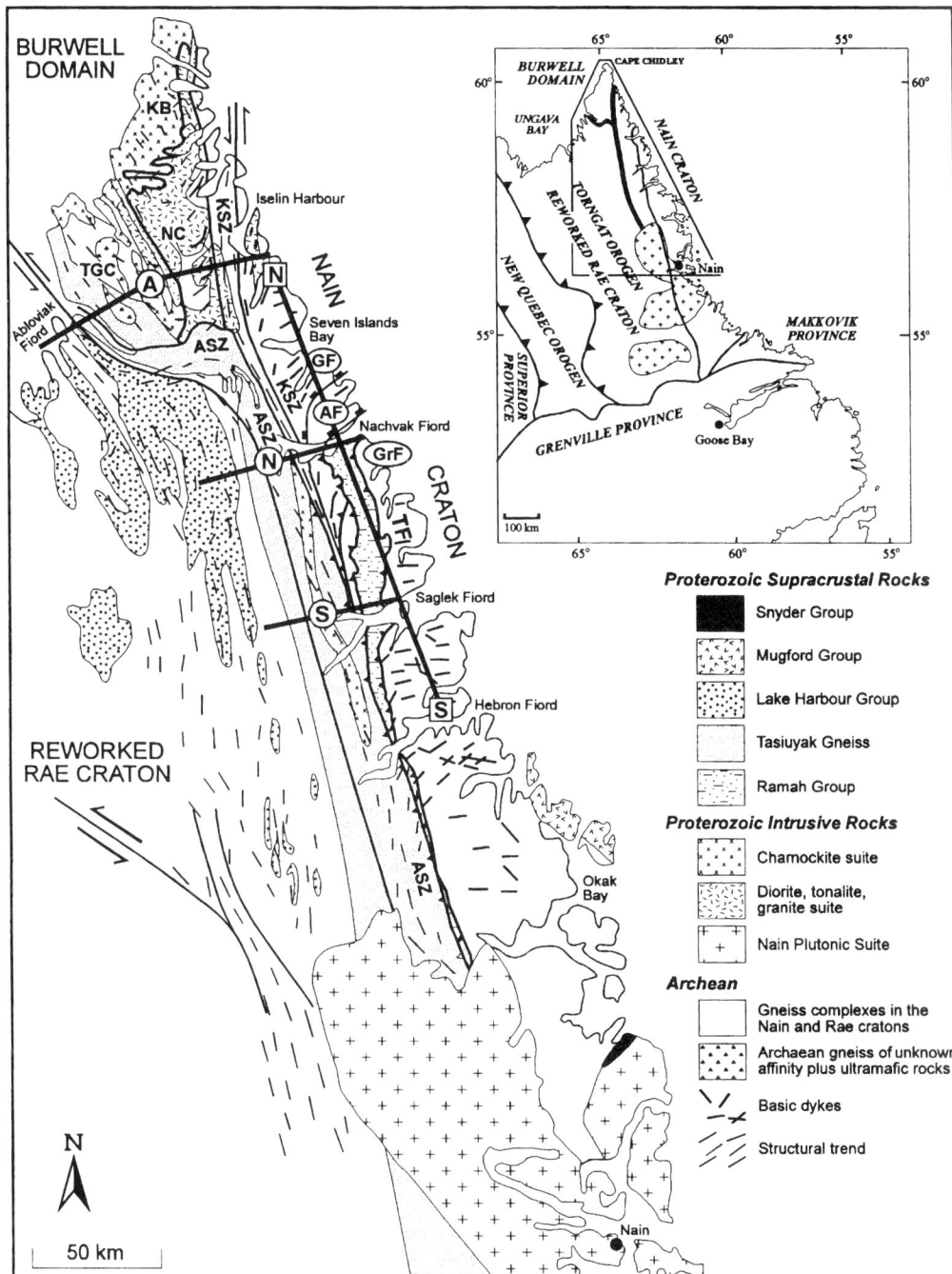

Fig. 1. Geological map showing the main lithotectonic elements of Torngat Orogen (from Van Kranendonk *et al.* 1993*b*). Thick black lines labelled A, N and S are the locations of the Abloviak, Nachvak and Saglek transects shown in Fig. 2. Thin line marked N–S is location of schematic longitudinal profile shown in Fig. 5. TGC, NC and KB in Burwell domain refer to Tasiuyak gneiss complex, Noodleook complex and Killinek charnockite batholith respectively. ASZ and KSZ refer to Abloviak and Komaktorvik shear zones. Inset map of northern Labrador and Québec shows location of Torngat Orogen with respect to other Archaean cratons and Proterozoic orogens in eastern Canada.

in the centre of the domain consists of the reworked Nain gneisses, slivers of Proterozoic metasediments and a large volume of Proterozoic intrusive rocks of the DTG suite. **Killinek batholith** in the north of Burwell domain consists predominantly of foliated orthopyroxene-bearing granodiorite of the charnockite suite. Contacts between the three units are interpreted to be thrusts by Van Kranendonk & Wardle (1994).

In this study, the transition from the marginal fold-thrust belts to the axial Abloviak shear zone has been investigated in two transects along Abloviak and Saglek fiords referred to as the **Abloviak** and **Saglek transects** (Fig. 1), and data from a third transect along Nachvak Fiord are taken from published reports by Wardle (1983, 1984) for comparison. Below we present the key structural features of the main lithotectonic units in each transect; for accounts of the regional geology we refer the reader to descriptions by Taylor (1979), Van Kranendonk *et al.* (1993*b*), Van Kranendonk & Wardle (1994) and Wardle *et al.* (1992, 1993).

Abloviak transect

From east to west, the Abloviak transect crosses little reworked Nain gneisses with intrusions of Avayalik dykes and plutons of the DTG suite, the Komaktorvik shear zone, the Noodleook and Tasiuyak gneiss complexes of Burwell domain, the Abloviak shear zone, and reworked gneisses of the Southeast Rae Province with their Lake Harbour Group cover. This summary is based on previous work by Goulet & Ciesielski (1990), Wardle *et al.* (1992), Mengel & Rivers (1993, 1994*a*), Rivers *et al.* (1993*a*), Rivers & Mengel (1994), Van Kranendonk *et al.* (1993*b*) and Van Kranendonk & Wardle (1994). The schematic cross-section is shown in Fig. 2a. Mineral abbreviations follow Kretz (1983).

Nain craton. In the east of the transect, Nain gneisses are cut by several suites of east- to northeast-trending mafic dykes, of which the northeast-trending Avayalik suite is the most prominent. East of the Komaktorvik shear zone, both gneisses and dykes are little affected by Proterozoic deformation, but are characterized by a static, high-pressure granulite facies metamorphism of Proterozoic age, best displayed by the presence of Grt–Cpx–Pla assemblages in the dykes (Mengel & Rivers 1993). This assemblage is variably overprinted by north-trending, amphibolite facies fabrics that are related to Komaktorvik shear zone.

Komaktorvik shear zone. At this latitude, KSZ is an anastomosing, variably penetrative, sinistral transcurrent shear zone about 5–10 km wide, characterized by sub-vertical protomylonitic to mylonitic foliations and associated sub-horizontal to moderately north- or south-plunging stretching lineations in areas of high strain. It is developed in reworked Nain gneisses with Avayalik dykes and in Proterozoic intrusive rocks of the DTG suite. Grade of metamorphism is predominantly in amphibolite facies, with widespread evidence of overprinting of the earlier high-pressure Proterozoic granulite facies assemblage (e.g. Grt–Cpx replaced by Hbl–Pla in Avayalik dykes and mafic supracrustal rocks).

Noodleook complex. On the west side of KSZ, within the Noodleook complex of Burwell domain, rotation of structures into the shear zone provides clear evidence of its sinistral sense of displacement. Noodleook complex consists of gneisses correlated with those in the Nain craton, and of Proterozoic intrusive and metasedimentary rocks that are interpreted to be in thrust contact with gneisses of the Tasiuyak gneiss complex (Van Kranendonk & Wardle 1994). Grade of metamorphism is uppermost amphibolite to granulite facies.

Tasiuyak gneiss complex. In southern Burwell domain, Tasiuyak gneiss complex lies outside the zone affected by shearing along ASZ, and its early structural development can be more easily studied. Within the metasedimentary Tasiuyak gneiss component of the complex, a sub-horizontal compositional layering (S_0) can be readily distinguished, defined by quartz-, mica- and graphite ± sulphide-rich layers. S_0 is sub-parallel to a penetrative foliation, referred to here as S_1, and outcrop-scale rootless, intrafolial, isoclinal F_1 folds with sub-horizontal axes are widespread, but sparse. Local evidence of refolding of F_1 and the combined $S_0//S_1$ fabric by a second generation of coaxial isoclinal folds (F_{1A}), together with the generation of a new axial planar fabric (S_{1A}) that is petrographically indistinguishable from S_1, is also considered part of the first deformation. L_1, the penetrative stretching lineation parallel to F_1 axes, plunges gently towards the southeast and northwest (Fig. 3; Goulet & Ciesielski 1990) and is defined by extremely elongate quartz aggregates and garnet grains (L >> S) and by the preferred orientation of sillimanite.

Within Burwell domain, Van Kranendonk & Wardle (1994) have mapped folded, structurally bound slices of Archaean gneisses (of unknown affiliation), ultramafic rocks and Proterozoic intrusives within the Tasiuyak gneiss complex, implying an important early thrust history to the complex.

The $S_0//S_1$ fabric in the imbricated Tasiuyak gneiss complex is refolded into decimetre-scale, upright, gently plunging to sub-horizontal, open to tight F_2 folds that are coaxial with F_1 (Fig. 3). Leucogranite sheets and charnockite bodies are clearly affected by D_2 structures, and locally have been seen to cut the S_1 fabric, implying intrusion was post D_1 - pre D_2. In Tasiuyak gneiss, the associated axial planar $L_2 \geq S_2$ fabric is defined by spaced quartz lenses that are elongate parallel to F_2 and by the preferred orientation of sillimanite and locally orthopyroxene. L_2 is coaxial with L_1.

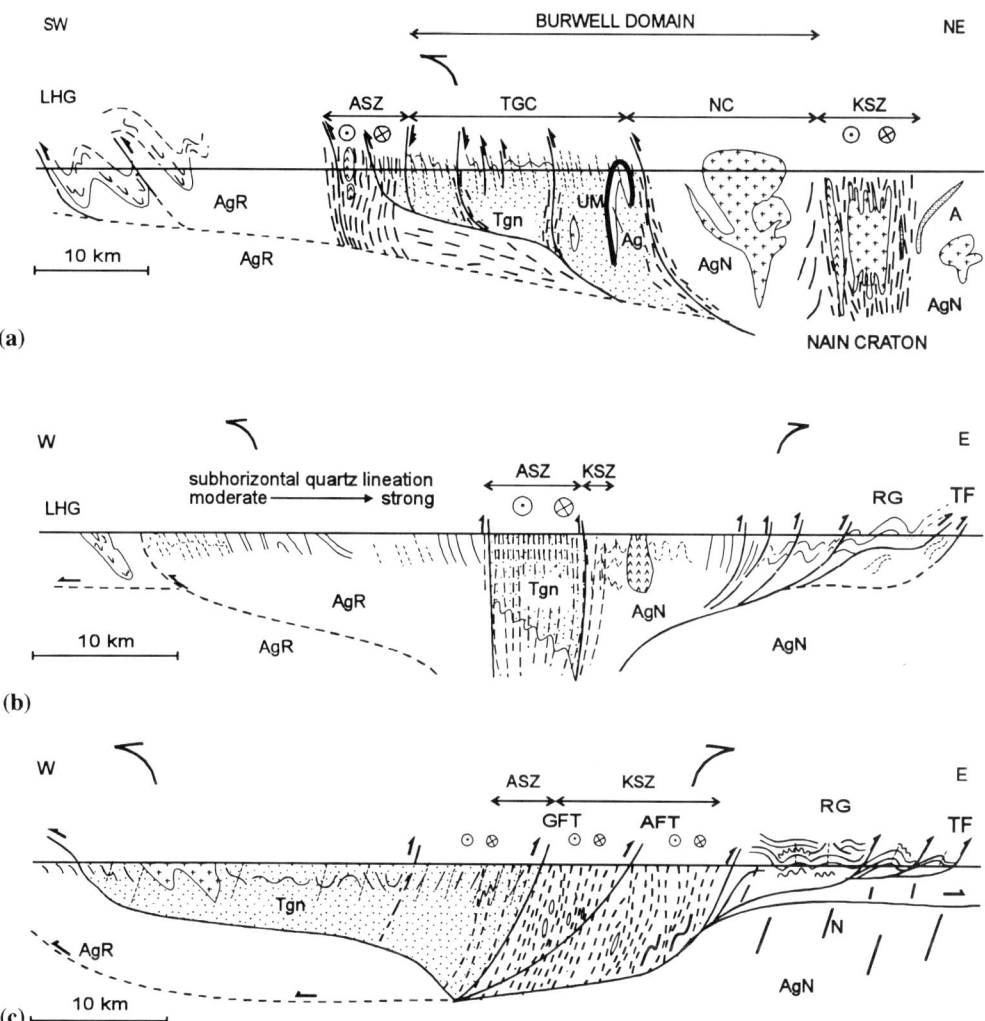

Fig. 2. Schematic cross-sections along the (**a**) Abloviak transect (after Rivers *et al.* 1993*a, b*), (**b**) Nachvak transect (after Wardle 1983, 1984) and (**c**) Saglek transect (after Rivers & Mengel 1994). Cross-sections are based on surface outcrops, and extrapolations to depth are poorly constrained. Shape and thickness of Tasiuyak gneiss (Tgn) are based on gravity modelling of Feininger & Ermanovics (1994). East-dipping structures in Southeast Rae craton are seen in seismic reflection profiles (Hall 1994). Ag, Archaean gneisses (affiliation unknown); AgN and AgR, Archaean gneisses of Nain and Southeast Rae cratons; dot pattern, Tasiuyak gneiss (Tgn); +, Proterozoic intrusions of DTG and charnockite suites; thick black line in (a), ultramafic rocks; inverted v pattern, anorthosite; mylonitic shear foliation in Abloviak shear zone (ASZ) and Komaktorvik shear zone (KSZ) is shown as bold dashed lines; dykes in Nain craton and KSZ are labelled A for Avayalik and N for Napaktok; light solid lines in Tasiuyak gneiss represent orientations of $S_0//S_1$ surfaces, light dashed lines represent orientations of S_2 foliations. GFT and AFT, granulite and amphibolite facies terranes of KSZ; geometry of the Ramah Group in (c) is after Calon & Jamison (1993, 1994). Large arrows at top of sections indicate overall directions of structural vergence.

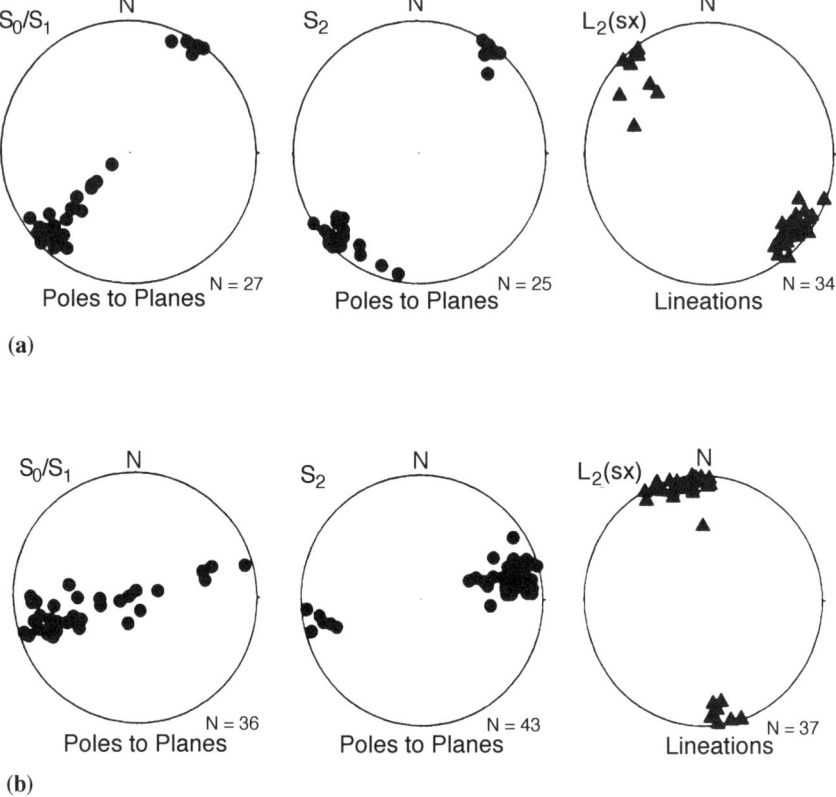

Fig. 3. Stereonets for structural data obtained from: (**a**) western part of Abloviak transect; and (**b**) western part of Saglek transect – (i) poles to $S_0//S_1$; (ii) poles to S_2; (iii) L_2 stretching lineations. Lower hemisphere, equal area projections.

Towards the southwest, F_2 folds become increasingly asymmetrical, with a consistent southwesterly vergence towards the Abloviak shear zone, and the angle between $S_0//S_1$ and S_2 is greatly reduced, such that within ASZ distinction between S_0, S_1 and S_2 in the Tasiuyak gneiss is not possible due to transposition and the formation of the penetrative mylonitic fabric.

Abloviak shear zone. Tasiuyak gneiss complex is in tectonic contact with the margin of the reworked Southeast Rae basement gneisses within the Abloviak shear zone, a major anastomosing transcurrent shear zone several kilometres wide in this transect. In Tasiuyak gneiss, the ASZ is characterized by sub-vertical mylonitic foliations and very strongly developed sub-horizontal stretching lineations (L >> S) (Fig. 3) defined by the shapes of deformed quartz aggregates and garnets and the preferred orientation of sillimanite and locally orthopyroxene. Widespread shear bands and local inclusion trails in rotated garnets show evidence of sinistral strike-slip displacement parallel to the sub-horizontal stretching lineations. In reworked Southeast Rae gneisses, ASZ is characterized by amphibolite facies mylonites, derived from granite and amphibolite, with a pinstripe mylonitic foliation that has locally been refolded into sheath folds, and by the lozenge shape and locally mylonitic character of a gabbro body incorporated in the shear zone (Rivers *et al.* 1993a).

The northern boundary of the ASZ is marked by a major brittle reverse fault zone up to 500 m wide, parallel to the layering in the gneisses, with extensive development of tectonic breccia and pseudotachylite, implying that uplift of high grade rocks in the ASZ continued into the brittle realm.

Reworked Southeast Rae gneisses and Lake Harbour Group. In the vicinity of Abloviak Fiord, the Southeast Rae basement consists of a variety of granitic (*sensu lato*) gneisses and subordinate anorthosite, gabbro and amphibolite cut by mafic dykes. Contacts and fabrics have been strongly reorientated into NW–SE directions, the local trend of Torngat Orogen, and narrow high strain zones are common within the basement and Lake Harbour Group cover. Prevailing metamorphic grade is in amphibolite facies, though relics of earlier granulite facies assemblages can be seen locally.

Near Abloviak Fiord, the Southeast Rae basement is disposed in an elongate dome and basin pattern between keels of its Lake Harbour Group cover. Lithologies of the Lake Harbour Group possess an S_1 foliation parallel to compositional layering (S_0) and to the foliation in the surrounding basement gneisses, and the dome and basin pattern is at least an F_2 feature. Lineations are sub-horizontal (Goulet & Ciesielski 1990), parallel to F_2 fold axes. Contacts between the Lake Harbour Group and its basement are sharp, and on the basis of evidence of imbrication of basement and cover (Goulet 1993) and the high degree of reworking of the Southeast Rae basement, are interpreted to be early thrusts. However their locations in map view are poorly constrained, and they are not shown in Fig. 1. F_2 folds in the Lake Harbour Group are overturned towards the southwest, away from Torngat Orogen.

Nachvak transect

From east to west, the Nachvak transect (Fig. 2b) crosses the Nain craton with east-trending Napaktok dykes and Ramah Group cover, the Komaktorvik shear zone developed in reworked Nain gneisses, and the Abloviak shear zone developed in Tasiuyak gneiss and reworked Southeast Rae gneisses with infolded keels of Lake Harbour Group cover. The description below is derived from the work of Wardle (1983, 1984).

Ramah Group and reworked Nain gneisses in Komaktorvik shear zone. In the eastern part of the transect, the Ramah Group unconformably overlies amphibolite facies Archaean gneisses of the Nain Province. Towards the west of its outcrop, the Ramah Group becomes increasingly deformed, and near its western limit is characterized by east-verging second-generation (F_2) folds. West-dipping thrusts are a characteristic feature of the eastern part of the Nachvak transect and Wardle (1983) described several thrusts in the Nain gneisses below the unconformity at the base of the Ramah Group, implying that the entire Ramah Group in this transect is transported and lies to the west of Torngat Front (the eastern margin of Torngat Orogen, Mengel *et al.* 1991) in this transect (Fig. 2c). To the west of, and structurally overlying the Ramah Group, are reworked Nain gneisses with deformed remnants of Napaktok dykes in the Komaktorvik shear zone, which is characterized by strong, sub-vertical north-trending fabrics and associated sub-horizontal stretching lineations, and evidence of amphibolite and granulite facies Proterozoic metamorphism. After strike-slip motion and formation of their transcurrent fabrics, rocks in the KSZ were thrust over the Ramah Group along brittle faults decorated with pseudotachylite.

Abloviak shear zone. The central part of the transect is the locus of the ASZ, which is in tectonic contact with the Nain gneisses in the KSZ along the North Arm thrust. In this transect, the ASZ affects the full width of the Tasiyak gneiss, and also extends into the reworked Southeast Rae gneisses to the west (Wardle 1983, 1984). As in the Abloviak transect, the structural style of the ASZ is characterized by steeply west-dipping to sub-vertical mylonitic foliations and sub-horizontal stretching lineations defined by quartz aggregates and the preferred orientation of sillimanite, with a variety of kinematic indicators implying sinistral transcurrent shear.

Reworked Southeast Rae gneisses. West of the Abloviak shear zone, the transect crosses into the reworked Southeast Rae basement, composed of migmatitic tonalitic to granodioritic gneisses with abundant amphibolite inclusions. The sub-horizontal quartz lineations that characterize the Abloviak shear zone extend in local high strain zones several kilometres into the Southeast Rae basement, and the unit is everywhere penetratively reworked. Metamorphic grade is largely in amphibolite facies, with widespread evidence in mafic lithologies of an earlier granulite facies event. Slivers and infolded keels of rusty, pelitic and psammitic Proterozoic supracrustal units, correlated with the Lake Harbour Group, occur throughout the western part of the transect, becoming more abundant towards the west. West of Abloviak shear zone, there is an overall change in orientation of D_2 structures from steep near ASZ to east-dipping, west-verging structures near the west end of the traverse. This is especially clear in the monocyclic supracrustal Lake Harbour Group, which is folded in large scale west-verging asymmetric F_2 folds.

Saglek transect

From east to west, the Saglek transect (Fig. 2c) crosses the Nain craton with east-trending Proterozoic Napaktok (Domes) dykes and Ramah Group cover, reworked Nain gneisses in Komaktorvik shear zone, Tasiuyak gneiss in Abloviak shear zone, and reworked Southeast Rae gneisses. The geology of the eastern part of the transect along Saglek Fiord was mapped by Mengel (1988) and has recently been the object of detailed structural studies by Calon & Jamison (1993, 1994). The western part of the transect was investigated by Rivers & Mengel (1994). Below, we review the transect from east to west.

Ramah Group and reworked Nain gneisses in Komaktorvik shear zone. In the vicinity of Saglek Fiord, the Torngat Front is located close to the eastern limit of outcrop of the Ramah Group (Figs 1 & 2). The Ramah Group forms part of an overall east-directed, metamorphic foreland fold-thrust belt (Mengel et al. 1991) in which deformation was principally driven by extrusion of high grade rocks from the interior of the orogen onto the foreland (Calon & Jamison 1993, 1994). Calon & Jamison (1993) showed that reworked amphibolite facies basement gneisses (amphibolite facies block – see below) were emplaced as a several kilometres thick wedge along a décollement surface (Lake Kiki thrust) located near the base of the Ramah Group, and that propagation of thrust-and-fold structures away from the front of the wedge gave rise to the easterly-vergent pattern of the belt. They inferred that the décollement breaches the erosion surface at Torngat Front (known locally as Branagin thrust), and thus that all the Ramah Group along this transect is transported. Pelitic lithologies in Ramah Group show evidence of two principal deformation events, D_1 involving formation of a sub-horizontal fabric with associated isoclinal folds, and D_2 resulting in more upright F_2 folds that are overturned towards the east. Peak metamorphism was attained between D_1 and D_2, and evidence of transposition of S_1 into S_2 is widespread in pelitic lithologies (Mengel et al. 1991; Calon & Jamison 1993).

Structurally overlying the Ramah Group along the Lake Kiki thrust are two fault-bound slices of reworked Nain gneisses with remnants of Napaktok dykes, referred to as the amphibolite and granulite facies blocks by Mengel et al. (1991), that constitute Komaktorvik shear zone in this transect. KSZ is characterized by a well-developed, steeply west-dipping, protomylonitic to mylonitic foliation and an associated penetrative sub-horizontal mineral lineation. These structures, together with map-scale evidence of sinistral rotation of the Napaktok dykes into the trend of the mylonitic foliation (Morgan 1978; Mengel et al. 1991) and other small-scale kinematic indicators of transcurrent shearing, define the location of Komaktorvik shear zone. Intensity of transcurrent shearing increases towards the west, and in the granulite facies block the original angular discordance between the Archaean gneissic fabric and Proterozoic Napaktok dykes is effectively obliterated by high strain. Uplift of the KSZ over the Ramah Group along the Lake Kiki and Nachvak Brook thrusts postdated the transcurrent shearing, as the mylonitic fabrics in the KSZ are truncated by the thrusts, both of which are locally decorated by pseudotachylite, indicating that contractional deformation continued to shallow levels.

Tasiuyak gneiss complex and Abloviak shear zone. At its western margin, the granulite facies block of the KSZ is in tectonic contact with Tasiuyak gneiss along the North Arm thrust. In the vicinity of the thrust and up to three or four kilometre farther west, the Tasiuyak gneiss is characterized by the steeply west-dipping mylonitic foliation and associated sub-horizontal lineation of the ASZ that is similar in character to that in the adjacent granulite facies block of KSZ. Within the Abloviak shear zone, the mylonitic fabric is generally the only planar feature discernible, although remnants of earlier layering and isoclinal folds can be discerned locally.

Within Tasiuyak gneiss complex, a few kilometres west of the Abloviak shear zone, there are narrow (10–100 m wide) zones of high strain, (proto-)mylonitic gneisses, generally located on the steep limbs of F_2 folds where the S_2 and $S_0//S_1$ surfaces are sub-parallel, implying that in detail the shear zone has an anastomosing form. However, overall, the western margin of the ASZ marks a rather abrupt transition into much less deformed rocks in which $S_0//S_1$ and S_2 can be distinguished, as in the Abloviak transect. In the western 20 km of the Saglek transect, Tasiuyak gneiss is characterized by open to tight F_2 folds (Figs 2c & 3b). S_2 is sub-vertical to steeply west-dipping across the western part of the section (Fig. 2c), and is sub-parallel to the mylonitic foliation in the Abloviak shear zone. These folds have sub-horizontal axes parallel to the regional stretching (L_1) lineations. In addition, there is no measurable change in the orientation of the lineation within the Abloviak shear zone (L_S in Fig. 3b), suggesting that the latter is a composite feature that developed during both D_1 and D_2 deformations. F_2 folds in the central part of the outcrop of the Tasiuyak gneiss are open, approximately upright structures, whereas at the west end of the transect, dips of $S_0//S_1$ are predominantly towards the east and minor folds indicate a westward sense of vergence. However, as noted previously, the S_2 fabric retains an approximately constant, steep westerly dip throughout the section. A large body of orthopyroxene-bearing tonalite in the western part of the transect contains a single (weak) fabric correlated with F_2 in the Tasiuyak gneiss.

In the western part of the section, the Southeast Rae gneisses are only known from reconnaissance. They consist of a tonalite gneiss complex that shows evidence of reworking under both ductile and brittle conditions. A superimposed foliation, oblique to the older (Archaean?) gneissosity, is subparallel to the foliation in the adjacent Tasiuyak gneiss, and is therefore assumed to have developed during Torngat Orogeny. Grade of metamorphism is in amphibolite facies.

Metamorphism

Details of the metamorphism in all three traverses are considered together.

Ramah Group and reworked Nain gneisses east of KSZ

Effects of Proterozoic metamorphism in the foreland to the east of the Torngat Front are localized, and are largely limited to static hydrous retrogression in most units (Ryan 1990). West of Torngat Front, within the Ramah Group, there is a variation in metamorphic grade from north to south as well as from east to west (Mengel 1988; Mengel & Rivers, 1994a). North of Saglek Fiord, chloritoid–chlorite, chloritoid–andalusite and chloritoid–kyanite sub-assemblages are present in pelitic units, suggesting upper greenschist to lower amphibolite facies conditions, and Mengel & Rivers (1994b) have estimated maximum P–T conditions in the range 3.5–4 kb and about 475°C. Further south, along the north shore of Saglek Fiord, andalusite is unstable and the sub-assemblage staurolite–chlorite replaces chloritoid–kyanite, indicating slightly higher temperatures and pressures. Approximately 15 km south of Saglek Fiord the sub-assemblage sillimanite–garnet–biotite is stable, for which Mengel & Rivers (1994b) have estimated P–T conditions of 6–7 kb and about 650°C. These are the maximum conditions recorded in the Ramah Group.

Reworked Nain gneisses in Komaktorvik shear zone

Maximum Proterozoic metamorphic conditions in reworked Nain gneisses in KSZ have been estimated at up to 800°C/10–12 kb in the granulite facies block and about 550°C/5–6 kb in the amphibolite facies block in the Saglek transect (Mengel & Rivers 1991). Decompression reactions between phases in granulite facies assemblages have been utilized to infer that up to 6 kb of syntectonic exhumation occurred during cooling and transcurrent motion on Komaktorvik shear zone (Fig. 4a; Mengel & Rivers 1991).

Further north, in the KSZ in the Abloviak transect, Mengel & Rivers (1994a) have recently documented comparable P–T–t paths from about 12 kb/800°C to 4 kb/500°C utilizing a variety of lithologies (Figs 4c, 4d). Highest pressures and temperatures occur in Proterozoic dykes and in Grt–Cpx–Hbl–Pla assemblages in Archaean metabasic supracrustal units with mylonitic fabrics (characteristic of KSZ) from the east of the area, implying that transcurrent movement and early retrogression along the shear zone occurred at near peak P–T conditions.

Tasiuyak gneiss in and adjacent to Abloviak shear zone

Mineral assemblages in pelitic parts of the Tasiuyak gneiss are similar in all three transects, both in and adjacent to ASZ, with the presence of Qtz–Kfs–Grt–Sil–Bt–Pl–L ± Gr ± Po being widespread. The unit is characteristically diatexitic, with garnet-bearing granitic leucosomes (L) interpreted to have been derived by *in situ* anatexis, being ubiquitous. Within the restites, orthopyroxene is rare and has not been observed in equilibrium with sillimanite, implying that it is restricted to Fe-rich compositions. This suggests that metamorphic conditions were in uppermost amphibolite facies, close to the amphibolite–granulite transition. The widespread presence of variable amounts of in situ granitic leucosome suggests that an H_2O-bearing fluid was available during metamorphism, and graphite is widespread, implying that f_{O_2} was at or below the graphite-CO_2 buffer. Magnetite is absent, giving rise to the distinctive low magnetic signature for the unit (Rivers *et al.* 1993a).

Thermobarometry on the assemblage Grt–Sil–Pla–Qtz–Bt from pelitic portions of the Tasiuyak gneiss within the Abloviak shear zone in the Saglek transect has yielded estimates that are in the range 6–7 kb and 650–750°C (Fig. 4b), whereas interlayered mafic rocks yielded P–T estimates of 10–12 kb and 800°C, similar to those in the adjacent granulite facies block of KSZ (Mengel 1988). This difference in estimated P–T conditions is interpreted to be a result of more pervasive post-peak re-equilibration in the pelitic rocks (Mengel & Rivers 1991). Analytical work on samples from low strain rocks outside the Abloviak shear zone is in progress. In this area, the presence of plagioclase haloes around large garnets, and a distinctly bimodal size distribution of garnet suggests that there were two generations of garnet growth, with large elongate early garnets with plagioclase haloes suggesting decompression (by reaction of grossular with sillimanite and quartz to produce anorthite).

126   T. RIVERS ET AL.

Reworked Southeast Rae gneisses and Lake Harbour Group

The metamorphic assemblages of reworked Southeast Rae gneisses have not been studied in detail, but preliminary examination suggests that the gneisses generally exhibit amphibolite facies assemblages. The age of relict orthopyroxene, which occurs in some mafic rocks is not known. Available data for supracrustal rocks of the Lake

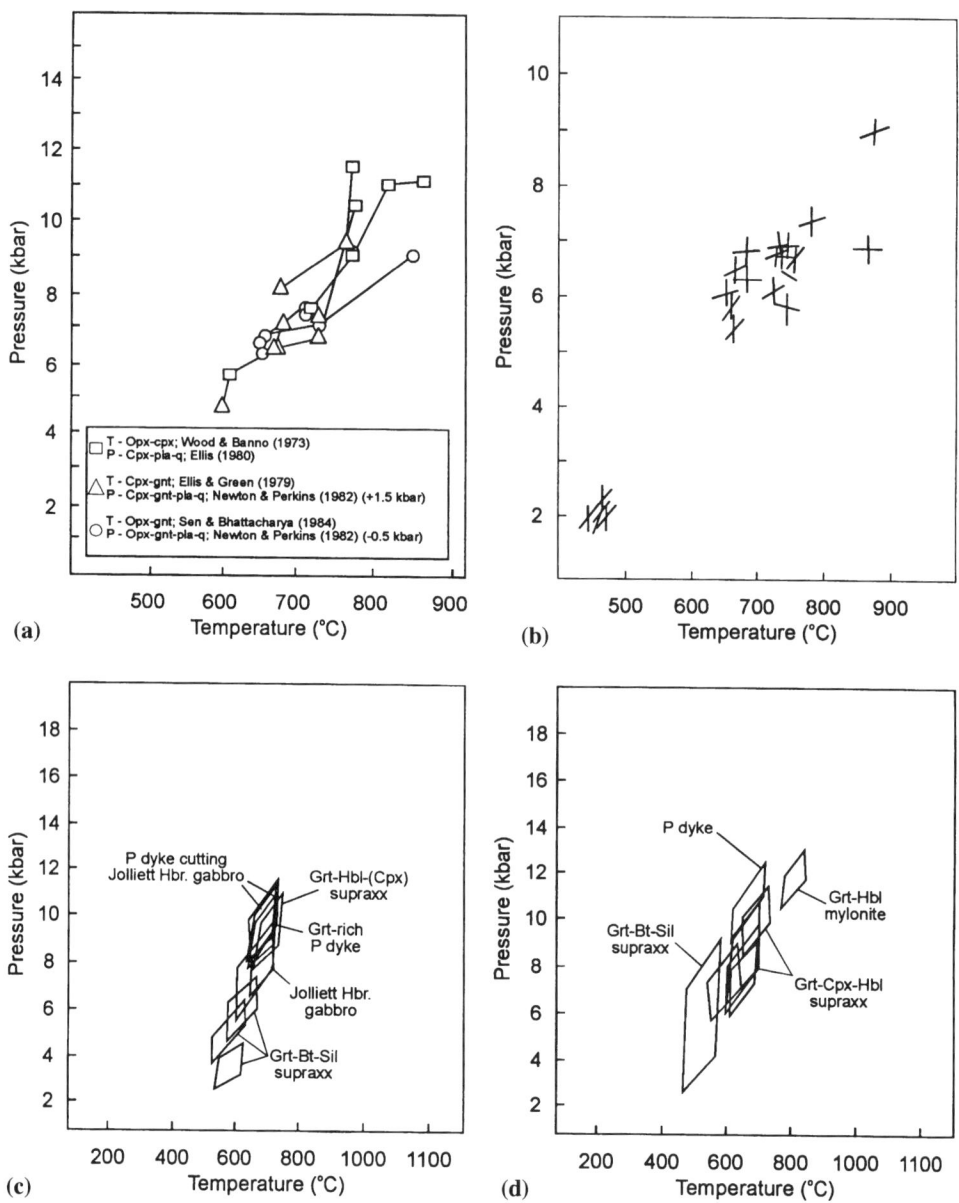

Fig. 4. P–T estimations for high grade rocks from Komaktorvik and Abloviak shear zones. (a) Mafic rocks (Opx–Cpx–Grt–Pla–Qtz) from granulite facies block of KSZ in Saglek transect; lines connect P–T estimates from cores, rims and rim symplectites in assemblages, yielding P–T–t paths from individual samples (from Mengel & Rivers 1991). (b) Tasiuyak gneiss (Grt–Sil–Pla–Qtz–Bt) from ASZ in Saglek transect (from Mengel 1988). (c–d) Various lithologies from western (c) and eastern (d) parts of northern KSZ in Abloviak transect (from Mengel & Rivers 1994a).

Harbour Group are also sparse, but suggest that the unit exhibits amphibolite facies assemblages.

Comparison between Abloviak, Nachvak and Saglek transects

The salient observations are summarised in point form.

- There is substantial along-strike continuity in the orogen, with the axial location of the Tasiuyak gneiss in an F_2 synclinorial structure and the presence of two bordering cratons occurring in each section. Opposing structural vergence towards the eastern and western forelands of the orogen, as discussed in the introduction, is clearly displayed in the Nachvak and Saglek transects, but foreland vergence occurs only at the western side of the Abloviak transect.
- In the Abloviak and Saglek transects, where there are suitable areas of low strain to determine the structural history, the development of the Tasiuyak gneiss is similar. D_1 gave rise to multiple generations of a sub-horizontal layer-parallel fabric ($S_0//S_1$) and an orogen-parallel stretching lineation (L_1), and D_2 resulted in the formation of open to tight F_2 folds with fold axes and a stretching lineation coaxial to L_1 and a steeply dipping S_2 fabric that is parallel to the mylonitic fabric in the Abloviak shear zone. We infer from these data that D_1 in Tasiuyak gneiss was a protracted, high strain event involving a continuum of isoclinal fold and fabric formation and a high degree of fabric transposition. The evidence of imbrication of slivers of ultramafic rocks and Archaean gneisses within the Tasiuyak gneiss complex (Van Kranendonk & Wardle 1994), together with recumbent isoclinal folding and the inference of tectonic transport about sub-horizontal axes, suggests that S_1 may be a recrystallized shear fabric associated with thrust imbrication of Tasiuyak gneiss complex. However, to date, insufficient continuous stratigraphic markers have been found in Tasiuyak gneiss to map out discrete internal thrust packages satisfactorily. D_2 involved shortening perpendicular to the length of the orogen, principally by upright folding and extension parallel to sub-horizontal fold axes. F_2 folds tighten towards Abloviak shear zone, implying that transcurrent shearing was synchronous with or postdated D_2.
- Sinistral rotation of Proterozoic high strain fabrics into the KSZ can be seen on both the east and west sides of the shear zone (in the Saglek and Abloviak transects respectively), implying that KSZ is a ductile transcurrent shear zone in the sense of Ramsay & Graham (1970). Although quantitative estimates of the amount of transcurrent displacement on the shear zone are not possible in the absence of a discrete cross-cutting marker unit, correlation of units on both sides of the shear zone suggests that total transcurrent displacement was small.
- In contrast, no rotation of foliations occurs adjacent to ASZ, which developed parallel to the S_2 fabric in Tasiuyak gneiss. ASZ crosses the Tasiuyak gneiss complex obliquely in the area covered by Fig. 1 (being adjacent to the Southeast Rae craton in the Abloviak transect and the Nain craton in the Saglek transect), but the observation that the ASZ remains within the Tasiuyak gneiss unit throughout the >250 km length of Torngat Orogen shown in the figure is evidence that it, too, cannot have been the locus of major transcurrent displacements.
- Southeast Rae craton and its Lake Harbour Group cover are pervasively reworked, even distal to the ASZ. In contrast, penetrative reworking of Nain craton is restricted to the KSZ and eastern portions of the Ramah Group have not been strongly deformed.
- Estimates of peak P–T conditions in Komaktorvik and Abloviak shear zones are similar, at about 800°C/12 kb. Although the mafic lithologies in the KSZ are better suited as mineralogical monitors of post-peak reactions (e.g. Hbl–Pla replacing Grt–Cpx) than are the assemblages in Tasiuyak pelites in the ASZ, estimates of the retrograde P–T–t paths are similar in the two shear zones, suggesting that both underwent quasi-isothermal decompression with lowest recorded P–T conditions at about 5 kbar/500°C in metabasic rocks and 2 kb/500°C in metapelites (Fig. 4).
- East of the KSZ in the northeast of Torngat Orogen, there is a static Proterozoic metamorphic overprint on Nain gneisses and Proterozoic dykes. Metamorphic isograds reflecting this event are approximately perpendicular to the strike of the orogen, such that in a north-northwest trending transect, granulite-, amphibolite- and greenschist-facies rocks are progressively encountered east of KSZ (Wardle et al. 1992; Fig. 1). This distribution, together with available estimates of the depth of peak metamorphism from the granulite and greenschist facies rocks, implies that the present erosion surface affords an oblique view through the eastern margin of Torngat Orogen and its adjacent foreland, with much greater uplift occurring in the north of the orogen near Iselin Harbour (about c. 40 km) than further south between Nachvak and Saglek fiords (< 20 km). This interpretation is also compatible with the absence of the Ramah Group in the north and with the north-to-south metamorphic zonation in the (transported) Ramah Group further south (Fig. 5). Wardle et al. (1992) interpreted the presence of Proterozoic high grade rocks northeast of the KSZ to be a result of oblique sinistral ramping of the Nain craton within the collisional

orogen. Three other observations are also consistent with this interpretation: (a) absence of easterly vergence at the eastern end of the Abloviak transect may be explained by erosion to below the level of a fold-thrust belt (Fig. 5); (b) restriction of the Proterozoic DTG and charnockite suites to the northern, high grade part of the orogen implies that they were emplaced into the lower crust, so their absence in the southern part of the orogen may be more apparent than real, due to the level of erosion; and (c) it is therefore likely that advection of heat into the lower crust, due to emplacement of the DTG and charnockite suites, was the cause of the early static Proterozoic metamorphism.

• Apart from the static metamorphism, highest metamorphic grades (uppermost amphibolite to granulite facies) are in the axial region of Torngat Orogen. There is insufficient evidence at present to indicate if the widespread amphibolite facies signature in Southeast Rae gneisses is a retrograde imprint on an earlier Proterozoic granulite facies event, although this appears likely. On the basis of mineral assemblages in Tasiuyak gneiss, grade of metamorphism does not vary significantly along the length of the axial zone of the orogen. Mineral assemblages in Tasiuyak gneiss are also similar within and outside the ASZ, implying that transcurrent shearing was coeval with peak metamorphism, with peak metamorphic minerals defining the stretching lineations characteristic of the shear zone.

• On the basis of the gently east-dipping pattern of seismic reflections in Southeast Rae craton (Hall 1994), we interpret the steep fabrics characteristic of the ASZ and KSZ to shallow out at depth, as shown schematically in Fig. 2. From these observations, we infer that there is a root zone located under the Tasiuyak gneiss complex, and that the deep crustal extensions of the Nain and Southeast Rae cratons are in tectonic contact along a suture in this region. On account of its oblique orientation with respect to the bounding cratons, Abloviak shear zone probably does not correspond precisely with the location of this suture.

• Uplift and extrusion of the high grade core of the orogen over the bounding forelands outlasted peak metamorphism and transcurrent shearing, as indicated by hinterland-dipping brittle faults decorated with pseudotachylite that cut transcurrent fabrics, and it involved a distinct late thrusting event prominently displayed in the Ramah Group (Calon & Jamison 1994).

Origin of the Tasiuyak gneiss complex

The role of the Tasiuyak gneiss complex in the evolution of Torngat Orogen is clearly crucial, but its origin has prompted much speculation on

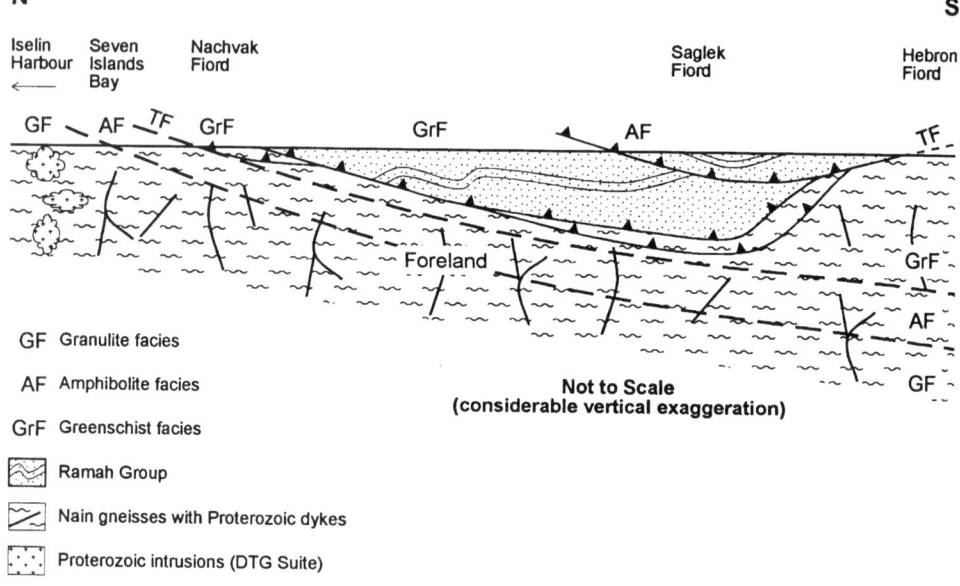

Fig. 5. Schematic longitudinal section along the western margin of the Nain craton, viewed in the direction of tectonic transport. The distribution of isograds for the static Proterozoic metamorphism in the Nain craton and in the transported Ramah Group are indicated, and imply that the Nain craton is tilted towards the south.

account of its relatively uniform composition, lack of continuous marker units, high state of strain, elevated metamorphic grade, extensive partial melting and magmatic injection, and its great longitudinal extent (>500 km) and thickness (up to 13 km, Feininger & Ermanovics 1994), which have all combined to make it a rather enigmatic unit. Additionally, early attempts at resolving its origin were hindered by the fact that the most accessible parts of the unit are within the ASZ, where original relationships are largely obscured. However, examination of the unit in areas of low strain leaves no doubt that it is sedimentary in origin, consisting principally of pelitic to quartzitic sediments with abundant thin rusty layers rich in graphite and sulphides. Pelitic portions of the unit have undergone extensive partial melting that has disrupted original layering, and in addition there are abundant subconcordant intrusions of leucogranite, probably derived from partial melting of the unit below the present level of erosion. Evidence of early isoclinal folding is sparse but widespread in the lower strain portions of the unit, and there is a pervasive strong amphibolite facies LS fabric that is parallel to the S_0 layering where the latter can be discerned. Due to the high state of strain it has not proven possible, even in areas of good exposure, to trace out marker units for more than a few decimetres, suggesting that the unit may be dissected by annealed faults associated with the early ductile isoclinal folding. Furthermore, Bertrand *et al.* (1993) reported the existence of isolated mafic to ultramafic bodies in the unit to the south of the area considered here, and more recently Van Kranendonk & Wardle (1994) have documented evidence of imbrication of Tasiuyak gneiss with ultramafic rocks, Archaean gneisses and Proterozoic intrusives of the charnockite suite. On the basis of this combined evidence, it is inferred that the early tectonic history of the Tasiuyak gneiss complex involved ductile folding and thrust imbrication of ocean floor and adjacent continental margin to account for the incorporation of ultramafic rocks and Archaean gneisses respectively.

The recent gravity study of Feininger & Ermanovics (1994) has shown that the Tasiuyak gneiss is wedge-shaped in cross-section in the Nutak area, attaining its maximum thickness of c. 13 km near the Nain craton, and thinning to a feather edge at its contact with the Southeast Rae craton.

Scott & Machado (1994) have recently reported U/Pb ages of detrital zircon from quartzite horizons in the Tasiuyak gneiss. Their results (2062 ± 2, 2042 ± 2, >2017 and 1940 ± 2 Ma) imply that the Tasiuyak gneiss was not derived from an Archaean source. This conclusion is supported by neodymium isotopic studies on pelitic portions of the unit that have yielded ε_{Nd} values (at 1.85 Ga) in the range −2.1 to −3.1 (Thériault & Ermanovics 1994) to −3.7 (Campbell 1994) that are distinctly more positive than values obtained for the platformal Ramah and Lake Harbour groups which were interpreted to have been largely derived from Archaean protoliths (Thériault & Ermanovics 1994). These data therefore suggest that much of the sedimentary protolith of Tasiuyak gneiss was derived from a juvenile source. The 1940 ± 2 Ma age for detrital zircon provides a minimum age for the sedimentation of the protolith of the Tasiuyak gneiss.

Since the Tasiuyak gneiss formed between the Nain and Southeast Rae cratons, but was not principally derived from them, we follow Scott & Machado (1994) in inferring that significant transport of sediment must have occurred parallel to the continental margins. Incorporating the evidence of an early thrust history discussed above, we suggest that the Tasiuyak gneiss complex formed as an accretionary complex, as proposed by Scott & Campbell (1993), first on ocean crust and later on continental crust. Such an origin could plausibly account for the remarkable thickness and along-strike continuity of the unit, while at the same time explaining the lack of coherent internal marker horizons. The thicker section of Tasiuyak gneiss against the Nain craton is compatible with the accretionary wedge developing against this cratonic margin.

Geochronology

Geochronological data for the main tectonic events are summarized below. Ages for each event appear to be approximately coeval throughout the orogen, suggesting a relatively simple history for the orogen as a whole.

1910–1864 Ma: intrusion of DTG and Charnockite suites

Scott & Machado (1993, 1994, 1995) have reported six intrusive ages, determined by the U/Pb method on zircon, for the full range of compositions of the DTG suite, that lie between 1910–1864 Ma with a central tendency at c. 1895–1885 Ma. These ages overlap with three U/Pb zircon dates obtained by the same authors for the proximal Killinek charnockite suite (range 1895–1886 Ma), implying the existence of a major intrusive event in the northern part of the orogen at this time (Scott & Machado 1994).

Nd isotopic studies show that the DTG suite exhibits a wide range in ε_{Nd} (back-calculated for the age of crystallization) from positive values of +2.69 and +1.45, just below those of contemporaneous depleted mantle, to negative values of −1.12 (Campbell 1994). Campbell interpreted these data to indicate that the DTG suite was derived from a mantle source with variable degrees of crustal contamination. ε_{Nd} for one sample from the charnockite suite is c. −6.71, overlapping with the range of values determined for Nain gneisses (Campbell 1994), and compatible with the hypothesis that the charnockites were derived, in whole or in part, by partial melting of Archaean crust.

Most intrusions of the DTG suite outside the KSZ carry igneous orthopyroxene, and this mineral can also be recognized locally as a relict phase in retrogressed bodies within KSZ. This implies that the DTG magmas were both hot and dry. We interpret the calc-alkaline compositions (Wardle et al. 1992) and ε_{Nd} signatures of the DTG suite to indicate a subduction origin for these magmas, and we infer that they caused substantial partial melting as they passed through the lower Nain crust, resulting in variable contamination of the DTG suite with crustally-derived Nd and the formation of the large volumes of the contemporaneous, crustally-derived magmas of the Killinek charnockite suite.

1870–1840 Ma: collision and crustal thickening

This stage of the orogenic process has not been dated directly, but the age range is interpreted from the following data of Bertrand et al. (1993) in the area around Nutak (Fig. 1) and of Scott & Machado (1994) from the north of the orogen. (i) Scott & Machado (1994) have reported a U/Pb monazite age of 1870 ± 3 Ma from the Tasiuyak gneiss in Burwell domain that is interpreted to date cooling from granulite facies conditions. (ii) Bertrand et al. (1993) have reported U/Pb zircon ages of 1858 ± 3 Ma (cores) and 1843.9 ± 1 Ma (overgrowths) and a U/Pb monazite age of 1857 ± 1.1 Ma from late tectonic, anatectic, garnet-bearing leucogranite in the Tasiuyak gneiss complex. (iii) Bertrand et al. (1993) dated a late tectonic, hypersthene-bearing anatectic vein in the Tasiuyak gneiss complex at 1856 ± 3.6 Ma. (iv) Scott & Machado (1994) dated the emplacement of an orthopyroxene-bearing tonalite into the Tasiuyak gneiss just south of Burwell domain at 1839 ± 2 Ma. (v) Bertrand et al. (1993) reported three U/Pb zircon ages between 1854–1849 Ma for pegmatites that cut the Proterozoic gneissosity in reworked Southeast Rae gneisses. From cross-cutting relations noted above, the 'late tectonic' leucogranites and orthopyroxene-bearing tonalite were emplaced post-D_1–pre-D_2 into the Tasiuyak gneiss, and so we infer that peak metamorphism, D_2 deformation, emplacement of minor 'charnockite' intrusions and anatexis in Tasiuyak gneiss occurred between about 1870 and 1840 Ma, the time span that we associate with the main collisional event in Torngat Orogen.

1845–1820 Ma: sinistral transcurrent motion on ASZ

Within the ASZ, the 1858 Ma leucogranites within the Tasiuyak gneiss show evidence of sinistral transcurrent shearing, and thus provide a maximum age limit this deformation. Bertrand et al. (1993) reported that the main transcurrent deformation in ASZ in the Nutak area is approximately dated by the 1844 ± 3.6, 1843.9 ± 1 and 1843.7 ± 1.6 Ma zircon and monazite age determinations from the Tasiuyak gneiss complex and adjacent reworked Southeast Rae gneisses. Scott & Machado (1994) reported that a syntectonic granitic vein, which exhibits intrusive relations into the Tasiuyak gneiss within the ASZ and also shows evidence of sinistral transcurrent deformation, has a U/Pb zircon crystallization age of 1824 ± 2 Ma. This date, therefore, provides a minimum age for shearing on ASZ. Bertrand et al. (1993) also reported evidence of continued zircon growth up to 1822 Ma in samples from reworked Southeast Rae gneisses, and we interpret these data to imply that sinistral transcurrent shearing occurred in the Tasiuyak gneiss and reworked Southeast Rae gneisses in the ASZ under an elevated temperature regime between about 1845–1820 Ma.

1798 Ma: flexural folding of the bend in ASZ

On the basis of quartz petrofabrics, Godin (1993) has shown that the L > S strain in the Tasiuyak gneisses in the bend in the Abloviak shear zone south of Burwell domain is unrelated to position on the bend, and predated formation of the amphibolite-facies, dextral Katherine River mylonite zone (Van Kranendonk et al. 1993b) located on the north side of the bend. The Katherine River mylonite zone is interpreted to have formed as a result of dextral strike-slip movement associated with minor flexural folding and tightening of a primary bend in the shear zone (Godin 1993; Van Kranendonk et al. 1993b). Zircon from a syntectonic granitic vein that crosscuts the mylonite zone, but is itself deformed, has yielded a crystallization age of 1798 ± 2 Ma (Scott & Machado 1994). This age is therefore considered to approximately date the time of formation of the dextral mylonites and the flexural tightening of the bend in the ASZ.

1790–1720 Ma: uplift and sinistral transcurrent motion on KSZ

Bertrand et al. (1993) reported U/Pb zircon ages of 1790.7 ± 1.3 and 1787.9 ± 1.1 Ma for an undeformed vein of leucogranite that cuts high grade mylonitic rocks of the Tasiuyak gneiss complex in the Nutak area, thus providing a minimum age for high grade deformation in that part of the complex, and they interpreted a monazite age of 1780 Ma from the same sample as the time of cooling through the closure temperature of Pb in monazite (c. 725°C). There is also considerable U/Pb evidence that KSZ was active under amphibolite-facies conditions after 1791 Ma, and remained hot until 1710 Ma. For instance, Scott & Machado (1995) reported U/Pb zircon ages of 1789 ± 1 and 1719 ± 2 Ma for two syntectonic granitic veins near Seven Islands Bay (Fig. 1) in this time interval, and Scott & Machado (1994) reported overlapping zircon and monazite ages of 1780 ± 2 Ma for a syntectonic granite pegmatite that they considered was emplaced during the amphibolite facies retrogression in KSZ in the Abloviak transect. In addition, Scott & Machado (1993, 1994) reported five U/Pb titanite ages from KSZ in the range 1775–1710 Ma which they interpreted to date the time of cooling through the closure temperature of Pb in titanite (c. 550–600°C), and therefore by extension also some stage of the amphibolite facies retrogression in KSZ. This age range has recently been extended by Scott & Machado (1995) to as young as 1632 ± 4 Ma, and since amphibolite-facies minerals (e.g. hornblende) define the main fabric in KSZ, these ages are also interpreted to date some stage of movement on the shear zone. The majority of the titanite ages overlap with $^{40}Ar/^{39}Ar$ cooling ages of 1790–1750 Ma determined on hornblendes (closure temperature of Ar in hornblende c. 500°C) from the KSZ in the Saglek transect (Mengel et al. 1991) and with a poorly resolved $^{40}Ar/^{39}Ar$ age of c. 1730–1740 Ma determined on hornblende from an 'enderbite' intrusive into reworked Southeast Rae gneisses by Bertrand et al. (1993) from the Nutak area. These ages appear to date the final cooling stages of the orogenic cycle, the wide spread of ages being due to sporadic uplift of hot rocks from the core of the orogen during the waning stages of contraction. The late reverse faults along which this uplift took place cut the transcurrent fabrics in the ASZ and KSZ and are decorated with pseudotachylite, indicating that they remained active into the brittle realm.

Tectonic model for Torngat Orogen

The tectonic model proposed for the development of Torngat Orogen is based on the conceptual numerical modelling experiments of Willett et al. (1993) for doubly vergent orogens. Their modelling was based on sand box experiments (Malavieille 1984) in which analogues of the crust and sub-continental mantle delaminate at a singularity, the sub-continental mantle of the down-going plate being subducted under the opposing plate, whereas the detached crust is cycled through the orogen (Fig. 6).

Five additional assumptions underlie our proposed tectonic model. These are:

- that convergence was oblique, not orthogonal as in the Willett et al. (1993) model;
- that the Tasiuyak gneiss was formed as an accretionary thrust complex, as argued above;
- that double crustal thicknesses were attained in the centre of Torngat Orogen, as suggested by Mengel (1988) and Mengel & Rivers (1991) on the basis of P–T–t paths determined from Tasiuyak gneiss and from reworked Nain gneisses;
- that subduction was towards the east, on the basis of the emplacement of the DTG and Killinek charnockite suites into the Nain craton;
- that the Tasiuyak gneiss complex tectonically overlies the western margin of the Nain craton, as

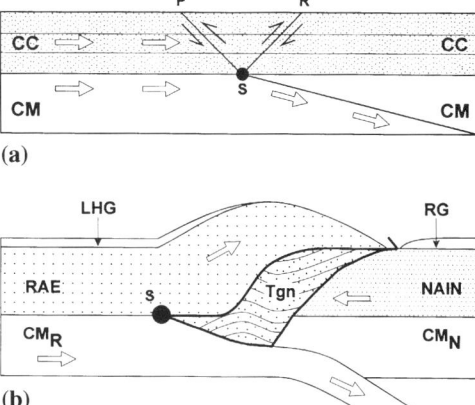

Fig. 6. (a). Schematic diagram of process modelled by Willett et al. (1993). CC, continental crust; CM, continental mantle; S, singularity; P, pro-wedge; R, retro-wedge. Continental crust and continental mantle of left-hand plate detach at singularity, mantle is subducted, crust is cycled through the orogen.
(b). Model applied to Torngat Orogen, showing relative positions and motions of Nain and Southeast Rae plates with their platformal Proterozoic supracrustal sequences (Ramah and Lake Harbour groups respectively), and the accretionary Tasiuyak gneiss complex. S, singularity; CM_R and CM_N, continental mantle beneath Southeast Rae and Nain cratons.

indicated by the presence of slivers of Tasiuyak gneiss and correlative metasedimentary rocks in eastern and southern Burwell domain (Godin 1993).

The model is presented as a series of panels in Fig. 7. Following Willett *et al.* (1993), the singularity is the point at which delamination of the down-going plate occurs. It is inferred that between *c.* 1910 and 1870 Ma, subduction of the subcontinental mantle of the Southeast Rae plate under the Nain plate gave rise to mantle- and crustally-derived magmatism of the DTG and charnockite suites, and to accretion of the Tasiuyak gneiss complex as a thrust wedge (early D_1 deformation) above the subduction zone at the edge of the Nain plate. Oblique collision of the Nain and Southeast Rae plates *(c.* 1870–1840 Ma) resulted in doubling of crustal thickness in the centre of the orogen as the Southeast Rae plate overrode the Nain plate. The Tasiuyak gneiss complex was also overridden by the Southeast Rae plate and accreted to its base (basal accretion in the thrust wedge, Rivers *et al.* 1993*b*; van Gool & Cawood 1994), causing additional thrust imbrication (late D_1 deformation) of the Tasiuyak gneiss complex. Coeval with and following crustal thickening, the metamorphic structure of the thickened orogenic crust was established, with advection of heat into the lower crust due to emplacement of leucogranite and small charnockite bodies into the Tasiuyak gneiss contributing to the elevated maximum temperatures attained in the core of the orogen. Continued shortening of the doubly thickened crust caused the formation of a large-scale (F_2) structural fan corresponding to the pro- and retro-wedges in the model of Willett *et al.* (1993). Axial planar fabrics (S_2) associated with the structural fan overprint the D_1 fabrics, and commonly display L > S shape fabrics, indicating that an important element of extension parallel to the length of the orogen occurred at this time.

Between 1845 and 1820 Ma, continued oblique convergence resulted in establishment of the ductile sinistral Abloviak shear zone system within the doubly thickened crust. Minor flexural folding of the pre-existing bend in the ASZ took place at about 1798 Ma. KSZ became active at about 1790 Ma, probably in response to internal lateral ramping and oblique uplift of Nain crust in the northeast of the orogen, with movement on KSZ giving rise to the tilted crustal section in this part of the orogen. Continued crustal shortening caused renewed thrusting and uplift of high grade rocks from the core of the orogen over the bounding forelands.

Such renewed thrusting and concomitant uplift, exhumation and rapid cooling are indicated by the quasi-isothermal decompression paths of high grade rocks in ASZ and KSZ, and by the late faults decorated with pseudotachylite that cut the transcurrent shear fabrics in ASZ and KSZ. Uplift of the high grade core of the orogen is a predicted consequence of collision in the Willett *et al.* model, and we interpret the age data to indicate that collision continued at least locally until 1790 Ma. Between 1790 and 1730 Ma, uplift probably became increasingly driven by isostasy rather than plate motions, and the exhumation process resulted in cooling of most of the rocks presently exposed at the surface through the closure temperatures of radiogenic Pb in monazite and titanite, Ar in hornblende, and of Fe, Mg, Al and Ca in minerals in the geothermobarometric assemblages. The long (*c.* 60 Ma, possibly extending as long as 160 Ma) duration of the uplift/exhumation event was probably a result of the sporadic uplift of hot rocks from the interior of the orogen along hinterland-dipping reverse faults, such as those cutting the mylonitic fabrics in ASZ and KSZ that are decorated with pseudotachylite. Second generation structures in the Ramah fold-thrust belt, including its macroscopic easterly vergence are inferred to have developed at this time.

Outstanding problems

Outstanding problems that require particular attention with respect to this model are: (i) the timing of formation of early structures (D_1) in the Ramah Group; (ii) the timing of thrusting and folding in the Lake Harbour Group, which is presently poorly constrained by geological evidence; (iii) the question of whether the KSZ was active prior to 1790 Ma; and (iv) the early structural history of the Tasiuyak gneiss complex, including interpretation of the kinematic significance of the orogen-parallel L_1 stretching lineations. In addition, understanding of the causes and timing of the ductile Proterozoic reworking of the Southeast Rae gneisses are currently inadequate; it is not clear whether the widespread ductile reworking was related to the Southeast Rae/Nain collision that resulted in Torngat Orogen, or was an effect of the slightly younger Southeast Rae/Superior collision that gave rise to the New Québec Orogen, or some combination of both.

Conclusions

Many aspects of the proposed model fit well with the published tectonic evolution scheme of Van Kranendonk & Ermanovics (1990), Van Kranendonk *et al.* (1993*b*) and Bertrand *et al.* (1993). However, the model differs significantly with respect to the origin of the Tasiuyak gneiss complex

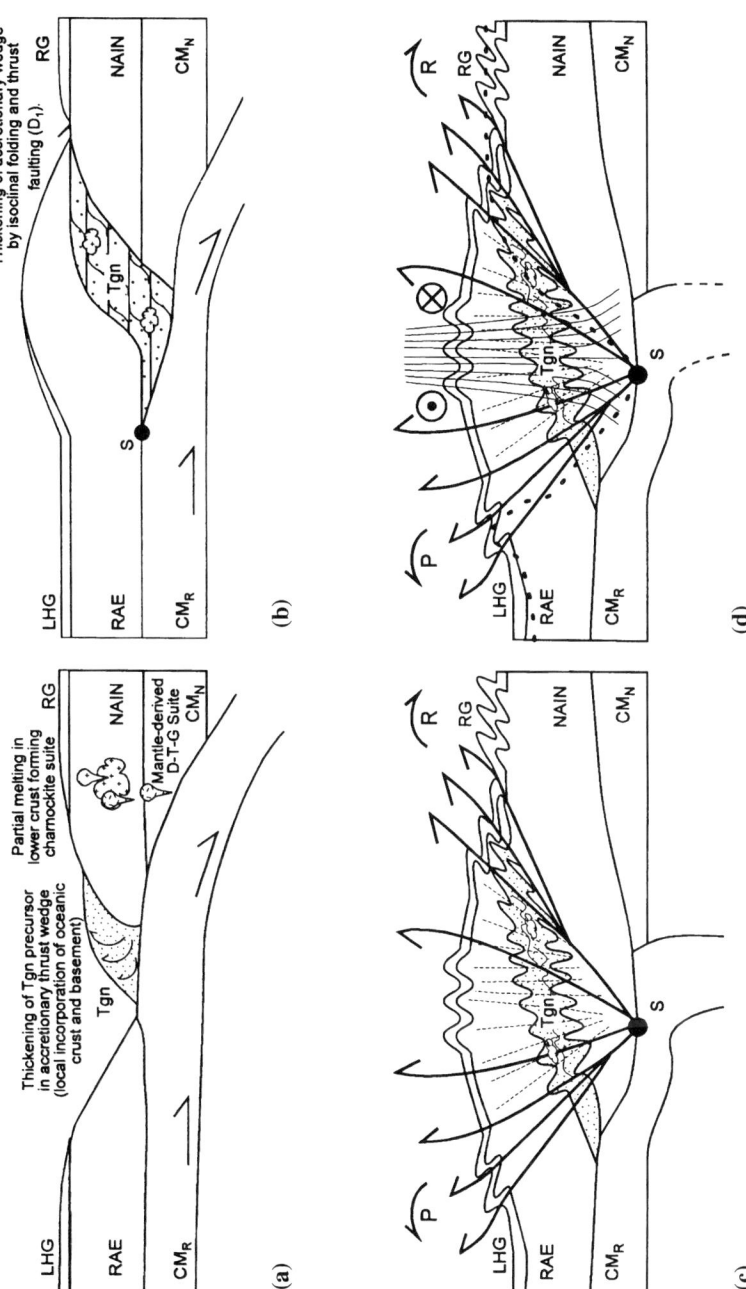

Fig. 7. Schematic evolution of Torngat Orogen utilizing the Willet et al. (1993) tectonic model for doubly vergent compressional orogens. (**a**) Precollisional configuration of plates, with subduction beneath the Nain plate (1910–1869 Ma) causing formation of the mantle-derived (and crustally contaminated) DTG suite, and indirectly causing melting of Nain crust to give rise to the charnockite suite. Tasiuyak gneiss complex accumulates as an accretionary prism above subduction zone. (**b**) Early configuration of collisional orogen showing attainment of double crustal thickness (D_1), with Tasiuyak gneiss complex accreted to base of Rae plate before being sandwiched between Nain and Southeast Rae plates. Partial melting in Tasiuyak gneiss gives rise to leucogranites (1870–1860 Ma). Possible early imbrication of Lake Harbour Group with Southeast Rae basement and accretion to base of Nain plate. Peak metamorphism 11 kb/850°C. S, singularity. (**c**) Collision continues, with formation of pro-wedge (in west), and retro-wedge (in east) giving rise to marginal fold belts with opposing vergence (D_2); establishment of peak metamorphic isotherms. (**d**) Formation of Abloviak and Komaktorvic shear zones and initiation of sinistral transcurrent motion (1845–1820 Ma) on ASZ parallel to S_2 in doubly thickened crust. Continued uplift and exhumation of hot core of orogen up pro- and retro-wedges (giving rise to measured P–T–t paths), and emplacement of hot rocks from core of orogen over cold rocks in foreland (giving rise to metamorphic fold-thrust belt in Ramah and Lake Harbour groups); initiation of KSZ, and continued uplift and exhumation into brittle realm along late faults cutting mylonitic fabric in ASZ and KSZ (1790–1730 Ma); final closure of geochronologic and geothermobarometric systems.

and the direction of subduction prior to collision. Reasons for preferring an east-dipping subduction zone model are discussed above and we believe they are well-founded. The implication that the Abloviak and Komaktorvik shear zones are late features, and perhaps not as fundamental as the preceding crustal thickening episode, has been made previously by both the current authors and other workers in the area. Uplift of the axial zone of the orogen along reverse faults is seen as a fundamental result of the collision, driven by the opposing displacement vectors of the converging plates. Late orogenic collapse along extensional faults, for which there is local evidence across the orogen, probably occurred at this time.

Finally, we note that subduction in this model is a fundamentally asymmetrical process, yet the crustal products, fold-thrust belts with opposing vergence at either side of the orogen, are symmetrical and by inference broadly coeval. This is a result of the bouyancy of continental crust.

Clearly many of these results build on the work of others, particularly the mapping project led by Martin Van Kranendonk and Dick Wardle. We gratefully acknowledge this debt, and also their hospitality and willingness to help with our logistics in the field. The insightful numerical modelling of Chris Beaumont and co-workers continues to stimulate novel, and we trust more realistic, interpretations of the multi-facetted datasets required to analyse ancient orogens. Rod Churchill drafted the figures and we are grateful to Simon Hanmer for a very inciteful review. TR acknowledges receipt of NSERC and *LITHOPROBE* grants; FM acknowledges fellowships from the Danish National Research Council; DJS and NG acknowledge support from *LITHOPROBE*; LMC acknowledges receipt of a Fulbright scholarship. This is *LITHOPROBE* contribution No. 677.

References

BERTRAND, J-M., RODDICK, J. C., VAN KRANENDONK, M. J. & ERMANOVICS, I. 1993. U–Pb geochronology of deformation and metamorphism across a central transect of the Early Proterozoic Torngat Orogen, North River map area, Labrador. *Canadian Journal of Earth Sciences*, **30**, 1470–1489.

BRIDGWATER, D. & SCHIØTTE, L. 1991. The Archean gneiss complex of northern Labrador. A review of current results, ideas and problems. *Bulletin of the Geological Society of Denmark*, **39**, 153–166.

CALON, T. & JAMISON, W. R. 1993. Structural evolution of the Eastern Borderland of the Torngat Orogen, Lake Kiki transect, Saglek Fiord area, northern Labrador. *In*: WARDLE, R. J. & HALL, J. (eds) *Eastern Canadian Shield Onshore–Offshore Transect (ECSOOT), Report of Transect Meeting*. Université du Québec à Montréal, Lithoprobe Report No. **32**, 90–112.

—— & —— 1994. Structural evolution of the Eastern Borderland of the Torngat orogen, Kiki Lake transect, Saglek Fiord area, northern Labrador. *In*: WARDLE, R. J. & HALL, J. (eds) *Eastern Canadian Shield Onshore–Offshore Transect (ECSOOT), Report of Transect Meeting*. Université du Québec à Montréal, Lithoprobe Report No. **36**, 89–99.

CAMPBELL, L. 1994. Preliminary results from a Sm–Nd isotopic study of Proterozoic crustal formation and accretion in the Torngat orogen, northern Labrador. *In*: WARDLE, R. J. & HALL, J. (eds) *Eastern Canadian Shield Onshore–Offshore Transect (ECSOOT), Report of Transect Meeting*. Université du Québec à Montréal, Lithoprobe Report No. **36**, 100–107.

FEININGER, T. & ERMANOVICS, I. 1994. Geophysical interpretation of the Torngat Orogen along the North River–Nutak transect, Labador. *Canadian Journal of Earth Sciences*, **31**, 722–727.

GODIN, L. 1993. *Structural analysis of the folded Abloviak shear zone, Paleoproterozoic Torngat Orogen, northeastern Québec*. MSc thesis, Université du Québec à Montréal, Montréal, QC.

GOULET, N. 1993. Structural pattern of the northern Torngat Orogen. *In*: WARDLE, R. J. & HALL, J. (eds) *Eastern Canadian Shield Onshore–Offshore Transect (ECSOOT), Report of Transect Meeting*. Université du Québec à Montréal, Lithoprobe Report No. **32**, 52–53.

—— & CIESIELSKI, A. 1990. The Abloviak shear zone at the NW Torngat Orogen, eastern Ungava Bay, Québec. *Geoscience Canada*, **17**, 269–272.

HALL, J. 1994. Interpretation of deep seismic reflection profiles with application to ECSOOT. *In*: WARDLE, R. J. & HALL, J. (eds) *Eastern Canadian Shield Onshore–Offshore Transect (ECSOOT), Report of Transect Meeting*. Université du Québec à Montréal, Lithoprobe Report No. **36**, 6–13.

HOFFMAN, P. F. 1990. Dynamics of the tectonic assembly of northeast Laurentia in geon 18 (1.9–1.8 Ga). *Geoscience Canada*, **17**, 222–226.

KORSTGÅRD, J., RYAN, B. & WARDLE, R. J. 1987. The boundary between Proterozoic and Archean rocks in central West Greenland and northern Labrador. *In*: PARK, R. G. & TARNEY, J. (eds) *Evolution of the Lewisian and Comparable High Grade Terrains*. Geological Society, London, Special Publication, **27**, 247–259.

KRETZ, R. 1983. Symbols for rock-forming minerals. *American Mineralogist*, **68**, 277–279.

MALAVIEILLE, J. 1984. Modélisation expérimentale des chevauchements imbriqués: Application aux chaînes des montagnes. *Bulletin du Société Géologique de France*, **26**, 129–138.

MENGEL, F. C. 1988. *Thermotectonic evolution of the Proterozoic–Archaean boundary in the Saglek area, northern Labrador*. PhD thesis, Memorial University of Newfoundland, St. John's, NF.

—— & RIVERS, T. 1991. Decompression reactions and

P–T conditions in high-grade rocks northern Labrador: P–T–t paths from individual samples and implications for Early Proterozoic tectonic evolution. *Journal of Petrology*, **32**, 139–167.

—— & —— 1993. 3-D architecture and thermal structure of deep crustal-scale transcurrent shear zones – Report 2: Preliminary results from the Komaktorvik shear zone north of 60°N. *In*: WARDLE, R. J. & HALL, J. (eds) *Eastern Canadian Shield Onshore–Offshore Transect (ECSOOT), Report of Transect Meeting*. Université du Québec à Montréal, Report No. **32**, 64–73.

—— & —— 1994a. 3-D architecture and thermal structure of a crustal-scale transcurrent shear zone – report 4: P–T results from the northern Torngat transect. *In*: WARDLE, R. J. & HALL, J. (eds) *Eastern Canadian Shield Onshore–Offshore Transect (ECSOOT), Report of Transect Meeting*. Université du Québec à Montréal, Lithoprobe Report No. **36**, 163–170.

—— & —— 1994b. Development of metamorphic assemblages in the Paleoproterozoic Ramah Group pelites, Saglek area, northern Labrador; mineral reactions, P–T conditions and influences of bulk composition. *Canadian Mineralogist*, **32**, 781–801.

——, —— & REYNOLDS, P. H. 1991. Lithotectonic elements and tectonic evolution of Torngat Orogen, Saglek Fiord, northern Labrador. *Canadian Journal of Earth Sciences*, **28**, 1407–1423.

MORGAN, W. C., 1978. *Bears Gut–Saglek Fiord*. Geological Survey of Canada, map 1478A.

RAMSAY, J. G. & GRAHAM, R. H. 1970. Strain variation in shear belts. *Canadian Journal of Earth Sciences*, **7**, 619–622.

RIVERS, T. & MENGEL, F. C. 1994. A cross-section of Abloviak shear zone at Saglek Fiord, and a preliminary tectonic model for Torngat Orogen. *In*: WARDLE, R. J. & HALL, J. (eds) *Eastern Canadian Shield Onshore–Offshore Transect (ECSOOT), Report of Transect Meeting*. Université du Québec à Montréal, Lithoprobe Report No. **36**, 171–184.

——, —— & GOULET, N. 1993a. A cross-section of the Abloviak shear zone at Abloviak Fiord – a preliminary report. *In*: WARDLE, R. J. & HALL, J. (eds), *Eastern Canadian Shield Onshore–Offshore Transect (ECSOOT), Report of Transect Meeting*. Université du Québec à Montréal, Lithoprobe Report No. **32**, 54–63.

——, VAN GOOL, J. A. M. & CONNELLY, J. N. 1993b. Contrasting tectonic styles in the northern Grenville Province: Implications for the dynamics of orogenic fronts. *Geology*, **21**, 1127–1130.

RYAN, B. 1990. Basement-cover relationships and metamorphic patterns in the foreland of Torngat Orogen in the Saglek-Hebron area, Labrador. *Geoscience Canada*, **17**, 276–279.

SCOTT, D. J. & CAMPBELL, L. 1993. Evolution of Paleoproterozoic Torngat Orogen, Labrador, Canada: recent advances using U–Pb geochronology and Nd isotope systematics. *Geological Society of America Meeting, Abstracts with Program*, **25**, A-237.

—— & MACHADO, N. 1993. U–Pb geochronology of the northern Torngat Orogen, Labrador: Report of progress to date. *In*: WARDLE, R. J. & HALL, J. (eds) *Eastern Canadian Shield Onshore–Offshore Transect (ECSOOT), Report of Transect Meeting*. Université du Québec à Montréal, Lithoprobe Report No. **32**, 32–41.

—— & —— 1994. U–Pb geochronology of the northern Torngat Orogen: results from work in 1993. *In*: WARDLE, R. J. & HALL, J. (eds) *Eastern Canadian Shield Onshore–Offshore Transect (ECSOOT), Report of Transect Meeting*. Université du Québec à Montréal, Lithoprobe Report No. **36**, 141–155.

—— & —— 1995. U–Pb geochronology of the northern Torngat Orogen, Labrador, Canada: A record of Paleoproterozoic magmatism and deformation. *Precambrian Research*, **70**, 169–190.

TAYLOR, F. C. 1979. *Reconnaissance geology of a part of the Precambrian Shield, northeastern Québec, northern Labrador and Northwest Territories*. Geological Survey of Canada, Memoir **393**.

THÉRIAULT, R. J. & ERMANOVICS, I. 1994. Nd isotopic transect across the Torngat orogen, North River–Nutak area. *In*: WARDLE, R. J. & HALL, J. (eds) *Eastern Canadian Shield Onshore–Offshore Transect (ECSOOT), Report of Transect Meeting*. Université du Québec à Montréal, Lithoprobe Report No. **36**, 119-125.

VAN GOOL, J. A. M. & CAWOOD, P. A. 1994. Frontal vs. basal accretion and contrasting particle paths in metamorphic thrust belts. *Geology*, **22**, 51–54.

VAN KRANENDONK, M. J. & ERMANOVICS, I. 1990. Structural evolution of the Hudsonian Torngat Orogen in the North River map area, Labrador: evidence for east–west transpressive collision of Nain and Rae continental blocks. *Geoscience Canada*, **17**, 283–288.

—— & WARDLE, R. J. 1994. Geological synthesis and musings on possible subduction-accretion models in the formation of northern Torngat Orogen. *In*: WARDLE, R. J. & HALL, J. (eds) Eastern Canadian Shield Onshore–Offshore Transect (ECSOOT), Report of Transect Meeting. Université du Québec à Montréal, Lithoprobe Report No. **36**, 62–80.

——, ST-ONGE, M. R. & HENDERSON, J. R. 1993a. Paleoproterozoic tectonic assembly of Northeast Laurentia through multiple indentations. *Precambrian Research*, **63**, 325–347.

——, WARDLE, R. J., MENGEL, F. C., SCOTT, D. J., BRIDGWATER, D., GODIN, L. & CAMPBELL, L. M. 1993b. Geology and tectonic evolution of the northernmost part of Torngat Orogen, Labrador, Québec and Northwest Territories. *In*: WARDLE, R. J. & HALL, J. (eds) *Eastern Canadian Shield Onshore–Offshore Transect (ECSOOT), Report of Transect Meeting*. Université du Québec à Montréal, Lithoprobe Report **32**, 6–20.

WARDLE, R. J. 1983. Nain–Churchill Province cross-section, Nachvak Fiord, northern Labrador. *In*: *Current Research*. Newfoundland Department of Mines and Energy, Mineral Development Division, Report **83-1,** 68–90.

—— 1984. Nain–Churchill cross-section: Rivière Beaudancourt–Nachvak Lake. *In*: *Current Research*. Newfoundland Department of Mines and Energy, Mineral Development Division, Report **84-1,** 1–11.

——, VAN KRANENDONK, M. J., MENGEL, F. & SCOTT, D. 1992. Geological mapping in the Torngat Orogen: Preliminary results. *In*: *Current Research*. Newfoundland Department of Mines and Energy, Geological Survey Branch, Report **92-1**, 413–429.

——, ——, ——, ——, SCHWARZ, S. & RYAN, B. 1993. Geological mapping in the Torngat orogen, northernmost Labrador: Report 2. *In*: *Current Research*. Newfoundland Department of Mines and Energy, Geological Survey Branch, Report **93-1**, 77–89.

WILLETT, S., BEAUMONT, C. & FULLSACK, P. 1993. Mechanical model for the tectonics of doubly vergent compressional orogens. *Geology*, **21**, 371–374.

The Palaeoproterozoic Southeastern Churchill Province of Labrador–Quebec, Canada: orogenic development as a consequence of oblique collision and indentation

R. J. WARDLE[1] & M. J. VAN KRANENDONK[2]

[1] *Geological Survey Branch, Newfoundland Department of Natural Resources, PO 8700, St John's, NF, A1B 4J6, Canada*
[2] *Continental Geoscience Division, Geological Survey of Canada, 601 Booth Street, Ottawa, ON, K1A 0E8, Canada (Present address: Department of Geology, University of Newcastle, Newcastle, NSW, 2308, Australia)*

Abstract: Palaeoproterozoic orogenic development in the northeastern Canadian Shield was controlled by the successive, oblique collision of the Archaean Nain (North Atlantic) and Superior cratons with a southwards projecting promontory of the Archaean Rae Province (part of the northern Churchill Province hinterland). By this process the Rae Province became sutured to the Nain craton by the Torngat Orogen and to the Superior craton by the New Quebec Orogen, resulting in the formation of a 400 km wide orogenic belt known as the Southeastern Churchill Province (SECP). Initial rifting at 2.45–2.1 Ga, and early (arc?) magmatism and deformation at 2.3–1.9 Ga, were restricted to the Rae Province. They were followed by arc magmatism in the Torngat Orogen at *c.* 1.91–1.88 Ga and Rae\Nain collision 1.87–1.86 Ga, which resulted in the formation of an orogen with east- and west-verging structures. Arc magmatism in the New Quebec Orogen commenced at *c.* 1.845 Ga and was succeeded by Rae\Superior collision and widespread deformation across the SECP at *c.* 1.83 Ga. Deformation at this time was dominated by west-vergent thrusting in the New Quebec Orogen and Rae Province, and by renewed east-vergent thrusting in Torngat Orogen. Deformation was then progressively transferred to major sinistral (1.845–1.82 Ga in the eastern SECP) and dextral (1.83–1.80 Ga? in the western SECP) shear systems that accommodated continued northwards motion of the Nain and Superior cratons relative to the Rae Province. Juvenile crust expelled by thrusting effectively doubled crustal thickness in parts of the Rae Province and the northwestern edge of the Nain craton. Late stage development (1.8–1.71 Ga), was restricted to the margins of the SECP, where deformation was associated with renewed outward-directed overthrusting and transcurrent shear in conjunction with uplift of the orogenic core.

The Churchill Province of Canada represents a broad belt of Palaeoproterozoic and reworked Archaean orogenic crust extending from western Canada, through the Arctic Islands into the Nagssugtoqidian mobile belt of northern Greenland. In northeastern Canada (Fig. 1), a splay of this belt, known as the Southeastern Churchill Province (SECP; Wardle *et al.* 1990a) projects southwards to separate the Archaean cratons of the Nain (part of North Atlantic Craton) and Superior provinces (Fig. 1).

Tectonic models for the origin of the Churchill–Nagssugtoqidian orogens have long been based on the concept of indentation tectonics. Watterson (1978) was the first to apply this concept to the North Atlantic Craton (NAC), arguing that it had indented northwards into a more ductile Churchill/Nagssugtoqidian hinterland in a manner analogous to the Tertiary collision of India with Asia. Evidence for indentation was provided by symmetrical sinistral and dextral shear belts on the northeastern and northwestern sides of the NAC. Significantly, the model predicted that the western margin of the Nain Province in Labrador should be dominated by sinistral shear, although corroboratory field evidence was lacking at the time. Gibb (1983), in a somewhat similar model, proposed that Palaeoproterozoic deformation in the Churchill Province of northern Canada resulted from indentation of the rigid Superior Province into a more ductile Churchill Province softened by northwest-dipping subduction. Hoffman's (1990b) regional synthesis of northeast Laurentia drew further on the southeast Asia analogy and modelled the evolution of the various components of the northern Churchill Province as the result of successive indentation of the Superior and then NAC cratons into a Churchill Province hinterland (Rae Province). Indentation

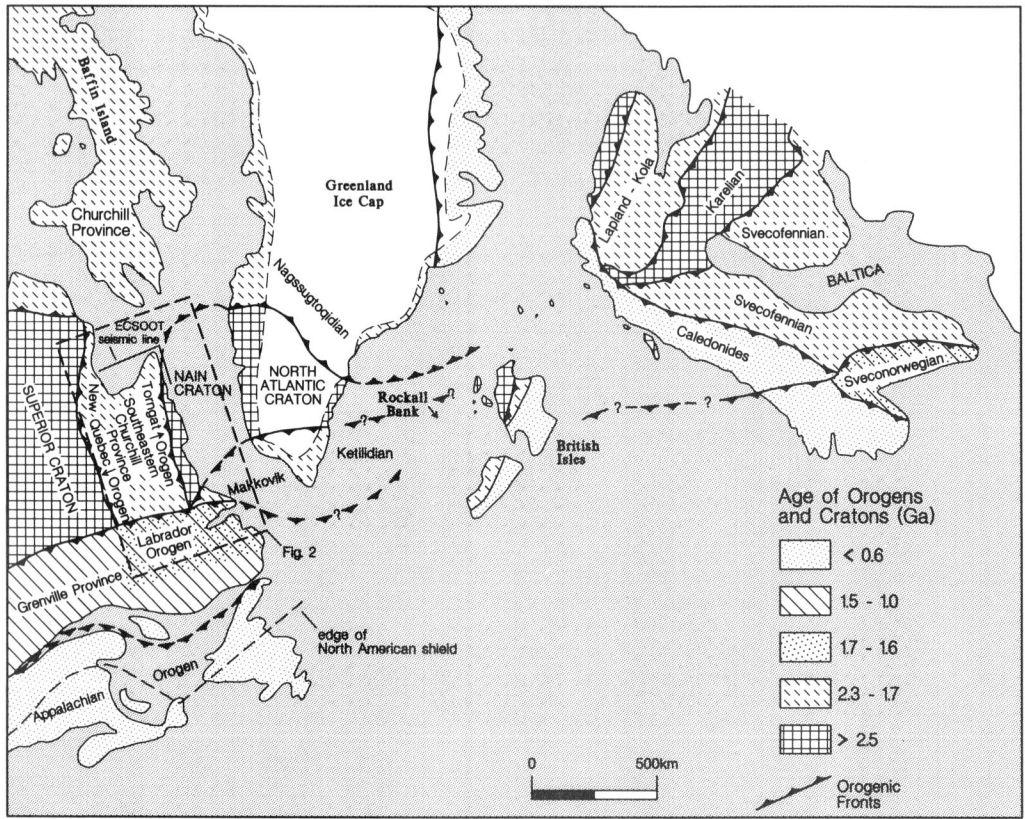

Fig. 1. Proterozoic orogenic belts of the North Atlantic region in a 600 Ma reconstruction (after Gower *et al.* 1990). The rectangle represents the area shown in Fig. 2.

was inferred to have forced the escape and extrusion of Rae Province crust into the gap between the two indenting cratons, with escape being accommodated along the major transcurrent shear systems that dominate the SECP. This model was substantially modified by Van Kranendonk *et al.* (1993a) who, with the aid of a much-improved U–Pb geochronological database, were able to test some of Hoffman's (1990b) predictions and more precisely define the sequence of indentations leading to the development of northeast Laurentia. A consequence of both Hoffman's (1990b) and Van Kranendonk *et al.*'s (1993a) models is that the SECP is assumed to represent crust that was trapped between two obliquely converging cratons and deformed in a transpressive environment. This paper seeks to describe the pre- and post-collisional history of the SECP in the light of recent mapping and U–Pb geochronology; with some additional insight derived from recent deep crustal seismic studies. All ages and age ranges are based upon U–Pb dating unless otherwise stated.

The Southeastern Churchill Province

The SECP has been divided (cf. Wardle *et al.* 1990a, b; Hoffman 1990a) into three principal components (Fig. 2). In the west, the New Quebec Orogen, which is a foreland fold-and-thrust belt composed of Palaeoproterozoic, predominantly low- to medium-grade metasedimentary and metavolcanic rocks, separates the Archaean Superior Province craton from a dominantly internal Archaean terrane thought (Hoffman 1990a) to form a southeast extension of the Rae Province (Fig. 2). In the east this same segment of the Rae Province is separated from the Nain Province by the Palaeoproterozoic Torngat Orogen (Van Kranendonk & Ermanovics, 1990).

New Quebec Orogen

The New Quebec Orogen is composed of a number of thrust-bounded sedimentary and volcanic packages, most of which have been previously described under the terms Labrador Trough or Labrador Geosyncline (e.g. Dimroth 1972) and most of which contain stratigraphic units that can be correlated between adjacent zones. Early rift-related activity is seen in the form of the c. 2.1 Ga Payne River dykes (Fahrig & West 1986). These do not cut, and are therefore probably older than, the metasedimentary rocks of the New Quebec Orogen. The western parts of the orogen are formed by the predominantly sedimentary rocks of the autochthonous **Chioak zone** and the parautochthonous **Melezes–Schefferville zone** (Clark & Avramtchev 1990; Fig. 2), each of which comprises two sedimentary/volcanic cycles. The lower cycle (cycle 1) comprises a sequence of arkose, shale and dolomite that rests unconformably upon the Superior Province basement and which represents the transition from a terrestrial to shallow-shelf environment during initial rifting of the Superior craton. Cycle 1 is overlain by a transgressive cycle 2 sequence of quartzite, iron formation, alkalic lavas (dated at c. 1879 Ma; Birkett et al., in Wardle et al. 1990b) turbiditic shale-sandstone and arkose. This lies disconformably upon cycle 1 rocks and oversteps them to rest unconformably upon Archaean gneisses of the Superior Province in the west.

This predominantly sedimentary package is overthrust in the east by the parautochthonous **Baby–Howse zone**. The zone is floored by cycle 1 arkose, shale and dolomite, identical in most respects to the equivalent units seen in the Melezes–Schefferville zone, interstratified with basalts and intruded by gabbro sills. One of the sills has been dated at 2169 ± 2 Ma (Rohon et al. 1993) and a thin rhyolite at 2142 +4/-2 Ma (Krogh & Dressler, in Clark 1984). These intrusions are associated with rocks near the base of cycle 1 (Seward Subgroup) and are interpreted to date its initiation. They are overlain, apparently conformably, by cycle 2 ironstone and a thick sequence of turbiditic sandstone and pillow basalt that is intruded by a gabbroic and ultramafic sill swarm locally dated at 1883–1874 Ma (Machado 1990; Birkett et al. cited in Wardle et al. 1990b). The basalts and sills of both cycles 1 and 2 are predominantly low-K tholeiites transitional between MORB and continental tholeiite in their composition (Boone & Hynes 1990; Skulski et al. 1993). Nd isotopic results indicate derivation of the volcanic rocks from a depleted mantle source whereas the sills contain evidence for what is either a slight crustal influence or derivation from a heterogeneous mantle (Rohon et al. 1993; Skulski et al. 1993). The significance of the voluminous basaltic rocks in both cycles 1 and 2 of the Baby–Howse zone has been the subject of different interpretations. Wardle & Bailey (1981) and Skulski et al. (1993) suggested an origin as transitional continent–oceanic crust formed by intense sill injection into a sediment-dominated rift margin in a manner analogous to models proposed for development of modern, oceanic pull-apart basins such as the Gulf of California (Einsele 1986). True ophiolite, however has not been recognized. A two-stage model for such development has been proposed by Skulski et al. (1993); an early, c. 2.17–2.14 Ga (cycle 1) stage, related to initial rifting and development of a Superior continent passive margin, and a later 1.88–1.87 Ga (cycle 2) stage, characterized by magmatism in pull-apart intra-oceanic basins developed as a result of dextral transtension along the Superior margin. This differs from an earlier model proposed by Hoffman (1987, 1990a) who envisaged the sedimentary and volcanic rocks of the entire cycle 2 sequence as the products of a westerly migrating foredeep produced in response to encroachment of hinterland thrust sheets from the east. The western, predominantly sedimentary, part of cycle 2 which records deepening to basinal conditions and then shallowing in concert with a westerly overstep over cycle 1 is compatible with such a model. A formidable drawback, however, is posed by the voluminous volcanic-sill swarm sequences in the eastern part of cycle 2; such rocks are generally atypical of foredeep environments. Hoffman (1990b) postulated that these might have been produced in a supra-subduction zone environment. However, this requires placing the sill swarms on the eastern side of such a subduction zone, an interpretation that is at odds with the clear stratigraphic ties between the host rocks to the sills and the Superior margin. The foredeep model may still be valid, however, if it is restricted to the turbidites and terrestrial arkoses which form the top of cycle 2. These rocks may be younger than the c. 1.88–1.87 Ga eastern sill swarms. If this is the case, the foredeep should be associated with a change in provenance from a western shield source during early rifting, to an eastern juvenile source as arc assemblages were thrust onto the Superior margin. This prediction can be tested through sedimentological and detrital zircon studies.

To the east of the Baby–Howse zone lies the **Rachel zone** (Moorhead & Hynes 1990) which is dominated by an amphibolite-facies volcano-sedimentary package interpreted to be contiguous with the **Laporte terrane** of Wardle et al. (1990b) to the south. Both sequences are dominated by biotite schist and gneiss, apparently derived from a turbiditic sequence, together with metabasalt and gabbro intrusions. The Rachel zone is further characterized by several complex domes and refolded nappes of Archaean basement dated by Machado et al. (1989) at 2883–2868 Ma. The ages

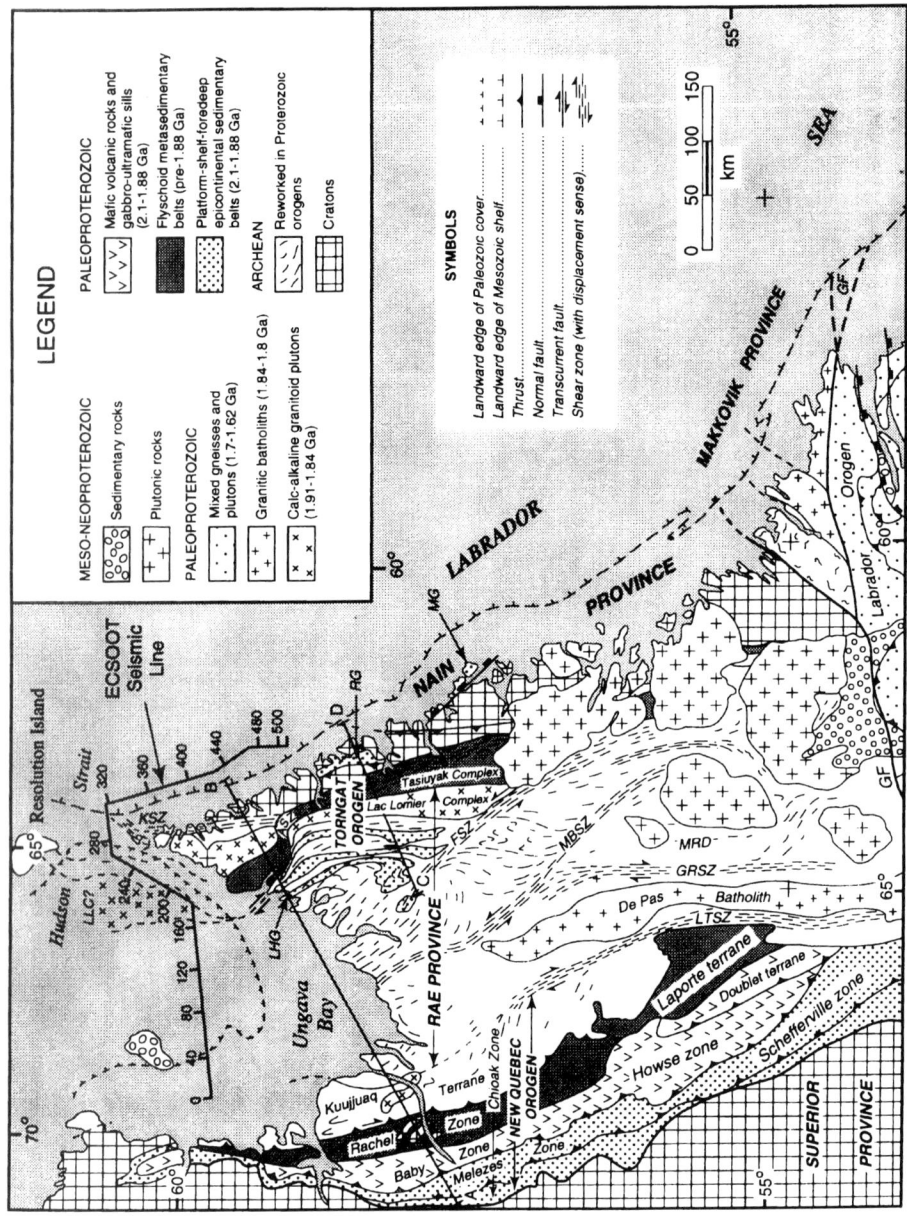

Fig. 2. The Southeastern Churchill Province: LHG, Lake Harbour Group; RG, Ramah Group; MG, Mugford Group; LLC?, seaward extent of Lac Lomier Complex based upon marine magnetic maps; MRD, Mistinibi–Raude domain; LTSZ, Lac Tudor shear zone; GRSZ, George River shear zone; MBSZ, Moonbase shear zone; FSZ, Falcoz shear zone; ASZ, Abloviak shear zone; KSZ, Komaktorvik shear zone; GF, Grenville front.

are somewhat older than the 2725–2688 Ma ages generally typical of the adjacent parts of the Superior Province (Percival et al. 1992). They are also older than ages in the adjacent Kuujjuaq zone (Machado et al. 1988), but fall within the range of 2922–2688 Ma ages determined from part of the northern Rae Province on Ungava Bay (see below). With regard to the origin of the Rachel zone Archaean gneisses, all that can be said at present is that they are of uncertain Superior or Rae province origin.

The origin of the Rachel–Laporte zone is speculative; Poirier et al. (1990) and Wardle et al. (1990b) proposed that it represents a distal part of the Superior passive margin sequence. Alternatively, van der Leeden et al. (1990) suggested an origin as a forearc accretionary wedge. Geochronological and isotopic geochemical work is required to discriminate between these models.

The easternmost component of the New Quebec Orogen is taken as the **Kuujjuaq terrane** (Perreault & Hynes 1990), which is given a terrane designation since it has no clear stratigraphic ties with either the Superior margin or the Rae Province. The terrane consists of reworked Archaean gneisses, locally dated at 2721 ± 4 and 2719 ± 7 Ma (Machado et al. 1989) and also of uncertain Superior or Rae affinity, infolded with Palaeoproterozoic metasedimentary rocks, including thick sequences of metasandstone-arkose, possible pyroclastic rocks, and amphibolites (Poirier et al. 1990). These are intruded by the axial Kuujjuaq batholith, dated between 1845 and 1833 Ma (Machado et al. 1988) which, on the basis of its calc-alkaline composition and chemistry, has been ascribed a magmatic arc origin (Poirier et al. 1990). The batholith has not been recognized south of its position in Fig. 2 but this may be due to the lack of detailed mapping in this area. The metasedimentary and metavolcanic rocks of the terrane are distinctly different from those of the Rachel zone and may represent remnants of arc superstructure. Subduction to form this arc was probably east-directed, since areas to the west are not affected by this type of activity, and it is likely that the arc was accreted along the Lac Turcotte thrust (Fig. 3A).

Rae Province

The Rae Province, which is thrust onto the Kuujjuaq terrane, is the least understood component of the SECP, particularly in its northern exposures around Ungava Bay. Where the province has been recently mapped and dated, an Archaean age has generally been established. Machado et al. (1989) have reported ages of $c.$ 2922–2688 Ma from a migmatite on the south shore of Ungava Bay. Nd isotopic studies (Emslie & Thériault 1991; Kerr et al. 1994) have also provided evidence of Archaean crust adjacent to the Torngat Orogen. In the south, Nunn et al. (1990) have documented extensive areas of well preserved 2.68–2.67 Ga Archaean crust with minimal Proterozoic overprint in areas adjacent to the Grenville Front; Ryan et al. (1991) reported large tracts of 2.8–2.6 Ga Archaean gneisses variably refoliated by Palaeoproterozoic amphibolite facies events in eastern parts of the province.

An Early Palaeoproterozoic tract of gabbro-norite, 'I'-type granitoids and intermediate to basaltic metavolcanics of apparent calc-alkaline character is represented in the **Mistinibi–Raude domain** of the southeastern Rae Province (Fig. 2). Plutonic rocks and amphibolite-facies orthogneisses have been dated at $c.$ 2.32 Ga in an event which appears to represent intrusion followed almost immediately by metamorphism (van der Leeden et al. 1990; Krogh 1992; R. Girard, pers. comm. 1992; B. Ryan pers. comm. 1992). These rocks are also characterized by $c.$ 1.94 Ga lower-intercept and concordant zircon ages apparently representing a later amphibolite-facies metamorphic overprint.

In the northeastern part of the Rae Province, the Archaean gneisses are infolded with quartzites, marbles and pelitic schists of the Lake Harbour Group (the type area for which is southern Baffin Island), which is interpreted as a Palaeoproterozoic cover sequence of shallow water origin (Jackson & Taylor, 1972). The only other Palaeoproterozoic feature of note is the linear De Pas batholith (Fig. 2), which comprises post-tectonic granitoids dominated by K-feldspar megacrystic granite, locally containing primary orthopyroxene. This has been reported by van der Leeden et al. (1990) to have an 'I'-type character and calc-alkaline trend. Interior parts of the batholith are generally only weakly deformed but the western margin has been strongly sheared (James et al. 1994). The batholith has yielded intrusion ages of $c.$ 1.84 Ga (Bowring, in van der Leeden et al. 1990), 1830 +6/–5 Ma (Connelly 1993) and 1811 ± 3 Ma (Krogh, 1986).

Torngat Orogen

The Torngat Orogen has four principal lithotectonic components, including, from west to east (Figs 2 & 3A): the Lac Lomier complex, the Tasiuyak complex, the Killinek/DTG (diorite–tonalite–granite) plutonic suites, and the part of the Nain Province craton lying within the Torngat foreland referred to as the Four Peaks domain (Fig. 3). The Tasiuyak complex and the Killinek/DTG suites are largely juvenile Palaeoproterozoic domains; the

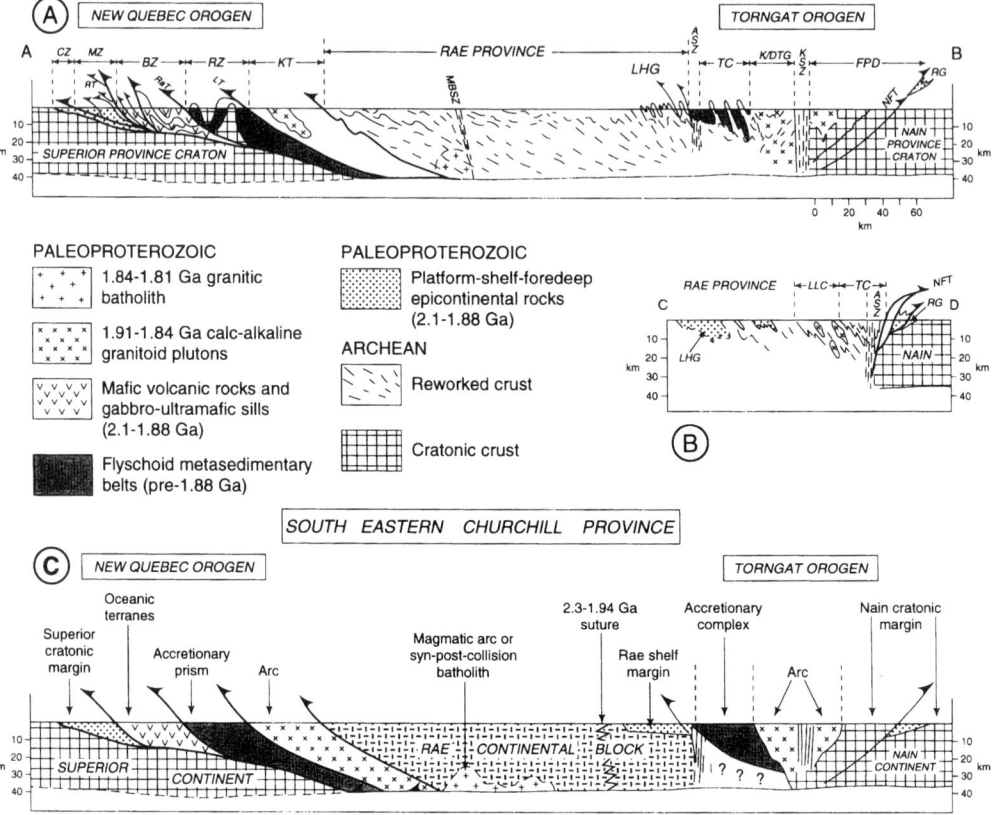

Fig. 3. Structural and interpretative cross-sections of the Southeastern Churchill Province. (A) Structural cross-section along line A–B (Fig. 2). New Quebec Orogen structure simplified from Wares & Goutier (1990), Moorhead & Hynes (1990) and Poirier et al. (1990). Vertical = horizontal scale. Rae Province crustal thickness and structure are constrained according to data from the ECSOOT 1992 seismic line (Fig. 2), assuming an average crustal velocity of 6 km s^{-1} (Hall et al. 1995). CZ, Chioak zone; MZ, Melezes zone; BZ, Baby zone; RZ, Rachel zone; KT, Kuujjuaq terrane; MBSZ, Moonbase shear zone; LHG, Lake Harbour Group; ASZ, Abloviak shear zone; TC,Tasiuyak complex; K/DTG, Killinek/DTG suites; KSZ, Komaktorvik shear zone; FPD, Four Peaks domain; NFT, Nachvak Fiord thrust; RG, Ramah Group; RT, Robelin thrust; RaT, Rachel thrust; LT, Lac Turcotte thrust. (B) Structural section of Torngat Orogen along section line C–D (Fig. 2), after Wardle (1983) and Rivers & Mengel (1994). Abbreviations as for (A). (C) Interpretative section along A–B showing inferred tectonic origin of major lithotectonic elements of Southeastern Churchill Province.

other components represent the reworked margins of the bounding Archaean provinces.

The Lac Lomier complex contains a mixture of Archaean granitoid gneisses, contiguous with those in the Rae Province, infolded with Palaeoproterozoic supracrustal gneisses and intruded by subsequently deformed orthopyroxene-bearing tonalite to granodiorite plutons. The supracrustal rocks are thought to be correlative with those in the Lake Harbour Group (Ermanovics & Van Kranendonk 1990). The plutonic rocks are thought to be similar in age and origin to those in the Tasiuyak complex (see below), which have been dated at 1877 ± 1 Ma.

The Tasiuyak complex consists of the metasedimentary Tasiuyak gneiss, a distinctive unit of garnet–sillimanite–biotite–graphite quartzofeldspathic gneiss that extends the length of the Torngat Orogen and which, by virtue of its distinctively low aeromagnetic signature, has been traced across Hudson Strait to Resolution and Baffin islands (Fig. 2) (Hoffman 1990a). The gneiss is intruded by numerous, linear, strongly deformed bodies of orthopyroxene-bearing tonalite, one of which has been dated at 1877 ± 1 Ma (Bertrand et al. 1993), and another at 1839 ± 2 Ma (Scott & Machado 1994). These have been proposed as the roots of a magmatic arc related to westwards subduction under the Rae Province (Van Kranendonk & Ermanovics 1990).

At its northwestern end, near Ungava Bay, the Tasiuyak gneiss contains tectonic slices of inferred Archaean crust that are partially bounded by ultramafic pods (Van Kranendonk et al. 1994) in a manner that suggests early interslicing with the protolith of the enveloping metasedimentary gneisses. The source and affinity of this Archaean crust has yet to be established.

The Tasiuyak gneiss is interpreted to be a metamorphosed flyschoid sequence that marks the boundary between rocks clearly associated with the Rae Province and those tied to the Nain craton. Nd isotopic studies of the gneiss (Kerr et al. 1994; Thériault & Ermanovics 1994) indicate derivation from either a late Archaean source or a mixed late Archaean–juvenile Palaeoproterozoic source. The Nain Province is generally too old to act as a significant source (although, as noted below it may have provided a small amount of Palaeoproterozoic detritus), moreover the Sm–Nd results reported by Thériault & Ermanovics (1994) show the Tasiuyak gneiss to have a significantly different source than the metasedimentary rocks within the Lac Lomier Complex, thereby ruling out the possibility that the Tasiuyak gneiss developed as a continental rise prism on the Rae margin. An accretionary wedge model is therefore inferred for the origin of the Tasiuyak gneiss and is supported by reports of c. 2.0 to 1.94 Ma detrital zircons (Scott & Machado 1994), which establish a maximum age of c. 1.94 Ga for deposition of the gneiss protolith. Potential local sources for these zircons may be c. 2.0 to 2.1 Ga granitoid plutons of the Nain craton (Emslie & Loveridge 1992) and/or the 2.3–1.94 Ga rocks of the Mistinibi–Raude domain but more distal sources may also be possible. The 2.0–1.9 Ga Thelon Orogen of northwestern Canada (Hoffman 1990a), which affects substantial areas of Rae Province crust (Bickford et al. 1994), might be a suitable candidate for such a source, although this would involve transport over considerable distances.

The Killinek/DTG suites are predominantly composed of Palaeoproterozoic plutonic rocks of calc-alkaline composition divided into suites of orthopyroxene-bearing ('charnockitic') rocks – the Killinek suite, and amphibolite-facies diorites, tonalites and granites – the DTG suite (Van Kranendonk et al. 1993b). These intrude pelitic gneisses similar to those of the Tasiuyak gneiss and contain variably sized enclaves of older mafic granulite and orthogneiss (Wardle et al. 1993) for which Sm–Nd data indicate an Archaean age (Campbell 1994). Dating of seven samples of plutonic rocks, including both amphibolite- and granulite-facies varieties, has given intrusion ages ranging between 1910 and 1885 Ma (Scott & Machado 1995). A younger suite of intrusions is represented by metatonalite (1869 +3/–2 Ma) and megacrystic granite (minimum age of 1864 ± 2 Ma) which intrude both the Killinek/DTG and Nain Province rocks in areas adjacent to, and within, the Komaktorvik shear zone (Scott & Machado 1995; Fig. 3A). They thus provide a minimum age for amalgamation of the Killinek/DTG suite and Nain Province, if indeed they were ever separate. The Killinek/DTG suite plutonic rocks have been proposed as the roots of a magmatic arc (Wardle et al. 1992) that must, in part at least, be founded upon Archaean crust. The parentage of this Archaean crust, however, is unclear; Pb–Pb isotopic evidence (Bridgwater & Wardle 1992) suggests that it lacks the radiogenically depleted signature characteristic of the northern Nain province, but there are insufficient data to allow any clear comparison with Rae Province crust.

A tectonically bounded belt of mafic gneiss derived from metavolcanic and intrusive rocks lies along the eastern margin of the Killinek/DTG suites adjacent to their contact along the Komaktorvik shear zone with the Four Peaks domain. The amphibolites are undated but, on the basis of their MORB-like chemistry (Van Kranendonk, unpublished information), they have been tentatively interpreted as the dislocated remnants of a backarc basin (Van Kranendonk & Wardle 1994) formed during attempted separation of the Killinek/DTG suite arc from the Nain craton.

The Four Peaks domain (northern Nain Province) is composed largely of Archaean gneisses (Wardle et al. 1994), intruded on their western margin by Palaeoproterozoic (?) anorthosite and by plutons similar to those of the DTG suite. It is separated from the Killinek/DTG suites by the Komaktorvik shear zone and from the remainder of the Nain Province by the Nachvak Fiord thrust (Fig. 3A, B). The Archaean gneisses are typically migmatitic, tonalitic orthogneisses that are intercalated with belts of sillimanite-bearing metasedimentary gneiss and abundant mafic and ultramafic material, much of which appears derived from supracrustal volcano-plutonic belts (Wardle et al. 1994), all intruded by late, foliated granites. The gneisses generally preserve a late Archaean granulite-facies assemblage that is variably overprinted by Palaeoproterozoic assemblages (see below). U–Pb zircon data from the orthogneisses indicate an intrusion age of 2834 ± 4 Ma (Scott & Machado 1994) and an age of migmatization of 2767 ± 6 Ma. An amphibolite-facies, late foliated granite has given monazite ages of c. 2650 Ma (Scott & Machado 1994). The age data confirm previous arguments (Wardle et al. 1993), based upon lithological comparisons, for correlation with gneisses in the remainder of the Nain Province to the south.

To the west, the Archaean gneisses become progressively redeformed within the amphibolite-facies Komaktorvik shear zone where the gneisses have been tectonically intersliced and infolded with the elongate Hutton anorthositic suite (not shown on Fig. 2). This suite is suspected to be of early Palaeoproterozoic (c. 2.4–1.91 Ga) age and related to early rifting of the Nain craton, however, a late Archaean age cannot be ruled out (Van Kranendonk et al. 1994).

Both the Archaean gneisses and Hutton anorthositic suite are intruded by a swarm of east–west to northeast–southwest trending mafic dykes that have been collectively termed the Avayalik dykes (Wardle et al. 1992). The age of these dykes is contentious. Field observations suggest that many predate intrusion of c. 1.91 Ga plutons (Van Kranendonk et al. 1994) and thus may correlate with c. 2.4–2.2 Ga Napaktok dykes in more southerly parts of the Nain Province. However, this conflicts with a single zircon age of 1834 +7/–3 Ma from an Avayalik dyke, which has been variably interpreted as the age of dyke intrusion (Scott & Machado 1994) and as the age of a static, granulite-facies metamorphic overprint (see discussion in Van Kranendonk et al. 1994). There is also evidence for younger dykes in the form of syntectonic dioritic dykes that intrude 1.91 Ga plutons, and syntectonic amphibolite dykes within the Komaktorvik shear zone that are bracketed by 1804–1779 Ma ages (D. Scott, pers. comm. 1993).

The western part of the Four Peaks Domain is dominated by a group of compositionally varied diorite, tonalite to granite plutons that intrude Archaean gneisses and the Hutton anorthositic suite within, and east of, the Komaktorvik shear zone. The plutons are compositionally similar to those in the DTG suite and have yielded comparable ages of 1891 ± 2 and 1885 ± 2 Ma, indicating that the magmatic arc-related activity that produced the Killinek/DTG suites also affected the northwestern margin of the Nain craton.

The Komaktorvik shear zone is a relatively young structure (see below) that postdates the Abloviak shear zone. It probably nucleated along a long-lived zone of crustal weakness that may originally have controlled location of the Hutton anorthositic suite and also the backarc rifting of the Killinek/DTG suite. It is not, however, envisaged as a fundamental suture.

The remainder of Nain craton within the Torngat Foreland zone, defined as that part of the craton lying south of the Nachvak Fiord thrust (Fig. 3A, B), is characterized by early Archaean crust of 3.8–3.3 Ga age, locally intruded by posttectonic granites of c. 2.8 Ga age. Palaeoproterozoic rocks are represented by the 2.45–2.2 Ga (K–Ar ages; Taylor 1979) Napaktok dykes (Ermanovics & Van Kranendonk 1990) and the unconformably overlying rocks of the Ramah and Mugford groups (Fig. 2). Neither of these units has been reliably dated and they are constrained in age only between 2.45 and 1.8 Ga. The Ramah Group consists of a lower package of dominantly clastic, shallow-water rocks, inferred to have been deposited in a shelf environment (Knight & Morgan 1981), overlain by an upper sequence of deep-water turbidites proposed by Hoffman (1987) as a foredeep sequence related to encroachment of Torngat thrust sheets from the west. The Mugford Group comprises a lower, largely clastic sequence comparable to that of the lower Ramah Group but in its upper part is dominated by subaerial mafic flows of uncertain age and significance.

A series of small granitoid plutons dated at 2.1 to 2.0 Ga (Emslie & Loveridge 1992; Connelly & Ryan 1994) has been recognized from the central regions of the Nain Province (not shown on Fig. 2) and may record the initial rifting of the craton; their relationship to the Ramah and Mugford groups, however, is unknown. The youngest units of the northern Nain Province are several small granitoid plutons exposed on islands along the Atlantic coast (not shown on Fig. 2) and dated at c. 1.77 Ga (Emslie & Loveridge 1992).

Crustal structure and tectonic evolution

New Quebec Orogen

Figures 3A & B illustrate simplified structural cross-sections of the SECP, which are keyed to Fig. 2. Crustal thickness and structure in parts of Fig. 3A are constrained by data from the 1992 Lithoprobe ECSOOT deep-reflection seismic profile carried out across the offshore extension of the Rae Province and Torngat Orogen in Ungava Bay (Fig. 2; Hall et al. 1995). Figure 3C is an interpretative section that summarizes the structural relationship of the various tectonic and lithologic elements described above.

The New Quebec Orogen is a west-verging fold-and-thrust belt characterized by a thin-skinned imbricate thrust belt in the west (Chioak and Melezes zones; Wares & Goutier 1990) whose roots extend down to the east into a thick-skinned belt that comprises the eastern part of the orogen and which is associated with basement slices. Metamorphic grade rises progressively from greenschist facies in the west to amphibolite, and locally granulite, facies in the east. Peak PT conditions for granulites in the Kuujjuaq terrane have been established at 800–850°C at 6–8 kb (Perreault & Hynes 1990). Initial thrusting, represented by westward transport of basement-cored nappes in the Rachel zone (Fig. 3A), was succeeded by more widespread low-angle and bedding-parallel thrusting and then by a series of high angle, out-of-sequence thrusts that define most of the present zone and terrane boundaries (Wares & Goutier 1990). These structures are interpreted to connect downward with a master decollement at the basement interface and are associated with refolding of earlier structures. Shortening within the New Quebec Orogen has been estimated at between 64% (Boone & Hynes 1990) and 78% (Wares & Goutier 1990).

The timing of metamorphism in western parts of the orogen is not well constrained but to the east in the Kuujjuaq zone, granulite-facies metamorphism has been dated at 1833–1829 Ma (Machado 1990). Deformation in the Rachel zone–Kuujjuaq terrane area appears to have taken place in the interval 1793–1783 Ma, the inferred time of movement on the Rachel and Lac Olmstead thrusts that bound the zone, and terminated by 1775 Ma, the age of post-tectonic pegmatites in the Rachel zone (Machado et al. 1989).

Rae Province

Structural style within the western Rae Province appears similar to that in the New Quebec Orogen. Data from the interior of the province are restricted to the regional mapping of Taylor (1979) and a coastal reconnaissance by Goulet & Ciesielski (1990). Analysis of the regional data indicates a predominance of easterly dips of gneissic (Archaean?) foliation at between 15 and 80°. Taken in conjunction with the more detailed sections of Goulet & Ciesielski (1990), this is interpreted in terms of a Archaean gneissosity which has been folded and reorientated by asymmetric west-verging Palaeoproterozoic D2 folds, as indicated in schematic fashion on Fig. 3A. Further evidence for the three-dimensional orientation of Palaeoproterozoic fabrics is provided by the Lithoprobe ECSOOT seismic profile (Hall et al. 1995), which reveals a pervasive east-dipping pattern of crustal-scale reflectors representing an easterly continuation of the structural style seen in the New Quebec Orogen. This pattern, indicated schematically in Fig. 3A, wanes to the east but pervades the entire crust and is compatible with surface evidence for westerly structural vergence.

There is little geochronological control on the age of fabric development in the Ungava Bay region; however, data from 250 km to the south indicate a widespread amphibolite-facies overprint(s) of c. 1860 to 1800 Ma age (Ryan et al. 1991; Krogh 1992), with most ages falling within the period 1830–1800 Ma. This is in addition to the more limited evidence for a c. 1.94 Ga event in the Mistinibi–Raude domain.

The prevailing grade of Palaeoproterozoic metamorphism in the interior Rae Province appears to have been amphibolite facies, although it should be noted that in the south, in the Grenville front region, there are areas that have experienced only a minimal overprint (Nunn et al. 1990). It is also significant that the metamorphic grade rises to granulite-facies in the bordering areas of the Rae Province represented by the Kuujjuaq terrane and Lac Lomier complex, reinforcing the concept that these areas mark collisional interfaces with the adjacent Superior and Nain cratons.

The Rae Province is also dissected by several, prominent, vertical transcurrent shear zones (Figs 2 & 3A), which are characterized by dextral sense in the west, (Lac Tudor and George River shear zones), and sinistral in the east (Moonbase, Mistastin, and Falcoz shear zones). Dextral shearing, as represented by the Lac Tudor shear zone, is probably constrained in age between 1830 (or possibly 1811 Ma, the youngest age for the De Pas batholith which it affects) and 1.8 Ga, the age of regional titanite crystallization (James & Dunning 1994). Sinistral shearing may have commenced as early as 1857 Ma, based upon a thorite age for the Moonbase shear zone (Ryan et al. 1991), which is also in approximate agreement with nearby 1859 and 1854 Ma metamorphic ages. Shearing is probably no younger than 1802 Ma on the basis of a titanite age from the Mistastin shear zone (Figs 3A & 4A; Ryan et al. 1991). The Falcoz shear zone (Figs 3A & 4A) is undated but is probably similar in age to the 1.84–1.82 Ga Abloviak shear zone of Torngat Orogen from which it splays (Fig. 4A).

Torngat Orogen

The orogen displays an overall doubly vergent profile (Figs 3A & B) formed by west-verging folds in the Lac Lomier complex and adjacent parts of Rae Province, and east-verging thrusts and folds in the Torngat foreland of Nain craton. The axial zone of Torngat Orogen, the Tasiuyak complex, is characterized by vertical structures and is also the locus for the sinistral Abloviak shear zone. In the northern part of the orogen, the vertical Komaktorvik and Abloviak shear zones isolate the triangular Burwell structural domain, including

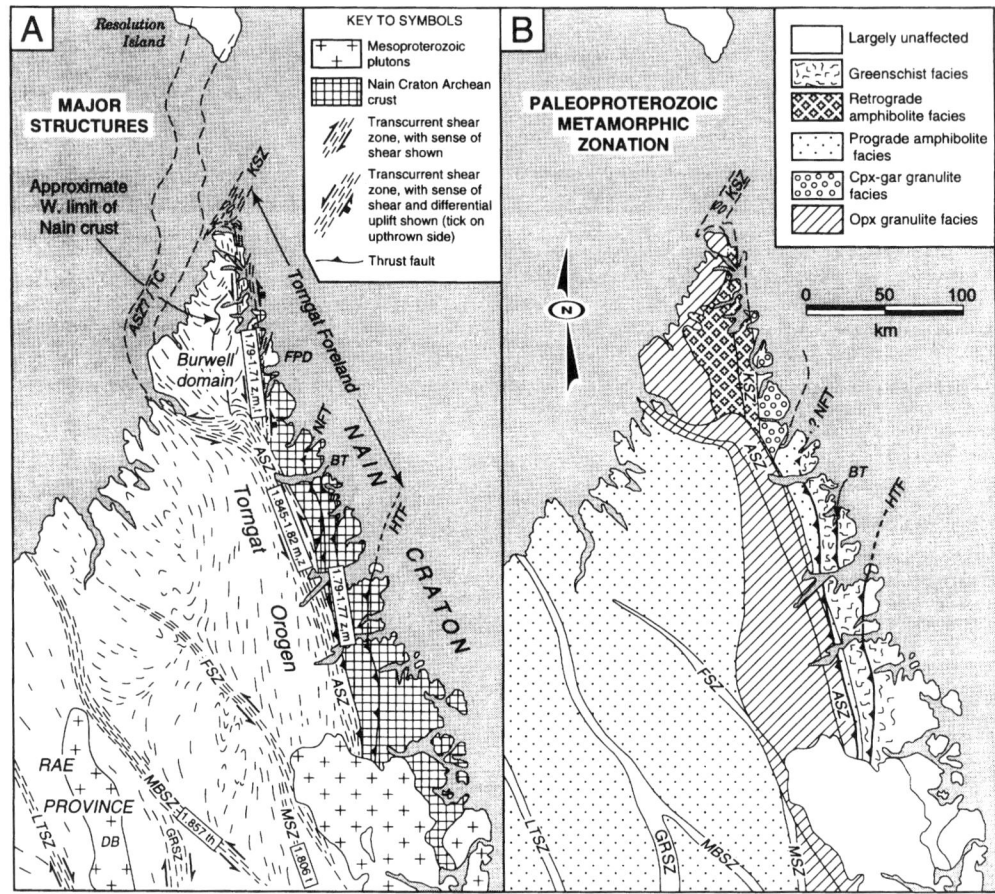

Fig. 4. Torngat Orogen. (**A**) Structural features and tectonic zonation: LTSZ, Lac Tudor shear zone; GRSZ, George River shear zone; MBSZ, Moonbase shear zone; MSZ, Mistastin shear zone; FSZ, Falcoz shear zone; ASZ, Abloviak shear zone; ASZ?TC, inferred offshore extent of ASZ and Tasiuyak complex based upon marine magnetic data; KSZ, Komaktorvik shear zone; NFT, Nachvak Fiord thrust; HTF, Handy thrust; BT, Brannigan thrust. U–Pb accessory mineral ages: z, zircon; m, monazite; t, titanite; th, thorite. (**B**) Palaeoproterozoic metamorphic zonation; abbreviations as in (A).

the Killinek/DTG suites and associated Archaean gneisses, from the Nain and Rae provinces (Fig. 4A).

The chronology of kinematic development in the Torngat Orogen has been deciphered in considerable detail by Bertrand *et al.* (1993) and Scott & Machado, (1994, 1995). The earliest metamorphic ages have been recorded from the Tasiuyak complex and record an initial event related to the formation of an axial belt of upper amphibolite- to granulite-facies migmatites c. 1870–1860 Ma ago (Bertrand *et al.* 1993; Scott & Machado 1994). These ages are believed related to initial Rae\Nain collision and to early thrusting of the Tasiuyak complex westwards over Rae crust (Van Kranendonk 1992; Rivers & Mengel 1994).

To the south, Van Kranendonk (1992) and Ryan (1990) have also recognized east-verging thrusts that interleave Nain gneisses and Ramah Group cover. These thrusts predate the Abloviak shear zone and represent the initial stage in the development of a doubly vergent orogen at c. 1.86 Ga.

This initial event was followed by a prolonged period of granulite-facies D_2 metamorphism and deformation that is bracketed at between 1845 and 1820 Ma in the Tasiuyak complex, where it is related principally to the development of the sinistral Abloviak shear zone (Bertrand *et al.* 1993). Similar metamorphic ages of 1843 and 1824 Ma in Burwell domain (Scott & Machado 1994, 1995), appear to be related to an equivalent period of folding and high-grade metamorphism.

A similar period of metamorphism may also be represented in the Torngat foreland where Archaean orthopyroxene-bearing granulite-facies gneisses of the Four Peaks domain have been overprinted by a static, high pressure clinopyroxene–garnet assemblage, formed at ≥750°C and 10 kb (Mengel & Rivers 1994), and thought to indicate burial under thrust sheets expelled eastward out of Torngat Orogen (Wardle *et al.* 1994). Since this overprint also affects the Avayalik dykes, it must be bracketed in age between 1834 Ma, which is the Avayalik dyke intrusion or metamorphic age, and 1.79 Ga, the oldest age of deformation in the superimposed Komaktorvik shear zone.

A distinctly younger period of deformation is represented by the Komaktorvik shear zone and the thrusts (e.g. Nachvak Fiord thrust) that splay eastward into the Nain craton (Fig. 4A). The Komaktorvik shear zone is a composite structure that shows evidence of sinistral translation that in the east is associated with east-side-up oblique uplift of the Nain craton, and in the west with oblique west-side-up uplift of the Burwell domain (Van Kranendonk *et al.* 1994). Scott & Machado (1995), on the basis of zircon and monazite ages from synkinematic granitoid veins, have constrained movement on the zone to between 1798 and 1719 Ma, though the main episode of amphibolite-facies metamorphism appears to have occurred between 1789 and 1740 Ma.

Sinistral motion on the Komaktorvik shear zone appears to have also coincided with folding of the Abloviak shear zone into the major flexure shown in Fig. 4A (Van Kranendonk *et al.* 1993*b*).

South of the Four Peaks domain, a rather different late-stage kinematic picture emerges. In this area the boundary of the Tasiuyak complex and Nain craton has been deformed in a narrow zone of steep, westerly dipping mylonite–ultramylonite, that overprints the Abloviak shear zone and also affects the western part of the Ramah Group. This zone, characterized by down-dip lineations, can be traced northwards into the Nachvak Fiord thrust, a major discontinuity along which Archaean crust of the Four Peaks domain has been translated eastward over the Ramah Group and the Nain craton (Figs 3B & 4A; Wardle 1983). The Ramah Group and Nain craton are cut by other faults such as the Brannigan and Handy thrusts (Calon & Jamison 1994; Fig. 4A) which splay eastward in a curvilinear fashion similar to that of the Nachvak Fiord thrust and which are presumed to be of similar age. The age of mylonitization adjacent to, and within, the western Ramah Group, and by inference the age of the Nachvak Fiord thrust, has been established by Bertrand *et al.* (1993) as between 1795 and 1740 Ma, a period that is broadly contemporaneous with development of the Komaktorvik shear zone to the north.

Metamorphic zonation in the Torngat Orogen (Fig. 4B) displays an asymmetric pattern in which the axial granulite-facies belt, centred on the Tasiuyak and Lac Lomier complexes, is flanked by contrasting assemblages. In the west these assemblages consist of amphibolite-facies rocks of the Rae Province, and to the east comprise a series of zones in the Nain Province that progress from high-pressure granulite facies in the north, to greenschist facies in the south. The axial granulites, where studied in detail in the Saglek and North River–Nutak areas, were formed *c.* 1.87 Ga at peak temperatures and pressures of 9.5–10 kbar at 800–50°C (Mengel & Rivers 1990, 1994; Van Kranendonk 1992). According to Van Kranendonk (1992), the peak pressures reflect early burial during initial Rae\Nain collision and were succeeded by lower pressures and temperatures of 7.3 kb and 750°C during formation of the Abloviak shear zone. The oblique pattern of isograds in the Four Peaks domain of the Nain Province is thought to be younger and reflect a combination of burial at *c.* 1.83 Ga, and subsequent scissor-like uplift (1.79–1.71 Ga) on the Komaktorvik shear zone to expose an oblique section of Nain crust (Van Kranendonk & Scott 1992).

In summary, the doubly vergent profile of Torngat Orogen is a composite structure formed by sequential deformation between 1.87 and 1.71 Ga. West-verging structures predate development of the Abloviak shear zone and are thereby constrained as pre-1.82 Ga in age. East-verging structures formed in three episodes; *c.* 1.86 Ga, *c.* 1.83–1.79 Ga and finally 1.79 and 1.74 Ga. The latter episode took place in conjunction with sinistral displacement along the Komaktorvik shear zone in association with refolding of the Abloviak shear zone and scissor-like uplift of the Nain Province.

Tectonic synthesis: a model for oblique collision and indentation

At present, there is no evidence to suggest that the Nain, Rae and Superior provinces were ever linked prior to Palaeoproterozoic ocean development. However, the Nain and Superior cratons do show evidence of early Palaeoproterozoic rifting *c.* 2.4–2.1 Ga, suggesting that they formed parts of larger cratons. In both cratons this is seen initially as dyke activity (2.4–2.2 Ga in Nain; 2.1 Ga in Superior) followed by formation of passive margin sequences, and (in the New Quebec Orogen at least) the development of basalts and gabbro sill swarms related to incipient ocean opening (Rohon *et al.* 1993). The Lake Harbour Group may represent the similar development of a Palaeoproterozoic passive margin sequence on the Rae Province.

An important, but incompletely understood, period of Early Palaeoproterozoic accretion may be represented by the igneous and metamorphic ages of 2.3–1.94 Ga from the Mistinibi–Raude domain. The tectonic setting of this event is far from established but is speculated to mark an early stage of arc magmatism (R. Girard, pers. comm. 1992) and accretion associated with early Palaeoproterozoic assembly of the Rae Province. Crocker *et al.* (1993) and Bickford *et al.* (1994) have described widespread growth of metamorphic zircons from *c.* 2.3–2.0 Ga, including a significant granulite-facies event, from the Rae and Hearne provinces of western Canada, indicating that such early Palaeoproterozoic ages may be more widespread than generally thought. The metamorphic ages of *c.* 1.94 Ga from Mistinibi-Raude domain also fall within the 1.96–1.91 Ga range of collisional events in the Thelon Orogen and suggest the possibility of contemporaneous accretionary events in the eastern Rae Province. These ages underline the distinctiveness of the Rae Province and suggest that it may have been a composite Archaean/Palaeoproterozoic terrane, rather than a pristine Archaean craton, prior to its incorporation into the SECP.

Van Kranendonk *et al.* (1993*a*) have recently proposed a scenario for the evolution of northeast Laurentia that is based upon the oblique convergence of the Nain and Superior cratons against a south-projecting promontory of the Rae Province. This idea is adapted in Fig. 5 to illustrate selected stages in the progressive development of the SECP. Figure 5a summarizes the situation at *c.* 1.91 Ga, by which time the Nain craton had commenced its northward advance towards the Disko and (amalgamated?) Rae provinces (Van Kranendonk *et al.* 1993*a*) and Killinek/DTG suite arc magmatism had commenced on the western flank of the Nain craton. The Tasiuyak domain accretionary prism, by virtue of its 2.0–1.94 Ga detrital zircon content, is speculated to have developed on the eastern margin of the southern Rae promontory at this time.

The early subduction history of the Torngat Orogen at around this time is complicated by the occurrence of magmatic arc sequences on both Nain (1.91–1.88 Ga) and Rae (Lac Lomier complex, 1.88–1.84 Ga) margins, a feature which can be explained either by flips of subduction with time or by double subduction (Fig. 5b; see Van Kranendonk & Wardle 1994 for discussion). Deformation in the Torngat Orogen may have commenced as early as 1.87 Ga, possibly in response to early collision of the Killinek/DTG suite arc, but was definitely underway by 1.86 Ga in response to collision of the Rae and Nain Provinces (Fig. 5c). This collision was probably associated with early thrusting (*c.* 1.86 Ga) of the Tasiuyak domain westwards over the Rae crust, (Rivers *et al.* 1993), and early east-directed thrusting to produce a doubly vergent profile (Fig. 5c). On the Superior margin, oceanic rifting was still under way at *c.* 1.87 Ga in a dextral transtensional environment (Skulski *et al.* 1993) (Fig. 5c) but, by 1.845 Ga, arc magmatism had commenced in the Kuujjuaq terrane and was followed *c.* 1.83 Ga ago by collision of Superior and Rae cratons (Fig. 5d). At about the same time, east-verging thrusting was probably in progress in the Torngat Orogen to produce the static, high-pressure overprint on the Nain craton (Fig. 5d). Penetrative ductile deformation in the interior of Torngat Orogen was probably more or less continuous between 1.86 and 1.82 Ga (Fig. 5d), during which time deformation was largely transferred to the Abloviak shear zone (1.845–1.82 Ga) and other sinistral shear zones within the Rae Province (Fig. 5d). The period from 1.845–1.82 Ga was a time of widespread deformation in both the NQO and the western Torngat Orogen. This is probably also the time at which the intervening areas of the Rae Province were deformed to produce both the east-dipping fabrics imaged on the ECSOOT seismic profile, and the major transcurrent shear zones that transect the Rae crust. From 1.845 Ga onwards, the SECP therefore acted as a single orogenic entity.

The peak pressures in the axial granulites of Torngat Orogen correspond to burial depths of 30–35 km and point to an early doubling of crustal thickness (Mengel & Rivers 1990; Rivers & Mengel 1994) which was probably maintained between 1.86 and 1.82 Ga. The most likely source for the load necessary to produce this burial is the middle and upper crustal juvenile material that must have been expelled from the axial parts of the orogen during closure. This load may also have covered most of the Rae Province, thus accounting for its pervasive amphibolite-facies overprint. Similar depths of burial probably affected the northwestern parts of the Nain craton to produce the static garnet-clinopyroxene overprint (Fig. 5d). An event of uncertain significance is the intrusion of the De Pas batholith between 1.84 and 1.81 Ga. As suggested by van der Leeden *et al.* (1990) and Machado *et al.* (1989), this may mark the site of another magmatic arc and suture zone; alternatively, it may be a syn- to post-collisional feature related to crustal thickening and subsequent crustal melting. Intrusion of the younger parts of the

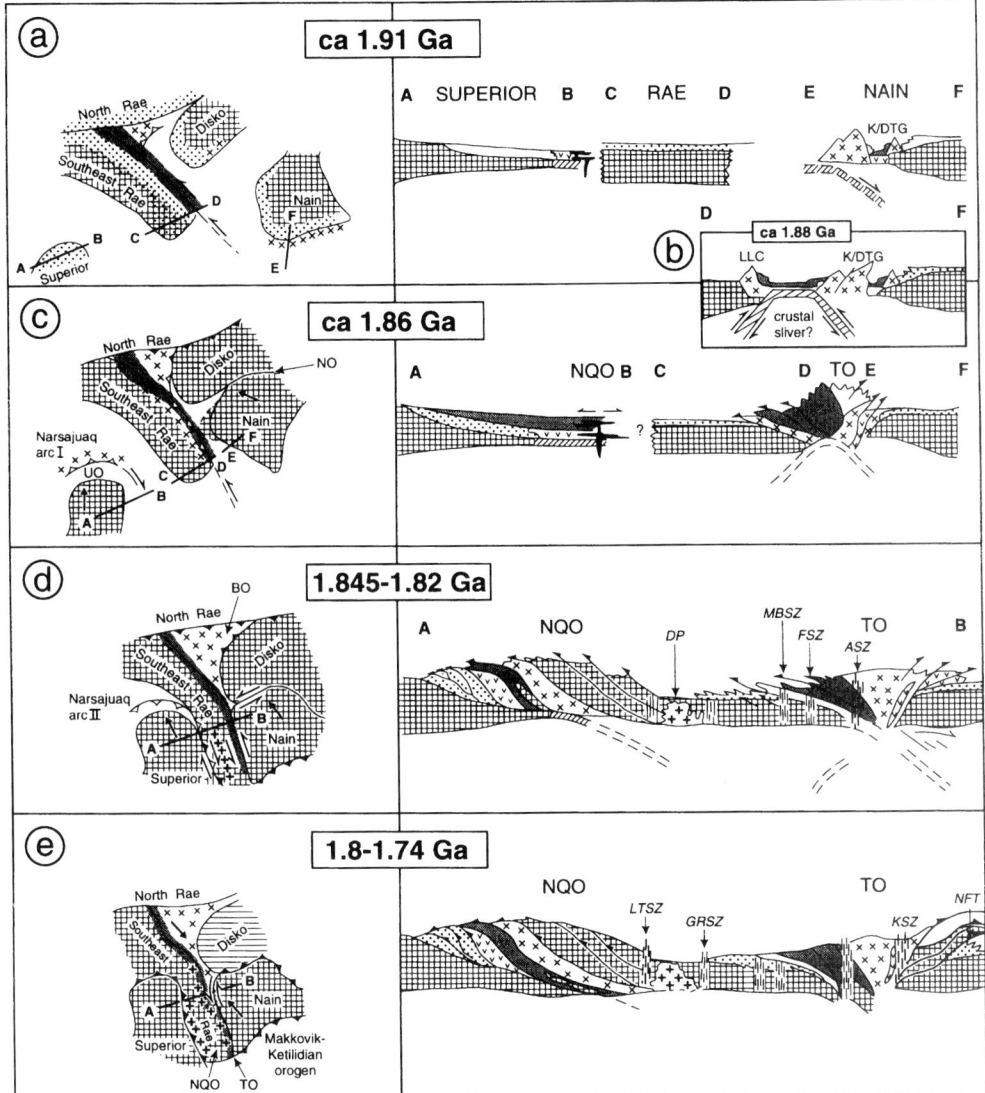

Fig. 5. Critical stages in the tectonic development of the Southeastern Churchill Province in the context of the evolution of northeastern Laurentia (after Van Kranendonk et al. 1993a). Cartoon cross-sections are referenced to section lines A–B etc. in the palinspastic reconstructions at left. Stages a–e are described in the text; shading patterns are as for Figs 2 & 3. K/DTG, Killinek/DTG suite; LLC, Lac Lomier complex; NQO, New Quebec Orogen; TO, Torngat Orogen; UO, Ungava Orogen; BO, Baffin Orogen; DP, De Pas batholith; ASZ, Abloviak shear zone; LTSZ, Lac Tudor shear zone; GRSZ, George River shear zone; MBSZ, Moonbase shear zone; FSZ, Falcoz shear zone; ASZ, Abloviak shear zone; KSZ, Komaktorvik shear zone; NFT, Nachvak Fiord thrust.

batholith was approximately coincident with cessation of deformation in the Rae Province at about 1.8 Ga.

From 1.8 Ga onwards, deformation appears to have been focused on the margins of the SECP (Fig. 5e). In the Torngat Orogen this was associated with continued sinistral motion on the Komaktorvik shear zone (1.79–1.719 Ga), folding of the ASZ and development of the Nachvak Fiord thrust and allied structures in the east-verging thrust fold-and-thrust belt of the Ramah Group (1.79–1.74 Ga; Fig. 5d). Deformation in New Quebec Orogen, probably in association with continued thrusting, appears to have terminated by 1.75 Ga.

As noted by Van Kranendonk et al. (1993a), these events mark late-stage contraction across the SECP, possibly in association with exhumation of the orogenic core, as the Superior and Nain cratons accomplished the final stages of their indentation into the hinterland of the northern Churchill Province.

Summary

The SECP represents the site of the oblique convergence and collision of the Archaean Nain and Superior cratons with a southern promontory of the Archaean Rae Province. Deformation occurred in an overall transpressive environment but the balance of contractional deformation versus transcurrent extension varied with time. Initial deformation was dominated by collision-induced westward-directed thrusting of Rae over Nain crust and doubly vergent thrusting at the Rae\Nain boundary. Later stages were dominated more by a combination of transcurrent shear, induced by sideslip of the Nain and Superior cratons against the Rae promontory, and contractional stress transferred outwards onto the margins of the cratons. The following are the salient points in the history of the amalgamation of this part of northeast Laurentia.

(1) Initial rifting occurred between c. 2.45 and 2.2 Ga in the Nain and at c. 2.1 Ga in the Superior craton. Early Palaeoproterozoic arc-type plutonism and deformation at c. 2.3–1.94 Ga in the Rae Province may signify an early terrane amalgamation event.

(2) Initial collision occurred at c. 1.87–1.86 Ga between the Nain and Rae provinces to form the Torngat Orogen and at c. 1.83 Ga to form the New Quebec Orogen. The sites of collision are marked by belts of granulite-facies crust exhumed from c. 6–10 kb (approximately 22–35 km) depth.

(3) Juvenile Palaeoproterozoic crust, including substantial amounts of material inferred to be derived from the continent–ocean transition, is preserved largely in the New Quebec Orogen, a feature which is probably due to the burial and structural protection of juvenile allochthons by overthrusting of Rae Province crust. Juvenile crust in Torngat Orogen was probably expelled from the axial parts of the orogen and produced a superincumbent thrust load that must have effectively doubled the crustal thickness and induced high grade metamorphism throughout the central and eastern parts of the SECP.

(4) Final closure of SECP between 1.845 and 1.80 Ga was accompanied by the continued northwards slip of the bordering cratons against the Rae promontory along major transcurrent shear systems, sinistral in the east (1.845–1.82 Ga) and dextral in the west (1.83–1.82 Ga).

(5) Deformation during the latest stage of the SECP, in the period 1.79–1.71 Ga, was largely restricted to the margins of the orogens adjacent to the rigid cratons. This probably occurred in conjunction with exhumation of the orogenic core and small-scale plate adjustments as the various components of the northeast Laurentian mosaic jostled into their final position.

We wish to thank our colleagues of the Torngat mapping project, namely David Scott, Flemming Mengel, David Bridgwater and Lisa Campbell for numerous stimulating and constructive discussions which have led directly and indirectly to the concepts proposed in this paper. Publication is with the permission of the Executive Director of the Newfoundland Geological Survey Branch. Garth Jackson and Jack Henderson are thanked for their critical reviews of earlier drafts. Lithoprobe contribution no. 682; GSC contribution no. 46794.

References

BERTRAND, J.-M., RODDICK, J. C., VAN KRANENDONK, M. J. & ERMANOVICS, I. 1993. U–Pb geochronology of deformation and metamorphism across a central transect of the Early Proterozoic Torngat Orogen, North River map area, Labrador. *Canadian Journal of Earth Sciences*, **30**, 1470–1489.

BICKFORD, M. E., COLLERSON, K. E. & LEWRY, J. F. 1994. Crustal history of the Rae and Hearne provinces, southwestern Canadian Shield, Saskatchewan: constraints from geochronologic and isotopic data. *Precambrian Research*, **68**, 1–21.

BOONE, E. & HYNES, A. 1990. A structural cross-section of the northern Labrador Trough, New Québec. *In*: LEWRY, J. F. & STAUFFER, M. R. (eds) *The Early Proterozoic Trans-Hudson Orogen of North America*. Geological Association of Canada, Special Paper **37**, 387–396.

BRIDGWATER, D. & WARDLE, R. J. 1992. Whole rock Pb isotopic compositions of Archaean and Proterozoic rocks from northern Labrador used to separate major units in the shield. *In*: WARDLE, R. J. & HALL, J. (eds) *Eastern Canadian Shield Onshore–Offshore Transect (ECSOOT), Report of Transect Meeting*. University of British Columbia, Lithoprobe Report, **32**, 42–51.

CALON, T. J. & JAMISON, W. R. 1994. Structural evolution of the Eastern Borderland of the Torngat Orogen, Kikki Lake transect, Saglek Fiord area, northern Labrador. *In*: WARDLE, R. J. & HALL, J. (eds) *Eastern Canadian Shield Onshore–Offshore Transect (ECSOOT), Report of Transect Meeting*.

University of British Columbia, Lithoprobe Report No. 36, 89–99.

CAMPBELL, L. M. 1994. Preliminary results from an Sm–Nd isotopic study of Proterozoic crustal formation and accretion in the Torngat Orogen, northern Labrador. *In*: WARDLE, R. J. & HALL, J. (eds) *Eastern Canadian Shield Onshore–Offshore Transect (ECSOOT), Report of Transect Meeting*. University of British Columbia, Lithoprobe Report No. 36, 100–107.

CLARK, T. 1984. *Géologie de la région du lac Cambrien, Territoire du Nouveau-Québec*. Ministère de l'Energie et des Ressources du Québec, ET 83-02.

—— & AVRAMTCHEV, L. 1990. *Gîtes Minéraux du Québec Région de la Fosse du Labrador, Feuille Rivière Arnaud, 25D*. Ministère de l'Energie et des Ressources du Québec, Direction générale de l'exploration géologique et minérale, Carte no M-371 du DV 84-01.

CONNELLY, J. N. 1993. *U–Pb geochronological research agreement: final report for the Newfoundland Department of Mines and Energy, Labrador Mapping Section*. Unpublished report. Newfoundland Department of Mines and Energy, Geological Survey Branch, File LAB 0978.

—— & RYAN, B. 1994. Late Archaean and Proterozoic events in the central Nain craton. *In*: WARDLE, R. J. & HALL, J. (eds) *Eastern Canadian Shield Onshore–Offshore Transect (ECSOOT), Report of Transect Meeting*. University of British Columbia, Lithoprobe Report No. 36, 53–61.

CROCKER, C. H., COLLERSON, K. D., LEWRY, J. F. & BICKFORD, M. E. 1993. Sm–Nd, U–Pb and Rb–Sr geochronology and lithostructural relationships in the southwestern Rae Province: constraints on crustal assembly in the western Canadian Shield. *Precambrian Research*, 61, 27–50.

DIMROTH, E. 1972. The Labrador Geosyncline revisited. *American Journal of Science*, 272, 487–506.

EINSELE, G. 1986. Interaction between sediments and basalt injections in young Gulf of California-type spreading centres. *Geologisches Rundschau*, 75, 197–208.

EMSLIE, R. F. & LOVERIDGE, W. D. 1992. Fluorite-bearing Early and Middle Proterozoic granites, Okak Bay area, Labrador, Geochronology and Petrogenesis. *Lithos*, 28, 87–109.

—— & THÉRIAULT, R. J. 1991. Sm–Nd and Rb–Sr isotopic characteristics of ferrodiorites and related rocks associated with anorthositic complexes, central Labrador. *Geological Association of Canada–Mineralogical Association of Canada, Joint Annual Meeting, Program with Abstracts*, 16, A34.

ERMANOVICS, I. & VAN KRANENDONK, M. J. 1990. The Torngat Orogen in the North River–Nutak transect area of Nain and Churchill provinces. *Geoscience Canada*, 17, 279–283.

FAHRIG, W. F. & WEST, T. D. 1986. *Diabase dyke swarms of the Canadian Shield*. Geological Survey of Canada, Map 1627A.

GIBB, R. A. 1983. Model for suturing of Superior and Churchill plates: an example of double indentation tectonics. *Geology*, 11, 413–417.

GOULET, N. & CIESIELSKI, A. 1990. The Abloviak shear zone and the NW Torngat Orogen, eastern Ungava Bay, Québec. *Geoscience Canada*, 17, 269–272.

GOWER, C. F., RYAN, A. B & RIVERS, T. 1990. Mid-Proterozoic Laurentia–Baltica: an overview of its geological evolution and a summary of the contributions made by this volume. *In*: GOWER, C. F., RYAN, A. B. & RIVERS, T. (eds) *Mid-Proterozoic Laurentia–Baltica*. Geological Association of Canada, Special Paper 38, 1–22.

HALL, J., WARDLE, R. J., GOWER, C. F., KERR, A., COFLIN, K., KEEN, C. E. & CARROLL, P. 1995. Proterozoic orogens of the northeastern Canadian Shield: new information from crustal reflection seismic surveys. *Canadian Journal of Earth Sciences*, 32, 1119–1131.

HOFFMAN, P. F. 1987. Early Proterozoic foredeeps, foredeep magmatism, and superior-type iron-formations of the Canadian Shield. *In*: KRÖNER, A. (ed.) *Proterozoic Lithospheric Evolution*. American Geophysical Union, Geodynamics Series, 17, 85–98.

—— 1990a. Subdivision of the Churchill Province and extent of the Trans-Hudson Orogen. *In*: LEWRY, J. F. & STAUFFER, M. R. (eds) *The Early Proterozoic Trans-Hudson Orogen*. Geological Association of Canada, Special Paper 37, 15–39.

—— 1990b. Dynamics of the assembly of northeast Laurentia in Geon 18 (1.9–1.8 Ga). *Geoscience Canada*, 17, 222–226.

JACKSON, G. D. & TAYLOR, F. C. 1972. Correlation of major Aphebian rock units in the northeastern Canadian Shield. *Canadian Journal of Earth Sciences*, 9, 1659–1669.

JAMES, D. T. & DUNNING, G. R. 1994. Preliminary Results of detailed U–Pb Geochronology from the Rae province, Western Labrador. *In*: WARDLE, R. J. & HALL, J. (compilers) *Eastern Canadian Shield Onshore–Offshore Transect (ECSOOT), Report of Transect Meeting*. University of British Columbia, Lithoprobe Report No. 36, 63–71.

——, KILFOIL, G. & NUNN, G. A. G. 1994. Structural, metamorphic and intrusive relationships in the Rae Province, western Labrador. *In*: WARDLE, R. J. & HALL, J. (eds) *Eastern Canadian Shield Onshore–Offshore Transect (ECSOOT), Report of Transect Meeting*. University of British Columbia, Lithoprobe Report No. 36, 81–88.

KERR, A., FRYER, B. J., WARDLE, R. J., RYAN, B. & BRIDGWATER, D. 1994. Nd isotopic and geochemical studies in the Labrador shield: progress report and preliminary data from Torngat Orogen. *In*: WARDLE, R. J. & HALL, J. (eds) *Eastern Canadian Shield Onshore–Offshore Transect, Report of Transect Meeting*. University of British Columbia, Lithoprobe Report No. 36, 108–118.

KNIGHT, I. & MORGAN, W. C. 1981. The Aphebian Ramah Group, northern Labrador. *In*: CAMPBELL, F. H. A. (ed.) *Proterozoic Basins of Canada*. Geological Survey of Canada, Paper 81-10, 313–330.

KROGH, T. E. 1986. *Report to Newfoundland Department of Mines and Energy on isotopic dating results from the 1985-1986 geological research agreement*. Newfoundland Department of Mines and Energy,

Mineral Development Division, Open File Report LAB 707.
—— 1992. *Report on Geochronological Contract Work; Labrador.* Unpublished report to Newfoundland Department of Mines and Energy, Geological Survey Branch, File LAB 944.

MACHADO, N. 1990. Timing of collisional events in the Trans-Hudson Orogen: evidence from U–Pb geochronology for the New Quebec Orogen, the Thompson Belt and the Reindeer Zone (Manitoba and Saskatchewan). *In*: LEWRY, J. F. & STAUFFER, M. R. (eds) *The Early Proterozoic Trans-Hudson Orogen of North America.* Geological Association of Canada, Special Paper **37**, 433–441.

——, GOULET, N. & GARIÉPY, C. 1989. U–Pb geochronology of reactivated Archaean basement and of Hudsonian metamorphism in the northern Labrador Trough. *Canadian Journal of Earth Sciences*, **26**, 1–15.

——, PERREAULT, S. & HYNES, A. 1988. Timing of continental collision in the northern Labrador Trough, Québec: Evidence from U–Pb geochronology. *In: Geological Association of Canada, Mineralogical Association of Canada and Canadian Society of Petroleum Geologists, Joint Annual Meeting, St. John's 1988. Program with Abstracts*, **13**, A76.

MENGEL, F. & RIVERS, T. 1990. The synmetamorphic P–T–t path of granulite-facies gneisses from the Torngat Orogen, and its bearing on their tectonic history. *Geoscience Canada*, **17**, 288–293.

—— & —— 1994. 3-D architecture and thermal structure of a crustal-scale transcurrent shear zone – report 4: P–T results from the northern Torngat transect. *In*: WARDLE, R. J. & HALL, J. (eds) *Eastern Canadian Shield Onshore–Offshore Transect (ECSOOT), Report of Transect Meeting.* University of British Columbia, Lithoprobe Report No. **36**, 163–170.

MOORHEAD, J. & HYNES, A. 1990. Nappes in the internal zone of the northern Labrador Trough: Evidence for major early, NW-vergent basement transport. *Geoscience Canada*, **17**, 241–244.

NUNN, G. A. G., HEAMAN, L. & KROGH, T. E. 1990. U–Pb geochronological evidence for Archaean crust in the continuation of the Rae Province (eastern Churchill Province), Grenville Front Tectonic Zone, Labrador. *Geoscience Canada*, **17**, 259–265.

PERCIVAL, J. A., MORTENSEN, J. K., STERN, R. A., CARD, K. D. & BÉGIN, N. J. 1992. Giant granulite terranes of northeastern Superior Province: the Ashuanipi complex and Minto block. *Canadian Journal of Earth Sciences*, **29**, 2287–2308.

PERREAULT, S. & HYNES, A. 1990. Tectonic evolution of the Kuujjuaq terrane, New Québec Orogen. *Geoscience Canada*, **17**, 238–240.

POIRIER, G., PERREAULT, S. & HYNES, A. 1990. Nature of the Eastern boundary of the Labrador Trough near Kuujjuaq, Quebec. *In*: LEWRY, J. F. & STAUFFER, M. R. (eds) *The Early Proterozoic Trans-Hudson Orogen of North America.* Geological Association of Canada, Special Paper **37**, 397–411.

RIVERS, T. & MENGEL, F. 1994. A cross-section of the Abloviak shear zone at Saglek Fiord and a preliminary model for Torngat Orogen. *In*: WARDLE, R. J. & HALL, J. (eds) *Eastern Canadian Shield Onshore–Offshore Transect (ECSOOT), Report of Transect Meeting.* University of British Columbia, Lithoprobe Report No. **36**, 171–184.

——, —— & Goulet, N. 1993. A cross-section of the Abloviak shear zone at Abloviak Fiord – a preliminary report. *In*: WARDLE, R. J. & HALL, J. (eds) *Eastern Canadian Shield Onshore–Offshore Transect (ECSOOT), Report of Transect Meeting.* University of British Columbia, Lithoprobe Report No. **32**, 54–63.

ROHON, M.-L., VIALETTE, Y., CLARK, T., RODGER, G. & OHNENSTETTER, D. 1993. Aphebian mafic-ultramafic magmatism in the Labrador Trough (New Québec): its age and the nature of its mantle source. *Canadian Journal of Earth Sciences*, **30**, 1582–1593.

RYAN, B. 1990. Basement-cover relationships and metamorphic patterns in the foreland of the Torngat Orogen in the Saglek–Hebron area, Labrador. *Geoscience Canada*, **17**, 276–279.

—— (COMPILER), KROGH, T. E., HEAMAN, L., SCHÄRER, U., PHILIPPE, S. & OLIVER, G. 1991. On recent geochronological studies in the Nain Province, Churchill Province and Nain Plutonic Suite, north-central Labrador. *In: Current Research 1991.* Newfoundland Department of Mines and Energy, Geological Survey Branch, Report 91-1, 257–261.

SCOTT, D. J. & MACHADO, N. 1994. U–Pb geochronology of the northern Torngat Orogen: results from work in 1993. *In*: WARDLE, R. J. & HALL, J. (eds) *Eastern Canadian Shield Onshore–Offshore Transect (ECSOOT), Report of Transect Meeting.* University of British Columbia, Report No. 36, Lithoprobe, 141–155.

—— & —— 1995. U–Pb geochronology of the northern Torngat Orogen, Labrador, Canada: a record of Palaeoproterozoic magmatism and deformation. *Precambrian Research*, **70**, 169–190.

SKULSKI, T., WARES, R. P. & SMITH, A. D. 1993. Early Proterozoic (1.88–1.87 Ga) tholeiitic magmatism in the New Québec orogen. *Canadian Journal of Earth Sciences*, **30**, 1505–1520.

TAYLOR, F. C. 1979. *Reconnaissance geology of a part of the Precambrian Shield, northeastern Quebec, northern Labrador and Northwest Territories.* Geological Survey of Canada, Memoir 393.

THÉRIAULT, R. J. & ERMANOVICS, I. 1994. Nd isotopic transect across the Torngat Orogen, North River–Nutak area. *In*: WARDLE, R. J. & HALL, J. (eds) *Eastern Canadian Shield Onshore–Offshore Transect (ECSOOT), Report of Transect Meeting.* University of British Columbia, Lithoprobe Report No. **36**, 119–125.

VAN DER LEEDEN, J., BÉLANGER, M., DANIS, D., GIRARD, R. & MARTELAIN, J. 1990. Lithotectonic domains in the high-grade terranes east of the Labrador Trough (Québec). *In*: LEWRY, J. F. & STAUFFER, M. R. (eds) *The Early Proterozoic Trans-Hudson Orogen of North America.* Geological Association of Canada, Special Paper **37**, 371–386.

VAN KRANENDONK, M. J. 1992. *Geological evolution of the Archaean Nain Province and the Early Proterozoic Torngat Orogen as seen along a transect in the North River–Nutak map area, northern*

Labrador, Canada. PhD thesis, Queen's University, Kingston, Ontario, Canada.

—— & ERMANOVICS, I. 1990. Structural evolution of the Hudsonian Torngat Orogen in the North River map area, Labrador: Evidence for east-west transpressive collision of Nain and Rae continental blocks. *Geoscience Canada,* **17,** 283–288.

—— & SCOTT, D. J. 1992. Preliminary report on the geology and structural evolution of the Komaktorvik Zone of the Early Proterozoic Torngat Orogen, Eclipse Harbour area, northern Labrador. *In: Current Research, Part C.* Geological Survey of Canada, Paper **92-1C,** 59–68.

—— & WARDLE, R. J. 1994. Geological synthesis and musings on possible subduction–accretion models in the formation of the northern Torngat Orogen. *In:* WARDLE, R. J. & HALL, J. (eds) *Eastern Canadian Shield Onshore–Offshore Transect (ECSOOT), Report of Transect Meeting.* University of British Columbia, Lithoprobe Report No. **36,** 62–80.

——, ST-ONGE, M. R. & HENDERSON, J. R. 1993a. Palaeoproterozoic tectonic assembly of Northeast Laurentia through multiple indentations. *Precambrian Research,* **63,** 325–347

——, GODIN, L., MENGEL, F. C., SCOTT, D. J., WARDLE, R. J., CAMPBELL, L. C. & BRIDGWATER, D. 1993b. Geology and structural development of the Archaean and Palaeoproterozoic Burwell domain, northern Torngat Orogen, Labrador and Quebec. *In: Current Research, Part C.* Geological Survey of Canada, Paper **93-1C,** 329–340.

——, WARDLE, R. J., MENGEL, F. C., CAMPBELL, L. M. & REID, L. 1994. New results and summary of the Archaean and Palaeoproterozoic geology of the Burwell domain, northern Torngat Orogen, Labrador, Quebec, and Northwest Territories. *In: Current Research 1994-C.* Geological Survey of Canada, 321–332.

WARDLE, R. J. 1983. Nain–Churchill Province cross-section, Nachvak Fiord, northern Labrador. *In: Current Research.* Newfoundland Department of Mines and Energy, Mineral Development Division, Report **83-1,** 68–89.

—— & BAILEY, D. G. 1981. Early Proterozoic sequences in Labrador. *In:* CAMPBELL, F. H. A. (ed.) *Proterozoic Basins of Canada.* Geological Survey of Canada, Paper **81-10** suppl., 331–358.

——, BRIDGWATER, D., MENGEL, F., CAMPBELL, L., VAN KRANENDONK, M. J. *ET AL.* 1994. Mapping in Torngat Orogen, northernmost Labrador: Report 3, the Nain Craton (including a note on ultramafic dyke occurrences in northernmost Labrador. *In: Current Research 1994.* Newfoundland Department of Mines and Energy, Geological Survey Branch, Report **94-1,** 399–407.

——, RYAN, B. & ERMANOVICS, I. 1990a. The Eastern Churchill Province, Torngat and New Québec orogens: An overview. *Geoscience Canada,* **17,** 217–222.

——, NUNN, G. A. G. & MENGEL, F. C. 1990b. Labrador segment of the Trans-Hudson Orogen: crustal development through oblique convergence and collision. *In:* LEWRY, J. F. & STAUFFER, M. A. (eds) *The Early Proterozoic Trans-Hudson Orogen of North America.* Geological Association of Canada, Special Paper **37,** 353–369.

——, VAN KRANENDONK, M. J., MENGEL, F. & SCOTT, D. 1992. Geological mapping in the Torngat Orogen, northernmost Labrador: preliminary results. *In: Current Research 1992.* Newfoundland Department of Mines and Energy, Geological Survey Branch, Report **92-1,** 413–429.

——, ——, ——, ——, SCHWARZ, S., RYAN, B. & BRIDGWATER, D. 1993. Geological mapping in the Torngat Orogen, northernmost Labrador: Report 2. *In: Current Research 1993.* Newfoundland Department of Mines and Energy, Geological Survey Branch, Report **93-1,** 77–89.

WARES, R. P. & GOUTIER, J. 1990. Deformational style in the foreland of the northern New Québec Orogen. *Geoscience Canada,* **17,** 244–249.

WATTERSON, J. 1978. Proterozoic intraplate deformation in the light of South-east Asia neotectonics. *Nature,* **273,** 636–640.

The Makkovik Province : extension of the Ketilidian Mobile Belt in mainland North America

ANDREW KERR[1], BRUCE RYAN, CHARLES F. GOWER
& RICHARD J. WARDLE

Geological Survey of Newfoundland and Labrador, Department of Mines and Energy,
PO Box 8700, St John's, NF, Canada A1B 4J6
[1] *also of the Department of Earth Sciences, Memorial University of Newfoundland,*
St John's, NF, A1B 3X5

Abstract: The Makkovik Province of Labrador represents the extension of the Ketilidian Mobile Belt of south Greenland into mainland North America; it exhibits a threefold division into a foreland region, a fold-and-thrust belt, and an interior magmatic zone. The **Kaipokok Domain** is dominated by Archaean basement rocks that form an extension of the North Atlantic Craton, but Proterozoic reworking is recorded by the reorientation of a *c.* 2230 Ma dyke swarm. Supracrustal rocks, consisting of shallow-marine sedimentary rocks overlain by greywackes and mafic volcanic rocks, rest unconformably upon Archaean basement, but towards the interior of the belt the basal unconformity is eradicated by northwest-directed thrusting. In the **Aillik Domain**, high-grade supracrustal rocks of similar aspect to those of the Kaipokok Domain are separated from the basement by mylonite zones, in a thick-skinned fold-and-thrust belt (**Kaipokok Bay Structural Zone**), believed to record significant northwest-directed translation. The Aillik Domain also contains abundant felsic volcanic rocks that lack typical arc-like geochemical signatures. The **Cape Harrison Domain** is dominated by plutonic rocks, including suites of 1840 Ma, 1800 Ma, 1720 Ma and 1650 Ma age, but gneissic inliers apparently represent 'juvenile' Proterozoic crust. The dominant 1800–1720 Ma plutonic suites are late-orogenic to post-orogenic, siliceous, potassic granitoid rocks, which resemble Phanerozoic post-collisional suites, rather than subduction-related arc batholiths. Nd isotopic variations position the eastern edge of the North Atlantic Craton close to the boundary between the Aillik and Cape Harrison Domains. The structural evolution of the Makkovik Province records a shift from pre-1800 Ma northwest-directed thrusting to post-1800 Ma tight upright folding, also northwest-verging. However, there is also evidence for earlier (pre-1890 Ma) events in the Kaipokok Domain. Major unresolved problems include the timing of early sub-horizontal deformation (perhaps related to collisional events), the age relations and setting of supracrustal sequences, the location of suture zones, the absence of clear arc-like magmatic assemblages, and the nature and antiquity of the eastern juvenile crustal block.

In the mid-1930s, long before the plate tectonics revolution, E. H. Kranck commented on geological similarities between the Labrador coast and southwest Greenland. In a wider discussion of possible correlations, his words were prophetic:

'...it will be shown that Labrador really can deliver some contributions towards an answer to these questions, and therefore also can contribute to the solving of the question regarding the connection between the Pre-Cambrian in Northern Europe and North America'

(Kranck 1939, p. 66).

Specific correlations between the Makkovik Province and the Ketilidian Mobile Belt of southern Greenland (e.g., Sutton *et al.* 1972; Bridgwater *et al.* 1973; Gower & Ryan, 1986) have now firmly established the pre-Cretaceous connection between these two areas (Fig. 1). This paper presents a synthesis of published work and new data from the Makkovik Province, emphasizing recent mapping, U–Pb geochronology and geochemical–isotopic studies, and discusses correlations across the Labrador Sea. This is timely in the light of new seismic reflection data from the Lithoprobe project (Hall *et al.* 1995), and re-appraisal of the Ketilidian Mobile Belt (Chadwick & Garde, this volume).

The Makkovik Province lies in east-central Labrador, between the Archaean Nain Province and the Early to Middle Proterozoic Grenville Province (Fig. 1). It was originally viewed as a subprovince of the Nain Province (Gandhi *et al.* 1969; Taylor 1971), but was elevated to full province status by

From Brewer, T. S. (ed.), 1996, *Precambrian Crustal Evolution in the North Atlantic Region,*
Geological Society Special Publication No. 112, pp. 155–177.

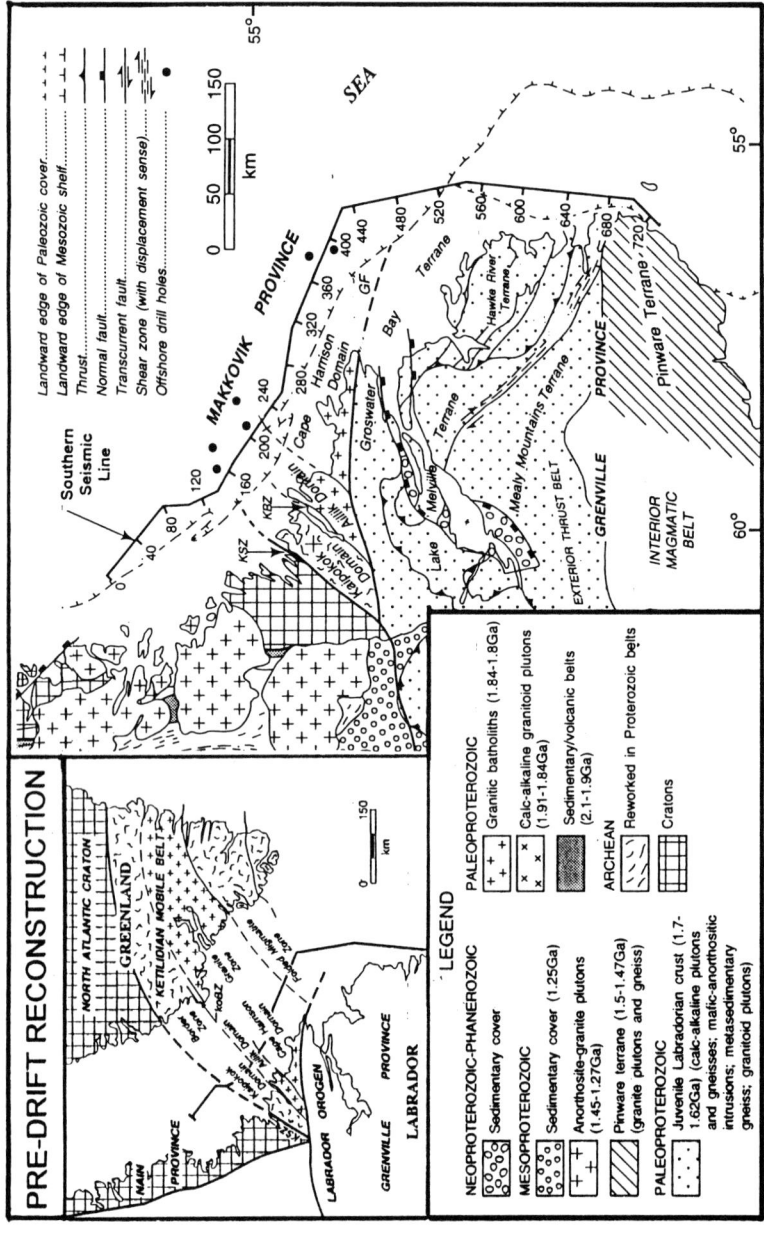

Fig. 1. Geology of the Labrador coast region, Canada, showing major subdivisions of the Makkovik Province and adjacent regions of the Nain and Grenville provinces. Map also shows the location of the Lithoprobe seismic-reflection line run in 1992 (Hall *et al.* 1995). Inset shows correlations between the Makkovik Province and Ketilidian Mobile Belt in Greenland. KSZ, Kanairiktok Shear Zone; KBZ, Kaipokok Bay Structural Zone; KoBZ, Kobbermine Bugt Zone.

Gower & Ryan (1986). The Makkovik Province and Ketilidian Mobile Belt are coeval with other parts of an extensive orogenic belt on the southern margin of Laurentia–Baltica, including the Penokean Belt of the mid-continent USA and the Svecofennian Belt of Scandinavia (Gower et al. 1990).

Regional setting and principal subdivisions

Eastern Labrador (Fig. 1) includes portions of the Archaean Nain Province (part of the North Atlantic Craton), the early Proterozoic Eastern Churchill and Makkovik provinces, and the Early to Middle Proterozoic Grenville Province. The northwestern boundary of the Makkovik Province (Fig. 1) is a shear zone that defines the limit of Proterozoic reworking of the Archaean rocks. The southern boundary of the Makkovik Province is the Grenville Front, marking the northern limit of Grenvillian (c. 1000 Ma) deformation (Gower et al. 1980), but it is likely that elements of the Makkovik Province persist within the northernmost Grenville Province. The Grenville Province in this area is dominated by rocks formed c. 1710–1630 Ma ago, and these components were probably initially assembled onto the southern margin of Laurentia during this period (Wardle et al. 1986; Gower et al. 1990).

The Makkovik Province is subdivided into three regions (after Gower & Ryan, 1986; Fig. 1). The **Kaipokok Domain** is dominated by an Archaean high-grade basement complex which is interpreted as a reworked extension of the Nain Province. The **Aillik Domain** is dominated by supracrustal rocks, including a mafic volcanic–psammitic assemblage and a thick sequence of felsic volcanic and volcaniclastic rocks. It is separated from the Kaipokok Domain by a zone of structural complexity and strong deformation here termed the **Kaipokok Bay Structural Zone**, which includes major mylonite zones along the basement-cover contact. The **Cape Harrison Domain** is dominated by syntectonic and post-tectonic plutonic suites, with scattered enclaves of supracrustal material and earlier orthogneisses. Its boundary with the Aillik Domain is largely intrusive, although it is locally defined by faults. In the following discussions, all ages quoted are U–Pb zircon determinations unless otherwise stated.

Kaipokok Domain

Reworked Archaean Basement Complex

This consists of grey, banded, tonalitic to granodioritic orthogneisses considered to be reworked equivalents of the southern Nain Province, which dates back to at least 3250 Ma (Loveridge et al. 1987; Wasteneys et al. 1992). A direct link between the Nain Province and the Kaipokok Domain is provided by the Kikkertavak Dykes, now dated precisely at c. 2230 Ma (Cadman et al. 1993). These dykes are undeformed in the Nain Province, where they truncate all structures in the older gneisses but, within the Kaipokok Domain, they are at amphibolite facies, and reorientated and folded into dominantly NE–SW structural trends (Ryan et al. 1983; Fig. 1). In some areas, Proterozoic reworking is pervasive and a 'straightened' gneiss is produced by transposition and attenuation of the dyke remnants (see later discussion). In the northeast, this straightened gneiss was migmatized and subsequently overprinted by mylonite zones associated with the Kaipokok Bay Structural Zone. However, Archaean structural and chronological relationships characteristic of the Nain Province remain recognizable in low-strain enclaves (Ryan et al. 1983). It is likely that Proterozoic reworking was multiphase, and its exact timing and duration in various parts of the area are not fully resolved (see later discussion).

Proterozoic supracrustal rocks (Moran Lake Group)

The Moran Lake Group (Smyth et al. 1978; Wardle & Bailey 1981; Ryan 1984) rests unconformably upon Archean rocks in which a regolith is developed (Ryan 1984). The Moran Lake Group is affected by at least two periods of folding, and its basal unconformity is disturbed by thrusting. In the east, amphibolite and schist remnants are surrounded by refoliated gneisses (Ryan 1984), that are cut by granites dated at c. 1895–1891 Ma (Kerr et al. 1992; see below) which (indirectly) provide a minimum age for the supracrustal rocks.

The Moran Lake Group is at least 1.6 km thick, and divided into two formations (Smyth et al. 1978; Ryan 1984; Fig. 3). The Warren Creek Formation comprises mature sandstones and quartzites overlain by thinly-bedded shale and silicate-oxide iron formation, in turn overlain by a shallow water shale–dolostone unit, passing upward into turbiditic siltstone and greywacke, locally with massive sulphides. The Joe Pond Formation consists of at least 1 km of massive and pillowed mafic volcanic rocks. There are stratigraphic similarities between the Moran Lake Group and the Lower Aillik Group of the Aillik Domain (Fig. 3; see below). The mafic volcanic rocks evolve stratigraphically from oceanic island-type basalts to mid-ocean ridge basalts (Ryan 1984; North 1988; Fig. 5).

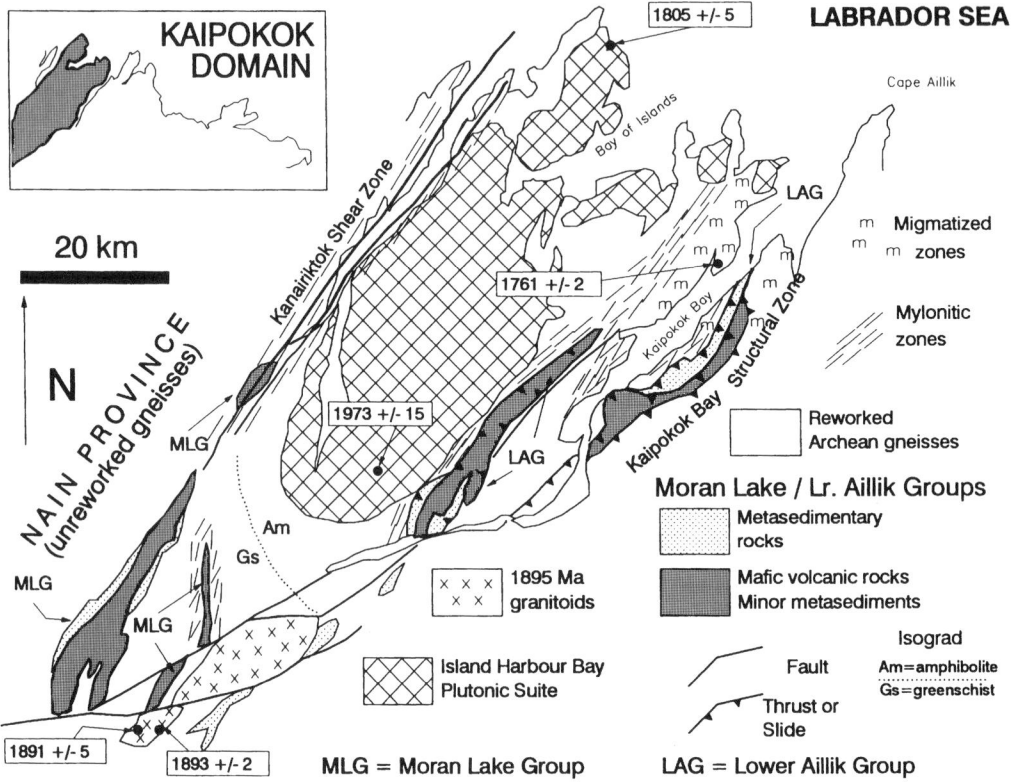

Fig. 2. Simplified geological map of the Kaipokok Domin of the Makkovik Province (after Ryan *et al.* 1983; Ryan 1984; Ermanovics 1992). Ages indicated on map are all U–Pb determinations (mostly on zircon); see text for sources and discussion.

Early Proterozoic intrusive rocks

Foliated potassic granites dated at 1893 ± 2 Ma and 1891 ± 5 Ma (Kerr *et al.* 1992) in the southwest (Fig. 2) cut foliations associated with interleaving of basement and cover, and thereby constrain the age of the Moran Lake Group. In the northwest, reworked gneisses are intruded by the Island Harbour Bay Plutonic Suite (IHBPS), a complex, polyphase body ranging from hornblende diorite and tonalite to granite (Ryan *et al.* 1983; Ermanovics 1992). The youngest phases are massive and undeformed, but magmatic fabrics and primary layering, parallel to pluton margins, are evident in earlier, marginal phases, which are locally foliated. The structural patterns within the intrusion and host gneisses have been interpreted in terms of diapiric emplacement, but it is possible that some early phases are syntectonic (Ryan 1984; Korstgård & Ermanovics 1985). A massive phase has been dated at 1805 ± 5 Ma (Loveridge *et al.* 1987), but there is no age control on foliated members, except for strongly discordant U–Pb zircon ages of 1973 ± 15 Ma and *c.* 1931 Ma (Brooks 1982; 1983). These early results must be interpreted with caution because of possible inheritance, but they are potentially significant in view of the *c.* 1895 Ma granites (Kerr *et al.* 1992) and similar U–Pb ages from offshore wells (Wasteneys *et al.* 1992). Major element geochemistry shows that the IHBPS has contrasting trondhjemitic and calc-alkaline evolutionary trends (Ermanovics 1992) which are inconsistent with a single, related, magma sequence. These tonalitic and trondhjemitic rocks of the IHBPS form the only truly 'primitive' calcic and sodic igneous suites in the western Makkovik Province (see below).

Aillik Domain

The Aillik Domain (Fig. 4; after Gower *et al.* 1982; Kerr 1989a) contains the most important supracrustal sequences and major structures in the Makkovik Province. Its boundary with the

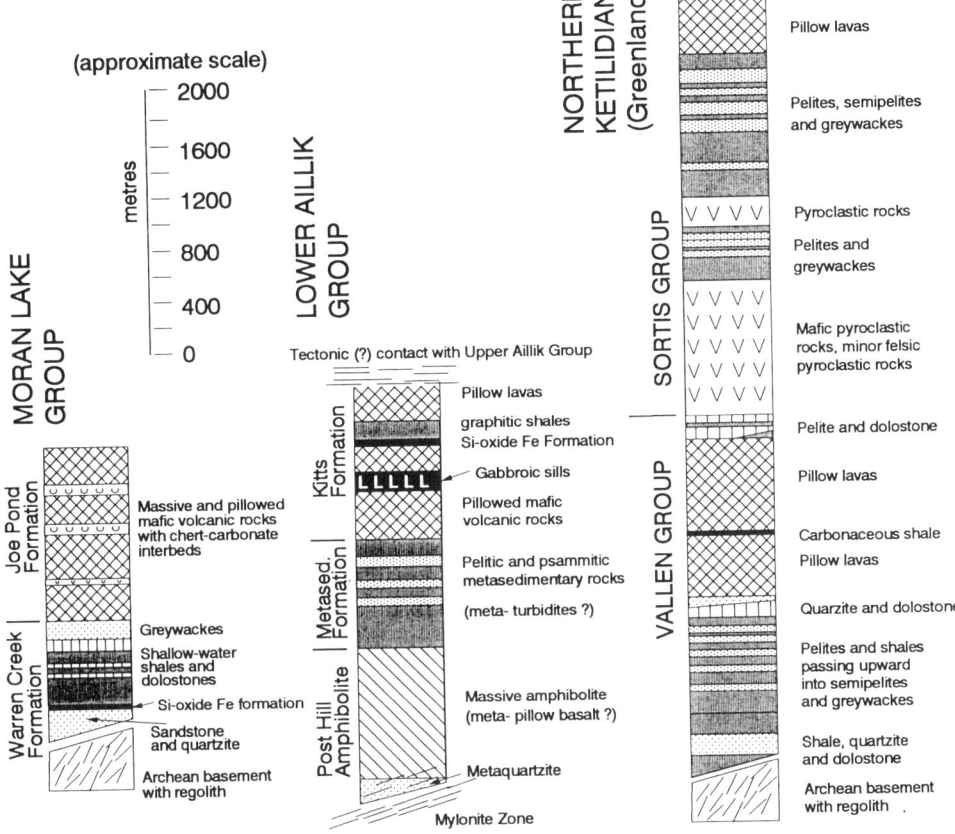

Fig. 3. Partly schematic illustration of the stratigraphy of Early Proterozoic cover sequences in the Makkovik Province (after Wardle & Bailey 1981) and on the northern margin of the Ketilidian Mobile Belt (after Allaart 1976).

Kaipokok Domain is here drawn at the mylonite zone(s) that separate(s) the reworked gneisses from high-grade supracrustal rocks of the Lower Aillik Group, but we stress that the boundary region is a wide zone of structural complexity and intense deformation (Kaipokok Bay Structural Zone). Outliers of amphibolite and metasedimentary rocks, surrounded by mylonitic gneisses, also occur within the Kaipokok Domain (Fig. 2), and are correlated with the rocks of the Aillik Domain. Also, the boundary between the Lower Aillik and Upper Aillik Groups (Fig. 4) may be an important internal structural break within the Aillik Domain.

Basement orthogneisses

Quartzofeldspathic gneisses immediately east of the mylonite zone, and within thrust slices in the Kaipokok Bay Structural Zone, have a complex early history that predates their interleaving with the supracrustal rocks (Marten 1977; Gower *et al.* 1982; Ryan *et al.* 1983). Low-strain enclaves resemble 'straightened' gneisses north of Kaipokok Bay (Ryan *et al.* 1983; see above). Proterozoic granitoid rocks in the basement have very low ε_{Nd} consistent with anatectic derivation from mid-Archaean sources (Kerr & Fryer 1994). A poorly known gneissic enclave southeast of Kaipokok Bay (Fig. 4) lies on the southeast side of the Kaipokok Bay Structural Zone, and does not necessarily represent Archaean basement (see later discussions).

Early Proterozoic supracrustal rocks (Lower Aillik Group)

The Lower Aillik Group forms several elongate panels bounded by ductile shear zones within the Kaipokok Bay Structural Zone (Fig. 4), and is

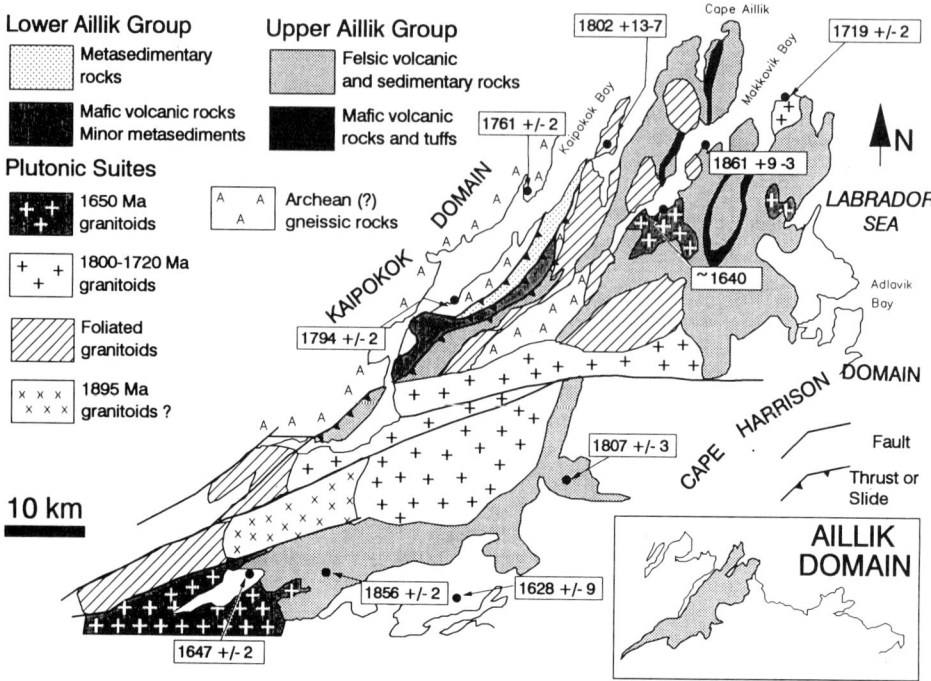

Fig. 4. Simplified geological map of the Aillik Domain of the Makkovik Province (after Gower *et al.* 1982; Kerr 1989*a*, in press). Ages indicated on map are all U–Pb determinations (mostly on zircon); see text for sources and discussion.

separated from the gneissic rocks of the Kaipokok Domain by a major mylonite zone or zones. It shows a steep metamorphic gradient from greenschist (southeast) to amphibolite (northwest), and the original stratigraphy is poorly preserved. The assemblage is similar to the Moran Lake Group, and the two have been loosely correlated (Smyth *et al.* 1978; Wardle & Bailey 1981; Fig. 3). A quartzitic unit is locally present within and immediately above the basal mylonite, which is elsewhere developed in a thick amphibolite unit. A middle unit of psammitic and pelitic metasedimentary rocks, locally with bedded sulphide-rich zones, is overlain by a 500 m–1000 m thick sequence of pillowed metabasalt, mafic tuffaceous rocks and gabbroic sills. The mafic volcanic rocks include graphitic schist and silicate-oxide iron formation, which host stratiform uranium mineralization (Gandhi 1978; Gower *et al.* 1982). Mafic volcanic rocks are subalkaline, low-K tholeiites, which could either represent MORB-like or immature arc settings (Evans 1980; Fig. 5).

The earliest phases of deformation that affected the Lower Aillik Group have a minimum age of 1802 + 13/−7 Ma (see later discussion). Correlation with the Moran Lake Group, if valid, would suggest a minimum age of 1895 Ma (see above). The original relationship between the Lower Aillik Group and the Archaean gneiss complex is also uncertain. On the basis of correlations with the Moran Lake Group, it has been viewed as originally unconformable (Sutton 1972; Gower *et al.* 1982). However, anatectic granites within Lower Aillik Group metasedimentary rocks have ε_{Nd} of +2 at 1800 Ma, suggesting that their sources were relatively juvenile (Kerr 1989*a*), and unlikely to represent local Archaean crust.

Early Proterozoic supracrustal rocks (Upper Aillik Group)

The term 'Upper Aillik Group' is a relic of the original stratigraphic term 'Aillik Series' (Kranck 1939), and may be misleading, as the relationship between the lower and upper Aillik Groups is controversial (see below). Marten (1977) contended that both shared a common deformational history in the Kaipokok Bay area, but Clark (1979) suggested that the Upper Aillik Group escaped early deformational events. Wardle (1984) reported that their mutual contact is intensely deformed.

There is no clear indication that it represents a tectonized unconformity, and the Upper Aillik Group may be allochthonous. There is also little consensus on the evolution or internal relationships of the Upper Aillik Group. It appears to contain two divisions, which are separated by a persistent unit of mafic volcanics and tuffs (Gower & Ryan 1987; Fig. 4); however, these divisions lack laterally

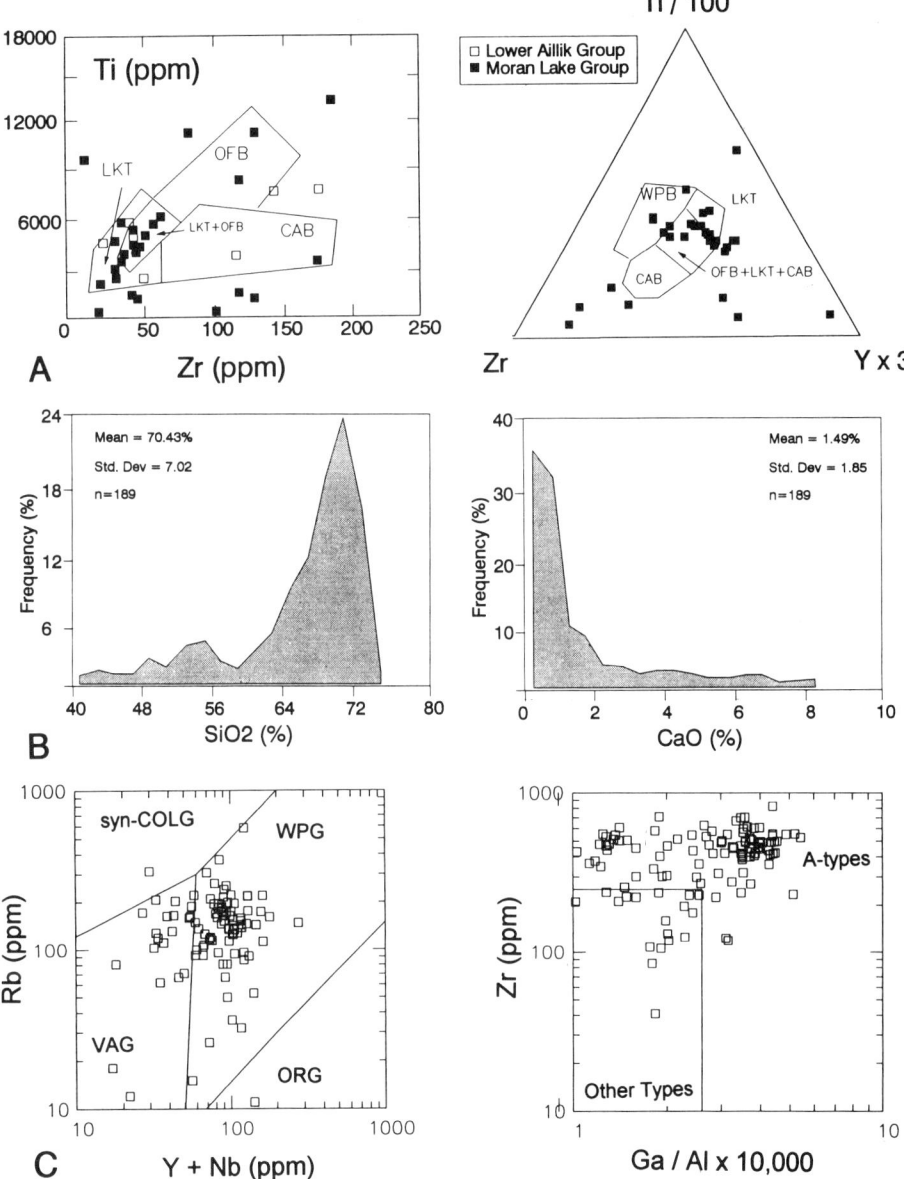

Fig. 5. Salient geochemical features of volcanic rocks within supracrustal sequences of the Makkovik Province. (A) Trace element discrimination diagrams (Pearce & Cann 1973) for mafic volcanic rocks of the Moran Lake Group and Lower Aillik Group; data from North (1988) and Wilton (1996). (B) SiO_2 and CaO frequency histograms for felsic volcanic rocks of the Upper Aillik Group, data from Kerr (1989a). (C) Trace element discrimination diagrams (Pearce et al. 1984; Whalen et al. 1987) for felsic volcanic rocks of the Upper Aillik Group, data from Kerr (1989a). OFB, ocean floor basalt; CAB, calc-alkaline basalt; WPB, within-plate basalt; LKT, low-K tholeiite; VAG, volcanic arc granite; WPG, within-plate granite; syn-COLG, syncollisional granite; ORG, ocean ridge granite.

continuous internal stratigraphy, due to rapid original facies variations, complicated by deformation and late faulting. The earlier division includes arkosic sandstones, felsic volcaniclastic rocks, conglomerate, siltstone and shales, with subordinate dacitic to rhyolitic flows and sills. The later division is dominated by variably fragmental siliceous ash-flow and ash-fall tuffs, rhyolite, agglomerates and breccias, with subvolcanic quartz-feldspar porphyry. Original sedimentary and volcanic textures are locally well-preserved, but the rocks are more commonly recrystallized, saccharoidal and locally silicified. The thickness of the group is tentatively estimated to be around 5000 m (Wardle & Bailey 1981). The age of the Upper Aillik Group is best defined by U–Pb ages of 1861 +9/–3 Ma, 1856 ± 2 Ma and 1807 ± 3 Ma, which indicate two discrete components (Schärer et al. 1988). The youngest age is from a quartz-feldspar porphyry, but the c. 1860 Ma ages from both 'early' and 'late' components underline the stratigraphic uncertainties. The c. 1860 Ma ages have been cited as a minimum age for the 'underlying' Lower Aillik Group (Schärer et al. 1988); given the uncertainty of their contact relations, this conclusion is equivocal, and their ages could overlap or (conceivably) be reversed from those implied by the nomenclature.

Most geochemical studies indicate that the sequence is dominated by high-SiO_2 rhyolites, with subordinate mafic rocks, and a dearth of intermediate compositions. White & Martin (1980) suggested an originally alkaline composition, but recognized effects from alkali metasomatism. Evans (1980) and Wardle & Bailey (1981) preferred a high-K calcalkaline affinity. Additional data, including areas well-removed from mineralization (Kerr 1989a), show that the Upper Aillik Group is strongly siliceous and potassic, and has trace element characteristics transitional between volcanic arc (VAG) and within-plate (WPG) or so-called 'A-type' granites (Fig. 5). Such compositions are not necessarily indicative of an 'anorogenic' setting, as rocks of this general affinity are known from postorogenic environments (e.g. the Appalachians) and also in distal, mature-arc settings. There are also geochemical differences between the earlier and later components of the Upper Aillik Group (Gower & Ryan 1987; Kerr 1989a) and the former is more sodic. Thus, the petrological and geochemical data from the Upper Aillik Group do not rule out an arc-type setting, but the case for equating these rocks with 'Andean' or island arcs is weak; a distal arc setting, far removed from the locus of subduction, is the only candidate. However, a setting unrelated to subduction (e.g. postcollisional extension) is a valid alternative.

Early Proterozoic deformed ('Syntectonic') plutonic suites

Deformed ('syntectonic') plutonic suites (Fig. 4) include melanocratic quartz monzonites with dioritic enclaves, evolved fluorite-bearing granites, and anatectic leucogranites developed within the Lower Aillik Group and adjacent basement rocks (Marten 1977; Gower et al. 1982; Kerr 1989a; 1994). These truncate early fabrics related to interleaving of basement and Lower Aillik Group, but contain a single fabric related to later upright folding (Marten 1977). The 1802 +13/–7 Ma (Kerr et al. 1992) Long Island quartz monzonite cuts early structures in the Kaipokok Bay Structural Zone and constrains the age of the Lower Aillik Group and the timing of this early deformation (see later discussion). The significance of a 1985 ± 15 Ma U–Pb age from an inland unit (Brooks 1982) is unclear; it may record inheritance and (or) point to the presence of pre-1890 Ma intrusive rocks (cf. Kerr et al. 1992; Wasteneys et al. 1992). Titanite ages of 1794 ± 2 Ma, 1761 ± 2 Ma and 1746 ± 2 Ma (Schärer et al. 1988; Kerr et al. 1992) suggest that the region around Kaipokok Bay remained thermally (and perhaps tectonically) active for up to 50 Ma following emplacement of these plutonic suites.

Early Proterozoic massive ('post-tectonic') plutonic suites

Undeformed plutonic suites of c. 1800 Ma, 1720 Ma and 1650 Ma age also occur in the Aillik Domain, where they mostly intrude the Upper Aillik Group. These rocks become dominant to the east, forming most of the Cape Harrison Domain, discussed below.

Cape Harrison Domain

The Cape Harrison Domain (Fig. 6) forms a narrow strip bounded by the Labrador Sea to the north and the Grenville Front to the south. The Cape Harrison Domain is dominated by plutonic suites of varied age, but contains enclaves of older gneissic rocks and supracrustal rocks. Its contact with the Aillik Domain is largely intrusive, but the transition from dominantly supracrustal to plutonic environments likely also indicates deeper crustal levels in the east.

Cape Harrison Metamorphic Suite

The Cape Harrison Metamorphic Suite (CHMS; Gower 1981) is the largest inlier of gneisses in this area, although smaller enclaves (not shown in Fig.

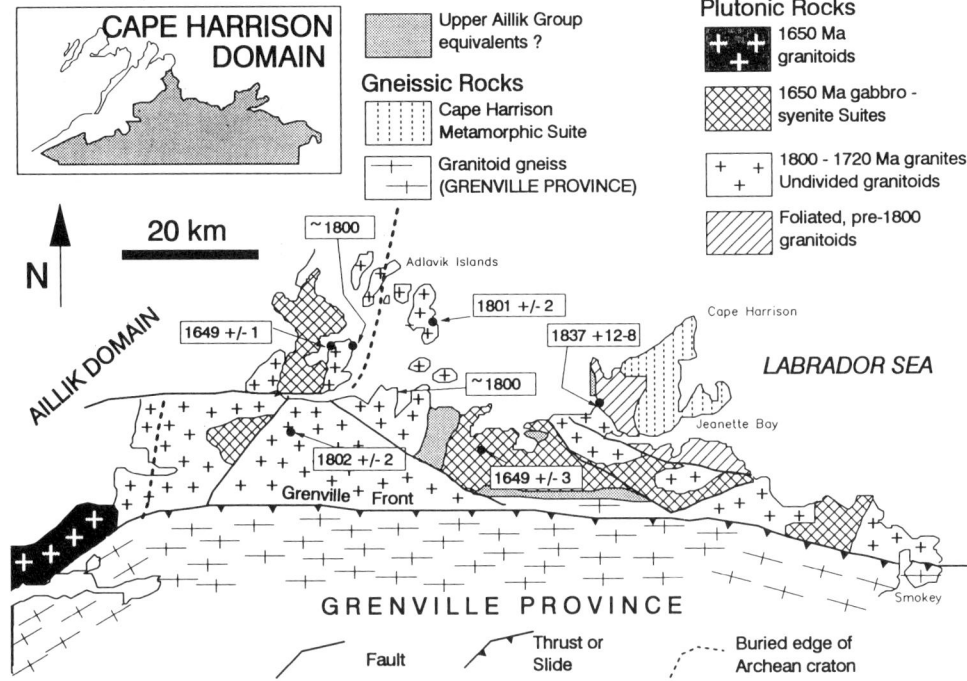

Fig. 6. Simplified geological map of the Cape Harrison Domain of the Makkovik Province (after Gower 1981; Gower *et al.* 1982; Kerr 1989a, in press). Ages indicated on map are all U–Pb determinations (mostly on zircon).

6) are present further east. The CHMS consists of variably banded gneissic tonalite, granodiorite and monzodiorite, with conformable units of amphibolite, calcsilicate and fine-grained siliceous rocks, possibly felsic metavolcanics (Gower 1981). Although these rocks superficially resemble the Archaean gneisses of the Kaipokok Domain), a Rb–Sr errorchron age of 1740 ± 85 Ma (Brooks 1982) gave a low initial Sr isotope ratio (0.7034 ± 6) inconsistent with long crustal residence. Sm–Nd isotopic studies gave ε_{Nd} of +1 to +2 at 1850 Ma, and depleted-mantle model ages of 2100–2000 Ma (Kerr & Fryer 1993; 1994). These 'juvenile' results indicate that the CHMS represents post-2100 Ma juvenile crust, but do not reliably date it; it could be as young as *c.* 1840 Ma, a minimum age provided by an adjacent granite with a simple fabric. In the easternmost Makkovik Province, similar gneissic rocks gave imprecise Rb–Sr errorchrons of 1899 ± 187 Ma and 1923 ± 148 Ma, also indicative of an Early Proterozoic age (Owen *et al.* 1986). The CHMS and similar gneissic enclaves attest to a discrete, post-Archaean crustal block that is 'foreign' to the reworked Nain Province rocks of the Kaipokok Domain. This block may represent a cratonic region of short (*c.* 200 Ma) crustal residence or (at the other extreme) pre- or syntectonic magmatic rocks only slightly older than the predominant massive suites.

Supracrustal rocks (Upper Aillik Group equivalents ?)

Isolated enclaves and fault-bounded slivers of supracrustal rocks in the Cape Harrison Domain are presently correlated with the Upper Aillik Group, but there are no good constraints on their ages. Most are metamorphosed and deformed, but a belt of fresh volcanic rocks immediately south of the Adlavik Islands (Fig. 6), includes well-preserved ignimbrite and spherulitic rhyolite, and may be related to nearby 1719 ± 2 Ma A-type granites (Kerr *et al.* 1992).

Early Proterozoic plutonic suites

Plutonic suites are the dominant component of the Cape Harrison Domain (Fig. 6). These were previously assumed to be of *c.* 1650 Ma age, and part of a major magmatic province termed the Trans-Labrador Batholith (e.g. Wardle *et al.* 1986). However, recent mapping, geochemical studies and geochronology (Kerr 1989a, b; Kerr & Fryer

1993, 1994; Kerr et al. 1992) show that these rocks include suites of c. 1840 Ma, c. 1800 Ma, c. 1720 Ma and c. 1650 Ma age.

The oldest suites contain pervasive NE-trending fabrics, and have been dated at 1837 +12/–8 Ma in the Cape Harrison area (Kerr et al. 1992); geological mapping in the easternmost Makkovik Province (Gower 1981; Owen et al. 1986) suggests that these are regionally extensive. The most abundant suites, based on lithological correlations of dated units, are defined by four U–Pb ages that cluster at 1801 ± 3 Ma (Kerr et al. 1992), and by Rb–Sr isochrons of 1798 ± 28 Ma and 1798 ± 48 Ma (Kerr 1989a). These include monzonite, quartz monzonite and (locally fayalitic) syenite, regionally extensive K-feldspar porphyritic hornblende–biotite granite with distinctive 'pseudorapakivi' (i.e. plagioclase mantled by K-feldspar) texture, and a variety of other, generally siliceous and potassic, granitoid rocks. These are massive and unfoliated, except locally at their margins, and in areas adjacent to the Grenville Front, where sporadic east-trending foliations are developed. The close correspondence between these ages and those reported from pervasively foliated granites in the Aillik Domain (Schärer et al. 1988; Kerr et al. 1992; see above) can be interpreted in two ways; either the terminal deformation was extremely short-lived at c. 1801 Ma, or it was regionally heterogeneous, and focused strongly in the Kaipokok Bay area. A subordinate group of potassic, fluorite-bearing, 'A-type' granites occur within both the Aillik and Cape Harrison domains, and have been dated at 1719 ± 2 Ma (Kerr et al. 1992). These represent either late postorogenic magmatism or a subsequent anorogenic episode.

Plutonic suites of c. 1650 Ma age appear to be subordinate in the Cape Harrison Domain. The Adlavik Intrusive Suite (Gower 1981) is a complex, multiphase, layered mafic intrusion with a hydrous biotite-rich primary mineral assemblage. Its most evolved phase is monzodiorite, dated at 1649 ± 1 Ma (Kerr et al. 1992). The nearby Mount Benedict Intrusive Suite is dominated by syenomonzonite (1647 ± 3 Ma; Kerr et al. 1992) and quartz syenite, with subordinate gabbro and diorite. Trace element modelling (Kerr 1989a) suggests that it was derived by extended fractionation of a mafic parental magma. Numerous gabbroic dykes and pipes that cut massive granitoid units are also viewed as part of this association; these commonly develop spectacular 'net-veined' structure, with cuspate mafic-felsic contacts and chilled margins on disrupted mafic 'pillows' indicating magma mingling. In the main body of the Adlavik Suite, identical net-veined dykes are interpreted as feeder zones to higher parts of the evolving magma chamber (Kerr 1989a). As noted initially by Kranck (1939), these net-veined dykes are closely similar to those described from southwest Greenland (e.g. Allaart 1976; Chadwick & Garde this volume). Granitoid rocks of similar age include an alaskitic granite in the Aillik Domain, dated at c. 1640 Ma, a 1628 ± 9 Ma leucogranite (Brooks 1982), and a regionally extensive granodiorite to monzogranite unit in the west, dated at 1647 ± 2 Ma (Kerr et al. 1992). The latter resembles typical megacrystic granitoid rocks of the Trans-Labrador Batholith in central Labrador (e.g. Wardle et al. 1986). The Bruce River Group in central Labrador (Ryan 1984; Fig. 2) has been dated at 1649 ± 1 Ma (Schärer et al. 1988), and provides evidence for 1650 Ma volcanism. These rocks lie within the Grenville Province, but are parautochthonous and little-deformed.

Offshore exploratory petroleum wells along-strike from the Kaipokok and Aillik Domains (Fig. 1) intersected foliated granitoid rocks with ages of 1895 ± 8 Ma and 1865–1839 Ma; wells offshore from the Cape Harrison Domain gave ages of 1801 ± 5 Ma, 1813 ± 3 Ma and 1806 +10/–6 Ma (Wasteneys et al. 1992). These results provide additional evidence for early events, as discussed above, and confirm the regional importance of the c. 1800 Ma magmatic pulse throughout the eastern Makkovik Province.

Granitoid gneisses of the northernmost Grenville Province

The region south of the Grenville Front, defined by the Benedict Fault Zone (Fig. 6), is dominated by poorly-exposed granitoid gneisses, with east-trending high-strain zones. In areas of relatively low-strain, augen gneisses resemble the K-feldspar megacrystic granites that dominate much of the Cape Harrison Domain, and they locally have vestiges of 'pseudorapakivi' texture. The granitoid gneisses are geochemically akin to the 1800–1720 Ma plutonic assemblage (see below). The field evidence suggests that parautochthonous Makkovik Province rocks extend for some distance into the northern Grenville Province. Their exact southern limit is uncertain, but it is certainly north of Groswater Bay (Fig. 1; Gower et al. 1992).

Geochemistry and affinities of plutonic suites

Large amounts of geochemical data are now available from the plutonic rocks of the Makkovik Province (Kerr 1989a, 1994). Earlier discussions (Gower et al. 1982; Kerr 1989b) were hampered by a lack of geochronological data, but their general

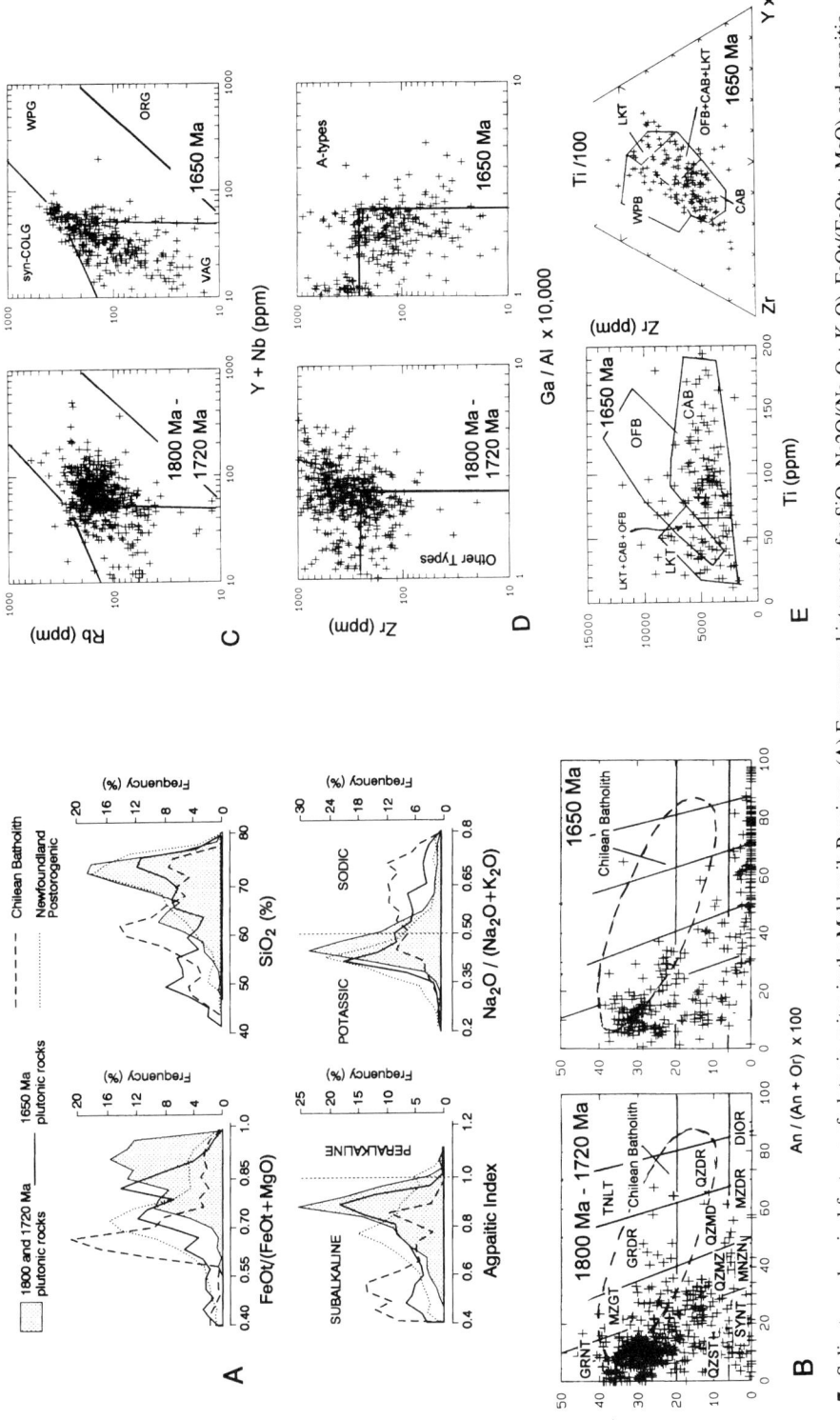

Fig. 7. Salient geochemical features of plutonic suites in the Makkovik Province. (**A**) Frequency histograms for SiO$_2$, Na$_2$O/(Na$_2$O + K$_2$O), FeOt/(FeOt + MgO) and agpaitic index values for 1800–1720 Ma and 1650 Ma plutonic suites, compared to the coastal batholith of Chile and the postorogenic granites of the Newfoundland Appalachians (after Kerr 1989a). (**B**) Normative equivalent to IUGS rock classification showing distribution of 1800–1720 Ma and 1650 Ma suites compared to the coastal batholith of Chile. (**C–D**) Trace element discrimination diagrams (Pearce et al. 1984; Whalen et al. 1987) for both 1800–1720 and 1650 Ma suites. (**E**) Trace element discrimination diagrams (Pearce and Cann 1973) for mafic and intermediate rocks assigned to 1650 Ma suites. All abbreviations as for Fig. 5.

conclusions remain valid (Fig. 7). The 1840–1800 Ma and 1720 Ma plutonic suites are dominated by high-SiO$_2$ granites, with subordinate monzonite to quartz syenite; tonalites and granodiorites are virtually absent. These granites are significantly less calcic, but more siliceous, potassic, agpaitic and Fe-enriched than typical arc-type suites such as the Sierra Nevada batholith or coastal batholiths in Peru or Chile (Kerr 1989a; Fig. 7). They are also enriched in HFS trace elements (e.g. Zr, Nb and Ga), and some LIL elements (e.g. Rb, U). In discrimination diagrams, they plot mostly in the WPG and A-type granite fields (Fig. 7). They are unlike typical continental arc magmatic suites in almost every respect, and their closest compositional analogues are the late- to postorogenic (Silurian and Devonian) plutonic suites of the Appalachians, which have been termed 'Caledonian I-type' granites (Pitcher 1983). The 1800–1720 Ma granites also have some affinities to so-called anorogenic granites, like those in some other Proterozoic orogenic belts (e.g. Wyborn et al. 1987). Amongst the 1800 Ma and older suites, the only rocks that exhibit strong similarities to arc magmas are the tonalitic and trondhjemitic components of the Island Harbour Bay plutonic suite (Ermanovics 1992).

Plutonic suites of c. 1650 Ma age have bimodal tendencies, and also do not closely correspond to arc-type suites in major element compositions (Fig. 7). They cluster more in the VAG and I-type granite fields, and their mafic to intermediate components range from low-K tholeiites to calc-alkaline basalts (Pearce & Cann 1973), but are broadly shoshonitic (Fig. 7). A similar affinity has been proposed for the c. 1650 Ma Bruce River Group (Ryan et al. 1987). A shoshonitic distal arc setting is thus possible for the 1650 Ma plutonic suites, and they may represent distal manifestations of subsequent 'Labradorian' events within the area now included in the Grenville Province.

Nd isotopic studies of igneous suites across the Makkovik Province (Kerr & Fryer 1993, 1994) yield crucial information about the distribution of lower crustal provinces. Both 1840–1800 Ma and 1720 Ma plutonic assemblages demonstrate a radical shift in ε_{Nd}, from –3 to –8 (locally –15) in the west, to +2 to +4 in the east (Fig. 8). The unradiogenic signatures in the west must reflect the influence of the Archaean North Atlantic Craton on granitoid magmas that were (in part) derived from lower crustal rocks. The inflection to positive values defines its eastern limit and boundary with a region of Early Proterozoic juvenile crust, perhaps represented by the CHMS and other gneissic enclaves. This boundary is located some 20–30 km east of the Kaipokok Bay Structural Zone, which is the most obvious tectonic boundary on surface (Fig. 6; Fig. 8). However, the ε_{Nd} of granites in the west is well above that expected for c. 3200 Ma Archean crust, which should have ε_{Nd} of –14 or less at 1800 Ma. Thus, this late- to postorogenic magmatism also involved a major contribution from more juvenile (presumably mantle) sources, and facilitated significant crustal growth (Kerr & Fryer 1994). Plutonic suites of c. 1650 Ma age display less systematic geographic variation, and some may record high-level contamination by Archaean crust (Fig. 8). Kerr & Fryer (1994) suggested that these record indirect crust-mantle mixing by sediment subduction, as documented in modern arc environments, rather than the direct lower-crustal interaction implied for older, non-subduction-related suites.

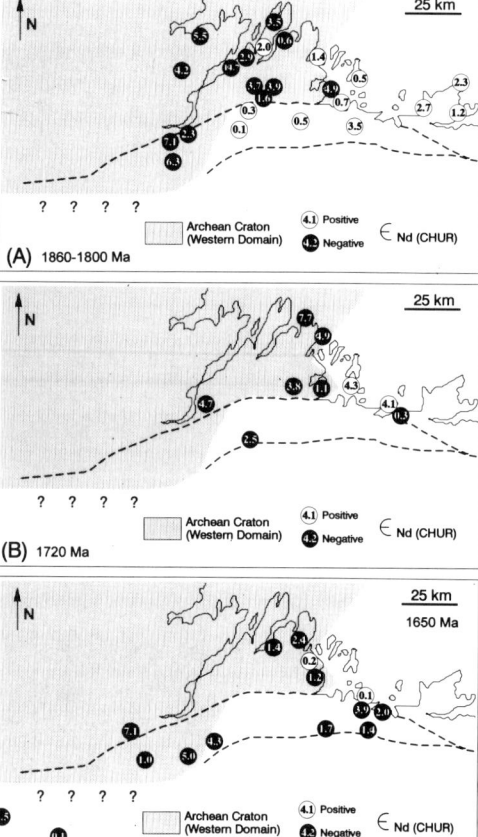

Fig. 8. Spatial Nd isotopic variations (as initial ε_{Nd} relative to CHUR) for plutonic suites of the Makkovik Province, showing the west to east shift from negative to positive values: (**A**) 1800 Ma and older suites; (**B**) 1720 Ma suites; (**C**) 1650 Ma suites. After Kerr & Fryer (1994).

Structural and metamorphic development

The structural development of the Makkovik Province is best understood in the Kaipokok Domain and the Kaipokok Bay Structural Zone, which escaped most Grenvillian effects. Elsewhere, localized Grenvillian fabrics occur to the north of the Grenville Front (Benedict Fault Zone), and it is locally difficult to discriminate Early and Middle Proterozoic effects. The discussion of tectonic evolution is treated geographically, from northwest to southeast, corresponding to a transect from the foreland to the internides of the belt.

Kanairiktok Shear Zone ('Makkovik Front')

The Kanairiktok Shear Zone (Korstgård & Ermanovics 1985; Ermanovics 1992), is a 2 km-wide belt of greenschist-facies mylonites, extending for 100 km from the coast towards the Moran Lake area (Fig. 2). The fold-and-thrust belt of the Moran Lake Group lies along-strike from this zone, but its relationship to it is unclear. In the northeast, the Kanairiktok Shear Zone defines a sharp break, from greenschist facies (northwest) to middle amphibolite facies (southeast). The mylonitic foliation dips steeply southeast, and lineations are sub-horizontal or gently northeast-plunging. Archaean gneisses, Kikkertavak dykes, and both trondhjemitic and late granite phases of the Island Harbour Bay Plutonic Suite (IHBPS) are locally recognizable through the intense deformation (Ermanovics 1992). Some shearing thus postdates the late granites, dated nearby at 1805 ± 5 Ma (Loveridge *et al.* 1987), but initiation of deformation is unconstrained. Kinematic indicators and metamorphic contrasts suggest a reverse motion with strong dextral shearing, possibly as sequential events; displacements may have been 100 km or more, based on strain measurements (Korstgård & Ermanovics 1985). The relationship between various phases of the IHBPS and shearing provides the best avenue to resolve early events via geochronology, and perhaps correlate with the Kaipokok Bay Structural Zone (see below).

Reworking and migmatization in the Kaipokok Domain

On a regional scale, the Kaipokok Domain can be divided into two sub-areas where Proterozoic metamorphism was at amphibolite and greenschist facies respectively (Ryan *et al.* 1983; Ryan 1984; Fig. 2). Common events can be recognized in both areas, which implies that they represent different crustal levels, but there are also important differences in their evolution. Deformational events are here designated as PD_1, PD_2, etc., indicating that they are Proterozoic in age; the complex Archaean tectonic history of parts of the area is not discussed.

In the low-grade southwestern region, Archaean gneisses are cut by Kikkertavak diabase dykes up to 5 m wide, and both rock types are overprinted by greenschist facies metamorphism. Areas of strong deformation show isoclinal folding of both gneisses and dykes (PD_1), with transposition on the limbs of structures. This deformation produced a 'layered' gneiss composed of retrogressed Archaean rocks alternating with greenschist-facies, actinolite-bearing, metadykes (Ryan 1984). In the Moran Lake Group, the earliest deformation involved small-scale isoclinal folding, sub-horizontal translation of the basement-cover interface, and local transposition, particularly in eastern equivalents of the Moran Lake Group, which are locally mylonitic (Ryan 1984). Northwest-verging isoclinal folding and layer-parallel shearing was suggested by Ryan (1984), but no large-scale nappe structures have been defined. The similar early tectonic histories of basement and cover indicate that both record PD_1 deformation. The PD_1 fabric in the gneisses is cut by the 1891 ± 5 Ma granite (Kerr *et al.* 1992), indicating that basement-cover interaction is pre-1890 Ma. A second period of deformation produced regional-scale tight upright folds with NE-trending axes, commonly overturned to the northwest. In terms of style, this folding resembles the PD_4 event recognized further east, which also established the dominant structural pattern (see below). Upright folding was followed by late, open folds with east-trending axes.

In the higher-grade northeastern region, Kikkertavak dykes were progressively reoriented and a new amphibolite-facies gneissosity developed (Ryan & Kay 1982; Ryan *et al.* 1983). Although ductile deformation is widespread, metadykes retain delicate intrusive apophyses in low-strain enclaves. Elsewhere, the dykes are intensely folded, and, together with the earlier gneissosity, are transposed into a 'new' layering, defined by alternations of attenuated dyke remnants and gneiss ('straightened gneiss' of Ryan *et al.* 1983). This event is logically correlated with the pre-1895 Ma PD_1 event in the southwest (see above). However, in the northeast, PD_1 was apparently followed by migmatization, involving at least three generations of leucosome veins, locally producing nebulitic gneisses (Sutton 1972; R. J. Wardle, unpublished data). There may be a link between this migmatization (termed Iggiuk Event by Schuärer *et al.* 1988) and the 1895–1891 Ma granites in lower-grade areas, i.e. these could represent magmas derived from Iggiuk 'migmas' at greater depth. This is also supported by the presence of inherited Archaean zircons in one granite sample (Kerr *et al.* 1992).

However, there is an important conflict in the **relative** timing of later events in these two sub-areas. In the southwest, the *c.* 1895 Ma granites intrude gneisses that had **already** been interleaved with their cover (Moran Lake Group), whereas the migmatized straightened gneisses in the northeast are **overprinted** by mylonite zones (here labelled as composite PD_2–PD_3 structures; see below) that interleave the gneiss complex and Lower Aillik Group. Thus, the early deformation and basement-cover interaction in the southwest was apparently earlier than the juxtaposition of the Lower Aillik Group and basement in the northeast. Also, there is no evidence that any of the Lower Aillik Group rocks experienced any **pre-mylonitization** migmatization; the metasedimentary rocks yielded local S-type granitoid rocks (Marten 1977; Kerr 1989*a*), but these 'migmas' overprint the mylonites, and are thus related to subsequent metamorphism (see below). A major question that arises from this conflict is the absence of the major PD_2–PD_3 event in the southwest, unless it is actually recorded by the northwest-verging tight folding that we have here correlated with PD_4 based on its general style. An alternative explanation is that the initial reorientation of the dyke swarm, and Iggiuk migmatization, were **earlier** events in the northeast that were not recorded in the southwest, where reorientation of the dyke swarm was first seen in association with basement-cover interaction, i.e. the earliest Proterozoic deformation in the southwest is actually the PD_2–PD_3 event. In this respect, dating this event around Kaipokok Bay is of great importance.

The mylonite zones associated with the basement-cover interface form the dominant structural elements in the northeast, and locally achieve widths up to 10 km, notably on the north side of Kaipokok Bay, where a 'keel' of high-grade amphibolite and paragneiss is enveloped by intensely mylonitized gneiss. A second mylonite zone defines the northwest edge of the fold-and-thrust belt along Kaipokok Bay, which contains most of the Lower Aillik Group (Fig. 9). The relationship between these two mylonite zones is uncertain. Lower Aillik Group rocks north of Kaipokok Bay may represent klippen, linked to a single basal mylonite zone that has been refolded;

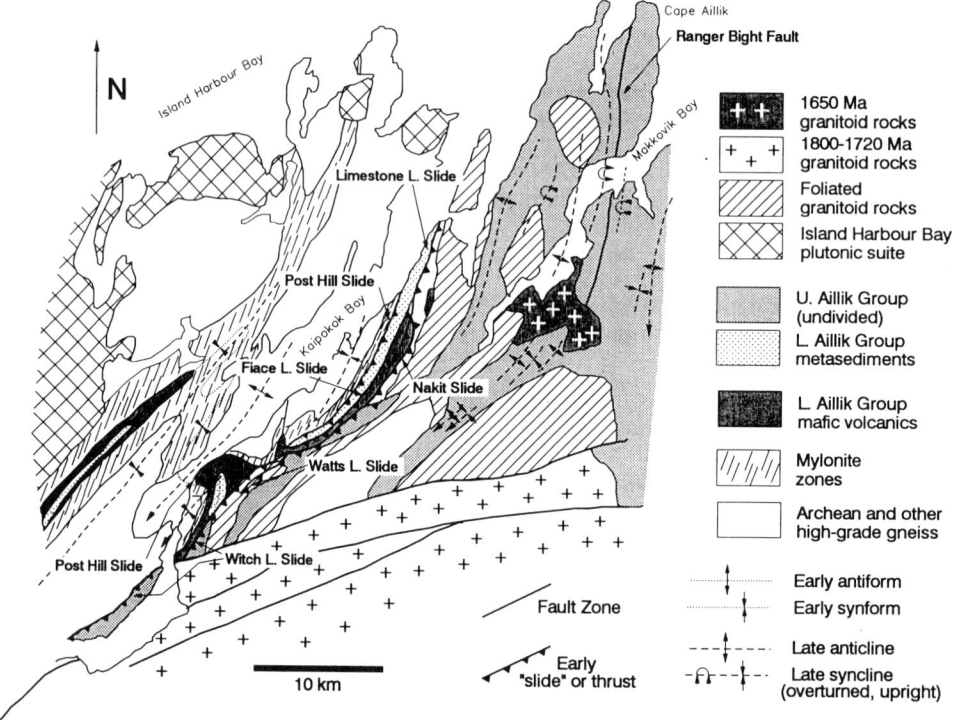

Fig. 9. Major structural elements and simplified geology of the Kaipokok Bay Structural Zone, along the boundary between the Kaipokok and Aillik domains. After Ryan & Kay (1982), Marten (1977) and Gower *et al.* (1982).

an alternative view is that the area around Kaipokok Bay represents a structural slice (thrust wedge ?) dominated by basement rocks, between two discrete mylonite zones. The mylonite zones are probably the most important structures within the Makkovik Province, but relatively little is known of their kinematic evolution. Preliminary indications (Marten 1977; N. Culshaw, pers. comm. 1994) suggest a dextral component, similar to that in the Kanairiktok Shear Zone (see above). North of Kaipokok Bay, lineations range from steep to shallow, and generally plunge southwest (Ryan et al. 1983), but interpretation is complicated by later refolding. Within the mylonite zone, there is evidence of polyphase deformation and isoclinal folding, and a variety of granites are caught up in the deformation (Sutton 1972; Ryan et al. 1983). Some of these may be of Proterozoic age, and they may provide a route towards understanding the history and longevity of the zone(s). The mylonite zones are interpreted as northwest-verging PD_2–PD_3 thrusts (see below), which were rotated into their present attitudes (Marten 1977; Ryan & Kay 1982). The deformation and attenuation within the mylonite is very intense; for example, the Post Hill amphibolite unit (Fig. 3; Fig. 4) is reduced from almost 1000 m thickness in the south, to a mere 30 m in the northeast (Marten 1977). If the intensity of deformation is an indication of the amount of translation, the Lower Aillik Group may be highly allochthonous with respect to the adjacent gneisses (Ryan & Kay 1982). Metamorphism accompanying this event was predominantly of upper greenschist to lower amphibolite facies, and is largely recorded by actinolite–plagioclase–epidote assemblages in mafic metavolcanic rocks, which locally retain vestiges of pillow structure (Ryan et al. 1983).

Kaipokok Bay Structural Zone and Aillik Domain

The Kaipokok Bay Structural Zone and western Aillik Domain (Figs 4, 9) retains the most complete sequence of Makkovikian tectonism. Marten (1977) recognized six main periods of deformation, including three events in the basement considered to be of Archaean age. Correlations across Kaipokok Bay (Ryan et al. 1983; see above) indicate that at least two of these are Proterozoic events, representing reorientation of Kikkertavak dykes (PD_1) followed by the Iggiuk Event migmatization. Marten (1977) provided a detailed description of the interleaving of basement and cover southeast of Kaipokok Bay, which we regard as a composite PD_2–PD_3 event, and correlate with the northeastern Kaipokok Domain (see above).

The PD_2–PD_3 episode also produced the major ductile structures associated with lithological contacts or incompetent units within the Lower Aillik Group rocks (Fig. 9). These were described as 'slides' (Marten 1977), implying that they originated as low-angle thrust structures; the 'Post Hill slide' (Fig. 9) is equivalent to the main mylonite zone described above. Other structures, notably the Fiace Lake and Nakit slides (Fig. 9), are also characterized by mylonites, although not to the same extent. These ductile structures separate Lower Aillik Group units, but some also involve the basement rocks, notably the dextral Limestone Lake slide, which juxtaposes Archaean basement against metasedimentary rocks (Marten 1977). As it is near-vertical, the original configuration is debatable, but it may be a steepened thrust that exhumed the basement rocks. Minor structures show that tectonically interleaved Lower Aillik Group was refolded **before** the late upright folding. Following Marten (1977), the early deformation is thus designated as a composite PD_2–PD_3 deformation, but it may represent a single semi-continuous episode of low-angle shearing, rather than two discrete events. The only constraint on timing is that it predates emplacement of syn- or pre-PD_4 granites dated at 1801 +13/–7 Ma and 1794 ± 2 Ma (Kerr et al. 1992; Schuärer et al. 1988).

Tight, upright folding (here assigned to PD_4) was the last major event throughout the Aillik Domain, and most map-scale structures (Fig. 4) are of this generation. The earlier PD_2–PD_3 'slides' were reactivated, and rotated into steeply-inclined or subvertical attitudes, and a number of tight, upright folds developed (Marten 1977; Gower et al. 1982). The so-called 'Ranger Bight slide' (Fig. 9) of Clark (1979) was interpreted as a PD_4 structure that disrupts an antiformal hinge (Gower et al. 1982), but it may represent a late high-angle reverse fault rather than a reactivated early thrust. A variety of minor structures were developed, varying in form and style according to the host rock type. PD_4 deformation affected most parts of the Upper Aillik Group, as evidenced by regional folding (Fig. 4); however, the presence of earlier (PD_2–PD_3) structures is controversial. Along the Lower Aillik–Upper Aillik contact southeast of Kaipokok Bay, both components are affected by strong ductile deformation of unknown age, but Marten (1977) contended that the adjacent Upper Aillik Group showed evidence of the earlier events. Further to the east, where the metamorphic grade is lower, much of the Upper Aillik Group contains only a single penetrative fabric of unknown age, implying that it did not experience PD_2–PD_3 events. Thus, the contact of Lower and Upper Aillik groups could be either a PD_2–PD_3 structure or a PD_4 structure (or both). The age of PD_4 events is constrained by the 1801 +13/−7 Ma age from the Long Island quartz monzonite, which was affected by this deformation, and (possibly) by the c. 1800 Ma ages obtained from some massive granitoid rocks that lack this fabric (Kerr et al. 1992). If this is true, it was a very short-lived pulse of deformation at c. 1800 Ma; however, if it was regionally heterogeneous, it may have continued beyond this time in the Kaipokok Bay–Makkovik Bay area, where the only local constraint is the c. 1640 Ma age from the Monkey Hill Granite (Kerr et al. 1992), which cuts the Ranger Bight fault (Fig. 4). As noted previously, titanite ages of 1761–1740 Ma in the Kaipokok Bay area (Schärer et al. 1988; Kerr et al. 1992) imply an extended history of activity here.

Later events (PD_5) include gentle, open folding around E–W trending axes, which created the tight dome-and-basin structures that dominate outcrop patterns (Fig. 4), and is also recognized in the Kaipokok Domain. Subsequent (PD_6 and PD_7) deformation events are trivial, forming minor kink bands, and late brittle fault structures, respectively. PD_5 and subsequent episodes are of unknown age. As pointed out by Gower et al. (1982) gentle folding around E–W axes (PD_5) may reflect distal Grenvillian deformation, rather than Early Proterozoic tectonism.

Marten (1977) recognized three periods of metamorphism in the Kaipokok Bay area, of lower to middle amphibolite facies, defined largely by assemblages in mafic and pelitic supracrustal rocks. Gower et al. (1982) examined regional metamorphic grade variations and concluded that there was a decrease from middle amphibolite facies along the northwest side of the Kaipokok Bay Structural Zone, to upper greenschist on its southeast side, and greenschist and lower through most of the Aillik Domain. This regional overprint is considered to be the last metamorphism, and is thus probably syn-PD_4. The presence of anatectic granitoid rocks in the Kaipokok Bay Structural Zone (Kerr 1989a), including the 1794 ± 2 Ma leucogranite of Schärer et al. (1988), indicates that at least upper amphibolite-facies conditions were achieved at deeper levels, and supports a temporal link to PD_4. Earlier metamorphic events in the Lower Aillik Group presumably correlate with the composite PD_2–PD_3 event, and there is no clear evidence of an earlier high-grade metamorphism in most of the Upper Aillik Group.

Cape Harrison Domain and areas proximal to Grenville Front

The structural chronology outlined in the Kaipokok and Aillik Domains cannot be extended with confidence to the Cape Harrison Domain, which is dominated by massive and weakly foliated plutonic rocks, locally affected by Grenvillian deformation. Early deformational events are preserved within the Cape Harrison Metamorphic Suite (Gower 1981), but these have not been outlined in detail. In the south, Grenvillian deformation becomes more significant, especially adjacent to the Grenville Front, mostly defined by the Benedict Fault Zone (Fig. 6). This is a 1–2 km wide zone of intense deformation, including several individual south-dipping mylonite zones, with steeply plunging lineations (Gower 1981; Kerr 1989a); contrasts in metamorphic grade indicate reverse motion. The major east-trending faults such as the Adlavik Brook and Benedict faults (Figs 4, 6) are of Grenvillian age (Gower et al. 1982), but they may have an earlier history, as recently documented by Krogh (1994) for parts of the Grenville Front.

Correlations with the Ketilidian Mobile Belt of Greenland

Correlations between coastal Labrador and southwest Greenland have been suggested by Kranck (1939), Bridgwater (1969), Sutton et al. (1972), Bridgwater et al. (1973) and Gower & Ryan (1986). Here, we review these main links, and comment on more detailed correlations suggested by the recent work in Labrador. Such considerations are of particular importance in the light of recent reappraisal of Ketilidian geology (Chadwick & Garde, this volume) and the proximity of southern Greenland to Lithoprobe seismic-reflection lines (Fig. 1). Our treatment of correlations is largely based on reviews (e.g. Bridgwater et al. 1973; Allaart 1976; Windley 1991) which divide the belt into four principal regions (Fig. 1, inset). Chadwick & Garde (this volume) employ a similar fourfold division, but suggest a revised terminology; in this paper we have retained the names of Allaart (1976).

The Kaipokok Domain of the Makkovik Province is equivalent to the **Border Zone**, where Archaean basement and diabase dykes (Iggavik Dykes) are similarly reworked and reorientated. However, the zone of reworking is wider, and there is no clear indication of a well-defined 'Ketilidian front' analogous to the Kanairiktok Shear Zone. Supracrustal rocks of the Vallen and Sortis groups rest unconformably on the basement (Allaart 1976), and are probable equivalents of the Moran Lake and Lower Aillik (?) groups (Gower & Ryan 1986). The stratigraphic sequence is similar (see Fig. 3), but the Greenland succession is thicker and better preserved. The basal unconformity is obliterated southward, where north-directed thrusts disrupt and repeat supracrustal units (Allaart 1976; Windley 1991), a progression similar to relationships north of Kaipokok Bay. The basement–cover contact is commonly defined by a unit termed 'gneissic schist' (mylonite ?).

The Aillik Domain and the spatially associated Kaipokok Bay Structural Zone have no explicit terminological equivalent in Greenland, but the **Kobbermine Bugt Shear Zone** (Bridgwater et al. 1973; Windley 1991) has an equivalent location, separating the reworked foreland from a major belt of granitoid rocks (Fig. 1, inset). This shear zone is an ENE-trending, sub-vertical, zone, up to 15 km wide, with a metamorphic gradient from upper amphibolite facies on its north side, to subgreenschist in the south, i.e. similar to the pattern in the northwestern Aillik Domain. The Kobbermine Bugt Zone is dominated by intensely deformed supracrustal rocks and granitoid gneisses. Major structures dip steeply to the south, and contain subhorizontal lineations. The latest motions are interpreted to be transcurrent, but the kinematic sense and early history of the zone is presently unclear, as in Labrador. Attenuation is intense (approaching 500:1), and 'autochthonous' granites were produced by anatexis of reworked basement rocks (Allaart 1976). All of these features have analogues around Kaipokok Bay (Fig. 9). Supracrustal rocks south of the Kobbermine Bugt Zone include mafic volcanic and volcaniclastic rocks, but also intermediate to felsic volcanic rocks, possibly analogous to the Upper Aillik Group. As in Labrador, their relationship to the amphibolites and pelitic–psammitic metasedimentary rocks of the Kobbermine Bugt Zone is uncertain. The Kobbermine Bugt Shear Zone represents an important break within the Ketilidian Mobile Belt, interpreted as a possible suture zone (Windley 1991).

The Cape Harrison Domain has an obvious counterpart in the **Granite Zone (Julianehåb Batholith** of Chadwick & Garde, this volume), which is dominated by complex, polyphase granitoid intrusions, with rare remnants of supracrustal rocks (mostly felsic volcanics) and enclaves of earlier foliated granites and gneisses (Allaart 1976). Extensive areas (up to 80%) are reported to be dominated by 'late' granites, which are only weakly foliated or undeformed (Allaart 1976). Mafic intrusive rocks are also widespread, including norites, gabbros, and pyroxene-bearing monzonites and syenites; many are hornblende-bearing and locally pegmatitic ('appinitic'), and show spectacular net-veined, agmatitic textures. As noted by Allaart (1976), the age relations of these are very hard to determine, but most are considered to postdate the first generation of 'late' granites.

Late, postorogenic 'rapakivi-type' granites also occur in the granite zone, but these are more abundant in the southernmost Ketilidian (see below). These represent possible equivalents of the c. 1720 Ma 'A-type' granites in the Makkovik Province (see below).

There are few geochemical, isotopic or geochronological constraints for Ketilidian granites. However, early U–Pb ages of 1845 ± 15 Ma for 'gneissose' granites, 1805 ± 25 Ma for 'early' granites and 1755–1740 Ma for 'late' rapakivi suites (van Breeman et al. 1974; Gulson & Krogh 1975) resemble the age groupings recorded from Makkovik Province granites (Kerr et al. 1992). Isotopic contrasts between granites in the Border Zone and the Granite Zone have also been reported (Kalsbeek & Taylor 1985). Most suites south of the Kobbermine Bugt Zone show neutral to positive ε_{Nd} suggesting that they represent relatively 'juvenile' crust (Patchett & Bridgwater 1984; Kalsbeek & Taylor 1985), as in the eastern Makkovik Province. Although the Ketilidian granites have been described as an 'Andean-type' assemblage (e.g. Bridgwater et al. 1973; Windley 1991), there are very few geochemical data to confirm this model. Based on findings in the Makkovik Province (Kerr 1989a; Kerr et al. 1992), we suspect that there may be a complex, episodic plutonic history in the Ketilidian, and that it may be premature to assume that all of its components are of arc affinity, or necessarily of the same age. Kerr et al. (1992) pointed out that the age distribution of Makkovik Province granites is similar to the Trans-Scandinavian Igneous Belt (TSIB) of Sweden (e.g. Johansson 1988), a conclusion strengthened by recent dating, which reveals c. 1700 and c. 1650 Ma plutonism (Brewer et al. this volume). The TSIB cuts obliquely across earlier Svecofennian structural trends, but is in part contemporaneous with Svecofennian magmatism; this possible link between Sweden and parts of the Makkovik Province suggests that TSIB-type plutonic rocks may occur in southern Greenland.

The southern part of the Ketilidian Mobile Belt has no on-land equivalent in North America, except perhaps some of the poorly known gneissic rocks and metasedimentary remnants at the eastern extremity of the Makkovik Province (Gower & Ryan 1986; Owen et al. 1986). However, plate reconstructions (Fig. 1; inset) imply that this domain should occur on the adjacent continental shelf, in regions imaged by the Lithoprobe seismic reflection line (Fig. 1). The **Folded Migmatite Zone** (Psammite Zone of Chadwick & Garde this volume) consists of low-P, high-T, amphibolite-facies granitoid gneisses, migmatitic paragneisses, quartzofeldspathic gneisses, and a wide variety of later (mostly granitic) plutonic rocks (Bridgwater et al. 1973). The **Flat-lying Migmatite Zone** (Pelite Zone of Chadwick & Garde, this volume) exhibits sub-horizontal gneissosity and compositional layering, and was described by Allaart (1976) as a gentle domal structure some 100 km across, intruded by rapakivi granites emplaced either as diapirs or as sheet-like bodies. No 'basement' orthogneisses have been recognized anywhere in this region, and the supracrustal rocks include calc-silicate gneisses and graphitic rocks in addition to pelitic and psammitic gneisses. Chadwick & Garde (this volume) present some reappraisal and reinterpretation of the earlier ideas about this region summarized by Allaart (1976), but confirm the dominantly low-angle structural trends. An important aspect of southern Greenland geology is the absence of an older basement complex that might represent a second continental block on the other side of the Mobile Belt. Nd isotopic studies (Patchett & Bridgwater 1984) suggest that this region is dominantly juvenile. However, results permit slightly older material with formation ages up to c. 2100 Ma.

Summary and discussion: outstanding problems in Makkovik Province geology

The Makkovik Province records a southeastward transition from a foreland fold-and-thrust belt into a zone of migmatization and ductile deformation, and eventually into a major plutonic belt. The tectonic history shows an overall temporal change from horizontal to vertical tectonics. These patterns resemble those documented in other orogenic belts in the Canadian Shield, which have been interpreted as convergent margin and/or collisional orogens (e.g. Hoffman 1988; van Kranendonk et al. 1993).

Previous tectonic models for the Makkovik Province and Ketilidian Mobile Belt vary considerably. Ryan (1984) discussed a collisional option, where the Moran Lake and Lower Aillik groups represent a continental margin cover sequence, overlain by parts of the Upper Aillik Group, suggested to be of backarc affinity and related to northward subduction. Deformation was attributed to collision with an undefined southern cratonic block, then followed by crustal thickening and anatexis. Kerr (1989a) suggested collision between the North Atlantic Craton and a composite arc terrane of juvenile affinities. In at least the latter stages, the polarity of subduction was to the southeast, away from the craton, and sub-horizontal deformation was interpreted to record the arc terrane–continent collision. For the Ketilidian Mobile Belt, Bridgwater et al. (1973) suggested an 'Andean' analogy, but also recognized similarities to the Hercynian of Europe. Windley (1991) interpreted the Ketilidian belt as a collisional orogen, analogous to the Himalayan Belt; this model implies south-directed subduction, with a magmatic arc (the Granite Zone) sandwiched between the North Atlantic Craton and a southern block of unknown heritage. More recently, Chadwick & Garde (this volume) have returned to a broadly 'Andean' analogy invoking north-directed subduction. We do not outline a new tectonic model for the Makkovik Province here, but we hope that integration of surface geology with deep-crustal structure (Hall et al. 1995) will permit this development. We focus instead on the constraints imposed on such models by the existing data, and on outstanding problems that require resolution to test existing and future models.

Early events in the Kaipokok Domain

Parts of the Kaipokok Domain have a complex history that apparently **predates** the ductile deformation that juxtaposed the basement and the Lower Aillik Group. This deformation involved the reorientation and transposition of c. 2230 Ma diabase dykes, followed by migmatization (Iggiuk Event). However, the ages of these events are unknown, and there is an apparent conflict in the relative timing of basement-cover interaction in high- and low-grade areas. A c. 1895 Ma age for Iggiuk migmatization would confirm a link to granites of this age, and support the presence of pre-1895 Ma PD_1 deformation events throughout the region. A substantially older age would indicate that early dyke reorientation and thermal events are a discrete episode unrelated to Makkovikian events. Most importantly, an age for Iggiuk migmatization would provide an older limit for PD_2–PD_3 deformation in the Kaipokok Bay Structural Zone (see below).

Chronology and character of plutonism in the Kaipokok Domain

Of particular importance in this context is the Island Harbour Bay Plutonic Suite (IHBPS), presently constrained only by a 1805 ± 5 Ma age from a fluorite-bearing granite (Loveridge et al. 1987). This age is similar to late- and postorogenic suites (Kerr et al. 1992), but may not represent the variably deformed dioritic, tonalitic and trondhjemitic members of the IHBPS. These complex early phases of the IHBPS are one of the few igneous suites in the Makkovik Province that have some affinities to those of magmatic arc environments (see below), and they may therefore document and date any north-directed subduction beneath the North Atlantic Craton.

Age(s) and provenance(s) of the Moran Lake and Lower Aillik groups

These important supracrustal sequences remain undated. Precise age information would assist in correlation, in assessing the relationship between the Lower and Upper Aillik groups, and in constraining early sub-horizontal deformation (see below). Provenance is also an important question for both sequences. The generalized stratigraphic sequences (Fig. 3) resemble those proposed for foredeep basins (e.g. Hoffman 1987), which show provenance reversals recording the approach of arc-related terranes and may constrain collisional events. U–Pb studies of detrital zircons and Nd isotopic studies offer a means of pinpointing such reversals.

Timing and kinematics of early deformation in the Kaipokok Bay Structural Zone

Early sub-horizontal translation in the Kaipokok Bay Structural Zone is the most obvious candidate for a 'collisional' tectonic event. At present, the only constraint is that it happened before c. 1800 Ma (Schärer et al. 1988; Kerr et al. 1992). Although the interpretation of the mylonites and related zones as rotated thrusts (Marten 1977) is generally accepted, there is little kinematic data to indicate the original sense(s) of motion and subsequent history. A related question concerns the history of motion on the Kanairiktok Shear Zone, which defines the reworking front, and affects several phases of the IHBPS, including the 1805 Ma granite (Ermanovics 1992; see above). The history of the Kanairiktok shear zone thus must be more protracted than those around Kaipokok Bay, which are cut by similar late phases of the IHBPS.

Location of potential suture zone(s)

No suture zones have yet been clearly defined within the Makkovik Province. Ryan (1984) felt that such a zone, if present, must lie well to the east, based on interpretation of the Upper Aillik Group as a backarc sequence. However, Nd isotope studies of plutonic rocks (Kerr 1989a; Kerr &Fryer 1994) show that an 'ancient' crustal signature is mostly restricted to the Kaipokok and Aillik Domains (Fig. 8). There is no recognizable surface structure that corresponds to the isotopic boundary, but the buried edge of the craton is roughly coincident with the eastern edge of the Aillik Domain supracrustal rocks (Fig. 6).

The Kaipokok Bay Structural Zone is the most obvious structural element in the entire Makkovik Province, and represents a fundamental break in its geology. The equivalent zone in southwest Greenland (Kobbermine Bugt Zone) has been interpreted as a suture (Windley 1991). These zones may represent the interface between the North Atlantic Craton and a second block, where cover (?) sequences have been imbricated and intensely deformed during collision. If this zone dips eastward beneath the Aillik Domain, Archaean rocks would be present at depth in the source regions of late- and postorogenic granitoid rocks some 20–30 km to the east, as suggested by isotopic data (Fig. 8). Regardless of the precise interpretation, the isotopic data of Kerr & Fryer (1994) demonstrate that the bulk of the plutonic terranes of the Makkovik Province were formed east (outboard) of the eastern limit of Archaean crust, although remnants of such material may have been present at high crustal levels (Kerr & Fryer 1994; Fig. 8).

Presence (or absence ?) of arc-type magmatic rocks

A fundamental problem in Makkovik Province geology is the almost complete absence of igneous rocks that resemble continental-arc or island-arc assemblages related to subduction of oceanic crust. The Upper Aillik Group occurs where such rocks might be expected (see above), but is virtually devoid of andesites, and dominated by high-SiO_2, variably alkaline rhyolites (Fig. 5; see above). A distal, backarc, extensional setting is possible (e.g. Ryan 1984) but conflicts with its position at the edge of the North Atlantic Craton (Kerr & Fryer 1994) and the virtual absence of more primitive arc-like rocks further to the east. Moreover, the age of parts of the sequence (c. 1807 Ma) is very close to the age of late- to postorogenic plutonic rocks (Kerr et al. 1992), which postdate most of the deformation in the area. Within the Cape Harrison Domain, most of the plutonic suites appear to be relatively late, and they are distinct from younger continental-arc magmas such as those of Peru or Chile (Fig. 7); they instead resemble late, largely postcollisional assemblages such as those that dominate eroded Phanerozoic mobile belts such as the Appalachians. Any model that uses these plutonic suites in an 'Andean' context must be questioned on geochemical grounds.

However, arc-type suites are not completely absent. The Cape Harrison Metamorphic Suite contains rocks described as 'tonalitic' (Gower

1981), which are the best candidates for early arc-type rocks in the east. At present, there are few geochemical data to confirm their affinities, and no precise geochronology; it is thus also possible that these rocks instead represent somewhat older crust (see below). In the western Makkovik Province, the undated tonalitic and trondhjemitic rocks within the IHBPS are possible candidates (see above).

Presence or absence of a second continental block

The isotopic data from granites (Kerr & Fryer 1994) indicate that Archaean crust is not present at depth beneath the Cape Harrison Domain, and that the gneisses of the Cape Harrison Metamorphic Suite are juvenile rocks that do not belong to the North Atlantic Craton. These could, therefore, represent a sample of a discrete post-Archaean crustal block involved in collisional orogenesis. However, virtually nothing is known about the age and evolution of these rocks, apart from c. 2100 Ma Sm–Nd model ages (Kerr & Fryer 1994), and it is also possible that the CHMS simply represents an earlier Makkovikian magmatic episode (see above).

The absence of evidence for an older continental block in the southern Ketilidian has been used to support a broadly 'Andean-type' setting for the belt, where oblique convergence of the North Atlantic Craton and an oceanic plate to the south resulted in a major magmatic arc over a north-dipping subduction zone (Chadwick & Garde, this volume). The arc (Granite Zone) is interpreted as the provenance for metasedimentary rocks of southernmost Greenland, and the supracrustal rocks of the Border Zone are interpreted to be of backarc affinity. This model does not involve collisional tectonics, in the strict sense of continent–continent or arc–continent collisions. However, absence of a second ancient cratonic block does not preclude collisional tectonism, which can result from the accretion of composite arc-type terranes, or involve continental blocks with short crustal residence periods. An excellent example of the latter is provided by the northern Appalachians, where the southeastern (Gondwanaland) foreland is formed by the Pan-African Avalon Composite Terrane, which was formed only 100–150 Ma prior to its Ordovician–Silurian collision with Laurentia. This 'completed', more deeply-eroded orogenic belt may provide a better analogy to the Makkovikian–Ketilidian system than either the Himalayan or Andean orogenic belts.

Relationship to coeval and subsequent orogenic events

The Torngat Orogen, on the western edge of the Nain Province, is in places only 100 km from the Makkovik Front, although details of its geology in central Labrador are obscured by Middle Proterozoic plutonism (Fig. 1). In northern Labrador, the Torngat Orogen was active from c. 1880 Ma to c. 1790 Ma (Bertrand et al. 1993), and it was approximately coeval with the Makkovik Province and the Ketilidian Belt. Links between these areas should be expected. As an example, the c. 1844 Ma change from convergent to sinistral transcurrent motion in the Torngat Orogen may be a manifestation of events in the Makkovikian–Ketilidian area resulting in northward motion of the North Atlantic Craton. There must also be links between the Makkovik Province and subsequent (Labradorian) events now preserved in the Grenville Province. The 1650 Ma plutonic rocks in the Makkovik Province may represent initiation of north-directed subduction (e.g. Kerr 1989a) or the accretion of slightly older (1680–1670 Ma) Labradorian arc terranes (e.g. Gower et al. 1992). Also, pelitic to psammitic metasedimentary rocks in the Labradorian terranes have Sm–Nd model ages of 2200 Ma or less (e.g. Kerr et al. 1993). These rocks cannot be derived from the Archaean rocks that form the present foreland to the Labradorian terranes in central Labrador, but may represent 'flysch' derived from the Makkovikian–Ketilidian belt, which was later accreted to Laurentia in conjunction with Labradorian arcs. These suggestions are pure speculation, but illustrate an important point, i.e. that any model for Makkovik–Ketilidian evolution must fit into a broader evolutionary scheme for the southern margin of Laurentia that spans several hundred million years.

The work discussed in this review paper was mostly conducted by the Geological Survey of Newfoundland and Labrador (Department of Mines and Energy), Memorial University of Newfoundland, and the Geological Survey of Canada. It was supported in large part by several cost-shared agreements on mineral development, funded by the Governments of Newfoundland–Labrador and Canada, with contributions from the National Sciences and Engineering Research Council (NSERC) to university research projects. We thank Tim Brewer for organizing the productive Nottingham meeting. Reviewers Brian Chadwick and Sunil Gandhi are thanked for their constructive input, and Adam Garde is thanked for helpful comments. This paper is published with the permission of the executive director of the Geological Survey of Newfoundland and Labrador. Lithoprobe Contribution no. 661.

References

ALLAART, J. H. 1976. Ketilidian mobile belt of West Greenland. *In*: ESCHER, A. & WATT, W. S. (eds) *Geology of Greenland*. Grønlands Geologiske Undersøgelse, Copenhagen, 120–150.

BERTRAND, J. M., RODDICK, J. C., VAN KRANENDONK, M. & ERMANOVICS, I. 1993. U–Pb geochronology of deformation and metamorphism across a central transect of the Early Proterozoic Torngat Orogen, North River map area, Labrador. *Canadian Journal of Earth Sciences*, **30**, 1470–1489.

BREWER, T. S., DARBYSHIRE, D. P. F. & LARSON, S. A. 1996. Paleoproterozoic granitoid magmatism within the southeastern Baltic Shield: Implications from whole rock geochemistry and Nd isotopic systematics. *This volume*.

BRIDGWATER, D. 1969. Observations on the Precambrian rocks of Scandinavia and Labrador and their implications for the interpretation of the Precambrian of Greenland. *Rapport Grønlands Geologiske Undersøgelse*, **28**, 43–47.

——, ESCHER, A. & WATTERSON, J. 1973. Tectonic displacments and thermal activity in two contrasting Proterozoic mobile belts from Greenland. *Philosophical Transactions of the Royal Society of London*, A **273**, 513–533.

BROOKS, C. 1982. *Third report on the geochronology of Labrador*. Newfoundland Department of Mines and Energy, Mineral Development Division, unpublished report.

BROOKS, C. 1983. *Fourth report on the geochronology of Labrador*. Newfoundland Department of Mines and Energy, Mineral Development Division, unpublished report.

CADMAN, A. C., HEAMAN, L., TARNEY, J., WARDLE, R. J. & KROGH, T. E. 1993. U–Pb geochronology and geochemical variation within two Proterozoic mafic dyke swarms, Labrador. *Canadian Journal of Earth Sciences*, **30**, 1490–1504.

CHADWICK, B. & GARDE, A. A. 1996. Paleoproterozoic oblique convergence in south Greenland: A reappaisal of the Ketilidian Orogen. *This volume*.

CLARK, A. M. S. 1979. Proterozoic deformation and igneous intrusion in part of the Makkovik subprovince, Labrador. *Precambrian Research*, **10**, 95–114.

ERMANOVICS, I. F. 1992. *Geology of Hopedale Block, southern Nain Province, and the adjacent Proterozoic terranes, Labrador, Newfoundland*. Geological Survey of Canada, Memoir **431**.

EVANS, D. 1980. *Geology and petrochemistry of the Kitts and Michelin Uranium deposits and related prospects, Central Mineral Belt, Labrador*. PhD thesis, Queen's University, Kingston, Ontario, Canada.

GANDHI, S. S. 1978. Geological setting and genetic aspects of uranium occurrences in the Kaipokok Bay–Big River area, Labrador. *Economic Geology*, **73**, 1492–1523.

——, GRASTY, R. L. & GRIEVE, R. A. F. 1969. The geology and geochronology of the Makkovik Bay area, Labrador. *Canadian Journal of Earth Sciences*, **6**, 1019–1035.

GOWER, C. F. 1981. *The Geology of the Benedict Mountains, Labrador*. Newfoundland Department of Mines and Energy, Mineral Development Division, Report **81-3**.

—— & RYAN, A. B. 1986. Proterozoic Evolution of the Grenville Province and adjacent Makkovik Province in east-central Labrador. *In*: MOORE, J. M., BAER, A. J. & DAVIDSON, A. (eds) *The Grenville Province*. Geological Association of Canada, Special Paper **31**, 281–295.

—— & —— 1987. Two stage felsic volcanism in the Lower Proterozoic Upper Aillik Group, Labrador, Canada: Its relationship to syn- and post- kinematic plutonism. *In*: PHAROAH, T. C., BECKINSALE, R. D. & RICKARD, D. (eds) *Geochemistry and Mineralization of Proterozoic Volcanic Suites*. Geological Society, London, Special Publication **33**, 201–210.

——, FLANAGAN, M. J., KERR, A. & BAILEY, D. G. 1982. *Geology of the Kaipokok Bay–Big River Area, Central Mineral Belt, Labrador*. Newfoundland Department of Mines and Energy, Mineral Development Division, Report **82-7**.

——, RYAN, A. B., BAILEY, D. G. & THOMAS, A. 1980. The position of the Grenville Front in eastern and central Labrador. *Canadian Journal of Earth Sciences*, **17**, 784–788.

——, —— & RIVERS, T. 1990. Mid-Proterozoic Laurentia–Baltica: an overview of its evolution and a summary of the contributions made by this volume. *In*: GOWER, C. F., RYAN, A. B. & RIVERS, T. (eds) *Mid-Proterozoic Laurentia–Baltica*. Geological Association of Canada, Special Paper **38**, 1–23.

——, SCHÄRER, U. & HEAMAN, L. M. 1992. The Labradorian Orogeny in the Grenville Province, eastern Labrador, Canada. *Canadian Journal of Earth Sciences*, **29**, 1944–1957.

GULSON, B. L. & KROGH, T. E. 1975. Evidence of multiple intrusion, possible resetting of U–Pb ages, and new crystallization of zircons in the post-tectonic intrusions (Rapakivi Granites) and gneisses from south Greenland. *Geochimica et Cosmochimica Acta*, **39**, 65–82.

HALL, J., WARDLE, R. J., GOWER, C. F., KERR, A., COFLIN, K., KEEN, C. E. & CARROLL, P. 1995. Proterozoic orogens of the northeastern Canadian Shield: new information from crustal reflection seismic surveys. *Canadian Journal of Earth Sciences*, **32**, 1119–1131.

HOFFMAN, P. F. 1987. Early Proterozoic foredeeps, foredeep magmatism, and Superior type iron formations of the Canadian Shield. *In*: KRONER, A. (ed.) *Proterozoic Lithospheric Evolution*. American Geophysical Union, Geodynamics Series, **17**, 85–98.

—— 1988. United plates of America, the birth of a craton: Early Proterozoic assembly and growth of Proto-Laurentia. *Annual Reviews of Earth and Planetary Sciences*, **16**, 543–603.

JOHANSSON, A. 1988. The age and geotectonic setting of the Småland-Vårmland granite-porphyry belt. *Geologiska Foreningens i Stockholm Forhandlingar*, **110**, 105–110.

KALSBEEK, F. & TAYLOR, P. N. 1985. Isotopic and chemical variation in granites across a Proterozoic continental margin – the Ketilidian mobile belt of South Greenland. *Earth and Planetary Science Letters*, **73**, 65–80.

KERR, A. 1989a. *Early Proterozoic Granitoid Magmatism and Crustal Evolution in the Makkovik Province of Labrador: A Geochemical and Isotopic Study*. PhD thesis, Memorial University of Newfoundland, St John's, Newfoundland, Canada.

—— 1989b. Geochemistry of the Trans-Labrador Granitoid Belt, Canada: A quantitative comparative study of a Proterozoic Batholith and possible Phanerozoic counterparts. *Precambrian Research*, **45**, 1–17.

—— 1994. Early Proterozoic magmatic suites of the eastern Central Mineral Belt (Makkovik Province), Labrador: Geology, geochemistry, petrogenesis and mineral potential. Newfoundland Department of Mines and Energy, Geological Survey Branch Report **94-3**.

—— & FRYER, B. J. 1993. Nd isotopic evidence for crust-mantle interaction in the genesis of A-type granitoid suites in Labrador, Canada. *Chemical Geology*, **104**, 39–60.

—— & —— 1994. The importance of late- and post-orogenic crustal growth in the Early Proterozoic: evidence from Sm–Nd studies of igneous rocks in the Makkovik Province, Canada. *Earth and Planetary Science Letters*, **125**, 71–88.

——, ——, WARDLE, R. J. & NUNN, G. A. G. 1993. Early and Middle Proterozoic crustal accretion on the southern margin of Laurentia: Evidence from Nd isotopic transects in central Labrador. *Geological Association of Canada. Program with Abstracts, 1993*, A51.

——, KROGH, T. E., CORFU, F., SCHARER, U., GANDHI, S. S. & KWOK, Y. Y 1992. Episodic Early Proterozoic granitoid plutonism in the Makkovik Province, Labrador: U–Pb geochronological data and geological implications. *Canadian Journal of Earth Sciences*, **29**, 1166–1179.

KORSTGÅRD, J. & ERMANOVICS, I. 1985. Tectonic evolution of the Hopedale Block and the adjacent Makkovik Subprovince, Labrador, Newfoundland. *In*: AYRES, L. D., THURSTON, P., CARD, K. D. & WEBER, W. (eds) *Evolution of Archean Supracrustal Sequences*. Geological Association of Canada, Special Paper **28**, 223–229.

KRANCK, E. N. 1939. The rock-ground of the coast of Labrador and the connection between the Pre-Cambrian of Greenland and North-America. *Bulletin Commission Geologique Finlande*, **125**, 65–87.

KROGH, T. E. 1994. Precise U–Pb ages for Grenvillian and pre-Grenvillian thrusting of Proterozoic and Archean metamorphic assemblages in the Grenville Front tectonic zone, Canada. *Tectonics*, **13**, 963–982.

LOVERIDGE, W. D., ERMANOVICS, I. F., & SULLIVAN, R. W. 1987. U–Pb ages on zircon from the Maggo Gneiss, the Kanairiktok Plutonic Suite and the Island Harbour Plutonic Suite, coast of Labrador, Newfoundland. *In*: *Radiogenic Age and Isotopic Studies; Report 1*. Geological Survey of Canada, Paper **87-2**, 59–65.

MARTEN, B. E. 1977. *The relationship between the Aillik Group and the Hopedale Gneiss Kaipokok Bay, Labrador*. PhD thesis, Memorial University of Newfoundland, St John's, Newfoundland, Canada.

NORTH, J. W. 1988. *The stratigraphy, structure, geochemistry and metallogeny of the Moran Lake Group, Central Mineral Belt, Labrador*. MSc thesis, Memorial University, St John's, Newfoundland, Canada.

OWEN, J. V., RIVERS, T. & GOWER, C. F. 1986. The Grenville Front on the Labrador coast. *In*: MOORE, J. M., BAER, A. J. & DAVIDSON, A. (eds) The Grenville Province. Geological Association of Canada, Special Paper **31**, 95–106.

PATCHETT, P. J. & BRIDGWATER, D. 1984. Origin of continental crust of 1.9–1.7 Ga age defined by Nd isotopes in the Ketilidian terrain of South Greenland. *Contributions to Mineralogy and Petrology*, **87**, 311–318.

PEARCE, J. A. & CANN, J. R. 1973. Tectonic setting of basic volcanic rocks investigated using trace element analyses. *Earth and Planetary Science Letters*, **19**, 290–300.

——, HARRIS, N. B. W. & TINDLE, A. G. 1984. Trace element discrimination diagrams for the tectonic interpretation of granitic rocks. *Journal of Petrology*, **25**, 956–974.

PITCHER, W. S. 1983. Granite type and tectonic environment. *In*: HSU, K. (ed.) *Mountain Building Processes*. Academic Press, London, 19–44.

RYAN, A. B. 1984. *Regional geology of the central part of the Central Mineral belt, Labrador*. Newfoundland Department of Mines and Energy, Mineral Development Division, Memoir **3**.

—— & KAY, A. 1982. *Basment-cover relationships and plutonic rocks in the Makkovik subprovince, north of Postville, coastal Labrador*. Newfoundland Department of Mines and Energy, Mineral Development Division, Report **82-1**, 109–121.

——, BARAGAR, W. R. A. & KONTAK, D. J. 1987. Geochemistry, tectonic setting and mineralization of high-potassium Middle Proterozoic rocks in central Labrador, Canada. *In*: PHAROAH, T. C., BECKINSALE, R. D. & RICKARD, D. (eds) *Geochemistry and Mineralization of Proterozoic Volcanic Suites*. Geological Society, London, Special Publication, **33**, 241–254.

——, KAY, A. & ERMANOVICS, I. 1983. *The geology of the Makkovik Subprovince between Kaipokok Bay and Bay of Islands, Labrador*. Newfoundland Department of Mines and Energy, Mineral Development Division, Maps 83–38 and 83–41, with descriptive notes.

SCHÄRER, U., KROGH, T. E., WARDLE, R. J., RYAN, A. B. & GANDHI, S. S. 1988. U–Pb ages of Early and Middle Proterozoic volcanism and metamorphism in the Makkovik Orogen, Labrador. *Canadian Journal of Earth Sciences*, **25**, 1098–1107.

SMYTH, W. R., MARTEN, B. E. & RYAN, A. B. 1978. A major Aphebian–Helikian unconformity within the Central Mineral Belt of Labrador: Definition of new groups and metallogenic implications.

Canadian Journal of Earth Sciences, **15**, 1954–1966.

SUTTON, J. S. 1972. The Precambrian gneisses and supracrustal rocks of the western shore of Kaipokok Bay, Labrador. *Canadian Journal of Earth Sciences*, **9**, 1677–1692.

——, MARTEN, B. E., CLARK, A. M. S. & KNIGHT, I. 1972. Correlation of Precambrian supracrustal rocks of coastal Labrador and southwestern Greenland. *Nature*, **238**, 122–123.

TAYLOR, F. C. 1971. A revision of Precambrian provinces in northeastern Quebec and northern Labrador. *Canadian Journal of Earth Sciences*, **8**, 579–585.

VAN BREEMAN, O., AFTALION, M. & Allaart, J. H. 1974. Isotopic and geochronologic studies on granites from the Ketilidian mobile belt of South Greenland. *Bulletin of the Geological Society of America*, **85**, 403–412.

VAN KRANENDONK, M. J., ST-ONGE, M. R. & HENDERSON, J. R. 1993. Paleoproterozoic tectonic assembly of northeast Laurentia through multiple indentations. *Precambrian Research*, **63**, 325–347.

WARDLE, R. J. 1984. *Geological fieldwork in the lower Proterozoic Aillik Group, eastern Labrador.* Newfoundland Department of Mines and Energy, Mineral Development Division, Report of activities for 1984, 14–18.

—— & BAILEY, D. G. 1981. Early Proterozoic sequences in Labrador. *In*: CAMPBELL, F. H. A. (ed.) *Proterozoic Basins of Canada.* Geological Survey of Canada, Special Paper **81-10**, 331–359.

——, RIVERS, T., GOWERS, C. F., NUNN, G. A. G. & THOMAS, A. 1986. The northeastern Grenville Province: New insights. *In*: MOORE, J. M., BAER, A. J. & DAVIDSON, A. (eds) *The Grenville Province.* Geological Association of Canada, Special Paper **31**, 13–29.

WASTENEYS, H., WARDLE, R. J. & KROGH, T. E. 1992. Extrapolation of tectonic boundaries across the Labrador Shelf: U–Pb geochronology of well samples. *In*: WARDLE, R. J. & HALL, J. (eds) *Eastern Canadian Shield Onshore–Offshore Transect (ECSOOT), Report of Transect Meeting.* University of British Columbia, Lithoprobe Report, **32**, 145–174.

WHALEN, J. B., CURRIE, K. L. & CHAPPELL, B. W. 1987. A-type granites: geochemical characteristics, discrimination and petrogenesis. *Contributions to Mineralogy and Petrology*, **95**, 407–419.

WHITE, M. V & MARTIN, R. F. 1980. The metasomatic changes that accompany mineralization in the non-orogenic rhyolites of the upper Aillik Group, Labrador. *Canadian Mineralogist*, **18**, 459–479.

WILTON, D. H. C. 1996. *Metallogeny of the Central Mineral Belt and adjacent Archean basement, Labrador.* Newfoundland Department of Natural Resources, Mineral Resource Report 8.

WINDLEY, B. F. 1991. Early Proterozoic collision tectonics, and rapakivi granites as intrusions in an extensional thrust-thickened crust: The Ketilidian Orogen, south Greenland. *Tectonophysics*, **195**, 1–10.

WYBORN, L. A., PAGE, R. W. & PARKER, A. J. 1987. Geochemical and geochronological signatures in Australian Proterozoic igneous rocks. *In*: PHAROAH, T. C., BECKINSALE, R. D. & RICKARD, D. (eds) *Geochemistry and Mineralization of Proterozoic Volcanic Suites.* Geological Society, London, Special Publication **33**, 377–394.

Palaeoproterozoic oblique plate convergence in South Greenland: a reappraisal of the Ketilidian Orogen

B. CHADWICK[1] & A. A. GARDE[2]

[1] *Earth Resources Centre, University, Exeter EX4 4QE, UK*
[2] *Grønlands Geologiske Undersøgelse, Østervoldgade 10, Copenhagen K, DK 1350, Denmark*

Abstract: New field and petrographic data are presented for a revised four-fold division of the Ketilidian Orogen into a Border Zone in the north, followed to the south by the Julianehåb batholith, the Psammite Zone and the Pelite Zone, as the basis for a new plate tectonic model. The Border Zone within the southern margin of the Archaean foreland contains shallow marine, volcano-sedimentary basins, granites and appinite dykes which are regarded as part of an incipient backarc. The wedge-shaped Julianehåb batholith comprises polyphase calc-alkaline granites, granodiorites, tonalites and hornblende diorites with swarms of appinite dykes. The batholith was emplaced during sinistral transpression as a result of oblique convergence between the Archaean continental plate to the north and an oceanic plate (no longer preserved) to the south. New Rb–Sr isotopic data from the southern part of the batholith confirm previous views that it consists of juvenile Palaeoproterozoic crust. The batholith is interpreted as the root of a volcanic arc that was the provenance of, and in part the basement to, polymict conglomerates, arkosic arenites, semi-pelites and pelites in the Psammite and Pelite Zones, deposited in intra-arc basins and the inner part of a forearc between the batholith and an ocean to the south. The sedimentary rocks and their subordinate intercalations of basalt and andesite were intensely migmatized and deformed in high temperature–low pressure conditions prior to the emplacement of broadly concordant sheets of granites and related rocks of the rapakivi suite. Appinite dykes were emplaced at various stages of the migmatization and deformation. We attribute the thermal effects which characterize the Ketilidian oblique convergent system to voluminous basaltic underplating in accord with recent work on the rapakivi suite and associated metamorphism.

The Ketilidian Orogen is an important component of the system of Palaeoproterozoic mobile belts in the North Atlantic region (Van Kranendonk *et al.* 1993; Park 1994), and it is one of four major orogens of similar age in Greenland (Fig. 1). Tracts of Palaeoproterozoic orthogneisses and high-grade supracrustal rocks elsewhere in Greenland are presumed to be parts of other, so far undefined, belts, for example within the Caledonides in the Dove Bugt region on the northeast coast (Kalsbeek *et al.* 1993; Chadwick & Friend 1994).

The Ketilidian Orogen lies to the south of its Archaean foreland which has remained stable since *c.* 2500 Ma, apart from intrusion of Proterozoic and younger dolerite dyke swarms, carbonatites, kimberlites and lamprophyres (Larsen & Rex 1992). The Palaeoproterozoic Rinkian (Escher & Pulvertaft 1976; Kalsbeek *et al.* 1988; Grocott & Pulvertaft, 1990), Nagssugtoqidian (Escher *et al.* 1976; Kalsbeek *et al.* 1987) and Ammassalik mobile belts (Chadwick *et al.* 1989) lie immediately north of the Archaean block (Fig. 1). The major involvement of Archaean crust in these belts contrasts strongly with its absence in the Ketilidian Orogen. The northwestern part of the orogen forms the basement to sedimentary and volcanic rocks of the Gardar Province, *c.* 1300–1120 Ma, and hosts a series of Gardar syenitic plutons (Emeleus & Upton 1976; Kalsbeek *et al.* 1990). Gardar dolerite dykes are relatively common throughout the Ketilidian Orogen and its foreland to the north.

The dramatic exposures of Ketilidian plutonic and high-grade metasedimentary rocks in the mountainous coastal tracts of South Greenland have attracted wide attention since Wegmann (1938) first applied the term Ketilidian to the orogen after Ketils Fjord (now Tasermiut; Fig. 1). Systematic mapping by the Geological Survey of Greenland in the west of the orogen in the 1960s and 1970s, and a reconnaisance survey of the eastern coastal tract, led to the establishment of a series of tectonic divisions (Allaart 1976). More recent work has been concerned with isotopic age studies (e.g. Kalsbeek & Taylor 1985), specific issues such as uranium mineralization (Steenfelt & Armour-Brown 1988), the mode of emplacement and metamorphic setting of the rapakivi granites (Hutton *et al.* 1990; Dempster *et al.* 1991) and the broad canvas of Ketilidian plate tectonic setting (Windley 1991). The wider role of the Ketilidian

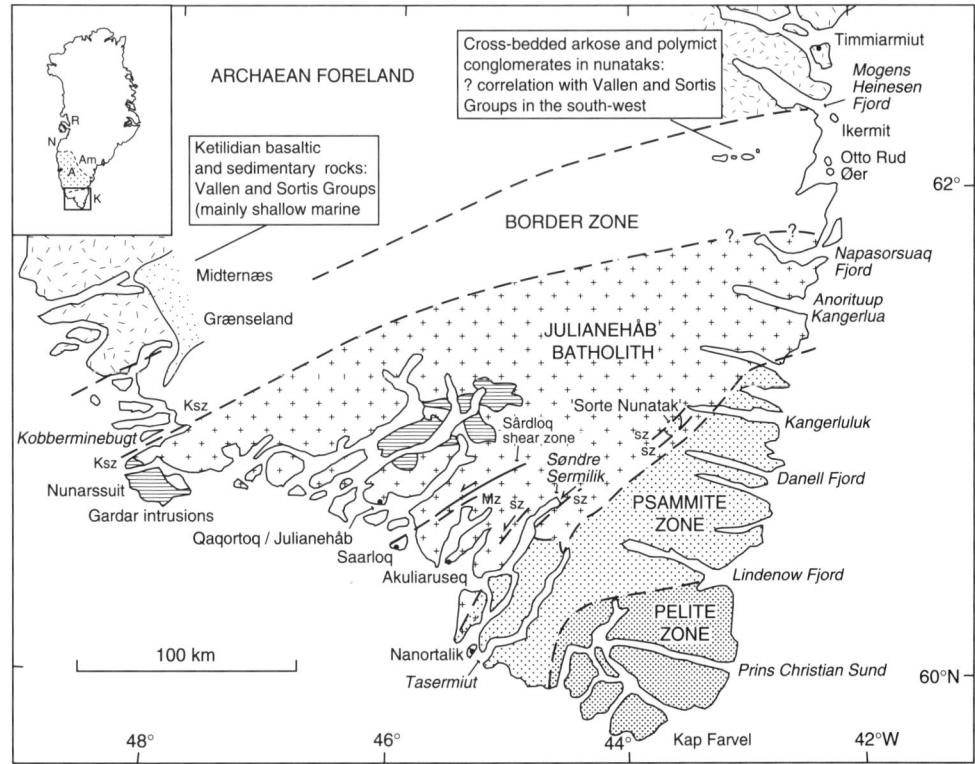

Fig. 1. Summary map of the Ketilidian Orogen, South Greenland, showing the principal divisions: Archaean foreland, Border Zone, Julianehåb batholith, Psammite Zone and Pelite Zone. Much of the orogen is covered by the Inland Ice. Ksz, Kobberminebugt shear zone; sz, shear zone; Mz, Matorssuaq zone of high-strain orthogneisses; horizontal ruling: Gardar sedimentary and volcanic rocks and intrusions. Inset map of Greenland shows the area of the summary map, the Archaean block (A), and the Palaeoproterozoic Ketilidian (K), Ammassalik (Am), Nagssugtoqidian (N) and Rinkian (R) orogenic belts.

Orogen in models of Palaeoproterozoic plate collision in the North Atlantic region has been addressed by Watterson (1978), Dickin (1992), Park (1994) and van Kranendonk *et al.* (1993). In this paper a new interpretation of the Ketilidian Orogen is presented, based on surveys between Mogens Heinesen Fjord and Prins Christian Sund on the east coast in 1992 and in the area of Søndre Sermilik in the west in 1993 (Fig. 1; Chadwick *et al*, 1994a).

Reviews of previous work by Allaart (1976) and Kalsbeek *et al.* (1990), the latter as the descriptive text to the 1:500 000 mapsheet of South Greenland (Allaart 1975), showed that the Ketilidian Orogen could be described in terms of four distinct divisions, namely:

- **Border Zone** in the north in which the orthogneiss complex of the Archaean foreland and its local cover of Palaeoproterozoic supracrustal rocks were variably affected by Ketilidian deformation and metamorphism;
- **Granite Zone** comprising variably deformed Ketilidian plutons and associated mafic hornblende-rich dykes;
- **Folded Migmatite Zone** of Ketilidian pelitic to arkosic metasedimentary rocks with subordinate metavolcanic suites; and
- **Flat-lying Migmatite Complex** of Ketilidian migmatitic paragneisses dominated by flat-lying structures and post-tectonic norites, monzonites and adamellites of the rapakivi suite.

Our surveys have shown that, apart from the Border Zone, these divisions are in need of revision and new definitions. Following Bridgwater *et al.* (1973) and Windley (1991), we propose that the Granite Zone be replaced by the term **Julianehåb batholith**. We also propose that the terms **Psammite Zone** and **Pelite Zone** replace, respectively, the Folded Migmatite Zone and Flat-lying Migmatite Complex in the terminology of Allaart (1976). Whereas Allaart incorporated the record of the late tectonic and thermal history (migmatization, porphyroblast growth and emplacement of the rapakivi suite) in his terminology, our new terms highlight the hitherto unrecognized significance of the metasedimentary rocks in the evolution of the Ketilidian Orogen.

The Border Zone

The Border Zone in the northwest of the orogen (Fig. 1) includes Ketilidian basaltic and shallow marine sedimentary rocks which rest unconformably on Archaean basement gneisses in Midternæs and Grænseland (Vallen and Sortis Groups; Fig. 1; Bondesen 1970; Higgins 1970). The basement gneisses are cut by Palaeoproterozoic dolerite dyke swarms. The effects of Ketilidian deformation and metamorphism increase southward from undeformed, low-grade cover and basement into intensely deformed, high-grade supracrustal rocks and modified basement gneisses (Berthelsen & Henriksen 1975). The Palaeoproterozoic dykes were transformed into deformed metadolerites (Bondesen & Henriksen 1965). The Ketilidian deformation was characterized by thrusting of the cover rocks from the southeast, although local thrusting and back-folding to the southeast affected the cover and basement in the southwest of the Border Zone. Ketilidian granites (c. 1750 Ma) were emplaced into the basement gneiss complex (Kalsbeek & Taylor 1985; Kalsbeek et al. 1990). The southern boundary of the Border Zone in the northwest is marked by the Kobberminebugt shear zone which Windley (1991) described as a steeply dipping belt of high strain. Watterson (1965) presumed that the principal displacements in the shear zone were transcurrent, but Windley (1991) noted that displacement directions have yet to be resolved.

The Border Zone on the east coast between Mogens Heinesen Fjord and Napasorsuaq Fjord comprises Archaean amphibolite facies orthogneisses with enclaves of metagabbro and amphibolite and local intercalations of supracrustal amphibolite and paragneiss. The orthogneisses have yielded an Rb–Sr whole-rock age of 2565 + 75 Ma (Pedersen et al. 1974; revised age after Kalsbeek et al. 1990). Andrews et al. (1971, 1973) believed that much of the orthogneisses were Ketilidian acid volcanic rocks, but the isotopic age, together with field relations and petrography described by Chadwick et al. (1994b), have shown that their belief was unfounded. The gneisses on Ikermit (Fig. 1) are deformed in small-scale shear zones of presumed Ketilidian age. Similar gneisses on Otte Rud Øer have a white 'bleached' aspect with blebby mica textures which may be a product of Ketilidian retrogression from granulite facies, although granulite facies gneisses with variable degrees of Archaean retrogression are abundant in the foreland north of Mogens Heinesen Fjord. The gneisses on Otte Rud Øer are intruded by plugs of hornblende diorite. These islands also include granodiorites which may be part of the Julianehåb batholith. Some of the steep shear zones in the islands are up to 25 m wide, many include mylonites and ultramylonites, and some form conjugate sets with NE and NW trends. The shear zones are also presumed to be of Ketilidian age like those on Ikermit. Some are cut by undeformed sheets of granite.

Metadolerite dykes between Mogens Heinesen Fjord and Napasorsuaq Fjord may be the equivalents of the pre-Ketilidian dykes in the west of the Border Zone, whereas numerous appinite dykes are similar to those in the Julianehåb batholith to the south. Mapping by the first author and B. J. Walton in 1987 showed that Ketilidian igneous effects in the form of relatively abundant dykes and low-angle sheets of appinite extend c. 80 km into the foreland north of Timmiarmiut (Fig. 1). The appinites cut dolerite dykes of presumed Palaeoproterozoic age, but are in turn cut by dolerite dykes presumed to be part of the Gardar system on the west coast.

Thick sequences of cross-bedded arkoses and polymict conglomerates in the isolated nunataks west of Otte Rud Øer (Fig. 1; Chadwick et al. 1994b) are intruded by Ketilidian granites (M. Rosing, pers. comm., 1995) and may be equivalents of the Ketilidian cover rocks in the southwest of the Border Zone.

The foregoing shows that the Border Zone in the east differs considerably from the western part. Although Ketilidian granites which intrude the arkoses in the nunataks west of Otte Rud Øer can be compared with the Ketilidian granites in the west of the Border Zone, no thrusting, back-folding or transitions to high-grade gneisses have been seen in the east. The Ketilidian effects in the east were restricted to emplacement of appinite dykes, concentrations of strain in ductile and brittle shear zones, and retrogression of Archaean gneisses and Palaeoproterozoic dolerite dykes.

The Julianehåb batholith

We use the term **Julianehåb batholith** to describe the polyphase granites, granodiorites, tonalites, diorites and subordinate metagabbros and amphibolite dykes (appinite suite *sensu* Pitcher 1993) which make up the wedge-shaped outcrop between

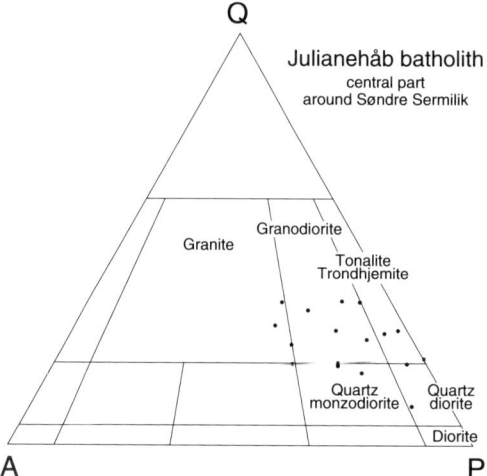

Fig. 2. Quartz–alkali feldspar–plagioclase (QAP) diagram showing normative mineralogy of samples from the southwestern part of the Julianehåb batholith in the Søndre Sermilik region. The diagram is based on chemical analyses using the cation norm calculation, modified so that biotite replaces hypersthene and K-feldspar where the latter two phases appear together in the norm (6hy + 5or = 8bi + 3qz).

the Border Zone and the Psammite Zone (Figs 1,2; Chadwick et al. 1994a). The batholith widens from c. 50 km between Napasorsuaq Fjord and Kangerluluk on the east coast to c. 150 km between Kobberminebugt and Søndre Sermilik in the southwest: it corresponds broadly with the Granite Zone of Allaart (1976). Its total area, including that part beneath the Inland Ice, is c. 30 000 km². Bridgwater et al. (1973) pointed out that xenoliths in the Julianehåb batholith had a general horizontal elongation which was consistent with transcurrent displacements suggested by fabrics in the Kobberminebugt shear zone (Watterson 1965, 1968) and steep shear zones in the Sârdloq (Windley 1966a, b) and Akuliaruseq areas (Persoz 1969).

Interpretation of the Ketilidian Orogen in the 1960s was strongly influenced by the theory of granitization, i.e. it was believed that much of the Julianehåb batholith and its associated mafic igneous rocks were formed by granitization of supracrustal rocks or pre-existing basement gneiss. For example, Persoz (1969) interpreted a large body of amphibolite within the granodiorite northwest of Søndre Sermilik as a mixture of metavolcanic and metagabbroic rocks. We have found, however, that this body and others are hornblende diorites s.l. of the batholith with variable degrees of net-veining.

Initial Sr ratios and the absence of old U–Pb zircon ages in a suite of granitic rocks and orthogneisses in the Julianehåb batholith showed that they were juvenile additions (van Breemen et al. 1974), not partial melts of Archaean crust as had been suggested by field evidence. The granites were regarded as partial melts of volcanic and sedimentary rocks that were only slightly older than the peak of Ketilidian metamorphism. Pb–Pb and Rb–Sr isotopic age data (Kalsbeek & Taylor 1985) showed that the orthogneisses and granites were emplaced episodically within the period c. 1850–1750 Ma and the juvenile component increases progressively southeast from the Archaean basement. New Rb–Sr whole-rock data from 14 samples (including high-strain gneisses) from the southern part of the batholith (Table 1; Fig. 3) further substantiate its juvenile affinity.

Chadwick et al. (1994a) showed that the batholith northwest of Søndre Sermilik comprises a complex range of porphyritic granites, granodiorites, tonalites, hornblende diorites and gabbro with major swarms of appinite dykes. Grey granodiorite and tonalite predominate, but details of boundary relationships between individual plutons have not been established so far. The granodiorites are dominated by plagioclase, microcline perthite and quartz with subordinate biotite and hornblende. Titanite, apatite, epidote and metamict allanite are common accessory minerals which are in accord with the I-type characteristics of the Julianehåb batholith indicated by its isotopic compositions (van Breemen et al. 1974; Kalsbeek & Taylor 1985).

Large bodies and small enclaves of aphyric, plagioclase- or hornblende-phyric, coarse- to medium-grained, mafic diorite s.l. are common in the granodiorites. The large bodies are a few hundred metres in size and commonly net-veined by the host granodiorite. Enclaves which resulted from mingling of pulses of immiscible mafic melts with the host granodiorite are 5–50 cm in size with smooth rounded boundaries. They are either widely dispersed or form swarms and clumps in the host: some of the swarms occur as well-defined steep sheets c. 2 m wide. Enclaves also formed by spalling of fragments from the margins of more massive net-veined dykes.

Schistosity and linear fabric

Much of the batholith is characterized by a steeply dipping or vertical schistosity trending NE–SW. An intense linear fabric with a shallow plunge in the plane of the schistosity prevails in many outcrops: L tectonites are relatively common. The Søndre Sermilik area includes several steep or vertical

Table 1. Rb–Sr isotope data for the Julianehåb batholith in the Søndre Sermilik area, South Greenland

GGU No	Locality		Rb (ppm)	Sr (ppm)	$^{87}Rb/^{86}Sr$	$^{87}Sr/^{86}Sr$
404504	45°10'W	60°50'N	94.8	406	0.67488	0.71985
404507	45°05'W	60°48'N	30.9	988	0.09019	0.70495
404514	44°33'W	60°50'N	34.3	623	0.15916	0.70718
404524	44°59'W	60°32'N	99.5	198	1.45378	0.74055
404530	44°59'W	60°32'N	96.1	353	0.78747	0.72411
404541	44°38'W	60°34'N	146	169	2.50093	0.77273
404552	45°13'W	60°54'N	38.7	1465	0.07633	0.70482
404902	45°15'W	60°49'N	65	843	0.22271	0.70830
404909	44°59'W	60°52'N	31.1	841	0.10666	0.70544
404925	44°57'W	60°37'N	25.4	892	0.8233	0.70529
404936	45°03'W	60°44'N	30.1	716	0.12120	0.70590
404954	45°17'W	60°44'N	35.1	711	0.14256	0.70627
404215	44°56'W	60°56'N	138	491	0.81607	0.72391
405216	44°58'W	60°56'N	81.5	668	0.35259	0.71170
405217	44°54'W	60°52'N	23.1	1177	0.05662	0.70412
405233	44°51'W	60°57'N	85	469	0.52413	0.71642
405234	44°49'W	60°47'N	35.3	565	0.18061	0.70746

Rb and Sr were analysed by XRF on pressed powder tablets (precision of Rb/Sr c. 1%, of ppm values c. 5%). Isotope determinations were performed on a VG 3540 multicollector mass-spectrometer and are reported relative to a $^{87}Sr/^{86}Sr$ value of 0.710250 ± 11 (1σ) for the Sr standard NBS 987 (all analyses at the Geological Institute, University of Copenhagen).

shear zones up to 1.5 km wide. Many are broadly parallel to the schistosity, but others are oblique. The shear zones are of two principal kinds: one, for example the **Sârdloq shear zone** (Fig. 1), is characterized by mylonitized and ultramylonitized plutonic rocks, whereas the other, for example the **Matorssuaq zone** (Fig. 1), is distinguished by high-strain orthogneisses.

Hutton (1988) and Paterson et al. (1989), among others, have emphasized the critical role of linear and planar fabrics and their microtextures, i.e. whether they are magmatic (pre-full crystallization) or crystal-plastic, solid state phenomena, in the interpretation of the tectonic setting and mechanism of emplacement of acid-intermediate plutonic rocks. Microtextures and fabrics in the Julianehåb batholith, specifically those in the granodiorites and shear zones in the Søndre Sermilik area, are reviewed below in the light of this recent work as a basis for the interpretation of the structure and intrusive setting.

The granodiorites are characterized by medium- to coarse-grained laths of plagioclase, commonly with oscillatory zoning. In isotropic non-foliated granodiorite, plagioclase occurs as coarse aggregates or as solitary subhedral laths with slightly sutured boundaries, bent twin lamellae and undulatory extinction. In the schistose granodiorites, many laths of plagioclase have a parallel preferred orientation which, with medium-grained flakes of biotite and rare prisms of hornblende and titanite, define the schistosity. We interpret this preferred orientation as the effect of magmatic or submagmatic flow as defined by Paterson et al. (1989). Whereas plagioclase twin lamellae may distort during submagmatic flow, much of the undulatory extinction, subgrain mosaics and

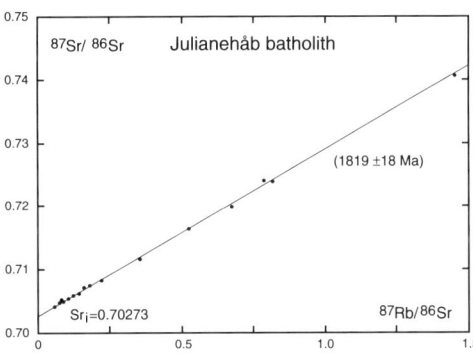

Fig. 3. Rb–Sr whole-rock isochron diagram of samples from the Julianehåb batholith in the Søndre Sermilik region (see Table 1 for localities and isotope data). The samples were collected from within an area of c. 1500 km² and plot close to a common c. 1820 Ma reference line with a low initial $^{87}Sr/^{86}Sr$ value. This part of the batholith thus comprises juvenile Palaeoproterozoic additions. The reference line (1819 + 18 Ma (2σ), MSWD 6.88; initial $^{87}Sr/^{86}Sr$ 0.70273) was calculated omitting the most radiogenic sample (404541, Table 1).

flexuring of the twin lamellae in the plagioclase in the isotropic and schistose granodiorites may be attributed to subsequent high temperature solid state crystal-plastic deformation.

Coarse aggregates of fine-grained, granoblastic-polygonal quartz with weakly sutured grain boundaries characterize granodiorite with an isotropic fabric. In schistose granodiorites, these aggregates appear as elongate lenticles coplanar with schistosity. Quartz in these lenticles has undulatory extinction and strongly sutured grain boundaries. These variations suggest that quartz crystallized as part of the magmatic to submagmatic isotropic fabric, but its texture was modified by submagmatic or high temperature solid state flow during the formation of the schistosity.

Microcline is a relatively minor, interstitial component of the granodiorites. Minor myrmekite in adjacent plagioclase suggests that high temperature solid state deformation was superimposed on the magmatic–submagmatic texture of microcline in accord with the review of the significance of myrmekite by Paterson *et al.* (1989).

Scattered enclaves of hornblende diorite with prolate ellipsoidal shapes, 5–50 cm long, are commonly elongated parallel to the shallow linear fabric. Paterson *et al.* (1989) interpreted the preferred alignment of elongate enclaves as an effect of magmatic flow if they show no evidence of plastic deformation or recrystallization. Coarse laths of plagioclase and prisms of hornblende in many of the strongly elongate enclaves are parallel to their long axes, but in weakly elongate enclaves there is a mixed, random and partial preferred orientation of these minerals. The plagioclase laths have oscillatory and patchy zoning, subgrain mosaics, slightly bent twin lamellae and undulatory extinction. Prisms of hornblende also have undulatory extinction and are partly replaced by neoblasts of biotite. The zoning in plagioclase indicates that they are magmatic crystals, but the other microtextures are indicative of crystal-plastic, submagmatic or high temperature solid state deformation. Whereas the shape and parallel preferred alignment of hornblende diorite enclaves suggest that the lineation and schistosity in the granodiorites were the products of predominantly submagmatic flow, crystal-plastic microtextures suggest high temperature solid state deformation outlasted the effects of submagmatic flow.

Whereas the bulk of the batholith is characterized by penetrative linear and planar fabrics, there are a few major bodies of younger, undeformed granites and syenites, for example on the Akuliaruseq peninsula northwest of Søndre Sermilik (Persoz 1969) and the north coast of Naparsorsuaq Fjord.

Shear zones with mylonites

The western part of the batholith is bounded to the north by the Kobberminebugt shear zone which was described by Windley (1991) as 15 km wide, although belts of mylonite are relatively thin (*c.* 100 m; Harry & Oen 1964). Bridgwater *et al.* (1973) reported that this shear zone extends northeast to Napasorsuaq Fjord on the east coast, but we were unable to locate it during our helicopter reconnaissance in 1992. The north coast of Napasorsuaq Fjord is dominated by undeformed granite and syenite which were intruded into fractured and mylonitized orthogneisses and metadolerite dykes. Comparable orthogneisses and local sillimanite–paragneisses west and north of Napasorsuaq Fjord appear to be part of the foreland in the Border Zone.

The Sârdloq shear zone (Fig. 1) is typical of the larger shear zones in the Julianehåb batholith. Based on descriptions by Bridgwater *et al.* (1973), Windley (1991) defined the Sârdloq shear zone as '... 20–30 km-wide high deformation zone (separating) totally different geological terranes – an igneous batholith to the north from a thrust stack of ... supracrustal rocks ... and rapakivi granites to the south. The borders of the shear zone consist of two 10–15 km-wide steep shear belts ...' We found that there are no shear belts as wide as 10–15 km, and the Sârdloq shear zone does not separate two different terranes. We propose that the term **Sârdloq shear zone** be applied only to the 1.5 km wide steep belt of mylonites and ultramylonites in the granodiorites northwest of Søndre Sermilik and their probable extension along-strike to the SW into the narrow belt of 'veined gneisses' trending NE–SW through Saarloq itself (Windley 1966*a*, Map 1).

Mylonites and ultramylonites in the Sârdloq and other prominent shear zones in the Julianehåb batholith were derived by intense deformation of the adjacent granodiorites and granites with the development of steeply dipping or vertical type I S-C fabrics (*sensu* Lister & Snoke 1984). Within *c.* 100 m of the Sârdloq shear zone the regional schistosity in the granodiorites is intensified, and within the shear zone itself it is intensified in the C fabric. In the ultramylonites the pre-existing schistosity appears to have been obliterated completely.

Mineral lineations with a shallow NE plunge in the Sârdloq and other shear zones indicate a transcurrent displacement. Lateral displacement of dykes and enclaves, textures of feldspar porphyroclasts, an extensional crenulation fabric (shear bands *sensu* White *et al.* 1980; normal slip crenulations *sensu* Dennis & Secor 1987) and rare asymmetric S folds indicate a consistent sinistral sense of shear. The total finite displacement in the shear zones is unknown, but in the Sârdloq shear zone alone it was probably many kilometres on the grounds of its width and the intensity of the mylonitization. Other small-scale, vertical or steeply dipping shear zones trending N–S and ESE–WNW with sinistral and dextral displacements appear to have been contemporaneous effects related to the major NE–SW shear zones.

Textures of feldspar and quartz indicate crystal-plastic solid state deformation, although the infilling of gaps between disrupted feldspar porphyroclasts by quartz and feldspar suggests the presence of small amounts of melt during mylonitization. Stable hornblende and biotite suggest that deformation took place in amphibolite facies PT conditions, whereas local replacement of biotite by chlorite indicates late greenschist facies conditions. Effects of hydrothermal alteration, mainly in the form of quartz veining, silicification, chloritization, epidotization and pyritization are common within and adjacent to the largest shear zones. These effects appear to be related to late stages of the evolution of the batholith. Gold anomalies appear to be closely tied to the hydrothermal phenomena (Chadwick et al. 1994a). Some of the shear zones were reworked as brittle fractures with development of cataclasites, including fault breccias with strontianite and hematite mineralization. Although they are cut by Gardar dykes, some of the brittle reworking was probably an effect of Gardar deformation.

Zones of high-strain orthogneisses

Another variety of shear zone which also trends NE–SW is represented by steeply dipping or vertical belts of high-strain orthogneisses, for example the **Matorssuaq zone** (Fig. 1) which is c. 500 m wide and traceable along-strike for c. 10 km. A mineral lineation with a shallow plunge to the NE is found on some schistosity surfaces. The high-strain gneisses were intruded by discordant sheets of pale granitic rocks, some of which are folded with axial surfaces coplanar with the schistosity of their host, whereas others share the same schistosity as their host gneisses or are not deformed. Thin concordant seams and intrafolial ptygmatically folded veins of pegmatite and aplite are restricted to the high-strain gneisses and appear to have been generated as partial melts contemporaneous with the ductile shearing.

The regional granodiorite and its enclaves of dark grey titanite-bearing granodiorite adjacent to the Matorssuaq zone can be traced into the high-strain gneisses across a transition zone c. 25 m wide. The titanite-bearing granodiorite is modified in the high-strain zone to a fine-grained schist with marked preferred planar orientation of biotite and linear orientation of subhedral prisms of hornblende. Plagioclase, quartz and subordinate microcline form a fine-grained equidimensional groundmass with weakly sutured grain boundaries: there are no quartz ribbons like those in the mylonites in the Sârdloq shear zone. We interpret the fabrics and microtextures as high temperature solid state flow phenomena consistent with localised anatexis and emplacement of various deformed and non-deformed sheets of granite s.l. as the zone evolved. The field relationships show that the high-strain gneisses were derived by deformation of the granodiorites and are not relics of a gneiss complex older than the Julianehåb batholith as indicated by previous workers. Some of the orthogneisses in the batholith with c. 1850 Ma isotopic ages reported by van Breemen et al. (1974) and Kalsbeek & Taylor (1985) may be relics of similar early high-strain zones.

Appinite suite in the Julianehåb batholith

Amphibolitic and mafic dioritic dykes typical of the appinite suite (sensu Pitcher 1993) are common in the Julianehåb batholith (Walton 1965; Allaart 1967). They range in composition from hornblendites to pale diorites s.l., and are commonly net-veined (Watterson 1965, 1968; Windley 1965). The similarity in composition between many of the dykes and the enclaves and larger bodies of diorite in the batholith suggests they were probably consanguineous. The dykes were intruded at various stages during the emplacement and deformation of the batholith. Some dykes have disrupted margins where elongate fragments a few metres long have floated off into the host granodiorite, some cut across the schistosity in the granodiorite host, some have an internal schistosity oblique to that in the host, and others curve into shear zones with NE–SW trends or they may be displaced by a few metres. This curvature is compatible with the sense of transcurrent displacement in the shear zones. The oblique internal schistosity, the consistent curvature and the commonly unbroken passage of dykes along shear zones until they link with the high-angle dykes outside indicate emplacement contemporaneous with the localized plastic yielding in the shear zones.

The Psammite Zone

This zone has a sharp boundary with the Julianehåb batholith in the Søndre Sermilik area, northeast of Tasermiut and in the Sorte Nunatak area, but it has not been fully defined on the east coast (Fig. 1). Large enclaves of supposed metavolcanic felsic rocks (Chadwick *et al.* 1994*a*) lie just within the batholith northwest of Søndre Sermilik. Much of the northwestern boundary of the Psammite Zone is marked by steep NE–SW shear zones or shallow contacts with underlying granodiorite, for example in the Sorte Nunatak area. The southeastern boundary of the Psammite Zone on the west coast is arbitrary because of the abundance of rapakivi granites. The southeastern boundary on the east coast is a transition zone south of Lindenow Fjord where intensely migmatized pelites with relatively shallow dips become predominant over psammites.

Lithological characteristics on the east coast

The Psammite Zone on the east coast comprises a major zone of migmatized, bedded psammites which extends northeast from Lindenow Fjord to Kangerdluluk. The psammites dip gently northwest in Danell Fjord, but dips are more variable north of Lindenow Fjord because of large-scale folding. The total thickness of this tract of psammites is unknown because of the reconnaissance scale of the mapping, but it is probably in the order of a few thousand metres. Bedding thickness in the psammites varies from a few centimetres to about one metre. Cross-bedding and other primary structures indicative of fluvial or shallow marine conditions are preserved in a few outcrops (Fig. 4). Pervasive recrystallization has obscured much of the primary detrital grains and textures, but the predominance of microcline and plagioclase over quartz indicates that the psammites were originally arkosic arenites. Brown biotite, bluish green hornblende and subordinate titanite, apatite, zircon and magnetite make up 10–15% of the mode. Concentrations of these accessory grains on foreset beds show that they are detrital. Calcareous psammites with microcline, plagioclase, quartz and subordinate diopside, garnet and calcite are common.

The psammites are interbedded with variably migmatized pelites and semi-pelites which at the head of Danell Fjord and in the nunataks to the north are characterized by graded bedding indicative of deposition by turbidity flows. Some of the pelites are graphitic and many include swathes of fibrolite and poikiloblasts of cordierite and andalusite. Basaltic and andesitic amphibolites are a volumetrically minor component of the Psammite Zone. Many are intensely deformed, but some areas preserve plagioclase-phyric flows, hyaloclastite breccias, bedded volcaniclastic rocks and rare pillow-structured flows.

The newly discovered supracrustal sequence in Sorte Nunatak (*c.* 70 km^2, Fig. 1) is dominated by feldsparphyric, amygdaloidal metabasalts and andesites with interbedded semi-pelites, psammites and polymict conglomerates. The sequence is folded by large-scale, tight to open, upright structures with a steep schistosity trending NE–SW parallel to the shear zones that bound the outcrop on the northwest and southeast (Figs 1,5). Clasts in the conglomerates are elongated parallel to a shallow NE–SW lineation within the plane of the schistosity. The schistosity is overprinted by

Fig. 4. Primary structures in psammites, north of Danell Fjord. Note slumped foreset beds top right. Seams of anatectic melt occur in the same outcrop, but are not seen in the photograph. Coin is *c.* 2.5 cm in diameter.

Fig. 5. Steep NE–SW shear zone marking the northwestern boundary between the Julianehåb batholith and the metalavas (dark) and interbedded psammites (pale) of Sorte Nunatak (see Fig. 1); the upright fold has a steep schistosity; cliff is c. 250 m high and the main summit is 1710 m.

randomly arranged prisms of hornblende. The conglomerates include unsorted, clast- or matrix-supported, rounded clasts of porphyritic granite, deformed and undeformed, equigranular granite or granodiorite, aplite and feldsparphyric amphibolite which range from a few centimetres to one metre in size. Trough cross-bedding, with soft-sediment slumping of foresets, is common in the conglomerates, psammites and semi-pelites (Fig. 6). Similar conglomerates underlie amphibolites in Kangerdluluk.

Most of the psammites and pelites in the east of the Psammite Zone are intensely migmatized. *Lit-par-lit* migmatization is so intense in some outcrops that the psammites look identical to banded orthogneisses. Small bodies of anatectic granite with enclaves of the host migmatized sedimentary rocks are common: the granites are cut by net-veined diorite dykes.

Lithological characteristics on the west coast

In the northeast of the peninsula between Søndre Sermilik and Tasermiut (Fig. 1) the supracrustal rocks of the Psammite Zone lie within a large-scale asymmetrical syncline at right angles to the trend of the Julianehåb batholith. The northeast limb of the syncline is in part overturned to the southwest, whereas the southwest limb is sub-horizontal. Strongly foliated, partly migmatized semi-pelites which immediately overlie the batholith are followed by c. 500 m of schistose amphibolites with local pillow structures and breccias with

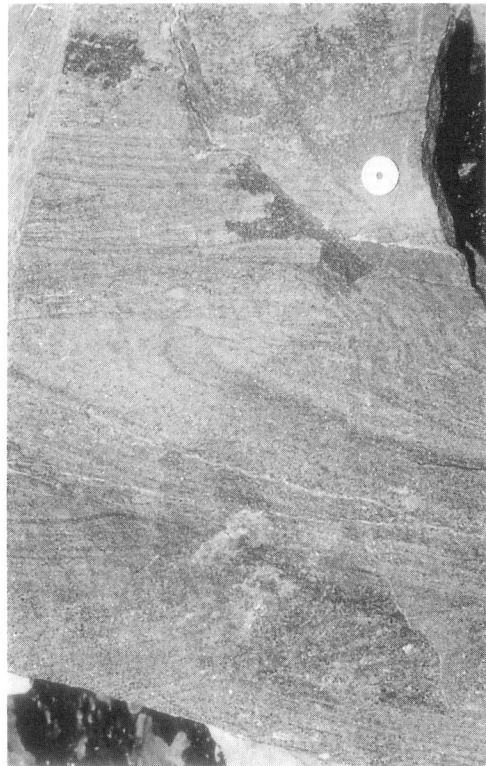

Fig. 6. Effects of soft-sediment slumping, including folding of foresets, in arenite; loose boulder in moraine from Sorte Nunatak (see Fig. 1). Coin is c. 2.5 cm in diameter.

deformed fragments of leucogabbro and mafic amphibolite. The amphibolites are overlain by a thin layer of black pelites and c. 1000 m of psammites. The psammites in the shallow limb are intensely migmatized, but those in the steep overturned limb are less migmatized and preserve trough cross-bedding and grading which show that the stratigraphy is right way up. Bedding is 1–2 m thick and the composition of the psammites is identical to those on the east coast.

Coarse polymict conglomerates up to a few metres thick with closely packed, well-rounded clasts of granodiorite up to 1 m in size and smaller clasts of pale, presumed metavolcanic rocks, vein quartz, metadolerite and a variety of granitic compositions, including aplite, are interbedded with the psammites. Other conglomerates at the base of graded psammite beds are matrix-supported. Their clasts are 5–20 cm in size, well-rounded and predominantly of granitic rocks with subordinate clasts of semi-pelite with detrital magnetite. The rounding of the clasts and the common trough cross-bedding indicate a high energy fluvial or marine deltaic environment.

Similar conglomerates and arkosic psammites, up to 700 m thick, east of Tasermiut (Wallis 1966; Dawes 1970) occur within an upper tectonic unit which is relatively undeformed, but variably migmatized. Migmatized pelitic rocks, which appear to be part of a lower tectonic unit, are abundant further south in the Tasermiut area. Wallis (1966) suggested that the conglomerates were either slumped sub-aqueous deposits or products of fluvio-glacial redeposition. He rejected the suggestion of Wegmann (1948) that the conglomerates were tillites. Wegmann (1939) also proposed that the conglomerates were desert deposits. Dawes showed that age relationships between the various units of metasedimentary rocks and basic metavolcanic rocks are uncertain because of the intersheeting of younger rapakivi granites.

Escher (1966) recognized five principal units of supracrustal rocks in the Nanortalik peninsula in the extreme southwest of the Psammite Zone (Fig. 1). They are dominated by variably migmatized pelites, semi-pelites, psammites (called quartzites by Escher) and metavolcanic amphibolites with local pillow structures and agglomerates.

Basement to the supracrustal rocks of the Psammite Zone

There are no unambiguous field indications of the age of the basement to the metasedimentary and metavolcanic rocks of the Psammite Zone, although three lines of circumstantial evidence point to the Julianehåb batholith as the basement and provenance.

First, the abundance of alkali feldspar in the psammites and the common clasts of granite *s.l.* in the conglomerates indicate a granitic provenance not far removed from the sedimentary basin. The clasts of volcanic and sedimentary rocks show that supracrustal sequences were also exposed to erosion within the provenance. The rounding of the clasts suggests either a fluvial or shallow marine environment or a two-stage depositional process in which rivers delivered rounded clasts to the basin margin before they were reworked in marine environments. The abundance of detrital titanite, apatite, zircon and magnetite in the psammites is also consistent with a provenance of granitic and dioritic rocks similar to those in the Julianehåb batholith. The mainly tonalitic composition and distance from the basin make the gneisses of the Archaean foreland an unlikely source of detrital alkali feldspar and boulders of granite *s.l.*.

The second thread of circumstantial evidence comes from relationships between the supracrustal rocks and the underlying batholith in Sorte Nunatak and south of Søndre Sermilik. In the northeast of Sorte Nunatak, psammites and a polymict conglomerate, c. 50 m thick with unsorted clasts up to 1 m in size, have a sub-horizontal contact with the underlying granite (Fig. 7). The conglomerate is overlain by mafic metalavas. The contact is inaccessible, but the overstep of the conglomerate above an underlying psammite suggests it is an unconformity. Moreover, two mafic dykes that cut the granite do not extend across the contact into the conglomerate (Fig. 7). Mafic metalavas have a sub-horizontal contact with the granite in the southwest of Sorte Nunatak. The contact is sheared with thin, concordant veins of quartz. Sheared psammite also intervenes locally between the metalavas and the underlying granite. A few large sub-horizontal apophyses of granite which extend into the metalavas suggest that part of the underlying granite is younger than the supracrustal rocks. In conclusion, the contacts in the Sorte Nunatak area may be interpreted both as unconformable and intrusive, but the presence of granitic clasts in the conglomerates strongly suggests that the Julianehåb batholith was the basement and provenance. The contact between the granodiorites and the overlying supracrustal rocks southeast of Søndre Sermilik suggests age relationships comparable to those in Sorte Nunatak, although sheared areas seen from the air may be the result of thrusting.

The third line of evidence, much more tenuous than the foregoing, is the general absence of xenoliths of supracrustal rocks in the batholith, apart from certain large bodies of pale quartzofeldspathic rocks (perhaps acid volcanic rocks, Chadwick *et al.* 1994*a*) in the granites west of Søndre Sermilik and minor occurrences of

Fig. 7. Probable unconformity (larger arrows) at the base of the psammites and conglomerates (pale) and metalavas (dark) in Sorte Nunatak (see Fig. 1); note: the granodiorite beneath the unconformity contains two mafic dykes (smaller arrows) that do not extend up into the sedimentary rock above the unconformity, and a thin dark bed overlying the granodiorite in the left of the photograph oversteps a wedge of pale grey psammite above the granodiorite in the right. The height of the cliff is c. 100 m.

cordierite–andalusite gneisses to the north which were interpreted as metasedimentary rocks by Persoz (1969).

On the grounds of these various lines of evidence we favour the evolving Julianehåb batholith as the basement and provenance to the sedimentary rocks of the Psammite Zone. This interpretation is supported by preliminary U–Pb ages of zircon in the granodiorites (M. A. Hamilton, pers. comm., 1995) and SHRIMP U–Pb ages of detrital zircon in the psammites (A. P. Nutman, pers. comm., 1995).

Structure of the Psammite Zone

The structure in the Psammite Zone on the east coast has not been fully established. The shallow dip of bedding in the east of the zone steepens adjacent to the boundary with the Julianehåb batholith at the head of Danell Fjord. The steepening appears to be related to folding of nappes which are indicated by widespread overturning of bedding in the psammites, but their structure is not known in detail. Top-to-the-southeast displacements are suggested by folds with abundant seams of neosome parallel to fold axial surfaces south of Danell Fjord. Large-scale tight-open folds with NE trend which deform bedding and neosome north of Lindenow Fjord (Fig. 8) appear to be related to the shallow structures further northwest: these folds also have coplanar seams of neosome. Folding younger than these NE-trending structures is indicated by steep crenulation cleavages trending E or SE in the more pelitic parts of the Psammite Zone. These cleavages are younger than poikiloblasts of cordierite and andalusite which grew over an early bedding-parallel schistosity, although andalusite appears to have continued growing after the microbuckling. These relationships show that high temperature conditions prevailed throughout the deformation.

The structure of the supracrustal rocks in the Psammite Zone northwest of the head of Tasermiut on the west coast is dominated by a large-scale upright syncline which trends NW–SE (see above). It correlates with the predominant second phase of fold structures of Escher (1966) further southwest in the Nanortalik peninsula. These NW–SE structures were superimposed on early folds with NNE-trending axes. Effects of the early phase were mostly restricted to the mica schists and gneisses in the lowest part of the supracrustal rocks, whereas the volcanic rocks and psammites were largely unaffected (Escher 1966). The third phase is characterized by large wavelength (up to 10 km), ENE- and NNE- trending folds with crenulation cleavages, faults and thrust-faults.

Large-scale recumbent folds resulted in repetition of the succession southeast of Tasermiut (Dawes 1970). They are confined mainly to the lower pelitic and hypersthene-bearing paragneisses, whereas the higher level psammites, conglomerates and volcanic rocks are largely unaffected. The east

Fig. 8. Refolded isoclines in psammite; loose boulder, head of Danell Fjord. Coin is c. 2.5 cm in diameter.

or southeast transport of the recumbent folds (nappes) was probably accompanied by thrusting in the psammites and conglomerates. The early folds were locally refolded by open NW-trending folds.

Appinite and related dykes in the Psammite Zone

Deformation in the Psammite Zone was accompanied throughout by emplacement of appinite dykes and granites *s.l.* Many of the appinite dykes were emplaced parallel to the axial surfaces of early folds and were then boudinaged as deformation continued. Others were intruded much later, but before the complete cessation of ductile deformation. Steep sheets of granite full of rounded inclusions of medium-grained diorite, metagabbro and metaperidotite 10 cm to 3 m in size were also emplaced during the later period of folding. These relationships show that deformation in the Psammite Zone was characterized not only by pervasive anatexis and high temperature–low pressure metamorphic conditions as indicated by poikiloblasts of cordierite and andalusite, but also intrusion of various melt compositions of the appinite suite.

Rapakivi granites

Sub-horizontal sheets of brown Fe-orthopyroxene-fayalite monzonite and grey biotite-hornblende monzonite (Harrison *et al.* 1990), both of the rapakivi granite suite, were emplaced c. 1750 Ma ago (van Breemen *et al.* 1974; Gulson & Krogh 1975) into the supracrustal rocks of the Psammite Zone after the cessation of the ductile deformation. The brown and grey facies intermingle on various scales. Intense migmatization of supracrustal gneisses immmediately beneath a thick sheet of brown granite northwest of Lindenow Fjord suggests localized contact metamorphism, although Dawes (1970) reported no contact effects adjacent to the sheets of rapakivi granite in the Tasermiut region. Brown *et al.* (1992) attributed the presence of aureoles adjacent to the rapakivi granites as an effect of the level of their emplacement, i.e. those no deeper than c. 6 km overprinted regional assemblages, whereas at deeper levels they were contemporaneous with regional metamorphism and have no aureoles. These relationships imply that the rapakivi suite was emplaced very soon after the close of the ductile deformation.

Bridgwater *et al.* (1974) believed that the rapakivi granites were sill-like masses with a mushroom shape which had been fed by a vertical stem. They believed that the steepening of the basal contacts of the sheets of rapakivi granite, for example that of the brown granite in the west of Lindenow Fjord, was an effect of the diapiric emplacement process. Hutton *et al.* (1990) proposed that emplacement of the sheets of granite was facilitated by late-stage, ductile extensional shear zones. In a related study, Brown *et al.* (1992) concluded that the genesis of the rapakivi granites and associated norites involved predominantly juvenile crust or a mantle component.

The Pelite Zone

The Pelite Zone (Fig. 1) comprises mainly flat-lying, intensely migmatized pelitic rocks. Relics of psammites, some with cross-bedding, occur as thin, boudinaged resistor layers or as xenoliths in the lenses of anatectic granite and the sheets of rapakivi granite. Garnet, cordierite and sillimanite are common in the palaeosome and anatectic neosome. Disrupted discordant amphibolites (appinites) and rare ultramfic rocks occur locally. Isoclinal folds of varying scales are deformed by more upright folds with E or SE trends, but the structure of the Pelite Zone is not known in any detail because of the lack of persistent markers. Dempster *et al.* (1991) showed that the peak of high-grade regional metamorphism in the Pelite Zone coincided with the emplacement of the rapakivi suite at pressures of 2–4 kb and temperature in the range 650–800°C. They attributed the high geothermal gradient of $>60°C$ km^{-1} to underplating by large volumes of basaltic melts in a regime of lithospheric extension. The extensional setting for the peak of metamorphism reported by Dempster *et al.* (1991) contrasts with the syntectonic compressional setting of peak metamorphic cordierite and andalusite growth and migmatization observed by us in the Psammite Zone to the north.

Many parts of the intensely migmatized paragneisses are transitional into sheets of anatectic, schlieric garnetiferous granites with foliation parallel to that in their flat-lying hosts. These granites, in turn, grade into larger and more homogeneous grey granites with scattered megacrysts of alkali feldspar and xenoliths of psammite (some with sheets of amphibolite) and variably digested semi-pelite.

Rapakivi granites *s.l.* and associated norites are more abundant in the Pelite Zone than the Psammite Zone. Their setting and compositions are similar to those in the Psammite Zone described above.

Plate tectonic setting of the Ketilidian Orogen

Bridgwater *et al.* (1973) were the first to compare the Ketilidian and Andean belts on the basis of petrological similarities and vertical and transcurrent movements. Our findings in the Julianehåb batholith are in accord with their views, but we can advance their comparison with Phanerozoic destructive plate boundaries a stage further with our new data.

Our field and microtextural data show that the steep schistosity and shallow linear fabrics which characterize much of the Julianehåb batholith formed during submagmatic flow and subsequent high temperature solid state deformation. Microtextures and fabrics in major shear and high-strain gneiss zones, such as Sârdloq and Matorssuaq, formed as part of the high temperature solid state deformation. Moreover, the S-C fabrics, feldspar porphyroclasts and dykes in the major shear zones indicate sinistral displacement. On the grounds of these various lines of evidence we propose that emplacement of the polyphase components of the Julianehåb batholith took place in a regime of sinistral transpression (Chadwick *et al.* 1994a) which was the consequence of oblique plate convergence as new Palaeoproterozoic crust was accreted in the form of the Julianehåb batholith onto the Archaean craton of South Greenland. This view is in accord with the isotopic age data of van Breemen *et al.* (1974), Patchett & Bridgwater (1984) and Kalsbeek & Taylor (1985), and the transpressional settings of younger orogenic magmatism described by D'Lemos *et al.* (1992) and Hutton & Reavy (1992).

Comparison of the Julianehåb batholith with those in younger zones of plate convergence suggests that its granitic and dioritic components can be regarded as parts of the root zone of a volcanic arc that developed partly on, and accreted adjacent to, a continental margin (Fig. 9). There are few inclusions of volcanic rocks within the batholith outcrop. Lavas and pyroclastic rocks are presumed to have been eroded more or less contemporaneously with their eruption to provide detritus for the psammites and other clastic rocks within the Psammite and Pelite Zones. Moreover, uplift during transpression and batholith emplacement is likely to have exposed part of the batholith itself to erosion. Consequently, we regard the batholith and its volcanic roof as the provenance of the psammites and related sedimentary rocks. We also presume that the batholith extends southeast as the basement beneath the northwestern part of the Psammite Zone.

The conglomerates, psammites and pelites of the Psammite and Pelite Zones are distributed asymmetrically: the conglomerates are largely restricted to the northeast of the Psammite Zone, and they are followed to the southeast by a tract dominated by psammites. Pelitic rocks are most abundant in the extreme southeast, i.e. in the Pelite Zone, although subordinate psammites, including some with cross-bedding, are preserved as xenoliths in the anatectic and rapakivi granites in this zone. Accumulation of the original sediments in the Psammite and Pelite zones took place as a gradational series with fluvial or mixed fluvial and shallow marine conglomerates and arenites proximal to the arc, followed by shallow marine arenites and more distal argillites, respectively. Cross-bedding in xenoliths of psammite in granites in the Pelite Zone suggests that

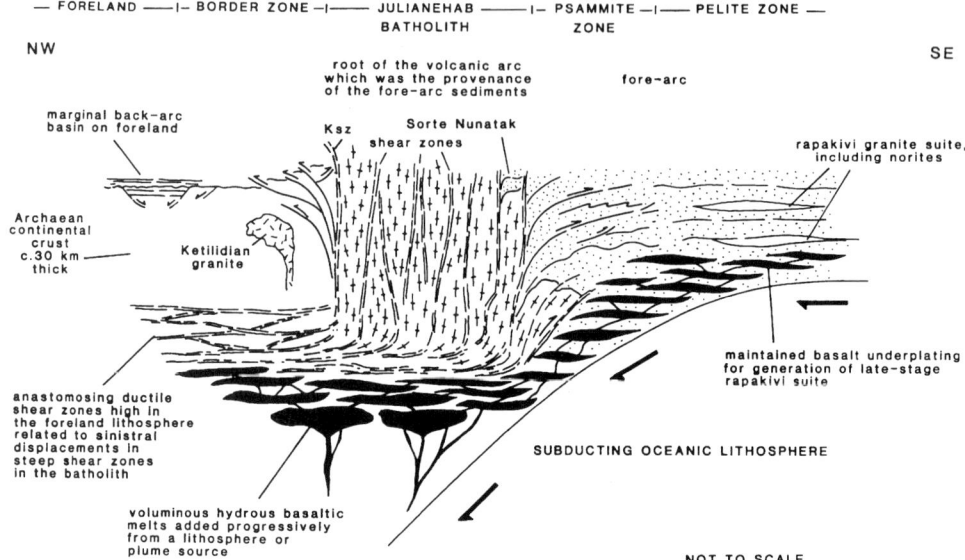

Fig. 9. Schematic representation of the Ketilidian Orogen in terms of oblique plate convergence; basalt underplating is regarded as having taken place throughout the period of accretion of the Julianehåb batholith, basin development, deformation, metamorphism and emplacement of the rapakivi suite.

water depths may not have been extreme. The asymmetric distribution of the sedimentary rocks in the Psammite and Pelite zones, together with the subordinate basaltic and andesitic components, suggest deposition and volcanism took place in intra-arc basins (e.g. Sorte Nunatak) that were transitional into the inner fluvial and shallow marine part of a forearc between the volcanic arc with its batholith and foreland to the north and an ocean to the south (Fig. 9).

Although the forearc pile was thickened by thrusts and nappes, the absence of high pressure mineral assemblages suggests that thickening was not nearly as great as that in Phanerozoic continent–continent collision zones. The high temperature–low pressure metamorphism and widespread anatexis in the Psammite and Pelite zones are consistent with basalt underplating beneath the forearc and its arc batholith. Underplating continued throughout the periods of ductile deformation of the Psammite and Pelite zones as indicated by the syntectonic dykes of the appinite suite. Partial melting of the sedimentary rocks at depth is presumed to have contributed to the granitic components of the rapakivi suite at a late stage in the evolution of the Ketilidian Orogen. This view is in accord with the views of Dempster et al. (1991) and Brown et al. (1992), although they proposed a backarc setting for the Psammite and Pelite zones.

Nd isotope data from supracrustal rocks and the rapakivi suite in the Pelite Zone and the west of the Psammite Zone led Patchett & Bridgwater (1984) to visualize the Ketilidian Orogen as a 'subduction related continental margin orogen' in which 5–17% of Archaean sediments were mixed with juvenile melt additions. In our model (Fig. 9), Archaean oceanic sediment suggested by Patchett & Bridgwater (1984) may have been added from the lithospheric plate south of the orogen.

Windley (1991) compared the Ketilidian Orogen with the plate tectonic setting of the Himalayas, especially with reference to the Karakoram region at the western uplifted end of the Tibetan plateau. He suggested that the Border Zone represented a shelf-foredeep which was thrust northward towards the foreland and back-folded and thrust southward to a suture which extended from Kobberminebugt to Napasorsuaq Fjord (Fig. 1). He regarded this suture zone as an association of basaltic lavas, gabbros, shear belts, mylonites and gold and copper mineralisation. We have found no evidence of this suture in the area of Napasorsuaq Fjord.

Windley (1991) regarded the Julianehåb batholith as an Andean-type batholith that had been intruded into an island arc of volcanic rocks, gneissose plutons and layered noritic rocks. He based his identification of volcanic rocks in the batholith on the few inclusions reported by Allaart (1967) and the erroneous interpretation of dioritic bodies as basic volcanic rocks by previous workers. He interpreted the 25 km wide tract southeast of the Sârdloq shear zone (our definition) as a steep backarc shear belt which extended northeast to just north of Kangerluluk on the east coast. Finally, he proposed that the Psammite and Pelite zones represent an area of 'thrust-thickened continental crust' where '...ensuing thermal relaxation led to partial dehydration melting of deep continental crust.' These proposals are open to question on the grounds of the lack of inclusions of supracrustal rocks in the batholith, the presence of intra-arc basins and the absence of a 25 km wide shear zone in the south of the batholith, and the lack of thrust-thickened continental crust in the Psammite and Pelite zones. We share the view of Brown *et al.* (1992) who noted that the igneous and metamorphic history of the southern part of the Ketilidian Orogen is inconsistent with the plate tectonic model of Windley (1991).

Conclusions

The Ketilidian Orogen in South Greenland formed as part of a Palaeoproterozoic oblique convergent system between an Archaean continent to the north and an oceanic plate to the south which is no longer preserved (Fig. 9). Shallow marine metabasalts and sedimentary rocks on the southern margin of the Archaean foreland in the west of the orogen and fluvial or shallow marine arkoses and conglomerates in the east indicate Ketilidian passive margin or backarc extension. Parts of these cover rocks and their basement in the west were deformed and metamorphosed in the southern margin of the foreland. Several granite plutons intruded this marginal zone in the west: granites which intrude the foreland in the east of the orogen are also presumed to be derivatives of partial melting of the foreland gneisses like those in the west. Swarms of appinite dykes which extend at least 80 km north into the foreland in the east are regarded as part of the Ketilidian igneous province. These thermal phenomena favour a backarc setting for this southern part of the foreland.

The wedge-shaped Julianehåb batholith immediately south of the foreland is a polyphase, calc-alkaline complex with swarms of appinite dykes. Microtextures and fabrics indicate that the components of the batholith were emplaced in a regime of sinistral transpression. The batholith formed the root of a volcanic arc which was the provenance of, and in part the basement to, the metasedimentary rocks to the south.

The sedimentary facies and distribution of polymict conglomerates, metamorphosed arkosic arenites, semi-pelites and pelites south of the batholith indicate that they were deposited in fluvial and shallow marine environments in intra-arc basins and the inner part of a forearc between the volcanic arc represented by the batholith and an oceanic area to the south.

The metasedimentary rocks and their intercalations of metabasaltic amphibolites were intensely migmatized in syntectonic, high temperature–low pressure conditions as indicated by the relationship between structures, tectonic fabrics and abundant andalusite, cordierite, *lit-par-lit* neosome and large lenses of anatectic granite. Overthrusting and recumbent folds propagating from the northwest may have led to limited thickening. Appinite dykes were emplaced during at least two major stages of deformation. Broadly concordant sheets of rapakivi granite with mixtures of pale grey biotite-hornblende and brown Fe-orthopyroxene–fayalite facies were emplaced into the metasedimentary rocks late in the history of the orogen.

The evolution of the Ketilidian Orogen was dominated by widespread thermal phenomena which included the emplacement of appinite dyke swarms and granites in the foreland, the varied suite of plutonic rocks and appinitic dyke swarms in the calc-alkaline Julianehåb batholith, and the migmatization and emplacement of appinite dykes and the rapakivi suite in the supracrustal rocks south of the batholith. Prolific basaltic underplating beneath the evolving convergent zone, perhaps as an effect of a mantle plume, appears to be the likely source of the necessary thermal energy. Our model of the Ketilidian Orogen differs from the Himalayan collision model of Windley (1991), but it is in accord with the isotopic age data of van Breemen *et al.* (1974), Patchett & Brigwater (1984) and Kalsbeek & Taylor (1985), the metamorphic PT conditions in the Pelite Zone determined by Dempster *et al.* (1991) and the petrogenetic model of the rapakivi suite proposed by Brown *et al.* (1992) which together indicate the predominance of juvenile additions to the bulk of the orogen and thermal effects dominated by major basaltic underplating.

The prolific generation of juvenile plutonic and sedimentary rocks, the high temperature–low pressure metamorphism and lack of Archaean crust in the Ketilidian Orogen are in marked contrast with the subordinate volumes of newly formed crust and abundant reworked Archaean crust in the thrust stacks and nappe complexes in the Palaeoproterozoic Rinkian, Nagssugtoqidian and Ammassalik orogens north of the stable Archaean block of southern Greenland. Although it is still uncertain whether Proterozoic sedimentary rocks are volumetrically important in the Nagssugtoqidian (Kalsbeek *et al.* 1987), the orogens north of the Archaean block have more in common with the Alpine–Himalayan system than the Ketilidian Orogen which compares more closely with the Andean collision zone. The differences between the Ketilidian Orogen and those to the north may be related to differences in the size of oceanic plates adjacent to the Archaean block of southern Greenland in the gross pattern of Palaeoproterozoic plates in the North Atlantic region.

We thank J. Bailey and S. Pedersen, Geological Institute, University of Copephagen, for the Rb–Sr isotope data, and we acknowledge the generous contribution of field data and discussion by P. W. U. Appel, P. Erfurt, T. Frisch, R. A. Frith, M. A. Hamilton, F. Kalsbeek, T. F. D. Nielsen, H. K. Schønwandt, H. Stendal and B. Thomassen. We are grateful to C. R. L. Friend, R. P. Hall and M. Stone for critical comments on the manuscript. We are indebted to T. F. D. Nielsen and P. Bay, Geological Survey of Greenland, and J. Gowen and B. Sieborg, Nuna Oil A/S, for their ready help with logistical support, and we also thank G. M. Spencer, University of Exeter, for the provision of high quality thin sections. This paper is published with permission of the Geological Survey of Greenland.

References

ALLAART, J. H. 1967. Basic and intermediate igneous activity and its relationship to the evolution of the Julianehåb Granite, South Greenland. *Bulletin Grønlands Geologiske Undersøgelse*, **69** [*Meddelelser om Grønland* **175**, 1].

—— 1975. *Geologiske kort over Grønland, 1:500 000, sheet 1, Sydgrønland*. Grønlands Geologiske Undersøgelse, Copenhagen.

—— 1976. Ketilidian mobile belt in South Greenland. *In*: ESCHER, A. & WATT, W. S. (eds) *Geology of Greenland*. Geological Survey of Greenland, Copenhagen, 120–151.

ANDREWS, J. R., BRIDGWATER, D., GORMSEN, K., GULSON, K., KETO, L. & WATTERSON, J. 1973. The Precambrian of South-East Greenland. *In*: PARK, R. G. & TARNEY, J. (eds) *The Early Precambrian of Scotland and related rocks of Greenland*. University of Birmingham Press, 143–156.

——, ——, GULSON, B. & WATTERSON, J. 1971. Reconnaissance mapping of south-east Greenland between 62°30'N and 60°30'N. *Rapport Grønlands Geologiske Undersøgelse*, **35**, 32–38.

BERTHELSEN, A. & HENRIKSEN, N. 1975. *Geologisk kort over Grønland 1:100 000*. Descriptive text Ivigtut 61 V.1 Syd. The orogenic and cratogenic geology of a Precambrian shield area. Copenhagen: Grønlands Geologiske Undersøgelse [*Meddelelser om Grønland* **186**, 1].

BONDESEN, E. 1970. The stratigraphy and deformation of the Precambrian rocks of the Grænseland area, South-West Greenland. *Bulletin Grønlands Geologiske Undersøgelse*, **86** [*Meddelelser om Grønland* **185**, 1]

—— & HENRIKSEN, N. 1965. On some Precambrian metadolerites from the Central Ivigtut region, SW Greenland. *Bulletin Grønlands Geologiske Undersøgelse*, **52**.

BRIDGWATER, D., ESCHER, A. & WATTERSON, J. 1973. Tectonic displacements and thermal activity in two contrasting Proterozoic mobile belts from Greenland. *Philosophical Transactions of the Royal Society, London, Series A*, **273**, 513–533.

——, SUTTON, J, & WATTERSON, J. 1974. Crustal downfolding associated with igneous activity. *Tectonophysics*, **21**, 57–77.

BROWN, P. E., DEMPSTER, T. J., HARRISON, T. N. & HUTTON, D. H. W. 1992. The rapakivi granites of S Greenland – crustal melting in response to extensional tectonics and magmatic underplating. *Transactions of the Royal Society of Edinburgh: Earth Sciences*, **83**, 173–178.

CHADWICK, B., DAWES, P. R., ESCHER, J. C., FRIEND, C. R. L. ET AL. 1989. The Proterozoic mobile belt in the Ammassalik region, South-East Greenland (Ammassalik mobile belt): an introduction and re-appraisal. *Rapport Grønlands Geologiske Undersøgelse*, **146**, 5–12.

——, ERFURT, P., FRISCH, T., FRITH, R. A., GARDE, A. A. ET AL. 1994a. Sinistral transpression and hydrothermal activity during emplacement of the Early Proterozoic Julianehåb batholith: field work in the area adjacent to Søndre Sermilik, South Greenland. *Rapport Grønlands Geologiske Undersøgelse*, **163**, 5–22.

——, ERFURT, P., FRITH, R. A., NIELSEN, T. F. D., SCHØNWANDT, H. K. & STENDAL, H. 1994b. Re-appraisal of the Ikermit supracrustal suite of the Ketilidian Border Zone in South-East Greenland. *Rapport Grønlands Geologiske Undersøgelse*, **163**, 23–31.

—— & FRIEND, C. R. L. 1994. Reaction of Precambrian high-grade gneisses to mid-crustal ductile deformation in western Dove Bugt, North-East Greenland. *Rapport Grønlands Geologiske Undersøgelse*, **162**, 53-70.

DAWES, P. R. 1970. The plutonic history of the Tasiussaq area, South Greenland, with special reference to a high-grade gneiss complex. *Bulletin Grønlands Geologiske Undersøgelse*, **88** [*Meddelelser om Grønland*, **189**, 3].

D'LEMOS, R. S., BROWN, M. & STRACHAN, R. A. 1992. Granite generation, ascent and emplacement within a transpressional orogen. *Journal of the Geological Society, London*, **149**, 487–490.

DEMPSTER, T. J., HARRISON, T. N., BROWN, P. E. & HUTTON, D. H. W. 1991. Low-pressure granulites from the Ketilidian mobile belt of southern Greenland. *Journal of Petrology*, **32**, 979–1004.

DENNIS, A. J. & SECOR, D. T. 1987. A model for the development of crenulations in shear zones with applications from the Southern Appalachian Piedmont. *Journal of Structural Geology*, **9**, 809–818.

DICKIN, A. P. 1992. Evidence for an Early Proterozoic

crustal province in the North Atlantic region. *Journal of the Geological Society, London*, **149**, 483–486.

EMELEUS, C. H. & UPTON, B. G. J. 1976. The Gardar period in southern Greenland. *In*: ESCHER, A. & WATT, W. S. (eds) *Geology of Greenland*. Geological Survey of Greenland, Copenhagen, 152–181.

ESCHER, A. 1966. The deformation and granitisation of Ketilidian rocks in the Nanortalik area, S. Greenland. *Bulletin Grønlands Geologiske Undersøgelse*, **59** [*Meddelelser om Grønland* **172**, 9].

—— & PULVERTAFT, T. C. R. 1976. Rinkian mobile belt of West Greenland. *In*: ESCHER, A. & WATT, W. S. (eds) *Geology of Greenland*. Geological Survey of Greenland, Copenhagen, 104–119.

——, SØRENSEN, K. & ZECK, H. P. 1976. Nagssugtoqidian mobile belt in West Greenland. *In*: ESCHER, A. & WATT, W. S. (eds) *Geology of Greenland*. Geological Survey of Greenland, Copenhagen, 76–95.

GROCOTT, J. & PULVERTAFT, T. C. R. 1990. The Early Proterozoic Rinkian Belt of central West Greenland. *In*: LEWRY, J. F. & STAUFFER, M. R. (eds) *The Early Proterozoic Trans-Hudson Orogen of North America*. Special Paper of the Geological Association of Canada, **37**, 443–463.

GULSON, B. L. & KROGH, T. E. 1975. Evidence of multiple intrusion, possible resetting of U–Pb ages, and new crystallisation of zircons in the post-tectonic intrusions ('Rapakivi granites') and gneisses from South Greenland. *Geochimica et Cosmochimica Acta*, **39**, 65–82.

HARRISON, T. N., PARSONS, I. & BROWN, P. E. 1990. Mineralogical evolution of fayalite-bearing rapakivi granites from Prins Christian Sund pluton, South Greenland. *Mineralogical Magazine*, **54**, 57–66.

HARRY, W. T. & OEN, Ing Soen. 1964. The pre-Cambrian basement of Alángorssuaq, South Greenland and its copper mineralization at Josvaminen. *Bulletin Grønlands Geologiske Undersøgelse*, **47**, [*Meddelelser om Grønland*, **179**, 1].

HIGGINS, A. K., 1970: The stratigraphy and structure of the Ketilidian rocks of Midternæs, South-West Greenland. *Bulletin Grønlands Geologiske Undersøgelse*, **87**, [*Meddelelser om Grønland*, **189**, 2].

HUTTON, D. H. W. 1988. Granite emplacement mechanisms and tectonic controls: inferences from deformation studies. *Transactions of the Royal Society of Edinburgh: Earth Sciences*, **79**, 245–255.

—— & REAVY, R. J. 1992. Strike-slip Tectonics and Granite Petrogenesis. *Tectonics*, **11**, 960–967.

——, DEMPSTER, T. J., BROWN, P. E. & BECKER, S. D. 1990. A new mechanism of granite emplacement: intrusion in active extensional shear zones. *Nature*, **343**, 452–455.

KALSBEEK, F. & TAYLOR, P. N. 1985. Isotopic and chemical variation in granites across a Proterozoic continental margin – the Ketilidian mobile belt of South Greenland. *Earth and Planetary Science Letters*, **73**, 65–80.

——, LARSEN, L. M. & BONDAM, J. 1990. *Geological Map of Greenland. 1:500 000*. Descriptive text. Sheet 1. Sydgrønland. Grønlands Geologiske Undersøgelse, Copenhagen.

——, NUTMAN, A. P. & TAYLOR, P. N. 1993. Palaeoproterozoic basement province in the Caledonian fold belt of North-East Greenland. *Precambrian Research*, **63**, 163–178.

——, PIDGEON, R. T. & TAYLOR, P. N. 1987. Nagssugtoqidian mobile belt of West Greenland: cryptic 1850 Ma suture between two Archaean continents – chemical and isotopic evidence. *Earth and Planetary Science Letters*, **85**, 365–385.

——, TAYLOR, P. N. & PIDGEON, R. T. 1988. Unreworked Archaean basement and Proterozoic crustal rocks from northwestern Disko Bugt, West Greenland: implications for the nature of Proterozoic mobile belts in Greenland. *Canadian Journal of Earth Sciences*, **25**, 773–782.

LARSEN, L. M. & REX, D. C. 1992. A review of the 2500 Ma span of alkaline-ultramafic, potassic and carbonatitic magmatism in West Greenland. *Lithos*, **28**, 367–402.

LISTER, G. S. & SNOKE, A. W. 1984. S-C Mylonites. *Journal of Structural Geology*, **6**, 617–638.

PARK, R. G., 1994. Early Proterozoic tectonic overview of the northern British Isles and neighbouring terrains in Laurentia and Baltica. *Precambrian Research*, **68**, 65–79.

PATCHETT, P. J. & BRIDGWATER, D. 1984. Origin of continental crust of 1.9–1.7 Ga age defined by Nd isotopes in the Ketilidian terrain of South Greenland. *Contributions to Mineralogy and Petrology*, **87**, 311–318.

PATERSON, S. R., VERNON, R. H. & TOBISCH, O. T. 1989. A review of criteria for the identification of magmatic and tectonic foliations in granitoids. *Journal of Structural Geology*, **11**, 349–363.

PEDERSEN, S., LARSEN, O., BRIDGWATER, D. & WATTERSON, J. 1974. Rb/Sr whole rock isochron age determinations on metamorphosed acid volcanic rocks and granitic gneisses from the Ketilidian mobile belt, South-East Greenland. *Rapport Grønlands Geologiske Undersøgelse*, **66**, 12–20.

PERSOZ, F. 1969. Évolution plutonique et structurale de la presqu'île d'Akuliaruseq, Groenland Méridional. *Bulletin Grønlands Geologiske Undersøgelse*, **72**, [*Meddelelser om Grønland*, **175**, 3].

PITCHER, W. S. 1993. *The Nature and Origin of Granite*. Blackie, London.

STEENFELT, A. & ARMOUR-BROWN, A. 1988. Characteristics of the South Greenland Uranium Province. *In*: *Recognition of Uranium Provinces*. International Atomic Energy Agency, Vienna, 305–335.

STRECKEISEN, A. L. 1976. To each plutonic rock its proper name. *Earth Science Reviews*, **12**, 1–33.

VAN BREEMEN, O., AFTALION, M., & ALLAART, J. 1974. Isotopic and geochronologic studies on granites from the Ketilidian mobile belt of South Greenland. *Bulletin of the Geological Society of America*, **85**, 403–412.

VAN KRANENDONK, M. J., ST-ONGE, M. R. & HENDERSON, J. R. 1993. Paleoproterozoic tectonic assembly of Northeast Laurentia through multiple indentations. *Precambrian Research*, **63**, 325–347.

WALLIS, R. H. 1966. *The Geology of North-east Tasermiut Fjord, South Greenland*. PhD thesis, University of Birmingham.

WALTON, B. J. 1965. Sanerutian appinitic rocks and Gardar dykes and diatremes, north of Narssarssuaq, South Greenland. *Bulletin Grønlands Geologiske Undersøgelse*, **57** [*Meddelelser om Grønland* **179**, 9].

WATTERSON, J. 1965. Plutonic development of the Ilordleq area, South Greenland. Part I: Chronology, and the occurrence and recognition of metamorphosed basic dykes. *Bulletin Grønlands Geologiske Undersøgelse*, **51**, [*Meddelelser om Grønland*, **172**, 7].

—— 1968. Plutonic development of the Ilordleq area, South Greenland. Part II. Late-kinematic basic dykes. *Bulletin Grønlands Geologiske Undersøgelse*, **70** [*Meddelelser om Grønland*, **185**, 3].

—— 1978. Proterozoic intraplate deformation in the light of Southeast Asian neotectonics. *Nature*, **273**, 636–640.

WEGMANN, C. E. 1938. Geological investigations in southern Greenland. Part I. On the structural divisions of southern Greenland. *Meddelelser om Grønland*, **113**, 2.

—— 1939. Übersicht über die Geologie Südgrönlands. *Mitteilungen der Naturforschenden Gesellschaft*, **16**, 188–212, Scaffhausen, Switzerland.

—— 1948. Note sur la chronologie des formations précambriennes du Groenland méridional. *Eclogae geologicae Helvetiae*, **40**, 7–14.

WHITE, S. H., BURROWS, S. E., CARRERAS, J., SHAW, N. D. & HUMPHREYS, F. J. 1980. On mylonites in ductile shear zones. *Journal of Structural Geology*, **2**, 175–187.

WINDLEY, B. F. 1965. The composite net-veined diorite intrusives of the Julianehåb district, South Greenland. *Bulletin Grønlands Geologiske Undersøgelse*, **58** [*Meddelelser om Grønland*, **172**, 8].

—— 1966a. The Precambrian geology of the Sârdloq area, South Greenland. *Rapport Grønlands Geologiske Undersøgelse*, **5**.

—— 1966b. Superposed deformations of the Ketilidian gneisses in the Sârdloq area, South Greenland. *Bulletin Grønlands Geologiske Undersøgelse*, **64** [*Meddelelser om Grønland*, **179**, 3].

—— 1991. Early Proterozoic collision tectonics, and rapakivi granites as intrusions in an extensional thrust-thickened crust: the Ketilidian orogen, South Greenland. *Tectonophysics*, **195**, 1–10.

The evolution of the Grenville Province in eastern Labrador, Canada

CHARLES F. GOWER

Newfoundland Department of Mines and Energy, PO Box 8700, St John's, Newfoundland A1B 4J6, Canada

Abstract: Rifting at *c.* 1.71 Ga separated the Mealy Mountains terrane from pre-Labradorian Laurentia and resulted in a short-lived (backarc?) basin, followed by southward subduction giving rise to 1.68–1.66 Ga Labradorian calc-alkaline arcs, partly built on older crust in the Mealy Mountains terrane. The arcs and subduction were also short lived, terminating when accretion against the Makkovikian and Eastern Churchill provinces occurred at 1.65 Ga. The Trans-Labrador batholith developed as a result of crustal thickening during suturing of the 1.68–1.66 Ga arc-related rocks to pre-Labradorian Laurentia. Further bimodal magmatism, in response to crustal thickening, continued until *c.* 1.62 Ga, followed by waning granitic magmatism until 1.60 Ga.

The Labrador Orogen and older rocks in the Mealy Mountains terrane are flanked to the south by Pinwarian (1.51–1.45 Ga) granitoid rocks, possibly the inboard manifestation of a northward-subducting continental-margin arc at the southern margin of Laurentia. Extension, mostly coincident with the 1.71 Ga zone of weakness, started at *c.* 1.45 Ga. Although rifting was successful further west, it is only represented by basaltic magmatism (Michael and Shabogamo gabbros) along a linear zone in Labrador. Similar tectonism characterized the same zone periodically for at least the next 150 million years during which time the 1.27 Ga Harp-Nutak dykes were emplaced and slightly younger rocks of the Seal Lake Group formed north of the Grenville Province.

After *c.* 1.23 Ga tectonic conditions changed, partly in response to Elzevirian accretion of arc terranes preserved in the southwest Grenville Province, that are speculated to have once also existed south of the exposed eastern Grenville Province. Alternating compressional and extensional conditions terminated at *c.* 1.0 Ga during collisional orogenesis associated with the final stages of development of the Grenville Orogen. Grenvillian deformation in Labrador (1.08–0.97 Ga) resulted in northwesterly thrusting, the effects of which are focused in the Exterior Thrust Belt, and widespread plutonism in the Interior Magmatic Belt. Thrusting in the Exterior Thrust Belt is sited along the zone of 1.71 and 1.43 Ga rifting. Uplift and cooling continued until 0.90 Ga as crustal stability was gradually achieved.

Waves of tectonism crashed on the crustal southern shore of Laurentia–Baltica from the time of amalgamation of Archaean crustal nuclei and early Palaeoproterozoic orogens at *c.* 1850 Ma to at least the time of Grenvillian orogenesis. Crust that accreted between 1850 and 970 Ma, therefore, is the debris from 850 million years of tectonic storms impinging on this crustal shoreline.

Implicit in the above metaphor are concepts used throughout this paper in an attempt to understand the geological evolution of a 150 km wide section across the full width of the Grenville Province in eastern Labrador. One result of 1:100,000 scale geological mapping over the last 15 years is the recognition that the crustal debris can be grouped into lithotectonic terranes. From supporting U–Pb geochronological studies it is established that these preserve histories showing both comparisons and contrasts with adjacent terranes.

The first half of the text below is devoted to a terrane-by-terrane review of their histories. The second part uses the review as a basis for drawing together various threads into a model for the Proterozoic tectonic evolution of the Grenville Province in eastern Labrador. As no tectonic model can be viable if it is not consistent with data from surrounding areas, the broader eastern-Laurentian context is considered at each stage of the history.

As a final opening comment it is important to clarify that the term 'terrane', although identifying regions differing from each other (e.g., in rock types, structural style and geological evolution), does not imply that they must have been widely separate from each other at some earlier stage (i.e., they are not necessarily 'suspect'). The terranes are considered to have most significance in a Grenvillian context, although believed to have existed in some form long before Grenvillian orogenesis.

Geological framework

Geological subdivision

The eastern Grenville Province (Fig. 1) can be divided into an Exterior Thrust Belt and an Interior Magmatic Belt (Gower et al. 1991). The Exterior Thrust Belt is almost entirely underlain by rocks having pre-Grenvillian protoliths, most of which were formed during the 1710–1600 Ma Labradorian orogeny (Schärer et al. 1986; Thomas et al. 1986; Schärer & Gower 1988). The Grenvillian characteristic of the Exterior Thrust Belt is that it forms a collage of generally south-dipping, thrust-bound terranes, which achieved their final configuration during Grenvillian (1080–970 Ma) orogenesis (Gower & Owen 1984; Thomas et al. 1986; Rivers et al. 1989). Grenvillian isotopic ages, almost entirely related to tectonic and metamorphic effects, are linked to telescoping and final assembly of the thrust-bound lithotectonic terranes. The Exterior Thrust Belt has been divided into (i) parautochthonous terranes, showing a marked Grenvillian overprint, and (ii) allochthonous terranes, in which Grenvillian effects are less evident (Rivers et al. 1989: Gower et al. 1991).

Parts of the Interior Magmatic Belt mapped so far lack the extensive tracts of high-grade pelitic gneiss and calc-alkaline plutonic rocks that characterize the terranes to the north. In contrast, abundant alkali-rich granitoid rocks (some of which are known to have post-Labradorian ages) and abundant AMCG-suite (anorthosite–mangerite–charnockite–granite) rocks are present. It is the numerous Grenvillian plutons (Gower et al. 1991), however, that give the Interior Magmatic Belt its name.

In eastern Labrador, the parautochthonous part of the Exterior thrust Belt is represented by the Groswater Bay terrane and bounded on its northern side by the Grenville front. The Lake Melville Terrane and Hawke River terranes show both parautochthonous and allochthonous features. The allochthonous Exterior Thrust Belt is represented by the Mealy Mountains terrane. The Interior Magmatic Belt is represented by the Pinware terrane. In the following sections, the geological

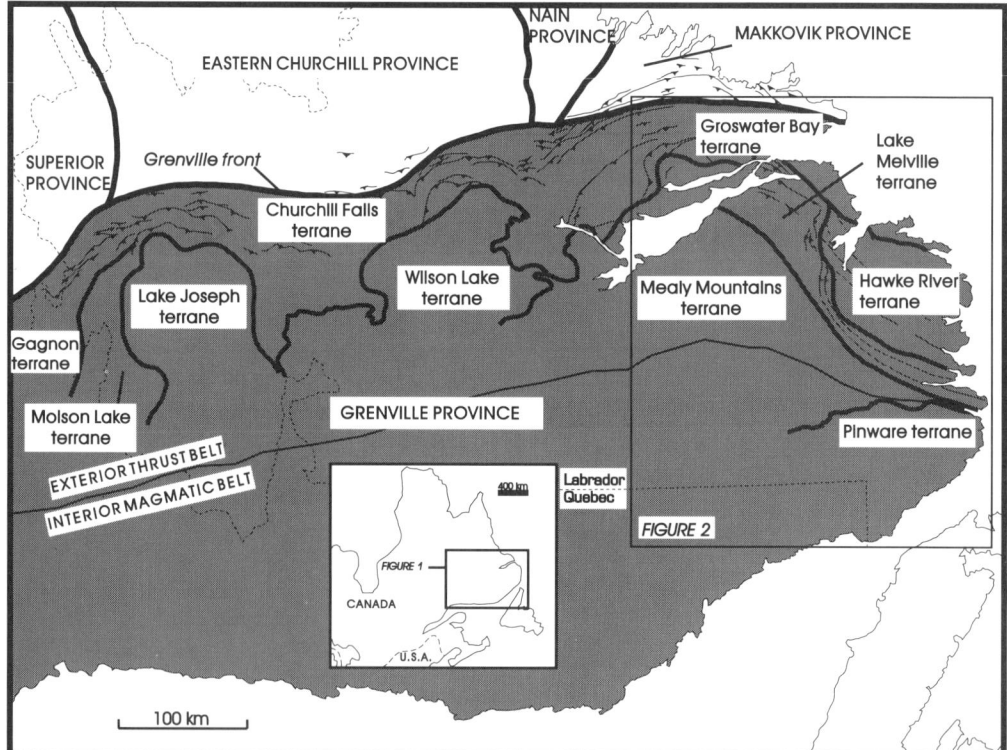

Fig. 1. The eastern Grenville Province showing the Exterior Thrust Belt, Interior Magmatic Belt, terrane subdivision and major structures (Fig. 2 is outlined).

attributes of each terrane and its boundaries are reviewed, starting in the north at the Grenville front. Regional geological features are shown in Fig. 2 and geological histories of each terrane depicted and correlated in Fig. 3.

The Grenville front

The position of the Grenville front in eastern Labrador has been debated for several decades (cf. Gower *et al.* 1980), but apart from suggestions

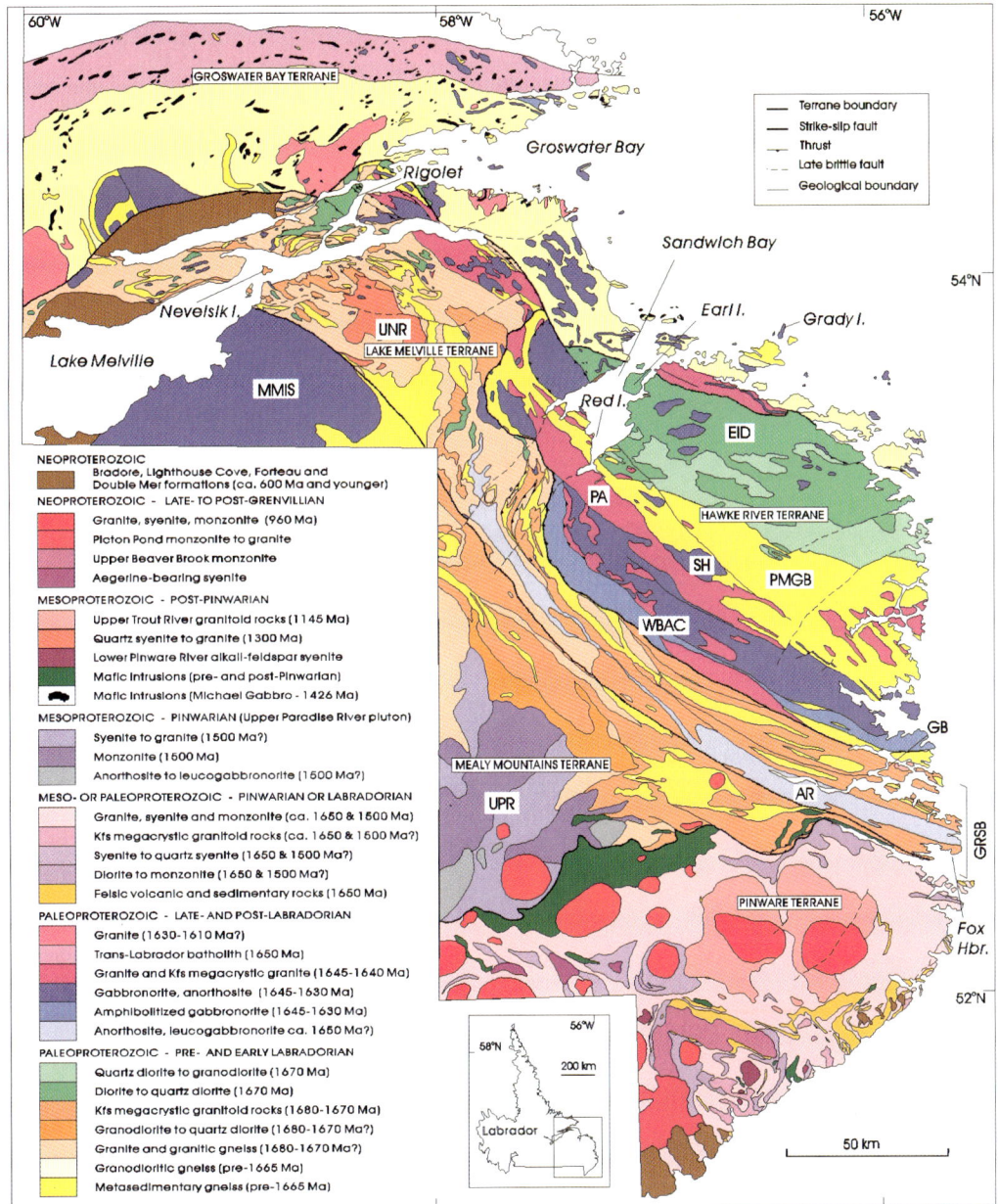

Fig. 2. Geological map of the Grenville Province in eastern Labrador. AR, Alexis River anorthosite; EID, Earl Island domain; GB, Gilbert Bay pluton; GRSB, Gilbert River shear belt; MMIS, Mealy Mountains Intrusive Suite; PA, Paradise Arm pluton; PMGB, Paradise Metasedimentary Gneiss Belt; SH, Sand Hill Big Pond gabbronorite; UNR, Upper North River pluton; UPR, Upper Paradise River pluton; WBAC, White Bear Arm complex.

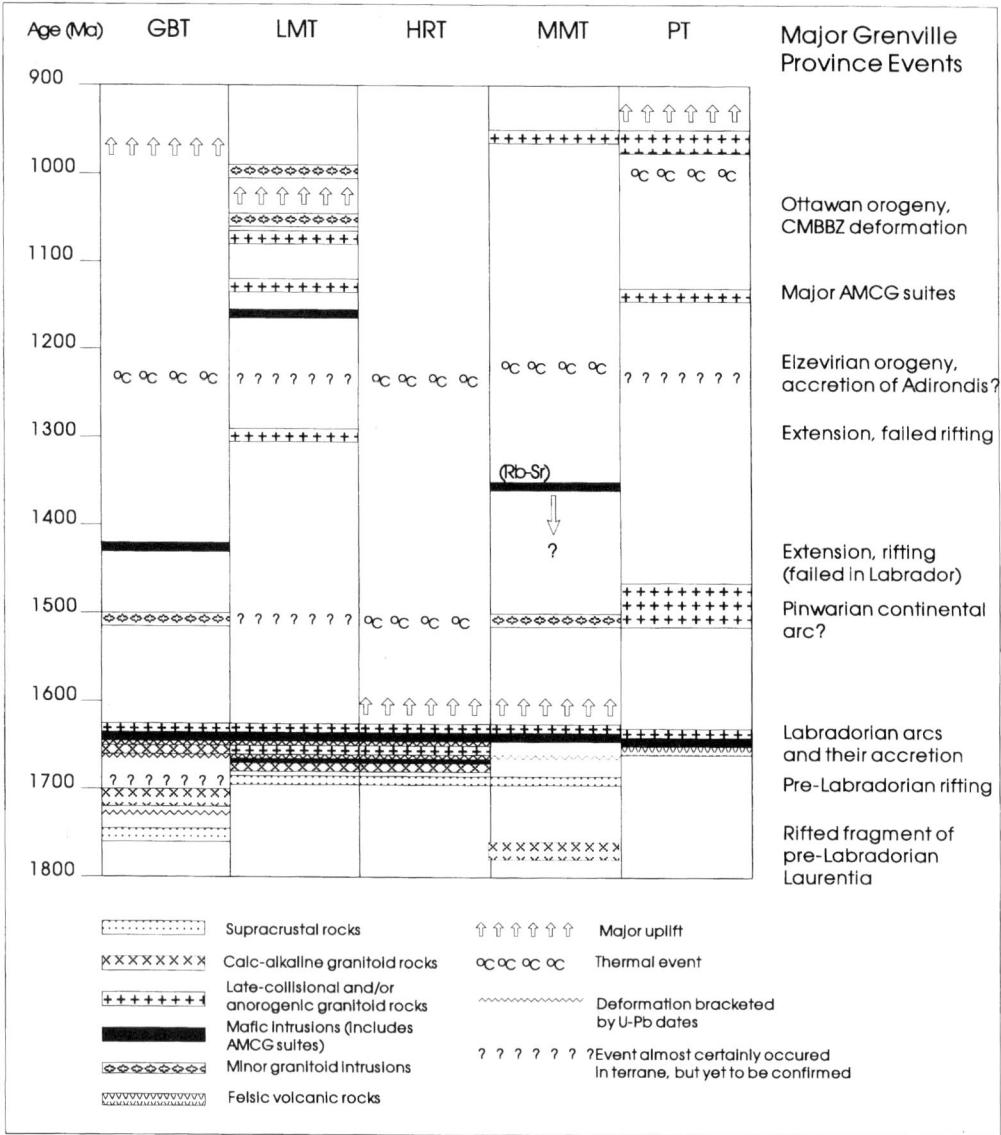

Fig. 3. Time–space relationships between terranes in the Grenville Province, eastern Labrador. All events defined by high-precision U–Pb dates, with the exception of the 1380 Ma mafic dykes (Mealy dykes) in the Mealy Mountains terrane. GBT, Groswater Bay terrane; LMT, Lake Melville terrane; HRT, Hawke River terrane; MMT, Mealy Mountains terrane; PT, Pinware terrane; CMBBZ, Central Metasedimentary Belt Boundary Zone (southern Ontario).

for minor shifts (generally to the north), the location of the Grenville front in eastern Labrador is normally depicted as shown in Fig. 1. The front is defined as the northern limit of widespread Grenvillian deformation, although what constitutes 'widespread' and 'Grenvillian' remain open to discussion. In any event, it is gradational in that some Makkovikian-aged rocks (>1800 Ma) occur south of the line and Grenvillian deformational and minor thermal effects extend into the Makkovik Province.

Detailed study of the Grenville front on the Labrador coast has been carried out by Owen (1985) and Owen *et al.* (1986; 1988), who described a transition from Makkovikian to Grenvillian structural zones over a 3 km wide zone that is bounded by major mylonite zones (the Benedict fault and the Cut Throat Island thrust). Kinematic indicators and jumps in metamorphic grade demonstrate that these shear zones are characterized by reverse displacement, such that Grenville Province rocks are uplifted relative to those in the Makkovik Province. The Benedict fault dips between 50 and 70°S and the Cut Throat Island thrust at about 30°S.

Although there is little doubt that Grenvillian deformation occurred along this zone, the debate about what constitutes the Grenville front is compounded by the fact that this line is also approximately the northern limit of the Labradorian (1710–1600 Ma) Orogen. Krogh (1994) from dating zircon overgrowths and titanite in post-deformation melt pods, has shown that the Cut Throat Island thrust largely predates 1662 ± 10 Ma, and that Grenvillian effects were mostly confined to reheating at 1037 ± 19 Ma.

Groswater Bay terrane

The Groswater Bay terrane (defined by Gower & Owen 1984) is an arcuate belt over 400 km long and about 50 km wide (Fig. 2). It is structurally overlain by the Wilson Lake terrane at its western end. The northern part of the Groswater Bay terrane comprises the 1654–1649 Ma Trans-Labrador batholith, which has an ill-defined boundary with gneisses to the south that is probably partly structural and partly intrusive.

The terrane consists mostly of calc-alkaline granitoid orthogneiss, which grades into less migmatized granitoid plutons (Gower & Owen 1984). Early orthogneisses are represented by a granodioritic gneiss dated at 1709 +7/−6 Ma and later orthogneisses have an age range between 1658 ± 5 and 1649 +4/−3 Ma (Schärer et al. 1986; Schärer & Gower, 1988). The later period of orthogneiss formation has been termed the Double Island magmatic event (Gower et al. 1992). It clearly overlaps with the formation of the Trans-Labrador batholith and the simplest interpretation is that orthogneiss coeval with the Trans-Labrador batholith is a deeper-level equivalent that was exhumed during Grenvillian orogenesis (Gower & Erdmer 1988; Gower 1990).

Supracrustal rocks are a minor, but important component of the Groswater Bay terrane. Gower & Owen (1984) suggested that there are two groups, namely (i) medium to high-grade pelitic paragneiss, and (ii) lower-grade metasediments with recognizable primary fabrics. As their suggestion has been neither confirmed nor denied, the options they raised concerning whether differences in appearance are due to age or grade of metamorphism variations still remain. The absolute age of any of the supracrustal rocks remains unknown. An enclave of a biotite-rich gneiss, probably of metasedimentary origin, in the granitoid rock dated at 1709 Ma (Schärer et al. 1986) suggests that some are pre-Labradorian, but all the high-grade metasedimentary gneisses need not necessarily have this minimum age (nevertheless, the current hypothesis is that the high-grade supracrustal rocks are pre- or syn-Labradorian).

Remnants of layered mafic intrusions are a subsidiary igneous component of the Groswater Bay terrane. One example, at Grady Island (Fig. 2), which was studied in detail by O'Flaherty (1986) and previously dated by an Rb–Sr whole-rock isochron to be 1610 ± 30 Ma (Brooks 1983), has yielded a preliminary U–Pb age of 1645 ± 3 Ma (Kamo et al. 1995).

Some less-migmatized or unmigmatized granitoid rocks within the Groswater Bay terrane are younger than the orthogneiss and postdate some of the mafic intrusions. A granite emplaced into the southern Groswater Bay terrane having an age of 1632 +10/−9 Ma (Schärer et al. 1986, revised from Krogh 1983) is one example, and two other intrusions, also in the southern Groswater Bay terrane, similarly may be late-Labradorian intrusions (cf. Gower & Owen 1984). On the basis of a pegmatite dated at 1499 +8/−7 Ma (Schärer et al. 1986), it can be inferred that there was some post-Labradorian granitoid magmatic activity, but its extent and duration remain uncertain.

Both the ortho- and metasedimentary gneiss and the 1499 Ma pegmatite were intruded by the 1426 ± 6 Ma Michael gabbro (Schärer et al. 1986). This is a widespread tholeiitic olivine gabbro suite (Gower et al. 1990a) that appears to be confined principally to the Groswater Bay terrane (but see comment below). None of the 1:100,000 geological maps of the Groswater Bay terrane (e.g. Gower et al. 1983a, b) discriminate between ages of mafic intrusions, although Gower et al. (1990a, Fig. 1) excluded mafic rocks that they considered to be unrelated to the Michael gabbro. The exclusion was done on the basis of largely unpublished geochemical data (especially Zr/Y values; Murthy et al. 1989) and is partly the basis on which mafic rocks are separated in Fig. 2.

Especially in the southern Groswater Bay terrane, folds are isoclinal and overturned to the north and commonly refolded. Axial surfaces are locally horizontal or dip shallowly to the south (Gower et al. 1982). As the Michael gabbro discordantly intrudes complexly deformed gneiss, but is itself folded and boudinaged, structures are the product of both Grenvillian and earlier deformation.

An increase in metamorphic grade across the Groswater Bay terrane, seen in both the Michael gabbro and its host rocks, has been interpreted in terms of a sheaf of thrust slices that exhumed progressively deeper crustal levels to the south (Gower 1986). The Michael gabbro shows a series of coronitic textures ranging from simple double orthopyroxene-amphibole+spinel coronas between olivine and plagioclase in igneous-textured rocks in the north to complex multiple coronas involving garnet in almost completely recrystallized rocks in the south (Emslie 1983; Gower 1986; Gower & Erdmer 1988). These have been interpreted as indicating gradually increasing pressures from north to south (Gower 1986; Gower & Erdmer 1988; van Nostrand 1988). The pattern is consistent with mineralogical variations seen in associated orthogneisses and pelitic metasedimentary gneisses. For example, the pelitic gneisses show systematic mineralogical variation that includes a transition from muscovite-bearing assemblages in the north to garnet- and kyanite-bearing assemblages in the south. Preliminary geothermobarometry indicates pressures to have been generally above 8 kb, and up to 14 kb (Gower 1986; van Nostrand 1988) in the southern part of the Groswater Bay terrane.

No entirely satisfactory conclusions have been drawn regarding the time of formation of the metamorphic/coronitic assemblages, and all three of the alternative models (incorporating Labradorian versus Grenvillian effects) presented by Gower & Erdmer (1988) remain viable. Their suggestion that much of the metamorphism may not be Grenvillian derives support from geochronological investigations by Schärer et al. (1986) and Krogh (1994). Nevertheless, it is clear from 980–970 Ma U–Pb titanite ages, which are characteristic of the whole of the Groswater Bay terrane (Schärer et al. 1986), that final uplift and the present structural configuration was achieved during Grenvillian orogenesis.

Lake Melville terrane

The Lake Melville terrane, named by Gower & Owen (1984), is an arcuate belt up to about 60 km wide, but tapering to a much narrower, more highly deformed region toward the southeast, where the name Gilbert River shear belt has also been applied (Gower et al. 1987). Hanmer & Scott (1990), after investigating the southeast Labrador coastal cross-section, concluded that it is not a crustal-scale shear zone, and revised the name to Gilbert River belt.

The main rock types in the Lake Melville terrane are similar to those of the Groswater Bay terrane, but proportions differ and not all rock types are common to both regions. As in the Groswater Bay terrane, granitoid orthogneisses grading into less migmatized equivalents are dominant, although K-feldspar megacrystic granitoid rocks are much more abundant. Available data indicate that the orthogneiss and associated megacrystic granitoid rocks are slightly older than similar rocks in the Groswater Bay terrane. Dates reported are 1678 ± 6 Ma for a megacrystic granitoid rock and 1677 +16/–15 Ma for a banded migmatitic orthogneiss (Schärer et al. 1986). This age group includes a gneissic tonalite dated at c. 1677 +18/–11 Ma (Schärer et al. 1986; revised from Krogh 1983). The area in which the tonalite occurs was originally mapped as part of the southwest Groswater Bay terrane. On the basis of the 1677 Ma age and the timing of Grenvillian metamorphism, Gower et al. (1992) reassigned the area to the Lake Melville terrane. This sleight of interpretation puts all the c. 1677 Ma orthogneisses into one terrane and prompted Gower et al. (1992) to coin the name Neveisik magmatic event for this plutonism. Accepting this revision means, however, that the previous statement that the Michael gabbro is restricted to the Groswater Bay terrane is no longer valid.

Two other Labradorian dates have been reported from igneous rocks in the Lake Melville terrane. One is from a granitic vein from which an age of 1664 +14/–9 Ma was obtained and the other is from a megacrystic granitoid rock that gave an age of 1644 +8/–6 Ma age (Scott et al. 1993). The 1664 Ma granitic vein transects amphibolite-facies mylonitic fabrics but is itself affected by ductile deformation thought by the authors to be part of the same on-going deformation. Its significance is discussed below in conjunction with coeval events in the Hawke River terrane. The age of the megacrystic granitoid rock is more enigmatic for the following reasons: (i) neither the 1644 Ma upper intercept nor the c. 890 Ma lower intercept of the megacrystic granitoid rock are typical of the Lake Melville terrane on the basis of present knowledge; (ii) the megacrystic granitoid rock is intruded by two phases of mafic dykes, a feature that elsewhere in eastern Labrador is consistently associated with dated granitoid rocks older than 1665 Ma; (iii) at the same outcrop, Scott et al. (1993) reported a Grenvillian minor intrusion showing 1790 +142/–122 Ma inheritance, which is analytically older than any dated rock in the Lake Melville terrane. Although it is tempting to regress the Labradorian megacrystic rock with the Grenvillian intrusion (arguing that the inheritance might have come from the immediate host rock and caused the Pb loss in the host rock and generating a result more consistent with existing data), D. Scott (pers. comm., 1994) noted that such a model leads to zero probability of fit. Two additional zircon fractions refined the upper intercept to 1644 +5/−3 Ma (D. Scott, pers. comm., 1995), suggesting that further modifications to Labradorian geological history in the Lake Melville terrane may be needed.

Metasedimentary gneiss forms a higher proportion of the Lake Melville terrane than the Groswater Bay terrane. Pelitic and psammitic gneisses are most common, but minor quartzite and calc-silicate units are present. Both of the latter are rare in the Groswater Bay terrane. The pelitic–psammitic gneisses commonly grade into metasedimentary diatexite. Gower et al. (1987) emphasized a close spatial association between K-feldspar megacrystic granitoid rocks and pelitic metasedimentary gneisses, advocating derivation of the granitoid rock from a pelitic protolith. This suggestion implies that the pelitic gneisses predate the granitoid rocks, consistent with field evidence, but presently unconfirmed by geochronological data.

Layered mafic intrusions are a key feature of the Lake Melville terrane and mostly concentrated close to the leading (north) edge of the terrane. The rocks include anorthosite, norite, gabbro and monzonite; none have been dated, but they are assumed to be mid- to late Labradorian (see below). A unit unique to the Lake Melville terrane is the Alexis River anorthosite. This severely deformed, distinctive layered intrusive body of anorthosite and leucogabbronorite forms an excellent marker unit, which, despite being less than 5 km wide, is traceable along-strike for over 150 km (Gower et al. 1985, 1987; van Nostrand 1992; van Nostrand et al. 1992). The unit is undated but is assumed to be Labradorian.

A large time interval separates the Labradorian rocks from the next two dated events in the Lake Melville terrane, both involving the emplacement of granitoid plutons. The first event was the intrusion of the Upper North River pluton at 1296 +13/−12 Ma (Schärer et al. 1986). This pyroxene-bearing syenite to granite is strongly deformed, presumably during Grenvillian deformation. The second event was the emplacement of the Gilbert Bay pluton, although its age of 1132 +7/−6 Ma is equivocal (Gower et al. 1991). The Gilbert Bay pluton is surrounded by an envelop of minor granitic dykes, very similar to the main body; a zircon age of 1113 +6/−5 Ma from a granitic vein 2 km west of the pluton (Scott et al. 1993) may provide confirmatory evidence that the date obtained from the Gilbert Bay pluton age is approximately correct. The Gilbert Bay pluton contains enclaves of mylonite (Gower et al. 1987), from which it may be inferred that major movement of the Hawke River terrane relative to the Lake Melville terrane occurred prior to c. 1132 Ma. A greenschist-facies fabric in the nearby 1113 Ma granitic vein dated by Scott et al. (1993) shows that there was also later deformation, however. The enclave is on-strike with mylonite intruded by the 1664 Ma granitic vein alluded to earlier, thus mylonitization in the enclave may be Labradorian.

Activity in the Lake Melville terrane continued intermittently for the next 100 million years. It was characterized by the emplacement of minor granitic intrusions and accompanied by widespread thermal effects. Geochronological evidence for granitic magmatism is provided by a zircon age for a granite southwest of Sandwich Bay (1079 ± 6 Ma; Schärer et al. 1986) and 1062 +5/−6 Ma for a granitic vein intruding Labradorian megacrystic granitoid rock further to the southeast (Scott et al. 1993). Thermal effects are recorded from monazite ages of 1078 ± 2 Ma from the previously noted 1113 Ma granitic vein and 1077 ± 3 Ma from a Labradorian granodioritic gneiss (Scott et al. 1993). It is tempting to relate a Rb–Sr whole-rock age of 1085 ± 110 Ma (Brooks 1984) from the 1296 Ma Upper North River pluton to the same event, but it must be remembered that the date is analytically indistinguishable from a 1038 Ma metamorphic overprint dated from the same locality.

Evidence for a 1038 Ma metamorphic overprint is based on an imprecise age from an amphibolitized mafic dyke (Schärer et al. 1986). The 1038 Ma metamorphic age is one of several zircon, monazite and titanite dates between 1040 and 1026 Ma obtained from the Lake Melville terrane (Schärer et al. 1986; Schärer & Gower 1988) and, like the c. 1077 Ma event, seems to be confined to this terrane. Beyond constraining its emplacement between c. 1296 and 1038 Ma the age of the dyke is unknown, but this bracket does exclude the possibility that the dyke is related to the Michael gabbro.

A pegmatite emplaced at 1003 ± 6 Ma (zircon and titanite; Gower et al. 1991) is the youngest Grenvillian age reported from the Lake Melville terrane. The near coincidence of dates from three minerals spanning a 200°C range in blocking temperatures suggests that geon 10 (the 100 million year period between 1100 and 1000 Ma) was a time of rapid uplift and cooling in the Lake Melville terrane.

The structural geometry of the Lake Melville terrane is complex and has yet to be fully unravelled, but the following points can be made.

(i) The structurally complex leading edge of the Lake Melville terrane may embody two

opposing transport directions and may represent an attenuated tectonic melange of the Hawke River terrane. This tectonic melange would be spatially equivalent to the Mount Gnat granulite belt of Gower & Owen (1984).
(ii) The Lake Melville terrane is distinguished from the Groswater Bay terrane by lower metamorphic pressures (6–9 kb; time of formation uncertain), but there is a pressure increase along the leading margin of the terrane to values close to those for the southern Groswater Bay terrane.
(iii) The southeast part of the Lake Melville terrane (the Gilbert River shear belt), although lithologically comparable to the remainder of the terrane, is structurally distinct. Fabrics throughout the region are steep to vertical, especially at its southeast end (further to the northwest there are broad areas where fabrics dip shallowly to the northeast on southwest-verging thrusts). Lineations and minor fold axes are sub-horizontal, but have a preferred plunge to the northwest. Kinematic indicators indicate that movement involved north-side-up movement on the north side of the Gilbert River shear belt but was dominantly sub-horizontal and dextral on the southern side (Gower et al. 1987; Hanmer & Scott 1990). Later faulting offset the northern and southern halves of the Gilbert Bay pluton in a 2.5 km apparent dextral sense (Gower et al. 1987).
(iv) Partly to explain regional sigmoidal folds within the Lake Melville terrane, Gower et al. (1985) presented a structural model involving overthrusting of the Lake Melville terrane onto the Groswater Bay terrane, dextral transport of the Mealy Mountains terrane relative to the Lake Melville terrane and overthrusting of the Lake Melville terrane by the Hawke River terrane.

Much remains to be done to evaluate the relative effects of Labradorian, Pinwarian (1510–1450 Ma) and Grenvillian deformation and metamorphism in generating these thrust relationships. Earlier studies in the Lake Melville terrane indicating that much of the deformation could not be Grenvillian have been extended by Scott et al. (1993), who have clearly demonstrated that some of the mylonitization was earlier than mid-Labradorian.

Hawke River terrane

The Hawke River terrane was first defined by Gower (1987) and geographically outlined by Gower et al. (1988). The terrane is bounded by thrusts and strike-slip faults on its west and southwest sides, respectively, and the northern boundary is probably also structural. U–Pb isotopic studies by Schärer et al. (1986), Schärer & Gower (1988) and Gower et al. (1992) and Sm–Nd isotopic studies by Schärer (1991) have indicated that the region is underlain by juvenile Labradorian (1710–1620 Ma) crust. A key feature of the terrane is that it largely escaped Grenvillian high-grade metamorphism and deformation.

In a regional geological sense, the Hawke River terrane is the simplest in the eastern Grenville Province. It comprises four consistently southeast-trending longitudinal segments (Gower et al. 1985, 1986, 1987), making it quite distinct from the structurally convoluted adjacent terranes. The four segments are, from northeast to southwest: (i) the Earl Island domain, (ii) the Paradise metasedimentary gneiss belt, (iii) the Paradise Arm pluton, and (iv) the White Bear Arm complex.

The Earl Island granodiorite–diorite domain consists of alternating zones of calc-alkaline diorite and granodiorite, the granodiorite becoming dominant progressing south. Mafic rocks, subsidiary ultramafic units, and remnants of pelitic metasedimentary gneiss are also present. The mafic rocks, occurring both as remnants of large layered gabbronorite intrusions and small mafic bodies are similar to those found in adjacent terranes. Pelitic metasedimentary gneiss is mostly sillimanite bearing and is found as enclaves in the calc-alkaline granitoid rocks.

Three ages obtained from the dioritic to granodioritic rocks of 1671 +4/−3 Ma, 1671 ± 4 Ma, 1668 +6/−4 Ma (Schärer et al. 1986; Schärer & Gower 1988) were taken by Gower et al. (1992) to indicate a discrete magmatic event in the Hawke River terrane, which they referred to as the Red Island magmatic event. Although, analytically, the 1671 Ma Red Island magmatic event in the Hawke River terrane overlaps with the 1677 Ma Neveisik Island magmatic event in the Lake Melville terrane, Gower et al. (1992) justified distinguishing two magmatic pulses because of linkage between age range and terrane.

Detailed geochronological investigations of outcrops showing unequivocal field relationships allowed Schärer & Gower (1988), Gower et al. (1992) and Kamo et al. (1995) to establish the following sequence of events: (i) the Red Island Magmatic event (1671 Ma), (ii) mafic dyke injection, (iii) a major deformational and migmatization event (c. 1665 Ma), (iv) granite emplacement (1663–1658 Ma), (vi) further mafic dyke injection and, (vi) declining metamorphism at c. 1645 Ma. That the major tectonism was at c. 1665 Ma is constrained by a granite dated at 1662 ± 3 Ma

(Kamo et al. 1995) discordantly truncating the fabric in a 1671 Ma tonalite–granodiorite gneiss, which was dated at the same locality (Schärer & Gower 1988). This result confirms the same conclusion based on a date of 1663 ± 3 Ma from a similar granite 2 km farther east intruding undated rocks showing comparable gneissic fabrics (Schärer & Gower 1988).

The 1663 Ma granite is intruded by unmigmatized amphibolite-facies mafic dykes. A concordant titanite age of 1649 ± 4 Ma, from the same granite, shows that the later mafic dykes must have been emplaced and metamorphosed between 1663 and 1649 Ma (unless it is insisted that the titanite was unaffected by a post-1649 Ma amphibolite-facies event). The 1663–1649 Ma bracket is supported by similar data from two other outcrops (Gower et al. 1992) and a cross-cutting pegmatite dated to be 1622 ± 3 Ma (Gower et al. 1992) provides an absolute minimum age for the unmigmatized mafic dykes.

Scott et al. (1993) observed that their 1664 Ma age from a granitic vein in the Lake Melville terrane, analytically overlaps with the 1663 Ma granite in the Hawke River terrane, and establishes a time of common history between the two regions. If the small age difference between the 1677 Ma Neveisik Island event and the 1671 Ma Red Island event is valid, the major 1665 Ma deformational and metamorphic event can be taken as the time at which the two terranes became linked (although granting that they may never have been widely separated in the first place).

There are reasonable grounds for accepting that the Hawke River and Groswater Bay terranes were also linked after the 1665 Ma event. A 1645 +7/–5 Ma age for a large body of alkali-feldspar granite (Kamo et al. 1995) intrudes the Earl Island domain–Groswater Bay terrane mutual boundary, whereas, further southeast, the same boundary is marked by well-banded gneiss, mylonite and ultramylonite dipping 45°S (Gower et al. 1986). This zone is interpreted as a thrust; kinematic indicators indicate that the hanging-wall has been transported to the north. Given that the 1645 Ma alkali-feldspar granite is unmylonitized, it seems likely that the deformation zone is early- to mid-Labradorian. The Hawke River and Groswater Bay terranes probably were also joined to pre-Labradorian Laurentia, as the 1665 Ma age of migmatization in the Earl Island domain was essentially coeval with the 1662 Ma minimum age for mylonitization at the Grenville front.

The Paradise metasedimentary gneiss belt consists mostly of pelitic and semipelitic gneiss. Calc-silicate units and quartzite are associated minor rock types. Mafic volcanic rocks (generally minor) include pillowform lavas (Gower & Swinden 1991). These are associated with mafic dykes, larger mafic intrusions and fine-grained quartz- and sulphide-rich sediments that were probably derived from chemical precipitates. A U–Pb zircon age of c. 1645 Ma from the pillow lava locality is interpreted as dating metamorphism in the area (Kamo et al. 1995). The supracrustal rocks are intruded by dioritic to granitic plutons and mafic dykes.

Previous geochronological data were interpreted in favour of deposition of sedimentary protoliths in the Paradise metasedimentary gneiss belt after 1627 Ma (Gower et al. 1992). Further work (Kamo et al. 1995), however, indicates that the rocks were also affected by metamorphism at c. 1645 Ma. Based on these results, coupled with the similarity of the rocks with enclaves of pelitic gneiss in the calc-alkaline granitoid rocks of the Earl Island domain, it is now inferred that the high-grade pelitic metasedimentary gneisses were deposited before 1670 Ma. Mylonitized metasedimentary gneiss samples collinear with unmylonitized samples on an Rb–Sr isochron collectively yield a date of 1629 ± 90 Ma and imply Labradorian mylonitization (Prevec et al. 1990), consistent with U–Pb data presented above for other parts of the Hawke River terrane and adjacent terranes.

Part of the Paradise metasedimentary gneiss belt was intruded by the Sand Hill Big Pond gabbronorite, one result of which was the formation of osumilite and sapphirine in a contact metamorphic aureole on the northwest and southeast sides of the gabbronorite (cf. Gower et al. 1986). This osumilite locality, one of only two known in Canada, has been studied in detail by Arima & Gower (1991), who show that the gabbro was emplaced at minimum P–T conditions of about 7 kb and 850°C and possibly up to 10 kb and 1000°C. Lack of a contact aureole on the northeast side of the Sand Hill Big Pond intrusion is due a major northwest-trending fault, downthrown to the north, which is one of many strike-parallel faults transecting the Paradise metasedimentary gneiss belt.

The Paradise Arm pluton was emplaced by wedging along a fault between the Paradise metasedimentary gneiss belt and the White Bear Arm complex, hence separating the Sand Hill Big Pond intrusion from the White Bear Arm complex. The dominant rock type in the Paradise Arm pluton is a massive to strongly foliated K-feldspar megacrystic granodiorite. It has a zircon age of 1639 ± 2 Ma and a monazite age of 1631 Ma (Kamo et al. 1995). In coastal areas, other plutons of similar composition within the Paradise metasedimentary gneiss belt may have been emplaced at the same time. One coastal pluton, containing an extraordinary variety of ultramafic and mafic enclaves, deserves detailed study.

The White Bear Arm complex comprises anorthosite, minor lherzolite, gabbronorite, monzonorite, and monzonite-syenite. Coronitic olivine gabbronorite is dominant. Gower *et al.* (1987) structurally divided the huge White Bear Arm complex into two parts separated by a fault, which they suggested could be a steep thrust. North of the fault, the White Bear Arm complex is composed mainly of massive mafic plutonic rocks. South of the fault, the rocks are intermixed amphibolite and metasedimentary gneiss in a zone of intense folding and transposition. Gower *et al.* (1987) interpreted these effects to be due to isoclinal folding and tectonic slicing during southwestward- or westward-directed thrusting. Kamo *et al.* (1995) have obtained an age of >1641 Ma from the western part of the body.

Apart from the 615 ± 2 Ma Long Range dykes (Kamo *et al.* 1989), the Hawke River terrane is not known to be intruded by younger Precambrian rocks. The preservation of mid- to late-Labradorian titanite ages demonstrates that at least parts of the Hawke River terrane were not affected by a severe Grenvillian overprint. This is supported by a well-defined $^{40}Ar-^{39}Ar$ hornblende plateau age of 1241 ± 3 Ma from a metamorphosed mafic dyke near the western end of the Earl Island domain (van Nostrand 1988).

The 1038–1026 Ma thermal and granitic magmatic event that affected the Lake Melville terrane is also documented on their mutual boundary at the west end of the Hawke River terrane. Here, an unfoliated microgranite dyke has a monazite and zircon lower intercept age of 1029 ± 2 Ma, which is interpreted as its time of intrusion (Schärer *et al.* 1986). The microgranite dyke discordantly intrudes granulite-facies mylonite, thereby providing a minimum age for the mylonitization. The upper intercept of 1566 ± 13 Ma cannot be linked to any established event in eastern Labrador, and, despite good precision, the age should be treated with caution as the data are very discordant.

Mealy Mountains terrane

The northern part of the Mealy Mountains terrane is dominated by the Mealy Mountains Intrusive Suite, which consists of an older group of anorthositic, leucogabbroic and leucotroctolitic rocks and a younger assemblage of rocks of variable compositions, dominated by pyroxene quartz monzonite (Emslie 1976). The rocks have genetic links with monzonite, granite and alkali-feldspar granite of the Dome Mountain intrusive suite to the west (Wardle *et al.* 1990). Emslie & Hunt (1990) have reported ages of 1646 ± 2 Ma and 1635 +22/–8 Ma from pyroxene monzonite and pyroxene granite, respectively.

The northeast flank of the Mealy Mountains Intrusive Suite is faulted against the Lake Melville terrane, but the southeast side intrudes sillimanite-bearing pelitic gneisses in which cordierite has formed as a contact metamorphic mineral (Gower *et al.* 1983c).

Although much of the Mealy Mountains terrane remains unmapped, some areas have been studied in its southeast part, where the terrane consists of tracts of sillimanite-bearing pelitic gneiss, with which granitoid and mafic plutonic rocks and orthogneiss are associated. The granitoid rocks include quartz diorite, quartz monzonite, granodiorite, granite and K-feldspar megacrystic variants. K-feldspar megacrystic granodiorite is the most common, forming ovoid plutons intruding metasedimentary gneiss. One hornblende granodiorite has a monazite age of 1631 ± 1 Ma, plus two discordant zircon fractions having minimum ages of 1735 and 1718 Ma. The 1631 Ma monazite age was interpreted to be the time of emplacement and the zircons to indicate older source components (Schärer & Gower 1988).

At the coast, the terrane is attenuated to a very narrow belt consisting of strips of granitoid orthogneiss, lenses of K-feldspar megacrystic rocks, remnants of layered mafic intrusions and slivers of metasedimentary gneiss. This attenuation coincides with a major structural break (obvious from regional geological information, potential field data and topographic expression) that separates the Mealy Mountains and Lake Melville terranes. Along much of its length, kinematic indicators demonstrate sub-horizontal, dextral movement. Extensive mylonite (Gower *et al.* 1988; Hanmer & Scott 1990) at Fox Harbour is taken as defining the Lake Melville–Mealy Mountains boundary on the southeast Labrador coast. Scott *et al.* (1993) presented U–Pb data from a deformed aplitic vein crosscutting the mylonite giving a lower intercept age of 1509 +11/–12 Ma and a 2067 ± 28 Ma upper intercept. The upper intercept was taken as evidence of an older component, and the lower intercept as the minimum time of emplacement of the vein into an actively deforming shear zone, but the possibility that the mylonitization was Labradorian and Grenvillian must not be dismissed.

A large AMCG intrusion named the Upper Paradise River pluton (van Nostrand *et al.* 1992; Gower *et al.* 1993), underlies at least 2000 km^2 of the southeast Mealy Mountains terrane and consists of anorthosite, leucogabbronorite, and pyroxene-bearing monzonite to quartz syenite.

Preliminary geochronological data indicate that it was emplaced during the 1510–1470 Ma Pinwarian event. (Krogh, unpublished; cf. Gower et al. 1993).

The next major addition to the Mealy Mountains terrane was the emplacement of the Mealy dykes. The dykes are northeast-trending, olivine tholeiitic gabbros and diabases emplaced into the Mealy Mountains intrusive suite. They have been dated by the Rb–Sr whole-rock method to be 1380 ± 54 Ma (Emslie et al. 1984), although, on the basis of an $^{40}Ar-^{39}Ar$ amphibole plateau age of c. 1215 Ma, Reynolds (1989) has suggested emplacement might be later. Given that the dykes are chemically similar (although not identical) to the Michael gabbro (Gower et al. 1990a), and that there is evidence for a c. 1240 Ma thermal resetting in the Hawke River terrane (see earlier), an alternative explanation (and offered by Reynolds) is that the $^{40}Ar-^{39}Ar$ age is related to later reheating.

The only firmly established younger event in the Mealy Mountains terrane is the emplacement of a granite at 962 ± 3 Ma, in the southeast part of the Mealy Mountains terrane. This granite is part of a widespread suite emplaced into the southern part of the eastern Grenville Province (see below).

Pinware terrane

The Pinware terrane, defined by Gower et al. (1988), is the most southerly currently identified terrane in eastern Labrador and its full extent is not yet known. Rocks in the terrane are subdivided as follows: (i) supracrustal units, (ii) foliated to gneissic granitoid rocks, (iii) layered mafic intrusions and mafic dykes, (iv) syn- to late-Grenvillian(?) granitoid rocks, (v) late- to post-Grenvillian granitoid rocks.

Units interpreted to be of supracrustal origin are dominated by fine-grained, recrystallized, commonly banded quartzofeldspathic rocks, locally showing inhomogeneous texture in detail. Although a granitoid protolith cannot be completely discounted, the rocks are suspected to be of felsic volcanic origin. This interpretation is made on the basis that the inhomogeneous texture most likely indicates a pyroclastic origin. The assertion that the rocks are supracrustal rests mainly upon their common association with units having obvious supracrustal parentage, such as quartzite, calc-silicate units and pelitic schist and gneiss. Associated banded mafic rocks are interpreted as having a mafic volcanic protolith. The age of these rocks is uncertain; the currently favoured hypothesis is that all are coeval and pre-1500 Ma (Gower et al. 1994).

U–Pb dating of recrystallized granitoid rocks in the Pinware terrane has provided ages of 1490 ± 5 Ma, 1479 ± 2 Ma, and 1472 ± 3 Ma (Pinwarian) from three coastal localities (Tucker & Gower, 1994). Inheritance in an inland Grenvillian granite of 1530 ± 30 Ma (Gower et al. 1991) and the 1509 Ma age reported from a granitoid vein by Scott et al. (1993) suggest that Pinwarian ages are widespread. Not all strongly deformed granitoid rocks in the Pinware terrane necessarily have this age range, and it is possible that Labradorian (or older) granitoid rocks and (or) younger post-1470 Ma granitoid rocks are included. Alkali-rich granitoid rocks dominate (granite, alkali-feldspar granite, syenite, alkali-feldspar syenite), grading into monzonite locally. Aegerine- and nepheline-bearing alkali-feldspar syenite also occur. Amphibolite, thought to represent remnants of extremely deformed mafic dykes, is common throughout the unit.

Mafic rocks in the Pinware terrane fall broadly into two groups, namely layered mafic intrusions and mafic dykes. The largest mafic plutonic body (110 km long and up to 12 km wide) is located at the boundary between the Mealy Mountains and Pinware terranes and is termed the Kyfanan Lake layered mafic intrusion. That it remained unmapped until 1992 (Gower et al. 1993), illustrates how poorly known some parts of the eastern Grenville Province still remain. The intrusion comprises ultramafic rocks (websterite and clinopyroxenite), gabbronorite, gabbro, leucogabbronorite, anorthositic gabbro and anorthosite. Layering is evident in several places and coronitic textures around olivine are common. Several smaller mafic bodies exist in other parts of the Pinware terrane and a wide range of discordant mafic dykes are also present (Gower et al. 1994). The mafic dykes clearly postdate deformation and metamorphism that affected the foliated to gneissic granitoid rocks.

Strongly to weakly foliated granitoid rocks, but apparently lacking mafic dykes, are assumed younger. The only indication of the age of these rocks comes from a unit dated to be c. 1145 Ma, but this is open to interpretation (Tucker & Gower 1994) and, in any case, need not be representative of the whole group. Granitoid rock types include granite, alkali-feldspar syenite, syenite and monzonite and quartz-bearing varieties. One of two deformational events in the Pinware terrane was interpreted to have coincided with metamorphism of an amphibolite, dated by zircon to be 1000 ± 2 Ma (Tucker & Gower 1994).

Late- to post-tectonic granitoid rocks form discrete monzonite, syenite and granite plutons that are generally circular in plan, homogeneous and undeformed. Some show mantled feldspar textures. Their emplacement age is between 966 and 956 Ma (Gower & Loveridge 1987; Gower et al. 1991; Tucker & Gower 1994). The boundary between the Pinware and Mealy Mountains terrane, excluding that part associated with the Kyfanan Lake layered mafic intrusion, is marked by well-banded gneiss and mylonite (Gower et al. 1988). Along northwest-trending segments of the boundary, foliations are steep (50–80°NE) and lineations plunge moderately to the northwest. Where the boundary is orientated northeast (correlating with particularly well developed mylonite) foliations dip moderately (20–50°NW) and lineations plunge down-dip. No unequivocal kinematic indicators were recorded during reconnaissance mapping, but if mafic plutonic rocks on the northern side of the boundary link up with similar mafic rocks on the south side, 20 km apparent dextral movement and southeast-directed overthrusting of the Pinware terrane by the Mealy Mountains terrane is implied.

Regional extrapolations and tectonic models

Each terrane has some unique geological features and also features in common with other terranes. Some comparisons and contrasts have been discussed above, and comments made concerning timing of terrane juxtaposition and subsequent history. The following section explores the wider significance of these data, in the context of known geological history in surrounding regions, and attempts to identify particular stages in the Proterozoic evolution of this part of Laurentia.

Post-Makkovikian–Pre-Labradorian crust (1790–1710 Ma)

In the preceding review, it was indicated that, in eastern Labrador, post-Makkovikian–pre-Labradorian rocks are confined to the Mealy Mountains terrane (post-Makkovikian being defined here as later than 1800 Ma). These data can be augmented with information from the Goose Bay area, where zircon dates of 1754 +37/−34 Ma from a foliated quartz monzonite, 1712 ± 23 Ma from a granitic leucosome, and 1775 ± 16 Ma from a quartz diorite gneiss have been reported, and interpreted as maximum ages for the formation of the rocks, or their early metamorphism (Wardle et al. 1990; Philippe et al. 1993).

The closest area where rocks of similar age are found is in the Makkovik Province, but they are separated from the Mealy Mountains terrane by Labradorian crust, lacking evidence of pre-1710 Ma rocks. In the Makkovik Province, post-Makkovikian–pre-Labradorian events are indicated by monazite and titanite ages of 1794 ± 2 Ma and 1761 ± 2 Ma (Schärer et al. 1988), and a titanite age of 1746 ± 2 Ma (Gandhi et al. pers.comm., in Loveridge et al. 1987; Kerr et al. 1992, 1994, this volume). Coupled to these data, Wilton & Longerich (1993) reported Pb–Pb data on uraninite, concluding that there was a U mineralization event at 1741 ± 23 Ma, which they linked to metamorphism, deformation and granitoid emplacement. The youngest pre-Labradorian event was emplacement of the Strawberry Intrusive Suite, a group of coarse grained, variably K-feldspar megacrystic, fluorite-bearing subsolvus biotite granites dated at 1719 ± 3 Ma (Kerr & Krogh 1990; Kerr et al. 1992). In southern Greenland, in the Ketilidian Mobile Belt, rapakivi granites have been dated at 1.74 Ga (Gulson & Krogh 1975). In Rockall and western Ireland, Daly (1994) have identified calc-alkaline crust having an age of 1.75 Ga, consistent with previous information for crust of this age forming the Rhinns complex in northwest Scotland (Muir et al. 1992).

In the central Grenville Province, there are hints for material derived from pre-Labradorian crustal sources in the Atikonak and Baie Comeau areas, based on Nd–Sm model ages (cf. Dickin & Higgins 1992; Emslie & Hegner 1993). The extent to which these data indicate the presence of pre-Labradorian crust, rather than merely younger rocks containing older inherited material, remains unclear.

To explain this growing but scattered collection of age data, it is postulated that pre-Labradorian crust was separated from the rest of pre-Labradorian Laurentia along its southern flank during the formation of a backarc basin or small ocean at 1.71 Ga, so that only a small fragment of this crust remains in the Makkovik Province (Fig. 4a). Thus, the 1794–1746 Ma titanite ages and the U-mineralizing event in the Makkovik Province may document an inboard manifestation of a post-Makkovikian continental-margin arc much further to the south. Remnants of the arc may be present in the southern part of the eastern Grenville Province and other fragments are preserved in northwest Britain and Ireland. Such an arc could provide an explanation for the elusive granitoid source for c. 1.74 Ga U-mineralized fluids in the Makkovik Province proposed by Wilton & Longerich (1993), as coeval granitoid rocks have yet to be found in the region.

The Labradorian orogenic cycle (1710–1600 Ma)

1. Formation of a backarc basin or small ocean (1710–1680 Ma). Overall, evidence points to deposition of broad tracts of the sedimentary precursors to pelitic to semi-pelitic gneisses in the

Grenville Province in eastern Labrador, either before the Labradorian Orogeny, or during its early stages (the exact timing of sediment deposition remains unclear). The pre-1709 Ma biotite-rich gneissic enclave in the Groswater Bay terrane, interpreted to have been derived from a sedimentary protolith provides one indication for pre-Labradorian deposition (and deformation) of some supracrustal rocks in the Groswater Bay terrane. Enclaves of sedimentary gneiss in 1668 Ma calc-alkaline rocks in the Hawke River terrane provides another indication, although it cannot yet be demonstrated that the two groups of supracrustal rocks correlate.

Further west, a zircon age of 1699 ± 3 Ma for a sillimanite-bearing gneiss from Wilson Lake was reported by Currie & Loveridge (1985). Pb/Pb ages of 1707 ± 2 Ma and 1709 ± 2 Ma were interpreted to indicate older cores in two zircons. With one exception from central Labrador, there is no geochronological evidence for the sediments having been deposited appreciably earlier than Labradorian orogenesis, or derived from rocks that were significantly older. The exception is a metasedimentary gneiss from central Labrador containing zircons having a minimum age of 2600 Ma.

A plausible tectonic depositional environment is a backarc basin or small ocean, the detritus having been derived from both the pre-Labradorian Laurentian hinterland to the north and pre-Labradorian rocks to the south (Fig. 4a). Rifting at c. 1.71 Ga is postulated, a suggestion that is consistent with Schärer's (1991) interpretation that the source materials for the Labradorian granitoid rocks separated from the mantle at 1712 ± 66 Ma (Sm–Nd isochron).

2. Formation of arc-related terranes (1680–1660 Ma). Widespread calc-alkaline plutonism between 1680 and 1650 Ma in eastern Labrador, and grouped by Gower *et al.* (1992) into three sequential magmatic events at c. 1677 Ma, 1671 Ma and 1658–1649 Ma, broadly indicate that the earliest Labradorian arc-related rocks are the most southerly, decreasing in age to the north and temporally merging into the 1654–1649 Ma age range for the Trans-Labrador batholith.

Data from the Goose Bay area are broadly compatible with this pattern, but with some complications. For example, the Susan River quartz diorite, dated at 1672 +11/–10 Ma (Philippe *et al.* 1993), most closely resembles quartz diorite belonging to the Red Island magmatic event in the Hawke River terrane, but is assigned to the Groswater Bay terrane (Philippe *et al.* 1993). Also, the region in which the tonalite dated to be 1677 +18/–11 Ma occurs was originally assigned to the Groswater Bay terrane, but could be re-assigned to the Lake Melville terrane (Gower *et al.* 1992).

Regardless of these details, the important point to be stressed is that **all** pre-1660 Ma Labradorian plutonic rocks dated so far occur south of the 1650 Ma Trans-Labrador batholith. This paper contends this is a key feature in assigning Labradorian subduction polarity. Had the polarity been uninterruptedly to the north, under pre-Labradorian Laurentia, a continuum of ages from Makkovikian through to Labradorian would be expected; such is not the case. No granitoid rocks having ages between the 1719 Ma Strawberry Intrusive Suite and the 1654–1649 Ma Trans-Labrador batholith are known to have been emplaced into Makkovikian or older crust. Coupling this observation with evidence presented by Schärer (1991) for lack of contamination from an appreciably older crustal source for Labradorian arc-related rocks, it is argued that subduction must have been to the south (Fig. 4a). Southward-dipping reflectors in the mantle in the southern Grenville Province (Hall *et al.* 1995) may be evidence of that subduction.

3. Collision of arc-related terranes (1660–1645 Ma). The Trans-Labrador batholith is interpreted here as a consequence of the accretion of Labradorian arc-related terranes to pre-Labradorian Laurentia. The accretion is inferred to have resulted in crustal thickening, which then triggered melting to generate the batholith. The timing of collision remains uncertain, but must have been before 1650 Ma, as this is the crystallization age for much of the batholith. Evidence for migmatization between 1668 and 1663 Ma in the Hawke River terrane and Labradorian mylonitization in both the Hawke River and Lake Melville terranes coupled with a 1662 Ma age for termination of major mylonitization at the Grenville (Labradorian) front suggests accretion of the arc-related terranes at c. 1665 Ma.

4. Post-collisional events (1645–1600 Ma). The period between 1645 and 1625 Ma is characterized by the emplacement of both mafic and felsic plutonic bodies (including AMCG suites). Waning granitoid magmatism and cooling continued until about 1600 Ma. Mafic-felsic magmatism has been dated in the Groswater Bay, Hawke River and Mealy Mountains terranes, and similar undated rocks occur in the Lake Melville terrane. Thus, with the possible exception of the Pinware terrane (excluded only because of inadequate data), all crustal blocks in eastern Labrador were bonded together and to pre-Labradorian Laurentia by 1645 Ma at the latest (probably 10 to 20 million years earlier).

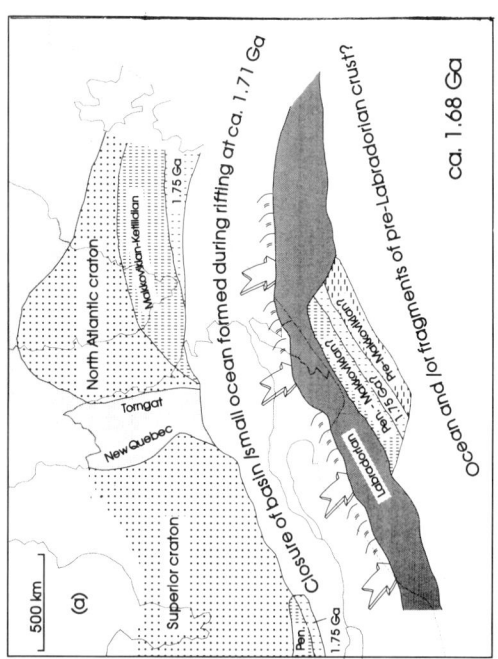
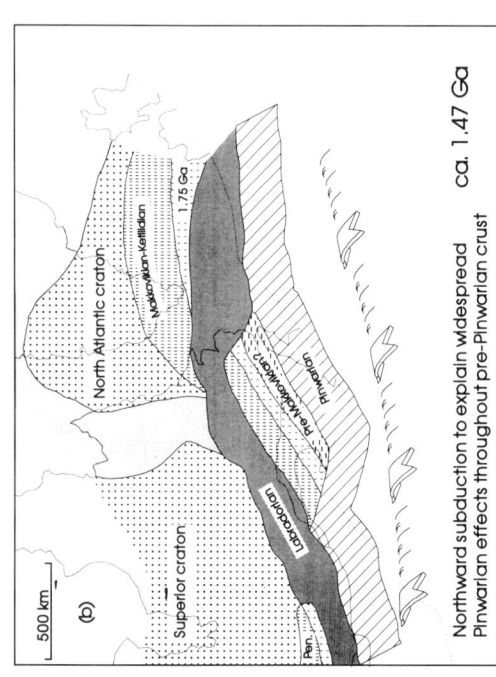
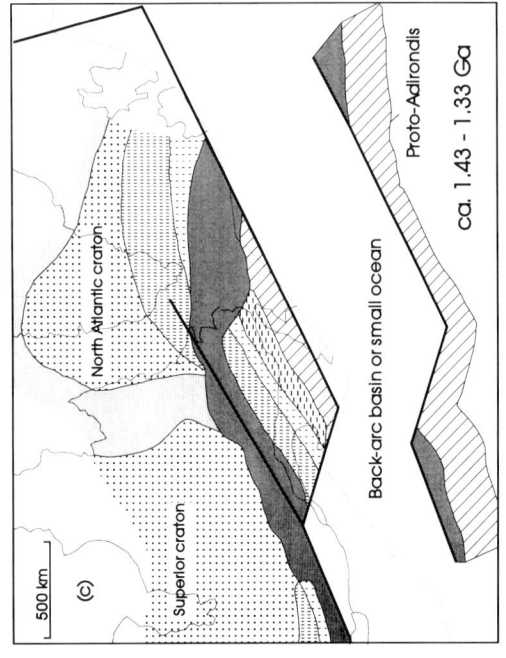
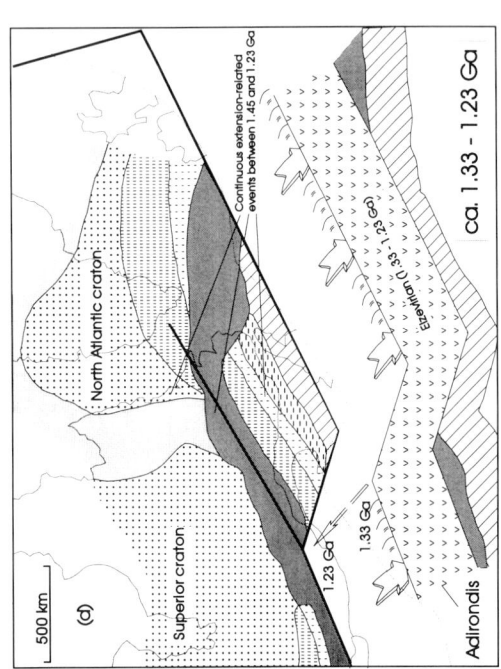

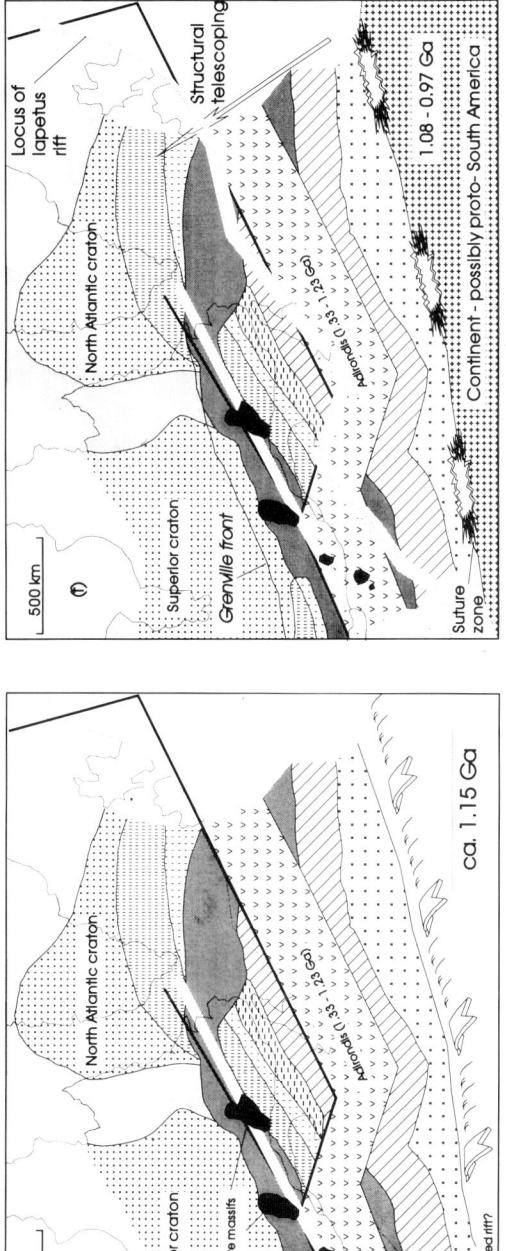

Fig. 4a–f. Schematic tectonic evolution diagrams for the evolution of the eastern Grenville Province (Pen., Penokian).

There is no clear cut temporal demarcation between the collisional magmatic stage of the Labradorian orogenic cycle and the post-collisional bimodal magmatism. For example, in the Makkovik Province, coeval mafic and felsic magmatism resulted in the emplacement of satellite plutons north of the main body of the Trans-Labrador batholith (the mafic Adlavik Intrusive suite dated to be 1649 ± 1 Ma and the felsic Mount Benedict Intrusive Suite at 1649 ± 3 Ma; Kerr et al. 1992). In the southern Groswater Bay terrane, similar rocks at Grady Island have an age of 1645 ± 3 Ma.

Gower et al. (1992) and Connelly et al. (1995) have both linked the Trans-Labrador batholith to northward-dipping subduction. Gower et al. envisaged a reversal of subduction from south to north following accretion of the arc terranes to pre-Labradorian Laurentia, whereas Connelly et al. considered that there were two northward-dipping subduction zones, one related to the arc terranes and a second one responsible for the Trans-Labrador batholith. An additional possibility to the north-dipping subduction models is that there was no northward-dipping subduction at any stage. Following accretion of the 1680–1660 Ma magmatic arcs, the Trans-Labrador batholith and post-1650 Ma magmatism could be generated simply as a result of the crustal thickening, which then merely dwindled away as isostatic equilibrium was regained. Such a model would explain the lack of magmatic products after 1600 Ma – the system simply 'shut down' and the Laurentian margin became passive.

Passive continental margin (1600–1530 Ma)

The period between 1600 and c. 1530 Ma is an enigmatic one for the whole of Laurentia, as there is very little indication of geological activity of any kind. If the lack of activity is related to the development of a passive continental margin as noted above, then evidence of continental-margin basins should be present south of Labradorian (and any pre-Labradorian) crust. Large tracts of undated metasedimentary gneiss in eastern Quebec, flanking the north shore of the Gulf of St Lawrence, are a potential candidate.

Pinwarian orogenesis (1530–1450 Ma)

In eastern Labrador, two Pinwarian tectonic components can be recognized, namely (i) extensive quartz monzonite to granite plutonism in the south (in the Pinware terrane), and (ii) widespread thermal effects and very minor felsic magmatism farther north. On the basis of Rb–Sr isochrons from rocks of Labradorian age, the thermal effects in the north bracket the period between 1530 and 1500 Ma (cf. Tucker & Gower 1994). It is suggested here that the 1530–1500 Ma Rb–Sr ages hint at thermal effects slightly earlier than the 1510 to 1470 Ma period presently identified as Pinwarian.

Recall that southward-dipping subduction during Labradorian arc formation was envisaged above because of lack of evidence for thermal and magmatic effects inboard of a continental-margin arc. In contrast, widespread Pinwarian isotopic resetting and minor granitoid magmatism throughout the northern part of the Grenville Province and the granitoid pluton emplacement into Labradorian-aged crust further south are the expected manifestations of a continental-margin arc over a north-dipping subduction zone (Fig. 4b). The arc is inferred to have existed between about 1510 and 1450 Ma. That such an arc existed (rather than, say, an extensional setting) is based on Nd isotopic evidence for a major crust-forming event at 1.53 ± 0.07 Ga in the Baie Comeau area (Dickin & Higgins 1992) and for c. 1.5 Ga juvenile crustal material in the mid-west USA (Bowring et al. 1992). Coeval activity included 1.52 Ga high-grade metamorphism in southwest Norway (Corfu 1980), high-grade metamorphism in the Hart Jaune terrane at 1469 ± 5 Ma (Scott & Hynes 1994) and metamorphism in the basement to the Wakeham Supergroup at 1495 ± 2 Ma (Clark & Machado 1994).

According to the continental-arc model, the bulk of the arc would lie to the south of presently investigated areas in the eastern Grenville Province (Fig. 4b). Granulite facies mafic to felsic gneisses of the Disappointment Hill complex in western Newfoundland dated to be 1498 +9/–8 Ma (Currie et al. 1992) support such a thesis. Pinwarian geochronological data establish beyond little doubt that terrane assembly in the eastern Grenville Province was accomplished prior to Pinwarian orogenesis: no support exists for sutures within the eastern Grenville Province at any later time, although telescoping of pre-existing crustal blocks certainly occurred.

Events between 1450 and 1230 Ma in eastern Laurentia – intermittent rifting of a passive continental margin

Gower & Tucker (1994) reviewed geochronological data for the period between 1500 and 1400 Ma, emphasizing a contrast in the age of the crust between the southwest part and the remainder of the Grenville Province. They explained this contrast as due to rifting (to produce either a small ocean or a backarc basin) that occurred in the latter part of geon 14 (Fig. 4c). Rifting was unsuccessful in the eastern Grenville Province but a tensional tectonic environment permitted the mafic magmatism that resulted in the Shabogamo (c. 1450 Ma) and Michael (1426 Ma) gabbros, and also the Mealy dykes, the latter currently dated by Rb–Sr methods to be c. 1380 Ma.

Geochronological documentation of events between 1400 and 1300 Ma in the eastern Grenville Province is rather scarce, and so indications of the prevailing tectonic environment partly come from outside the Grenville Province, or from its fringes. These give little reason to suppose that much change occurred overall throughout the period, consistently pointing to a stable but intermittently extensional tectonic environment. Relevant units include the AMCG Nain Plutonic Suite (1350–1290 Ma), the alkaline Red Wine and Arc Lake intrusive suites (1337 +10/–8 Ma) and Letitia Lake Group (1327 ± 4 Ma). A few plutons within the eastern Grenville Province, such as the Upper North River body (1296 Ma), the Arrowhead Lake intrusion (Rb–Sr age of 1307 ± 28 Ma; Fryer 1983), a monzonite in the Havre–St Pierre area of eastern Quebec (1332 ± 7 Ma; Martignole et al. 1994), and the Rivière Pentecôte anorthosite (1354 ± 3 Ma; Martignole et al. 1993: 1365 +7/–4 Ma; Emslie & Hunt 1990), show that similar magmatic activity was also on-going farther south.

In contrast to the predominantly felsic magmatic activity between 1400 and 1300 Ma, the next 60 to 100 million years show increasing evidence of mafic magmatism, perhaps due to progressive thinning of continental crust hence allowing deep-source mafic magmas access to upper-crustal regions. Again, evidence of this activity comes partly from outside the Grenville Province, for example the Nutak and Harp dykes were emplaced at 1268 Ma and 1273 ± 1 Ma, respectively (Cadman et al. 1993) and gabbro sills emplaced into the Seal Lake Group between 1250 and 1225 Ma (Romer et al. 1994).

Meanwhile, the alkaline Flowers River Igneous Suite was emplaced at 1271 ± 15 Ma (Hill 1982), and within the eastern Grenville Province, bimodal felsic and mafic volcanic activity were almost coeval in the Wakeham Supergroup at 1271 +13/–3 Ma (age on rhyolite from Aguanus Group; Martignole et al. 1994). A granulite facies meta-basic gneiss that yielded a titanite age of 1281 Ma and linked to an amphibolite facies retrogression in the Lac Joseph terrane (Connelly 1991) shows that, in contrast to deep-level rocks exhumed from the northern Grenville Province, the Flowers River and Wakeham areas have undergone relatively little uplift since c. 1270 Ma.

Limited evidence for a thermal event in eastern Labrador comes from c. 1220 Ma K–Ar and Ar–Ar ages from the Mealy Mountains and Hawke River terranes. These dates are similar to an age of 1225 +47/–33 Ma (Thomas et al. 1986) from a granulite tonalite to quartz diorite gneiss of the Lac Long Igneous Suite in southwest Labrador. Two granitoid bodies dated at 1245 ± 3 Ma and 1239 ± 3 Ma marginal to the Wakeham Supergroup (Martignole et al. 1994) and a 1253 to 1246 Ma felsic volcanic unit in the upper part of Wakeham Supergroup (Clark & Machado 1994) indicate that magmatism, only slightly older, was widespread further south. Neither the extent nor the tectonic setting of this magmatic and thermal activity are yet established, except that c. 1800 Ma inheritance in the 1245 Ma granite suggests emplacement through older crust. The thermal effects and magmatism may be a distal manifestation of accretion of Adirondis (new name; Fig. 4d) south of the exposed eastern Grenville Province, but 10 to 20 million years earlier. This suggestion is consistent with lack of evidence for extensional conditions in the eastern Grenville Province after 1225 Ma.

1150–1110 Ma granitoid activity in southeast Labrador – magmatism inboard of a continental-margin subduction zone?

In southeast Labrador, two granitoid plutons have equivocal ages of c. 1145 and 1132 Ma and a granitic vein an age of 1113 Ma. If these dates are representative of the type of activity in eastern Labrador during geon 11, then it would seem that conditions differed from those prevailing at the time throughout most of the Grenville Province and southwest Greenland, as this was mainly a period of mafic and AMCG magmatism, especially between 1160 and 1120 Ma (Emslie & Hunt 1990; Gower et al. 1990b). For example, two bodies in the eastern Grenville Province are the Lac Allard pyroxene monzonite (1126 +7/–6 Ma) and the Atikonak River massif (1123 ± 4 Ma for pyroxene quartz monzonite and 1133 +10/–5 Ma for rapakivi granite), both dated by Emslie & Hunt (1990). Recent dates of 1177 +5/–4 Ma from mafic sills emplaced into arenites of the Davy Lake Group of the Wakeham Supergroup (Martignole et al. 1994) and an age of 1169 ± 2 Ma from the Brien troctolite–anorthosite (Scott & Hynes 1994), however, suggest that the 'mafic–anorthositic event' started 20 million years earlier in some areas than Gower et al. (1990b) suggested.

It is proposed here that following the accretion between 1230 and 1190 Ma (Elzevirian Orogeny) of geon 12 arc-related rocks, north-dipping subduction may have been initiated well south of the presently exposed Grenville Province and that the

1180–1120 Ma AMCG magmatism and mafic magmatism is a distal expression of that subduction and subsequent backarc rifting (Fig. 4e), analogous to the tectonic environment proposed for geon 14 when similar rocks were emplaced in eastern Laurentia (cf. Gower & Tucker 1994).

Magmatism at c. 1078 Ma and 1040–1026 Ma in the Lake Melville terrane

Despite only the Lake Melville terrane in eastern Labrador recording events at 1078 Ma and 1040–1026 Ma, it is clear that magmatism and thermal effects of this age in the eastern Grenville Province were not confined to this terrane. In the central-south part of the Grenville Province in Quebec, reported dates include 1079 ± 5 Ma from the monzogranitic margin of the Romaine River anorthosite (Loveridge 1986), 1082 ± 2 Ma for the Chicoutimi mangerite, 1079 ± 22 Ma for the St Urbain anorthosite (Higgins & van Breemen 1992).

Two of the 1040–1026 Ma dates in the Lake Melville terrane are crystallization ages from minor intrusions discordant to gneissic/mylonitic fabrics defining parts of the boundary of the Lake Melville terrane with the Mealy Mountains and Hawke River terranes. These suggest that displacement of the Lake Melville terrane relative to the Hawke River and the Mealy Mountains terrane had ceased by c. 1035 Ma. A 1038 ± 2 Ma monazite date (Philippe et al. 1993) from the Cape Caribou allochthon provides support for this conclusion inasmuch as it shows similar activity in the Mealy Mountains terrane.

As a final point, it should be noted that the 1078 Ma event is documented by zircon and monazite, whereas the later event also includes some titanite ages, suggesting declining temperatures with time, perhaps related to tectonically driven uplift (Fig. 4f).

Termination of orogenesis

Throughout much of the eastern Grenville Province, a metamorphic event between 1010 Ma and 990 Ma seems to be dominant. In easternmost Labrador, only two ages in that time interval are known – dating a pegmatite at 1003 Ma at the Mealy Mountains–Lake Melville terrane boundary and dating metamorphism of an amphibolite at 1000 Ma in the Pinware terrane. In other parts of the eastern Grenville Province this c. 1000 Ma event was linked to northwest-directed thrusting (e.g. Connelly et al. 1995). If the Mealy Mountains–Lake Melville terrane boundary, the locus of dextral transcurrent movement, was the northeastern flank for this thrusting event, this explains the lack of c. 1000 Ma metamorphism in the Hawke River and Groswater Bay terranes further northeast, and prompts the prediction that it will be found to be extensive in the Mealy Mountains and Pinware terranes.

The Groswater Bay terrane was involved in a yet younger Grenvillian event, as indicated by several titanite ages between 980 and 968 Ma obtained from Labradorian gneisses in the region. In addition, many granitoid plutons have been dated between 966 and 956 Ma in the Pinware terrane and one at 962 Ma in the southeastern Mealy Mountains terrane. These data reflect contrasting coeval tectonic settings; one in the north in which very little magmatic activity occurred, as opposed to the south where granitoid plutonism was extensive. The ages are interpreted to be related to syn- to post-Grenvillian collisional tectonism (Fig. 4f), explaining the 980–968 Ma ages in the north as reflecting titanite closure after exhumation and cooling related to thrusting, and the 966–956 Ma ages in the south as indicating magmatism following crustal thickening (cf. Gower et al. 1991).

In both the Groswater Bay and Pinware terranes, late- to post-Grenvillian uplift and cooling is indicated by U–Pb 930–920 Ma rutile data (Schärer et al. 1986) and K–Ar hornblende and biotite data (Gower et al. 1991), respectively. The K–Ar data, taken in conjunction with U–Pb zircon and titanite data for the same samples indicate a cooling rate for granitoid plutons of 11°C per million years between 960 and 910 Ma (Gower et al. 1993).

Discussion

The plate tectonic evolutionary model presented in the previous section embodies several suggestions that are not part of Grenvillian lore and therefore deserve some discussion.

Perhaps the greatest departure from established concepts is the concept that there was a rifting event at c. 1.71 Ga, at which time a large block of pre-Labradorian Laurentia was torn out of its crustal context, but cryptically remains in the southern part of the Grenville Province. Some of the isotopic hints for pre-Labradorian crust in the southern Grenville Province have been noted in the preceding text. An argument against rifting at 1.71 Ga, however, is the apparent lack of evidence for mafic dykes and typical rift-related supracrustal rocks of this age in the Grenville Province or adjacent areas to the north.

The polarity of Labradorian subduction remains controversial. The thesis for southerly polarity advanced here relies heavily on lack of evidence for pre-1660 Ma Labradorian intrusions north of the Trans-Labrador batholith; if this is proved incorrect, the argument is seriously weakened. Although the model calls for a short-lived basin, it must be remembered that even with a modest rate of spreading of 5 cm a^{-1}, a 1000 km wide ocean could form over a 20-million-year period.

A second departure from prevailing concepts is the idea that a continental-margin arc existed during the early part of geon 14. The evidence for this proposal is not (yet?) compelling and alternative models are possible. Certainly, crust of this age having calc-alkaline signatures has not been identified in the Grenville Province, but the model argues that most of it was removed during late-geon 14 rifting and that any remaining crust of this age is situated in the region about which very little is known.

Although the proposal that extensional conditions dominated during the latter half of geon 14 is easier to justify than 1.71 Ga rifting, the thesis must still be regarded as tenuous. Nevertheless, longitudinal correlations within the Grenville Province, especially the northeast extrapolation of the Central Metasedimentary Belt, have consistently encountered problems and the model is attractive in that it offers a way of initiating the crustal differences that exist between the southwest part and the remainder of the Grenville Province. The schematic Laurentian continental margin would have been masked during subsequent orogeneses, of course, and would not appear as depicted in Fig. 4c.

Implicit in the model is that not all AMCG suites are necessarily anorogenic. For example, some are regarded here as distal intrusions inboard of continental arcs, others could be post collisional. As with genuinely anorogenic bodies, such suites would have been emplaced into stabilized crust and, from a field perspective, might show few differences.

The 1180–1120 Ma AMCG magmatism in the Grenville Province is of interest from another perspective. These intrusions extend diagonally across almost the full width of the Grenville Province and are coeval with the 1.14 Ga Abitibi dykes in the Superior Province. This implies a common (largely extensional?) tectonic setting for all regions at that time. It follows that the last major accretionary event must have been during 1.22–1.19 Ga Elzevirian orogenesis and that no sutures postdating this accretion are present anywhere within the exposed part of the Grenville Province. The suture related to the major continent–continent collisional event postulated to have occurred between 1.08 and 0.97 Ga would have to be further south. In this context, the concept of a Grenvillian orogenic cycle (extending from c. 1.30 Ga to 1.00 Ga) requires re-evaluation, at least as it applies to the eastern Grenville Province. Evidence points to an anorogenic setting continuing until 1.23 Ga for this region; it is only after 1.08 Ga, at the very earliest, that 'Grenvillian' tectonism began.

Tectonic environments must be either extensional, compressional or neutral. For the Grenville Province, these seem to have alternated in the following fashion:

pre-1.74 Ga Compression (re-amalgamation of Archaean cratons)
1.74–1.69 Ga Extensional (formation of pre-Labradorian basin)
1.69–1.60 Ga Compression (Labradorian Orogeny – arc accretion)
1.60–1.51 Ga Uncertain (possibly neutral to mildly extensional)
1.51–1.45 Ga Compression (Pinwarian continental-margin arc)
1.45–1.23 Ga Extensional (Elsonian anorogenic plutonism)
1.23–1.18 Ga Compression (Elzevirian Orogeny – arc accretion)
1.18–1.08 Ga Mixed extension and compression (backarc?)
1.08–0.97 Ga Compression (Grenvillian Orogeny – continent–continent collision).

Even given such a model, the extent to which extension was adequate to form new ocean floor versus merely backarc basins or intra-continental rifts remains uncertain, and addressing this issue may be one of the more fruitful avenues for future research.

This review would not be possible without the dedicated contributions made by U–Pb geochronologists over the last decade, many of whom initiated their work at the Royal Ontario Museum under the direction of T. E. Krogh. The manuscript benefitted from the comments of J. M. McLelland and an anonymous reviewer. This paper is published with the permission of Bryan Greene, Executive Director, Geological Survey of Newfoundland and Labrador.

References

ARIMA, M. & GOWER, C. F. 1991. Osumilite-bearing granulites in the eastern Grenville Province, eastern Labrador, Canada: mineral parageneses and metamorphic conditions. *Journal of Petrology*, **32**, 29–61.

BOWRING, S. A., HOUSH, T. B., VAN SCHMUS, W. R. & PODOSEK, F. A. 1992. A major Nd isotopic boundary along the southern margin of Laurentia (abs.). *American Geophysical Union Spring Meeting, Montreal, Program with Abstracts*, 333.

BROOKS, C. 1983. Fourth report on the geochronology of Labrador. U–Pb ages. Unpublished report to Newfoundland Department of Mines and Energy, Mineral Development Division, Open File Lab 519.
—— 1984. Determination of Rb–Sr and Sm–Nd geochronological ages of Labrador and insular Newfoundland rock samples; part 1, Rb–Sr determinations. Unpublished report to Energy Mines and Resources concerning contract serial no. OSQ83-00200. Newfoundland Department of Mines and Energy, Open File Lab 745.
CADMAN, A. C., HEAMAN, L., TARNEY, J., WARDLE, R. & KROGH, T. E. 1993. U–Pb geochronology and geochemical variation within two Proterozoic mafic dyke swarms, Labrador. *Canadian Journal of Earth Sciences*, **30**, 1490–1504.
CLARK, T. & MACHADO, N. 1994. New U–Pb dates from the Wakeham Terrane and its basement (abs.). *In*: An Overview of the Grenville orogen, from the Great Lakes to the Labrador Sea. *Joint Abitibi-Grenville–ECSOOT LITHOPROBE meeting*. École Polytechnique, November, 1994.
CONNELLY, J. N. 1991. *The thermotectonic history of the Grenville Province of western Labrador*. PhD thesis, Memorial University of Newfoundland, St John's.
——, RIVERS, T. & JAMES, D. T. 1994. Thermotectonic evolution of the Grenville Province of western Labrador. *Tectonics*, **14**, 202–217.
CORFU, F. 1980. U–Pb and Rb–Sr systematics in a polyorogenic segment of the Precambrian shield, central southern Norway. *Lithos*, **13**, 305–323.
CURRIE, K. L. & LOVERIDGE, W. D. 1985. Geochronology of retrogressed granulites from Wilson Lake, Labrador, Current Research, Part B. *Geological Survey of Canada*, Paper **85-1B**, 191–197.
——, VAN BREEMAN, O., HUNT, P. A. & VAN BERKELL, J. T. 1992. Age of high-grade gneisses south of Grand Lake, Newfoundland. *Atlantic Geology*, **28**, 153–161.
DALY, J. S. 1994. Proterozoic geology in western Ireland. Post-conference field guide. *Precambrian Crustal Evolution in the North Atlantic Regions, IGCP-275/371*.
DICKIN, A. P. & HIGGINS, M. D. 1992. Sm/Nd evidence for a major 1.5 Ga crust-forming event in the central Grenville province. *Geology*, **20**, 137–140.
EMSLIE, R. F. 1976. Mealy Mountains Complex, Grenville Province, southern Labrador. *In*: Report of Activities, Part A. *Geological Survey of Canada*, Paper **76-1A**, 165–170.
—— 1983. The coronitic Michael gabbros, Labrador: assessment of Grenvillian metamorphism in northeastern Grenville Province. *Current Research, Part A. Geological Survey of Canada, Paper* 83-1A, 139–145.
—— & HEGNER, E. 1993. Reconnaissance isotopic geochemistry of anorthosite–mangerite–charnockite–granite (AMCG) complexes, Grenville Province, Canada. *Chemical Geology*, **106**, 279–298.
—— & HUNT, P. A. 1990. Ages and petrogenetic significance of igneous mangerite–charnockite suites associated with massif anorthosites, Grenville Province. *Journal of Geology*, **98**, 213–231.
——, LOVERIDGE, W. D. & STEVENS, R. D. 1984. The Mealy dykes, Labrador: petrology, age, and tectonic significance. *Canadian Journal of Earth Sciences*, **21**, 437–446.
FRYER, B. J. 1983. Report of geochronology – Labrador mapping. Unpublished report to the Newfoundland Department of Mines and Energy, Open File Lab 617.
GOWER, C. F. 1986. Geology of the Double Mer White Hills and surrounding region, Grenville Province, eastern Labrador. *Geological Survey of Canada*, Paper **86-15**.
—— 1987. The Hawke River Terrane – a newly defined lithotectonic entity in the Grenville Province of eastern Labrador (abs.). *GAC–MAC Annual Meeting*, **12**, 48.
—— 1990. Mid-Proterozoic evolution of the eastern Grenville Province, Canada. *Geologiska Föreningens i Stockholm Förhandlingar*, **112**, 127–139.
—— & ERDMER, P. 1988. Proterozoic metamorphism in the Grenville Province: a study in the Double Mer–Lake Melville area, eastern Labrador. *Canadian Journal of Earth Sciences*, **25**, 1895–1905.
—— & LOVERIDGE, W. D. 1987. Grenvillian plutonism in the eastern Grenville Province. Radiogenic age and isotopic studies: Report 1. *Geological Survey of Canada*, Paper **87-2**, 55–58.
—— & OWEN, V, 1984. Pre-Grenvillian and Grenvillian lithotectonic terranes in eastern Labrador – correlations with the Sveconorwegian Orogenic Belt in Sweden. *Canadian Journal of Earth Sciences*, **21**, 678–693.
—— & SWINDON, H. S. 1991. Pillow lavas in the Dead Islands area, Grenville Province, southeast Labrador. *Current Research*. Newfoundland Department of Mines and Energy, Mineral Development Division, Report **91-1**, 205–215.
—— & TUCKER, R. D. 1994. Distribution of pre-1400 Ma crust in the Grenville province: Implications for rifting in Laurentia–Baltica during geon 14. *Geology*, **22**, 827–830.
——, BAILEY, D. G., DOHERTY, R. A., NOEL, N. & GILLESPIE, R. T. 1983*a*. *Rigolet Map Region*. Newfoundland Department of Mines and Energy, Mineral Development Division Map **8342**.
——, HEAMAN, L. M., LOVERIDGE, W. D., SCHÄRER, U. & TUCKER, R. D. 1991. Grenvillian granitoid plutonism in the eastern Grenville Province, Canada. *Precambrian Research*, **51**, 315–336.
——, NEULAND, S., NEWMAN, M. & SMYTH, J. 1987. Geology of the Port Hope Simpson map region, Grenville Province, eastern Labrador. *Current Research*. Newfoundland Department of Mines, Mineral Development Division, Report **87-1**, 183–199.
——, NOEL, N. & GILLESPIE, R. T. 1983*b*. *Groswater Bay Map Region*. Newfoundland Department of Mines and Energy, Mineral Development Division Map **8343**.
——, ——, FINN, G. & EMSLIE, R. 1983*c*. *English River Map Region*. Newfoundland Department of Mines and Energy, Mineral Development Division Map **8344**.
——, —— & VAN NOSTRAND, T. 1985. Geology of the

Paradise River region, Grenville Province, eastern Labrador. *Current Research.* Newfoundland Department of Mines and Energy, Mineral Development Division, Report **85-1**, 19–32.

——, OWEN, V. & FINN, G. 1982. The geology of the Cartwright region, Labrador. *Current Research.* Newfoundland Department of Mines and Energy, Mineral Development Division, Report **82-1**, 122–130.

——, RIVERS, T. & BREWER, T. S. 1990a. Middle Proterozoic mafic magmatism in Labrador, eastern Canada. *In*: GOWER, C. F., RIVERS, T. & RYAN, A. B. (eds) *Mid-Proterozoic Laurentia–Baltica.* Geological Association of Canada Special Paper **38**, 485–506.

——, RYAN, A. B., BAILEY, D. G. & THOMAS, A. 1980. The position of the Grenville Front in eastern and central Labrador. *Canadian Journal of Earth Sciences*, **17**, 784–788.

——, & RIVERS, T. 1990b. Mid-Proterozoic Laurentia–Baltica: an overview of its geological evolution and a summary of the contributions made by this volume. *In*: GOWER, C. F., RIVERS, T. & RYAN, A. B. (eds) *Mid-Proterozoic Laurentia–Baltica.* Geological Association of Canada Special Paper **38**, 1–20.

——, SCHÄRER, U. & HEAMAN, L. M. 1992. The Labradorian Orogeny in the Grenville Province, eastern Labrador. *Canadian Journal of Earth Sciences*, **29**, 1944–1957.

——, VAN NOSTRAND, T. & EVANS-LAMSWOOD, D. 1994. Geology of the Pinware River region, southeast Labrador. *Current Research.* Newfoundland Department of Mines and Energy, Geological Survey Branch, Report **94-1**, 347–369.

——, VAN NOSTRAND, T., McROBERTS, G., CRISBY, L. & PREVEC, S. 1986. Geology of the Sand Hill River–Batteau map region, Grenville Province, eastern Labrador. *Current Research.* Newfoundland Department of Mines and Energy, Mineral Development Division, Report **86-1**, 101–111.

——, ——, PECKHAM, V. & ANDERSON, J. 1993. Geology of the upper St Lewis River map region, southeast Labrador. *Current Research.* Newfoundland Department of Mines and Energy, Geological Survey Branch, Report **93-1**, 17–34.

——, ——. & SMYTH, J. 1988. Geology of the St. Lewis River map region, Grenville Province, eastern Labrador. *Current Research.* Newfoundland Department of Mines, Mineral Development Division, Report **88-1**, 59–73.

GULSON, B. L. & KROGH, T. E. 1975. Evidence of multiple intrusion, possible resetting of U–Pb ages, and new crystallization of zircons in the post-tectonic intrusions ('Rapakivi granites') and gneisses from South Greenland. *Geochimica et Cosmochimica Acta*, **39**, 65–82.

HALL, J., WARDLE, R. J., GOWER, C. F., KERR, A., COFLIN, K., KEEN, C. E., & CARROLL, P. 1995. Proterozoic orogens of the northeast Canadian Shield: new information from the Lithoprobe ECSOOT crustal reflection seismic survey. *Canadian Journal of Earth Sciences*, **32**, 1119–1131.

HANMER, S. & SCOTT, D. J. 1990. Structural observations in the Gilbert River belt, Grenville Province, southeastern Labrador. *Current Research, Part C.* Geological Survey of Canada, Paper **90-1C**, 1–11.

HIGGINS, M. D. & VAN BREEMAN, O. 1992. The age of the Lac-Saint-Jean Anorthosite Complex and associated mafic rocks, Grenville Province, Canada. *Canadian Journal of Earth Sciences*, **29**, 1412–1423.

HILL, J. D. 1982. *Geology of the Flowers River–Notakwanon River area, Labrador.* Newfoundland Department of Mines and Energy, Mineral Development Division, Report **82-6**.

KAMO, S. L., GOWER, C. F. & KROGH, T. E. 1989. Birth date for the Iapetus Ocean? A precise U–Pb zircon and baddeleyite age for the Long Range dikes, southeast Labrador. *Geology*, **17**, 602–605.

——, WASTENEYS, H., GOWER, C. F. & KROGH, T. E. 1995. U–Pb geochronological constraints on Labradorian events in the eastern Grenville Province of Labrador. *GAC-MAC Program with Abstracts*, **20**, A-51.

KERR, A. & KROGH, T. E. 1990. The Trans-Labrador granitoid belt in the Makkovik Province: new geochronological and isotopic data and their geological implications. *Current Research.* Newfoundland Department of Mines and Energy, Report **90-1**, 237–249.

——, CORFU, F., SCHÄRER, U., GANDHI, S. S. & KWOK, Y. Y. 1992. Episodic Early Proterozoic granitoid plutonism in the Makkovik Province, Labrador: U–Pb geochronological data and geological implications. *Canadian Journal of Earth Sciences*, **29**, 1166–1179.

——, RYAN, B., GOWER, C. F. & WARDLE, R. J. 1996. The Makkovik Province: extension of the Ketilidian Mobile Belt in mainland North America. *This volume.*

——, ——, ——, —— & HALL, J. 1994. The Makkovik Province: geological overview, unanswered questions and unexplained reflectors. *In*: WARDLE, R. J. & HALL, J. (eds) *Eastern Canadian Shield Onshore–Offshore Transect (ECSOOT), Transect Meeting December 10–11, 1993.* Université du Québec à Montréal, Report **36**, 35–52.

KROGH, T. E. 1983. Report on U–Pb zircon geochronology, Labrador and Newfoundland. Unpublished report to the Newfoundland Department of Mines and Energy, Open File Lab 708.

—— 1994. Precise U–Pb ages for Grenville and Pre-Grenville thrusting of Proterozoic and Archean metamorphic assemblages on the Grenville front tectonic zone, Canada. *Tectonics*, **13**, 963–982.

LOVERIDGE, W. D. 1986. U–Pb ages on zircon from rocks of the Lac de Morhiban map area, Quebec. *Current Research, Part A.* Geological Survey of Canada, Paper **86-1A**, 523–530.

MARTIGNOLE, J., MACHADO, N. & INDARES, A. 1994. The Wakeham terrane: a Mesoproterozoic terrestrial rift in the eastern part of the Grenville Province. *Precambrian Research*, **68**, 291–306.

MARTIGNOLE, J., MACHADO, N. & NANTEL, S. 1993. Timing of intrusion and deformation of the Rivière–Pentecôte anorthosite (Grenville Province). *Journal of Geology*, **101**, 652–658.

MUIR, R. J., FITCHES, W. R. & MALTMAN, A. J. 1992. Rhinns complex: A missing link in the Proterozoic basement of the North Atlantic region. *Geology*, **20**, 1043–1046.

MURTHY, G., GOWER, C. F., TUBRETT, M. & PÄTZOLD, R. 1889. Paleomagnetism of pre-Grenvillian mafic intrusions from the Grenville Province, southeast Labrador. *Canadian Journal of Earth Sciences*, **26**, 2544–2455.

O'FLAHERTY, C. A. 1986. *The geology of the Layered mafic intrusion on Little Grady Island, south-eastern Labrador*. BSc thesis, Memorial University of Newfoundland, St John's, Newfoundland.

OWEN, J. V. 1985. *Tectono-metamorphic evolution of the Grenville front zone, Smokey archipelago, Labrador*. PhD thesis, Memorial University of Newfoundland, St John's, Newfoundland.

——, DALLMEYER, R. D., GOWER, C. F. & RIVERS, T. 1988. Metamorphic conditions and $^{40}Ar/^{39}Ar$ geochronologic contrasts across the Grenville front zone, coastal Labrador, Canada. *Lithos*, **21**, 13–35.

——, RIVERS, T. & GOWER, C. F. 1986. The Grenville front on the Labrador coast. *In*: MOORE, J. M., DAVIDSON, A. & BAER, A. J. (eds) *The Grenville Province*. Geological Association of Canada Special Paper **31**, 95–106.

PHILIPPE, S., WARDLE, R. J. & SCHÄRER, U. 1993. Labradorian and Grenvillian crustal evolution of the Goose Bay region, Labrador: new U–Pb geochronological constraints. *Canadian Journal of Earth Sciences*, **30**, 2315–2327.

PREVEC, S. A., MCNUTT, R. H. & DICKIN, A. P. 1990. Sr and Nd isotopic and petrological evidence for the age and origin of the White Bear Arm Complex and associated units from the Grenville Province in eastern Labrador. *In*: GOWER, C. F., RIVERS, T. & RYAN, B. (eds) *Mid-Proterozoic Laurentia–Baltica*. Geological Association of Canada Special Paper **38**, 65–78.

REYNOLDS, P. H. 1989. $^{40}Ar/^{39}Ar$ dating of the Mealy dykes of Labrador: paleomagnetic implications. *Canadian Journal of Earth Sciences*, **26**, 1567–1573.

RIVERS, T. & CHOWN, E. H. 1986. The Grenville Orogen in eastern Quebec and western Labrador – definition, identification and tectonometamorphic relationships of autochthonous, parautochthonous and allochthonous terranes. *In*: MOORE, J. M., DAVIDSON, A. & BAER, A. J. (eds) *The Grenville Province*. Geological Association of Canada Special Paper **31**, 31–50.

——, MARTIGNOLE, J., GOWER, C. F. & DACIDSON, A. 1989. New tectonic divisions of the Grenville Province, southeast Canadian Shield. *Tectonics*, **8**, 63–84.

ROMER, R. L., SCHÄRER, U., WARDLE, R. J. & WILTON, D. 1994. Middle Proterozoic intraplate rifting-related magmatism: baddeleyite and zircon U–Pb ages from the Seal Lake Group gabbros (abs.). *21st Nordic Geologic Winter Meeting*, Luleå, January, 1994.

SCHÄRER, U. 1991. Rapid continental crust formation at 1.7 Ga from a reservoir with chondritic isotope signatures, eastern Labrador. *Earth and Planetary Science Letters*, **102**, 110–133.

—— & GOWER, C. F. 1988. Crustal evolution in eastern Labrador; constraints from precise U–Pb ages. *Precambrian Research*, **38**, 405–421.

——, KROGH, T. E. & GOWER, C. F. 1986. Age and evolution of the Grenville Province in eastern Labrador from U–Pb systematics in accessory minerals. *Contributions to Mineralogy and Petrology*, **94**, 438–451.

SCOTT, D. J. & HYNES, A. 1994. U–Pb geochronology along the Manicouagan corridor, preliminary results: evidence for c. 1.47 Ga metamorphism. *Abitibi-Grenville Lithoprobe workshop report*.

——, MACHADO, N., HANMER, S. & GARIÉPY, C. 1993. Dating ductile deformation using U–Pb geochronology:examples form the Gilbert River Belt, Grenville Province, Labrador, Canada. *Canadian Journal of Earth Sciences*, **30**, 1458–1469

THOMAS, A., NUNN, G. A. G. & KROGH, T. E. 1986. The Labradorian Orogeny: evidence for a newly identified 1600 and 1700 Ma orogenic event in Grenville Province crystalline rocks from central Labrador. *In*: MOORE, J. M, DAVIDSON, A. & BAER, A. J. (eds) *The Grenville Province*. Geological Association of Canada Special Paper **31**, 175–189.

TUCKER, R. D. & GOWER, C. F. 1994. A U–Pb geochronological framework for the Pinware terrane, Grenville Province, southeast Labrador. *Journal of Geology*, **102**, 67–78.

VAN NOSTRAND, T. 1988. *Geothermometry–geobarometry and $^{40}Ar/^{39}Ar$ incremental release dating in the Sandwich Bay area, Grenville Province, eastern Labrador*. MSc thesis, Memorial University of Newfoundland, St John's.

—— 1992. *Geology of the Alexis River region, Grenville Province, southeastern Labrador*. Newfoundland Department of Mines and Energy, Report **92-3**.

——, DUNPHY, D. & EDDY, D. 1992. *Geology of the Alexis River map region, Grenville Province, southeastern Labrador*. Newfoundland Department of Mines and Energy, Report **92-1**, 399–412.

WARDLE, R. J., RYAN, B., PHILIPPE, S. & SCHÄRER, U. 1990. Proterozoic crustal development, Goose Bay region, Grenville Province, Labrador, Canada. *In*: GOWER, C. F., RIVERS, T. & RYAN, B. (eds) *Mid-Proterozoic Laurentia–Baltica*. Geological Association of Canada Special Paper **38**, 197–214.

WILTON, D. H. C. & LONGERICH, H. P. 1993. Metallogenic significance, of trace element and U–Pb isotope data for uraninite-rich mineral separates from the Labrador Central Mineral Belt. *Canadian Journal of Earth Sciences*, **30**, 2352–2365.

Accretion, rifting, rotation and collision in the North Atlantic supercontinent, 1700–950 Ma

IAN C. STARMER

*Department of Geological Sciences, University College London,
Gower Street, London WC1E 6BT, UK*

Abstract: Events in the North Atlantic region, from *c.* 1700 to 950 Ma, are unified in a model compatible with palaeomagnetic results and Neoproterozoic supercontinent reconstructions. Following Labradorian–Gothian accretion, largely between 1700 and 1550 Ma, post-collisional effects continued until *c.* 1500 Ma. Continental extension from *c.* 1500 Ma onwards, related to ocean opening, developed granite–rhyolite provinces: the Central Metasedimentary Belt (Grenville Province) started to form at *c.* 1425 Ma, with arcs outboard at *c.* 1350–1300 Ma. Baltica started to separate from Laurentia and rotate at *c.* 1240 Ma, causing crustal delamination of its leading edge, in South Norway. Here, subduction-related gabbroids intruded at *c.* 1230–1225 Ma, as rotation proceeded. The Elzevirian Orogeny occurred at *c.* 1190 Ma in the Southwest Grenville Province, marking the start of convergence of Baltica, in which backarc extension was accompanied by granites and bimodal volcanics. Collision at *c.* 1080 Ma caused the Ottawan Orogeny which thrust accreted terranes northwestwards and closed the Mid-Continental Rift. In southern Baltica, complex deformation occurred. Post-collisional convergence and sinistral slip of Baltica against Laurentia (possibly related to convergence of Amazonia) caused thrusting, sinistral strike-slip faulting, folding and granitic intrusion until *c.* 950 Ma. Post-tectonic granites intruded until *c.* 900 Ma.

Bridgwater & Windley (1973) recognized a continuity of Mid-Proterozoic (1800–1000 Ma) features across a reconstructed North Atlantic Shield, stretching from Laurentia (North America, Greenland, and the British Isles), through Baltica to eastern Europe. In recent years, knowledge of these regions has increased greatly and modern syntheses include that of Gower *et al.* (1990*a*). It is now widely accepted that a late Palaeoproterozoic supercontinent fragmented after *c.* 1500 Ma, in the Mesoproterozoic, and re-combined during worldwide, Grenvillian age orogenies) at *c.* 1100–1000 Ma) to form the first Neoproterozoic supercontinent (Bond *et al.* 1984; Hoffman 1991; Moores 1991; Dalziel 1992*a* & *b*).

Palaeomagnetic studies by Patchett *et al.* (1978) and Stearn & Piper (1984) suggested a clockwise rotation of at least 90°, of Baltica relative to Laurentia, between 1190 and 1000 Ma. Amongst the rocks used were 'hyperite' gabbroids from South Norway, then thought to have intruded at *c.* 1090–1010 Ma, but now believed to have been emplaced at *c.* 1230–1225 Ma. Recent studies by Elming *et al.* (1993) have shown a clockwise rotation of *c.* 60°, between 1250 and 1050 Ma, after separation from Laurentia at *c.* 1250 Ma.

In South Norway, Starmer (1993) showed that tectonism, consistent with clockwise rotation, separated several terranes by a crustal delamination due to oblique thrusting on listric, east-dipping planes. Granites which divided two pulses of this tectonism were probably emplaced *c.* 1240 Ma, but their intrusion ages were poorly-constrained and have been described as 'after ca. 1250 Ma' (Starmer 1990) or 'ca. 1250 Ma' (Starmer 1993). However, the granites and the tectonic fabrics of both pulses were cut by gabbroids intruded at *c.* 1230–1225 Ma.

The present study reviews the evolution of Baltica from *c.* 1700 to 950 Ma, relative to general events in Laurentia. It attempts to correlate contemporary events in a number of stages (I–IX), each illustrated by a figure, and to link them to possible plate-tectonic scenarios. The latter are necessarily idealized, particularly with respect to the oceanic systems. Movements of Baltica are depicted relative to a constant position for Laurentia, which was also moving during this period.

Earlier models have proposed separation of Baltica from Laurentia, rotation and later collision of the Scandinavian Sveconorwegian belt with the Grenville Province at *c.* 1100–950 Ma (Gower 1985, 1992; Park *et al.* 1991). Johansson *et al.* (1993) attributed the Sveconorwegian Orogeny in Sweden to collision with North America between *c.* 1100 and 900 Ma. Another model was proposed by Park (1992) in which Baltica detached and rotated *c.* 80° clockwise between *c.* 1200 and 1050 Ma, when it collided with South America, that had docked earlier with Laurentia at

c. 1100 Ma, forming the Grenville Belt. Baltica then moved along the margin of South America to collide with the British Isles and East Greenland at c. 1000 Ma. It underwent sinistral transform motion from c. 950 to 750 Ma, enabling it to reach the position off East Greenland shown on the Neoproterozoic supercontinent reconstruction of Bond *et al.* (1984) for c. 600 Ma.

The model suggested in the present study is similar to the earlier proposals of Gower (1985, 1992) and Park *et al.* (1991), involving collision of western Baltica with the Grenville Province of Laurentia. It is shown to be compatible with the magmatism, tectonism and subduction patterns recognized throughout Baltica and Laurentia during the Mesoproterozoic.

The main areas

By c. 1800 Ma, the supercontinent in the North Atlantic provinces (Fig. 2) consisted of an Archaean–Early Proterozoic craton bordered by Penokean–Makkovikian–Ketilidian–Svecofennian accreted terranes. By c. 1750 Ma, new Labradorian–Gothian terranes were forming and they were largely accreted to the supercontinent between 1700 and 1550 Ma and stabilized by c. 1500 Ma. The newly-accreted terranes then

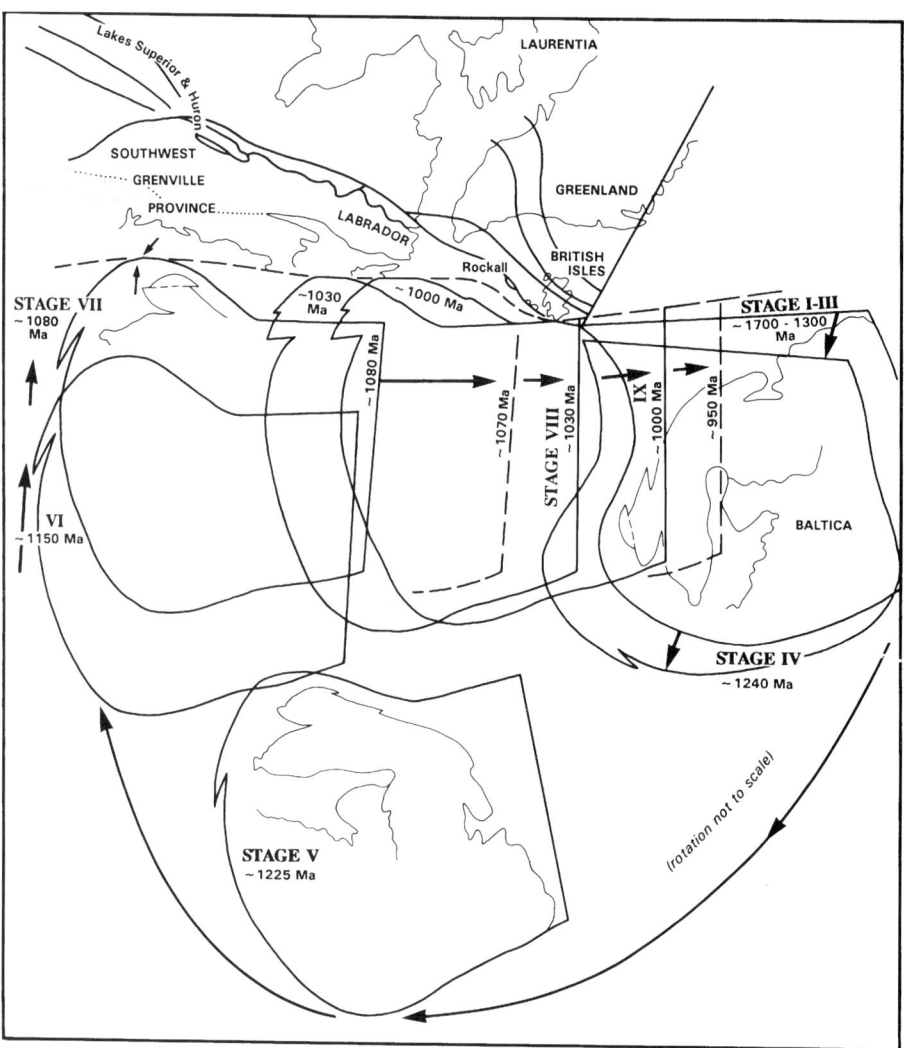

Fig. 1. The main regions and the motion of Baltica relative to Laurentia, 1700–950 Ma.

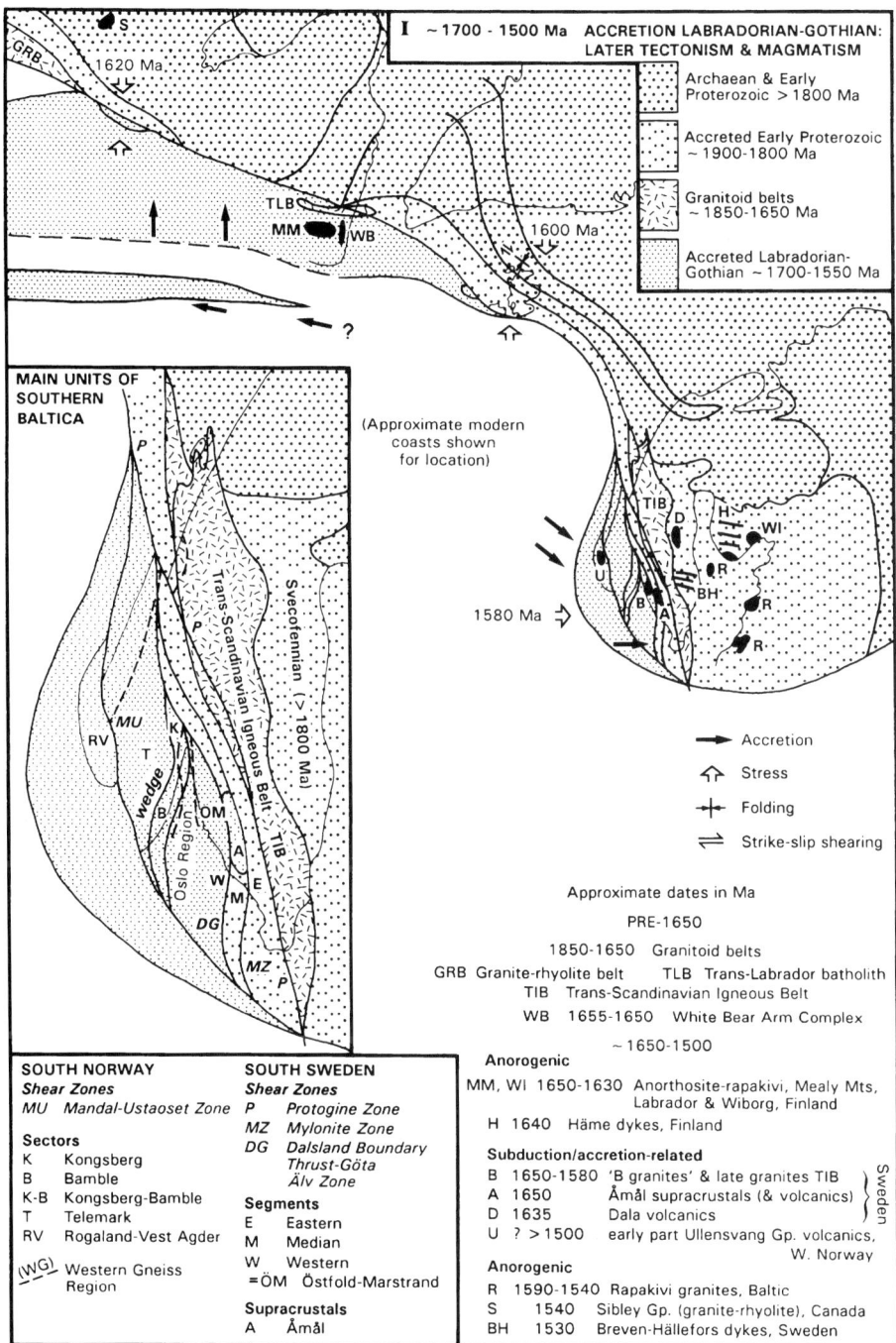

Fig. 2. Stage I: *c.* 1700–1500 Ma. Labradorian–Gothian accretion: later tectonism and magmatism.

underwent a different evolution in each of three sectors, namely Baltica, the British Isles–South Greenland–Labrador, and the Southwest Grenville Province outboard of the Great Lakes (Fig. 1).

Baltica

The eastern part of Baltica is represented by the East European craton, now largely covered by later rocks. Recent work has established the existence of Svecofennian terranes from the Baltic through Estonia, Belorussia and Poland to the western Ukraine (Puura & Huhma 1993; Gorbatschev & Bogdanova 1993; Bogdanova et al. 1994). Preliminary palaeomagnetic studies by Elming et al. (1993) have suggested that the Ukrainian Shield may have separated from Fennoscandia at c. 1300 Ma.

Central and southeastern Sweden consists of Svecofennian crust, formed before c. 1800 Ma and intruded by the Trans-Scandinavian Igneous Belt (TIB) in the west (Fig. 2). The TIB is a large calc-alkaline granitoid belt, emplaced in three stages, termed TIB1 (1810–1770 Ma), TIB2 (c. 1700 Ma), and TIB3 (1680–1650 Ma) by Larson & Berglund (1992). It intruded Svecofennian gneisses to the east and to the west where it became foliated and invaded by Gothian plutons. Near the western margin of the TIB, and partly along it, the 'Protogine Zone' (P. Fig. 2) was a continually reactivated shear zone.

West of the Protogine Zone, southwestern Sweden comprises Gothian terranes, formed from c. 1760 Ma onwards and accreted between 1700 and 1600 Ma. These terranes were intensively intruded by plutons (particularly granitoids), but probably were originally underlain by Svecofennian crust. This region is divided into Eastern, Median and Western Segments by two other c. N–S shear belts, namely the Mylonite Zone and the Dalsland Boundary Thrust–Göta Älv Zone (Fig. 2). The Mylonite Zone is now a west-dipping shear zone which cuts an earlier east-dipping shear zone further north. The Dalsland Boundary Thrust is another west-dipping shear zone which is replaced southwards by a number of shears, including the Göta Älv Zone.

The Eastern Segment comprises deformed granitoid gneisses with some supracrustals and amphibolites. In the Median Segment, these granitoids are overlain by the Åmål-Horred Belt supracrustals, deposited c. 1640–1610 Ma (Åhäll et al. 1995) and the later Dal Group supracrustals (possibly deposited c. 1150 Ma). The 'Western' or 'Östfold-Marstrand' Segment contains the Stora Le-Marstrand Formation, representing island-arc rocks formed c. 1760 Ma (Åhäll & Daly 1989) and intruded by Gothian and Sveconorwegian plutons.

Four periods of magmatism have been separated and termed 'Groups A, B, C, and D' (Samuelsson & Åhäll 1985; Åhäll et al. 1990). Calc-alkaline 'Group A' granites (intruded c. 1720–1700 Ma) and 'Group B' granites (c. 1680–1580 Ma) both underwent Gothian metamorphism and deformation. 'C1' alkaline diorite–trondjhemite bodies intruded at c. 1570–1530 Ma, in the closing stages of the Gothian Orogeny, and were followed by anorogenic 'C2' basic–intermediate–acid magmatism (c. 1508–1416 Ma). The 'C3' granites intruded at c. 1220 Ma and underwent the Sveconorwegian Orogeny, starting at c. 1090 Ma. Post-orogenic 'Group D' pegmatites and granites were intruded c. 890 Ma.

The Phanerozoic Oslo Region Rocks, around the Oslo–Skagerrak Graben, now separate the Östfold–Marstrand Segment from the Proterozoic in South Norway, where terrane boundaries were accentuated by the major crustal delamination at c. 1240 Ma (mentioned previously). In all terranes, the oldest exposed rocks are supracrustals with a maximum age of c. 1700 Ma (Falkum 1985; Starmer 1985). They show effects of both the Gothian and Sveconorwegian orogenies, including widespread plutonism, and now comprise variable (pelitic and semi-pelitic) gneisses with subordinate quartzites and carbonates. Some gneisses are probably andesitic–dacitic metavolcanics, which Jacobsen & Heier (1978) related to island-arc regimes.

Supracrustals in the basement of the separate terranes (the Kongsberg–Bamble Sector, the Telemark Sector, and the Rogaland–Vest Agder Sector of Fig. 2) show minor differences, but the Kongsberg–Bamble Sector is unique in also containing major units of massive quartzite and sillimanite gneiss, which have been thrust NW into the variable gneisses. The N–S Kongsberg Sector and the NE–SW Bamble Sector then formed a single curved belt, but are now separated by the Phanerozoic Oslo Region Rocks.

The Telemark Sector basement to the west was overlain by the Telemark Supergroup, comprising the Rjukan Group rhyolites and basalts (extruded c. 1500 Ma), the Seljord Group quartzites, and the Bandak Group rhyolites, basalts and sediments (deposited c. 1150 Ma).

The Rogaland–Vest Agder basement, west of the Mandal–Ustaoset Shear Zone, was also overlain by the Bandak Group and was intruded in the extreme southwest by the Rogaland anorthosite province, between c. 1050 and 935 Ma (Demaiffe & Michot 1985; Menuge 1988).

The Western Gneiss Region (WG, Fig. 2) is now separated from the Proterozoic of South Norway and South Sweden by the Caledonian belt, but Tucker et al. (1990) suggested that it contained a continuation of these terranes. The northeastern part had juvenile Gothian crust, but to the southwest (of a NNW–SSE boundary compared to the Protogine Zone), heterogeneous rocks had undergone Sveconorwegian metamorphism and granitic intrusion.

The British Isles and Greenland

In Scotland, the Archaean Lewisian underwent Early Proterozoic Inverian and Laxfordian deformations before c. 1870 Ma, with Late Laxfordian deformations at c. 1600 Ma (Coward & Park 1987). From Colonsay and Islay in western Scotland to Malin Head at the northern tip of Ireland, the Rhinns Complex of syenites and gabbros represented juvenile crust formed c. 1800 Ma (Marcantonio et al. 1988; Daly et al. 1991; Muir et al. 1992). On the northwest coast of Ireland, crust in the Annagh Division formed c. 1900 Ma (Menuge & Daly 1990). To the northwest, juvenile Labradorian crust formed in the Rockall Bank (Fig. 1) at c. 1670 Ma (Roberts et al. 1973).

Greenland consisted of Archaean and Early Proterozoic belts, including the accreted Ketilidian terrane at its southern tip, which located the Mesoproterozoic Gardar Province. An Early Gardar event (c. 1330–1290 Ma), a Mid Gardar event (c. 1280–1220 Ma) and a Late Gardar event (c. 1175–1120 Ma) all involved rifting and alkaline intrusive activity (Upton & Emeleus 1987). Central East Greenland was affected by Grenvillian metamorphism between 1250 and 1000 Ma (Hansen et al. 1981; Rex & Gledhill 1981), preserved within and just west of the Caledonian belt (see Fig. 10). In contrast, in northeast Greenland, Archaean rocks were reworked in an Early Proterozoic mobile belt, but there were no Grenvillian effects (Tucker et al. 1993). Similarly, along the northern coast, there was no Grenvillian-age metamorphism (Surlyk 1991).

North America

Granitoid belts developed around the junction of the craton and the accreted Labradorian terranes, in a position similar to that of the TIB in Baltica, and with a similar range of ages (c. 1800–1650 Ma). The Trans-Labrador Batholith (TLB), like the TIB, probably contained intrusions of 1800, 1720 and 1650 Ma age (Kerr & Fryer 1990). The granite–rhyolite (–basalt) belt (GRB) was emplaced at c. 1770–1740 Ma (Greenberg 1990) and extended from the west of the present area through Wisconsin. Smaller provinces also developed, like the granitoid intrusions at c. 1750 Ma in the Killarney Igneous Complex (Clifford 1990), just east of the GRB, in the Early Proterozoic (Huronian) margin of the craton.

The Archaean and Early Proterozoic were bordered by the Early Proterozoic terranes of the Makkovikian in Labrador and the Penokean in the Great Lakes region. The Labradorian terranes were accreted to this craton between 1710 and 1620 Ma (Gower et al. 1990a). In Labrador, these terranes underwent mainly extensional tectonism and anorogenic magmatism before being reworked during the Grenvillian Orogeny.

In contrast, to the southwest in the Southwest Grenville Province (Fig. 1), new, subduction-related terranes were accreted to the Labradorian crust before and during the Grenvillian Orogeny. This region was divided by Wynne-Edwards (1972) into a 'Central Gneiss Belt' in the northwest, a 'Central Metasedimentary Belt' (CMB), and a 'Central Granulite Terrane' in the southeast.

The (inboard) Central Gneiss Belt comprises imbricated Labradorian domains. The CMB (previously termed the 'Grenville Supergroup') was sub-divided by Davidson (1986) into the Bancroft, Elzevir, and Frontenac terranes. The Bancroft terrane and the Frontenac (–Adirondack Lowlands–Mont Laurier) terrane contain shallow marine sediments deposited c. 1425–1350 Ma. These were metamorphosed at mid amphibolite to granulite facies grade, whereas the Elzevir terrane contains bimodal volcanics with some shallow marine sediments, formed c. 1300–1240 Ma and metamorphosed at amphibolite and greenschist facies grade (Lumbers et al. 1990). Outboard of these, the 'Central Granulite Belt', or the Adirondack Highlands (–Morin) terrane, contains some metasediments, but seems to have been dominantly magmatic-arc tonalities formed c. 1350–1300 Ma (McLelland & Chiarenzelli 1990; Daly & McLelland 1991). These later terranes are shown on Figs 4–9. The small Sharbot Lake terrane has been recognized recently, between the Elzevir and Frontenac terranes (Easton 1992).

The Labradorian terranes of Labrador and the South West Grenville Province are now bounded inboard, against the craton, by the Grenville Front tectonic zone (see Fig. 8). To the southwest, the Grenville Front turns due south and then southwest around the Eastern Granite–Rhyolite Province (the St Francois terrane of Fig. 3). It continues southwestwards (outside the area of the figures), through midcontinental and southwestern USA, along the margins of the Western Granite–Rhyolite Province (the Spavinaw terrane mentioned on Fig. 3) and the Mazatzal Orogen (of Labradorian age). South of the front, the Grenvillian Llano Province extends through Texas into northern Mexico.

Stage I (*c.* 1700–1500 Ma): Labradorian–Gothian accretion, with later tectonism and magmatism (Fig. 2)

Stage I in Baltica

East of the Trans-Scandinavian Igneous Belt (TIB), anorogenic regimes dominated in the Svecofennian craton, but in the TIB and to the west, effects reflected accretion.

Within the Svecofennian craton, anorthosite-rapakivi granite complexes (e.g. the Wiborg Complex) and the associated Häme dykes were emplaced between 1650 and 1630 Ma in Finland (Rämo & Haapala 1990). In Sweden, the Sub-Jotnian Dala volcanics were emplaced at *c.* 1635 Ma (Welin & Lundqvist 1970) and comprised felsic volcanics with some basalts and granitoids. Later rapakivi complexes were emplaced at *c.* 1590–1540 Ma around the Baltic Sea and, at *c.* 1530 Ma, the NW–SE Breven-Hällefors dyke swarm intruded southeast Sweden (Gower *et al.* 1990a).

The TIB (with TIB1 and TIB2 granitoids emplaced at 1810–1770 and *c.* 1700 Ma) was intruded by TIB3 granites at *c.* 1680–1650 Ma (Larson & Berglund 1992). This latest phase (TIB3) post-dated the second ('D2') deformation of eastward thrusting around the Protogine Zone, at the western margin of the TIB, and a later ('D3') deformation, produced E–W axis folds, probably before 1500 Ma (Larson *et al.* 1990).

West of the TIB and the Protogine Zone, there must have been Svecofennian basement at this stage. Although Johannsson *et al.* (1993) recorded accreted Gothian terranes, formed between 1650 and 1610 Ma, next to the Protogine Zone at the southern tip of Sweden, they suggested that these were juxtaposed during late Sveconorwegian thrusting, just before 900 Ma.

To the west, the Median Segment basement was overlain by the Åmål-Horred Belt supracrustals, dominated by volcanics and formed *c.* 1640–1610 Ma (Åhäll *et al.* 1995). Further west, in the 'Western', or 'Östfold–Marstrand' Segment, the Stora Le–Marstrand Formation (of metamorphosed greywackes and basic volcanics) was thought to have formed as an arc at *c.* 1760 Ma (Åhäll & Daly 1989).

Within the Western and Median Segments, calc-alkaline 'Group B' granites (of Åhäll *et al.* 1990) intruded between 1680 and 1580 Ma, and in the closing stages of the Gothian Orogeny, 'C1' diorite–trondhjemite bodies intruded between 1570 and 1530 Ma.

In the Kongsberg–Bamble Sector of South Norway (KB, Fig. 2), there were similar intrusions of granite–charnockite at *c.* 1600–1580 Ma and gabbro–diorite–tonalite–trondhjemite before *c.* 1535 Ma (Starmer 1990). Here, the earliest deformation ('D1') occurred between 1650 and 1600 Ma, before the granite–charnockite intrusions: it resulted from quartzite–sillimanite gneiss units being thrust NW into variable gneisses which continued northwest in the Telemark Sector. The quartzites developed NW vergent recumbent folds above a mylonitized basal shear and beneath a roof decollement in sillimanitic rocks, in which shearing produced quartz–sillimanite lensoids and rods (Starmer 1990). The ductile basal shear dissipated towards the ends of the belt, in the north of the Kongsberg Sector and the southwest of the Bamble Sector, and these structures reflected a NW-directed tectonic wedge.

The Kongsberg–Bamble Sector had a bend of *c.* 40° after the 'D1' deformation, but the resulting structures suggested that the original basin may have had some curvature (Starmer 1991). The combination of the increased curvature and the tectonic wedge suggest that 'D1' resulted from oblique lateral terrane assembly from the northwest (as shown on Fig. 2). Later deformation ('D2') refolded the F1 recumbent folds and caused thrusting of the NE–SW Bamble Sector to the northwest and of the N–S Kongsberg Sector to the southwest. This tightening of the whole segment around the Telemark Sector occurred after the granite–charnockite intrusions at 1600–1580 Ma, but before *c.* 1535 Ma, and was possibly related to post-collisional convergence. A later deformation ('D3'), between 1535 and 1500 Ma, caused WSW thrusting of the N–S Kongsberg Sector and dextral strike-slip of the NE–SW Bamble Sector against the bounding Telemark Sector: it was related to an ENE–WSW compression, radial to the supercontinent margin.

In the Telemark and the Rogaland–Vest Agder sectors, Falkum (1985) recorded migmatitic banded gneisses yielding dates 'around 1650 Ma'.

The northern part of the Western Gneiss Region (WG, Fig. 2) contains juvenile tonalitic–granitic gneisses formed *c.* 1690–1650 Ma and correlated with the TIB (Tucker *et al.* 1990). At the western margin of the Caledonides at Hardanger, the lowermost, Kinsarvik formation of the Ullensvang Group comprised (basalt–) andesite–dacite and possibly represented a continental volcanic arc (Torske 1985; Torske *et al.* 1988): it formed before bimodal basalt–rhyolite volcanism which was correlated with the (*c.* 1500 Ma) Rjukan Group of Telemark. Within the Caledonian belt, in the Bergen Arcs, acidic gneisses were intruded (at *c.* 1750 Ma) by jotunite–mangerite suites (Emmett 1994).

Stage I in the British Isles and Greenland

In Britain, there is no evidence of Labradorian age crust, nor of subduction and collisional sutures: the tectonism appears to have been intraplate. In northwest Scotland, Late Laxfordian deformation, at c. 1600 Ma, produced NW–SE dextral shear zones and NW–SE axis upright folds. This has been related to a dextral transpression caused by a N–S compression of the NW–SE Nagssugtoqidian–Lewisian belt (Coward & Park 1987).

In contrast, in northwest Ireland, Daly et al. (1994) reported crust formed c. 1750 Ma and further west, juvenile crust formed in the Rockall Bank at c. 1670 Ma (Roberts et al. 1973), and was deformed and metamorphosed between c. 1670 and 1566 Ma (Morton & Taylor 1991).

Stage I in North America

In Labrador, the Labradorian terranes were accreted at c. 1710–1620 Ma (Wardle et al. 1990). In the White Bear Arm Complex, Prevec et al. (1990) recorded high-grade metamorphism with gabbro-norite and then granitoid intrusions at c. 1655–1650 Ma. The Trans-Labrador Batholith (TLB), containing earlier intrusions, was invaded by granitoids at c. 1650 Ma (Kerr & Fryer 1990) and by bimodal suites in the Makkovik Province (Kerr & Fryer 1994). The Mealy Mountains Intrusive Suite of anorthosite and monzonite was emplaced c. 1640 Ma (Emslie & Hunt 1990).

In the Southwest Grenville Province, Labradorian terranes formed the Central Gneiss Belt, possibly overlying some Penokean basement (Dickin & McNutt 1990). Metamorphism and deformation occurred before c. 1685 Ma (Corrigan 1990). Further outboard, the later Frontenac terrane had Nd model ages (Marcantonio et al. 1990), suggesting it may have had a Penokean basedment, unlike the rest of the Central Metasedimentary Belt inboard. Dickin & McNutt (1990) suggested that this might indicate that the Frontenac basement was a displaced terrane.

Much later, at c. 1540 Ma, the Sibley Group granite–rhyolite suite formed in the Archaean craton, just north of Lake Superior.

Stage I: Summary of events across the Supercontinent

In the early part of this period, the Trans-Labrador Batholith and the Trans-Scandinavian Igneous Belt, which were existing granitoid belts parallel to the craton margin, were intruded again by granitoids, at c. 1680–1650 Ma.

At c. 1650–1630 Ma, anorogenic anorthosite-rapakivi complexes intruded the Svecofennian craton and the Mealy Mountains anorthosite–monzonite suite intruded in Labrador, almost contemporaneously with the bimodal magmatism just to the north in the Makkovik Province.

The Labradorian terranes were formed and accreted to North America between c. 1710 and 1620 Ma. In the Southwest Grenville Province, there may have been some lateral displacement of the outer terranes, but inboard, NW-directed compressions were recorded at c. 1620 Ma, in the margin of the Archaean craton (Clifford 1990) and in the adjacent Central Gneiss Belt (Dickin & McNutt 1990). To the west, the coeval Mazatzal Orogen formed in southwestern USA.

In Baltica, the Stora Le–Marstrand formation had developed c. 1760 Ma as an island arc. Later subduction eastwards under the Svecofennian craton probably produced the TIB2 granites (c. 1700 Ma) and the 'Group A' calc-alkaline magmatism (c. 1720–1700 Ma).

Outboard, the earliest crust yet recognized in southern Norway, was formed between c. 1700 and 1650 Ma. In the extreme west, in the Rogaland–Vest Agder Sector (RV), Ragnhildstveit et al. (1994) recorded inherited zircons with discordant $^{207}Pb/^{206}Pb$ ages between 2503 and 1658 Ma, suggesting the possibility of older crust beneath. The later (c. 1240 Ma) crustal delamination of South Norway separated terranes with boundaries inherited from the initial Gothian assembly.

Inboard of the accreting terranes, in the Protogine Zone at the western margin of the TIB, Larson et al. (1990) recognized 'D2' eastward thrusting after the TIB2 intrusions (c. 1700 Ma) and before the TIB3 intrusions (c. 1680–1650 Ma) which were believed to be intruded during post-D2 N–S tension. Early 'Group B' calc-alkaline granites were thought to have intruded the adjacent accreting terranes from c. 1680 Ma onwards (Åhäll et al. 1990).

Subduction generated the Åmål–Horred volcanics at c. 1640–1610 Ma, near the Svecofennian continental margin (Åhäll et al. 1995) and possibly the continental Dala volcanics further north at c. 1635 Ma. The early (pre-1500 Ma) calc-alkaline andesite–dacite volcanics of the Ullensvang Group in West Norway were probably generated in a continental volcanic arc (Torske et al. 1988).

The collision of terranes is reflected by the Kongsberg–Bamble Sector being driven into the Telemark Sector as a 'D1' tectonic wedge before granite–charnockite bodies intruded between 1600 and 1580 Ma (Starmer 1990), coeval with the main 'Group B' magmatism in Sweden. To the east in

Östfold, recumbent folds with N–S axes, recorded by Graversen (1984), may have been contemporaneous with the 'D1' deformation. The tectonism in the Kongsberg–Bamble Sector suggests that it was involved in oblique terrane collision, from the present northwest. This would have facilitated trapping of the older Östfold–Marstrand arc.

Later, between 1600 and 1535 Ma, the 'D2' deformation refolded the tectonic wedge structures and caused some increased curvature of the Kongsberg–Bamble Sector. It was possibly synchronous with the deformation of the Stora Le–Marstrand Formation at c. 1580 Ma, recorded by Åhäll & Daly (1989), and the NNE–SSW folding recorded just to the north in Östfold by Graversen (1984). These structures probably resulted from post-collisional convergence. Far within the craton, rapakivi complexes were intruded at 1590–1540 Ma.

In the closing stages of the Labradorian–Gothian Orogeny, between 1570 and 1530 Ma, (gabbro)–diorite–trondhjemite ('C1') magmatism occurred in SW Sweden and directly to the north in the Kongsberg Sector, with basic sheets intruded in the Bamble Sector. Anatectic red granites intruded the extreme south of Sweden at c. 1550 Ma (Johansson et al. 1993) and were thought to have followed the final compressional thickening. Deformations and high grade metamorphism occurred in South Norway at c. 1535–1530 Ma. In the craton, the NW–SE Breven–Hällefors dyke swarm intruded SE Sweden at c. 1530 Ma.

After the Gothian Orogeny, anorogenic 'C2' magmas intruded SW Sweden from 1508 onwards (probably until 1416 Ma). South Norway was affected by a 'D3' compression radial to the supercontinent margin: this may have caused the NW–SE axis folding at c. 1500 Ma in Östfold, recorded by Graversen (1984).

In summary, accretion in Laurentia and Baltica was probably diachronous between c. 1710 and 1600 Ma. Inboard terranes may have stabilized first (e.g. around the Mealy Mountains Intrusive Suite). Outer terranes (e..g. Frontenac and South Norway) show some evidence of lateral terrane displacement. The curvature of the supercontinent margin and the obliquity of North America and Baltica to the spreading directions (whatever their precise orientation) made these regions susceptible to lateral terrane displacements. In Britain, the lack of Labradorian crust may have resulted from its orientation relative to the (generalized) spreading system: this region was susceptible to strike-slip removal of slivers, which may have formed older basement beneath the outboard terranes in South Norway. Compressional effects radial to the supercontinent margin occurred between c. 1620 and 1580 Ma (in Laurentia, Scotland and southern Baltica) and later deformations until c. 1500 Ma may have resulted from post-collisional convergence.

Stage II (c. 1500–1350 Ma); continental extension (Fig. 3)

In southwest Sweden, from c. 1508 Ma onwards (and possibly until c. 1416 Ma), anorogenic 'C2' tonalite–granite, gabbro-dolerite, bimodal augen granite–dolerite and discrete augen granites were intruded (Åhäll et al. 1990). Gabbroic rocks of the Värmland Hyperite Suite intruded the Eastern Segment at c. 1500 Ma (Zeck & Willadsen, 1990). The Varberg charnockite was probably emplaced at c. 1420 Ma (Welin & Gorbatschev 1978), as a granite which was metamorphosed at granulite facies grade at c. 900 Ma (Johansson et al. 1993) and just to the east in southeast Sweden, the Karlshamn granite intrusions at c. 1400 Ma (Åberg 1988) may have been related. The Askim Granite intruded southwest Sweden at c. 1360 Ma (Welin & Samuelsson 1987). In the Oslofjord region, between Sweden and Norway, the Kattsund–Koster dykes intruded at c. 1420 Ma (Hageskov 1987), or possibly at c. 1460 Ma, coeval with the Orust dykes just to the south (Åhäll et al. 1994).

In South Norway, anorogenic magmatism occurred in the Kongsberg–Bamble Sector between c. 1500 and 1340 Ma (Starmer 1993). Between 1500 and 1395 Ma, 'early hyperite' alkaline gabbroids and diorites intruded. The were accompanied, in the Bamble Sector, by complexes of alkali- and iron-enriched, basic–intermediate–acid suites (showing anorthositic affinities) with associated granitic and charnockitic augen gneisses (Milne & Starmer 1982). They intruded as subconcordant (NE–SW) bodies in the Bamble Sector, which was undergoing NE–SW dextral transtension (Starmer 1990): their restriction to a NW–SE zone suggested possible links with the 'C2' magmatism in South Sweden. Like the Varberg intrusion, one augen gneiss body in Bamble seems to have acquired its granulite facies mineralogy later, at c. 1150 Ma (Hagelia 1984). The precise age of these complexes is not known, but they were emplaced during this anorogenic period: the present author considers that deformational fabrics within them are coeval with structures elsewhere, which are cut by granites intruded between 1395 and 1340 Ma.

To the northwest in the Telemark Sector, the Rjukan Group rhyolites and later basalts were extruded at c. 1500 Ma (Dahlgren et al. 1990) on continental crust, which was further extended

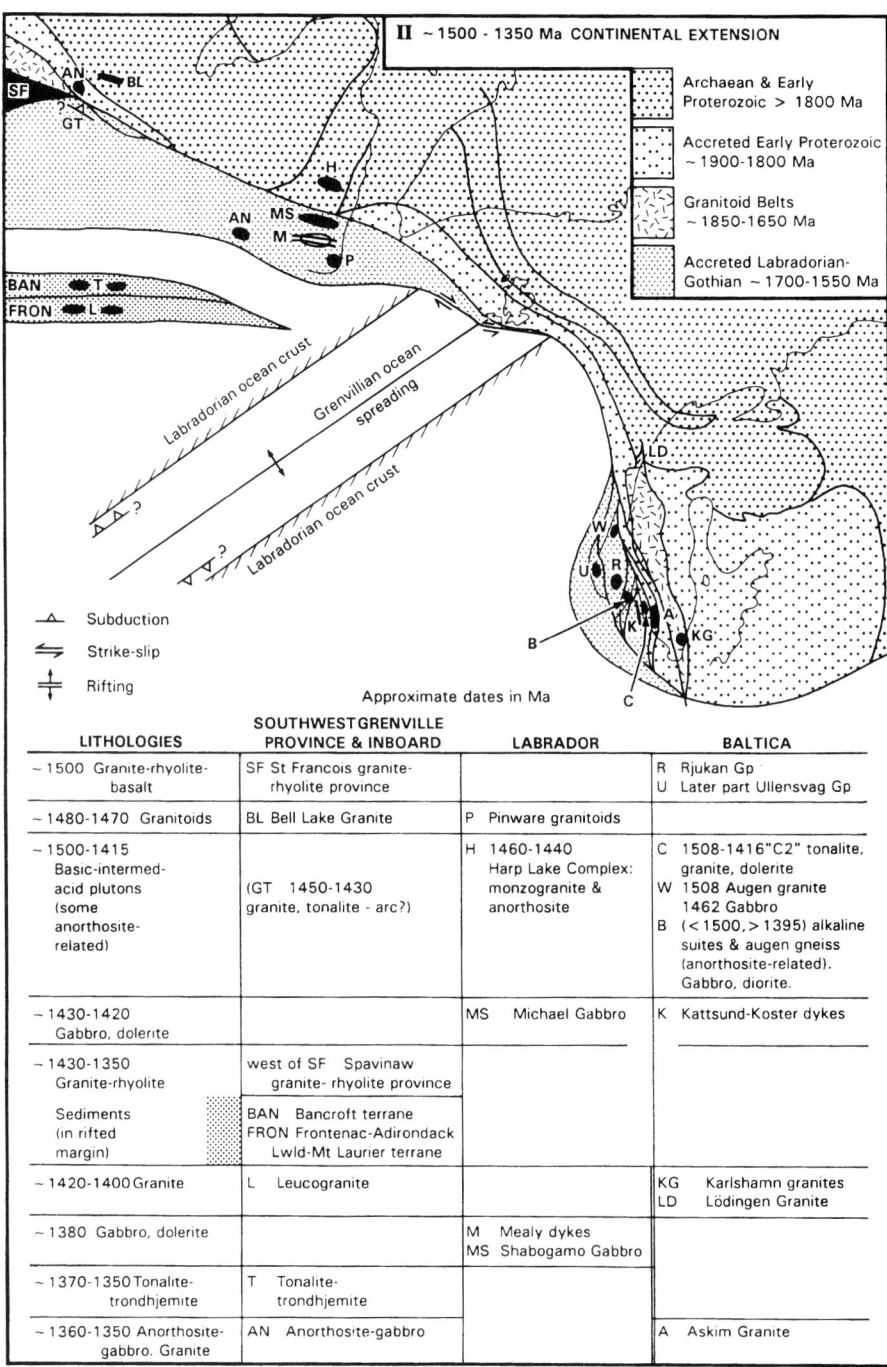

Fig. 3. Stage II: c. 1500–1350 Ma. Continental extension.

during the subsequent deposition of the Seljord Group quartzites. In West Norway, the later part of the Ullensvang Group, the bimodal rhyolite-basalt Jåstad Formation, was also extruded at this time (Torske et al. 1988).

In the Western Gneiss Region, augen granites intruded at c. 1508 Ma and gabbros at c. 1462 Ma (Tucker et al. 1990). Just to the southeast, basic magmas intruded at c. 1400 Ma into basement rocks which are now within the Caledonian belt (Emmett 1994). To the north, the Lödingen granite intruded in the Lofoten–Vesteralen islands at c. 1400 Ma (Griffin et al. 1978).

In North Greenland (north of the present area), the Independence Fjord Group of sandstones and red siltstones, developed in an intra-cratonic sag at c. 1380 Ma (Sønderholm & Jepsen 1991).

In Labrador, magmatism included the intrusion of granitoids in the Pinware Terrane at c. 1480 Ma (Gower 1992), the Harp Lake Complex of monzogranites and anorthosites at c. 1460–1440 Ma, the Michael gabbro at c. 1426 Ma, the Shabogamo Gabbro at c. 1379 Ma, and the doleritic Mealy Dykes at c. 1380 Ma (Gower et al. 1990b).

West of the Southwest Grenville Province, within the margin of the craton, granite–rhyolite provinces developed at c. 1510–1430 Ma in the St Francois terrane (Kisvarsanyi & Kisvarsanyi 1990) and further west at c. 1430–1350 Ma in the Spavinaw terrane (Sims 1990).

In the Central Gneiss Belt bordering the craton, inboard of the Central Metasedimentary Belt, there was high grade metamorphism and widespread granite-tonalite plutonism at c. 1450–1430 Ma (Ketchum et al. 1994). Although the widely accepted view was that this reflected anorogenic extension, these authors preferred the suggestion of Dickin & Higgins (1992) that it represented accretion of island arcs to the continent between 1500 and 1400 Ma. Just inboard, the anorogenic Bell Lake granite intruded at c. 1470 Ma (Clifford 1990) into the earlier Killarney Igneous Complex.

Outboard, the Central Metasedimentary Belt started to form between c. 1425 and 1350 Ma, with the sedimentation of the Bancroft terrane and the Frontenac (–Adirondack Lowlands–Mont Laurier) terrane. The rocks included quartzite, sandstone, arkose, siltstone and some dolomite and limestone (Lumbers et al. 1990). Leucogranites intruded the Frontenac–Adirondack Lowlands terrane at c. 1420–1415 Ma (McLelland & Chiarenzelli 1990) and a trondjhemite–granite–tonalite suite intruded the Bancroft terrane at c. 1370–1350 Ma (Lumbers et al. 1990). Inboard, some anorthosites were intruded at c. 1360–1350 Ma in the Superior region and in Quebec (Emslie & Hunt 1990).

Stage II: summary of events

Stage II, from c. 1500 to 1350 Ma, was an extensional period of high heat flows, with basic and acid magmatism throughout the margins of the supercontinent. In South Norway, upper amphibolite and granulite facies conditions prevailed throughout this period (Starmer 1990) and Kullerud & Machado (1991) recorded the growth of metamorphic zircons in charnockitic gneiss at c. 1437 Ma. Around 1400 Ma, K–Ar ages were reset in southeast Sweden (Åberg 1978) and U–Pb ages were disturbed in southwest Sweden (Hansen et al. 1989).

The initial continental extension at c. 1500 Ma was reflected by the granite–rhyolite (–basalt) provinces of the St Francois terrane in Laurentia and the Rjukan Group and the Jåstad Formation in Baltica. These magmas were probably generated by melting of thickened crust.

In Laurentia, Dickin & McNutt (1990) suggested the possibility of an Andean type, arc–backarc association developed in the Central Gneiss Belt of the Southwest Grenville Province between c. 1550 and 1350 Ma. If present, it could have resulted from closure of incomplete Labradorian suturing, or from the rifting discussed below.

Outboard, the Bancroft and Frontenac terranes of the Southwest Grenville Province formed at c. 1425–1350 Ma in a rift environment (Lumbers et al. 1990) and possibly in an ensialic rift (McLelland & Chiarenzelli 1990). Extension of the continental margin was also indicated, in the craton to the west, by the coeval production of the Spavinaw granite–rhyolite terrane. McLelland & Chiarenzelli (1990) suggested that the 1480–1340 Ma anorogenic magmatism of midcontinental USA extended into the Grenville Province, causing rifting and fragmentation of the continental margin, with related leucogranite intrusions in the Frontenac terrane (c. 1420–1415 Ma) and anorthosites in Labrador (c. 1460–1440 Ma). However, Nd model ages in the Frontenac terrane (Marcantonio et al. 1990) suggested the possibility of older (Penokean) basement and Dickin & McNutt (1990) therefore proposed that it might be a displaced terrane, already forming an outboard microcontinent at this time.

Whilst the supercontinent was undergoing extension, there was a protracted period of basic–intermediate–acid intrusive activity, in some cases intimately mixed, and in some cases anorthosite-related. The lack of compressional effects at the continental margin suggest that the new ocean crust may have been subducted beneath older, Labradorian ocean crust (Fig. 3).

Stage III (c. 1350–1300 Ma): continental extension, arcs in the Southwest Grenville Province (Fig. 4)

In Baltica, the Moss-Filtvet granite intruded the east shore of Oslofjorden at c. 1320 Ma (Hageskov & Pedersen 1980) and, to the north, the small Vatnås granite intruded the Kongsberg Sector at c. 1340 Ma (Jacobsen & Heier 1978).

In South Greenland, the Early Gardar event occurred at c. 1330–1290 Ma. Red sandstones and quartzites were deposited in an E–W graben, together with alkaline to sub-alkaline basalts and trachytes, and alkaline complexes (Upton & Emeleus 1987).

Just to the west, in Labrador, magmatism from c. 1330 to 1315 Ma produced the felsic Red Wine Intrusive Suite (peralkaline undersaturated and oversaturated granites and syenites), the Letitia Lake Group (quartz-feldspar porphyries, rhyolites and trachytes), the Nain Plutonic Suite (gabbros, diorites and granites), the Seal Lake Group (mafic volcanics and sills with sediments), and the Harp doleritic dykes (Hill & Miller 1990; Gower et al. 1990b).

The Southwest Grenville Province differed from the other areas between 1350 and 1300 Ma, when the Adirondack Highlands terrane (and probably part of the basement to the Appalachians) formed as juvenile arc tonalites, outboard of the continent (McLelland & Chiarenzelli 1990; Daly & McLelland 1991). Tonalites formed inboard, in the

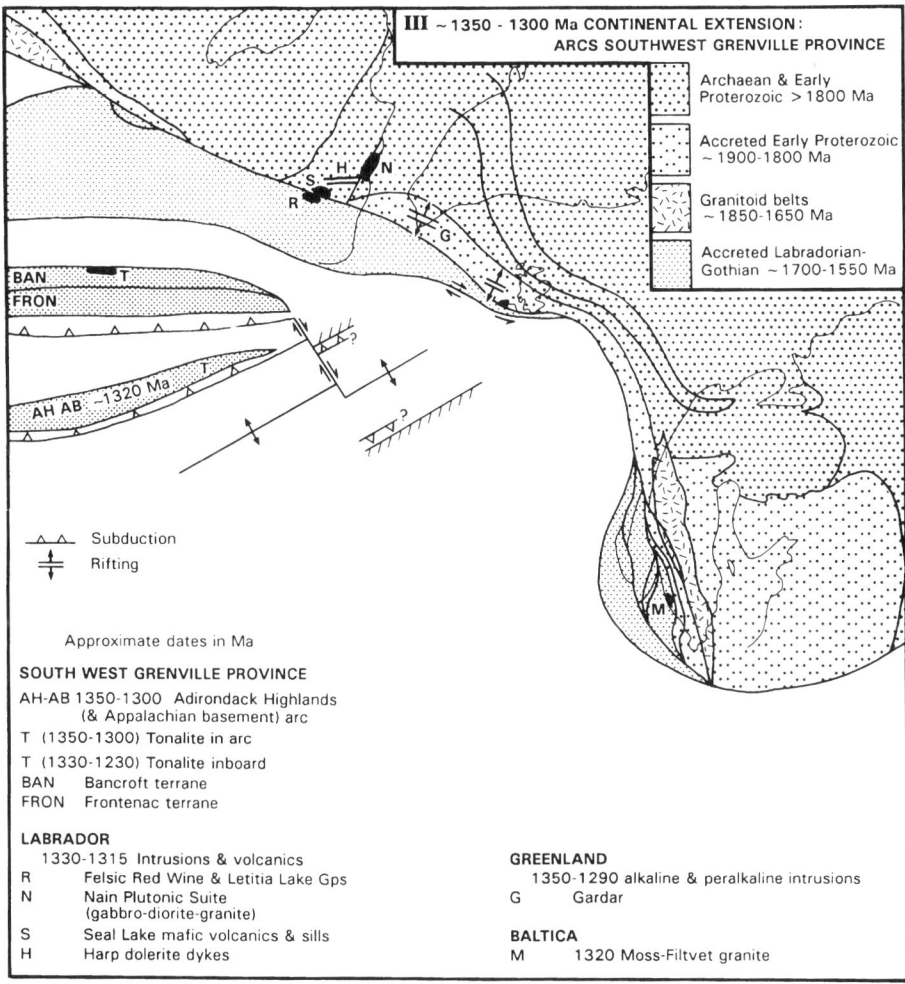

Fig. 4. Stage III: c. 1350–1300 Ma. Continental extension, arcs in the Southwest Grenville Province.

Bancroft terrane, as part of a protracted period of calc-alkaline, arc-related magmatism from 1330 to 1230 Ma.

Stage III: summary of events

Stage III, from c. 1350 to 1300 Ma, reflected continued extension in Labrador, Greenland and the British Isles: this regime was probably separated from the subducting margin of the Southwest Grenville Province by a major transform fault.

The Adironack Highlands terrane could have originated as an intra-oceanic arc, where the new ocean crust was being subducted beneath older, Labradorian crust. The arc-related magmatism in the Bancroft terrane was possibly related to a second subduction zone, produced partly because of the delayed break-up of the supercontinent.

Stage IV (c. 1300–1240 Ma): continental extension, start of the separation and rotation of Baltica (Fig. 5)

In Baltica, the Jotnian graben formed in Central Sweden and SW Finland and were filled with supracrustals before c. 1270 Ma, and probably just before 1300 Ma (Rämo & Haapala 1990; Pesonen 1991). Jotnian dolerite dykes cut these graben at c. 1270–1250 Ma (Suominen 1987) and just to the north, the NE–SW Central Scandinavian Dolerites intruded at c. 1270 Ma.

'Early-Sveconorwegian' granites intruded at c. 1250 Ma in southwest Sweden (Lindh 1987) and probably at c. 1240 Ma in South Norway (as discussed in the introduction). In South Norway, the granites separated two deformation phases ('D4' and 'D5') which produced fabrics that were cut by 'hyperite' gabbroids intruded from c. 1230–1225 Ma (Starmer 1993). The D4 and D5 deformations both produced a NW-directed, oblique sinistral thrusting at the terrane boundaries in South Norway and caused a major crustal delamination. The tectonism was consistent with clockwise rotational effects (Starmer 1993) and the northern end of the Kongsberg Sector was bent during D4, before the granites intruded. Whilst sinistral slip occurred between the continental **margin** and the ocean floor (becoming oblique subduction, in the west, as the rotation progressed), the continental block developed **internal** dextral transpression, particularly in its segmented leading edge of South Norway. In the extreme southwest of Norway, in Rogaland, there was recumbent folding at 1250–1200 Ma with 'synkinematic' granites (Demaiffe & Michot 1985).

In Ireland, Aftalion & Max (1987) recorded a tectonothermal event with emplacement of trondhjemitic and granodioritic migmatites between c. 1300 and 1225 Ma. In the Annagh Complex of northwest Ireland, Daly et al. (1994) dated the intrusion of syenite and alkali granite protoliths of gneisses at 1280–1270 Ma. In southern Greenland, crustal extension started the Mid-Gardar Event (c. 1280–1220 Ma) and lamprophyres and dolerites intruded (Upton & Emeleus 1987).

In Labrador and in the craton to the northwest, basic intrusions and volcanics were emplaced between c. 1275 and 1270 Ma. In Labrador, the Nain dolerite dykes intruded (c. 1276 Ma) (Gower et al. 1990b) and in the craton, the Mackenzie dyke swarm, the Muskox intrusion and the Coppermine River basalts were emplaced c. 1270 Ma (LeCheminant & Heaman 1989).

In the Southwest Grenville Province, the Elzevir terrane developed between 1300 and 1240 Ma, with the 'Hastings sequence' (of Lumbers et al. 1990) which seems to have overlapped into the southeast of the Bancroft terrane. This 'rift-related' sequence comprised early tholeiites (c. 1300 Ma) with later andesite-dacite and subordinate shallow water sediments, now metamorphosed at amphibolite to greenschist facies grade. An associated calc-alkaline suite intruded at c. 1290–1230 Ma (Lumbers et al. 1990; McLelland & Chiarenzelli 1990) and comprised nepheline-syenites, anorthosites, trondhjemites, diorites and alaskites: they were emplaced in the Bancroft, Elzevir and Adirondack Lowlands (–Frontenac) terranes.

To the east in eastern Quebec, the Wakeham Supergroup (WAK, Fig. 5) started to form in a continental rift, at c. 1270 Ma: it comprised arenites, bimodal rhyolite–basalt volcanics and gabbro sills (with some alkaline affinities): late granites and monzonites intruded at c. 1245 Ma (Martignole et al. 1994). Subsequent compression was followed by rifting again at c. 1180 Ma.

Rifting also occurred at c. 1250 Ma, inboard of the Southwest Grenville Province, in the Parry Sound Domain (Clifford 1990).

Far to the southwest, in the Llano uplift of Texas, igneous protoliths of gneisses were intruded between c. 1250 and 1232 Ma into tonalitic orthogneisses formed c. 1300 Ma (Walker 1992) and arc rocks were accreted northwestwards (Garrison 1981).

Stage IV: summary of events

Stage IV, from c. 1300 to 1240 Ma, spanned the transition from discontinuous, localized rifting (before c. 1250 Ma) to the development of two major rifts, which probably emanated from a triple point at the supercontinent margin. The East Greenland arm probably failed as an aulacogen in

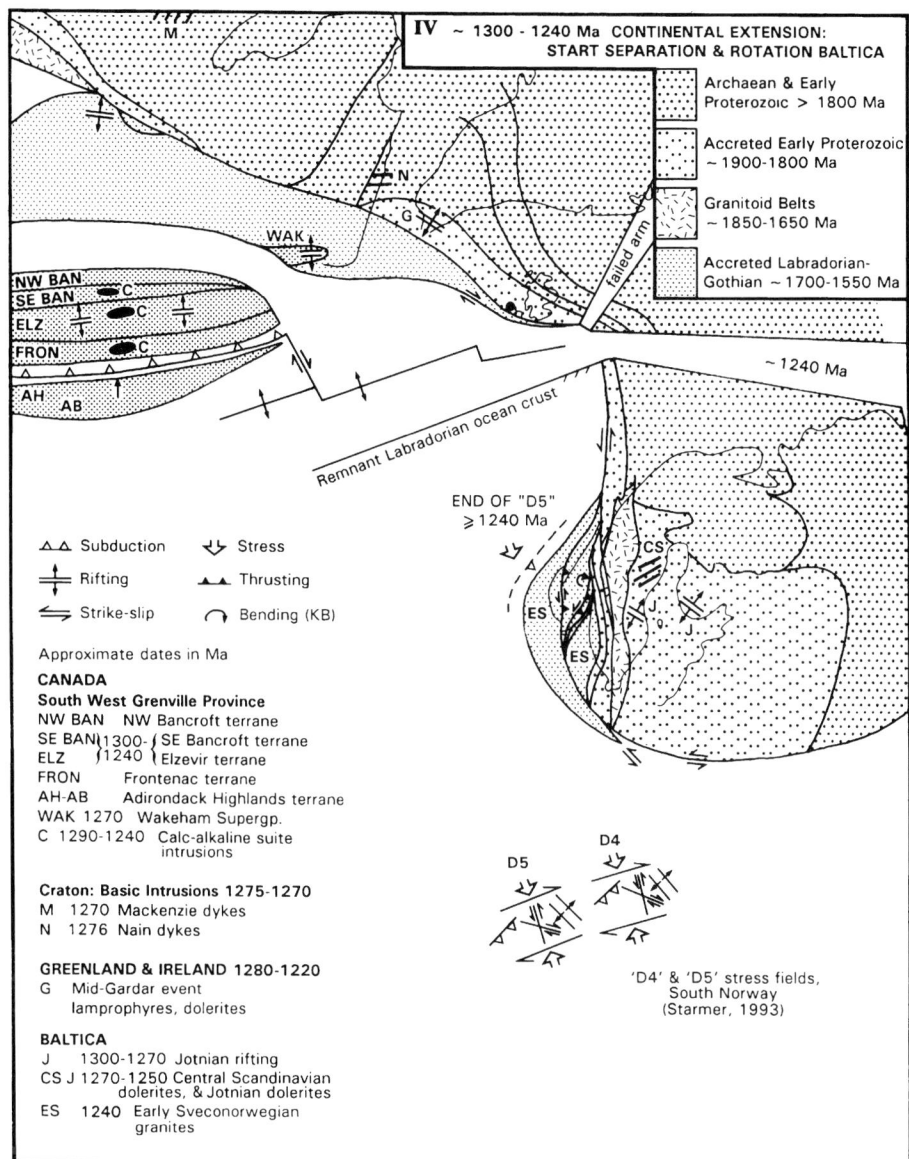

Fig. 5. Stage IV: c. 1300–1240 Ma. Continental extension, start of the separation and rotation of Baltica.

which the sediments of the Krummedal sequence were deposited.

From palaeomagnetic studies, Elming *et al.* (1993) suggested that Baltica separated at *c.* 1250 Ma and rotated *c.* 60° clockwise between 1250–1230 and 1050 Ma.

The rotation of Baltica seems to have started as it separated (*c.* 1240 Ma), developing relevant tectonism in its already-segmented leading edge in South Norway. Here, the 'D4' and 'D5' oblique sinistral thrusting, on east and southeast-dipping listric shears, followed shear zones along older terrane boundaries (Starmer 1993). Stresses were probably related to compression at the passive margin during the initiation of rotation. A slight change in orientation of stresses between D4 and D5 (Fig. 5) probably reflected a small amount of clockwise rotation. The basal shear along the

Mandal–Ustaoset Zone, the westernmost terrane junction, had less-regular movement directions and may have continued in the later stages.

At the eastern side of this shear regime, a major N–S lineament may have already existed in the Fjordzone (of Oslofjorden), where the Kattsund–Koster and Orust dykes had intruded at c. 1460–1420 Ma. To the east, southwest Sweden acted as a neutral prism between the shear regime and the craton.

Extension continued in Labrador and in Greenland, where the Mid-Gardar rifting event occurred.

In the southwest Grenville Province, it is suggested that the Elzevir terrane developed (c. 1300–1240 Ma) in an opening backarc basin, between the Bancroft and Adirondack Lowlands (–Frontenac) terranes and above a NW-dipping subduction zone, which produced the calc-alkaline magmatism in all three terranes. This subduction was probably associated with the closure of the outboard Adirondack Highlands terrane. The Wakeham Supergroup formed in a pull-apart basin on the continent.

The supercontinent probably changed its position relative to the spreading system. This might explain the initial opening and subsequent failure of the East Greenland aulacogen and the (possibly later) successful separation of Baltica. The subduction regime in the Southwest Grenville Province was probably still separated from the extensional regime in Labrador and Greenland by a major transform fault.

Stage V (c. 1240–1190 Ma): separation and rotation of Baltica, the Elzevirian Orogeny (Fig. 6)

The 'Inter-Sveconorwegian Extensional Period' started in South Norway. The curved Kongsberg–Bamble Sector developed a transtensional pull-apart due to dextral shear during rotation, but this was enhanced by a developing E–W tension as 'hyperite' gabbroids were intruded at c. 1230–1225 Ma (across 'D4' and 'D5' fabrics). The gabbroids had subduction-related chemical characteristics (Atkin & Brewer 1990) and developed corona growths by c. 1175 Ma (Starmer 1993). The cooling magnetizations of these rocks were among those used by Stearn & Piper (1984) to illustrate a rotation of Baltica. A few isolated basic intrusions occurred to the west of the Kongsberg–Bamble Sector (Falkum 1985). Further north, in the Western Gneiss Region, coronitic gabbroids intruded between 1300 and 1200 Ma (Mork & Mearns 1986) and basic magmas also intruded the basement of the Caledonian belt at c. 1200 Ma (Emmett 1994).

Further east (i.e. inboard), acidic plutons intruded in southern Sweden. The Protogine Zone underwent E–W tension as syenites and granites intruded it at c. 1230 Ma (Johansson 1990). The 'C3 Group' Hästefjorden granites intruded southwest Sweden at c. 1220 Ma (Åhäll et al. 1990).

In southern Greenland, the Mid-Gardar event (1280–1220 Ma) continued, with sinistral E–W and dextral N–S faulting before the intrusion of alkali syenite–granite complexes at c. 1220 Ma (Upton & Emeleus 1987). In central East Greenland, sediments of the Krummedal supracrustal sequence, which had been deposited on reworked Archaean basement, were metamorphosed and deformed between 1250 and 1000 Ma (Hansen et al. 1981; Rex & Gledhill 1981). Along the coast of northern Greenland (north of the present area), the rift-related, Midsommersø–Zig Zag volcanic event occurred at c. 1230 Ma (Surlyk 1991).

In the Southwest Grenville Province, early events included metamorphism and deformation in the Adirondack Highlands terrane, with an age of 1245–1215 Ma, reported by Daly & McLelland (1991). Inland, the Central Gneiss Belt was shortened by northwestward thrust stacking at c. 1240 Ma, with intense effects at the proto-Grenville Front (Bethune & Davidson 1988). In the Archaean craton, the Sudbury Dykes intruded (at c. 1240 Ma), adjacent to the proto-Grenville Front and at a high angle to it. The main Elzevirian Orogeny occurred from c. 1190 Ma onwards (1190–1060 Ma), according to McEachern & van Breemen (1993). It telescoped the Bancroft, Elzevir and Frontenac terranes of the Central Metasedimentary Belt by thrusting them NW onto the Central Gneiss Belt basement. This closed the rift in the Elzevir terrane.

Stage V: summary of events

The main rotation of Baltica was associated with subduction under the western continental margin in Norway. In South Norway, the subduction direction was orthogonal to the N–S regional grain: backarc extension in the upper plate was accompanied by high heat flows and subduction-related gabbroid intrusions at c. 1230–1225 Ma.

In South Greenland, strike-slip faulting affected the earlier rifting. It may have been related to the start of closure of the East Greenland aulacogen and the associated metamorphism and deformation of the Krummedal sequence (between 1250 and 1000 Ma).

In the Southwest Grenville Province, early metamorphism and deformation (at c. 1240 Ma) occurred as the Central Gneiss Belt was thrust NW towards the proto-Grenville Front, and orthogonal

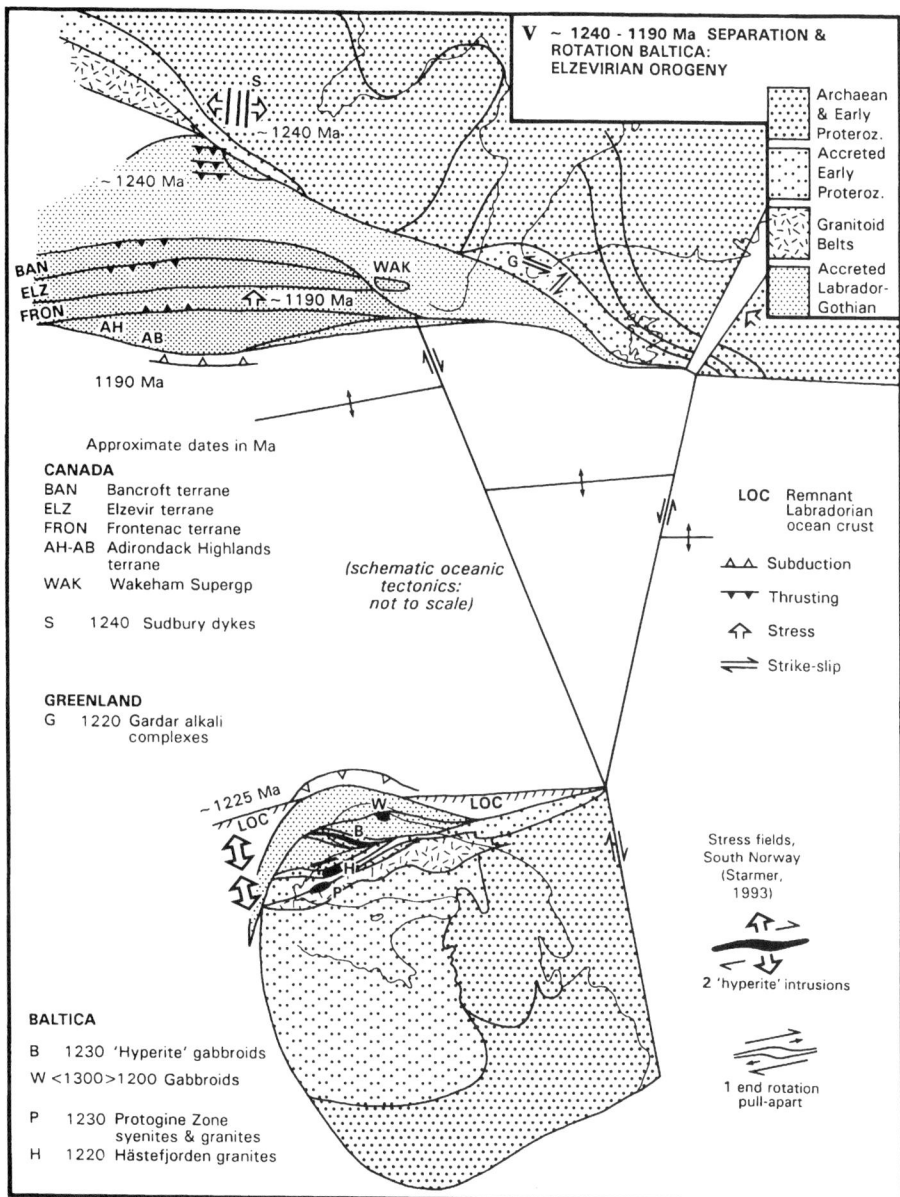

Fig. 6. Stage V: c. 1240–1190 Ma. Separation and rotation of Baltica, the Elzevirian Orogeny.

extension occurred in the adjacent Archaean craton. The early deformation in the (outboard) Adirondack Highlands terrane (at 1245–1215 Ma) is suggested to reflect initial docking with the Central Metasedimentary Belt. Subduction then switched outboard and convergence was renewed again during the main Elzevirian Orogeny (at c. 1190 Ma), after Baltica had rotated and was starting to close on the Southwest Grenville Province. The collision of the Adirondack Highlands terrane with the CMB was suggested, by McEachern & van Breemen (1993), to have occurred in the Ottawan Orogeny at c. 1080–1060 Ma. However, the AMCG intrusions at 1155–1120 Ma (discussed later) occurred across all the terranes, which must have been contiguous by that time. The Ottawan Orogeny is suggested to have resulted from collision with Baltica.

The differing tectonism in the three sectors (Baltica, Greenland–Labrador, and the Southwest Grenville Province) suggests that they may have been separated by major transform faults (shown schematically on Fig. 6). Such a system would facilitate rotation of Baltica, which could start to close on the Southwest Grenville Province in the next stage (Stage VI), when its northern, rifted margin lined-up with the major transform fault at the margin of the latter.

Stage VI (c. 1190–1100 Ma): closure of the Grenvillian Ocean (Fig. 7)

The 'Inter-Sveconorwegian Extensional Period' continued in South Norway, with high heat flows and medium-grained granites intruded in the west between c. 1200 and 1140 Ma, but becoming less common eastwards (Starmer 1993). Later, between c. 1150 and 1140 Ma, doleritic sheets were intruded and the Bandak Group bimodal volcanics were extruded in the west, with both subduction-related components and some MORB-like features (Atkin & Brewer 1990). Some terrestrial andesites occurred in the extreme west (Sigmond 1978). In the Bamble Sector, Kullerud & Dahlgren (1993) recorded local granulite facies metamorphism at c. 1150–1100 Ma, which they related to extension.

Further east, in southern Sweden, the Dal Group (sandstones, pelites, impure carbonates and basic volcanics) were possibly deposited during this period in a backarc position. To the east, mafic dykes of the Blekinge–Dalarna swarm and the Protogine Group intruded at c. 1180 Ma (Johansson 1990): in the Protogine Zone, they intruded and

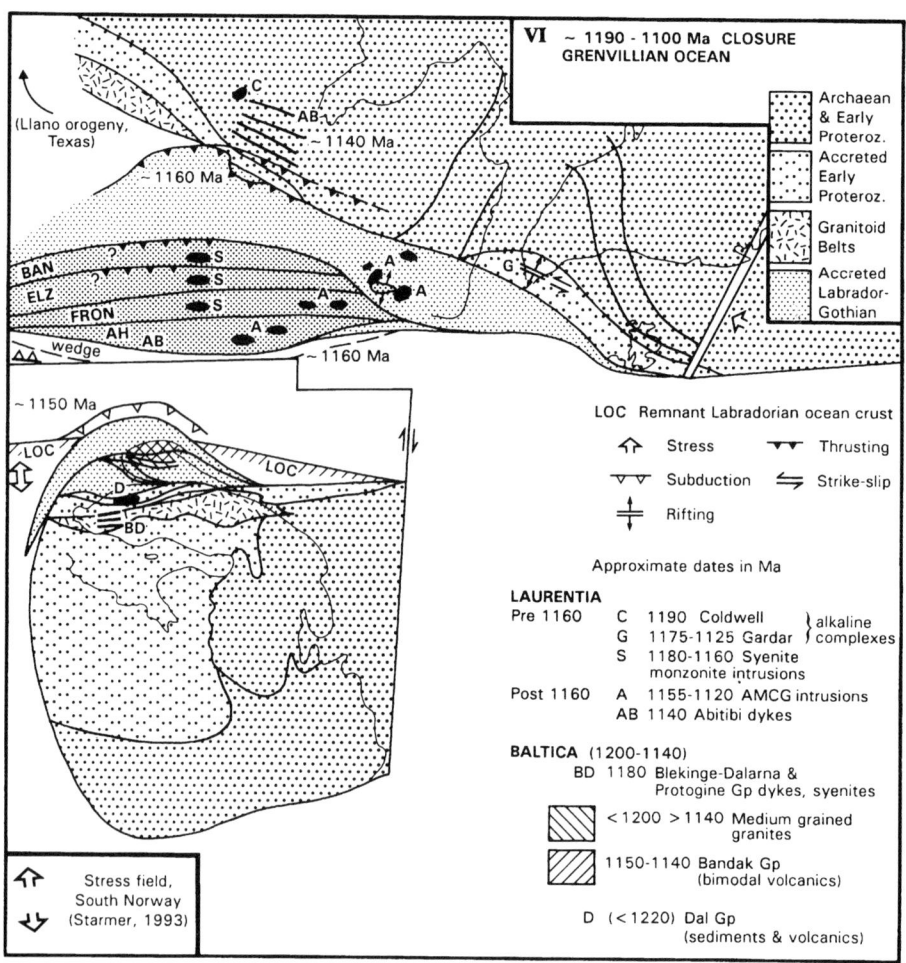

Fig. 7. Stage VI: c. 1190–1100 Ma. Closure of the Grenvillian ocean.

were intruded by syenites (emplaced between 1200 and 1150 Ma).

In South Greenland, the Late Gardar rifting event occurred between 1175 and 1125 Ma (Upton & Emeleus 1987). Crustal extension accompanied intrusion (at c. 1175 Ma) of major ENE–WSW alkaline dolerite dykes and later alkaline syenites and granites: then sinistral E–W faulting occurred before the intrusion of alkaline central complexes (at c. 1125 Ma) and late dolerite, lamprophyre, trachyte and phonolite dykes. In central East Greenland, the metamorphism continued from 1250 to 1000 Ma.

In Canada, alkaline magmatism occurred at c. 1190 Ma, in the Archaean craton at Coldwell (Gower et al. 1990a) and, at c. 1180–1160 Ma in the Southwest Grenville Province, where anorogenic potassic-alkaline suites of (gabbro-) syenite-monzonite intruded the Bancroft, Elzevir and Frontenac terranes (Lumbers et al. 1990; McLelland & Chiarenzelli 1990). To the east, conglomerates, arenites and continental tholeiitic gabbro sills developed in the Wakeham Supergroup during a second rifting event at c. 1180 Ma (Martignole et al. 1994). The Flinton Group (comprising continental volcanics and sediments) was deposited in a rift in the Elzevir terrane and was later folded in the Ottawan Orogeny (Rivers et al. 1989). Metamorphism occurred in the Southwest Grenville Province between 1170 and 1000 Ma (Mezger et al. 1993).

Inland, local compression caused NW-directed thrusting around the Grenville Front (Clifford 1990), with accompanying granulite facies metamorphism and syntectonic pegmatites emplaced at c. 1160 Ma (van Breeman et al. 1986).

Later, the widespread, anorogenic 'AMCG' (anorthosite–monzonite–charnockite–granite) suites intruded between 1155 and 1120 Ma from the Appalachians basement through the Adirondack Highlands and Frontenac terranes and into the Central Grenville Province (Emslie & Hunt 1989; McLelland & Chiarenzelli 1990). This suggests contiguity of all these terranes by c. 1155 Ma. At c. 1140 Ma, the Abitibi dyke swarm intruded the craton north of, and parallel to, the Grenville Front.

To the south, within the basement of the Appalachians in New England, the Blue Ridge terrane formed, between 1130 and 1030 Ma, as an island-arc outboard of the Southwest Grenville Province (Bartholomew & Lewis 1992). Further south (in Carolina), the older (1800–1200 Ma) Mars Hill sedimentary terrane was accreted on a west-dipping subduction zone.

In central Texas, the Llano Orogeny occurred between 1230 and 1116 Ma, with complex deformations, but an overall tectonic transport to the northeast: in west Texas, rhyolites (extruded at c. 1350 Ma) were thrust northwards onto granites and rhyolites formed at c. 1150–1130 Ma (Walker 1992). Post-tectonic granites intruded between 1116 and 1070 Ma. In central Arizona, dolerite sills intruded the Apache Group at c. 1120–1100 Ma (Silver 1978).

Stage VI: summary of events

Stage VI, between 1190 and 1100 Ma, reflected closure of the Grenvillian ocean and of the East Greenland aulacogen. In the early stages, between c. 1200 and 1160 Ma, subduction under the western part of South Norway generated medium-grained granites and caused metamorphism. Further east, away from the subduction, there were intrusions of mafic dykes and syenites in South Sweden. In Laurentia, alkaline complexes intruded at Coldwell and Gardar and anorogenic syenite–monzonites intruded the Southwest Grenville Province.

At c. 1160 Ma, granulite facies metamorphism and NW-directed thrusting occurred inland from the Central Metasedimentary Belt, which was also metamorphosed. It is suggested that this temporary compression may have been caused by post-collisional convergence in the basement, or by collision of the inactive spreading ridge with the trench and the Southwest Grenville salient (now with a Central Gneiss Belt **basement** and an accretionary wedge). When the ridge was over-ridden, tensional conditions returned, with intrusion of the AMCG suite and the Abitibi dykes, which suggest a front-parallel extension at c. 1140 Ma.

Subduction was enhanced under Baltica, producing the bimodal Bandak Group volcanics (at c. 1150 Ma) with both subduction and MORB-like characteristics.

The tectonism in the Gardar Province, particularly the sinistral E–W faulting, may have been related to the continued closure of the East Greenland aulacogen, where metamorphism continued from 1250 to 1000 Ma.

Stage VII (c. 1100–1070 Ma): collision of Baltica and Laurentia: the 'Ottawan' and 'Main Sveconorwegian' orogenies (Fig. 8)

It is suggested that collision of Baltica and Lautentia (at c. 1080 Ma) caused the Ottawan Orogeny and the Main Sveconorwegian Orogeny.

In South Norway, a protracted 'D6' deformation, poorly-constrained at c. 1100 Ma, produced complicated effects (discussed in detail by Starmer 1993). A complex system of folding was accompanied by sinistral strike-slip on the Mandal–Ustaoset Zone and NW-directed thrusting of the

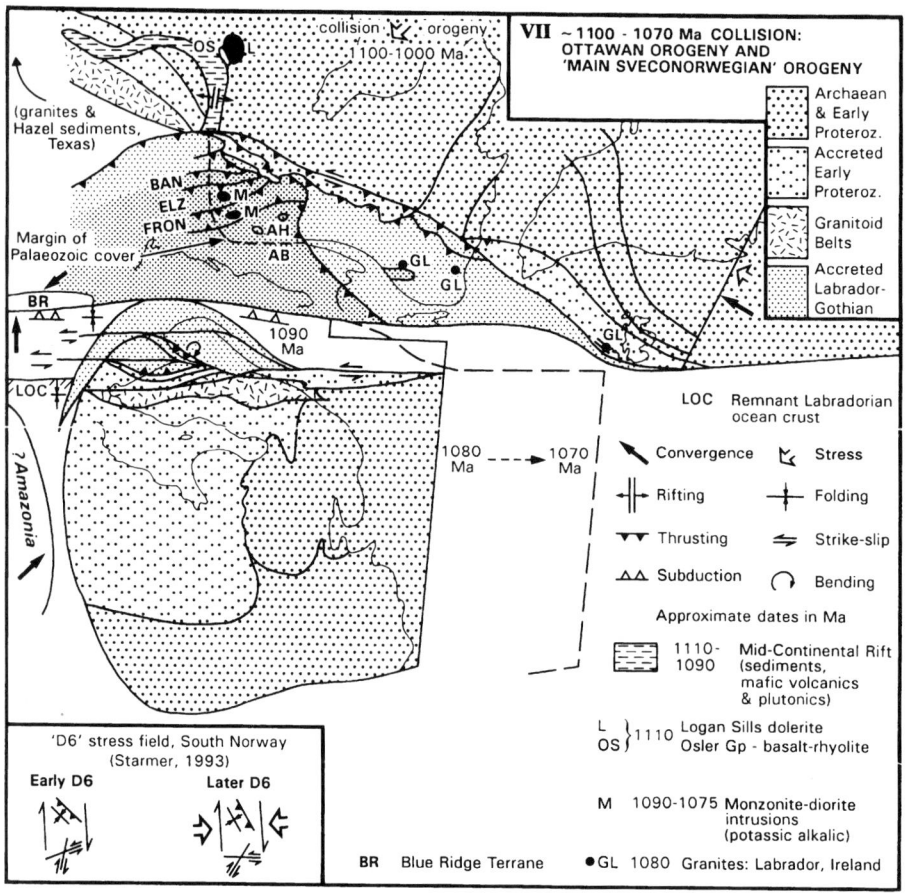

Fig. 8. Stage VII: c. 1100–1070 Ma. Collision of Baltica and Laurentia: the 'Ottawan' and 'Main Sveconorwegian' Orogenies.

Kongsberg–Bamble Sector, the curvature of which was increased by a clockwise rotation. The deformation was caused by regional E–W compression developing an E–W dextral shear couple which became increasingly transpressional, with a supplementary N–S compression. The complex 'D6' domain was limited by sinistral shears on the N–S Mandal–Ustaoset Zone in the west and on the N–S Fjordzone (in Oslofjorden) in the east. West of the Mandal–Ustaoset Zone, in southwest Norway, reclined folds formed on N–S axes (Falkum 1985).

In southwest Sweden, east of the Fjordzone, Sveconorwegian amphibolite facies metamorphism started at c. 1090 Ma (Åhäll et al. 1990). At this time in the N–S Fjordzone, or the 'Östfold–Marstrand boundary zone', Park et al. (1991) recorded a major sinistral shear, but to the east, SE-directed thrusting changed gradually to E-directed thrusting. Further east, in the Protogine Zone, Larson et al. (1990) recorded an ENE–WSW compression.

In Ireland, granites intruded at c. 1070 Ma (van Breemen et al. 1978). In East Greenland, Grenvillian metamorphism continued, probably around the closed aulacogen. In eastern Labrador, isolated granites intruded at c. 1080 Ma (Schärer & Gower 1988; Gower et al. 1991).

In the Southwest Grenville Province, Corriveau (1990) recorded late-stage (1090–1080 Ma), arc-related felsic to ultramafic rocks in the Elzevir terrane. Lumbers et al. (1990) described intrusions of an alkalic monzonite–diorite suite, with subordinate gabbro, granite, monzogranite and syenite, which occurred throughout the Central Metasedimentary Belt between 1090 and 1075 Ma.

The Ottawan Orogeny, in the Southwest Grenville Province, occurred between 1080 and 1060 Ma (McEachern & van Breemen 1993). It continued the NW-directed thrusting and imbrication started in the Elzevirian Orogeny at c. 1190 Ma and continued at c. 1160 Ma. At the northwest margin of the Central Metasedimentary Belt (CMB), stacking of thrust sheets continued until c. 1060 Ma, and probably until c. 1030 Ma (as discussed later). A SE-directed thrusting of the CMB over the Adirondack Highlands terrane developed the Carthage–Colton Mylonite Zone (at c. 1100 Ma), but it subsequently underwent Late-Ottawan normal faulting (Mezger et al. 1993).

The Adirondack Highlands terrane reached granulite facies grade at c. 1070–1050 Ma, with accompanying alaskite intrusions (McLelland & Chiarenzelli 1990). The grade was lower in the adjacent Adirondack Lowlands terrane and the rest of the Central Metasedimentary Belt, but this contrast may have been caused by Late-Ottawan down-faulting of these terranes on the Carthage–Colton Mylonite Zone.

The Central Gneiss Belt basement of the CMB was shortened. This reactivated older east-dipping thrusts including the Grenville Front and, south of it, formed the Allochthon Boundary Thrust of Rivers et al. (1989). Sinistral shearing was recorded along the Grenville Front by McWilliams & Dunlop (1978) and by Rivers et al. (1989), who commented that strike-slip was apparent, as in central Quebec 'where kilometer scale offsets are consistently sinistral'. In the Central Gneiss Belt, mildly deformed late-syntectonic pegmatites were dated at 1121–1097 Ma (van Breeman & Davidson 1990).

Inboard of this area and the Grenville Front, the curved Mid-Continental Rift formed between 1110 and 1090 Ma and was filled with continental sediments, together with voluminous mafic volcanics and plutonics. The rift curved in the earlier (c. 1140 Ma) extensional zone of the Abitibi dykes, possibly due to curving stress trajectories, or to the effects of a mantle plume. The southwestern arm was partially closed by thrusting between 1080 and 1040 Ma as strike-slip occurred along the southeastern arm (Cannon 1994). On the eastern margin of the Rift, the Osler Group mafic lavas and rhyolite porphyries and the doleritic Logan Sills were emplaced at c. 1100 Ma (Palmer & Davis 1987).

Outboard of the Southwest Grenville Province and to the south, in the Blue Ridge terrane in the basement of the New England Appalachians, there were translational tectonics (Bartholomew & Lewis 1992).

Far to the southwest, in the Llano Province of Texas, granites intruded between 1116 and 1070 Ma and, in west Texas, the Hazel Formation clastic wedge developed at the craton margin, between 1100 and 1080 Ma, during active strike-slip faulting (Soegaard & Callahan 1994). Dolerite sills intruded the Pahrump Group in California at 1087–1069 Ma (Heaman & Grotzinger 1992) and the Unkar Group of the Grand Canyon at c. 1070 Ma (Elston & McKee 1982).

Stage VII: summary of events

Stage VII, between 1100 and 1070 Ma, involved compressional tectonism of folding and thrusting in both Laurentia and Baltica, with major sinistral shearing along the Grenville Front, the Fjordzone and the Mandal–Ustaoset Zone.

This tectonism and the dextral transpression in South Norway suggest that Baltica converged sinistrally on Laurentia. The original collision at c. 1080 Ma closed the Mid-Continental Rift, but oblique convergence continued, producing major sinistral shears in Baltica. Contributory factors may have been the collision orogeny along the north of the Canadian Shield at 1100–1000 Ma (Trettin 1987) and the possible convergence of Amazonia from the south (discussed later).

The collision of Baltica with the Grenville basement expanded the Grenville Front as a continuous, through-crust detachment and produced the Allochthon Boundary Thrust as a major outboard shear (Fig. 8). The continued sinistral convergence progressively pushed the Central Metasedimentary Belt terranes over the basement, developing a NW-directed thrust stack between c. 1080 and 1060 Ma. The outer, Adirondack Highlands terrane was more deformed and metamorphosed than the inboard terranes, although this may result partly from the juxtaposition of different levels during subsequent Late-Ottawan extensional faulting.

Stage VIII (c. 1070–1000 Ma): post-collisional convergence and lateral slip (Fig. 9)

Across South Norway, 'Mid-Sveconorwegian' coarse- and medium-grained granite intrusions at 1060–1050 Ma were accompanied by high grade metamorphism (Starmer 1993). Later (at c. 1025 Ma), westward thrusting developed around the Mandal–Ustaoset zone and '$D7_1$' N–S axis folding occurred and decreased in intensity eastwards. In the south of the Western Gneiss Region, Sveconorwegian deformation occurred at c. 1050 Ma (Tucker et al. 1990) and 'Late-Sveconorwegian' thrusting was reported by Skjerlie & Pringle (1978).

In South Sweden, eastward thrusting occurred in the Fjordzone (of Oslofjorden) at c. 1015 Ma (Hageskov 1985) and in the Protogine Zone between 1100 and 1000 Ma (Larson et al. 1990).

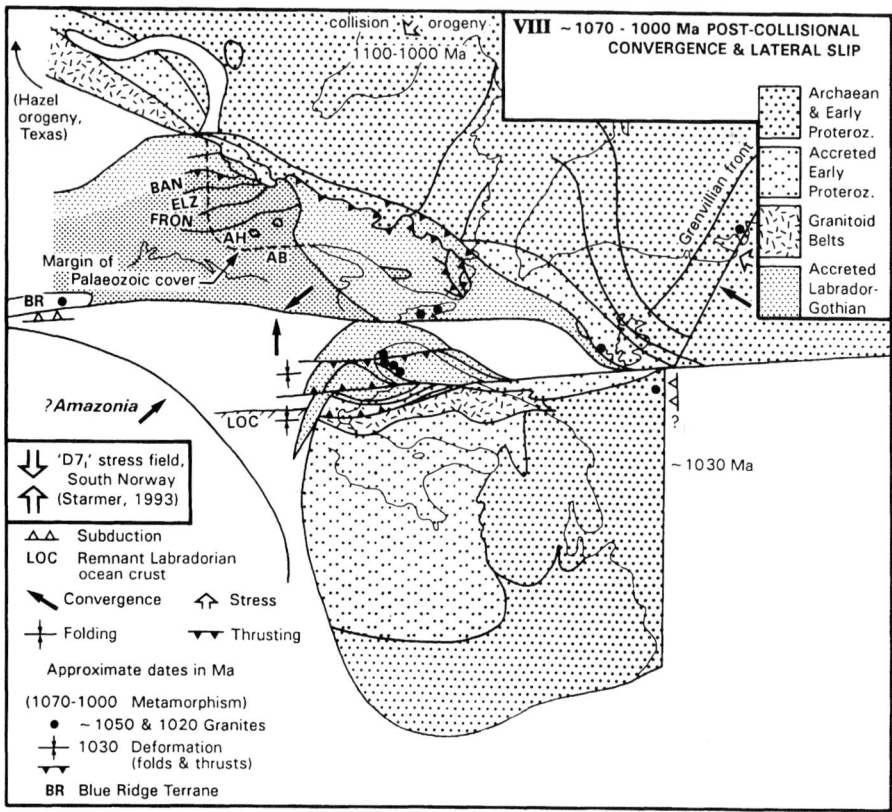

Fig. 9. Stage VIII: c. 1070–1000 Ma. Post-collisional convergence and lateral slip.

Deformation along the N–S belt of shear zones in southern Sweden, at c. 1000 Ma, was attributed to sinistral transpression by Park (1992). N–S folding of the Dal Group supracrustals occurred at c. 1030 Ma (Skiöld, 1976).

In Ireland, metamorphism, deformation and granitoid intrusions continued between 1070 and 1000 Ma (van Breemen et al. 1978; Winchester & Max 1990). The peralkaline Doolough granite intruded at c. 1015 Ma (Daly et al. 1994).

In East Greenland, metamorphism of the Krummedal supracrustal sequence continued until c. 1000 Ma (Hansen et al. 1981; Rex & Gledhill 1981) and its eastern side was migmatized and intruded by granites between 1058 and 977 Ma (Steiger & Meier 1990).

In Newfoundland, charnockite–granites intruded at c. 1050 Ma (Owen & Erdmer 1990). At Potato Hill, a potassic monzodiorite and granite intrusion at c. 1020 Ma was considered to have formed in a transpressional regime during collisional crustal thickening (Owen et al. 1992).

In Labrador, minor granitoid intrusions were emplaced at c. 1030 Ma (Schärer et al. 1986) and Grenvillian metamorphism and deformation occurred between 1050 and 990 (Wardle et al. 1990), with northward directed thrusting (and SW-directed thrusting, which may have been earlier and possibly pre-Grenvillian in age). Schärer & Gower (1988) noted that deformation and metamorphism generally finished at c. 1030 Ma, but that there was a later event at 980–960 Ma in the northernmost (Groswater Bay) terrane.

In the Central Metasedimentary Belt of the Southwest Grenville Province, a fenite-carbonatite suite was emplaced at c. 1070–1040 Ma (Lumbers et al. 1990): at its northwestern margin, thrust-stacking continued until c. 1060 Ma, accompanied by intrusion of granitic sheets (van Breemen & Hanmer 1986) and sheared pegmatites showed movements until c. 1030 Ma. Lumbers et al. (1990) recorded a thermal decline from 1030 Ma. Late-Ottawan extensional slip down-faulted the Central Metasedimentary Belt relative to the Adirondack Highland terrane and probably represented extensional collapse of the orogen.

Inboard, in the Central Gneiss Belt on the

east side of Lake Huron, Corrigan et al. (1994) recorded cessation of metamorphism by c. 1035 Ma, cooling below 600°C by 1003 Ma, and post-tectonic pegmatites at 990 Ma, probably during extensional collapse. In the northwest of the Belt, extensional effects were reported at c. 1020 Ma by Ketchum et al. (1994).

Plutons intruded the Blue Ridge island arc, in the basement of the Appalachians in New England, between 1045 and 1030 Ma (Bartholomew & Lewis 1992).

In west Texas, the Hazel Orogeny occurred after 1070 Ma, deforming the clastic wedge of the Hazel Formation in a transpressive regime (Soegaard & Callahan 1994).

Stage VIII: summary of events

Stage VIII, from c. 1070–1000 Ma, included very similar events in the contiguous areas of Laurentia and Baltica. Metamorphism occurred during much of this period, granites intruded mainly at 1050 (±20) and c. 1020 Ma and deformation occurred at c. 1030 (±20) Ma, with northward thrusting in Labrador and E–W compression in Baltica giving thrusting and N–S folding. In the Southwest Grenville Province, deformation generally ceased by c. 1030 Ma, apart from late, extensional faulting. Metamorphism and deformation continued around the closed aulacogen in East Greenland.

In Laurentia, the main focus of deformation had moved northeastwards, into Labrador. This is suggested to result from continued oblique convergence and the relative sinistral motion of Baltica (which underwent some sinistral transpression). At the northern margin of Baltica, granitic augen gneisses with an age of c. 1050 Ma have been recorded in Svalbard by Johansson et al. (1994) and might be subduction-related. The sinistral slip of Baltica might have been related to convergence of Amazonia from the south (discussed later) and to the collisional orogeny along the northern margin of the Canadian Shield at 1100–1000 Ma (Trettin 1987).

Stage IX (*c.* 1000–950 Ma): continued post-collisional convergence and lateral slip: late granites and deformations (Fig. 10)

In the extreme southwest of Norway, high grade metamorphism accompanied intrusion of the Rogaland anorthosite province and related rocks, mainly betwen c. 1000 and 935 Ma, although some bodies were emplaced earlier (Demaiffe & Michot, 1985). The Håland anorthosite-leuconorite was the earliest intrusion, at 1050–1020 Ma (Menuge 1988). The enigma of anorthosites emplaced in apparently orogenic settings was explained by Owens et al. (1994) as resulting from crustal thickening causing partial melting of mafic lower levels, followed by intrusion during local extension. Such a regime could have developed in southwest Norway during the sinistral transpressive slip.

Across South Norway, medium-grained 'Late-Sveconorwegian' granites intruded between 1000 and 990 Ma, followed by 'D7$_{II}$' N–S and 'D7$_{III}$' E–W folding and then medium-grained 'Post-Sveconorwegian' granites at c. 945 Ma (Starmer 1993). Events in the south of the Western Gneiss Region show many correlations, including late Sveconorwegian E–W folding (Abdel-Monem & Bryhni 1978) and metamorphism, migmatization and injection of granodioritic and pegmatite sheets between c. 950 and 940 Ma (Tucker et al. 1990).

In Oslofjorden, between Norway and Sweden, sinistral strike-slip movements occurred in the Fjordzone after 1015 Ma (Hageskov 1985).

In Ireland, late pegmatites intruded at c. 1000 Ma and cooling was prevalent by c. 985 Ma (Max & Sonet 1979). In East Greenland, granite intrusions continued until c. 977 Ma (Steiger & Meier 1990).

In Labrador, minor pegmatites were emplaced at c. 1003 Ma and a monzogranite intruded in eastern Quebec at 993 Ma: later granites intruded Labrador and eastern Quebec at 966–948 Ma and, in the northern Labradorian terranes next to the Grenville Front, NW-directed thrusting occurred at c. 960 Ma (Gower et al. 1991): later, NE–SW axis folds overprinted the thrusts (Connelly et al. 1995).

In Newfoundland, metamorphism at c. 978 Ma was possibly accompanied by deformation (Owen & Erdmer 1990) and granitic intrusions between 970 and 960 Ma may have been related to crustal thickening or to subsequent extensional collapse (Owen et al. 1992).

In the Southwest Grenville Province, deformation had ceased. In the Central Gneiss Belt of Ontario, Corrigan et al. (1994) considered that post-tectonic pegmatites, intruded at c. 990 Ma, were probably emplaced during extensional collapse.

Stage IX: summary of events

Stage IX, between c. 1000 and 950 Ma, shows the continued effects of oblique convergence and lateral slip of Baltica against Laurentia, until the two locked at c. 950 Ma, causing the E–W folding in Norway and effectively ending the Grenvillian Orogeny. There seems to have been intermittent compression, possibly related to the irregular margins, and some partitioning of the sinistral shear and compressional effects. The sinistral N–S shear

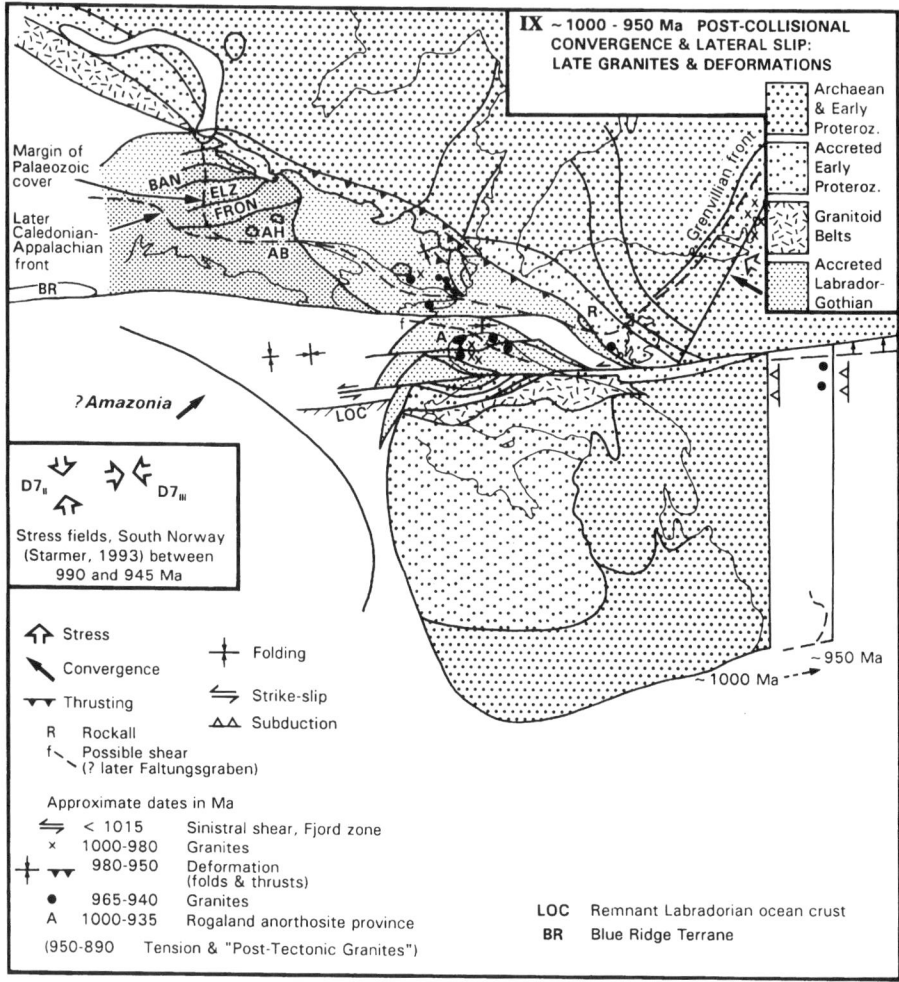

Fig. 10. Stage IX: c. 1000–950 Ma. Continued post-collisional convergence and lateral slip: late granites and deformations.

on the Fjordzone in southern Baltica reflected the lateral slip. Subduction may have occurred along the northern margin of Baltica, where Johansson et al. (1994) reported andesitic volcanics erupted at c. 1000 Ma and granites intruded at c. 950 Ma in Svalbard.

Deformation and granitoid intrusions affected the contiguous parts of Labrador and southern Norway, and Gower et al. (1991) noted the apparent continuity of granitic bodies. Granitoids intruded at c. 1000–980 Ma. Deformation and metamorphism occurred between c. 980 and 950 Ma, with N–S folding and then E–W folding in Norway, and late NW-directed thrusting in Labrador near the Grenville Front. Granites intruded again at c. 965–940 Ma.

The renewed shearing around the Grenville Front probably caused it to become the northern limit of Grenvillian K–Ar ages. Krogh (1994) detected a metamorphic disturbance along the entire Grenville Front tectonic zone between 995 and 975 Ma, resulting from thrusting or crustal thickening.

After c. 950 Ma, tensional conditions prevailed, with high heat flows. The 'Post-Tectonic Granites' intruded across South Norway and South Sweden (between 950 and 880 Ma). In South Sweden, local granulite facies metamorphism developed at c. 900 Ma in the Eastern Segment and extensional faulting occurred on the Mylonite Zone and the Protogine Zone (Johansson & Kullerud 1993). In the Rogaland anorthosite province of South

Norway, late charnockites and mangerites intruded (Demaiffe & Michot 1985).

Discussion

This study presents a possible model to link the Mid-Proterozoic evolution of Baltica and Laurentia. Its relative simplicity reflects that of the supercontinent and the consequent development of spreading systems. The model suggests that there may have been some lateral terrane displacement during Labradorian–Gothian accretion in both Baltica and the Southwest Grenville Province, but any subsequent effects of this type were on a smaller scale, considering the consanguineity of intrusions across terranes. Similarly, large-scale terrane transfers between Baltica and Laurentia seem unlikely, but Demaiffe & Michot (1985) suggested that the Rogaland anorthosite province **might** have been obducted onto the extreme southwestern tip of Baltica.

The events in central East Greenland correlate with the suggestion of a failed-arm aulacogen. The sediments of the Krummedal sequence were deposited on reworked Archaean crust: they were metamorphosed and deformed between 1250 and 1000 Ma and migmatized in the east between 1058 and 977 Ma, correlating with closure of the aulacogen. The apparent lack of Grenvillian effects in the north is also explained by the aulacogen.

The oblique sinistral collision of Baltica and Laurentia in the Ottawan Orogeny (at c. 1080 Ma) and the subsequent sinistral slip are suggested by the major N–S sinistral shearing (and N–S compression) in southern Scandinavia. Examples mentioned include sinistral shearing on the Mandal–Ustaoset Zone and Fjordzone at c. 1080 Ma with E–W dextral transpression in the intervening area of South Norway (Starmer 1993), sinistral slip on the Fjordzone at 1015 Ma (Hageskov 1985), N–S sinistral transpression in southern Sweden at c. 1000 Ma (Park 1992), and the E–W folding in Norway between c. 980 and 950 Ma (Starmer 1993). The Sveconorwegian orogen in southern Sweden was suggested to result from 'oblique collisional processes' by Stephens & Wahlgren (1994). Several references to sinistral slip and/or transpression in Laurentia have already been mentioned. They include sinistral shearing along the Grenville Front (McWilliams & Dunlop, 1978; Rivers et al. 1989), translational tectonism in the Grenvillian basement of the Appalachians between 1130 and 1030 Ma (Bartholomew & Lewis, 1992), and transpression at Potato Hill, Newfoundland at c. 1020 Ma (Owen et al. 1992). In addition, where the Grenville Front turns almost due south from the Southwest Grenville Province, Hauser (1993) demonstrated east-dipping thrusts in Ohio, beneath the Eastern (St Francois) Granite–Rhyolite Province. These thrusts were thought to reflect sinistral oblique convergence, consistent with 'southward tectonic escape' during the Ottawan Orogeny. Far to the southwest in Texas, transpression after 1070 Ma occurred during the Hazel Orogeny (Soegaard & Callahan 1994), although this may have been related to collisions with South America and Africa (see below). The convergence of the northern parts of South America may have contributed to the sinistral slip of Baltica.

At the inboard margin of the Grenville Province, the Grenville Front was a polyphase shear zone that varied along its length. It may have inherited shears from the original Labradorian accretion, but certainly formed the approximate northwestern front of thrusting in the Elzevirian Orogeny (c. 1190 Ma) and probably became a continuous, through-crust dislocation in the Ottawan Orogeny, during the collision with Baltica at c. 1080 Ma. Subsequent reactivations lasted until c. 960 Ma. Where the Grenville Front turned south into Ohio (see above), the east-dipping thrusts and sinistral oblique convergence (Hauser, 1993) were compatible with effects in the Southwest Grenville Province.

The Grenville Front then turned southwest through midcontinental and southwestern USA. In Texas, the Llano Orogeny involved overall tectonic transport to the northeast between 1232 and 1116 Ma, and probably between 1130 and 1116 Ma (Walker 1992) and the Hazel Orogeny occurred in a transpressive regime after 1070 Ma (Soegaard & Callahan 1994). Between these events, anorogenic conditions were recorded by the intrusion of post-tectonic granites (between 1116 and 1070 Ma), coeval with undeformed granites intruded into northern Mexico at c. 1078 Ma (Walker 1992): at this time, dolerite sills intruded in Arizona at 1120–1100 Ma, in California at 1087–1069 Ma and the Grand Canyon at c. 1070 Ma. During this period, the Ottawan Orogeny was occurring in eastern Laurentia, with overall tectonic transport to the northwest. The Texan orogenies may therefore have resulted from collision with South America and Africa rather than Baltica, particularly in the light of Neoproterozoic reconstructions. This would have compounded the composite nature of the Grenville Front.

Recent work (e.g. Bond et al. 1984; Moores 1991; Hoffman 1991; Dalziel 1992a & b) suggests that world-wide collisional orogenies of Grenvillian age welded cratons into the first Neoproterozoic supercontinent (at c. 1000 Ma). This fragmented at c. 750 Ma, affecting mainly the southern (Gondwana) continents, and re-combined during the Pan-African Orogeny (at c. 600–570 Ma) to form a second Neoproterozoic supercontinent (which broke-up between c. 600 and 550 Ma to form the Iapetus Ocean). The relative positions of cratons vary slightly on the different reconstructions: the variation ranges from South America adjacent to the Southwest Grenville Province with Baltica off the northeastern tip of Labrador and the southern tip of Greenland (similar to the positions in Stage IX, Fig. 10), to South America adjacent to the whole Grenville Belt and southern Greenland with Baltica off northern East Greenland.

The model proposed in the present study for the Mesoproterozoic is compatible with these Neoproterozoic reconstructions for 750–600 Ma. The collision of Baltica with Laurentia (at c. 1080 Ma) and the subsequent sinistral slip until (c. 950 Ma) would place both cratons in the appropriate positions in the Neoproterozoic (see Fig. 10). The model is also supported by some palaeomagnetic results. The conclusions of Elming et al. (1993), that Baltica separated c. 1250 Ma and rotated between 1250–1230 and 1050 Ma, have already been mentioned. These authors also showed that Baltica drifted south (>30° of latitude) from 1000 to 960 Ma. Greenland and North America have been suggested by Piper (1992) to have drifted north (c. 40° of latitude) on the 'Gardar Track' between 1200 and 1100 Ma and then returned south on the 'Keweenawan Track' between 1100 and 1050 Ma: the latter would encompass the time of oblique sinistral collision with Baltica and the subsequent sinistral slip. Later stages may be reflected by the southward drift (of c. 100° of latitude) of Laurentia between 980 and 920 Ma, shown by Irving & McGlynn (1981).

This study has been concerned principally with horizontal movements, but there were also considerable vertical movements of blocks. This is shown most convincingly by repeated instances of high grade metamorphic rocks being overlain by supracrustal sequences which, in many cases, were subsequently metamorphosed. The Svecofennian craton was overlain by the Dala volcanics at c. 1635 Ma and the Jotnian supracrustals at c. 1300 Ma. Parts of southwest Sweden were overlain by the Åmål–Horred Belt supracrustals at c. 1640–1610 Ma and the Dal Group between 1220 and 1030 Ma (possibly c. 1150 Ma). The Telemark Sector of Norway was overlain by the Rjukan Group volcanics at c. 1500 Ma and the Bandak Group volcanics at c. 1150 Ma.

A number of Phanerozoic structures may have nucleated on the Proterozoic shear zones and terrane boundaries. These include the Tornquist Lineament, the Oslo–Skagerrak Graben, the Caledonian Faltungsgraben (Fig. 10), and (to some extent) the entire Caledonian–Appalachian Orogen, especially in East Greenland.

The author thanks Janet Baker for interminable patience whilst drawing the diagrams. Thanks to Lennart Samuelsson are long overdue for his time, patience and hospitality, whilst introducing the geology of southern Sweden. Helpful reviews of the original manuscript by Chris Talbot and Graham Park are also acknowledged.

References

ABDEL-MONEM, A. A. & BRYHNI, I. 1978. A Rb/Sr date from anorthosite-suite rocks of the Gloppen-Eikefjord area, western Norway. *Norsk Geologisk Tidsskrift*, **58**, 229–232.

ÅBERG, G. 1978. Precambrian geochronology of southeastern Sweden. *Geologiska Föreningens i Stockholm Förhandlingar*, **100**, 125–154.

—— 1988. Middle Proterozoic anorogenic magmatism in Sweden and worldwide. *Lithos*, **21**, 279–289.

AFTALION, M. & MAX, M. D. 1987. U–Pb zircon geochronology from the Precambrian Annagh Division gneisses and the Termon Granite, NW County Mayo, Ireland. *Journal of the Geological Society, London*, **144**, 401–406.

ÅHÄLL, K-I. & DALY, J. S. 1989. Age, tectonic setting and provenance of Östfold-Marstrand Belt supracrustals: westward growth of the Baltic Shield at 1760 Ma. *Precambrian Research*, **45**, 45–61.

——, CONNELLY, J., BREWER, T. S. & SAMUELSSON, L. 1994. Constraints for the Mid-Proterozoic orogenic–anorogenic boundary and subsequent evolution in SW Sweden. *Terra Nova*, **6**, Abstract Supplement 2, 1.

——, DALY, J. S. & SCHÖBERG, H. 1990. Geochronological constraints on Mid-Proterozoic magmatism in the Östfold–Marstrand belt: implications for crustal evolution in SW Sweden. *In*: GOWER, C. F., RIVERS, T. & RYAN, B. (eds) *Mid-Proterozic Laurentia–Baltica*. Geological Association of Canada, Special Paper **38**, 97–115.

——, PERSSON, P-O. & SKIÖLD, T. 1995. Westward accretion of the Baltic Shield: implications from the 1.6 Ga Åmål-Horred Belt, SW Sweden. *Precambrian Research*, **70**, 235–251.

ATKIN, B. P. & BREWER, T. S. 1990. The tectonic setting of basaltic magmatism in the Kongsberg, Bamble and Telemark Sectors, southern Norway. *In*: GOWER, C. F., RIVERS, T. & RYAN, B. (eds) *Mid-Proterozic Laurentia-Baltica*. Geological Association of Canada, Special Paper **38**, 471–483.

BARTHOLOMEW, M. J. & LEWIS, S. E. 1992. Appalachian Grenville massifs: pre-Appalachian translational tectonics. *In*: MASON, R. (ed) *Basement Tectonics*, **7**, 363–374.

BETHUNE, K. M. & DAVIDSON, A. 1988. Diabase dykes and the Grenville Front, southwest of Sudbury,

Ontario. *In*: *Current Research, part C.* Geological Survey of Canada, paper **88-1C**, 151–159.

BOGDANOVA, S. V., BIBIKOVA, E. V. & GORBATSCHEV, R. 1994. Palaeoproterozoic U–Pb ages from Belorussia: new geodynamic implications for the East European Craton. *Precambrian Research*, **68**, 231–240.

BOND, G. C., NICKESON, P. A. & KOMINZ, M. A. 1984. Break-up of a supercontinent between 625 Ma and 555 Ma: new evidence and implications for continental histories. *Earth & Planetary Science Letters*, **70**, 325–345.

BRIDGWATER, D. & WINDLEY, B. F. 1973: Anorthosites, post-orogenic granites, acid volcanic rocks, and crustal development in the North Atlantic Shield during the mid-Proterozoic. *In*: LISTER, L. A. (ed.) *Symposium on Granites, Gneisses and Related Rocks.* Geological Society of South Africa Special Publication **3**, 307–318.

CANNON, W. F. 1994. Closing of the Midcontinent rift – a far-field effect of Grenvillian compression. *Geology*, **22**, 159–162.

CLIFFORD, P. M. 1990. Mid-Proterozoic deformational and intrusive events along the Grenville Front in the Sudbury–Killarney area, Ontario, and their implications. *In*: GOWER, C. F., RIVERS, T. & RYAN, B. (eds) *Mid-Proterozic Laurentia–Baltica.* Geological Association of Canada, Special Paper **38**, 335–350.

CONNELLY, J. N., RIVERS, T. & JAMES, D. 1995. Thermotectonic evolution of the Grenville Province in western Labrador. *Tectonics,* **14**, 202–217.

CORRIGAN, D. 1990. *Geology and U–Pb geochronology of the Key Harbour area, Britt domain, southwest Grenville Province.* MSc thesis, Dalhousie University.

——, CULSHAW, N. G. & MORTENSEN, J. K. 1994. Pre-Grenvillian evolution and Grenvillian overprinting of the Parautochthonous Belt in Key Harbour, Ontario: U-Pb and field constraints. *Canadian Journal of Earth Science*, **31**, 583–596.

CORRIVEAU, L. 1990. Proterozoic subduction and terrane amalgamation in the southwestern Grenville Province, Canada: evidence for ultrapotassic to shoshonitic plutonism. *Geology*, **18**, 614–617.

COWARD, M. P. & PARK, R. G. 1987. The role of mid-crustal shear zones in the Early Proterozoic evolution of the Lewisian. *In*: PARK, R. G. & TARNEY, J. (eds) *Evolution of the Lewisian Complex and Comparable Precambrian High-Grade Terrains.* Geological Society, London, Special Publications, **27**, 127–138.

DAHLGREN, S., HEAMAN, L. & KROGH, T. 1990. Geological evolution and U–Pb geochronology of the Proterozoic Central Telemark area, Norway. *Geonytt*, **17**, 38–39.

DALY, J. S. & MCLELLAND, J. M. 1991. Juvenile Middle Proterozoic crust in the Adirondack Highlands, Grenville Province, northeastern North America. *Geology*, **19**, 119–122.

——, FITZGERALD, R. C., MENUGE, J. F., BREWER, T. S., HEAMAN, L. M. & MORTON, A. C. 1994. Proterozoic crustal history of western Ireland and Rockall: Precambrian tectonic reconstructions in the North Atlantic. *Abstracts 8th International Conference on Geochronology, Cosmochronology & Isotope Geology*, 74.

——, MUIR, R. J & CLIFF, R. A. 1991. A precise U–Pb zircon age for the Inishtrahull syenitic gneiss, County Donegal, Ireland. *Journal of the Geological Society, London*, **148**, 639–642.

DALZIEL, I. W. D. 1992*a*. Antarctica; a tale of two supercontinents? *Annual Review of Earth & Planetary Sciences*, **20**, 501–526.

—— 1992*b*. On the organization of American plates in the Neoproterozoic and the breakout of Laurentia. *GSA Today*, **2**, 240–241.

DAVIDSON, A. 1986. New interpretations in the southwestern Grenville Province. *In*: MOORE, J. M, DAVIDSON, A. & BAER, A. J. (eds) *The Grenville Province.* Geological Association of Canada Special Paper **31**, 61–74.

DEMAIFFE, D. & MICHOT, J. 1985. Isotope geochronology of the Proterozoic crustal segment of southern Norway: a review. *In*: TOBI, A. C. & TOURET, J. L. R. (eds) *The Deep Proterozoic Crust in the North Atlantic Provinces.* NATO Advanced Science Institutes Series, series C, **158**, 411–433.

DICKIN, A. P. & HIGGINS, M. D. 1992. Sm/Nd evidence for a major 1.5 Ga crust-forming event in the central Grenville Province. *Geology*, **20**, 137–140.

—— & MCNUTT, R. H. 1990. Nd model-age mapping of Grenville lithotectonic domains: mid-Proterozoic crustal evolution in Ontario. *In*: GOWER, C. F., RIVERS, T. & RYAN, B. (eds) *Mid-Proterozic Laurentia–Baltica.* Geological Association of Canada, Special Paper **38**, 79–94.

EASTON, R. M. 1992. The Grenville Province and the Proterozoic history of central and southern Ontario. *In*: THURSTON P. C., WILLIAMS, H. R., SUTCLIFFE, R. H. & STOTT, G. M. (eds) *Geology of Ontario.* Ontario Geological Survey Special Volume **4**, part 2, 715–904.

ELMING, S-Å., PESONEN, L. J., LEINO, M. A. H., KHRAMOV, A. N., MIKHAILOVA, N. P. ET AL. 1993. The drift of the Fennoscandian and Ukrainian Shields during the Precambrian: a palaeomagnetic analysis. *Tectonophysics*, **223**, 177–198.

ELSTON, D. P & MCKEE, E. H. 1982. Age and correlation of the late Proterozoic Grand Canyon disturbance, northern Arizona. *Geological Society of America Bulletin*, **93**, 681–699.

EMMETT, T. F. 1994. Pre-Scandian continental flakes within the Scandian orogen of south central Norway. *Terra Nova*, **6**, Abstract Supplement 2, 6.

EMSLIE, R. F. & HUNT, P. A. 1989. The Grenvillian event: magmatism and high grade metamorphism. *In*: *Current Research, Part C.* Geological Survey of Canada, Paper **89-1C**, 11–17.

—— & —— 1990. Ages and petrogenetic significance of igneous charnockite suites associated with massif anorthosites, Grenville Province. *Journal of Geology*, **98**, 213–232.

FALKUM, T. 1985. Geotectonic evolution of southern Scandinavia in light of a Late-Precambrian plate collision. *In*: TOBI, A. C. & TOURET, J. L. R. (eds) *The Deep Proterozoic Crust in Atlantic Provinces.*

NATO Advanced Science Institutes Series, series C, **158**, 309–322.

GARRISON, J. R. 1981. Coal Creek serpentinite, Llano Uplift, Texas: a fragment of an incomplete Precambrian ophiolite. *Geology*, **9**, 225–230.

GORBATSCHEV, R. & BOGDANOVA, S. 1993. Frontiers in the Baltic Shield. *Precambrian Research*, **64**, 3–21.

GOWER, C. F. 1985. Correlations between the Grenville Province and Scevonorwegian Orogenic Belt – implications for Proterozoic Evolution of the Southern Margins of the Canadian and Baltic Shields. *In*: TOBI, A. C. & TOURET, J. L. R. (eds) *The Deep Proterozoic Crust in the North Atlantic Provinces*. NATO Advanced Science Institutes Series, series C, **158**, 247–257.

—— . 1992. The relevance of Baltic Shield metallogeny to mineral exploration in Labrador. *In*: *Current Research (1992)*. Newfoundland Department of Mines and Energy, report **92-1**, 331–366.

——, HEAMAN, L. M., LOVERIDGE, W. D., SCHÄRER, U. & TUCKER, R. D. 1991. Grenvillian magmatism in the eastern Grenville Province, Canada. *Precambrian Research*, **51**, 315–336.

——, RYAN, A. B. & RIVERS, T. 1990a. Mid-Proterozoic Laurentia-Baltica: an overview of its geological evolution and a summary of the contributions made by this volume. *In*: GOWER, C. F., RIVERS, T. & RYAN, B. (eds) *Mid-Proterozic Laurentia–Baltica*. Geological Association of Canada, Special Paper **38**, 1–20.

——, RIVERS, T. & BREWER, T. S. 1990b. Middle Proterozoic mafic magmatism in Labrador, eastern Canada. *In*: GOWER, C. F., RIVERS, T. & RYAN, B. (eds) *Mid-Proterozic Laurentia–Baltica*. Geological Association of Canada, Special Paper **38**, 485–506.

GRAVERSEN, O. 1984. Geology and structural evolution of the Precambrian of the Oslofjord–Öyeren area, S.E. Norway. *Norges Geologiske Undersøkelse*, **398**, 1–50.

GREENBERG, J. K. 1990. Anorogenic granite associations as products of progressive continental evolution. *In*: GOWER, C. F., RIVERS, T. & RYAN, B. (eds) *Mid-Proterozic Laurentia–Baltica*. Geological Association of Canada, Special Paper **38**, 447–457.

GRIFFIN, W. L., TAYLOR, P. N., HAKKINEN, J. W., HEIER, K. S., IDEN, I. K. ET AL. 1978. Archaean and Proterozoic crustal evolution in Lofoten–Vesterålen, N Norway. *Journal of the Geological Society, London*, **135**, 629–647.

HAGELIA, P. 1984. Rb–Sr dates on the Ubergsmoen augen gneiss and its country rock, Bamble, South Norway: evidence for a Sveconorwegian high grade event. NATO Advanced Study Institute, Norway 1984, Abstracts volume, 15.

HAGESKOV, B. 1985. Constrictional deformation of the Koster dyke swarm in a ductile sinistral shear zone, Koster islands, SW Sweden. *Bulletin of the Geological Society of Denmark*, **34**, 151–197.

—— 1987. Tholeiitic dykes and their chemical alteration during amphibolite facies metamorphism: The Kattsund–Koster dyke swarm, SE Norway–W Sweden. *Sveriges Geologiska Undersökning*, **C 817**, 1–61.

—— & PEDERSEN, S. 1980. Rb–Sr whole rock age determinations from the western part of the Östfold basement complex, SE Norway. *Bulletin of the Geological Society of Denmark*, **29**, 119–128.

HANSEN, B. T., PERSSON, P.-O., SOLLNER, F. & LINDH, A. 1989. The influence of recent lead loss on the interpretation of disturbed U–Pb systems in zircons from metamorphic rocks in southwest Sweden. *Lithos*, **23**, 123–136.

——, STEIGER, R. H. & HIGGINS, A. K. 1981. Isotopic evidence for a Precambrian metamorphic event within the Charcot Land window, East Greenland Caledonian fold belt. *Bulletin of the Geological Society of Denmark*, **29**, 151–160.

HAUSER, E. C. 1993. Grenville foreland thrust belts hidden beneath the eastern U.S. midcontinent. *Geology*, **21**, 61–64.

HEAMAN, L. M. & GROTZINGER, J. P. 1992. 1.08 Ga diabase sills in the Pahrump Group, California: implications for development of the Cordilleran miogeocline. *Geology*, **20**, 637–640.

HILL, J. D. & MILLER, R. R. 1990. A review of Middle Proterozoic epigenic felsic magmatism in Labrador. *In*: GOWER, C. F., RIVERS, T. & RYAN, B. (eds) *Mid-Proterozic Laurentia–Baltica*. Geological Association of Canada, Special Paper **38**, 417–431.

HOFFMAN, P. F. 1991. Did the breakout of Laurentia turn Gondwanaland inside-out? *Science*, **252**, 1409–1412.

IRVING, E. & MCGLYNN, J. C. 1981. On the coherence, rotation and palaeolatitude of Laurentia in the Proterozoic. *In*: KRÖNER, A. (ed.) *Precambrian Plate Tectonics*, Elsevier, Amsterdam, 561–598.

JACOBSEN, S. B. & HEIER, K. S. 1978. Rb–Sr isotope systematics in metamorphic rocks, Kongsberg Sector, South Norway. *Lithos*, **11**, 257–276.

JOHANSSON, Å. 1990. Age of the Önnestad syenite and some gneissic granites along the southern part of the Protogine Zone, southern Sweden. *In* GOWER, C. F., RIVERS, T. & RYAN, B. (eds) *Mid-Proterozic Laurentia–Baltica*. Geological Association of Canada, Special Paper **38**, 131–148.

—— & KULLERUD, L. 1993. Late Sveconorwegian metamorphism and deformation in southwestern Sweden. *Precambrian Research*, **64**, 347–360.

——, GEE, D. G. & LARIONOV, A. 1994. Precambrian basement within the eastern terrane of the Svalbard Caledonides. *Terra Nova*, **6**, Abstract Supplement 2, 8.

——, LINDH, A. & MÖLLER, C. 1991. Late Sveconorwegian (Grenville) high-pressure granulite facies metamorphism in southwest Sweden. *Journal of Metamorphic Petrology*, **9**, 283–292.

——, MEIER, M., OBERLI, F. & WIKMAN, H. 1993. The early evolution of the Southwest Swedish Gneiss Province: geochronological and isotopic evidence from southernmost Sweden. *Precambrian Research*, **64**, 361–388.

KERR, A. & FRYER, B. J. 1990. Sources of Early and Middle Proterozoic magmas in the Makkovik Province, Labrador: evidence from Nd isotope data. *In*: GOWER, C. F., RIVERS, T. & RYAN, B. (eds) *Mid-Proterozic Laurentia–Baltica*. Geological Association of Canada, Special Paper **38**, 53–64.

—— & —— 1994. The importance of late- and post-orogenic crustal growth in the early Proterozoic: evidence from Sm–Nd isotopic studies of igneous rocks in the Makkovik Province, Canada. *Earth & Planetary Science Letters*, **125**, 71–88.

KETCHUM, J. W. F., JAMIESON, R. A., HEAMAN, L. M., CULSHAW, N. G. & KROGH, T. E. 1994. 1.45 Ga granulites in the southwestern Grenville province: geologic setting, P–T conditions, and U–Pb geochronology. *Geology*, **22**, 215–218.

KISVARSANYI, E. B. & KISVARSANYI, G. 1990. Alkaline granite ring complexes and metallogeny in the Middle Proterozoic St. Francois terrane, southeastern Missouri, USA. *In*: GOWER, C. F., RIVERS, T. & RYAN, B. (eds) *Mid-Proterozic Laurentia–Baltica*. Geological Association of Canada, Special Paper **38**, 433–446.

KROGH, T. E. 1994. Precise U-Pb ages for Grenvillian and Pre-Grenvillian thrusting of Proterozoic and Archaean metamorphic assemblages in the Grenville Front tectonic zone, Canada. *Tectonics*, **13**, 963–982.

KULLERUD, L. & DAHLGREN, S. H. 1993. Sm–Nd geochronology of Sveconorwegian granulite facies mineral assemblages in the Bamble Shear Belt, South Norway. *Precambrian Research*, **64**, 389–402.

—— & MACHADO, N. 1991. End of a controversy: U–Pb geochronological evidence for significant Grenvillian activity in the Bamble area, Norway. *Terra Abstract*, **4**, 543.

LARSON, S-Å. & BERGLUND, J. 1992. A chronological subdivision of the Transscandinavian Igneous Belt – three magmatic episodes. *Geologiska Föreningens i Stockholm Förhandlingar*, **14**, 459–461.

——, ——, STIGH, J. & TULLBORG, E-L. 1990. The Protogine Zone, southwest Sweden: a new model – an old issue. *In*: GOWER, C. F., RIVERS, T. & RYAN, B. (eds) *Mid-Proterozic Laurentia–Baltica*. Geological Association of Canada, Special Paper **38**, 317–333.

LECHEMINANT, A. N. & HEAMAN, L. M. 1989. Mackenzie igneous events, Canada: Middle Proterozoic hotspot magmatism associated with ocean opening. *Earth & Planetary Science Letters*, **96**, 38–48.

LINDH, A. 1987. Westward growth of the Baltic Shield. *Precambrian Research*, **35**, 53–70.

LUMBERS, S. B., HEAMAN, L. M., VERTOLLI, V. M. & WU, T-W. 1990. Nature and timing of Middle Proterozoic magmatism in the Central Metasedimentary Belt, Grenville Province, Ontario. *In*: GOWER, C. F., RIVERS, T. & RYAN, B. (eds) *Mid-Proterozic Laurentia-Baltica*. Geological Association of Canada, Special Paper **38**, 243–276.

MARCANTONIO, F., DICKIN, A. P., MCNUTT, R. H. & HEAMAN, L. M. 1988. A 1,800 million-year-old Proterozoic gneiss terrane in Islay, with implications for the crustal structure and evolution of Britain. *Nature*, **335**, 62–64.

——, NCNUTT, R. H., DICKIN, A. P. & HEAMAN, L. 1990. Geochemical, isotopic and geochronological studies in the Frontenac Axis of the Grenville Province. *Chemical Geology*, **38**, 297–314.

MARTIGNOLE, J., MACHADO, N. & INDARES, A. 1994. The Wakeham Terrane: a Mesoproterozoic terrestrial rift in the eastern part of the Grenville Province. *Precambrian Research*, **68**, 291–306.

MAX, M. D. & SONET, J. 1979. A Grenville age for the pre-Caledonian rocks in NW Co. Mayo, Ireland. *Journal of the Geological Society, London*, **136**, 379–382.

MCEACHERN, S. J & VAN BREEMEN, O. 1993. Age of deformation within the Central Metasedimentary Belt boundary thrust zone, southwest Grenville orogen: constraints on the collision of the Mid-Proterozoic Elzevir terrane. *Canadian Journal of Earth Science*, **30**, 1155–1165.

MCLELLAND, J. M. & CHIARENZELLI, J. R. 1990. Geochronological studies in the Adirondack Mountains and the implications of a Middle Proterozoic tonalite suite. *In*: GOWER, C. F., RIVERS, T. & RYAN, B. (eds) *Mid-Proterozic Laurentia–Baltica*. Geological Association of Canada, Special Paper **38**, 175–194.

MCWILLIAMS, M. O. & DUNLOP, D. J. 1978: Grenville palaeomagnetism and tectonics. *Canadian Journal of Earth Science*, **15**, 687–695.

MENUGE, J. F. 1988. The petrogenesis of massif anorthosites: a Nd and Sr isotopic investigation of the Proterozoic of Rogaland/Vest Agder, SW Norway. *Contributions to Mineralogy & Petrology*, **98**, 363–373.

—— & DALY, J. S. 1990. Proterozoic evolution of the Erris Complex, northwest Mayo, Ireland: neodymium isotope evidence. *In*: GOWER, C. F., RIVERS, T. & RYAN, B. (eds) *Mid-Proterozic Laurentia–Baltica*. Geological Association of Canada, Special Paper **38**, 41–51.

MEZGER, K., ESSENE, E. J., VAN DER PLUIJM, B. A. & HALLIDAY, A. N. 1993. U–Pb geochronology of the Grenville orogen of Ontario and New York: constraints on ancient crustal tectonics. *Contributions to Mineralogy & Petrology*, **114**, 13–26.

MILNE, K. P. & STARMER, I. C. 1982. Extreme differentiation in the Proterozoic Gjerstad-Morkheia complex of South Norway. *Contributions to Mineralogy & Petrology*, **79**, 381–393.

MOORES, E. M. 1991. Southwest U.S. – East Antarctica (SWEAT) connection: A hypothesis. *Geology*, **19**, 425–428.

MORK, M. B. E. & MEARNS, E. W. 1986. Sm–Nd isotopic systematics of a gabbro-eclogite transition. *Lithos*, **19**, 255–267.

MORTON, A. C. & TAYLOR, P. N. 1991. Geochemical and isotopic constraints on the nature and age of basement rocks from Rockall Bank, NE Atlantic. *Journal of the Geological Society, London*, **148**, 631–634.

MUIR, R. J., FITCHES, W. R. & MALTMAN, A. J. 1992. Rhinns Complex: a missing link in the Proterozoic basement of the North Atlantic region. *Geology*, **20**, 1043–1046.

OWEN, J. V. & ERDMER, P. 1990. Middle Proterozoic geology of the Long Range Inlier, Newfoundland; regional significance and tectonic implications. *In*: GOWER, C. F., RIVERS, T. & RYAN, B. (eds) *Mid-Proterozic Laurentia–Baltica*. Geological

Association of Canada, Special Paper **38**, 215–231.

——, GREENOUGH, J. D., FRYER, B. J. & LONGSTAFFE, F. J. 1992. Petrogenesis of the Potato Hill pluton, Newfoundland: transpression during the Grenvillian orogenic cycle? *Journal of the Geological Society, London*, **149**, 923–935.

OWENS, B. E., DYMEK, R. F., TUCKER, R. D., BRANNON, J. C. & PODOSEK, F. A. 1994. Age and radiogenic isotopic composition of a late- to post-tectonic anorthosite in the Grenville Province: the Labrieville massif, Quebec. *Lithos*, **31**, 189–206.

PALMER, H. C. & DAVIS, D. W. 1987. Palaeomagnetism and U–Pb geochronology of volcanic rocks from Michipicoten island, Lake Superior, Canada: precise calibration of the Keweenawan polar wander track. *Precambrian Research*, **37**, 157–171.

PARK, R. G. 1992. Plate kinematic history of Baltica during the Middle to Late Proterozoic: A model. *Geology*, **20**, 725–728.

——, ÅHÄLL, K-I, & BOLAND, M. P. 1991. The Sveconorwegian shear-zone network of SW Sweden in relation to mid-Proterozoic plate movements. *Precambrian Research*, **49**, 245–260.

PATCHETT, P. J., BYLUND, G. & UPTON, B. G. J. 1978: Palaeomagnetism and the Grenville Orogeny: New Rb–Sr ages from dolerites in Canada and Greenland. *Earth & Planetary Science Letters*, **40**, 349–364.

PESONEN, L. J. 1990: The drift of Fennoscandia during the Proterozoic with special reference to the Bergslagen province, south-central Sweden. *Geologiska Föreningens i Stockholm Förhandlingar* **112**, 190–196.

—— 1991: The drift of Fennoscandia during the Proterozoic with special reference to the Bergslagen province, south-central Sweden; a reply. *Geologiska Föreningens i Stockholm Förhandlingar*, **113**, 251–253.

PIPER, J. D. A. 1976: Palaeomagnetic evidence for a Proterozoic Supercontinent. *Philosophical Transactions of the Royal Society, London*, **A280**, 469–490.

—— 1992. The palaeomagnetism of major (Middle Proterozoic) igneous complexes, South Greenland and the Gardar apparent polar wander track. *Precambrian Research*, **54**, 153–172.

PREVEC, S. A., McNUTT, R. H. & DICKIN, A. P. 1990. Sr and Nd isotopic and petrological evidence for the age and origin of the White Bear Arm complex and associated units from the Grenville Province in eastern Labrador. *In*: GOWER, C. F., RIVERS, T. & RYAN, B. (eds) *Mid-Proterozic Laurentia–Baltica*. Geological Association of Canada, Special Paper **38**, 65–78.

PUURA, V. & HUHMA, H. 1993. Palaeoproterozoic age of the East Baltic granulitic crust. *Precambrian Research*, **64**, 289–294.

RAGNHILDSTVEIT, J., SIGMOND, E. M. & TUCKER, R. D. 1994. Early Proterozoic supracrustal rocks west of the Mandal–Ustaoset Fault Zone, Hardangervidda, South Norway. *Terra Nova*, **6**, Abstract Supplement 2, 15–16.

RÄMO, O. T. & HAAPALA, I. 1990. The rapakivi granites of eastern Fennoscandia: a review with insights into their origin in the light of new Sm–Nd isotopic data. *In*: GOWER, C. F., RIVERS, T. & RYAN, B. (eds) *Mid-Proterozic Laurentia–Baltica*. Geological Association of Canada, Special Paper **38**, 401–415.

REX, D. C. & GLEDHILL, A. 1981. Isotopic studies in the East Greenland Caledonides ($72°–74°$ N) – Precambrian and Caledonian ages. *Rapport Grønlands Geologiske Undersøgelse*, **104**, 47–72.

RIVERS, T., MARTIGNOLE, J., GOWER, C. F. & DAVIDSON, A. 1989. New tectonic divisions of the Grenville Province, southeast Canadian Shield. *Tectonics*, **8**, 63–84.

ROBERTS, D. G., ARDUS, D. A. & DEARNLEY, R. 1973. Precambrian drilled rocks on the Rockall Bank. *Nature, Physical Science*, **244**, 21–23.

SAMUELSSON, L. & ÅHÄLL, K-I. 1985. Proterozoic development of Bohuslån, S.W. Sweden. *In*: TOBI, A. C. & TOURET, J. L. R. (eds) *The Deep Proterozoic Crust in the North Atlantic Provinces*. NATO Advanced Science Institutes Series, series C, **158**, 345–357.

SCHÄRER, U. & GOWER, C. F. 1988. Crustal evolution in eastern Labrador: constraints from precise U–Pb ages. *Precambrian Research*, **38**, 405–421.

——, KROGH, T. E. & GOWER, C. F. 1986. Age and evolution of the Grenville Province in eastern Labrador from U–Pb systematics in accessory minerals. *Contributions to Mineralogy & Petrology*, **94**, 438–451.

SIGMOND, E. M. O. 1978. Beskrivelse til det berggrunnsgeologiske kartbladet Sauda 1:250,000. *Norges Geologiske Undersøkelse, Skrifter*, **23**.

SILVER, L. T. 1978. *Precambrian formations and Precambrian history on Cochise County, southeastern Arizona*. New Mexico Geological Society Field Conference Guidebook, **29**, 157–163.

SIMS, P. K. 1990. *Precambrian basement map of the northern midcontinent, U.S.A.* United States Geological Survey Map 1-1853-A, 1 : 1,000,000.

SKIÖLD, T. 1976. The interpretation of Rb–Sr and K–Ar ages of Late Precambrian rocks in S.W. Sweden. *Geologiska Föreningens i Stockholm Förhandlingar*, **98**, 3–29.

SKJERLIE, F. J. & PRINGLE, I. R. 1978. Rb–Sr whole-rock isochron date from the lowermost gneiss complex of the Gaular area, west Norway and its regional implications. *Norsk Geologisk Tidsskrift*, **58**, 259–265.

SOEGAARD, K. & CALLAHAN, D. M. 1994. Late Middle Proterozoic Hazel Formation near Van Horn, Trans-Pecos Texas: evidence of transpressive deformation in the Grenville basement. *Geological Society of America Bulletin*, **106**, 413–423.

SØNDERHOLM, M. & JEPSEN, H. F. 1991. Proterozoic basins of North Greenland. *Bulletin Grønlands Geologiske Undersøgelse*, **160**, 49–69.

STARMER, I. C. 1985. The evolution of the South Norwegian Proterozoic as revealed by the major and mega-tectonics of the Kongsberg and Bamble Sectors. *In*: TOBI, A. C. & TOURET, J. L. R. (eds) *The Deep Proterozoic Crust in the North Atlantic*

Provinces. NATO Advanced Science Institutes Series, series C, **158**, 259–290.

—— 1990. Mid-Proterozoic evolution of the Kongsberg–Bamble belt and adjacent areas, southern Norway. *In*: GOWER, C. F., RIVERS, T. & RYAN, B. (eds) *Mid-Proterozic Laurentia–Baltica*. Geological Association of Canada, Special Paper **38**, 279–305.

—— 1991. The Proterozoic evolution of the Bamble Sector shear belt, southern Norway: correlations across southern Scandinavia and the Grenvillian controversy. *Precambrian Research*, **49**, 107–139.

—— 1993: The Sveconorwegian Orogeny in southern Norway, relative to deep crustal structures and events in the North Atlantic Proterozoic Supercontinent. *Norsk Geologisk Tidsskrift*, **73**, 109–132.

STEARN, J. E. F. & PIPER, J. D. A. 1984: Palaeomagnetism of the Sveconorwegian Mobile Belt of the Fennoscandian Shield. *Precambrian Research*, **23**, 201–246.

STEIGER, R. H. & MEIER, M. 1990. U-Pb isotope dilution analysis of zircon fragments for petrogenetic studies II. Analytical Results. *Abstracts Seventh International Conference on Geochronology, Cosmochronology and Isotope Geology*. Geological Society of Australia, **27**, 95.

STEPHENS, M. B. & WAHLGREN, C-H. 1994. Geometry and kinematics of ductile deformation zones in two Proterozoic orogens, south-central Sweden. *Terra Nova*, **6**, Abstract Supplement 2, 18–19.

SUOMINEN, V. 1987. Mafic dyke rocks in southwestern Finland. *In*: ARÖ, K. & LAITAKARI, I. (eds) *Diabases and other mafic rocks in Finland*. Geological Survey of Finland Report of Investigation **76**, 151–172.

SURLYK, F. 1991: Tectonostratigraphy of North Greenland. *Bulletin Grønlands Geologiske Undersøgelse*, **160**, 25–47.

THOMAS, J. J., SHUSTER, R. D. & BICKFORD, M. E. 1984. A terrane of 1,350 to 1,400 m.y.-old silicic volcanic and plutonic rocks in the buried Proterozoic of the mid-continent and in the Wet mountains, Colorado. *Geological Society of America Bulletin*, **95**, 1150–1157.

TORSKE, T. 1982. Structural effects on the Proterozoic Ullensvang Group (West Norway) relatable to forceful emplacement of expanding plutons. *Geologische Rundschau*, **71**, 104–119.

—— 1985. Terrane displacement and Sveconorwegian rotation of the Baltic Shield: a working hypothesis. *In*: TOBI, A. C. & TOURET, J. L. R. (eds) *The Deep Proterozoic Crust in the North Atlantic Provinces*. NATO Advanced Science Institutes Series, series C, **158**, 333–343.

——, BJÖRNSETH, H. M. & NILSEN, K. T. 1988. Calc-alkaline and associated metavolcanic rocks of the Ullensvang Group, South Norway Precambrian region. Geological Association of Canada-Mineralogical Association of Canada, Joint Annual Meeting, Program with Abstracts **13**, A125.

TRETTIN, H. P. 1987: Pearya:a composite terrane with Caledonian affinities in northern Ellesmere Island. *Canadian Journal of Earth Science*, **24**, 224–245.

TUCKER, R. D., DALLMEYER, R. D. & STRACHAN, R. A. 1993. Age and tectonothermal record of Laurentian basement, Caledonides of NE Greenland. *Journal of the Geological Society, London*, **150**, 371–379.

——, KROGH, T. E. & RÅHEIM, A. 1990. Proterozoic evolution and age-province boundaries in the central part of the Western Gneiss region, Norway: results of U–Pb dating of accessory minerals from Trondheimsfjord to Geiranger. *In*: GOWER, C. F., RIVERS, T. & RYAN, B. (eds) *Mid-Proterozic Laurentia–Baltica*. Geological Association of Canada, Special Paper **38**, 149–173.

UPTON, B. G. J. & EMELEUS, C. H. 1987. Mid-Proterozoic alkaline magmatism in southern Greenland: the Gardar Province. *In*: FITTON, J. & UPTON, B. G. J. (eds) *Alkaline Igneous Rocks*. Geological Society, London, Special Publications, **30**, 449–471.

VAN BREEMEN, O. & DAVIDSON, A. 1990. U–Pb zircon and baddeleyite ages from the Central Gneiss Belt, Ontario. *Geological Association of Canada Paper*, **89-2**, 85–92.

—— & HANMER, S. 1986. Zircon morphology and U–Pb geochronology in active shear zones: studies on syntectonic inclusions along the northwest boundary of the Central Metasedimentary Belt, Grenville Province, Ontario. *In: Current Research, Part B*. Geological Survey of Canada, Paper **86-1B**, 775–784.

——, DAVIDSON, A., LOVERIDGE, W. D. & SULLIVAN, R. W. 1986. U–Pb zircon geochronology of Grenville tectonites, granulites and igneous precursors, Parry Sound, Ontario. *In*: MOORE, J. M., DAVIDSON, A. & BAER, A. J. (eds) *The Grenville Province*. Geological Association of Canada Special Paper **31**, 191–207.

——, HALLIDAY, A. N., JOHNSON, M. R. W. & BOWES, D. R. 1978. Crustal additions in late Precambrian times. *In*: BOWES, D. R. & LEAKE, B. E. (eds) *Crustal Evolution in Northwestern Britain and Adjacent Regions*. Geological Journal Special Issue, **10**, 81–106.

WALKER, N. 1992. Middle Proterozoic geologic evolution of Llano uplift, Texas: evidence from U–Pb zircon geochronometry. *Geological Society of America Bulletin*, **104**, 494–504.

WARDLE, R. J., RYAN, B., PHILIPPE, S. & SCHÄRER U. 1990. Proterozoic crustal development, Goose Bay region, Grenville Province, Labrador, Canada. *In*: GOWER, C. F., RIVERS, T. & RYAN, B. (eds) *Mid-Proterozic Laurentia–Baltica*. Geological Association of Canada, Special Paper **38**, 197–214.

WELIN, E. & GORBATSCHEV, R. 1978. The Rb–Sr age of the Varberg charnockite, Sweden. *Geologiska Föreningens i Stockholm Förhandlingar*, **100**, 225–227.

—— & LUNDQVIST, T. 1970. New Rb–Sr age data for the Sub-Jotnian volcanics (Dala porphyries) in the Los-Hamra region. *Geologiska Föreningens i Stockholm Förhandlingar*, **92**, 35–39.

—— & SAMUELSSON, L. 1987. Rb–Sr and U–Pb isotope studies of granitoid plutons in the Göteborg region, southwestern Sweden. *Geologiska Föreningens i Stockholm Förhandlingar*, **109**, 39–45.

WINCHESTER, J. A. & MAX, M. D. 1990. A review of the Middle Proterozoic Annagh Division of northwest County Mayo, Ireland: a detached fragment of the Proterozoic North American Continent. *In*: GOWER, C. F., RIVERS, T. & RYAN, B. (eds) *Mid-Proterozic Laurentia-Baltica*. Geological Association of Canada, Special Paper **38**, 233–241.

WYNNE-EDWARDS, H. R. 1972. The Grenville Province. *In*: PRICE, R. A. & DOUGLAS, R. J. W. (eds) *Variations in Tectonic Styles in Canada*. Geological Association of Canada, Special Paper **11**, 263–334.

ZECK, H. P. & WILLADSEN, K. 1990. The ca. 1500 Ma Värmland hyperite suite, southwest Sweden – Petrography, magma chemistry and metasomatic changes of a series of partly recrystallised gabbroic intrusions. *In*: GOWER, C. F., RIVERS, T. & RYAN, B. (eds) *Mid-Proterozoic Laurentia–Baltica*. Geological Association of Canada, Special Paper **38**, 461–470.

The relationship between 1.88 Ga old magmatism and the Baltic–Bothnian shear zone in northern Sweden

A. WIKSTRÖM[1], T. SKIÖLD[2] & B. ÖHLANDER[3]

[1] *Geological Survey of Sweden, Box 670, S 751 28 Uppsala, Sweden*
[2] *Laboratory for Isotope Geology, Museum of Natural History, S 104 05 Stockholm, Sweden*
[3] *Division of Applied Geology, Luleå University of Technology, S 971 87 Luleå, Sweden*

Abstract: New geochemical data and U–Pb ages for Proterozoic mafic and felsic dykes and their granitic host rocks in the Bläsberget Massif, proximal to the Baltic–Bothnian shear zone, indicate that these rocks are part of a terrane distinct from Svecofennian terranes in Central and Southern Sweden. The dykes comprise a NNE-trending swarm which can be traced over a strike length of about 20 km. Most of the mafic dykes are internally deformed and have systematic internal structures showing a sinistral horizontal component and an eastern-side-up sense of movement. This sinistral shearing is also observed within the nearby Baltic–Bothnian shear zone.

The mafic dykes have a uniformly basaltic composition. High content of elements such as Ti and Zr illustrate the typical 'within-plate' character. The felsic dykes are chemically homogeneous and have a SiO_2 content of about 77 wt%. High content of Y, Yb and Nb confirms the 'within-plate' character of the dyke swarm.

Zircons extracted from one of the felsic dykes indicate a U–Pb age of *c.* 1.88 Ga. Approximately the same age has been obtained from the host rock granite and from a nearby occurrence of a similar late-orogenic granite. Since the dykes cross-cut most deformation and migmatite structures and most likely represent the waning stages of orogenic deformation in the area, it is evident that the timing of the orogenic events in the area is different and older than the Svecofennian development in central and southern Sweden.

A genetic relationship is suggested between the magmatism and the development of the Baltic–Bothnian shear zone.

A dyke swarm at Bläsberget, northern Sweden (Fig. 1) consisting of both mafic and felsic dykes was first observed by Ödman (1957). He also noted that the dykes cross-cut the major deformation and migmatitic structures, and in his stratigraphic scheme (Ödman 1957, table p. 133) these dykes were regarded by him to belong to the youngest group of magmatic rocks in the area. Further study of the dykes was motivated by their proximity with, and parallel trend to, the Baltic–Bothnian shear zone (Berthelsen & Marker 1986). This paper is a contribution to the on-going research in characterizing the geological development on both sides of the Archaean–Proterozoic palaeo-boundary in northern Sweden as defined by Öhlander *et al.* (1993). See also inset figure in Fig. 1.

Geology

The Baltic–Bothnian shear zone

This shear zone was described by Berthelsen & Marker (1986). It is mainly visible on regional, aeromagnetic maps of northern Sweden where it can be defined as a *c.* 10 km wide zone (Fig. 2). However, its continuation from the Gulf of Bothnia and southwards into the Baltic sea is very uncertain based on geophysical constraints (Aaro *et al.* 1994). In the field it is much more difficult to recognize the shear zone as a prominent structure. The major structures within the western Kalix map-sheets, where the zone has been investigated in more detail (Wikström 1993, 1995), are mostly from combined geological and geophysical data, defined as lithological breaks along a number of mainly north–south trending, unexposed lineaments. In part, the lithologies can be defined as distinct palaeosomes in different migmatite-dominated segments, all of which are permiated by a common potassium-rich granite. Some lineaments also separate segments with different metamorphic grades. The amphibolite facies foliations within the zone have in general a regular north–south trend, while they are more variable in the surrounding bedrock. In the migmatites, some local high-grade deformation structures with a dextral sense of shear have been observed (Wikström 1995). On the other hand one can find outcrops, especially in the coastal areas, where small-scale shear zones in lower amphibolite facies display a sinistral sense of shear. A regional, sinistral sense of shear can also be deduced from the aeromagnetic map (originally

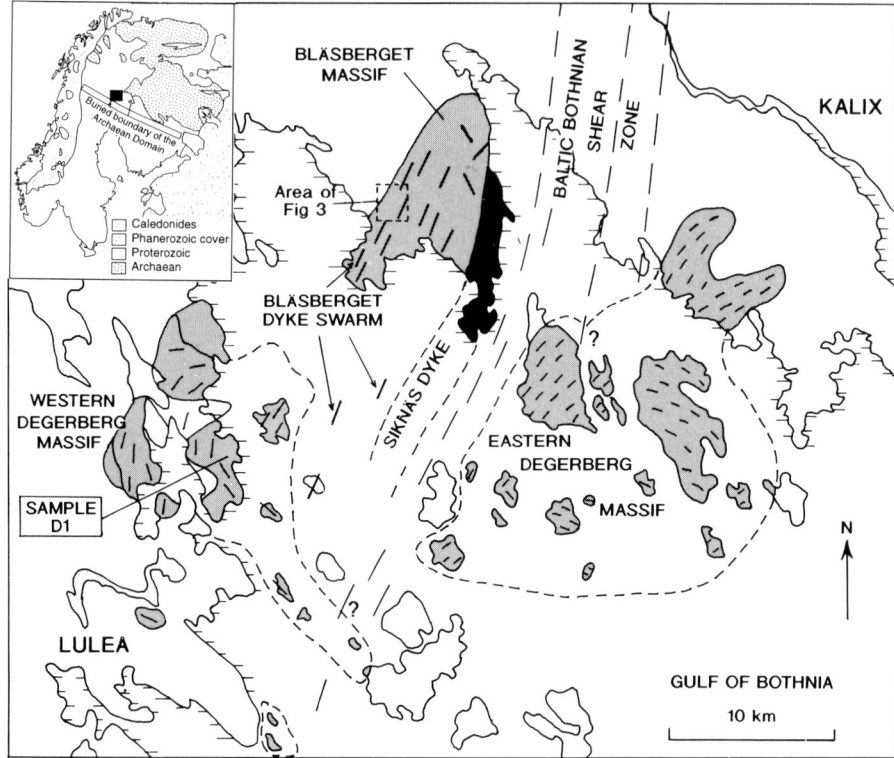

– Magmatic foliations within granite
— Magmatic and "solid state" foliations within granite

Fig. 1. The Bläsberget dyke swarm and other features discussed in the text.

1: 500 000) over the area (Fig. 2). In short, Berthelsen & Marker (1986) suggested an early, large-scale, dextral sense of shearing along this zone and a later, sinistral type of movement with even greater amounts of displacements. The field observations support this history of structural development in general terms although no quantification of the displacements in the area has been possible.

The Bläsberget dyke swarm

The Bläsberget dyke swarm runs in a NNE–SSW direction, to the west of and parallel with the Baltic–Bothnian shear zone (Fig. 1). The dyke swarm has been identified over a strike length of c. 20 km. Figure 3 illustrates the detailed geological relationships from the Bläsberget hill.

The **mafic dykes** are slightly retrograded and uralitized, fine-grained dolerites. Locally some elongated phenocrysts of andesine with combined Carlsbad-albite twins have been observed. The dyke width, in general, is within 1–3 m. All dykes have more or less well developed internal, incompetent deformation structures. As can be deduced from these, the dykes have experienced a sinistral, horizontal component of shear and a vertical component of eastern-side-up with a rotational axis around 335; 55 (azimuth; dip). Locally, a weak stretching lineation in 110; 15 has been developed. A narrow, marginal zone of vertical foliations due to contact strain is generally present. Close to the Baltic–Bothnian shear zone, thin dykes have been observed in a Riedel shear pattern with the same sense of sinistral shear as in the major shear zone (Wikström 1993, Fig. 5C).

Abundant **felsic dykes** occur together with the mafic dykes on Bläsberget Hill. Both groups of dykes have the same general structural trend. Cross-cutting relationships suggest the mafic dykes are most often the youngest, but the reverse relationship has also been observed. Apart from a hybridized dyke described below, the various dykes apparently did not chemically or physically interact. Microscopically the felsic dykes contain minor amounts of slightly corroded phenocrysts of have acted tectonically more competent and few deformation structures have been macroscopically observed in them.

A **hybrid (mixed) dyke** has also been found (Fig. 3). It is a heterogeneous, generally biotite-rich rock with local signs of magma mixing-textures like quartz ocelli (Fig. 4). The more biotite-rich parts in this dyke are relatively strongly deformed.

The Bläsberget and Degerberg granites

The country rock to the dyke swarm is the **granite** of the Bläsberget massif (Fig. 1). It is a grey to red, medium- to coarse-grained, partly feldspar phyric granite with greyish blue quartz and subordinate biotite. Macroscopically, the granite is undeformed. Along the eastern contact towards the Siknäs dyke (Fig. 1.), which is an intrusive breccia which will be described elsewhere, the grain size is reduced to fine grained in a zone approximately 400 m wide. The other contacts, which probably are towards 1.89 Ga (?) calc-alkaline tonalites, are unexposed.

The Degerberg granite (Skiöld 1977; Romer & Öhlander 1991) is found in the area of two major massifs (Fig. 1). It differs from the Bläsberget granite in being more homogeneously feldspar phyric.

The **Eastern Degerberg massif** (Fig. 1) consists of a biotite granite with characteristic, rectangular microcline phenocrysts (Wikström 1993, Fig. 4A). The orientation of the phenocrysts most likely reflects the magmatic flow during the intrusion. Few 'solid-state' deformation structures can be seen within the granite in spite of the fact that some of the Baltic–Bothnian shear zone structures trend towards the massif. There is a tendency for the phenocrysts to follow the general shape of the massif and this trend is not significantly disturbed when different water-covered, magnetic lineaments are crossed. Samples from the eastern massif have been dated at 1772+/−39 Ma (recalculated to the constants of Steiger & Jäger 1977) by the Rb–Sr method by Skiöld (1977).

The **Western Degerberg massif** (Fig. 1) has the same mineralogy and texture as the eastern massif but in the former one can also locally find 'solid-state' deformation structures mainly sub-parallel with the magmatic flow orientation of the rectangular, feldspar phenocrysts. A sample from the western massif has been radiometrically dated as described below.

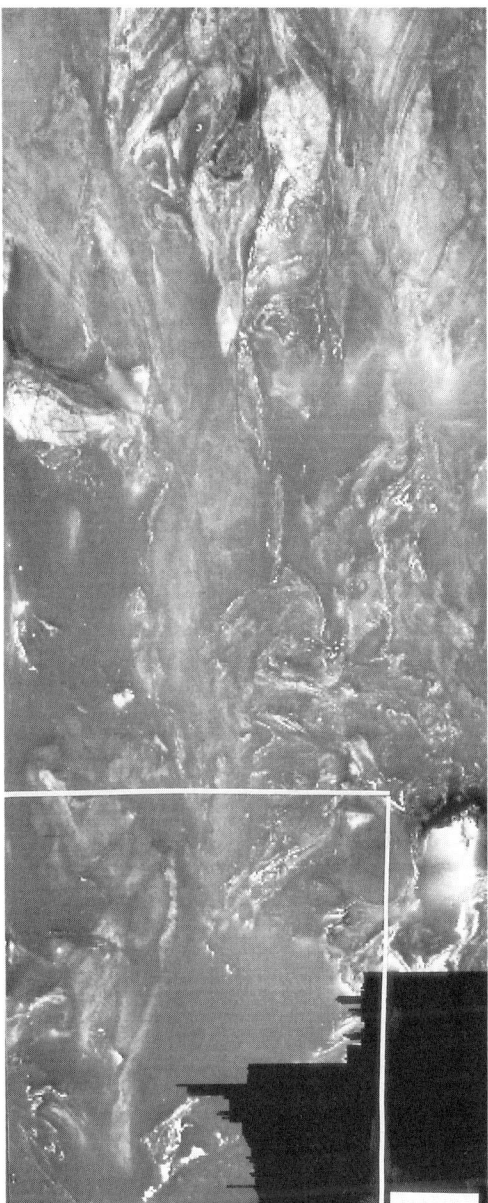

Fig. 2. Aeromagnetic relief map of the south-eastern part of the Norrbotten county. A sinistral sense of movement can be anticipated from the pattern over Baltic-Bothnian shear zone. Central and right-hand part of Fig. 1 is indicated. Scale bar is 10 km.

microcline and quartz in a very fine-grained matrix with quartz, feldspar and some biotite. The quartz is often granulated. Fluorite has been observed in accessory amounts.

Compared with the mafic dykes, the felsic dykes

Geochemistry

In 10 samples of the mafic dykes at Bläsberget, 5 samples of the felsic dykes, 3 samples of the host granite and 8 samples of the Degerberg granite from the western massif, major elements and Ba,

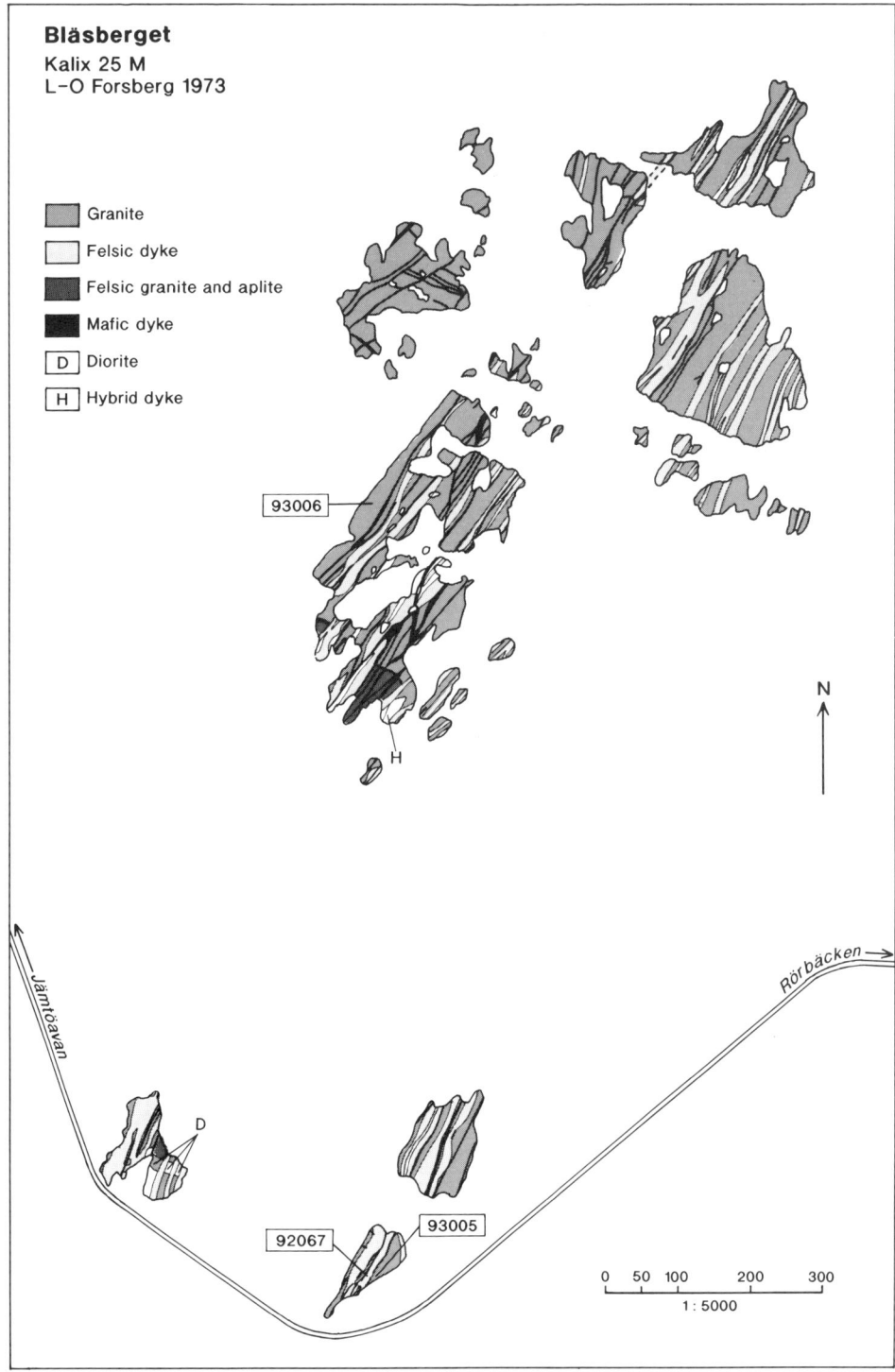

Fig. 3. A map of the dyke swarm on the Bläsberget hill. Sites for the radiometrically dated samples are indicated.

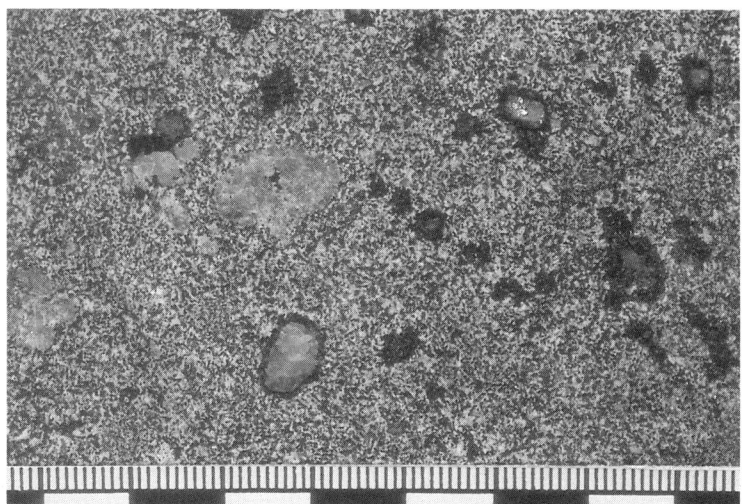

Fig. 4. Quartz ocelli (quartz grains with rims of hornblende) indicative of magma mixing in hybrid dyke on Bläsberget.

Be, Co, Cr, Cu, La, Ni, Nb, Sc, Sr, V, Y, Yb, Zn and Zr were analysed by SGAB Analys in Luleå by ICP-AES. In addition to the new chemical analyses, we have also used 17 unpublished major element XRF analyses of the Degerberg granite from the eastern massif obtained from the Geological Survey of Sweden.

Chemical analyses of typical samples of the various rock types are shown in Table 1. The mafic dykes at Bläsberget have probably to a large extent preserved their primary chemical composition. This is indicated by their restricted compositional range, illustrated by the well-defined cluster of samples in the R_1-R_2 diagram (La Roche et al. 1980) shown in Fig. 5A. Most samples plot in the monzo-gabbro and gabbro fields. All samples plot in the normal igneous field of the Hughes Igneous Spectrum (Fig. 5B). The SiO_2 content ranges in a narrow interval from 48.0 to 50.0 wt%. The mafic dykes at Bläsberget are rather rich in Fe (Fe_2O_3-t content between 12.1 and 14.6 wt%) and low in Mg (MgO between 4.80 and 6.81 wt%).

The obvious within-plate setting of the Bläsberget dyke swarm is reflected by the trace element composition of the mafic dykes. The relatively high Zr contents of the mafic dykes indicate within-plate field character, shown in Figs 5C and D. Other trace element dicrimination diagrams give the same result.

The felsic dykes are also chemically homogeneous with a very restricted compositional range. They are potassic (K_2O content between 4.84 and 4.92 wt%) and the SiO_2 content is high (from 76.9 to 77.3 wt%). They plot close to the dividing line between the granite and alkali granite fields in Fig. 5A. The low Ba, Sr and V contents attest to their leucocratic character. The content of elements such as Y, Yb and Nb is high. The Y and Nb content ranges from 70 to 85 ppm, typical of granitoids intruded in within-plate settings (Pearce et al. 1984).

The samples of the Degerberg granite form a rather well-defined trend in the R_1-R_2 diagram, with most samples plotting close to the dividing line between the granite and granodiorite fields. The trend continues into the quartz monzonite field. The samples of the host granite at Bläsberget plot on the same trend as the Degerberg granite samples. Also the trace element characteristics of the Bläsberget host granite are the same as those of the typical Degerberg granite. Most Degerberg granite samples are peraluminous, reflecting the high biotite content. The content of elements such as Y and Nb is too low for the Degerberg granitoid (including the Bläsberget granite) to be classified as a 'within-plate' granitoid according to Pearce et al. (1984). It is unlikely that the magma of the felsic dykes at Bläsberget was formed by differentiation of the magma that crystallized to the granite at Bläsberget. There is no evidence that felsic dykes are associated with the Degerberg granite elsewhere and there is no trend with increasing content of elements such as Y, Yb and Nb correlated with increasing SiO_2 content within the Degerberg suite. Although the felsic dykes and the host granite at Bläsberget have almost the same

Table 1. Chemical analyses of representative samples. SiO_2–LOI in wt%, Ba–Zr in ppm

	Bläsberget dyke swarm					Degerberg granitoids				
	mafic dykes			felsic dykes		Bläsberget		Western massif		
Sample no.	92401	92408	92411	92402	92406	93001	93002	D13	D17	D21
SiO_2	50.0	49.1	48.0	76.9	77.3	69.9	69.3	61.6	67.9	67.1
Al_2O_3	15.1	15.7	15.4	12.3	12.3	14.2	13.9	15.3	14.6	14.8
Fe_2O_3-tot	14.6	13.4	14.2	1.69	1.75	4.62	4.28	8.32	4.39	4.67
MnO	0.21	0.19	0.20	0.02	0.02	0.06	0.06	0.11	0.06	0.07
TiO_2	2.05	1.60	1.72	0.11	0.11	0.57	0.66	1.12	0.67	0.73
MgO	4.80	6.22	6.46	0.09	0.09	1.39	1.13	1.86	0.96	1.15
CaO	7.90	8.89	9.15	0.78	0.75	1.67	2.05	3.24	2.40	2.36
K_2O	1.69	1.43	1.85	4.90	4.89	4.10	4.31	3.95	4.55	4.40
Na_2O	2.96	2.84	2.76	3.49	3.46	3.09	3.44	3.29	3.43	3.53
P_2O_5	0.49	0.40	0.41	0.03	0.03	0.17	0.22	0.35	0.24	0.25
LOI	0.4	1.0	0.7	0.5	0.3	0.4	0.3	0.5	0.4	0.6
Ba	570	356	447	67	67	870	950	2770	1089	1072
Be	2.6	1.2	<1	3.6	3.8	2.8	2.9	<1	1.2	1.6
Co	38	47	44	<6	<6	8	<6	13	<6	6
Cr	55	84	102	<10	26	62	28	50	18	20
Cu	42	50	32	11	11	25	22	18	24	14
La	32	24	24	54	56	45	54	92	55	38
Nb	25	24	26	25	24	11	14	19	10	10
Ni	38	62	62	9	7	30	15	25	10	9
Sc	24	26	27	2.9	3.0	7.6	6.8	18	9.6	7.9
Sr	482	409	355	18	16	183	226	486	263	293
V	199	193	219	<6	<6	61	52	78	44	47
Y	30.3	24.7	25.8	60.3	59.9	17.7	24.5	30.0	21.9	21.1
Yb	4.28	3.61	3.73	6.43	6.27	1.52	2.11	2.91	1.78	2.12
Zn	252	136	162	96	88	77	74	94	48	36
Zr	162	137	135	188	190	223	330	514	231	397

U–Pb zircon age (see below), the conclusion would be that they were not formed by the same processes, and they do not stem from the same magma systems.

Geochronology

U–Pb zircon dating has been carried out on four rock samples close to the Baltic–Bothnian shear zone (Figs 1 & 3). The granite sample (D1) of the Western Degerberg massif was investigated by Dr Matti Vaasjoki at the Unit for Isotope Geology of the Geological Survey of Finland using large-scale standard zircon technique, while the Bläsberget granite and the transecting felsic dyke were investigated by one of the authors (T.S) at the Museum of Natural History in Sweden. Although the number of zircon crystals used for each determination as well as spiking, blank levels and data reduction differ between the laboratories, the chemical techniques are basically similar and stemming from the one presented by Krogh (1973). Therefore, the reported data and age results (Table 2 and Fig. 6) are considered comparable for the kind of relatively uncomplicated zircon analyses presented here.

Bläsberget, felsic dyke

This sample (92067) has stubby, transparent, short-prismatic zircons with simple pyramidal terminations. The colour of the zircons is in general somewhat brownish and the crystal faces give an impression of having been corroded at some later stage. Zircon fraction number 3 composes less regularly crystallized grains. Before decomposition, all zircons were severely abraded which reduced their weight (volume) to less than half of the original. A regression analysis of the dyke zircons yields upper and lower concordia intercepts at 1881+/−8 and 564 Ma, respectively (Fig. 6a). The MSWD (mean square of weighted deviates) value is less than unit and there are no reasons to suspect influence from more than one stage of zircon crystallization.

The very high U content of the dyke zircons, about 1200 ppm U in comparison to c. 200 ppm U for the country rock granite zircons, excludes the

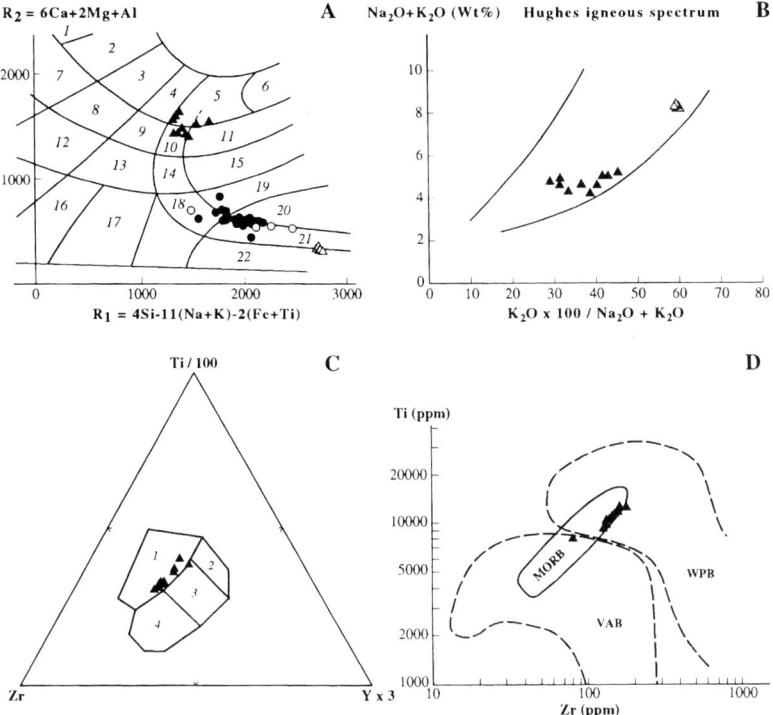

Fig. 5. Geochemical characteristics of the various rock types. (**A**). R_1–R_2 diagram after La Roche *et al.* (1980). Molar proportions multiplied by 1000. 2 = melteigite, 3 = theralite, 4 = alkali gabbro, 5 = (olivine) gabbro, 6 = gabbro–norite, 7 = ijolite, 8 and 12 = essexite, 9 = syeno-gabbro, 10 = monzo-gabbro, 11 = gabbro–diorite, 13 = syeno-diorite, 14 = monzonite, 15 = monzo-diorite and diorite, 16 = nepheline syenite, 17 = syenite, 18 = quartz monzonite, 19 = tonalite, 20 = granodiorite, 21 = granite, 22 = alkali granite. Filled triangles = mafic dykes from Bläsberget, open triangles = felsic dykes from Bläsberget, filled circles = Degerberg granite, open circles = Bläsberget granite. (**B**). Hughes Igneous Spectrum (Hughes 1972). Symbols as in (A). (**C**). Ti–Zr–Y diagram (Pearce & Cann 1971). Symbols as in (A). 1 = within-plate basalt, 2 = low-K tholeiite, 3 = ocean floor basalt + low-K tholeiite + calc-alkali basalt, 4 = calc-alkali basalt. (**D**). Ti–Zr diagram (Pearce 1982). Symbols as in (A). MORB = mid-ocean ridge basalt, VAB = volcanic arc basalt and WPB = within-plate basalt.

influence of granitic zircons among those analysed from the porphyry dyke. Thus, we state with quite some confidence that the present zircon dating of the dyke, which points at an age very similar to that of the country rock granite, reflects the crystallization of the dyke zircons at a time considered to approximate the dyke intrusion.

The Bläsberget massif

Two granite samples from this massif have been analysed. The amount of zircon crystals in each of the analysed fractions were between 7 and 18 grains of a size varying from 74 to 150 µm, thus permitting extreme selection and examination. In general, the selected zircons exhibit short-prismatic morphology with no visible inclusions, like the zircons described from the western Degerberg massif, although long-prismatic prisms also appear. The best quality crystals (fractions 3, 5 and 7) are completely transparent and almost devoid of internal textures or fracturing, while the others are more or less turbid from micro-fractures which make screening of the internal features difficult. Some of the latter crystals were abraded (Krogh 1982) in order to remove possible metamict outer portions, but also to get an indication whether or not these zircons are heterogeneous. We believe that fractions 2 and 8 exhibit memories of isotopically older lead which implies that these granites were to some extent formed from older sialic crust. A regression of the remaining 6 fractions indicates an age of crystallization of 1880 +/–7 Ma (Fig. 6b). The MSWD value of 5.6 is slightly high which indicates scatter in excess of analytical errors.

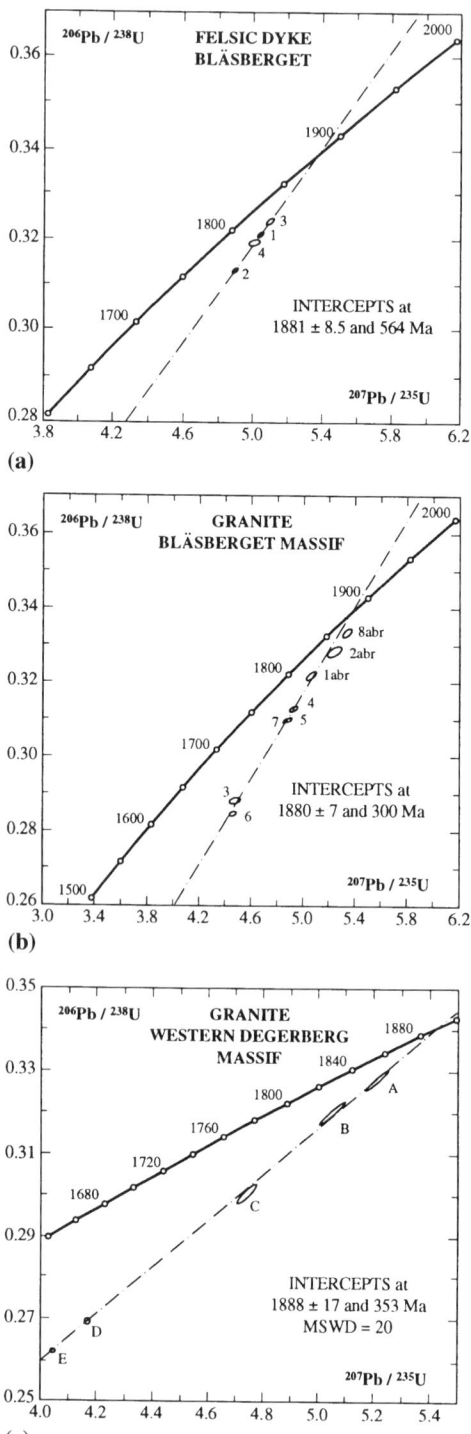

Fig. 6. Concordia diagrams of zircon fractions illustrating the isotope data in Table 2. In (b), fractions from two different samples have been included.

The Western Degerberg massif

The zircons of this sample are relatively coarse, about 80% of the grains being >100 μm. They make up stubby, euhedral prisms with simple pyramidal terminations. The length to width ratio is 2–4 with a median of about 2.5. The crystals are generally turbid and neither cores nor zonation are visible. The high-density zircon fraction (B in Table 2) is split and a part of the zircons has been abraded (fraction A). This fraction is dominated by colourless (light) zircon crystals, while metamict brown (dark) zircons make up about 30% of the 4.2–4.3 low-density fraction. It has been possible to separate these lighter zircons into fractions of light (D) and dark (E) crystals by hand-picking under microscope.

Because of the relatively large sample sizes used (1 to 3 mg), the composition of the common lead correction in the analysis of these zircons has no significant imprint on the results. The abraded zircons (A in Fig. 6c) are the most concordant, and since the light (D) as well as the dark (E) zircons plot close to the best-fit discordia line, colour difference seems to be insignificant in this respect. A discordia line with upper and lower intercept ages of 1888 +/–17 Ma and 353 Ma, respectively, has been calculated.

However, the high MSWD value of 20 makes the regression line of all data into an errorchron. If the technically poor analysis C is omitted, both the age and the MSWD value stay almost the same. The elevated MSWD value may, in the present case, be taken to indicate heterogeneities in the zircon population and/or differentiates in the impact of episodic lead loss and lead diffusion from large crystals. Thus, although there are potential heterogeneities relating to an inherited lead component, internal cores have not been identified. The significance of the errorchron age is further strengthened by the fact that it is close to the $^{207}Pb/^{206}Pb$ age of the most concordant plotting and abraded zircons.

Discussion

The felsic and mafic dykes are regarded as approximately contemporaneous (but derived from different, nearby magma chambers) due to their mutually cross-cutting field relationships and from the presence of the hybridized dyke in the central part of the complex.

The fracturing of the bedrock connected with the intrusion of the Bläsberget dyke swarm runs parallel to the Baltic–Bothnian shear zone. It is proposed that these features are genetically related to each other and that the dyke fractures were

Table 2. *Analytical data on zircons*

No.	Fraction[a] (μm)	Weight (μg)	U (ppm)[b]	Pb rad (ppm)[b]	$^{206}Pb/^{204}Pb$ meas.	Radiogenic (at. %)[c] ^{206}Pb	^{207}Pb	^{208}Pb	Atomic ratios[d] $^{206}Pb/^{238}U$	$^{207}Pb/^{235}U$	$^{207}Pb/^{206}Pb$ (Ma)
92067, Dyke porphyry at Bläsberget sampled at the same locality as 93005											
1	>106, spr, abr	23	1208	409.7	8050	81.5	9.28	9.25	0.3212 ± 7	5.045 ± 12	1863 ± 3
2	<74, spr, abr	29	1317	435.0	6850	81.6	9.25	9.19	0.3131 ± 7	4.894 ± 13	1854 ± 3
3	an, abr	24	1206	412.8	3780	81.5	9.30	9.24	0.3241 ± 7	5.100 ± 15	1866 ± 4
4	74–106, spr, abr	17	1285	435.9	638	81.0	9.21	9.79	0.3193 ± 8	5.009 ± 25	1861 ± 7
93006, Bläsberget granitoid sampled at Swedish grid 731800N 180810E											
1	>106, abr	27	174	61.9	3050	78.0	8.91	13.06	0.3217 ± 10	5.064 ± 27	1867 ± 7
2	>106, cr, abr	22	221	78.8	2356	79.3	9.19	11.51	0.3284 ± 10	5.247 ± 34	1894 ± 9
3	<106, bQ	50	188	59.9	1020	77.6	8.74	13.62	0.2883 ± 6	4.476 ± 32	1842 ± 11
4	<106, cr	58	182	62.1	3800	79.1	9.02	11.92	0.3129 ± 6	4.924 ± 20	1867 ± 6
5	>106, bQ	60	150	51.4	2270	77.7	8.80	13.43	0.3101 ± 6	4.887 ± 16	1869 ± 4
6	>106, cr	65	180	56.6	2805	77.9	8.85	13.25	0.2847 ± 5	4.461 ± 17	1859 ± 6
93005, Bläsberget granitoid sampled at Swedish grid 731710N 180810E											
7	<106, bQ	51	243	81.5	8800	79.5	9.05	11.45	0.3096 ± 6	4.861 ± 12	1863 ± 3
8	>106, cr, abr	36	171	62.2	2100	78.7	9.41	12.18	0.3334 ± 8	5.342 ± 24	1899 ± 6
D1, Granitoid of the Western Degerberg massif, sampled at Swedish grid 730410N 180025E (Analysed by Dr Matti Vaasjoki, Geological Survey of Finland)											
A	d > 4.5, abr	340	321	111.8	2113	83.0	9.62	7.35	0.3281 ± 19	5.216 ± 30	1885 ± 2
B	d > 4.5	370	331	115.1	1201	82.7	9.52	7.78	0.3202 ± 19	5.055 ± 30	1872 ± 2
C	d = 4.3–4.5	350	495	155.9	2330	83.9	9.65	6.42	0.3005 ± 17	4.742 ± 30	1871 ± 4
D	d = 4.2–4.3, light	260	760	219.7	1210	83.8	9.45	6.79	0.2693 ± 16	4.171 ± 20	1837 ± 1
E	d = 4.2–4.3, dark	335	784	216.1	1751	84.3	9.48	6.18	0.2621 ± 15	4.044 ± 20	1831 ± 1

[a] Size fractions pf colourless to pale brownish, clear, transparent and non-magnetic (Frantz isodynamic separator at 1.6 A and 2° tilt) zircons unless otherwise indicated; 74, 106 size in microns; spr = short-prismatic (length/width, 3); an = anhedral, sub-rounded; abr = abraded (Krogh 1982); cr = cracks; bQ = best quality crystals, d = density of stubby, euhedral zircon prisms in g cm^{-3}.
[b] Concentrations in pg/μg of U and Pb are known to about ± 5% for sample weights of about 40 μg.
[c] Corrected for lead blank and 1.88 Ga common lead (Stacey & Kramers 1975).
[d] Corrected for fractionation, spike, blank and initial common Pb; errors are at the 2σ level and refer to the last significant digits of the isotopic ratios.

formed by a stress-release mechanism in the same way as the dykes in the Riedel pattern (Wikström 1993, Fig. 5c) close to the shear zone. Although the felsic dykes and the country rock granite have overlapping ages of approximately 1.88 Ga, the dykes do not have a synplutonic character. The granite must have been wholly crystallized before the dyke intrusions.

These dykes clearly cross-cut the orogenic deformation structures and migmatised sedimentary gneisses on the islands south of the Bläsberget massif. The received age of 1881+/-8 Ma was therefore surprising for the felsic dykes of the combined felsic–mafic dyke swarm. In southern Sweden a major Svecofennian orogenic deformation and metamorphic episode have taken place c. 1.84–1.85 Ga (i.e. Persson & Wikström 1993). An age of c. 1.90 Ga is also generally associated with calc-alkaline, tonalite-dominated intrusives which have probably intruded in a compressive, orogenic regime.

In addition, the Degerberg granite, which has a late-orogenic character and was probably emplaced in a tectonic regime transitional from compressive to extensional (e.g. group H_{LO} of Barbarin 1990), is surprisingly old in relation to these tonalites.

It is not likely that the structural pattern (as seen, for example, in the aeromagnetic map) within the Baltic–Bothnian shear zone, is exclusively as old as 1.88 Ga, however. The intrusion of large volumes of potassium-rich granites within the shear zone, with a probable age c. 1.80 Ga, would hardly have left the old structures undisturbed. Some of the structures most likely post-date this age.

The geochemical results show that both the mafic and felsic dykes at Bläsberget have characteristics typical of rocks formed within tensional zones in continental crust (Fig. 5). The Bläsberget granite, on the other hand, as well as the whole Degerberg granite suite, has a content of elements such as Y, Yb and Nb which are too low to classify it as a typical 'within-plate' granitoid (Pearce *et al.* 1984). This would indicate that the formation of the Bläsberget granite was not directly associated with a tensional setting. Since the felsic dykes and the granite have the same U–Pb zircon age within analytical uncertainties, the geochemical results indicate that the tectonic conditions changed rapidly from the time of genesis of the granite to the time of formation of the dyke swarm. Another possibility is that the dyke swarm magmas and the granite magma were formed simultaneously but at different depths in the continental crust, and that the granite, which perhaps formed at a shallower depth than the dyke swarm magmas, crystallized earlier than the dykes. The most probable explanation for the Bläsberget intrusions is that contrasting rock types formed in a short time interval, when rapidly changing tectonic conditions prevailed at the waning orogenic stages in the area.

Romer & Öhlander (1991) suggested that the Degerberg granite could have a genetic relationship to the Baltic–Bothnian shear zone. The results of this study, which indicate that the shear zone was active around 1.88 Ga, support this conclusion.

Further support for the interpretation of combined tectonic and magmatic activity in this zone is found in the investigation by Björlykke & Ettner (1994) who concluded that magmatic activity at c. 1.88 Ga could be linked to the part of the Baltic–Bothnian shear zone occurring on the Norwegian side of the border.

Conclusions

- Felsic and mafic dykes and possibly the Degerberg igneous suite were emplaced in connection with tectonic movements along the Baltic–Bothnian mega-shear around, c. 1.88 Ga.
- The termination of major orogenic events took place earlier in the investigated area compared with the southern and central parts of the Svecofennian province. A possible explanation for this difference is the existance of a terrane boundary in the vicinity of Luleå and towards the northwest as suggested by Öhlander *et. al.* (1993).

The analysis D1 from the Western Degerberg massif was performed at the Unit for Isotope Geology of the Geological Survey of Finland under the leadership of Dr Matti Vaasjoki and was financed by the Mid-Norden project. A part of the investigation was also financed by the Research and Development program within the Geological Survey of Sweden. Dr Steve Noble and one anonymous referee made valuable comments on an earlier version of the manuscript. We gratefully acknowledge all support during the investigation and preparation of the manuscript.

References

AARO, S., ELO, S., GUSTAVSSON, N., HULT, K., KIHLE, O., ET AL. 1994. Mid-Norden: Gravity and magnetic maps. *21st Nordic Geological Wintermeeting*, Luleå.

BARBARIN, B. 1990. Granitoids: main petrogenetic classifications in relation to origin and tectonic setting. *Geological Journal*, **25**, 227–238.

BERTHELSEN, A. & MARKER, M. 1986. 1.9–1.8 Ga old

strike-slip megashears in the Baltic shield, and their plate tectonic implications. *Tectonophysics*, **128**, 163–181.

BJÖRLYKKE, A. & ETTNER, D. 1994. A model of gold transport and deposition from the Bidjovagge Au–Cu mine in Finnmark, Norway. *21st Nordic Geological Wintermeeting, Luleå*.

HUGHES, C. J. 1972. Spilites, keratophyres and the igenous spectrum. *Geological Magazine*, **109**, 513–527.

KROGH, T. E. 1973. A low contamination method for hydrothermal decomposion of zircon and extraction of U and Pb for isotopic age determinations. *Geochimica et Cosmochimica Acta*, **37**, 485–494.

—— 1982. Improved accuracy of U–Pb zircon ages by the creation of more concordant systems using an air abrasion technique. *Geochimica et Cosmochimica Acta*, **46**, 637–649.

LA ROCHE, H. DE, LETERRIER, J., GRANDCLAUDE, P. & MARCHAL, M. 1980. A classification of volcanic and plutonic rocks using R1R2 diagram and major element analyses. Its relationships with current nomenclature. *Chemical Geology*, **29**, 183–210.

ÖDMAN, O. 1957. Beskrivning till berggrundskarta över urberget i Norrbottens län. *Sveriges geologiska undersökning*, Ca **41**, 1–151. (English summary).

ÖHLANDER, B., SKIÖLD, T., ELMING, S.-Å., BABEL WORKING GROUP, CLAESSON, S. & NISCA, D. 1993. Delineation and character of the Archaean–Proterozoic boundary in northern Sweden. *Precambrian Research*, **64**, 67–84.

PEARCE, J. A. 1982. Trace element characteristics of lavas from destructive plate boundaries. *In*: THORPE, R. S. (ed.) *Andesites, Orogenic Andesites and Related Rocks*. Wiley and Sons, Chichester, 525–548.

—— & CANN, J. R. 1971. Ophiolite origin investigated by discriminant analysis using Ti, Zr and Y. *Earth and Planetary Science Letters*, **12**, 339–349.

——, HARRIS, N. B. W. & TINDLE, A. G. 1984. Trace element discrimination diagrams for the tectonic interpretation of granitic rocks. *Journal of Petrology*, **25**, 956–974.

PERSSON, P.-O. & WIKSTRÖM, A. 1993. A U–Pb dating of the Askersund granite and its marginal augen gneiss. *Geologiska Föreningens i Stockhom Förh.*, **115**, 321–329.

ROMER, R. & ÖHLANDER, B. 1991. The occurrence of the Degerberg migmatite granite and its constraints on the geological development of the Luleå area, northern Sweden. *Geologiska Föreningens i Stockholm Förh.*, **113**, 121–129.

SKIÖLD, T. 1977. Granite ages in the Kalix area, northern Sweden. *Geologiska Föreningens i Stockholm Förh.*, **99**, 75–79.

STEIGER, R. H. & JÄGER, E. 1977. Convention on the use of decay constants in geo- and cosmochronology. *Earth and Planetary Science Letters* **36**, 359–362.

WIKSTRÖM, A. 1993. Berggrundskartan Kalix SV. *Sveriges geologiska undersökning Ai* **81** (English summary).

—— 1995. Berggrundskartan Kalix NV. *Sveriges geologiska undersökning Ai* **79** (English summary).

The Mesoproterozoic cratonization of Baltica – new age constraints from SW Sweden

J. N. CONNELLY[1] & K-I. ÅHÄLL[2]

[1] *Department of Geological Sciences, University of Texas at Austin, 78712, USA*
[2] *Department of Geology, Earth Sciences Centre, University of Göteborg, S-413 81 Göteborg, Sweden*

Abstract: Comparable early Mesoproterozoic tectonic histories resulted in the formation of 1.71–1.62 Ga Labradorian and 1.75–1.55 Ga Gothian crust in Laurentia and Baltica respectively, along a postulated contiguous margin. New U–Pb geochronology in southwest Sweden has constrained the timing of three consecutive stages of the Gothian Orogeny. Calc-alkaline magmatism, presumably related to subduction, occurred as late as 1587 ± 3 Ma, indicating convergence and consumption of oceanic crust until this time. These magmas intruded terranes of previously accreted arcs, thus providing a minimum age for their juxtaposition and accretion. The age of the latest stage of Gothian migmatization is constrained between 1555 ± 2 and 1553 ± 2 Ma, synchronous with the transition from ductile to brittle conditions at the presently exposed erosional level. This transition permitted the emplacement of a suite of mafic–felsic 1555–53 Ma intrusions during an extensional regime that may have coincided with the post-deformational collapse of the Gothian Orogen. These age refinements in southwest Sweden indicate that the final stage of the Gothian Orogeny in Baltica occurred considerably later than that of the Labradorian Orogeny in Laurentia. The preservation of Gothian-aged titanite from southwest Sweden indicates that post-Gothian metamorphism, including that of Sveconorwegian–Grenvillian age, was not penetrative in this region.

The amalgamation of Archaean and Palaeoproterozoic microcontinents into proto-Laurentia–Baltica is thought to have been followed by stages of crustal growth along its western margin, relative to present-day Baltica (e.g. Gower *et al.* 1990). Although this margin appears to have experienced a grossly similar evolution during the Mesoproterozoic, it is becoming clear that important evolutionary stages were not synchronous. This paper reports new U–Pb geochronological data from SW Sweden that constrain the timing of successive stages of crustal growth and stabilization of Mesoproterozoic Baltica. These results bear on the diachroneity of events along the western Laurentia–Baltica margin (present-day Scandinavian frame of reference) and constrain the absolute ages of well established relative chronologies in SW Sweden. These constraints permit more accurate correlation of events between terranes and thus allow for the construction of more detailed tectonic models for the evolving geometry of the Mesoproterozoic Laurentia–Baltica margin.

Geological setting

The Southwest Scandinavian Domain (SSD) is located in the southwestern part of the Baltic/Fennoscandian Shield (Fig. 1) and was formed during the 1.75–1.55 Ga Gothian Orogeny (Gaál & Gorbatschev 1987; Gorbatschev & Bogdanova 1993). Since this same area was reworked by the 1.2–0.9 Ga Sveconorwegian–Grenvillian Orogeny it has also been termed the Sveconorwegian Province (Berthelsen 1980). Recognizing the poly-deformed nature of this domain, the less specific term SSD is used throughout this paper. East of the SSD, 1.85–1.65 Ga granitoid rocks of the Transscandinavian Igneous Belt (TIB; Lindh & Gorbatschev 1984; Larson & Berglund 1992; Persson & Wikström 1993) separate the SSD and the *c.* 1.9 Ga Svecofennian Domain (Fig. 1).

The SSD comprises several crustal segments that are separated by major N–S trending shear zones (Fig. 1), previously documented as terrane boundaries (Berthelsen 1980). It should be noted that these shear zones, at least those to the east of the Oslo Graben (Fig. 1), were active during the Sveconorwegian Orogeny (Larson *et al.* 1990; Park *et al.* 1991) and therefore may be unrelated to the formation and amalgamation of Gothian crust.

The Eastern Segment comprises the easternmost unit of the SSD adjacent to the TIB and is bounded on the east and west by the Protogine and Mylonite zones, respectively (Fig. 1). This segment comprises *c.* 1.70–1.66 Ga orthogneisses (Larson *et al.*

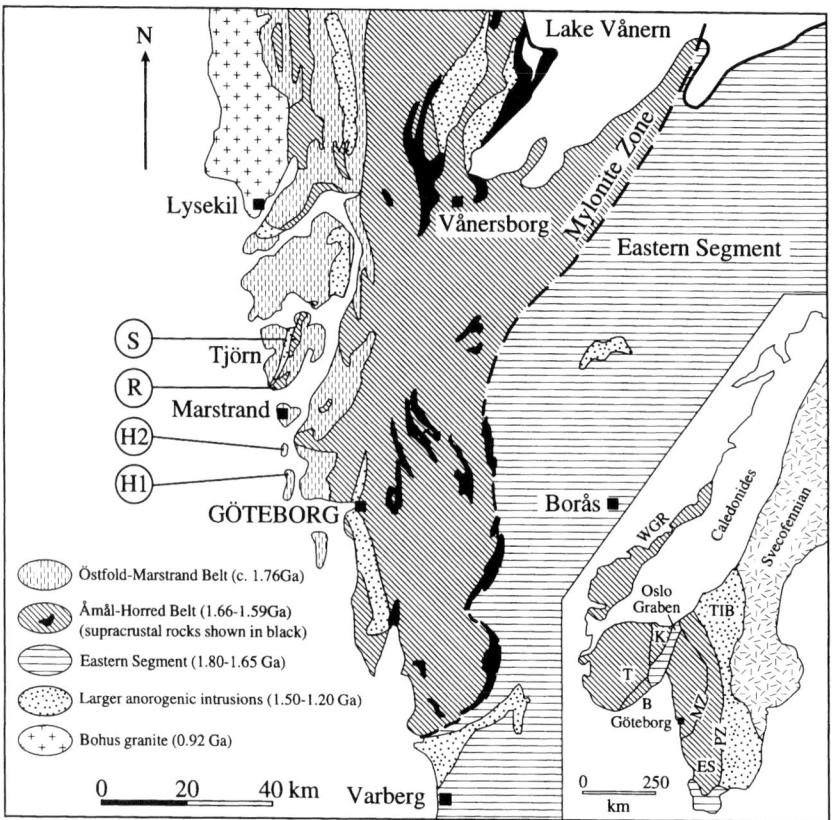

Fig. 1. Simplified map of the Göteborg region in southwestern Sweden. The four dated rocks are located to the NW of Göteborg: H1 = Hållsö diorite, H2 = Röseskär dyke, S = Stenkyrka granite, R = Rönnäng tonalite. The inset map shows major units of the southwestern part of the Baltic Shield. TIB = Transscandinavian Igneous Belt; light grey ornament = SSD, including the Eastern Segment (ES), the Kongsberg (K) and Bamble (B) sectors, the Telemark region (T) and the Western Gneiss Region (WGR); dark grey ornament = Phanerozoic rocks. MZ and PZ denote the Mylonite Zone and Protogine Zone respectively.

1993; Connelly *et al.* this volume) and *c.* 1.8 and 1.66 Ga granitoid rocks that are coeval with the early and late TIB units (Larson & Berglund 1992 and references therein; Welin 1994). Successively west of the Eastern Segment, the 1.6 Ga Åmål–Horred Belt and the *c.* 1.76 Ga Östfold–Marstrand Belt form two N–S trending crustal units that were juxtaposed by 1.59 Ga (Fig. 1, Åhäll *et al.* 1995).

Crustal growth of the SSD east of the Oslo Graben (Fig. 1 inset) included the accretion of outboard arcs, formation of a major calc-alkaline plutonic belt (the 1.59 Ga Göteborg Batholith, Table 1) and the subsequent deformation during the final orogenic episode of the Gothian period (Åhäll *et al.* 1995). There is no evidence for substantial addition of juvenile crust after 1.55 Ga (Lindh & Persson 1990; Johansson *et al.* 1993; Welin 1994).

The widespread intrusion of anorogenic gabbro-dolerite rocks at 1.51 Ga on both sides of the Mylonite Zone (Åhäll *et al.* 1990; Johansson & Johansson 1990) represents the earliest isotopic evidence for the existence of cratonized post-Gothian crust. The anorogenic development of Laurentia–Baltica is summarized by Gower *et al.* (1990), the evolution in the Kongsberg–Bamble sectors of Norway is reviewed by Starmer (1990) and the Östfold–Marstrand and Åmål–Horred belts are described by Åhäll *et al.* (1990).

The subsequent *c.* 1.2–0.9 Ga Sveconorwegian Orogeny reworked earlier-formed crust and ranged from high-pressure granulite facies metamorphism in the southern Eastern Segment (Johansson *et al.* 1991) to inhomogeneous deformation in the Östfold–Marstrand and Åmål–Horred belts. Post-kinematic intrusions include the 0.92 Ga Bohus granite (Fig. 1, Eliasson & Schöberg 1991).

Table 1. *Simplified sequence of events in the Östfold–Marstrand and Åmål–Horred belts in SW Sweden*

Tectono-Stratigraphic Group	Event (deformation–metamorphism–magmatism–deposition)	Age (Ma)
Group D	Post-kinematic magmatism – Bohus Granite (1)	920 ± 5
	D_3 – Sveconorwegian M_3: regional amph. facies metamorphism/deformation Deposition of continental sed. and extrusives (Dal Group)	c. 1090
Group C3	Bimodal magmatism; mainly granite (see references in 4) Bimodal magmatism; dolerite–granite	1249-03
	Askim granite (3)	1362 ± 9
	Torpa granite (4)	1380 ± 6
	Mafic dyke magmatism Orust and Kattsund–Koster dykes (5) <Rb–Sr> Subordinate deformation, probably local	c. 1420
Group C2	Bimodal magmatism; gabbro-dolerite–granite Brevik gabbro (6)	1508 + 16/–12
	Felsic dyke magmatism **Röseskär dyke** M_{2b}: mainly local, static amph. facies metamorphism	**1553 ± 2**
Group C1	Mafic–felsic magmatism; mainly complex intrusions **Hällsö diorite**	**1555 ± 2**
	Vallhamn trondhjemite (6) <Rb–Sr>	c. 1550
	Late-orogenic magmatism ('red granites')	c. 1560
Group B	D_2 – second Gothian orogenic episode M_{2a}: regional amph. facies metamorphism/deformation Calc-alkaline plutonism (Göteborg Batholith)	
	Rönnäng tonalite	**1587 ± 3**
	Stenkyrka granite	**1588 ± 5**
	Calc-alkaline volcanism/sedimentation Åmål Formation (continental margin) (7)	1614 + 7
Group A	D_1 – first Gothian orogenic episode M_1: regional amph. facies metamorphism/deformation Calc-alkaline volcanism (?plutonism) Horred Formation (island arc) (8)	1659 + 8/–6
	Deposition of greywacke and volcanics Stora Le-Marstrand Formation (9) <Sm–Nd>	c. 1760

All ages, except those especially noted, are U–Pb zircon ages. Dated rocks in this study are marked with bold letters. References: 1 = Eliasson & Schöberg (1991), 2 = Daly *et al.* (1983), 3 = Welin & Samuelsson (1987), 4 = Åhäll *et al.* (this volume), 5 = Åhäll *et al.* (1990), 6 = Åhäll *et al.* (1990), 7 = Lundqvist & Skiöld (1992), 8 = Connelly (unpublished data), 9 = Åhäll & Daly (1989).

Objectives of this study

The study area is located in the well-exposed archipelago northwest of Göteborg in the Östfold–Marstrand Belt (Fig. 1). Recent 1:10 000 scale mapping by the Swedish Geological Survey (Sveriges Geologiska Undersökning, SGU) has resulted in the publication of two 1:50 000 map-sheets (Samuelsson & Åhäll 1985a, 1990). This mapping and detailed studies (Samuelsson & Åhäll 1985b; Park *et al.* 1987; Åhäll *et al.* 1990) have resulted in a well-constrained relative lithostratigraphic chronology that differentiates four magmatic periods (Groups A, B, C, D), each separated by major periods of deformation and metamorphism. An updated version of this previously published chronology (Åhäll *et al.* 1990) is presented in Table 1 and includes the results of this study and Åhäll *et al.* (this volume).

The purpose of this study is to refine tectonic models by integrating complex field relationships and precise U–Pb ages for intrusive rocks belonging to Groups B and C. Granitoid rocks of Group B intruded during a magmatic interval between D_1 and D_2 deformation events. D_2 is divided into two amphibolite facies stages, M_{2a} and M_{2b} (Table 1), which represent the main fabric-forming stage and a later, mainly static stage of D_2

respectively (Åhäll et al. 1990). Although the subsequent Group C intrusive rocks cut across features attributed to the fabric-forming M_{2a} stage, field relations show that the oldest Group C rocks began to intrude before the M_{2b} stage. The intrusion of the oldest Group C rocks thus overlap the post-Gothian crustal stabilization such that they are subdivided into Groups C1 and C2 (Table 1). The earlier Group C1 rocks are disrupted by granitic M_{2b} neosomes that are mainly confined to contacts or interboudin sites in the Groups C1 rocks (Fig. 2a). In contrast, Group C2 rocks cut across all observed M_{2b} features (Fig 2b; cf. fig. 2 in Åhäll et al. 1990).

More specifically, this work was intended to determine the intrusive ages of Groups B, C1 and C2 and thus constrain the timing and duration of the two intervening stages during the final orogenic episode of the 1.75–1.55 Ga Gothian period (Åhäll et al. 1990, 1995). The Rönnäng tonalite (Group B) was re-analysed to refine its intrusion age by reducing the large errors associated with a previous U–Pb age of 1658 ± 68 Ma (Welin et al. 1982,

recalculated and quoting 2σ errors). This age allowed for a distinct plutonic phase older than the known 1.59 Ga Group B calc-alkaline magmatism of the Göteborg Batholith (Table 1). Furthermore, the presence of veined and folded xenoliths of host Stora Le-Marstrand gneiss in the Rönnäng tonalite (Åhäll 1990) requires that its age represents a minimum age for the early-Gothian D_1 episode (Table 1). The age difference between Groups C1 and C2 addresses whether the M_{2a} and M_{2b} stages were episodic or continuous events.

U–Pb geochronology

Three rock samples between 7 and 12 kg were crushed and zircon and titanite mineral separates were obtained by standard mineral separation techniques. Mineral fractions were selected for analysis on the basis of microscopic optical properties to ensure that only the highest quality grains were analysed. Mineral fractions from re-analysed samples were selected from residual mineral separates from the Laboratory of Isotope

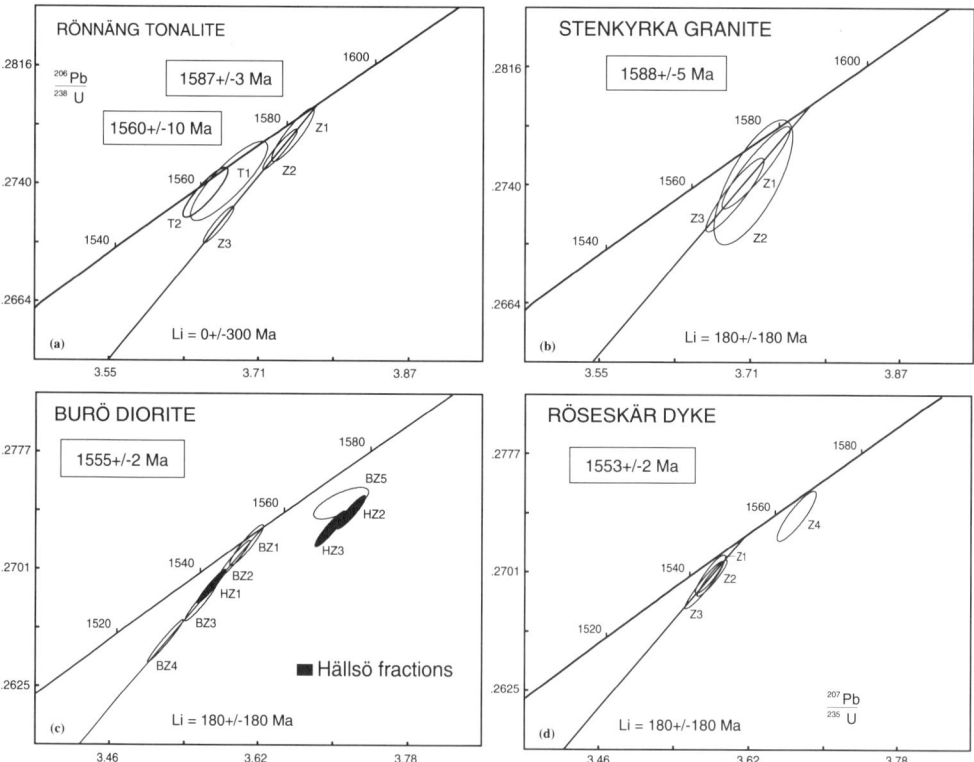

Fig. 2. U–Pb zircon diagrams: (**a**) = Rönnäng tonalite; (**b**) = Stenkyrka granite; (**c**) = Burö and Hällsö samples of the Hällsö diorite; (**d**) = Röseskär dyke. The 2σ analytical errors are indicated by the sizes of the data ellipses. Dashed discordia lines indicate pinned lower intercepts. See text for discussion.

Geology in Stockholm. Fraction sizes were small enough to maintain high sample integrity without compromising analytical precision. The outer surfaces of all zircons and titanite fractions were removed by air abrasion (Krogh 1982) prior to analysis (Appendix 1).

Group B intrusions

The Rönnäng tonalite. This medium-grained, predominantly tonalitic granitoid rock crops out in southern Tjörn (Fig. 1). The sample for age dating was collected in a moderately foliated portion of the intrusion, well away from the leucocratic phases associated with this body. The sample yielded zircons that are typically euhedral and elongate, with aspect ratios about 2.5:1 and titanite fragments, which ranged in colour from dark brown to light yellow.

Three clear, colourless fractions of inclusion-free zircons were analysed and define a discordia line with an upper intercept age of 1587 ± 3 Ma with an 83% probability of fit (Fig. 2a). Due to the near concordancy of these points, the lower intercept is poorly defined (0 ± 300 Ma). The euhedral zircon morphologies and the very consistent $^{208}Pb/^{206}Pb$ ratios (0.127 to 0.128, Table 2) indicate a common igneous source for all analysed fractions, such that the upper intercept is interpreted to represent the igneous crystallization age.

Two titanite analyses plot very near concordia with an average $^{206}Pb/^{238}U$ age of 1560 ± 10 Ma (Fig. 2a). This result establishes that titanite was not pervasively recrystallized everywhere in this terrane after 1560 ± 10 Ma and that subsequent temperatures did not exceed the closure temperature of titanite in these regions (approximately 550–600°C; Mattinson 1978; Tucker et al. 1987) for any significant interval.

The Stenkyrka granite. The field classification of this medium-grained granite from central Tjörn (Fig. 1) as a Group B intrusion (Åhäll et al. 1990) is consistent with its previously published U–Pb age of 1580 ± 18 Ma. As the youngest dated Group B rock, three new zircon fractions of this sample were analysed in an attempt to improve the precision, and thus to place a more accurate upper age constraint on the subsequent Gothian D_2 deformation and M_{2a} metamorphism (Table 1).

The fractions, consisting of clear, euhedral, short-prismatic zircons, all plot less than 2% discordant (Fig. 2b). Although their average $^{207}Pb/^{206}Pb$ age of 1587 ± 4 Ma might have been used to define the crystallization age, a lower intercept of 180 ± 180 Ma was used for the age calculation (cf. the Burö sample below). This estimate incorporates the full range of known lower intercepts for igneous rocks from this region and includes 0 Ma as a possible lower intercept. This method of regression yields an upper intercept of 1588 ± 5 Ma, which is interpreted to represent the age of igneous crystallization.

Group C1 intrusions

The Hällsö diorite. This medium-grained, dominantly dioritic intrusion from the archipelago NW of Göteborg (Fig. 1) is part of a weakly foliated, Group C1 complex comprising dyke-like sheets that range in composition from gabbro to leucogranite. The diorite was previously sampled at Hällsö and yielded a U–Pb age of 1570 +21/–7 Ma (Åhäll et al. 1990). In an attempt to refine the Group C1 intrusive suite age, three new analyses were made from the remaining zircon separate from this sample. However, they contained a significant component of inheritance (Fig. 2c) and a new dioritic sample was collected in the same dyke-like intrusion at Burö (800 m WNW of the Hällsö locality) in the hope that a sample from a different locality might contain less inherited zircon.

The Burö sample yielded a range of zircon morphologies including (1) round without crystal form, (2) euhedral to subhedral, subequant prisms, and (3) elongated to a maximum elongation of 5:1 with faceted to rounded terminations. The zircons were clear and colourless but the more elongated grains were typically fractured parallel to their length.

One fraction of euhedral prisms, two fractions of elongated prismatic grains and one fraction of round grains define a discordia line with upper and lower intercepts of 1555 ± 2 Ma and 180 ± 180 Ma (Fig. 2c) respectively, with a 63% probability of fit. The upper intercept age is interpreted to date the igneous crystallization. The large range of $^{208}Pb/^{206}Pb$ ratios (0.153–0.098) from zircon fractions of different morphologies of a common age suggests that zircon crystallized as the magma evolved, likely due to fractionation and/or modification by digested country rocks.

One fraction of the best zircon needles (5:1 aspect ratio) from the Hällsö sample (fraction HZ1; see Table 1 for fraction notation) falls on the discordia line defined by zircons from the Burö sample. Fractions of imperfect needles from both the Hällsö (HZ3) and Burö (BZ5) samples and rounded zircons (HZ4) from the former sample all yield $^{207}Pb/^{206}Pb$ ages older than the upper intercept age indicating the presence of inherited cores. The large sample sizes used in obtaining the previously published result of 1570 +21/–7 Ma are thought to have contained a significant component of inheritance, resulting in an older age than that obtained in this study. This interpretation is

Table 2. *U–Pb data*

Fraction	Weight (mg)	Concentration U (ppm)	Concentration Pb rad (ppm)	Measured total common Pb (pg)	Measured $\frac{^{206}Pb}{^{204}Pb}$	Measured $\frac{^{208}Pb}{^{206}Pb}$	Corrected atomic ratios* $\frac{^{206}Pb}{^{238}U}$		Corrected atomic ratios* $\frac{^{207}Pb}{^{235}U}$		Corrected atomic ratios* $\frac{^{207}Pb}{^{206}Pb}$		Ages (Ma) $\frac{^{206}Pb}{^{238}U}$	Ages (Ma) $\frac{^{207}Pb}{^{235}U}$	Ages (Ma) $\frac{^{207}Pb}{^{206}Pb}$
Rönnäng Tonalite															
Z1 2nd b from Z2	0.060	217	63.5	41	5464	0.1277	0.27741	142	3.7500	186	0.09804	22	1578	1582	1587
Z2 elong with xls	0.044	193	56.2	14	10087	0.1266	0.27647	108	3.7362	150	0.09801	12	1574	1579	1587
Z3 elong needles	0.104	195	56.0	69	5022	0.1274	0.27161	96	3.6708	136	0.09802	14	1549	1565	1587
T1 dk, med	0.085	136	36.7	154	1306	0.0472	0.27441	210	3.6815	338	0.09730	46	1563	1567	1573
T2 lgt, w abr	0.113	56	14.9	176	632	0.0379	0.27367	132	3.6565	200	0.09690	26	1559	1562	1565
Stenkyrka Granite															
Z1 sm frags equant	0.015	234	67.4	103	589	0.1201	0.27515	218	3.7157	294	0.09794	30	1567	1575	1585
Z2 b equant xls	0.006	157	45.2	5	3252	0.1220	0.27421	330	3.7129	340	0.09820	80	1562	1574	1590
Z3 equant xls	0.008	261	74.3	8	4149	0.1137	0.27338	194	3.6933	252	0.09798	28	1558	1570	1586
Burö Diorite															
BZ1 b needles	0.101	130	37.9	24	9473	0.1530	0.27190	108	3.6107	148	0.09631	14	1550	1552	1554
BZ2 rnd, clr	0.072	167	47.5	6	32252	0.1238	0.27096	102	3.5974	142	0.09629	10	1546	1549	1553
BZ3 b needles	0.097	191	53.8	6	52152	0.1218	0.26824	104	3.5615	144	0.09630	10	1532	1541	1554
BZ4 prisms, clr	0.122	213	58.2	7	63462	0.0980	0.26569	112	3.5232	156	0.09618	8	1519	1532	1551
BZ5 2nd b needles	0.158	46	13.4	1222	118	0.1425	0.27465	84	3.7118	240	0.09802	48	1564	1574	1587
Hällsö Diorite															
HZ1 b needles	0.049	157	44.6	30	4359	0.1314	0.26923	88	3.5728	128	0.09625	10	1537	1544	1553
HZ2 rnd, equant	0.092	208	58.9	33	10163	0.1031	0.27407	88	3.7225	126	0.09851	14	1561	1576	1596
HZ3 needles	0.051	156	45.4	29	4666	0.1414	0.27299	94	3.7004	134	0.09831	14	1556	1571	1592
Röseskär Dyke															
Z1 needles	0.066	178	50.8	8	24679	0.1312	0.27022	100	3.5822	134	0.09614	16	1542	1546	1551
Z2 elong-needles	0.059	185	53.0	10	18203	0.1369	0.26996	96	3.5840	132	0.09629	12	1541	1546	1553
Z3 elong clr, v abr	0.041	159	45.8	10	10583	0.1474	0.26953	126	3.5756	168	0.09622	14	1538	1544	1552
Z4 rnd weakly abr	0.018	214	62.3	7	9420	0.1353	0.27398	130	3.6738	170	0.09725	22	1561	1566	1572

*Ratios corrected for fractionation, 5–10 pg laboratory Pb blank, initial common Pb calculated using Pb isotopic compositions of Stacey & Kramers (1975) and 2 pg U laboratory blank. Two-sigma uncertainties on isotopic ratios, calculated with a modified unpublished error propagation program written by L. Heaman, are reported after the ratios and refer to the final digits.

Abbreviations are: abr = abraded; b = best; clr = clear; dk = dark; elong = elongate; frags = fragments; lgt = light colour; med = medium size (25–50 μm); rnd = round; sm = small size (50–100 μm); v = very; w = well; xls = crystals.

Mineral Codes: T, titanite; Z, zircon.

supported by older $^{207}Pb/^{206}Pb$ ages for fractions from the Hällsö sample that were abraded (Åhäll et al. 1990, p. 105). There is no correlation of $^{208}Pb/^{206}Pb$ ratios (Table 2) and apparent degree of inheritance.

Group C2 intrusions

The Röseskär dyke. This tonalitic dyke is part of an extensive Group C2 suite that is especially common in the southern part of the Östfold–Marstrand Belt. The sample was collected at Röseskär a few kilometres to the north of Hällsö (Fig. 1).

The zircons ranged from round to very elongated with a majority having aspect ratios of about 4:1. Except for the rounded grains, these zircons exhibited euhedral to subhedral form. The longest zircons were typically penetratively fractured parallel to the long axis and provided a means to ensure inherited cores were not included in this zircon type. The restricted range of $^{208}Pb/^{206}Pb$ ratios (Table 2) suggests that the analysed fractions are dominated by igneous zircons of a common source.

Three fractions of well abraded, clear, elongated zircons plot about 1% discordant (Fig. 2d) and yield an upper intercept age of 1553 ± 2 Ma when the lower intercept is pinned at 180 ± 180 Ma (for the same reason described above for Stenkyrka granite). A discordant fourth fraction of rounded grains (Z4) yields an older $^{207}Pb/^{206}Pb$ age confirming the presence of older zircon core in this intrusive suite (cf. the Group C2 Hällevik dyke in Åhäll et al. 1990). The upper intercept is interpreted to represent the age of crystallization of this prominent Group C2 dyke suite.

Implications for the geological evolution in the SSD

Timing of the calc-alkaline magmatism

The Rönnäng and Stenkyrka granitoids have been mapped as members of the 1.59 Ga calc-alkaline Göteborg Batholith (Åhäll et al. 1995; Table 1) belonging to the Group B intrusive phase. The new 1587 ± 3 Ma and 1588 ± 5 Ma crystallization ages represent the most precise ages for intrusions of the Göteborg Batholith and thus improve the constraints for the timing of this widespread calc-alkaline event (cf. Table 2 in Åhäll et al. 1995). The coincident ages for the tonalitic and granitic phases of the Göteborg Batholith dated in this study indicate that the progression of magmatism from intermediate to more felsic members, as observed in the field (cf. Samuelsson & Åhäll 1985b; Åhäll et al. 1990), occurred over a short time period.

Although still older or younger members may exist (for example the 1620 +35/-25 Ma Höjen tonalite located to the north of Lake Vänern (Fig. 1); Persson et al. 1983; Persson 1986), the available dataset is consistent with a single calc-alkaline intrusive phase at 1.59 ± 0.01 Ga (cf. Table 2 in Åhäll et al. 1995).

The revised age for the Rönnäng tonalite does not support the suggestion from earlier U–Pb data of an older calc-alkaline episode at c. 1.65 Ga (cf. Welin et al. 1982; Gaál & Gorbatschev 1987; Lindh 1991; Welin 1994) east of the Oslo Graben. Previous models envisioning calc-alkaline magmatism in the SSD contemporaneous with the c. 1.65 Ga alkali-calcic magmatism in the TIB may be still prove correct, but such magmatism cannot include the 1.59 Ga Göteborg Batholith (Åhäll et al. 1995). Although older calc-alkaline granitoids may exist to the east of the Mylonite Zone (Brewer & Åhäll unpublished data), and further north in the Western Gneiss Region (WGR, Fig. 1 inset; Tucker et al. 1987), the only geochronological indication now of such plutonism west of the Mylonite Zone comes from the 1.70 ± 0.08 Ga Rb–Sr whole-rock isochron age of the Grästorp granitoid gneiss located southeast of Vänersborg (Fig. 1) (Welin & Gorbatschev 1976).

The 1.59 Ga Göteborg Batholith, together with subordinate supracrustal rocks (the 1.66 Ga Horred and 1.61 Ga Åmål Formations), form a coherent calc-alkaline magmatic belt, the Åmål–Horred Belt (Åhäll et al. 1995). Since western plutons of the Göteborg Batholith, such as the Rönnäng and Stenkyrka granitoids (Fig. 1), have intruded into the Östfold–Marstrand Belt, it is clear that the two belts must have been juxtaposed by 1.59 Ga. Consequently, after welding during the Gothian episode, these two belts formed one single terrane, which was first intruded by the 1.5–1.2 Ga Group C intrusions and then transected by various N–S trending shear zones during the Sveconorwegian.

Timing of the Gothian deformation

The presence of veined and folded Stora Le-Marstrand xenoliths in the Rönnäng tonalite (Åhäll 1990) requires that emplacement of the Göteborg Batholith post-dates the D_1 episode of deformation. The 1587 ± 3 Ma age thus provides a minimum age for the first deformation in the Östfold–Marstrand Belt (D_1, Table 1).

The emplacement of the Göteborg Batholith was followed by the D_2 deformation event, which was associated with two amphibolite facies metamorphic stages (M_{2a} and M_{2b}; Samuelsson & Åhäll 1985b; Åhäll et al. 1990). The 1587 ± 3 Ma age for the Rönnäng tonalite thus provides a maximum age for the D_2 event. As a consequence of the tectonostratigraphic subdivision of Group B, C1 and C2 rocks (Åhäll et al. 1990 and Table 1), the new ages bracket the M_{2a} stage between 1587 ± 3 and 1555 ± 2 Ma and the M_{2b} stage between 1555 ± 2 and 1553 ± 2 Ma.

The timing of the Gothian metamorphism is also constrained by the 1560 ± 10 Ma titanite age of the Rönnäng tonalite since the age is interpreted to reflect D_2 metamorphism rather than igneous cooling. Although the titanite is related to the F_2/M_{2a} metamorphic fabric (Table 1), it is not certain whether titanite was affected by the later M_{2b} stage. Therefore, the 1560 ± 10 Ma titanite age only constrains the late-Gothian cooling through approximately 550–600°C and cannot independently constrain the M_{2a} and M_{2b} stages.

Post-Gothian crustal stabilization

Although localized melting around Group C1 dykes may have been induced by the dykes themselves (controlled by either spatial and/or minor temporal variations as indicated above) the timing of the static, waning stage of the last orogenic episode of the Gothian period is established by the age of the M_{2b} stage (i.e. between 1555 ± 2 and 1553 ± 2 Ma). This requires that conditions in the Östfold–Marstrand Belt changed from ductile to brittle at this time and marks a final crustal stabilization after a prolonged period of Gothian crustal growth.

The results from this study thus document that deformation and metamorphism did not end before 1560 ± 10 Ma and that fully stabilized crust did not exist, at least in the Östfold–Marstrand Belt, until 1555–1553 Ma. Orogenic conditions must therefore have prevailed much later in the SSD of SW Sweden than in other known segments along the proposed contiguous margin of Mesoproterozoic Laurentia–Baltica.

Post-kinematic c. 1.56–1.53 Ga magmatism

The oldest, unequivocally anorogenic event in the SSD is the magmatism represented by the 1.51 Ga Brevik gabbro (Åhäll et al. 1990). Still older post-kinematic events in the SSD are represented by the intrusion of c. 1.56 Ga 'red granites' (see below) and the mafic–felsic 1555–1553 Ma magmatism represented by the now dated Group C1 and C2 rocks. The geodynamic significance of these events is summarized in this section.

The c. 1.56 Ga 'red granites'. This widespread suite comprises c. 1.56 Ga granites *sensu strictu* that are described as less deformed and migmatized than adjacent Göteborg Batholith granitoids. In the Östfold–Marstrand and Åmål–Horred belts, the suite includes the 1564 ± 28 Ma Pinan granite (Welin 1994) and five more U–Pb dated rocks (Skiöld 1980; Welin et al. 1982; Persson et al. 1983, 1987) which are not as well constrained.

In the southern part of the Eastern Segment, three similar gneissic granites (at Skäralid, Mölle and Glimåkra) and an aplitic gneiss at Stenberget have all yielded U–Pb zircon ages between 1.58 and 1.50 Ga (Johansson 1990, Johansson et al. 1993). However, the presence of zircons with inherited cores in two samples, large uncertainties on three ages and high MSWDs for all discordia lines inhibit confident correlation.

While the 'red granites' may represent late-orogenic crustal melts (Lindh & Persson 1990), those characterized by low initial Sr values (e.g. the 1566 ± 39 Ma Sundsta granite, $Sr(i) = 0.705 + 0.003$; Persson & Hansen 1982) could also represent late fractionates of the 1.59 Ga Göteborg Batholith magmas. In addition, mapping in the Östfold–Marstrand and Åmål–Horred belts has shown that some granites that appear post-kinematic are actually recrystallized pre-kinematic Göteborg Batholith granitoids (Samuelsson & Åhäll 1985b, 1990).

The field relationships and geochemistry of the 'red granite' suite are, in general, insufficiently described to allow discrimination between these alternatives. Nevertheless, the timing requires that most of these rocks formed during the late-orogenic phase that corresponded to a period of thickened Gothian crust. Although available geochronology precludes detailed tectonic correlations, post-kinematic granitoid rocks of approximately the same age and composition are present in all terranes of the SSD east of the Oslo Graben (Fig. 1). This strongly suggests that these terranes followed a common late- to post-orogenic development.

The mafic–felsic 1555–1553 Ma magmatism. The overlapping ages of the 1555 ± 2 Ma Hällsö diorite (Group C1) and the intermediate 1553 ± 2 Ma Röseskär dyke (Group C2) suggest that these two rock types intruded as a continuous magmatic event. The earlier separation of these two units into distinct tectonostratigraphic groups (Åhäll et al. 1990) was based on their disparate relationships to the M_{2b} structures observed in the country rock gneisses. These differences can be explained by increased ambient temperatures associated with the emplacement of the larger, and commonly more mafic, Group C1 intrusions during the fading M_{2a} stage. This thermal pulse may have caused local mobilization of the country rocks, forming the M_{2b} neosomes, while the smaller, and probably slightly younger, Group C2 dykes intruded without associated mobilization (Figs 2a, b; cf. fig. 2 in Åhäll et al. 1990).

These two U–Pb age results are thus interpreted to reflect a single, continuous magmatic episode, involving rocks from both Groups C1 and C2 (Table 1). It should be noted, however, that this 1555–1553 Ma suite only includes a distinct rock type of the Group C1 and C2 (Table 1). The above interpretation is consistent with observations that Group C1 intrusions of this suite are positively identified in only a few restricted areas where this type of Group C2 dykes is sparse. In addition, such Group C2 dykes are not observed to unequivocally cut across any of these Group C1 intrusions, whereas intermediate Group C2 dykes are commonly cut by more evolved dykes of this suite.

The mafic–felsic 1555–1553 Ma magmas were emplaced as dykes and sills of varying size and complexity. Early magmas commonly formed complex intrusions up to 200 m wide (Group C1) in contrast to the later, more evolved magmas, that formed numerous smaller dykes (Group C2). Compositions range from mafic, with subordinate ultramafic pods, to granitic. Metre wide, plagioclase-rich granodioritic dykes form a dominant rock type. The progression from mafic to progressively more felsic members observed in the field is thought to be a consequence of magma fractionation. This interpretation is supported by the similarity in fractionation trends between the intrusions of the dated rock types (Brewer & Åhäll unpublished data). The significance of this magmatism, that occurred during the waning stage of the final orogenic episode of the Gothian period and thus far recognized with certainty only in the Östfold–Marstrand Belt, is discussed below.

Constraints for the Sveconorwegian orogenesis

The large contrast in Sveconorwegian metamorphism between different domains in the SSD is clearly evidenced by granulite facies assemblages in the southern part of the Eastern Segment (Johansson et al. 1991) and greenschist parageneses in an area of the Åmål–Horred Belt west of Lake Vänern (Fig. 1) (Lundqvist & Skiöld 1992). The concordant 1560 ± 10 Ma titanite age from the Rönnäng tonalite is the first firm U–Pb isotopic evidence to indicate that Sveconorwegian recrystallization was not pervasive throughout the SSD. This observation is consistent with the preservation of igneous features in several of the larger 1.51 Ga mafic bodies in the Östfold–Marstrand Belt. Nevertheless, the presence of garnet–amphibole assemblages in many of the anorogenic 1.51–1.25 Ga mafic rocks indicate that Sveconorwegian metamorphism at least locally reached middle amphibolite facies in the Östfold–Marstrand Belt (cf. Park et al. 1987, 1991).

Tectonics and trans-Atlantic correlation

In a model for the growth of Gothian crust in the SSD east of the Oslo Graben, the westernmost rim of Baltica is interpreted to represent an accreting continental margin (Åhäll et al. 1995). The model involves eastward subduction of oceanic crust that leads to (1) 1.61–1.59 Ga continental margin-type, calc-alkaline magmatism (the Åmål Formation and Göteborg Batholith) (2) amalgamation and accretion of outboard arcs by 1.59 Ga; (3) post-1.59 Ga orogenesis (D2, Table 1); and (4) 1.58–1.53 Ga rapakivi massifs (Haapala & Rämö 1992) further east in the Svecofennian Domain.

The presence of post-kinematic c. 1.56 Ga 'red granites', derived from late-orogenic crustal melts, is consistent with the model, whereas the largely coeval intrusions (the mafic–felsic 1555–1553 Ma magmatism) and the possibly related c. 1.55 Ga Vallhamn trondhjemite (Åhäll et al. 1990) suggest a more complex late-orogenic evolution, at least in the Östfold–Marstrand Belt. Since the 1555–53 Ma magmas are characterized by their dyke-like emplacement and include mafic components, they are interpreted to indicate an extensional regime that followed the compressional phase of the final Gothian orogenic episode. The calc-alkaline trends for some of the 1555–1553 Ma magmas may be derived from partial melting of the calc-alkaline 1.59 Ga Göteborg Batholith. This latter mechanism would explain the presence of the older zircon cores in all Group C1 and C2 samples in this study and the Hällevik dyke (Åhäll et al. 1990). Alternatively, the calc-alkaline trends may reflect that subduction had not ceased by 1.59 Ga everywhere in the SSD.

Outside of the SSD, there are few events so far dated that fall between 1.60 and 1.55 Ga. The 1576 ± 10 Ma age for North Pole Intrusive Suite, a quartz syenite in the Trans Labrador Batholith of south-central Labrador (Brooks 1983), represents one of the few well-constrained ages from this period in northeast Laurentia. It thus appears that magmatism, metamorphism and deformation associated with orogenic activity had ceased in most other regions of the North Atlantic Craton by 1.60 Ga, distinctly earlier than in the SSD of southwestern Sweden.

Conclusions

U–Pb zircon crystallization ages are reported for the 1587 ± 3 Ma Rönnäng tonalite, 1588 ± 5 Ma

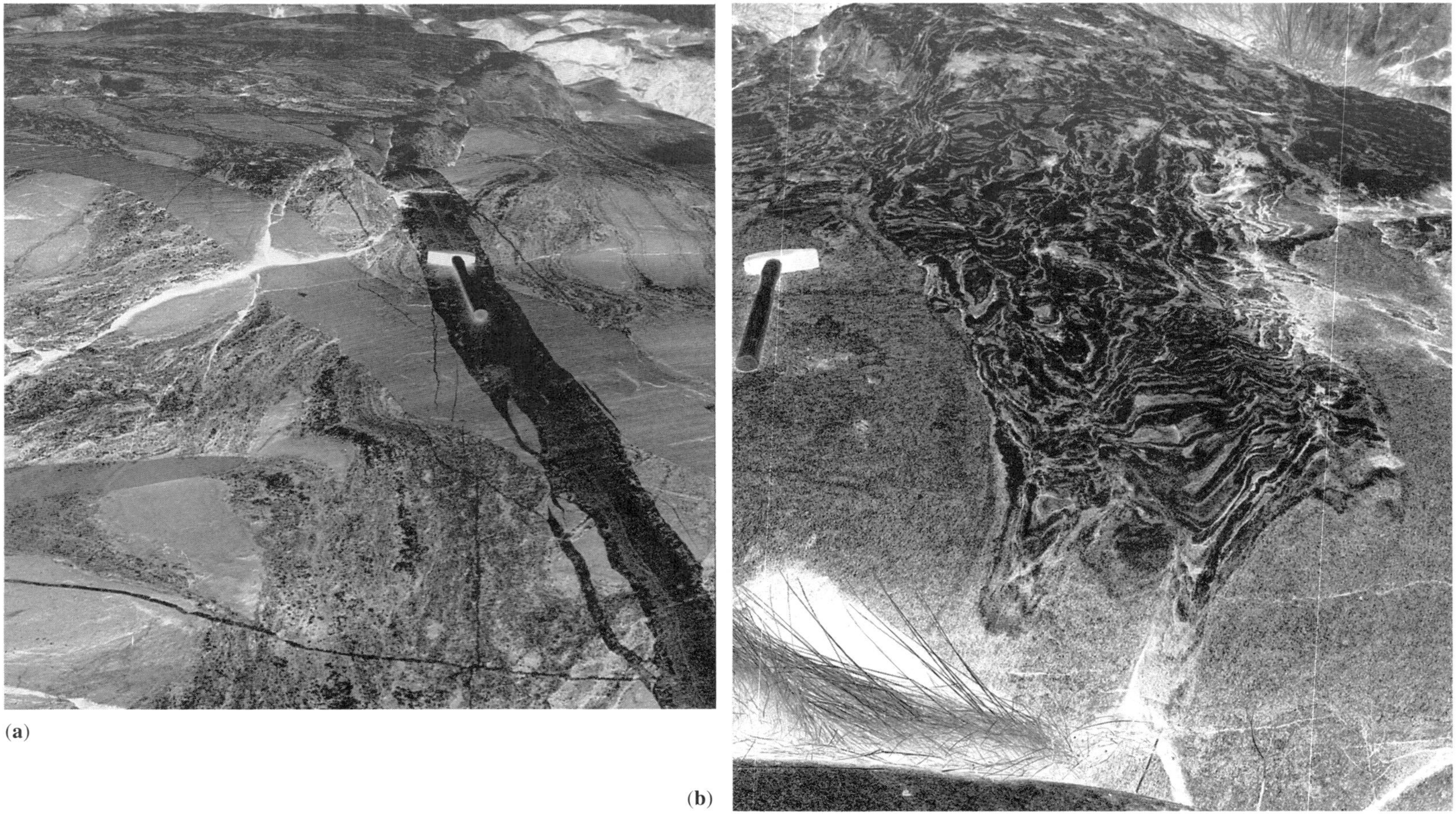

Fig. 3. (a) Disrupted tonatitic portion of the Hällsö diorite at Burö cutting migmatitic M_{2a} leucosomes; (b) typical granodioritic and granitic C2 dykes cross-cutting migmatitic M_{2a} structures in an area of insignificant Sveconorwegian deformation.

Stenkyrka granite, 1555 ± 2 Ma Hällsö diorite and 1553 ± 2 Ma Röseskär dyke in the Östfold–Marstrand Belt of the SSD. Since these rocks belong to three consecutive tectonostratigraphic groups, their ages constrain the intervening stages of the final orogenic episode of the Gothian period. These new ages bracket the F_2/M_{2a} stage between 1587 ± 3 and 1555 ± 2 Ma and the M_{2b} stage between 1555 ± 2 and 1553 ± 2 Ma (cf. Table 1). In addition, the Rönnäng tonalite age provides a minimum age for the D_1 metamorphic event at 1587 ± 3 Ma.

Although still older or younger calc-alkaline intrusions may exist, the ages of 1587 ± 3 Ma and 1588 ± 5 Ma for the Rönnäng tonalite and Stenkyrka granite, respectively, represent the most precise ages for members of the calc-alkaline Göteborg Batholith which is now constrained to 1.59 ± 0.01 Ga. The revised age for the Rönnäng tonalite does not support earlier suggestions of an older calc-alkaline episode at c. 1.65 Ga in the Swedish part of the SSD, as suggested by previous U–Pb data.

The presence of post-kinematic c. 1.56 Ga 'red granites', derived from crustal melts due to a thickened crust during the final Gothian orogenic episode, is consistent with the model presented by Åhäll et al. (1995), whereas the largely coeval intrusions (the mafic–felsic 1555–1553 Ma magmatism) and the possibly related c. 1.55 Ga Vallhamn trondhjemite (Åhäll et al. 1990) suggest a more complex late-orogenic evolution, at least in the Östfold–Marstrand Belt. This suite of 1555–1553 Ma magmas may reflect an extensional regime related to post-contraction movements in the collapsing Gothian Orogen. The partially calc-alkaline trends of these magmas suggest a significant contribution from older calc-alkaline rocks like those of the 1.59 Ga Göteborg Batholith. Alternatively, the calc-alkaline trends may reflect continued, areally restricted subduction of oceanic crust such that subduction had not ceased by 1.59 Ga everywhere in the SSD.

It is now documented that stabilized crust did not exist everywhere in the SSD until 1555–1553 Ma.

Orogenic conditions must therefore have prevailed much later in the SSD of SW Sweden than in other known segments along the postulated contiguous margin of the Mesoproterozoic Laurentia–Baltica.

This work has significantly benefitted from the personal involvement and support of Lennart Samuelsson, Geological Survey of Sweden. Discussions with Greg Dunning regarding the U–Pb component of this project and Tim Brewer have been most helpful. A critical review by Stephen Daly improved the manuscript. The study was financially supported by the Swedish Natural Research Council, grant G-GU 10286-300 (K-IÅ).

Appendix 1

U–Pb techniques

U–Pb analytical work was performed at Memorial University of Newfoundland, St John's, Canada. All fractions were abraded, re-evaluated and then washed in distilled 4N nitric acid, water and acetone. They were loaded into TEFLON capsules with a mixed ^{205}Pb/^{235}U isotopic tracer and dissolved with HF and HNO_3 acids. Chemical separation of U and Pb from zircon using minicolumns (0.055 ml resin volume) after Krogh (1973) resulted in a total Pb procedural blank between 2 and 5 picograms. Chemical separation from titanite on 0.5 ml columns resulted in a total procedural Pb blank of 10 picograms. Pb and U were loaded together with silica gel and phosphoric acid onto an outgassed rhenium ribbon and analysed on a multi-collector MAT 262 thermal ionization mass spectrometer, operating in static and/or peak jumping mode with ^{204}Pb measured in the axial secondary electron multiplier (SEM)/ion counter. Small samples were measured in peak jumping mode in the axial SEM/ion counter. Ages were calculated using the decay constants of Jaffey et al. (1971). Isotopic results are presented in Table 2 where errors on the isotopic ratios are reported at 2 σ. They were calculated by propagating uncertainties in measurement of isotopic ratios, fractionation and amount of blank with a program modified after an unpublished error propagation program written by L. Heaman (Royal Ontario Museum, Toronto, Canada). Linear regressions were calculated using the procedure of Davis (1982). The goodness-of-fit of a regressed line is represented as a probability of fit, where 10% corresponds to an MSWD of 2.

References

ÅHÄLL, K-I. 1990. An investigation of the Proterozoic Stenungssund granitoid, SW Sweden; conflicting geochronological and field evidence. In: GOWER, C. F., RIVERS, T. & RYAN, B. (eds) Mid-Proterozoic Laurentica–Baltica. Geological Association of Canada Special Paper 38, 117–129.

—— & DALY, J. S. 1989. Age, tectonic setting and provenance of the Östfold–Marstrand supracrustals: westward growth of the Baltic Shield at 1760 Ma. Precambrian Research, 45, 45–61.

——, DALY, J. S. & SCHÖBERG, H. 1990. Geochronological constraints on Mid-Proterozoic magmatism in the Östfold–Marstrand belt: implications for crustal evolution, SW Sweden. In: GOWER, C. F., RIVERS, T. & RYAN, B. (eds) Mid-Proterozoic Laurentica–Baltica. Geological Association of Canada Special Paper 38, 97–115.

——, PERSSON, P-O. & SKIÖLD, T. 1995. Westward accretion of the Baltic Shield: implications from the 1.6 Ga Åmål–Horred Belt, SW Sweden. Precambrian Research, 70, 235–251.

——, SAMUELSSON, L. & PERSSON, P-O. 1996. Geochronology and structural setting of the 1.38 Ga

Torpa granite: implications for charnockite formation in SW Sweden. *This volume.*

BERTHELSEN, A. 1980. Towards a palinspastic analysis of the Baltic shield. *International Geological Congress, Colloquium C6, Paris,* 5–21.

BROOKS, C. 1983. Supplement to the 4th report on the geochronology of Labrador (pertaining to subproject 1:09). Unpublished report to the Newfoundland Department of Mines and Energy, Open File No. Lab. 519.

CONNELLY, J. N., BERGLUND, J. & LARSON, S. Å. 1996. Thermotectonic evolution of the Eastern Segment of SW Sweden; tectonic constraints from U–Pb geochronology. *This volume.*

DALY, J. S., PARK, R. G. & CLIFF, R. A. 1983. Rb–Sr isotopic equilibrium during Sveconorwegian (= Grenville) deformation and metamorphism of the Orust dykes, S.W. Sweden. *Lithos*, **16**, 307–318.

DAVIS, D. W. 1982. Optimum linear regression and error estimation applied to U–Pb data, *Canadian Journal of Earth Sciences*, **23**, 2141–2149.

ELIASSON, T. & SCHÖBERG, H. 1991. U–Pb dating of the post-kinematic Sveconorwegian Bohus granite, SW Sweden: Evidence of restitic zircon. *Precambrian Research*, **51**, 337–350.

GAÁL, G. & GORBATSCHEV, R. 1987. An outline of the Precambrian Evolution of the Baltic Shield. *Precambrian Research*, **35**, 15–52.

GORBATSCHEV, R. & BOGDANOVA, S. 1993. Frontiers in the Baltic Shield. *Precambrian Research*, **64**, 3–21.

GOWER, C. F., RYAN, A. B. & RIVERS, T. 1990. Mid-Proterozoic Laurentia–Baltica: An Overview of its geological evolution and a summary of the contributions made by this volume. *In*: GOWER, C. F., RIVERS, T. & RYAN, B. (eds) *Mid-Proterozoic Laurentica–Baltica.* Geological Association of Canada Special Paper **38**, 1–20.

HAAPALA, I. & RÄMÖ, T. O. 1992. Tectonic setting and origin of Proterozoic rapakivi granites in southeastern Fennoscandia. *In*: BROWN, P. E. & CHAPPEL, B. W. (eds) *Proceedings of the 2nd Hutton Symposium: Origin of Granites and Related Rocks.* Geological Society of America, Special Paper, **272**, 165–171.

JAFFEY, A. H., FLYNN, K. F., GLENDENIN, L. E. BENTLEY, W. C. & ESSLING, A. M. 1971. Precision measurements of halflives and specific activities of ^{235}U and ^{238}U. *Physical Reviews.* **C4**, 1889–1906.

JOHANSSON, Å. 1990. Age of the Önnestad syenite and some gneissic granites along the southern part of the Protogine Zone, southern Sweden. *In*: GOWER, C. F., RIVERS, T. & RYAN, B. (eds) *Mid-Proterozoic Laurentica–Baltica.* Geolical Association of Canada Special Paper **38**, 131–148.

——, MEIER, M., OBERLI, F. & WIKMAN, H. 1993. The early evolution of the Southwest Swedish Gneiss Province: geochronological and isotopic evidence from southernmost Sweden. *Precambrian Research*, **64**, 361–368.

JOHANSSON, L. & JOHANSSON, Å. 1990. Isotope geochemistry and age relationships of mafic intrusions along the Protogine Zone, southern Sweden. *Precambrian Research*, **48**, 395–414.

——, LINDH, A. & MÖLLER, C. 1991. Late Sveconorwegian (Grenville) high-pressure granulite facies metamorphism in southwest Sweden. *Journal of Metamorphic Geology*, **9**, 293–292.

KROGH, T. E. 1973. A low-contamination method for hydrothermal decomposition of zircon and extraction of U and Pb for isotopic age determination. *Geochimica et Cosmochimica Acta*, **37**, 485–494.

—— 1982. Improved accuracy of U–Pb zircon ages by the creation of more concordant systems using an abrasion technique. *Geochimica et Cosmochimica Acta*, **46**, 637–649.

LARSON, S. Å. & BERGLUND, J. 1992. A chronological subdivision of the Transscandinavian Igneous Belt – three magmatic episodes. *Geologiska Föreningens i Stockholm Förhandlingar*, **114**, 459–461.

——, ——, STIGH, J. & TULLBORG, E.-L. 1990. The Protogine Zone, SW Sweden: a new model – an old issue. *In*: GOWER, C. F., RIVERS, T. & RYAN, B. (eds) *Mid-Proterozoic Laurentica–Baltica.* Geological Association of Canada Special Paper **38**, 317–333.

——, CONNELLY, J. & BERGLUND, J. 1993. Precise U–Pb dating of zircon and sphene from the Eastern Gneiss Segment, SW Sweden. *Symposium on 'the Svecofennian Domain' and annual meeting of the IGCP-275, Turku, Finland August 23–25, 1993.* Abstracts.

LINDH, A. 1991. Trace element variation in a suite of Proterozoic calc-alkaline granitoids, Sweden. *Geologiska Föreningens i Stockholm Förhandlingar*, **113**, 145–158.

—— & GORBATSCHEV, R. 1984. Chemical variation in a Proterozoic suite of granitoids extending across a mobile belt-craton boundary. *Geologische Rundschau*, **73**, 881–893.

—— & PERSSON, P.-O. 1990. Proterozoic granitoid rocks of the Baltic Shield – trends of development. *In*: Gower, C. F., Rivers, T. & Ryan, B. (eds) *Mid-Proterozoic Laurentica–Baltica.* Geological Association of Canada Special Paper **38**, 23–40.

LUNDQVIST, I. & SKIÖLD, T. 1992. Preliminary age-dating of the Åmål Formation, SW Sweden. *Geologiska Föreningens i Stockholm Förhandlingar*, **114**, 461–462.

MATTINSON, J. M. 1978. Age, origin and thermal histories of some plutonic rocks from the Salinian Block of California. *Contribrutions to Mineralogy and Petrology*, **67**, 233–245.

PARK, R. G., ÅHÄLL, K.I. & BOLAND, M. P. 1991. The Sveconorwegian shear-zone network of SW Sweden in relation to Mid-Proterozoic plate movements. *Precambrian Research*, **49**, 245–260.

——, ——, CRANE, A. & DALY, S. 1987. The structure and kinematic evolution of the Lysekil–Marstrand area, Östfold–Marstrand Belt, Southwestern Sweden. *Sveriges Geologiska Undersökning*, **C 797**.

PERSSON, P.-O. 1986. *A geochronological study of Proterozoic granitoids in the Gneiss Complex of southwestern Sweden.* Phd thesis, Lund University.

—— & HANSEN, B. T. 1982. The Rb–Sr age of the Sundsta granite in the Western Pregothian tectonic mega-unit, south-western Sweden. *Geologiska Föreningens i Stockholm Förhandlingar*, **104**, 17–21.

—— & WIKSTRÖM, A. 1993. U–Pb dating of the Askersund granite and its marginal augen gneiss. *Geologiska Föreningens i Stockholm Förhandlingar,* **115**, 321–329.

——, MALMSTRÖM, L. & HANSEN, B. T. 1987. Isotopic dating of reddish granitoids in southern Värmland, south-western Sweden. *Geologische Rundschau*, **76**, 389–406.

——, WAHLGREN, C. H. & HANSEN, B. T. 1983. U–Pb ages of Proterozoic metaplutonics in the gneiss complex of southern Värmland, south-western Sweden. *Geologiska Föreningens i Stockholm Förhandlingar*, **105**, 1–8.

SAMUELSSON, L. & ÅHÄLL, K.-I. 1985*a*. *Map of solid rocks Marstrand NO/Göteborg NV*. Sveriges Geologiska Undersökning, Af 146.

—— & —— 1985*b*. Proterozoic development of Bohuslän, south-western Sweden. *In*: TOBI, A. C. & TOURET, J. L. R. (eds) *The Deep Proterozoic Crust in the North Atlantic Provinces*, D. Reidel Publishing, 345–357.

—— & —— 1990. *Map of solid rocks Lysekil SO/Vänersborg SV*. Sveriges Geologiska Undersökning, Af 173.

SKIÖLD, T. 1980. Granite intrusions in the Proterozoic supracrustals of the Ellenö area, south-western Sweden. *Geologiska Föreningens i Stockholm Förhandlingar*, **102**, 201–205.

STACEY, J. S. & KRAMERS, J. D. 1975. Approximation of terrestrial lead isotopic evolution by a two-stage model. *Earth and Planetary Science Letters*, **26**, 207–221.

STARMER, I. C. 1990. Mid-Proterozoic evolution of the Kongsberg–Bamble Belt and adjacent areas, southern Norway. *In*: GOWER, C. F., RIVERS, T. & RYAN, B. (eds) *Mid-Proterozoic Laurentica–Baltica*. Geological Association of Canada Special Paper **38**, 279–305.

TUCKER, R. D., RÅHEIM, A., KROGH, T. E. & CORFU, F. 1987. Uranium–lead zircon and titanite ages from the northern portion of the Western Gneiss Region, south-central Norway. *Earth and Planetary Science Letters*, **81**, 203–211.

WELIN, E. 1994. Isotopic investigations of Proterozoic igneous rocks in south-western Sweden. *Geologiska Föreningens i Stockholm Förhandlingar*, **116**, 75–86.

—— & GORBATSCHEV, R. 1976. Rb–Sr age of granitoid gneisses in the 'Pregothian' area of south-western Sweden. *Geologiska Föreningens i Stockholm Förhandlingar*, **98**, 378–381.

—— & SAMUELSSON, L. 1987. Rb–Sr and U–Pb isotope studies of granitoid plutons in the Göteborg region, southwestern Sweden. *Geologiska Föreningens i Stockholm Förhandlingar*, **109**, 39–45.

——, GORBATSCHEV, R. & KÄHR, A.-M. 1982. Zircon dating of polymetamorphic rocks in south-western Sweden. *Sveriges Geologiska Undersökning*, **C797**, 1–34.

Mesoproterozoic anorogenic magmatism in southern Norway

J. F. MENUGE[1] & T. S. BREWER[2]

[1] *Department of Geology, University College Dublin, Belfield, Dublin 4, Ireland*
[2] *Department of Geology, University of Leicester, Leicester LE1 7RH, UK*

Abstract: The 1500 Ma Rjukan Group of the Telemark Supracrustal Suite, south Norway, consists of a metamorphosed sequence of acid volcanic rocks (Tuddal Formation) overlain by a comparable thickness of metabasalts (Vemork Formation). Both acid and basic volcanic rocks have within-plate chemical compositions. The acid rocks have initial ε_{Nd} values of 1.0 to 4.5, similar to the isotopic signature of some components of the Telemark gneiss complex, which is the probable basement to the Telemark Supracrustal Series. The metabasalts have initial ε_{Nd} values of 3.0 to 4.3, indicating a depleted mantle source region. A mantle plume origin for the Rjukan Group is unlikely given the absence of high temperature igneous rocks such as picrites. Magma genesis by lithospheric extension following orogeny is therefore proposed, with the Tuddal Formation acid magmas derived by anatexis of arc-generated crustal rocks 1500–2000 Ma old, whilst the Vemork Formation basaltic magmas were derived by decompression mantle melting. The Rjukan Group is correlated with the granite–rhyolite province of the US midcontinent, implying large-scale continental rifting. The substantial volume of basic volcanic rocks suggests greater or more prolonged extension in south Norway than in North America.

The tectonic setting of Proterozoic volcanic sequences is frequently obscure, especially where they have been subjected to later, unrelated orogeny. In this paper we present Sm–Nd isotopic and geochemical data from a Mesoproterozoic metamorphosed volcanic sequence, the Rjukan Group of south Norway. This sequence contains both acid and basic volcanic rocks which were subject to Sveconorwegian metamorphism and deformation c. 400 Ma after eruption. Geochemical data are integrated with field evidence and used to constrain the magmagenesis and tectonic setting of these volcanic rocks. Recent high-precision U–Pb zircon dating (Dahlgren *et al.* 1990) allows this sequence to be considered in the context of other dated tectonothermal events in the Baltic Shield and compared to similar volcanic sequences elsewhere in Laurentia–Baltica. This paper argues that the Rjukan Group may be correlated with the granite–rhyolite province of the US midcontinent and that these rocks represent continental rifting of Laurentia–Baltica.

Geological setting

The central Telemark region can be subdivided into three structural units (Fig. 1):

(a) a complex of foliated, predominantly granitic gneisses;
(b) the Telemark Supracrustal Suite (TSS), a metamorphosed and deformed sequence of acid and basic volcanic rocks, volcanogenic and arenaceous sedimentary rocks, with a maximum thickness of at least 6 km;
(c) late post-tectonic granites, which intrude both the gneisses and the TSS.

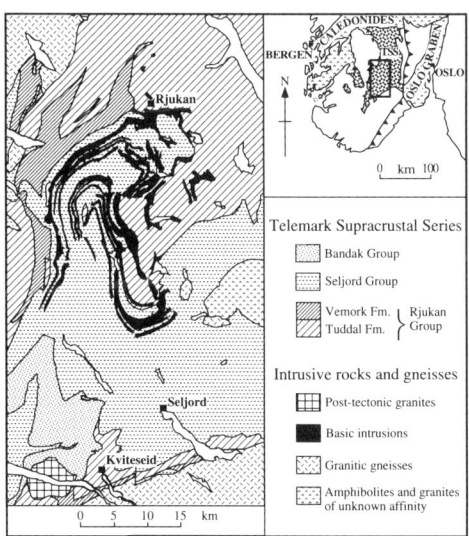

Fig. 1. Simplified geology and stratigraphy for the central Telemark region, modified from Dons (1972) and Dons & Jorde (1978). On the inset map, TSS indicates the extent of the Telemark Supracrustal Series including possible correlatives.

The gneiss complex consists of rocks mainly of granodioritic to granitic composition (Barth & Reitan 1963), which in western Norway appear to form the basement onto which correlative equivalents of the TSS (Bandak Group) rest unconformably (Sigmond 1978; Brewer & Atkin 1987). In central Telemark, however, the relationship is unclear, which has resulted in some authors regarding the gneisses as remobilized supracrustals (Werenskiold 1910; Mitchell 1967; Avila Martins 1969; Cramez 1970; Venugopal 1970; Dons 1972; Stout 1972), whilst others regard the gneisses as basement to the TSS (Brewer & Field 1985; Menuge 1985; Brewer & Atkin 1987; Starmer 1993). The relative age of the gneisses and the TSS remains uncertain, reflecting the lack of reliable supracrustal-basement contacts. The problem is further compounded by the lack of reliable isotopic ages for the gneisses.

The TSS was divided by Dons (1960a, b) into the oldest Rjukan Group, the Seljord Group and the youngest Bandak Group (Fig. 2). Dons (1972) considered that the number and intensity of deformations recorded by each of the groups differed and therefore that the groups were separated by angular unconformities. However, recent studies (Brewer & Atkin 1989; Dahlgren *et al.* 1990) have demonstrated that within the supracrustal pile only two folding episodes are preserved, with an early E–W orientated fold phase deforming both the Rjukan and Seljord groups prior to the deposition of the Bandak Group. Following deposition of the Bandak Group, the supracrustal pile was then deformed by approximately N–S orientated folds, which are related to the main Sveconorwegian orogeny (Starmer 1990). According to this interpretation, the Rjukan and Seljord groups represent one package, within which the discordance of individual units is a function of the mode of deposition, whilst the Bandak Group represents a separate, younger package.

The Rjukan Group has been subdivided into the Tuddal Formation and the overlying Vemork Formation (Figs 1 and 2). The Tuddal Formation is lithologically diverse, composed of metamorphosed crystal tuffs, tuffs, ignimbrites, rhyolites and volcanogenic sediments (Werenskiold 1910; Dons 1960a; Brewer 1985). Numerous xenoliths of quartzite within these acid volcanic rocks suggest eruption onto mature continental crust (Starmer 1993). Where metamorphic recrystallization is minimal, volcanic textures are preserved including glass shards in tuffs and rhyolites showing spherulites, flow banding and autobrecciation. Preserved phenocrysts are of quartz, plagioclase and K-feldspar.

The Vemork Formation rests disconformably upon the Tuddal Formation, and is composed of metamorphosed basaltic lavas and sedimentary rocks. Individual lava flows, frequently with vesicular flow tops, range in thickness from 0.5 to 5 m and are intercalated with cross-bedded sandstones, conglomerates and shales. The conglomerates are intraformational and include

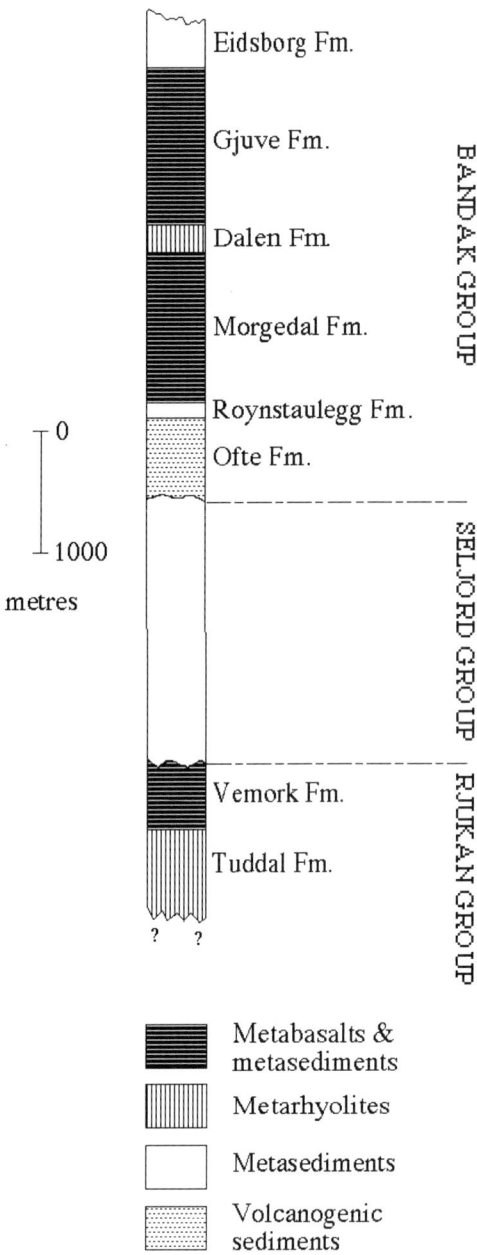

Fig. 2. Generalized stratigraphical column of the Telemark Supracrustal Series in central Telemark.

locally derived basalt clasts as well as clasts of acid volcanic rocks. Where deformation and recrystallization is slight, the basalts are plagioclase-phyric, with the groundmass containing microphenocrysts of plagioclase. Intercalated sedimentary rocks suggest deposition in shallow water conditions, though some volcanism may have been subaerial.

The Seljord Group was deposited disconformably on an irregular volcanic topography of the Rjukan Group. This has an erosion surface, commonly with metre-scale hollows, suggesting only short-term exposure at the surface (Starmer 1993). It consists mainly of quartzites, with subordinate conglomerates. The latter contain clasts primarily of quartzite but also of basic and acid volcanic rocks, schists and calcareous schists. Sedimentary structures including abundant ripple marks and cross-bedded units indicate a shallow water, probably intertidal, depositional environment. Raindrop imprints and mud cracks (Dons 1963) provide evidence of occasional subaerial conditions (Singh 1969). This implies that subsidence of the Seljord sedimentary basin was matched by accumulation of clastic sediment.

The Bandak Group, which rests with angular unconformity on the Rjukan and Seljord groups, is composed of metamorphosed basic and acid volcanic rocks, intercalated with sedimentary rocks (Fig. 2). Both the Morgedal and Gjuve formations are composed of a series of basaltic lava flows intercalated with cross-bedded sandstones and minor conglomeratic units. The Dalen Formation is an acid volcanic unit composed of flow-banded rhyolites which vary from moderately to sparsely phyric. Basic sills, intruding both the Rjukan and Seljord groups, have been interpreted as being contemporaneous with Bandak Group volcanism (Dahlgren et al. 1990). Late to post-tectonic coarse grained granites up to 10 km in diameter intrude both the gneisses and the TSS. They represent the last significant magmatic phase in the central Telemark region (Dons 1972; Sigmond 1978). The granites appear to have been forcefully emplaced and contain rafts of supracrustals.

Timing of eruption and geochemical effects of metamorphism

The geological evolution of the central Telemark area is summarized in Table 1. This section focuses specifically on the age and metamorphic history of the Rjukan Group. A recent U–Pb zircon study (Dahlgren et al. 1990) dated magmatic zircons from Tuddal Formation rhyolite at c. 1500 Ma. The major discrepancy between this age and a previously reported Sm–Nd whole-rock age of 1190 ± 37 Ma (also interpreted as the age of eruption of the Tuddal Formation (Menuge 1985)) led Brewer & Menuge (pers. comm.) to examine the relationship between metamorphism and element mobility in the Telemark volcanic rocks.

From the town of Rjukan (Fig. 1) to the village of Tuddal, an excellent section (1000 m) through the Tuddal Formation is exposed. In the lower parts of the section, metamorphic recrystallization is intense and the volcanic rocks have been transformed to quartz–feldspar–biotite–muscovite schists. In these lithologies, all primary textures and mineral assemblages have been replaced by a granoblastic quartz–feldspar mosaic within which muscovite and biotite define a pervasive metamorphic fabric. There is a gradation through to the upper parts of the section, where metamorphic recrystallization is extremely limited and excellent volcanic textures and mineral assemblages are preserved, with only a limited amount of muscovite developed defining a weak and non-pervasive fabric. Metamorphic recrystallization of the Tuddal Formation caused geochemical disturbance of trace elements, including the REE, and Nd

Table 1. *Geological history of the Telemark Supracrustal Suite in central Telemark*

	Date (Ma)	References
Intrusion of late kinematic granites and thermal metamorphism	c. 890	Priem et al. (1973)
N–S folding and regional main Sveconorwegian metamorphism	c. 1025	Starmer (1993)
Deposition of Bandak Group; early burial metamorphism	c. 1160	Dahlgren et al. (1990)
Uplift and erosion		
E–W folding		
Deposition of Seljord Group		
Deposition of Rjukan Group; early burial metamorphism	c. 1500	Dahlgren et al. (1990)

isotopes, so that only where recrystallization is slight can igneous geochemical signatures be retrieved (Brewer & Menuge, pers. comm.). Metamorphism caused partial resetting of the Sm–Nd isotopic system within the lower part of this section, from which the samples yielding the Sm–Nd isochron age of 1190 ± 37 Ma were collected (Menuge 1985). Therefore the U–Pb zircon date of c. 1500 Ma (Dahlgren et al. 1990) is considered to be the date of eruption of the Tuddal Formation, whilst the Sm–Nd date may have no geological significance.

The overlying Vemork Formation basalts record two metamorphic events (Table 1), an early burial type metamorphism and a later regional greenschist facies event as seen in the Tuddal Formation, the combination of which has overprinted the majority of primary textures and mineral assemblages. During the early burial event, anhedral inequigranular epidote developed near to the flow margins. In extreme cases, the flow margins have been transformed to epidote–quartz–chlorite (epidosite) assemblages (Brewer 1985; Brewer & Atkin 1987). Whole-rock geochemistry, including REE, Y, Zr, Ti and Nd isotope ratios, has been severely altered by the burial metamorphism. Samples showing petrographic evidence for such metamorphic effects have not been used for modelling of igneous petrogenesis. In contrast, the regional greenschist facies metamorphism has affected only the more mobile elements (e.g. K, Rb, Ba, Sr) (Brewer & Atkin 1989; Atkin & Brewer 1990; Brewer & Menuge, pers. comm.).

A U–Pb zircon age of c. 1160 Ma from an acid volcanic sample from the Bandak Group (Dahlgren et al. 1990) demonstrates a major break in the deposition of the TSS, with an older sequence (based on field evidence) probably composed of the Rjukan and Seljord groups, whilst the younger sequence is represented by the Bandak Group. However, since neither the Vemork Formation nor the Seljord Group has been directly dated, the timespan of Rjukan Group plus Seljord Group deposition remains unknown. Although the large age difference of c. 340 Ma between the Tuddal Formation and Bandak Group precludes a petrogenetic relationship between them, the term Telemark Supracrustal Series is retained for two reasons. Firstly, the position of the major break in time in the TSS is not known with certainty, and secondly, the stratigraphical position within the TSS of some outlying areas of supracrustal rocks in south Norway is also uncertain.

Sampling and analytical methods

Samples, 2–5 kg in weight, were collected in central Telemark. At each sample site individual samples were selected so as to be free of the effects of veining and recent weathering. All samples were crushed and then powdered in agate; the major and trace elements (Tables 2–5) were determined by X-ray fluorescence spectrometry at Nottingham University following the procedure of Harvey & Atkin (1982). REE were determined by inductively coupled plasma optical emission spectrometry following the procedure of Walsh et al. (1981).

Sm and Nd were determined by isotope dilution mass spectrometry and are presented with Nd isotopic data in Table 6. These data were obtained in four laboratories. The published data of Menuge (1985) were determined at Cambridge University by methods similar to those of Hooker et al. (1981). Further datasets were obtained at University College Dublin following the methods of Menuge (1988); at the British Geological Survey following the techniques described by Milne & Miller (1991); and at Leeds University using the techniques described by Chaffey et al. (1989). Data from all four laboratories have been standardized to a ^{143}Nd/^{144}Nd ratio of 0.511860 for the La Jolla Nd standard. Due to analytical uncertainties in the value for this standard in the different laboratories, there are possible systematic inter-laboratory differences in sample ^{143}Nd/^{144}Nd ratios of up to 0.00002.

Acid volcanic samples from the Tuddal Formation which exhibited petrographic evidence of moderate to extreme metamorphic recrystallization were excluded from the dataset presented here because of the likelihood of extensive element mobility (Brewer & Menuge, pers. comm.). For the same reason, Vemork Formation metabasalts showing petrographic evidence for burial metamorphism were similarly excluded. In Tables 2–6 and in the results and discussion which follow, only the screened datasets are considered.

Results

Tuddal Formation

The Tuddal Formation volcanic rocks sampled are all rhyolitic as defined by the alkali–silica plot of Le Maitre (1989) (Fig. 3). They are subalkaline and peraluminous to marginally metaluminous; none are peralkaline. The high-silica rhyolites have high Fe_2O_3/MgO ratios and low CaO and MgO concentrations. Harker diagrams (Fig. 4) show little or no correlation of most major elements with SiO_2. There is a good anticorrelation of Al_2O_3 with SiO_2, presumably reflecting variable proportions of quartz and feldspars, but otherwise fractionation trends are not apparent. In the case of CaO, Na_2O and K_2O, such trends might have been obliterated by subsequent metamorphism. Crystal fractionation

Table 2. Tuddal Formation major element analyses

Sample	SiO$_2$	TiO$_2$	Al$_2$O$_3$	Fe$_2$O$_3$	MnO	MgO	CaO	Na$_2$O	K$_2$O	P$_2$O$_5$	LOI	Total
RT2	70.80	0.48	14.04	4.30	0.07	0.54	1.57	3.26	3.80	0.04	0.65	99.55
RT3	77.01	0.13	12.34	1.49	0.03	bd	0.74	3.96	3.68	0.01	0.23	99.62
RT4	74.74	0.13	13.18	1.60	0.03	bd	0.40	3.38	6.07	bd	0.10	99.63
RT5	75.59	0.12	11.91	1.40	0.03	bd	0.30	2.86	5.90	0.01	0.50	98.62
RT6	76.00	0.28	12.07	2.48	0.05	0.21	1.27	2.91	3.96	0.05	0.45	99.73
RT9	73.23	0.26	13.94	2.59	0.06	0.39	0.86	2.91	3.78	0.03	1.55	99.60
RT11	71.07	0.50	14.56	3.97	0.05	0.55	1.44	1.91	3.91	0.07	1.63	99.66
RT16	73.32	0.37	12.91	3.06	0.05	0.72	0.22	3.43	4.46	0.04	1.29	99.87
RT19	74.41	0.36	12.44	3.13	0.06	0.67	0.13	2.85	5.09	0.04	0.44	99.62
RT20	78.35	0.17	11.03	1.76	0.03	0.01	bd	2.46	5.46	0.03	0.45	99.75
RT21	77.12	0.19	11.54	1.97	0.03	0.10	bd	1.79	6.27	0.01	0.55	99.57
RT22	75.63	0.31	12.27	1.99	0.04	0.11	0.63	2.62	5.69	0.03	0.66	99.98
RT23	75.96	0.14	12.55	1.58	0.02	bd	0.04	2.64	5.93	0.01	0.67	99.54
RT24	76.50	0.15	12.94	2.34	0.03	0.17	0.01	1.63	4.05	0.03	1.68	99.53
RT25	71.19	0.30	14.20	3.79	0.07	0.34	1.52	4.02	2.95	0.03	1.32	99.73
RT27	74.27	0.31	12.37	2.37	0.04	0.04	0.50	2.84	6.02	0.03	0.79	99.58
RT28	73.62	0.32	12.79	2.43	0.02	0.04	0.46	2.93	6.20	0.03	0.85	99.69
RT29	69.72	0.41	14.27	2.79	0.02	0.36	0.85	3.32	6.62	0.06	1.17	99.59
RT30	70.84	0.42	14.40	2.51	0.05	0.45	0.83	3.45	5.67	0.06	1.09	99.77
RT31	71.09	0.44	14.32	2.84	0.06	0.73	0.25	3.77	6.07	0.05	0.65	100.27
RT32	73.42	0.38	12.62	2.86	0.05	0.91	0.08	3.07	5.06	0.05	1.30	99.80
RT36	69.02	0.58	14.48	4.13	0.09	1.44	0.55	5.71	3.37	0.08	0.79	100.24
RT39	79.20	0.13	10.65	1.04	0.02	0.15	0.22	0.93	6.55	0.02	0.93	99.84
RT40	75.92	0.19	12.63	1.96	0.02	0.48	bd	0.34	6.76	0.02	1.32	99.64
RT41	78.36	0.17	11.18	1.40	0.01	0.07	0.04	2.55	5.34	0.03	0.49	99.64
RT42	74.34	0.31	13.45	2.15	0.02	0.69	bd	2.89	4.36	0.03	1.28	99.52
RT43	74.81	0.43	12.07	2.90	0.02	0.59	0.16	1.72	5.67	0.07	1.57	100.01
RT44	75.28	0.35	12.88	2.12	0.03	0.74	0.05	2.12	5.02	0.05	1.41	100.05
RT45	78.07	0.25	11.87	2.26	0.03	0.19	0.29	3.96	3.03	0.02	0.31	100.28
RT47	71.30	0.25	13.44	2.73	0.03	0.35	0.01	3.00	5.99	0.02	3.03	100.15
RT49	75.64	0.27	12.80	2.66	0.01	0.20	0.22	5.64	1.39	0.02	0.70	99.55
RT50	76.17	0.22	11.86	2.46	bd	0.25	0.09	2.28	6.21	0.02	0.54	100.10
RT51	75.77	0.23	12.36	2.49	0.01	0.34	0.04	3.44	4.83	0.03	0.54	100.08
RT52	75.50	0.24	11.78	2.60	0.02	0.31	0.04	2.20	5.88	0.02	1.06	99.65
RT53	76.10	0.22	12.19	2.04	0.02	0.33	0.18	2.92	5.23	0.05	0.45	99.73
RT54	77.65	0.22	11.43	2.48	0.01	0.26	0.04	3.84	3.82	0.03	0.38	100.16
RT55	75.40	0.23	12.13	2.57	0.03	0.16	0.01	3.03	5.33	0.02	0.63	99.54
RT56	75.10	0.24	12.31	2.42	0.01	0.34	0.07	2.22	6.22	bd	0.57	99.50
RT57	75.72	0.25	12.23	2.53	0.03	0.34	0.03	3.52	4.85	0.01	0.54	100.05
RT58	76.30	0.23	11.65	2.45	0.03	bd	0.01	4.18	4.03	0.01	0.75	99.64
RT59	75.87	0.22	11.83	2.56	0.03	0.17	0.10	2.82	5.37	0.02	0.54	99.53
RT60	76.48	0.22	11.82	2.57	0.01	0.10	0.08	5.25	3.06	0.04	0.22	99.85
RT61	75.95	0.21	11.86	2.34	0.01	bd	0.02	4.79	3.11	0.03	0.31	98.63
RT62	74.92	0.24	12.43	2.47	0.03	0.39	0.14	3.63	4.76	bd	0.70	99.71
RT63	74.79	0.23	13.20	2.62	0.02	0.16	0.01	3.35	5.44	0.01	0.18	100.01
RT64	75.29	0.23	12.49	2.52	0.01	0.25	0.05	4.12	4.29	0.02	0.32	99.59
RT65	71.84	0.27	13.90	3.19	0.04	0.48	0.06	2.14	7.23	0.03	0.80	99.98
RT66	73.34	0.27	13.63	2.61	0.03	0.40	0.01	3.72	5.16	0.02	0.70	99.89
RT67	75.99	0.22	11.81	2.35	0.02	0.14	0.05	3.16	5.04	0.03	0.88	99.69
RT68	77.28	0.21	11.52	2.41	0.02	0.34	bd	1.80	5.61	0.01	0.78	99.98
RT69	73.98	0.24	13.23	2.69	0.05	0.39	0.01	3.47	4.79	0.02	1.25	100.12
RT70	75.06	0.38	12.45	2.35	0.02	0.50	0.05	1.77	6.22	0.07	0.86	99.73
RT71	73.75	0.38	13.36	2.37	0.02	0.83	0.04	1.72	6.13	0.08	1.28	99.96
RT72	74.74	0.39	12.76	2.53	0.02	0.74	0.05	1.59	6.05	0.07	1.08	100.02
RT73	75.50	0.34	12.00	2.21	0.02	0.45	0.04	1.52	6.33	0.07	1.25	99.73
RT74	74.03	0.39	13.21	2.37	0.03	0.64	0.05	1.55	6.75	0.06	0.94	100.02
RT75	75.76	0.35	12.16	2.20	0.03	0.46	0.05	1.71	5.97	0.06	1.20	99.95
RT76	76.06	0.35	11.98	2.21	0.02	0.56	0.05	1.42	6.31	0.06	0.94	99.96
RT77	72.35	0.41	13.55	2.60	0.03	1.28	0.09	1.19	6.55	0.07	1.75	99.87
RT78	75.15	0.37	11.80	2.11	0.04	0.58	0.16	1.75	6.21	0.07	1.36	99.60
RT79	76.27	0.35	11.88	2.08	0.03	0.55	0.04	1.60	6.15	0.05	0.78	99.78
RT80	77.95	0.13	11.41	1.51	0.01	bd	0.03	2.81	5.35	0.01	0.32	99.53
RT82	75.44	0.13	13.04	1.59	0.02	0.25	0.08	3.82	5.26	bd	0.35	99.98
RT83	76.87	0.12	11.73	1.57	0.04	0.14	0.02	2.60	5.72	0.01	0.72	99.54
RT84	75.58	0.15	12.74	1.80	0.02	0.17	0.03	3.07	5.76	0.01	0.68	100.01
RT85	77.89	0.12	11.69	1.63	0.02	0.18	0.05	3.64	4.41	0.02	0.49	100.14
RT86	78.71	0.12	10.84	1.57	0.03	0.13	0.02	2.67	4.84	0.01	0.86	99.80
RT87	78.66	0.13	11.05	1.47	0.02	0.26	0.14	3.31	4.28	0.01	0.46	99.79

bd: below detection

Table 3. *Tuddal Formation trace element analyses*

Sample	Ba	Ce	Co	Cr	Cu	La	Ni	Nb	Pb	Rb	Sr	Th	U	V	Y	Zn	Zr
RT2	757	107	7	21	3	51	6	16	22	168	134	12	3	39	51	88	300
RT3	274	117	bd	7	1	53	3	16	16	110	54	5	7	2	78	26	336
RT4	323	134	4	7	5	71	8	15	10	215	36	7	5	6	82	44	335
RT5	258	104	4	4	7	44	4	15	10	153	42	5	6	8	75	22	323
RT6	417	78	7	10	28	31	10	15	26	198	96	7	2	23	67	99	243
RT9	347	57	6	17	3	25	10	22	24	262	44	8	6	21	77	150	257
RT11	843	107	12	28	3	58	6	14	15	162	148	15	6	41	51	81	275
RT16	425	18	7	7	bd	8	3	18	5	146	22	17	8	14	61	62	503
RT19	521	47	4	6	2	18	6	16	8	163	28	20	6	8	69	66	430
RT20	530	23	bd	bd	3	15	2	12	14	142	27	10	8	7	42	22	267
RT21	627	97	bd	8	3	35	4	15	bd	214	32	8	7	8	55	27	282
RT22	607	111	4	10	1	51	6	15	10	161	67	23	9	12	68	35	356
RT23	280	55	bd	8	3	18	4	25	8	220	29	9	6	4	69	28	275
RT24	212	110	3	5	7	51	6	27	5	260	13	11	5	7	55	38	302
RT25	317	288	3	7	19	142	16	41	2	189	39	27	11	24	128	54	591
RT27	614	119	bd	4	5	56	4	16	12	186	40	19	8	11	69	39	369
RT28	600	116	5	4	5	60	5	16	11	187	39	25	9	7	68	39	379
RT29	696	133	4	6	4	60	9	22	8	185	43	20	6	16	81	28	520
RT30	650	59	8	5	4	22	8	18	10	162	49	19	7	13	97	32	535
RT31	638	86	3	9	3	44	4	19	11	145	35	12	8	13	74	47	491
RT32	561	88	8	8	2	37	5	17	9	190	37	23	8	9	63	46	471
RT36	316	56	8	7	7	21	4	18	9	161	40	14	5	36	74	60	581
RT39	631	83	4	4	5	35	2	18	8	210	19	12	7	5	62	21	220
RT40	718	55	3	3	2	21	5	15	8	295	12	7	5	10	47	36	300
RT41	535	82	bd	3	bd	36	bd	13	2	154	20	10	4	6	52	24	252
RT42	414	60	6	7	3	26	2	12	4	223	27	19	8	18	41	24	246
RT43	465	46	5	19	6	22	4	13	7	229	24	14	11	15	39	26	252
RT44	460	48	7	12	2	20	bd	13	8	200	23	20	8	40	41	25	248
RT45	448	101	6	3	1	33	3	16	9	75	35	7	6	5	59	30	342
RT47	690	59	4	6	1	24	6	21	6	169	39	11	5	9	69	26	448
RT49	290	54	4	9	5	27	3	17	10	159	28	6	5	9	66	20	373
RT50	559	66	4	9	5	27	3	17	10	159	28	6	5	9	66	20	373
RT51	564	93	3	3	3	38	3	16	6	143	32	3	5	8	52	28	373
RT52	634	77	3	5	2	33	5	18	5	177	29	11	6	11	72	30	391
RT53	547	53	3	5	1	27	4	17	7	154	34	7	6	10	52	25	371
RT54	494	63	4	4	1	27	bd	16	5	101	41	2	6	7	62	19	354
RT55	662	106	bd	5	3	44	4	18	4	155	31	5	4	6	72	35	392
RT56	782	79	4	bd	3	30	4	16	11	190	27	8	3	8	58	43	380
RT57	645	78	6	4	7	33	4	17	8	159	28	10	5	4	60	44	408
RT58	656	42	bd	4	4	16	5	16	6	165	30	11	2	7	66	37	412
RT59	703	53	6	3	2	27	6	15	14	153	31	9	5	7	67	31	380
RT60	378	40	5	4	3	14	4	16	8	76	25	7	3	bd	62	19	353
RT61	540	95	5	8	bd	39	bd	15	5	116	31	5	5	6	52	18	368
RT62	589	41	4	bd	5	12	4	19	6	158	30	7	8	7	64	40	423
RT63	436	75	6	3	3	37	bd	17	3	104	29	16	3	5	50	17	340
RT64	551	82	8	9	bd	32	bd	17	6	126	28	6	4	8	62	34	384
RT65	874	20	bd	bd	3	8	6	17	10	236	28	17	7	10	56	50	427
RT66	609	80	bd	bd	1	27	3	20	5	168	30	5	5	8	61	37	408
RT67	588	54	3	5	2	21	bd	19	7	150	29	15	3	7	54	17	360
RT68	609	52	bd	5	bd	28	3	17	6	189	18	8	5	6	44	39	368
RT69	566	15	6	bd	4	7	3	17	10	169	30	11	6	7	63	43	382
RT70	433	93	9	7	bd	38	6	14	7	256	22	22	4	16	44	28	252
RT71	466	100	5	5	2	57	5	15	4	294	18	18	6	25	52	33	255
RT72	458	95	9	3	2	42	4	14	10	288	22	26	6	21	42	28	262
RT73	461	85	6	11	1	35	4	12	11	269	18	25	3	18	51	27	242
RT74	489	95	3	3	3	44	4	12	13	292	20	22	5	22	38	23	266
RT75	449	82	8	10	3	37	3	11	12	250	17	19	6	21	39	24	241
RT76	467	66	7	4	3	33	5	11	8	260	17	25	7	23	46	23	235
RT77	484	98	14	11	4	39	5	13	4	340	19	25	6	29	29	32	275
RT78	443	91	7	8	1	34	4	12	7	237	29	16	7	17	43	26	230
RT79	437	88	6	5	6	39	5	12	3	246	19	24	6	17	48	24	225
RT80	242	99	bd	5	2	38	3	22	7	146	24	11	3	5	64	16	261
RT82	227	95	4	4	6	31	5	24	9	156	23	11	6	6	64	23	260
RT83	270	89	bd	3	7	36	3	23	5	178	22	10	3	4	60	27	266
RT84	245	75	6	5	2	28	5	27	8	176	22	8	1	5	83	28	293
RT85	231	65	3	4	5	17	bd	22	6	129	19	9	3	5	57	18	262
RT86	240	82	bd	5	5	34	3	23	9	152	22	5	5	7	52	19	258
RT87	204	100	4	5	10	42	3	20	6	131	27	8	2	5	62	21	244

bd: below detection

Table 4. *Vemork Formation major element analyses*

Sample	SiO_2	TiO_2	Al_2O_3	Fe_2O_3	MnO	MgO	CaO	Na_2O	K_2O	P_2O_5	LOI	Total
RV1	47.90	2.07	15.22	12.89	0.18	8.25	5.29	4.40	0.93	0.41	2.21	99.75
RV2	47.77	1.93	15.03	11.82	0.16	6.45	7.05	3.38	0.70	0.38	5.30	99.97
RV6	49.09	1.91	15.34	11.56	0.21	7.00	7.62	3.68	0.41	0.42	2.49	99.73
RV7	50.46	2.09	14.58	11.35	0.20	7.00	7.28	4.10	0.37	0.44	2.31	100.18
RV8	49.47	1.96	14.71	11.90	0.17	7.12	7.92	3.73	0.68	0.43	2.01	100.10
RV39	47.23	2.11	15.66	12.80	0.17	8.05	4.86	2.79	0.81	0.32	4.99	99.79
RV40	44.81	1.86	14.89	13.80	0.17	6.61	8.16	3.74	0.32	0.28	5.51	100.15
RV44	47.26	2.21	15.31	13.59	0.19	7.91	5.80	3.44	0.52	0.39	3.51	100.13
RV45	51.43	1.86	13.88	12.45	0.18	6.61	6.70	2.84	0.27	0.37	3.31	99.90
RV46	52.78	1.72	13.77	12.73	0.17	6.34	5.42	2.94	0.56	0.29	3.19	99.91
RV47	51.45	1.83	13.90	11.62	0.17	6.57	7.79	2.64	0.61	0.32	2.89	99.79
RV48	42.72	2.15	17.08	14.39	0.26	8.59	7.49	3.27	0.24	0.40	3.07	99.66
RV49	48.30	2.28	14.08	13.21	0.17	7.18	7.62	2.84	0.99	0.37	3.05	100.09
RV50	54.49	2.00	12.07	11.29	0.16	5.78	8.44	2.30	0.78	0.34	2.39	100.04

of feldspars was therefore probably the main influence on major element compositions. K_2O/Na_2O ratios are quite variable, again probably as a result of metamorphic disturbance (Brewer & Menuge, pers. comm.), but most samples have ratios > 1.

A spiderdiagram plot of the 68 Tuddal Formation samples (Fig. 5) shows that the vast majority display rather similar element/element ratios. (The much greater variation in U and Th may be because concentrations of these elements are near detection limits.) The main exception is that La/Nb ratios exhibit considerable variation due to a wide range of La concentrations, in spite of rather uniform K/Nb and La/Ce ratios. This effect probably reflects control by one or more LREE-rich accessory minerals. A minority of samples have higher Sr concentrations and higher Sr/Zr ratios than the bulk of the sample set, suggesting control by plagioclase fractionation. Allowing for the effect of these igneous fractionation processes, the otherwise similar element patterns suggest that only minor metamorphic disturbance has affected these samples. Initial ε_{Nd} values for the Tuddal Formation range from 1.0 to 4.5, with T_{DM} ages ranging from 1510 to 1950 Ma. These are likely to be misleadingly high due to the high $^{147}Sm/^{144}Nd$ ratios of the rhyolites, which range from 0.121–0.158.

Vemork Formation

Harker diagrams of the Vemork Formation metabasalts (Fig. 6) reveal anticorrelation of both MgO and Al_2O_3, and perhaps Fe_2O_3, with SiO_2. CaO, Na_2O and K_2O show no correlation with SiO_2. As

Table 5. *Vemork Formation trace element analyses*

Sample	Ba	Ce	Co	Cr	Cu	La	Ni	Nb	Pb	Rb	Sr	V	Y	Zn	Zr
RV1	343	31	bd	379	37	11	76	19	13	23	153	188	37	168	174
RV2	248	32	47	380	16	bd	79	18	23	20	220	194	33	152	155
RV6	114	38	42	401	6	13	48	18	14	15	289	210	38	217	174
RV7	123	41	45	349	6	23	97	21	13	11	185	224	42	178	198
RV8	209	56	51	431	13	10	105	19	22	20	276	228	40	182	187
RV39	243	26	52	317	31	9	70	18	bd	20	334	193	37	201	151
RV40	138	30	60	357	14	17	96	18	bd	13	232	183	41	267	190
RV44	146	23	48	352	24	15	82	20	bd	13	247	197	40	190	185
RV45	76	27	47	298	41	12	69	20	bd	6	281	192	37	154	170
RV46	115	21	40	237	30	11	63	17	10	8	374	199	33	147	155
RV47	95	27	40	297	55	8	69	18	bd	6	295	193	34	143	165
RV48	78	37	47	291	20	12	70	21	11	8	284	207	42	217	165
RV49	205	40	44	380	22	13	75	19	bd	16	269	206	38	147	179
RV50	217	33	35	315	64	8	62	18	12	11	421	197	33	117	168

bd: below detection

Table 6. Sm–Nd isotopic analyses

Sample	Sm	Nd	^{143}Nd/^{144}Nd		^{147}Sm/^{144}Nd	$\varepsilon_{Nd}(T)$	T_{DM}	T_{DM2}
Rjukan Group – Tuddal Formation acid volcanic rocks								
§S-12	9.394	44.44	0.512189	12	0.1278	4.5	1507	1506
*RT45	10.62	50.72	0.512170	21	0.1265	4.4	1518	1515
*RT47	9.221	38.57	0.512306	30	0.1445	3.5	1613	1579
*RT53	10.78	53.79	0.512108	22	0.1212	4.2	1533	1527
*RT77	13.76	54.58	0.512363	20	0.1524	3.1	1679	1610
*RT84	13.44	51.59	0.512381	14	0.1575	2.5	1783	1661
*RT85	7.149	27.67	0.512379	12	0.1562	2.7	1748	1644
¶RT50	8.270	32.88	0.512250	20	0.1521	1.0	1952	1779
¶RT51	10.00	44.79	0.512214	16	0.1350	3.6	1596	1576
¶RT54	10.16	41.82	0.512225	14	0.1469	1.5	1856	1739
Rjukan Group – Vemork Formation basalts								
§S-3	6.783	28.25	0.512327	22	0.1451	3.8	1581	
§S-4	5.940	25.18	0.512296	10	0.1426	3.7	1591	
§S-5	5.714	24.25	0.512295	20	0.1424	3.7	1589	
†RV43	7.36	28.96	0.512390	8	0.1536	3.4	1644	
†RV44	8.00	34.88	0.512286	7	0.1386	4.3	1527	
†RV46	7.81	32.43	0.512327	10	0.1456	3.7	1592	
†RV48	8.75	36.00	0.512303	7	0.1469	3.0	1677	
Seljord Group – quartzite								
§S-14	10.21	47.56	0.512187	18	0.1298	4.1	1547	
Telemark gneisses								
§G-5	15.76	100.1	0.512185	18	0.09518	10.7	1103	
§G-7	7.614	37.28	0.512183	20	0.1235	5.2	1446	
§G-10	5.046	31.98	0.511904	16	0.09536	5.2	1461	

Sources of analysis: * University College Dublin; † British Geological Survey; § Cambridge University; ¶ Leeds University. Analyses of samples exhibiting metamorphic mobility of the high field strength elements (Brewer & Menuge, in press) have been excluded. Reproducibility of ^{147}Sm/^{144}Nd ratios is ± 0.2%. ^{143}Nd/^{144}Nd ratios were normalized to ^{146}Nd/^{144}Nd = 0.7219 and are relative to a value of 0.511860 for the La Jolla Nd standard. Within-run precision of ^{143}Nd/^{144}Nd ratios is given as $2\sigma_m$. Reproducibility of ^{143}Nd/^{144}Nd ratios is ± 0.00002 or the within-run precision, whichever is the greater. Sm and Nd spike solutions were calibrated against the CIT mixed normal Sm–Nd solution (Wasserburg et al. 1981). $\varepsilon_{Nd}(T)$ values calculated for T = 1500 Ma and using present day values for CHUR of ^{143}Nd/^{144}Nd = 0.512638 and ^{147}Sm/^{144}Nd = 0.1966. For the Tuddal Formation, T_{DM2} ages have been calculated assuming a crustal source with ^{147}Sm/^{144}Nd = 0.115 before eruption 1500 Ma ago.

previously discussed, regional metamorphism has pervasively disturbed CaO, Na$_2$O and K$_2$O in the metabasalts, so that any igneous fractionation trends originally present are likely to have been destroyed. The correlations of MgO and Al$_2$O$_3$ with SiO$_2$ suggest that fractional crystallization of both plagioclase and either olivine or pyroxene, or both, is responsible for some of the chemical variation of the metabasalts. However, the very low SiO$_2$ concentrations of some samples (two samples contain < 45% SiO$_2$) in spite of relatively low MgO concentrations, strongly suggest that SiO$_2$ was also mobile during regional metamorphism and that this is therefore responsible for part of the SiO$_2$ variation.

On a spiderdiagram plot (Fig. 7) the data are tightly clustered for the high field strength elements (HFSE) Y, Ti, Zr and Nb, suggesting that these elements were essentially immobile during regional metamorphism. La/Ce and La/Nb ratios are very variable and unrelated to La concentration, indicating metamorphic mobility of the LREE, whilst greater variation in Sr, Rb, Ba and K concentrations indicates that these elements were highly mobile during metamorphism. In addition, part of the variation in Sr and Ba abundances is probably due to fractional crystallization of plagioclase and to variation in plagioclase phenocryst abundance between samples, but the data provide no evidence for significant fraction-ation of other minerals.

Given the mobility of the alkali elements, conventional alkali–silica classifications (e.g. Le Maitre 1989) may be unreliable for characterization. Such diagrams (not illustrated) show considerable scatter but the samples nevertheless plot

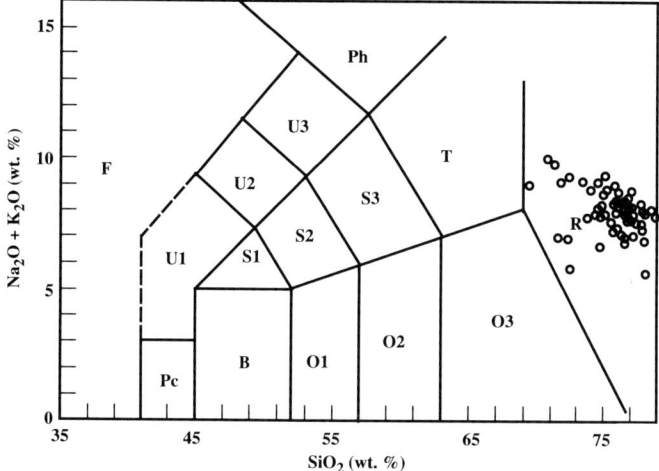

Fig. 3. Total alkali versus SiO_2 plot (Le Maitre 1989) for Tuddal Fm acid volcanic rocks.

mainly in the subalkaline field. As already noted, some of the variation in Si content may reflect metamorphic mobility. Sm/Nd ratios show little variation between samples, ranging from 0.139 to 0.154, with Nd concentrations ranging from 24 to 36 ppm. Initial ε_{Nd} values for the Vemork Formation metabasalts are tightly clustered in the range 3.0 to 4.3 (Fig. 8), indicating an isotopically depleted mantle source for the basic magmas. The low variation in both ε_{Nd} values, and in HFSE ratios and abundances, suggests that the Vemork Formation metabasalts were all comagmatic.

Petrogenesis of the volcanic rocks of the Tuddal and Vemork formations

The range of initial ε_{Nd} values is similar for the Tuddal and Vemork formations, but a comagmatic origin for the acid and basic volcanic rocks is unlikely for two reasons. Firstly, the volumes of acid and basic rocks presently exposed are comparable, rather than showing a high basic to acid ratio expected for a fractional crystallization origin, and intermediate compositions are rare. Secondly, the range of initial ε_{Nd} values for the Tuddal Formation acid volcanics extends to significantly lower values than those seen in the Vemork Formation metabasalts.

T_{DM} ages for the Tuddal Formation (Table 6) are unlikely to be reliable indicators of the mean crustal residence age of the source of the Tuddal Formation magmas because the rhyolites have $^{147}Sm/^{144}Nd$ ratios which are significantly higher than average continental crust. This may be attributed to fractional crystallization during ascent and ponding of magma at intracrustal levels. Better estimates of the mean crustal age of the source (T_{DM2} in Table 6) are calculated by assuming a typical crustal $^{147}Sm/^{144}Nd$ ratio of 0.115 prior to magma genesis at 1500 Ma. Such T_{DM2} ages range from 1510 to 1780 Ma compared to T_{DM} ages of 1100, 1450 and 1460 Ma for three samples of Telemark gneiss (Menuge 1985).

The proposed isotopic evolution of the Tuddal Formation samples and their crustal precursors is shown in Fig. 8. This comparison suggests that if the Tuddal volcanic rocks formed by melting of Telemark gneiss, the gneiss complex must contain elements of a greater T_{DM} age than those identified so far, similar to rocks of the Transscandinavian Igneous Batholith (TIB) (Brewer, unpublished data). The TIB consists of a major belt of granitoid magmatism post-dating the Svecofennian orogeny, with granite emplacement ages of c. 1800–1650 Ma. Chemically, the granites are predominantly I-type, with characteristics transitional to A-type. The Sm–Nd isotopic systematics of the TIB yield T_{DM} ages of 1820–2070 Ma (Brewer, unpublished data). Partial melting of continental crust whose T_{DM} ages range from those of the TIB to those known in the Telemark gneisses would therefore be a plausible source for Tuddal Formation acid magmas, in terms of their Nd isotopic composition (Fig. 8).

Elucidation of the Nd isotopic characteristics of the mantle source of the Vemork Formation metabasalts requires an assessment of the extent of crustal contamination. The most sensitive geochemical indicator is probably Rb, which ranges

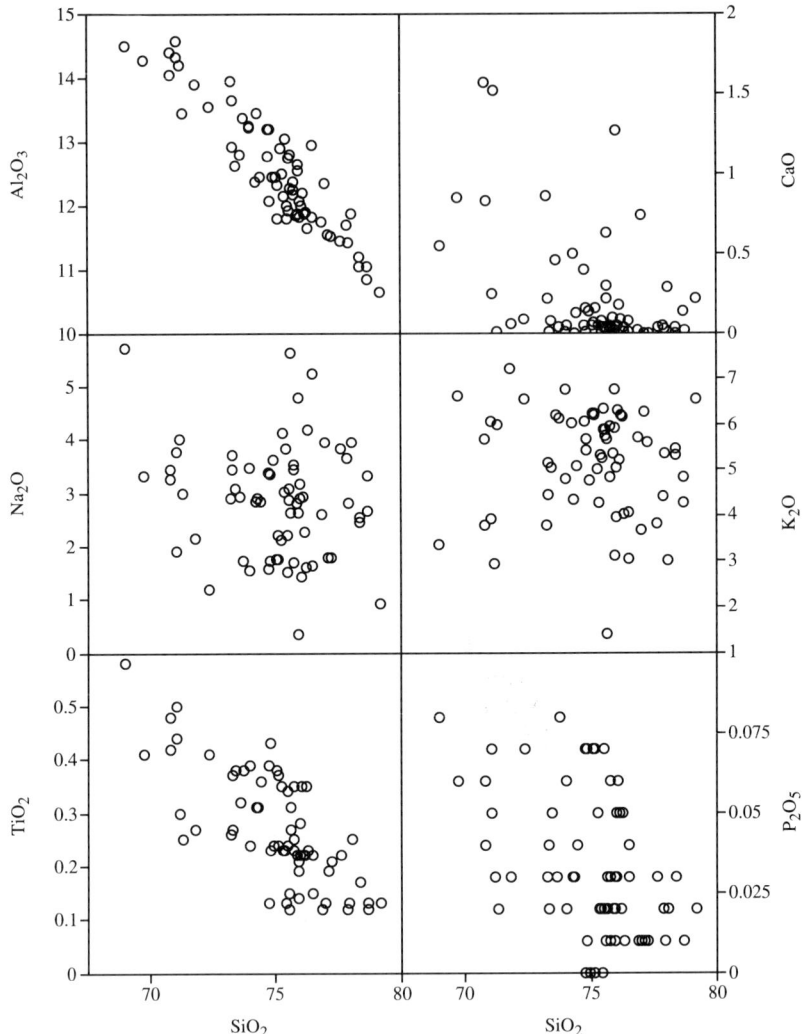

Fig. 4. Harker diagrams illustrating the chemical variation of the Tuddal Fm acid volcanic rocks. All axes are labelled in wt %.

from 6 to 23 ppm. Known potential crustal contaminants include the underlying Tuddal Formation, and the Telemark gneisses (assuming the latter formed the basement to the Rjukan Group). Tuddal Formation rocks contain 75–340 ppm Rb, with a mean of 184 ppm ($n = 68$). The Telemark gneisses include rocks of varied age and lithology, but are mostly of granitic chemical composition. In the Kviteseid region, immediately south of the central Telemark Rjukan Group outcrops, the Telemark gneisses contain 140–370 ppm Rb with a mean of 214 ppm ($n = 14$) (Priem *et al.* 1973). Simple mass balance considerations preclude bulk assimilation of more than about 10% of such gneissic material. In addition, the examination of samples for which both Rb concentration and initial ε_{Nd} are available reveals no correlation between these variables.

The possible effect of crustal contamination on ε_{Nd} is very limited because the potential contaminants have ε_{Nd} signatures at 1500 Ma which are very similar to that of the Vemork Formation (Fig. 8). The Tuddal Formation acid volcanic rocks have ε_{Nd} of 1.0 to 4.5, whilst three samples of Telemark granitic gneisses have $\varepsilon_{Nd}(1500)$ of 5.1, 5.2 and 10.7 (Menuge 1985). (The value of 10.7

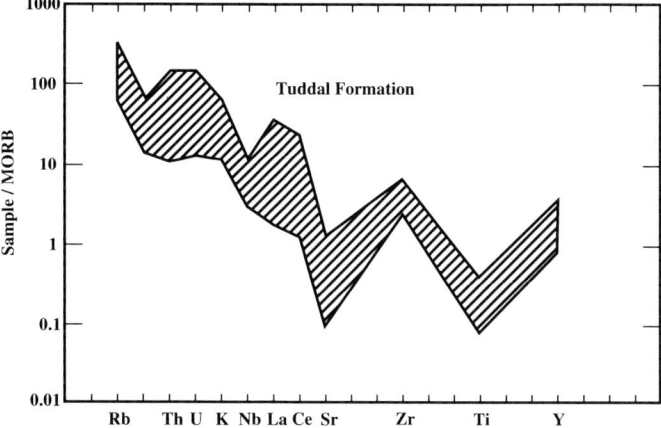

Fig. 5. Spiderdiagram of Tuddal Fm acid volcanic rocks.

for one sample almost certainly indicates that it represents crust much less than 1500 Ma old and cannot therefore have been a crustal contaminant of the Vemork Formation.) Given these constraints, the initial ε_{Nd} values of the Vemork Formation metabasalts (3.0–4.3) are unlikely to be significantly different to that of their mantle source, whose ε_{Nd} value is concluded to be slightly less depleted than the DePaolo (1981) depleted mantle value for 1500 Ma of 4.6.

Tectonic setting of Rjukan Group volcanism

Most published tectonic discrimination diagrams indicate A-type chemistry with a within-plate or anorogenic setting for the Tuddal Formation (e.g. Fig. 9a–c)). Eby (1992) has subdivided A-type granites into two chemical groups. Using this classification, the Tuddal Formation acid volcanic rocks are A_2-type, having relatively high Y/Nb and Rb/Nb ratios compared to A_1-type granites (Fig. 10a, b). A_2-type chemical compositions may indicate derivation of these magmas by partial melting of continental crust which has been through a cycle of subduction zone or continental collision magmatism, as opposed to differentiation of magma from an ocean island basalt source which may be related to a mantle plume (Eby 1992). Support for such an interpretation comes from experimental results. Dehydration melting of F-rich, biotite–hornblende–tonalitic gneiss at pressures of 6–14 kb can produce F-rich granitic liquids chemically similar to the Tuddal Formation and to A-type granites in general, leaving a granulitic residue dominated by orthopyroxene, quartz and plagioclase (Skjerlie & Johnston 1993).

For the Vemork Formation metabasalts, tectonic discrimination using only the HFSE is likely to be the most reliable approach because of the possibility of metamorphic disturbance of other trace elements, as previously discussed; a within-plate setting unrelated to subduction is indicated on the Zr/Y vs. Z diagram (Fig. 11). MgO concentrations of 5.8–8.6 % are typical of within-plate basalts and suggest substantial fractionation of olivine and/or pyroxene prior to the fractionation of plagioclase, which is present as phenocrysts. The low Mg/Fe ratios and low Cr and Ni concentrations of the Vemork Formation metabasalts are consistent with this interpretation.

Major and trace element patterns of the Vemork Formation metabasalts are similar to those of Phanerozoic continental flood basalts (CFBs). Their initial ε_{Nd} values of 3.0 to 4.3 indicate a slightly less depleted source than that of contemporary MORB (Fig. 8), a characterisitc also exhibited by many Phanerozoic CFBs. Whilst metamorphic disturbance has significantly affected the ratios of LILE and LREE, the concentrations of these elements are generally high for subalkaline basalts. One way in which these observations may be reconciled is to postulate enrichment of the mantle source in these elements no more than a few hundred Ma prior to eruption of the Vemork Formation metabasalts, perhaps by the addition of a component of metasomatized mantle lithosphere or mantle plume material which has sampled deeply subducted oceanic lithosphere. The absence of negative Nb anomalies meanwhile (Fig. 7), argues against the involvement of a mantle source which

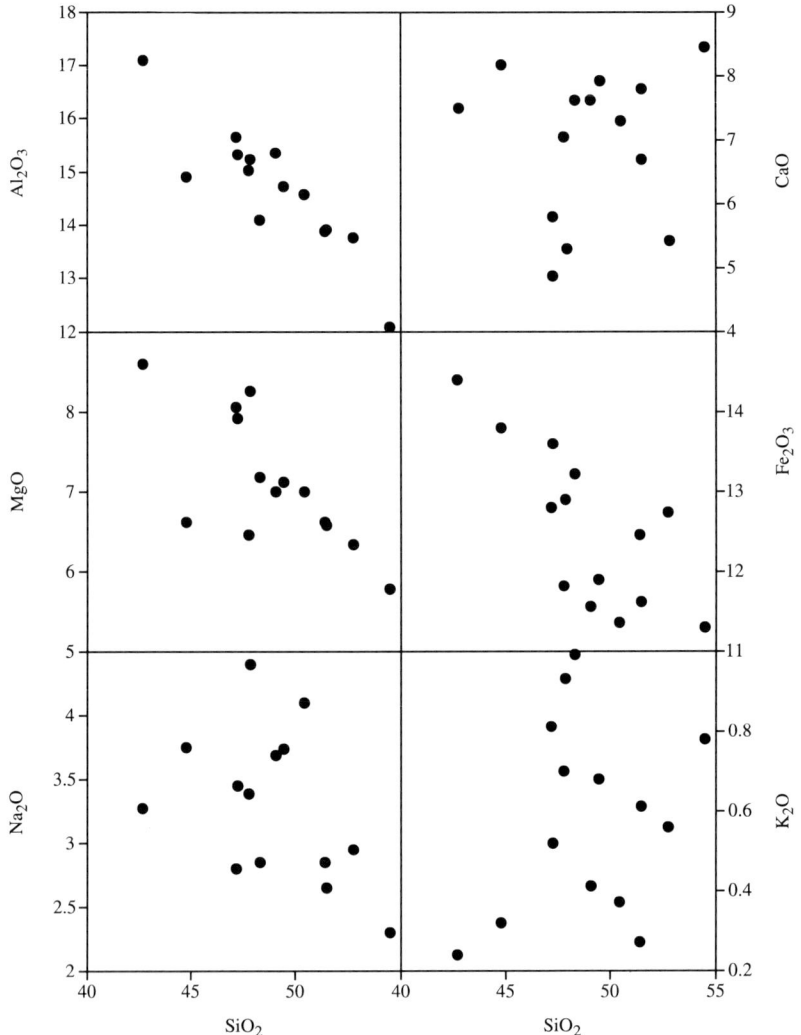

Fig. 6. Harker diagrams illustrating the chemical variation of the Vemork Fm metabasalts. All axes are labelled in wt %.

has been enriched by subduction-derived fluids. Consequently supra-subduction zone continental extension is an unlikely tectonic setting for the Rjukan Group.

The volume of erupted volcanic rocks in the Tuddal and Vemork formations is difficult to estimate due to subsequent deformation and erosion. In addition, the uncertainty of possible correlation with sequences outside the central Telemark area also hinders calculations. A conservative estimate, however, would be a mean thickness of 500 m for each formation over an aerial extent of 2500 km² for central Telemark. This corresponds to a volume per formation of 1250 km³. This is one or two orders of magnitude smaller than Phanerozoic CFB provinces, but the original extent of the Tuddal and Vemork formations may have been many times greater than this estimate for central Telemark and may have included metavolcanic rocks in the Fyresdal (Stout 1972) and Nisserdal (Mitchell 1967) areas, to the south of central Telemark. However one feature commonly associated with CFBs is absent in the Rjukan Group; there is no evidence for picritic basalts or other rocks with unusually high liquidus temperatures. Furthermore, in contrast to the Rjukan Group, CFBs are normally associated with subordinate (< 10%) amounts of acid volcanic

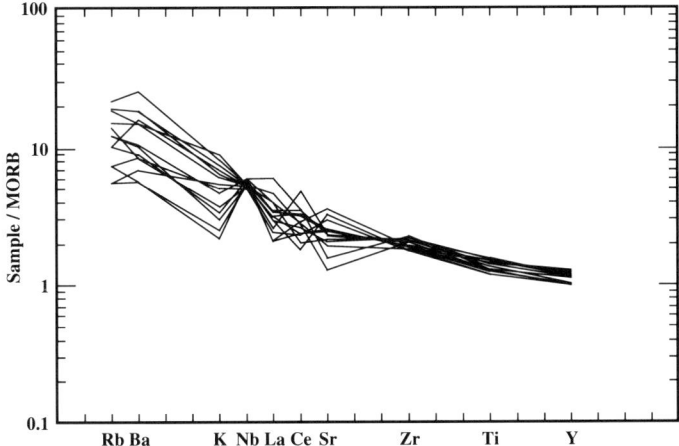

Fig. 7. Spiderdiagram of Vemork Fm metabasalts.

rocks erupted late in the volcanic succession. For these reasons, a **mantle plume** CFB origin is unlikely.

An alternative tectonic setting for the Rjukan Group is a slow-spreading continental rift zone similar to the East African Rift. Bimodal volcanism with comparable volumes of acid and basic volcanic rocks are a common feature in such settings. However, in this tectonic setting alkaline volcanic rocks are also usually present but these are apparently absent from the Rjukan Group. In the East African Rift, the rate of rifting is probably related to volcanic chemistry, with subalkaline rocks occurring where the rate of separation is relatively high.

Correlation of the Rjukan Group with the US midcontinent granite–rhyolite province

Gower *et al.* (1990) noted the broad similarity in age and structural setting of the Rjukan Group and the granite–rhyolite province of the US midcontinent. Most parts of this province are overlain by sedimentary cover and are known only from drilled core. The best known outcrops are those of the St Francois Mountains, Missouri, whose geology has been summarized by Kisvarsanyi & Kisvarsanyi (1990). These consist of granite ring complexes, intruding an overlying thin veneer of mainly pyroclastic acid volcanic rocks which are up to at least 1700 m thick locally, with occasional interbedded basalt flows. Minor intermediate rocks, mainly Fe-rich trachytes and trachyandesites, occur as ring intrusions.

The subsurface extent of the St Francois terrane is known from boreholes to be at least 60 000 km² (Kisvarsanyi & Kisvarsanyi 1990). Rocks of similar type and age are known from subsurface samples over a wide area of the US midcontinent (Fig. 12), with U–Pb zircon crystallization ages of 1380–1480 Ma. Ages are generally greater in the east and younger in the west (Bickford *et al.* 1981*b*; Van Schmus & Bickford 1981; Hoppe *et al.* 1983; Bickford *et al.* 1986).

The acid rocks have A-type chemistry, characterized by high Sn, Nb, Y, Be, Li, Rb, Ba and F, with greater enrichment of these elements in later

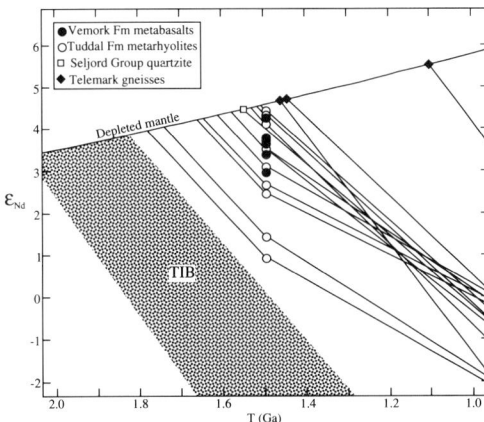

Fig. 8. ε_{Nd} vs. time diagram for the Rjukan Gp volcanic rocks. TIB = Transscandinavian Igneous Batholith (Brewer, unpublished data).

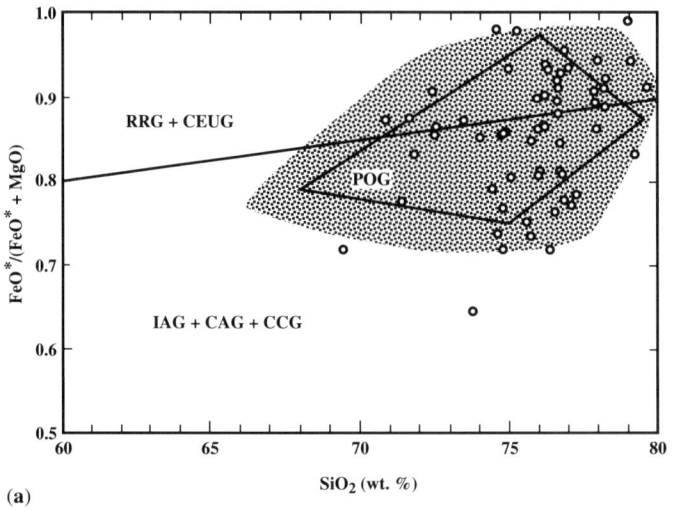

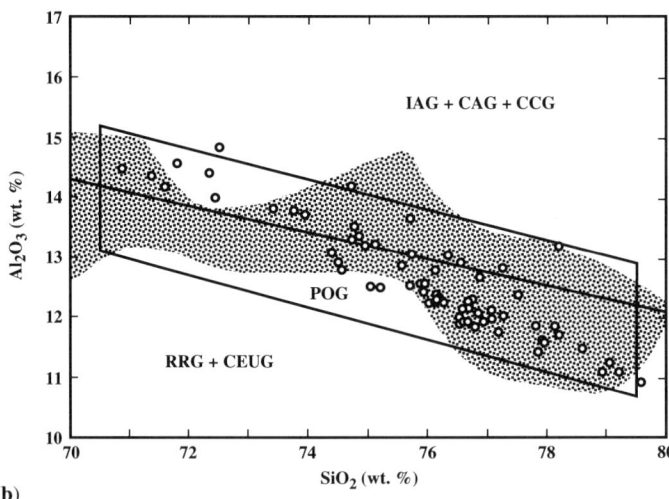

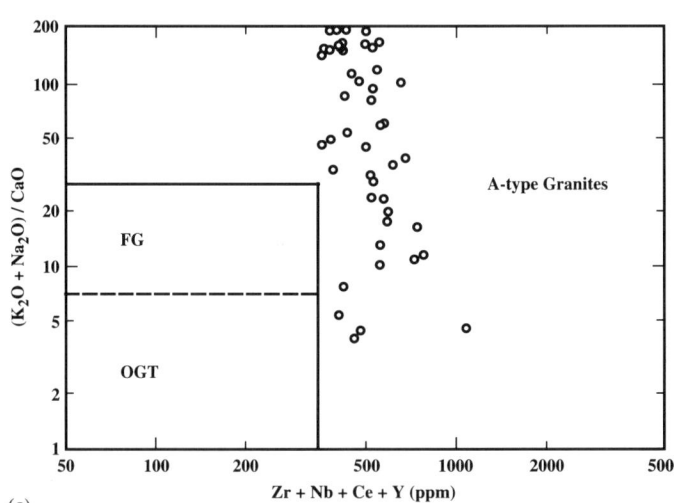

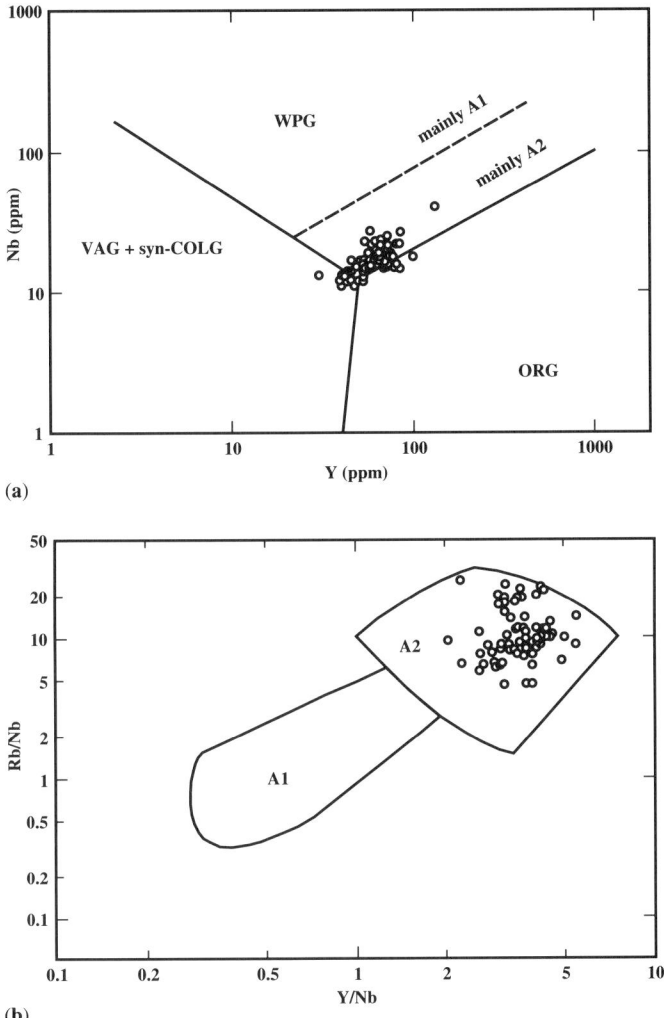

Fig. 10. Discrimination diagrams for Tuddal Fm acid volcanic rocks distinguishing A1- from A2-type magmas (Eby 1992). (**a**) Nb vs. Y (modified from Pearce *et al.* 1984); (**b**) Rb/Nb vs. Y/Nb (Eby 1992). WPG = within-plate granites, VAG = volcanic arc granites, syn-COLG = syn-collisional granites, ORG = ocean ridge granites.

Fig.9. Tectonic chemical discrimination diagrams for Tuddal Fm acid volcanic rocks, showing the range of published analyses from the St Francois Mountains, Missouri (shaded), where data are available (Kisvarsanyi 1980; Bickford *et al.* 1981*a*; Cullers *et al.* 1981). (**a**) FeO*/(FeO* + MgO) vs. SiO_2 (Maniar & Piccoli 1989); (**b**) Al_2O_3 vs. SiO_2 (Maniar & Piccoli 1989); (**c**) ($K_2O + Na_2O$)/CaO vs. Zr + Nb + Ce + Y (Whalen *et al.* 1987). POG = post-orogenic granitoids, IAG = island arc granitoids, CAG = continental arc granitoids, CCG = continental collision granitoids, RRG = rift-related granitoids, CEUG = continental epeirogenic uplift granitoids, FG = fractionated felsic granites, OGT = M-, I- and S-type granites.

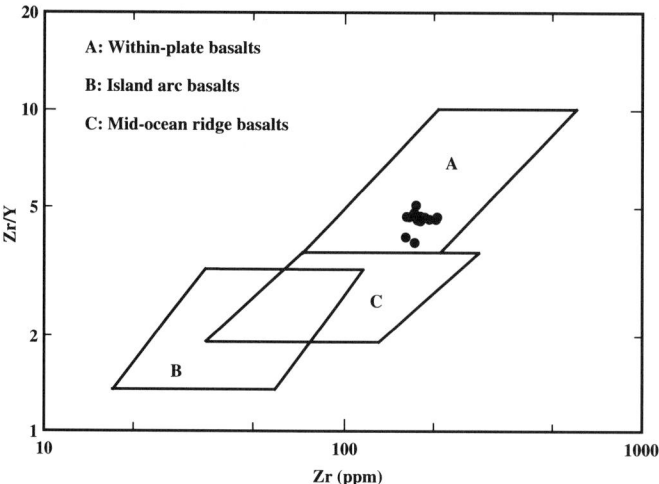

Fig. 11. Zr/Y vs. Zr tectonic discrimination diagram (Pearce & Norry 1979) for Vemork Fm metabasalts.

plutons. Some of the late-stage granites have abnormally high U and Th. Nearly all the acid rocks are subalkaline, peraluminous to metaluminous, and have major element compositions similar to the Tuddal Formation rhyolites. None are peralkaline.

When allowance is made for the slightly different range of elements analysed, the similarity of the trace element compositions of these rocks with those of the Tuddal Formation is apparent (Figs 9 & 13).

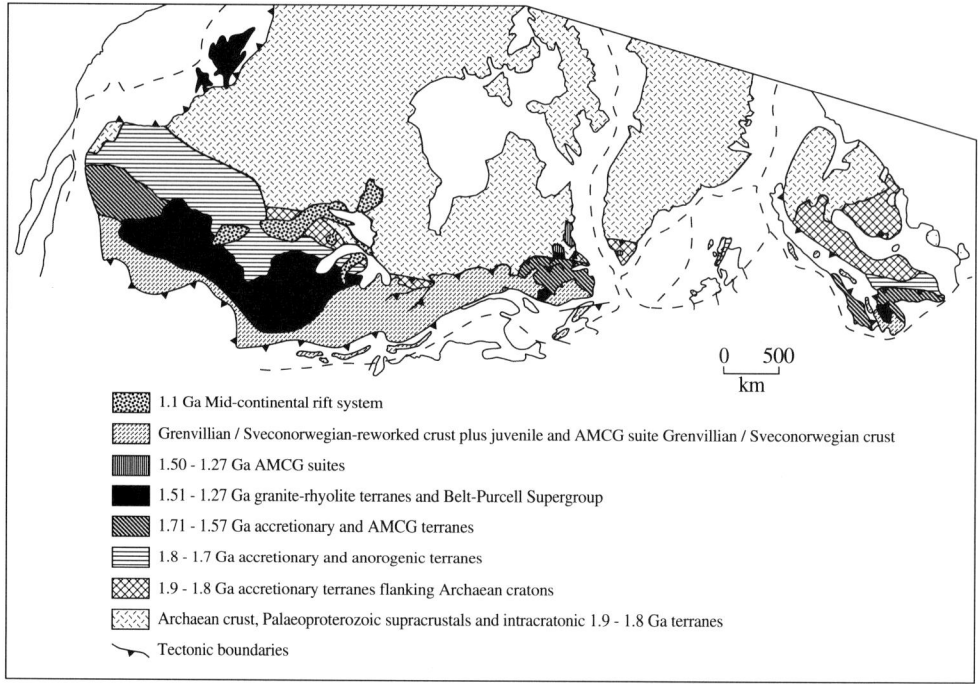

Fig. 12. Reconstruction of Laurentia–Baltica during the Mesoproterozoic, showing disposition of the main crustal age provinces (modified from Gower *et al.* 1990).

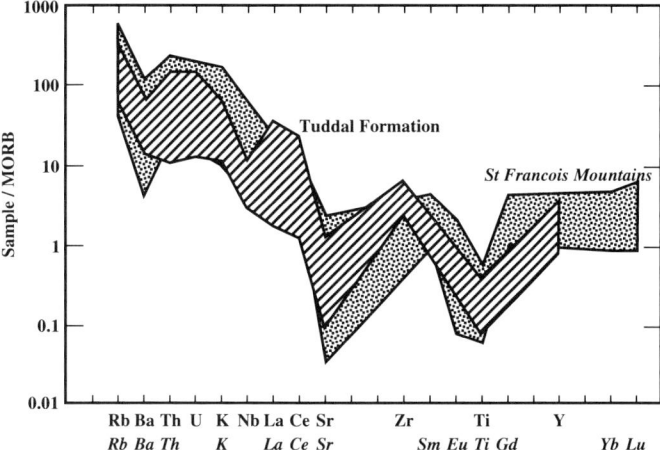

Fig. 13. Spiderdiagram illustrating the variation in composition of the Tuddal Fm acid volcanic rocks, compared to the range exhibited by published analyses of acid volcanic rocks from the St Francois Mountains, Missouri (data from Kisvarsanyi (1980), Bickford *et al.* (1981*a*) and Cullers *et al.* (1981)). Elements analysed in Tuddal Fm samples are shown in plain lettering; elements analysed in St Francois Mountains samples are shown in italics.

In both the US midcontinent and southern Scandinavia, the oldest crust is composed of c. 1900–1800 Ma old rocks. This represents mainly juvenile crust, which was affected soon after crustal accretion by the Penokean Orogeny in US and the Svecofennian Orogeny in Scandinavia. These deformed rocks were intruded by 1800–1600 Ma granites, which also represent juvenile crust, and include the TIB in Scandinavia and various units in the US. These were deformed prior to the eruption of rhyolites in the US midcontinent and of the Tuddal Formation rhyolites in south Norway. Anorogenic acid volcanic rocks in both Telemark and the US midcontinent, plus high level granites in the latter, appear to lie on what is now the southern margin of this composite, 1900–1600 Ma continental crust. In the US, midcontinent granites are seen to intrude these older rocks, whereas the relationship between the Tuddal Formation and pre-existing rocks in Telemark has yet to be satisfactorily demonstrated.

Nelson & DePaolo (1985) presented T_{DM} ages of US midcontinent granite and rhyolite samples, which range from 1500–2300 Ma, with most in the range 1700–2000 Ma, similar to the 1600–1900 Ma old gneisses which they intrude. Like the Tuddal Formation, the US midcontinent granites and rhyolites therefore seem to have been derived primarily by partial melting of pre-existing continental crust of Palaeoproterozoic age. The eastern extent of this crustal province, where it is intersected by the Grenville orogenic belt, is unknown. A prediction of the proposed correlation of the Rjukan Group with the US midcontinent granite–rhyolite province is that similar rocks should occur in the intervening region (Fig. 12) which was affected by the Grenville Orogeny. That such rocks probably exist is indicated by the recognition of granites intruded c. 1470 Ma ago (U–Pb zircon dating) lying north of Lake Huron and immediately to the west of the Grenville Front (Van Breemen & Davidson 1988).

Tectonic significance of 1400–1500 Ma bimodal volcanism in Laurentia–Baltica

A number of models have been proposed to account for the US midcontinent granite–rhyolite province.

(1) A mantle plume-related setting was suggested by Kisvarsanyi (1980) for the St Francois Mountains region, and for the US midcontinent in general by Hill *et al.* (1992) and Hill (1993). Windley (1993) suggested that post-1500 Ma anorogenic rocks in Laurentia–Baltica formed by plume impact, leading to tensional stress and in some cases continental break-up. In this model, ponding of mantle-derived magmas, which fractionated to produce massif anorthosites, plus extensive melting of lower continental crustal rocks, which may have been previously depleted by granulite facies metamorphism, then gave rise to extensive felsic magmatism. However, correlation of the US midcontinent granite–rhyolite province with the Rjukan Group results in a roughly linear

magmatic belt (Fig. 12). This provides further evidence against mantle plume interpretations in addition to the evidence discussed previously.

(2) Van Schmus & Bickford (1981) proposed that the US midcontinent felsic magmas resulted from anorogenic melting of tectonically thickened older crust, which had formed during a series of subduction-related orogenies. A similar origin for some Mesoproterozoic anorogenic magmatism was suggested by Windley (1993). The time lapse between cessation of orogeny and production of anorogenic granites in the US midcontinent and in south Norway is poorly constrained, so that the plausibility of this model is difficult to assess.

(3) Nelson & DePaolo (1985) proposed that US midcontinent granite–rhyolite magmatism was related to the supra-subduction zone production of Llano crust south of the midcontinent crustal province c. 1500–1300 Ma ago. They suggest an analogy between the US midcontinent and the Mesozoic–Cenozoic orogenies of the western margin of North America, which produced plutonism that extended far inland. The scarcity of basic magmatism, and the lack of evidence for a mantle-derived component in the felsic rocks is not, however, explained well by this model.

(4) A rift-related setting, proposed by Emslie (1978) and Anderson (1983), seems to provide the best explanation for the available data, with rifting of the US midcontinent and Rjukan Group possibly related to the opening of the Grenville ocean. The scarcity of basic volcanism in the US midcontinent province is not necessarily evidence against this interpretation, since basalts may have been scarce during Phanerozoic ocean opening settings. This was illustrated by Kay *et al.* (1989) who compared Gondwanaland margin volcanic rocks with those of the US midcontinent. They suggested that basic magmas were scarce due to the ponding of magma at the base of a pre-heated crust, formed 200–300 Ma earlier. This led to extensive crustal melting and the eruption of rhyolites, as well as granite emplacement. The depleted mantle isotopic signature of the Vemork Formation basalts, which do not have a subduction-related chemical signature, is consistent with this interpretation. It remains unclear however, whether the anorthosite–mangerite–charnockite–granite (AMCG) association was produced at deeper crustal levels within the same tectonic setting. An alternative explanation for the formation of the AMCG suite would be within an unrelated mantle plume-related setting.

Conclusions

Within-plate lithospheric extension associated with consequent crustal thinning and mantle melting due to decompression was the cause of Rjukan Group magmatism. The Tuddal Formation acid volcanic rocks represent anorogenic partial melts of subduction-generated continental crustal rocks, which were probably located within what is now the Telemark gneiss complex. The Vemork Formation basalts were derived by partial melting of depleted mantle. The Rjukan Group is correlated with the US midcontinent granite–rhyolite province. The age, chemical composition and structural relationships to pre-existing rocks of the Rjukan Group and the US midcontinent granite–rhyolite province are similar, with the notable exception that basic magmatism was volumetrically far more important in the former than in the latter. The roughly linear belt defined by these rocks in Mesoproterozoic continental reconstructions suggests that the intervening area, within the Grenville Province of the US and eastern Canada, may contain the deformed and metamorphosed extension of the US midcontinent granite–rhyolite province.

Widespread rifting and bimodal volcanism, including extensive melting of continental crust accreted 1600–1900 Ma ago, occurred along what is now the margin of Laurentia–Baltica. Extension and magmatism were not related either to subduction or to a mantle plume, with continental rifting, perhaps related to the opening of the Grenville ocean, the preferred model. In this model, magmatic activity was initiated progressively later from east to west, commencing at 1500 Ma in south Norway, 1460 Ma in the St Francois Mountains and 1430 Ma in the southwestern US.

The Seljord Group may represent an intracratonic basin which formed immediately after eruption of the Rjukan Group, as a result of crustal thinning associated with continuing crustal extension. Deposition of clastic sedimentary rocks in the much larger Belt–Purcell Supergroup of the western US and Canada commenced while the US midcontinent province was magmatically active (Zartman *et al.* 1982). These rocks may have a similar origin to the Seljord Group.

Two major obstacles hinder further understanding of the origin of the Rjukan Group. Firstly, the relationship of the Rjukan Group to the Telemark gneisses must be determined. If it can be shown that these gneisses formed the basement of the Rjukan Group, then the pre-1500 Ma magmatic, metamorphic and structural evolution of the gneisses, and the crustal residence ages of their protoliths, must be established. Origin of the Tuddal Formation rhyolites by partial melting of the Telemark gneisses can then be modelled. Secondly, the timing of Vemork Formation volcanism and Seljord Group deposition are required in order to establish whether they form part of the same tectonomagmatic cycle as the Tuddal Formation, as suggested by field evidence. Improved understanding of the Rjukan Group depends primarily on progress in these two areas.

We thank Arlene Hunter and Catrin Ellis Jones whose reviews have substantially improved the manuscript. JFM thanks Michael Murphy for assistance in the UCD mass spectrometry laboratory.

References

ANDERSON, J. L. 1983. Proterozoic anorogenic granite plutonism of North America. In: MEDARIS,L. G. Jr, BYERS, C. W., MICKELSON, D. M. & SHANKS, W. C. (eds) *Proterozoic Geology: Selected Papers from an International Proterozoic Symposium.*. Geological Society of America Memoir, **161**, 133–154.

ATKIN, B. P. & BREWER, T. S. 1990. The tectonic setting of basaltic magmatism in the Kongsberg, Bamble and Telemark Sectors, Southern Norway. In: GOWER, C., RIVERS, T. & RYAN, B. (eds) *Mid-Proterozoic Laurentia–Baltica*. Geological Association of Canada Special Paper **38**, 471–483.

AVILA MARTINS, J. 1969. The Precambrian rocks of the Telemark area in the South Central Norway, No. VII, the Vrådal area. *Norges Geologiske Undersøkelse*, **258**, 267–301.

BARTH, T. F. W. & REITAN, P. H. 1963. The Precambrian of Norway. In: RANKAMA, K. (ed.) *The Precambrian* (Volume 1). Wiley Interscience, New York, 27–80.

BICKFORD, M. E., SIDES, J. R. & CULLERS, R. L. 1981a. Chemical evolution of magmas in the Proterozoic terrane of the St. Francois Mountains, southeastern Missouri 1. Field, petrographic, and major element data. *Journal of Geophysical Research*, **86**, 10365–10386.

——, HARROWER, K. L., HOPPE, W. J., NELSON, B. K., NUSBAUM, R. L. & THOMAS, J. J. 1981b. Rb–Sr and U–Pb geochronology and distribution of rock types in the Precambrian basement of Missouri and Kansas. *Geological Society of America Bulletin*, **92**, 323–341.

——, VAN SCHMUS, W. R. & ZIETZ, I. 1986. Proterozoic history of the midcontinent region of North America. *Geology*, **14**, 492–496.

BREWER, T. S. 1985. *Geochemistry, Geochronology and Tectonic Setting of the Proterozoic Telemark supracrustals, Southern Norway.* PhD thesis, University of Nottingham.

—— & ATKIN, B. P. 1987. Geochemistry and tectonic setting of the Proterozoic Telemark supracrustal rocks, southern Norway. In: PHARAOH, T. C., BECKINSALE, R. D. & RICKARD, D. (eds) *Geochemistry and Mineralization of Proterozoic Volcanic Suites.* Geological Society, London, Special Publications **33**, 471–487.

—— & —— 1989. Elemental mobility produced by low-grade metamorphic events: A case study from the Proterozoic Telemark supracrustal rocks of Southern Norway. *Precambrian Research*, **45**, 143–158.

—— & FIELD, D. 1985. Tectonic environment and age relationships of the Telemark Supracrustals, southern Norway. In: TOBI, A. C. & TOURET, J. L. R. (eds) *The Deep Proterozoic Crust in the North Atlantic Provinces.* Reidel, Dordrecht, 291–307.

CHAFFEY, D. J., CLIFF, R. A. & WILSON, B. M. 1989. Characterization of the St. Helena magma source. In: SAUNDERS, A. D. & NORRY, M. J. (eds) *Magmatism in the Ocean Basins.* Geological Society, London, Special Publications **42**, 257–276.

CRAMEZ, C. 1970. The Precambrian rocks of the Telemark area in South Central Norway, No. VIII, Evolution structurale de la region Nisser - Vråvatn. *Norges Geologiske Undersøkelse*, **266**, 5–35.

CULLERS, R. L., KOCH, R. J. & BICKFORD, M. E. 1981. Chemical evolution of magmas in the Proterozoic terrane of the St. Francois Mountains, southeastern Missouri 2. Trace element data. *Journal of Geophysical Research*, **86**, 10388–10401.

DAHLGREN, S., HEAMAN, L. & KROGH, T. 1990. Geological evolution and U–Pb geochronology of the Proterozoic Telemark area, southern Norway. *Geonytt*, **17**, 38.

DEPAOLO, D. J. 1981. Neodymium isotopes in the Colorado Front Range and crust-mantle evolution in the Proterozoic. *Nature*, **291**, 193–196.

DONS, J. A. 1960a. Telemark supracrustals and associated rocks. In: HOLTEDAHL, O. (ed.) *Geology of Norway.* Norges Geologiske Undersøkelse, **208**, 49–58.

——1960b. The stratigraphy of supracrustal rocks, granitization and tectonics in the Precambrian Telemark area, Southern Norway. *Norges Geologiske Undersøkelse*, **212**.

—— 1963. Precambrian rocks of central Telemark, Norway: II. Ripple marks and mud cracks. *Norsk Geologisk Tidsskrift*, **43**, 477–495.

—— 1972. The Telemark area, a brief presentation. *Sciences de la Terre*, **17**, 25–29.

—— & JORDE, K. 1978. *Berggrunnskart Skien 1:250 000.* Norges Geologiske Undersøkelse.

EBY, G. N. 1992. Chemical subdivision of the A-type granitoids: Petrogenetic and tectonic implications. *Geology*, **20**, 641–644.

EMSLIE, R. F. 1978. Anorthosite massifs, rapakivi granites, and late Proterozoic rifting of North America. *Precambrian Research*, **7**, 61–98.

GOWER, C. F., RYAN, A. B. & RIVERS, T. 1990. Mid-Proterozoic Laurentia–Baltica: an overview of its geological evolution and a summary of the contributions made by this volume. In: GOWER, C.,

RIVERS, T. & RYAN, B. (eds) *Mid-Proterozoic Laurentia-Baltica*. Geological Association of Canada Special Paper **38**, 1–20.

HARVEY, P. K. & ATKIN, B. P. 1982. Automated X-ray Fluorescence analysis. *In: Sampling and Analysis for the Minerals Industry*. Special Publication of the Institution of Mining and Metallurgy, 17–26.

HILL, R. I. 1993. Mantle plumes and continental tectonics. *Lithos*, **30**, 193–206.

——, CAMPBELL, I. H., DAVIES, G. F. & GRIFFITHS, R. W. 1992. Mantle plumes and continental tectonics. *Science*, **256**, 186–193.

HOOKER, P. J., HAMILTON, P. J. & O'NIONS, R. K. 1981. An estimate of the Nd isotopic composition of Iapetus seawater from ca. 490 Ma metalliferous sediments. *Earth and Planetary Science Letters*, **56**, 180–188.

HOPPE, W. J., MONTGOMERY, C. W. & VAN SCHMUS, W. R. 1983. Age and significance of Precambrian basement samples from northern Illinois and adjacent states. *Journal of Geophysical Research*, **88**, 7276–7286.

KAY, S. M., RAMOS, V. A., MPODOZIS, C. & SRUOGA, P. 1989. Late Paleozoic to Jurassic silicic magmatism at the Gondwana margin: Analogy to the Middle Proterozoic in North America? *Geology*, **17**, 324–328.

KISVARSANYI, E. B. 1980. Granitic ring complexes and Precambrian hot-spot activity in the St. Francois terrane, Midcontinent region, United States. *Geology*, **8**, 43–47.

—— & KISVARSANYI, G. 1990. Alkaline granite ring complexes and metallogeny in the Middle Proterozoic St. Francois Terrane, southeastern Missouri, U.S.A. *In*: GOWER, C., RIVERS, T. & RYAN, B. (eds) *Mid-Proterozoic Laurentia–Baltica*. Geological Association of Canada Special Paper **38**, 433–446.

LE MAITRE, R. W. 1989. *A Classification of Igneous Rocks and Glossary of Terms*. Blackwell Scientific Publications, Oxford.

MANIAR, P. D. & PICCOLLI, P. M. 1989. Tectonic discrimination of granitoids. *Geological Society of America Bulletin*, **101**, 635–643.

MENUGE, J. F. 1985. Neodymium isotope evidence for the age and origin of the Proterozoic of Telemark, South Norway. *In*: TOBI, A. C. & TOURET, J. L. R. (eds) *The Deep Proterozoic Crust in the North Atlantic Provinces*. Reidel, Dordrecht, 435–448.

—— 1988. The petrogenesis of massif anorthosites: a Nd and Sr isotopic investigation of the Proterozoic of Rogaland/Vest-Agder, SW Norway. *Contributions to Mineralogy and Petrology*, **98**, 363–373.

MILNE, A. J. & MILLER, I. L. 1991. Mid-Palaeozoic basement in eastern Graham Land and its relation to the Pacific margin of Gondwana. *In*: THOMSON, M. R. A., CRAME, J. A. & THOMSON, J. W. (eds) *Geological Evolution of Antarctica*. Cambridge University Press, 335–340.

MITCHELL, R. H. 1967. The Precambrian rocks of the Telemark area in South Central Norway, No. V, The Nissedal supracrustal series. *Norsk Geologisk Tidsskrift*, **47**, 295–332.

NELSON, B. K. & DEPAOLO, D. J. 1985. Rapid production of continental crust 1.7–1.9 b.y. ago: Nd and Sr isotopic evidence form the basement of the North American midcontinent. *Geological Society of America Bulletin*, **96**, 746–754.

PEARCE, J. A. & NORRY, M. J. 1979. Petrogenetic implications of Ti, Zr, Y, and Nb variations in volcanic rocks. *Contributions to Mineralogy and Petrology*, **69**, 33–47.

——, HARRIS, N. B. W. & TINDLE, A .G. 1984. Trace element discrimination diagrams for the tectonic interpretation of granitic rocks. *Journal of Petrology*, **25**, 956–983.

PRIEM, H. N. A., BOELRIJK, N. A. I. M., HEBEDA, E. H., VERDURMEN, E. A. TH. & VERSCHURE, R. H. 1973. Rb–Sr investigations on Precambrian granites, granitic gneisses and acidic metavolcanics in central Telemark: metamorphic resetting of Rb–Sr whole-rock systems. *Norges Geologiske Undersøkelse*, **289**, 37–53.

SIGMOND, E. M. O. 1978. Beskrivelse til det berggrunsgeologiske kartbladet Sauda 1:250000 (Med fargetrykt kart). *Norges Geologiske Undersøkelse*, **341**, 1–94.

SINGH, I. B. 1969. The Precambrian rocks of the Telemark area in South Central Norway VI, Primary sedimentary structures in Precambrian quartzites of Telemark, Southern Norway, and their environmental significance. *Norsk Geologisk Tidsskrift*, **49**, 1–31.

SKJERLIE, K. P. & JOHNSTON, A. D. 1993. Fluid-absent melting behavior of an F-rich tonalitic gneiss at mid-crustal pressures: implications for the generation of anorogenic granites. *Journal of Petrology*, **34**, 785–815.

STARMER, I. C. 1990. Mid-Proterozoic evolution of Kongsberg Bamble Belt and adjacent areas, southern Norway. *In*: GOWER, C., RIVERS, T. & RYAN, B. (eds) *Mid-Proterozoic Laurentia–Baltica*. Geological Association of Canada Special Paper **38**, 279–305.

—— 1993. The Sveconorwegian Orogeny in southern Norway, relative to deep crustal structures and events in the North Atlantic Proterozoic Supercontinent. *Norsk Geologisk Tidsskrift*, **73**, 109–132.

STOUT, J. H. 1972. Stratigraphic studies of high-grade metamorphic rocks east of Fyresdal. *Norsk Geologisk Tiddskrift*, **52**, 23–41.

VAN BREEMEN, O. & DAVIDSON, A. 1988. Northeast extension of Proterozoic terranes of midcontinental North America. *Geological Society of America Bulletin*, **100**, 630–638.

VAN SCHMUS, W. R. & BICKFORD, M. E. 1981. Proterozoic chronology and evolution of the midcontinent region, North America. *In*: KRÖNER, A. (ed.) *Precambrian Plate Tectonics*. Elsevier, Amsterdam, 261–296.

VENUGOPAL, D. V. 1970. The Precambrian rocks of the Telemark area in south central Norway, No. X, Geology and structure of the area west of Fyresdal, Telemark, southern Norway. *Norges Geologiske Undersøkelse*, **268**, 1–57.

WALSH, J. N., BUCKLEY, F. & BARKER, J. 1981. The simultaneous determination of the rare earth elements in rocks using inductively coupled plasma spectrometry. *Chemical Geology*, **33**, 141–153.

WASSERBURG, G. J., JACOBSEN, S. B., DEPAOLO, D. J., MCCULLOCH, M. T. & WEN, T. 1981. Precise determination of Sm/Nd ratios, Sm and Nd isotopic abundances in standard solutions. *Geochimica et Cosmochimica Acta*, **45**, 2311–2324.

WERENSKIOLD, W. 1910. Om Ost-Telemark. *Norges Geologisk Undersøkelse*, **53**, 3–71.

WHALEN, J. B., CURRIE, K. L. & CHAPPELL, B. W. 1987. A-type granites: geochemical characteristics, discrimination and petrogenesis. *Contributions to Mineralogy and Petrology*, **95**, 407–419.

WINDLEY, B. F. 1993. Proterozoic anorogenic magmatism and its orogenic connections. *Journal of the Geological Society, London*, **150**, 39–50.

ZARTMAN, R. E., PETERMAN, Z. E., OBRADOVICH, J. D., GALLEGO, M. D. & BISHOP, D. T. 1982. Age of the Crossport C sill near Eastport, Idaho. *In*: READ, R. R. & WILLIAMS, G. A. (eds) *Society of Economic Geologists' Coeur d'Alène Field Conference, Idaho, 1977*. Idaho Bureau of Mines and Geology Bulletin **24**, 61–69.

Thermotectonic evolution of the Eastern Segment of southwestern Sweden: tectonic constraints from U–Pb geochronology

J. N. CONNELLY[1], J. BERGLUND[2] & S. Å. LARSON[2]

[1] *Department of Geological Sciences, University of Texas at Austin, Austin, TX 78712, USA*
[2] *Department of Geology, Earth Sciences Centre, Göteborg University, S-413 81 Göteborg, Sweden*

Abstract: The Eastern Segment forms a complex crustal terrane that occupies the southeastern part of the Southwest Scandinavian Domain adjacent to the less deformed 1.65–1.85 Ga Transscandinavian Igneous Belt (TIB). New U–Pb data from the Eastern Segment indicate that orthogneisses and granitoid rocks have protolith ages equivalent to the later magmatic phases of the TIB and that the earliest thermotectonic episodes in the Eastern Segment occurred between 1.70 and 1.61 Ga during the Gothian Orogeny. The comparable ages for Eastern Segment gneisses and TIB rocks represent permissive evidence for the hypothesis that the orthogneiss protoliths intruded into the western margin of Fennoscandia and are not exotic with respect to the pre-Gothian craton.

The eastern limit of Gothian deformation is thus interpreted to be an intracratonic deformation front that coincides with the later Sveconorwegian deformation front along the Protogine Zone south of Lake Vänern. The post-Gothian anorogenic period was interrupted, in at least the eastern part of the Eastern Segment, by a thermal and magmatic event at 1.47 Ga that generated granitic dykes and (re)crystallized titanite. Sveconorwegian U–Pb ages and field observations are compatible with a thrusting and exhumation of western supracrustal terranes over the Eastern Segment during the Sveconorwegian Orogeny, followed by isostatic unroofing and cooling of the lower deck at 950–930 Ma. Available, albeit limited, Sveconorwegian titanite ages in the Eastern Segment young from east to west, ranging from 950 to 930 Ma, and are substantially younger than those dated in the western terranes of SSD.

Proterozoic crustal growth in the Fennoscandian (Baltic) Shield resulted in progressively younger crust away from the Archaean nuclei in northwestern Russia and northeastern Finland (Gaál & Gorbatschev 1987). Geological mapping and geochronology have permitted correlation of major orogenic events associated with this growth in Fennoscandia (the *c.* 1.9–1.8 Ga Svecofennian Orogen and *c.* 1.6 Ga Gothian Orogen) with those in the North Atlantic craton (the *c.* 1.8 Ga Makkovikian Orogen and *c.* 1.65 Ga Labradorian Orogen). This work has led to models suggesting that Fennoscandia and Laurentia were together as part of a single continent until a mid-Proterozoic break-up (Gower & Owen 1984; Gower *et al.* 1990), which resulted in rift-related magmatism and sedimentation between ca 1.45 and 1.20 Ga in both crustal blocks (e.g. Gorbatschev *et al.* 1987; Starmer 1991; Gower & Tucker 1994). This break-up and subsequent rotation of Fennoscandia with respect to Laurentia eventually led to re-working during the *c.* 1.2–0.9 Ga Sveconorwegian–Grenvillian Orogeny along the margins of the two cratons (Gower 1990; Hoffman 1992).

While this emerging large-scale tectonic model offers a broad framework within which to conceptualize the development of Laurentia and Fennoscandia during the Proterozoic, much work is required if these models are to be refined further. This paper concentrates on Proterozoic tectonic development of the Southwest Scandinavian Domain (SSD, Gaál & Gorbatschev 1987) in southwestern Sweden. New U–Pb data are presented in conjunction with carefully defined field relationships, in order to better constrain the absolute timing of thermotectonic events, tectonic models for the SSD and correlations with the North Atlantic craton.

Regional geology of the Southwest Scandinavian Domain

The SSD occupies most of southern Norway and southwestern Sweden and lies adjacent to the 1.85–1.65 Ga Transscandinavian Igneous Belt (TIB; Fig. 1; Larson & Berglund 1992; Persson & Wikström 1993). The SSD comprises north–south trending lithotectonic domains that are separated by ductile shear zones. In southwestern Sweden, the SSD comprises crustal segments (Fig. 1) that were predominantly formed during the Gothian Orogeny between c. 1.75–1.55 Ga (Gaál & Gorbatchev 1987; Connelly & Åhäll, this volume). The westernmost Östfold–Marstrand Belt contains c. 1.76 Ga arc-related supracrustal rocks (Åhäll & Daly 1989) which lie adjacent to the Åmål–Horred Belt comprising younger, c. 1.65–1.61 Ga, supracrustal rocks (Lundquist & Skiöld 1992; Åhäll et al. 1995). The Östfold–Marstrand and Åmål–Horred belts are both intruded by 1.59 Ga calc-alkaline granitoid rocks of the Göteborg Batholith (Connelly & Åhäll, this volume), indicating a tectonic link by this time. The gneiss-dominated Eastern Segment (ES) lies between the Åmål–Horred Belt and Transscandinavian Igneous Belt (TIB) (Fig. 1) and is bounded in the west and east by kinematically complex zones of deformation known as the Mylonite (MZ) and Protogine (PZ) zones, respectively. These are mainly Sveconorwegian in age (e.g. Larson et al. 1990; Berglund & Connelly 1994). Although the exact timing of amalgamation of the Åmål–Horred Belt and ES is uncertain, the SSD of southwestern Sweden was mainly assembled by the end of the Gothian Orogeny (Gaál & Gorbatschev 1987)

Magmatism occurred episodically in the interval between the Gothian and Sveconorwegian orogenies, resulting in felsic to mafic 'anorogenic' intrusions. The Sveconorwegian Orogeny caused variable reworking in the SSD but did not result in significant new crustal growth (Starmer 1993; Connelly & Åhäll, this volume).

The Eastern Segment

The ES south of Lake Vänern lies between the known pre-Gothian Fennoscandian craton and accreted supracrustal terranes and may represent either a reworked margin or an accreted terrane itself (e.g. Lind & Gorbatchev 1980; Larson et al. 1990; Wahlgren et al., 1994). It thus occupies a critical intermediate position such that accurate tectonic models for the westward growth of the Fennoscandian craton will not be complete until the role of the ES is better understood. This paper reports results from an integrated mapping and U–Pb geochronology programme that constrain the tectonic history of the ES and thus its involvement in the development of the SSD as a whole.

The south-central ES is dominated by penetratively migmatized and banded calc-alkaline to alkali-calcic (T. Brewer and J. Berglund, unpublished data) orthogneisses with tonalitic, granodioritic, quartz-monzonitic and granitic compositions (Fig. 2; Larson et al. 1986). Informally, these orthogneisses are herein collectively referred to as the ES orthogneisses, although it is acknowledged that this suite may include rocks of disparate protolith ages. Gradational compositional contacts in this complex unit are interpreted to reflect primary magmatic differentiation, whereas sharp contacts are thought to represent intrusive relations, metamorphic segregations or enclaves in an intrusive phase. Demonstrable tectonic contacts are not common.

K-feldspar-rich granitoid rocks comprise a second lithologic unit in the ES that is subordinate to the ES orthogneisses. These rocks lack penetrative banding and have a uniform syenogranitic to alkaligranitic composition, close to that of a minimum melt, with a fine to medium-grained, equigranular texture. The term Hestra Suite will refer to any such homogeneous granitic rock (dykes excluded) of approximate Gothian age in the ES that lacks a high degree of migmatization. Where the contact can be observed, the Hestra Suite constitutes decimetre to metre thick bands interlayered with ES orthogneisses.

The ES orthogneisses and Hestra Suite are cut by several generations of post-Gothian granitoid and mafic dykes ranging from 1.61 Ga (this paper) to 1.22 Ga (Berglund & Connelly 1994), most of which have not been dated or studied in great detail. Post-Gothian granitoid rocks exhibit anorogenic, within-plate character and are variously overprinted by the Sveconorwegian metamorphism (e.g. Berglund & Connelly 1994; Åhäll et al. 1996)

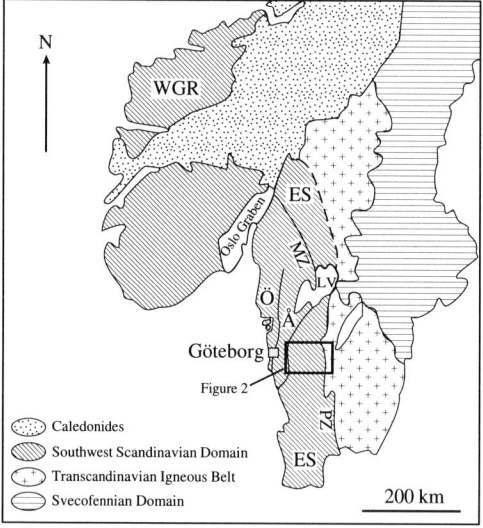

Fig. 1. Lithotectonic domains of southern Scandinavia. WGR = Western Gneiss Region; Ö = Östfold–Marstrand Belt; Å = Åmål–Horred Belt; ES = Eastern Segment; MZ = Mylonite Zone; PZ = Protogine Zone; LV = Lake Vänern.

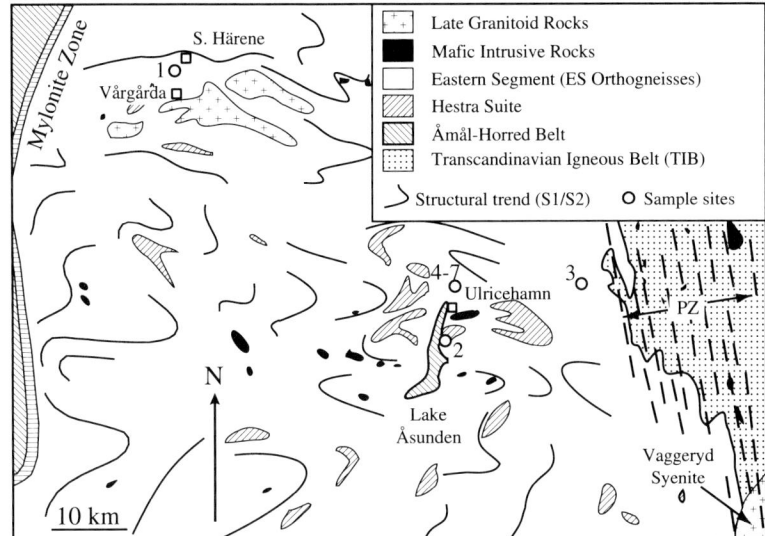

Fig. 2. Simplified geological map of the Eastern Segment east of Göteborg. Sveconorwegian shearing along the Protogine Zone (PZ) occurs over a wide zone. Location of geochronological samples 1–7 is indicated with open dots. Lithological map redrawn from SGU's map sheets Ulricehamn SV and SO (Larson 1988 and Larson & Berglund 1994) and PÖB Borås (Samuelsson et al. 1988).

Structure of the south-central Eastern Segment

The correlation of tectonic fabrics across the ES has been hampered by sporadic outcrop, complex structural and metamorphic relationships and the lack of correlative tectonostratigraphic markers. Larson et al. (1986) utilized detailed mapping and the existing geochronological database to establish an absolute chronology of events in the ES and initiated a regional structural framework which the geochronology in this study aims to test and refine. Although the following discussion summarizes the structural history of Larson et al. (1990), the reader is referred to the original paper for details. The nomenclature presented here is modified from Larson et al. (1990) to accommodate the new geochronological data presented in this paper (Table 1).

Field evidence indicates that the ES orthogneisses experienced several phases of deformation and metamorphism (Larson et al. 1986; 1990; Table 1). The main banding in the ES gneisses resulted from early, penetrative M1 (migmatitic veining), that was transposed during D_2 to define compositional layering parallel to a mineral alignment, collectively defining the dominant S1/S2 gneissosity/foliation. Only minor metamorphic segregation accompanied D_2. The migmatization associated with D_1 and D_2 suggest amphibolite-facies (or higher) conditions (Larson et al. 1990). In spite of the apparent simple history of the Hestra

Table 1. *Summary of ages of major intrusive, metamorphic and deformation events in the Eastern Segment and adjacent Transscandinavian Igneous Belt of south-central Sweden*

Approximate age (Ga)	Eastern Segment events	Transscandinavian Igneous Belt
1.85–1.75	—	TIB 1
1.7	Granitoids and	TIB 2,3
(bracketed only)	main Gothian event D_1–D_2, M1	
1.66–1.61	Late-to post- orogenic granitoids	
(bracketed only)	D_3, M3	
1.46–1.41	aplite dykes	aplite dykes
1.22–1.20	anorogenic granitoids	
	Main Sveconorwegian event D_4–D_5, M4	
0.95–0.93	Late Sveconorwegian–D_6	

Table 2. *U–Pb isotopic data*

Fraction	Weight mg	Concentration ppm		Measured			Corrected Atomic Ratios*						Age, Ma		
		U	Pb, rad	Total Common Pb, pg	$\frac{^{206}Pb}{^{204}Pb}$	$\frac{^{208}Pb}{^{206}Pb}$	$\frac{^{206}Pb}{^{238}U}$		$\frac{^{207}Pb}{^{235}U}$		$\frac{^{207}Pb}{^{206}Pb}$		$\frac{^{206}Pb}{^{238}U}$	$\frac{^{207}Pb}{^{235}U}$	$\frac{^{207}Pb}{^{206}Pb}$
Palaeosome Eastern Segment orthogneisses: South Härene (sample 1)															
Z1 30 equant; w abr	0.133	154	49.8	19	20231	0.1714	0.29411	138	4.1785	200	0.10304	16	1662	1670	1680
Z2 lg rnd; v w abr	0.093	159	48.2	11	23618	0.1378	0.28373	82	3.9817	128	0.10167	10	1610	1630	1657
Z3 rnd to equant; clr	0.103	186	55.9	14	23765	0.1409	0.28163	118	3.9480	168	0.10167	14	1600	1624	1655
T1 lgt yellow	0.181	209	30.4	280	1339	0.0117	0.15626	52	1.5163	58	0.07038	10	929	927	923
T2 dk brn; w abr	0.249	298	42.9	387	1879	0.0099	0.15505	50	1.4927	52	0.06983	10	929	927	923
Hestra Suite granitoid: Lake Åsunden (sample 2)															
Z1 sm rnd; w abr	0.047	288	94.4	25	9752	0.2277	0.28607	124	4.0132	162	0.10174	24	1622	1637	1656
Z2 lgt; w abr	0.095	172	53.7	7	39935	0.1862	0.28196	96	3.9393	142	0.10133	10	1601	1622	1649
Z3 sm rnd; clr	0.018	297	92.4	3	28684	0.1907	0.28006	112	3.9117	158	0.10130	16	1592	1616	1648
Z4 need xls	0.023	445	131.7	54	3210	0.1728	0.27000	116	3.7076	162	0.09959	24	1541	1573	1616
Z5 elong; fract	0.041	345	98.3	13	17716	0.1543	0.26404	88	3.5847	126	0.09846	10	1510	1546	1595
T1 dk brn; w abr	0.161	212	34.9	508	699	0.1091	0.16196	58	1.6038	68	0.07182	14	968	972	981
T2 lgt yellow; sm rnd	0.134	208	33.9	433	671	0.1015	0.16144	54	1.5930	64	0.07157	12	965	967	974
T3 dk brn; lg frags	0.196	216	35.1	626	698	0.1050	0.16112	78	1.5893	108	0.07154	36	963	966	973
Aplitic dyke: Vråna (sample 3)															
Z1 elong-rnd; best	0.020	246	65.3	3	22377	0.1563	0.24722	128	3.0848	148	0.09050	22	1424	1429	1436
Z2 need	0.010	213	56.2	7	4460	0.1760	0.24245	110	3.0098	130	0.09004	20	1399	1410	1426
Z3 need	0.012	274	71.5	24	2188	0.1617	0.24236	120	3.0022	160	0.08984	18	1399	1408	1422
Z4 sm clr xls	0.034	174	54.0	26	3684	0.2908	0.26025	92	3.3947	126	0.09460	12	1491	1503	1520
Z5 rnd; w abr	0.032	172	46.6	5	19572	0.1253	0.25809	86	3.3506	118	0.09416	10	1480	1493	1511
Z6 lg irreg	0.071	180	47.4	12	16268	0.1465	0.24664	78	3.0967	102	0.09106	12	1421	1432	1448
T1 dk brn; w abr	0.184	135	25.5	426	692	0.1095	0.18522	98	2.0013	122	0.07837	28	1095	1116	1156
T2 dk brn; abr	0.090	185	34.3	392	484	0.1410	0.17656	64	1.8548	78	0.07619	20	1048	1065	1100
T3 lgt yellow; vw abr	0.095	46	6.9	198	237	0.0271	0.16000	72	1.5788	118	0.07157	42	957	962	974
Palaeosome: Vistbergen (sample 4)															
Z1 elong; no xls	0.016	84	26.0	3	9514	0.1825	0.28205	190	3.9108	258	0.10056	22	1602	1616	1635
Z2 med rnd w fract	0.027	92	28.2	5	9650	0.1805	0.27645	158	3.8044	212	0.09981	22	1573	1594	1620
Z3 15 xls incl	0.017	113	33.6	3	9926	0.1673	0.27236	144	3.7114	190	0.09883	22	1553	1574	1602
Z4 rnd	0.045	99	29.1	4	17133	0.1666	0.27033	120	3.6778	170	0.09867	14	1542	1567	1599
T1 med color	0.040	126	21.7	104	559	0.0378	0.17937	88	1.9087	98	0.07718	20	1064	1084	1126
T2 dk brn	0.069	133	22.6	168	624	0.0414	0.17684	100	1.8670	116	0.07657	18	1050	1069	1110
T3 med; vw abr	0.065	105	17.7	155	501	0.0361	0.17629	58	1.8517	80	0.07618	18	1047	1064	1100
T4 lgt yellow	0.092	131	22.1	228	597	0.0432	0.17552	62	1.8416	76	0.07610	16	1042	1060	1098

THE EASTERN SEGMENT, SW SWEDEN

Fraction															
T5 lgt yellow xls	0.057	119	20.0	159	485	0.0400	0.17593	102	1.8592	146	0.07665	42	1045	1067	1112
T6 dk xls; rnd-flat	0.101	138	23.3	257	610	0.0470	0.17420	58	1.8231	82	0.07590	18	1035	1054	1093
First leucosome: Vistbergen (sample 5)															
Z1 elong xls	0.022	329	60.8	19	4579	0.0564	0.18862	72	2.1327	74	0.08200	14	1114	1159	1246
Z2 best xls	0.031	256	44.9	33	2785	0.0468	0.18108	64	1.9925	74	0.07980	12	1073	1113	1192
Z3 rnd	0.015	498	83.5	9	9211	0.0387	0.17484	62	1.8645	68	0.07735	10	1039	1069	1130
Z4 med rnd	0.017	574	92.4	12	8995	0.0303	0.16920	74	1.7679	78	0.07578	12	1008	1034	1089
Z5 best xls	0.018	778	119.6	15	9623	0.0193	0.16351	68	1.6690	72	0.07403	12	976	997	1042
Z6 equant xls	0.014	1023	154.0	17	8198	0.0142	0.16115	68	1.6223	72	0.07301	8	963	979	1014
T1 med color	0.071	187	30.6	184	797	0.0277	0.17283	66	1.7918	78	0.07519	22	1028	1042	1074
T2 light yellow xls	0.132	186	30.2	309	875	0.0260	0.17203	70	1.7816	82	0.07511	14	1023	1039	1071
T3 med color wi facets	0.113	251	39.9	268	1131	0.0299	0.16768	48	1.7054	60	0.07377	10	999	1011	1035
T4 dk brn rnd	0.083	420	64.3	210	1696	0.0300	0.16169	78	1.5981	80	0.07168	12	966	969	977
T5 dk brn frags	0.079	388	58.9	193	1625	0.0305	0.16023	56	1.5758	60	0.07133	10	958	961	967
Aplitic dyke: Vistbergen (sample 6)															
Z1 med some wi facets	0.007	255	71.1	6	4893	0.1326	0.26368	134	3.5112	164	0.09658	28	1509	1530	1559
Z2 rnd irreg	0.019	365	92.6	22	4903	0.1076	0.24538	80	3.1668	106	0.09360	16	1415	1449	1500
Z3 elong xls	0.012	318	77.9	11	5163	0.1161	0.23520	100	2.9948	122	0.09235	20	1362	1406	1475
Z4 elong smooth	0.019	445	104.5	12	10191	0.0829	0.23258	82	2.9256	106	0.09123	14	1348	1389	1451
Z5 med some wi facets	0.007	256	59.1	7	4183	0.0926	0.22594	118	2.9920	148	0.09251	22	1313	1377	1478
T1 lgt color xls	0.079	207	34.1	219	776	0.1036	0.16268	68	1.6162	74	0.07205	14	972	977	987
T2 dk xls	0.221	219	35.3	596	834	0.0968	0.16063	54	1.5758	64	0.07115	12	960	961	962
T3 dk rnd	0.171	211	33.9	436	845	0.0967	0.16009	52	1.5737	60	0.07129	14	957	960	966
T4 med rnd beige	0.195	204	32.8	516	788	0.1007	0.15947	58	1.5706	66	0.07143	14	954	959	970
Interboudin pegmatite: Vistbergen (sample 7)															
Z1 sm clr rnd	0.006	623	105.4	12	3498	0.0377	0.17631	84	1.8824	86	0.07743	20	1047	1075	1132
Z2 sm balls	0.008	477	79.4	12	3436	0.0339	0.17428	64	1.8474	70	0.07688	18	1036	1063	1118
Z3 0 mag & nm xls	0.032	461	75.6	13	12102	0.0305	0.17230	58	1.8196	64	0.07660	10	1025	1053	1111
Z4 xls; elong & need	0.010	562	88.1	232	259	0.0220	0.16649	60	1.6920	92	0.07370	26	993	1006	1033
Z5 med rnd	0.015	570	86.7	2	3932	0.0130	0.16321	80	1.6346	84	0.07264	8	975	984	1004
Z6 10 lg xls	0.016	734	107.7	37	3071	0.0075	0.15834	60	1.5637	60	0.07162	12	948	956	975
T1 lgt xls	0.040	165	25.8	145	472	0.0509	0.16153	54	1.6012	72	0.07189	18	965	971	983
T2 dk rnd frags	0.383	288	44.9	949	1180	0.0638	0.15984	74	1.5637	78	0.07095	10	956	956	956
T3 lgt yellow xls	0.081	232	36.4	212	907	0.0687	0.15983	86	1.5648	88	0.07101	22	956	956	958
T4 dk brn frags & xls	0.131	321	49.7	317	1339	0.0622	0.15908	64	1.5546	64	0.07087	16	952	952	954

Codes in fraction numbers are: T, titanite; Z, zircon.
Abbreviations are: abr, abraded; brm, brown; clr, clear; dk, dark; elong, elongate; fract, fractures; frags, fragments; incl, inclusions; inter, intermediate; irreg, irregular; lgt, light; lg, large; Mag, magnetic; med, medium; need, needles; nmag, non-magnetic; rnd, round; sm, small; v, very; w, well; wi, with; xls, crystals; +/–, with or without.
* values are corrected for fractionation, spike, laboratory blank of 5 to 25 pg, and initial common Pb at the time of mineral crystallization calculated from the model of Stacey & Kramers (1975) and 2 pg U blank. Two sigma uncertainties on isotopic ratios, calculated with a modified unpublished error propagation programme written by L. Heaman, are reported after the ratios and refer to the final digits.

Suite, evidence at several localities indicate that it was affected by the earliest deformation recognized in the ES. Larson et al. (1986) interpreted the massive texture to indicate that an older foliation in these rocks was obliterated during penetrative recrystallization. D_1–D_2 fabrics are tightly refolded around F3 folds that are associated with axial planar M3 migmatitic leucosomes and S3 mineral fabrics that are sub-parallel to S1/S2 fabrics on a regional scale.

D_1 to D_3 are folded in large-scale, east-west trending (Fig. 2), sub-vertical F4 fold structures that are only locally associated with pegmatitic M4 leucosomes. Although a mineral lineation (L4) is regionally present in the ES, axial planar foliations are not well developed. D_4 affects the 1224 ± 8 Ma Vårgårda granite and is thus presumed to be Sveconorwegian (Berglund & Connelly 1994). Deformation became progressively inhomogeneous towards the end of D_4 as deformation concentrated in ductile shear zones such as the Hammarö Shear Zone on the northern shore of Lake Vänern (Fig. 1; Vinnefors et al. 1993). D_5 created horizontal to gently dipping open to tight F5 folds that are locally associated with discontinuous, axial planar M5 pegmatitic leucosomes. The spatially discontinuous nature of this fold phase has necessitated regional correlation by means of style and orientation.

E–W trending D_4 fabrics are deflected by the MZ and PZ boundary shear zones on either side of the ES. Although the discontinuous nature of D_5 fabrics precludes definitive correlations with the boundary deformation zones, the latter are thought to post-date D_5 and are therefore labeled D_6. The boundary zones appear to have had a complex history and may have operated independently in time. D_6 in the PZ affects the 1.20 Ga old Vaggeryd syenite (Fig. 2; Larson et al. 1986; Jarl 1992) and thus provides a maximum age.

U–Pb Geochronology

Confidence in the regional correlations of thermotectonic events in the SSD, and thus tectonic models, will improve as the ages of the major lithotectonic units and thermotectonic events become better known. Towards this end, this study reports new U–Pb geochronological results that directly date or constrain protolith ages of the ES orthogneisses and Hestra Suite and the timing of metamorphism and deformation in the ES. Three samples were collected from widely spaced localities to determine the protolith age for the ES orthogneisses (sample 1), the intrusive age of the Hestra Suite (sample 2) and to constrain the age of D_1–D_3 fabrics (sample 3). Another four samples (samples 4–7) were collected at a well-exposed locality north of Ulricehamn (Fig. 2), at the Vistbergen ski centre. The objective was to determine a protolith age for the ES orthogneisses at a second locality, and to determine the ages of the first and second leucosome. The fourth sample (sample 7) from this locality was taken in a cross-cutting aplitic dyke in order to constrain the timing of D_1 and D_2. Titanite analyses from all seven samples address the timing and extent of post-Gothian thermotectonic events in this part of the ES. The results from the Vistbergen samples are collectively discussed after presentation of the results from the three far-field samples. Isotopic results are reported in Table 2.

S. Härene palaeosome: sample 1

A sample of granodiorite palaeosome from the ES orthogneiss was collected near S. Härene (Fig. 2) to determine a protolith age for this unit. This sample yielded equant-euhedral, large-irregular and rounded zircon grains that did not exhibit cores when examined optically. Dark and light brown titanite fragments were also separated. Three zircon fractions, one of each morphology, and two titanite fractions plot on a single discordia line with an upper and lower intercept of 1699 ± 3 and 932 ± 4 Ma respectively (Fig. 3a; probability of fit 39%). Both titanite fractions plot within 1% of the lower intercept.

This result from analyses of large, multigrain fractions (93–133 µg) representing all morphological types and the high probability of fit for the discordia line suggests that the upper intercept age represents the magmatic crystallization age of the orthogneiss. This is consistent with the narrow range of $^{208}Pb/^{206}Pb$ ratios between 0.14–0.17 (Table 2), indicating typical Th/U ratios for Proterozoic granitoid rocks. The lower intercept age reflects an age of titanite crystalliz-ation and Pb loss from older zircon and thus records a metamorphic event that affected this area at 932±4 Ma. The modest discordance of zircon (maximum 14%) indicates that, while Pb was lost at this time, new zircon did not crystallize in this palaeosome during the Sveconorwegian Orogeny.

The Hestra Suite: sample 2

A homogeneous sample representative of the Hestra Suite was collected 2 km south of Ulricehamn (Fig. 2) to determine the timing of intrusion and metamorphism of this unit. The sample was alkali-rich, fine- to medium-grained granite, dominated by quartz and alkali feldspar and yielded a range of zircon morphologies including grains that were large and irregular, elongate with terminations and needles. In addition,

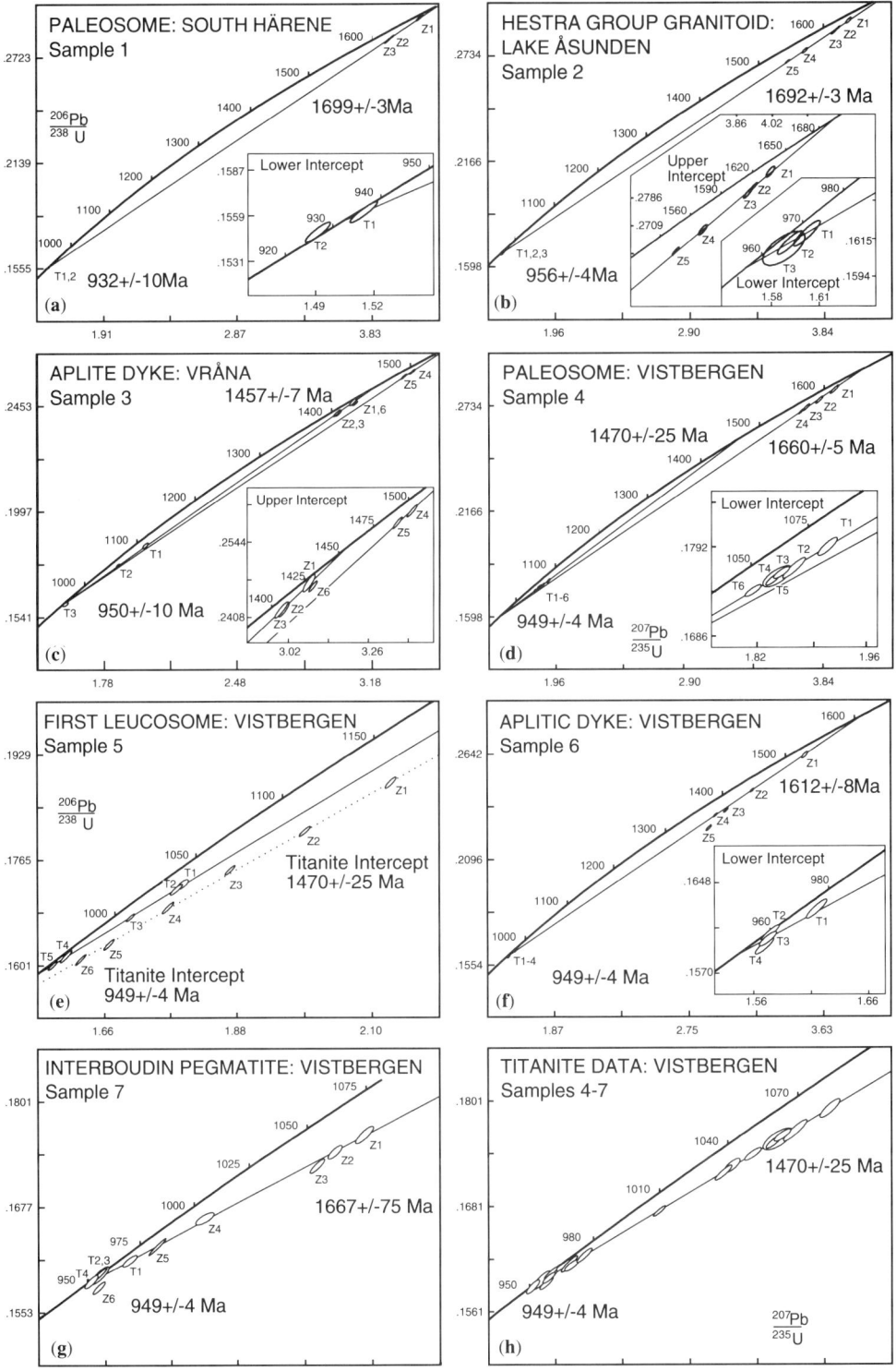

Fig. 3. (a–g) Concordia diagrams for samples 1–7. Intercept ages for Vistbergen samples are calculated utilizing the lower intercept age from all titanite analyses from this locality. See text for discussion. (h) Composite plot of 19 titanite fractions from the Vistbergen samples.

large fragments of titanite ranged in colour from dark to light brown. Five zircon and three titanite fractions define a simple, single discordia line with upper and lower intercepts of 1692 ± 3 and 956 ± 4 Ma respectively (Fig. 3b), with a 24% probability of fit.

The coincidence of five multigrain zircon fractions on a single line with an acceptable probability of fit indicates that zircons older than the upper intercept cannot exist in appreciable amounts in this sample and that no thermal events between 1692 Ma and 956 Ma disturbed the zircon U–Pb systematics. Given the simplicity of this data array and the restricted range of $^{208}Pb/^{206}Pb$ ratios between 0.15–0.22, the 1692 ± 3 Ma upper intercept is interpreted to represent the age of magmatic crystallization for the Hestra Suite. The lower intercept indicates that this sample was metamorphosed at 956 ± 4 Ma when temperatures were sufficiently high to (re)crystallize titanite and cause Pb-loss in zircons.

Cross-cutting aplitic dyke at Vråna: sample 3

An aplitic dyke that cuts across D_1–D_3 structures (including F3 folds, S3 and M3 leucosomes; Fig. 4) was collected at Vråna, c. 20 km east of Ulricehamn (Fig. 2). The crystallization age of this dyke thus provides a minimum age for the D_3–M_3 event in this part of the ES. A 4 kg sample of the dyke yielded a small amount of zircon that ranged from round, through elongate (with or without terminations), to needles with fractures parallel to the long axis. Titanite from the sample ranged from dark brown to light yellow.

Six zircon fractions, representing each morphological type (critical ones in duplicate), indicate a complex population including inherited grains and/or cores. However, two fractions of zircon needles (Z2 and Z3), one fraction of elongate zircons (Z1) and three fractions of titanite (T1–T3) define a discordia line with an upper and lower intercept of 1457 ± 7 Ma and 950 ± 10 Ma respectively (Fig. 3c), with an acceptable 15% probability of fit. Two dark brown fractions of titanite contain a pre-Sveconorwegian component while the light yellow fraction plots at the lower intercept and is apparently free of this older component.

The needles and elongate zircon fractions are interpreted to represent the only morphologies that can be confidently interpreted to have formed during the crystallization of the Vråna dyke and do not contain inheritance. The longitudinal fractures in the needle fractions (Z2 and Z3), a morphology typical of granitoid dykes, provided a criteria that ensured that this morphological type was free of older cores. The upper intercept age of 1457 ± 7 Ma is thus interpreted to represent the crystallization age of the Vräna dyke. The rounded grains (Z5) and the small zircons that exhibited crystal faces (Z4) both contain significant inheritance of older zircon while large irregular zircons (Z6) plot very close to the discordia line. Although the larger and rounded grains might have contained inherited grains missed during careful

Fig. 4. Photo showing the structural relations at the Vråna locality. Sample 3 was extracted from the aplitic dyke that cuts across S1–S2 planar fabrics shown in the photo. Planar fabrics are folded by F3 folds. Dyke contains a xenolith of foliated host rock with foliation oblique to fabrics in the country rock.

picking, the smaller crystals of Z4 contained a surprising component of inheritance. The distinct nature of this latter morphological type is also evidenced by the high $^{208}Pb/^{206}Pb$ ratio of 0.291 for this fraction in comparison to 0.156 to 0.176 for the needles and elongate zircons. This contrast indicates a distinct U/Th ratio, consistent with an interpretation that some component of this population was inherited. This chemical distinction further justifies the exclusion of this analysis in determining the crystallization age of this dyke utilizing the characteristic zircon needles. The lack of field evidence to suggest a complicated metamorphic and structural history for this dyke would suggest that the older, dark titanite is igneous in origin. The crystallization of light yellow titanite and Pb loss in dark titanite and zircon at 950 ± 10 Ma indicates that temperatures were elevated, at least above the 600°C closure temperature for titanite (Tucker et al. 1987), at this locality during the Sveconorwegian Orogeny.

Vistbergen samples

Samples of the palaeosome, first and second leucosomes and a cross-cutting aplitic dyke were collected at Vistbergen in order to establish the protolith age and the timing of migmatization and deformation at this locality. The results from each sample at this locality are presented individually but are discussed and interpreted collectively below.

Vistbergen palaeosome: sample 4

A sample of palaeosome from the ES orthogneisses at Vistbergen (Fig. 5) was collected to determine the protolith age and to complement the age determination of the palaeosome sample from S. Härene. This medium-grained, granoblastic granitoid rock yielded zircon morphologies that included both small round and elongate grains, with and without crystal faces. In addition, this sample yielded titanite crystals and fragments that ranged from dark brown to light yellow. Four zircon fractions representing each morphology (Table 2) plot in the upper 20% of a discordia line between the upper and lower intercepts of 1680 + 50/ −23 Ma and 1078 ± 160 Ma, respectively (>50% probability of fit) (Fig. 3d). The titanite defines a poorly constrained discordia line between 1499 ± 200 Ma and 955 + 3/ −6 Ma. The intercepts and precision of both lines are discussed and recalculated below in the context of the other samples from this locality. Figure 3d shows recalculated intercepts.

First leucosome from Vistbergen: sample 5

The palaeosome at the Vistbergen locality is interlayered on a decimeter scale with semi-continuous veins (Fig. 5) containing plagioclase–quartz–K-feldspar ± biotite that are interpreted to represent the first migmatitic leucosome in these gneisses. The veining is correlated with M1 that is parallel to the D_1–D_2 banding on a regional scale. A sample of

Fig. 5. Photo of the lithologies and structural style at the main outcrop at the Vistbergen locality. Samples 4,5 and 7 were extracted from the homogeneous, foliated palaeosome, layer parallel continuous leucosomes and interboudin leucosomes, respectively. Samples were collected by drill with care to avoid including adjacent rock type.

this material was collected to directly determine the age of this first metamorphic segregation event and yielded (1) small euhedral, equigranular, (2) angular, elongate, (3) small round grains and (4) large (>75 micron) irregular zircon morphologies. Dark brown to light yellow titanite was also separated from this sample. Six zircon fractions, representing at least one of each morphology, are highly discordant and define a discordia line between 1620 Ma and 916 Ma (Fig. 3e), with an unacceptably low probability of fit of 0.4%. The least discordant point (Z1) plots 73% down this discordia line and U concentrations range between 256–1023 ppm for the six fractions. Titanite analyses define a discordia line between 1459 ± 50 Ma and 947 ± 7 Ma. The intercepts shown on Fig. 3e reflect a regression of all titanite data from this locality, as discussed below.

Aplitic dyke from Vistbergen: sample 6

A weakly-foliated, non-migmatitic aplitic dyke (K-feldspar–plagioclase–quartz) is exposed in a second outcrop, about 15 m from the main migmatite locality at Vistbergen, where it cuts across the S2 foliation in the host. The dyke trends oblique (20°) to the gneissosity in the main outcrop. Its crystallization age thus provides a minimum age for the D_1–D_2 events at Vistbergen.

This sample yielded zircons with morphologies that included round, elongated and euhedral populations. Titanite separated from the sample ranged in colour from dark brown to light yellow. Three zircon fractions comprising euhedral faceted (Z1), anhedral elongate (Z4) and anhedral round (Z2) grains define a discordia line between 1612 ± 20 Ma and 963 ± 45 Ma, with a 98% probability of fit (Fig. 3f). A fraction of elongate anhedral zircons (Z3) touches this line whereas a fraction of large euhedral zircons (Z5) cannot be successfully regressed with the other fractions. This fraction is interpreted to contain inheritance or was insufficiently abraded and thus falls off the line due to late Pb-loss. Substantial pervasive Pb loss at 963 ± 45 Ma from zircon is indicated by the high degree of discordance (up to 42% for Z4) and the ineffectiveness of aggressive abrasion in creating more concordant analyses. Titanite analyses cluster near the lower intercept of the zircon discordia line, all plotting less than 2% discordant on $^{207}Pb/^{206}Pb$ lines.

Interboudin pegmatite at Vistbergen: sample 7

Late extension of the gneissic layering resulted in boudinage of the more competent mafic layers in the migmatites at the Vistbergen locality (Figs 5 & 6). Although this stretching is regionally interpreted to represent a D_4 feature, no conclusive correlation with F4 folding can be demonstrated in this outcrop. The interboudin spaces are filled with a coarse-grained, undeformed plagioclase–K-feldspar–quartz pegmatite a sample of which was collected in order to determine the timing of this late stretching. It yielded rounded to irregular zircons that were commonly cloudy, very fractured and/or inclusion filled and a small number of euhedral zircons were also recovered. Dark brown

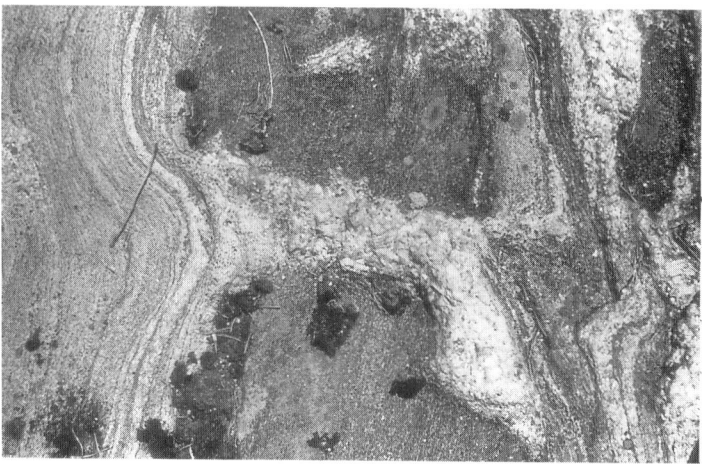

Fig. 6. Photo of interboudin leucsome at Vistbergen locality (sample 7). Photo was taken at the site shown in Fig. 5. Horizontal field of view is approximately 1.2 m.

to light yellow titanite crystals and fragments were also separated. Four of six zircon fractions define a discordia line with upper and lower intercepts of 1667 ± 70 Ma and 949 ± 9 Ma respectively (Fig. 3g with intercepts recalculated to include all titanite data from this locality; see below), with a 51% probability of fit. In contrast to the first leucosome, there is no apparent relationship between zircon morphology and degree of discordance. The titanite analyses all plot at the lower intercept defined by zircon and plot less than 1% discordant on $^{207}Pb/^{206}Pb$ lines.

Interpretation of Vistbergen results

Seventeen of nineteen abraded titanite fractions from the four Vistbergen samples define a single mixing line with upper and lower intercepts of 1470 ± 25 and 949 ± 4 Ma respectively (Fig. 3h), with an 87% probability of fit. The collective regression of titanite analyses from this locality thus yields better precision for both upper and lower intercepts than any individual sample. Considering the proximity of the sample sites and the high probability of fit for the mixing line, these intercepts are interpreted as the best age estimates for separate Mesoproterozoic and Sveconorwegian thermal recrystallization events.

The concordance at the lower intercept of a fraction of dark titanite (Sample 2-T3) that overgrows older, lighter cores, indicates that new titanite must have crystallized at 949 ± 4 Ma. Conversely, that analyses of titanite cores are not less than 78% discordant, in spite of avoiding rim material during mineral selection and extensive air abrasion, suggests that Pb was lost from the older titanite at 949 ± 4 Ma. Although amphibolite-facies assemblages suggest that temperatures exceeded the closure temperature for titanite during the Sveconorwegian Orogeny, this phase was not completely recrystallized.

The improved precision for the timing of Sveconorwegion recrystallization from the collective titanite regression necessitates re-evaluation of the zircon data for the Vistbergen samples. The lower intercepts of zircon discordia for the palaeosome, aplite and interboudin pegmatite samples overlap with the 949 ± 4 Ma lower intercept for titanite, suggesting that zircon and titanite responded to a common Sveconorwegian thermal event. Assuming reasonable cooling rates, it is practical to use a conservative Sveconorwegian age of 949 ± 10 Ma to constrain the lower intercepts of zircon discordia. This larger error estimate allows for differences in the timing of Pb closure of zircon relative to that of titanite.

Constraining the lower intercept at 949 ± 10 Ma, the regression of zircon analyses from the palaeosome yields an upper intercept age of 1660 ± 5 Ma, with a 25% probability of fit (Fig. 3d). The presence of faceted zircon, $^{208}Pb/^{206}Pb$ ratios between 0.167–0.183 and colinearity of all four zircon fractions are collectively consistent with an igneous origin such that the upper intercept is interpreted to represent the intrusive age of the protolith of this orthogneiss.

The definition of a single discordia by four multigrain fractions (16–45 µg) representing all zircon morphologies suggests that no zircon grains are older than the upper intercept. Presuming that a granitoid rock would contain primary zircons, it is interpreted that the upper intercept represents the crystallization of the igneous protolith rather than a metamorphic age. This is consistent with the $^{208}Pb/^{206}Pb$ ratios for these zircons between 0.14–0.18, reflecting typical Th/U ratios for Mesoproterozoic intrusive granitoid rocks.

Utilizing 949 ± 10 Ma to constrain the lower intercept of the aplite zircon data, a regression of four zircon analyses yields an upper intercept of 1612 ± 8 Ma with a probability of fit of 85% (Fig. 3f). This age is interpreted to represent the crystallization age of this aplitic granite and thus provides a minimum age for the D_1–D_2 features, including the first leucosome (M1).

The complexity of the zircon data from the first and second leucosomes precludes defining their crystallization age by zircon since both the upper and lower intercept might represent the age of leucosome crystallization. The presence of a 1459 ± 50 Ma titanite component in the first leucosome indicates preservation of pre-Sveconorwegian titanite in this leucosome. As it is unlikely that titanite would have survived a Sveconorwegian melt phase, this leucosome is interpreted to be pre-Sveconorwegian, most probably Gothian.

Zircon analyses from the second leucosome cluster nearer the lower intercept (minimum discordance 87%) than analyses from the first leucosome sample in spite of comparable U contents. The proximity to the lower intercept is compatible with crystallization of this leucosome at 949 ± 4 Ma, a process resulting in new zircon growth. This interpretation is consistent with the general lack of inherited pre-Sveconorwegian titanite (in contrast to the palaeosome and first leucosome). Although we favour this interpretation over Gothian crystallization, its speculative nature is emphasized.

Discussion of results

The protolith ages of 1699 ± 3, and 1660 ± 6 Ma for the ES orthogneiss indicates that these rocks were intruded contemporaneous with the later magmatic pulses of the TIB (i.e. TIB 2 and TIB 3 of Larson & Berglund 1992).These ages and field relationships indicate a major crust-forming event between 1.70–1.66 Ga, probably during the initial phase of the Gothian Orogeny in this part of the ES. Geochronological data are presently insufficient to determine whether early Gothian magmatism in the ES was continuous or episodic.

The 1612 ± 8 Ma age for the aplitic dyke at the Vistbergen locality suggests that the earliest migmatization and fabric transposition (D_1–D_2) in the ES occurred during the Gothian Orogeny between 1.70 Ga and 1.61 Ga. The migmatization of the 1.66 Ga palaeosome at Vistbergen suggests that D_1 and D_2 must have occurred, at least locally, after 1.66 Ga. It is possible that Gothian magmatism and migmatization overlapped in time. Attempts to date the age of the earliest leucosomes by zircon has yielded inconclusive results.

The 1692 ± 3 Ma igneous age for the Hestra Suite indicates that some phases of this suite intruded early in the Gothian magmatic interval as presently defined, synchronous with the emplacement of the ES orthogneiss protoliths. This is consistent with field observations that the earliest deformational events affected this suite. However, the early intrusive age raises questions about why this suite did not become layered during subsequent migmatization of the ES orthogneisses. The lack of high strain zones along contacts and its intercalation with ES orthogneisses rules out subsequent juxtapositioning by thrusting and folding. This would require that rocks of the Hestra Suite were not penetratively viened during D_1–D_2 in spite of their proximity to the ES orthogneisses. The lack of migmatization may be explained by variations in primary compositions, water availability and local temperature variations such that Gothian migmatization was not penetrative at the present erosional level. The non-penetrative migmatization may have been accentuated by post-Gothian broad-scale folding that created structural basins of previously higher crustal levels in parts of the ES.

Titanite from the Vistbergen samples shows evidence for a post-Gothian, Mesoproterozoic metamorphic event that affected these gneisses at 1470 ± 25 Ma synchronous (within error) with the 1457 ± 7 Ma emplacement of the Vråna aplitic dyke. Although D_3–M_3 fabrics are cross-cut by this dyke and are thus pre-1457 Ma in age, more data are needed to determine the precise age of the D_3 event. This Mesoproterozoic thermal event cannot be definitively correlated with any specific structural features observed in the field. The preservation of a significant component of c. 1470 Ma titanite in three samples from the ES indicates that later recrystallization was insufficient to completely reset U–Pb isotopic systematics in titanite. Titanite in samples 1, 3, and 7 do not show compelling evidence for the 1470 Ma crystallization recorded by other samples and may indicate: (1) older titanite did not exist in these samples, (2) older titanite was totally recrystallized or reset at 950 Ma, or (3) the rock sampled is younger than 1470 Ma. The latter explanation may account for the lack of significant inheritance in the interboudin pegmatite at Vistbergen (sample 7).

The 1224 ± 8 Ma Vårgårda granite near the town of Vårgårda (Fig. 2) (Berglund & Connelly 1994) records D_4 deformation and therefore requires that D_4 and subsequent events occurred during the Sveconorwegian Orogeny. However, precise age constraints on each deformation event are still lacking and therefore the onset of this orogenic cycle is not certain.

Titanite ages between 956 ± 4 Ma and 932 ± 4 Ma from all seven samples indicates that this region of the ES was penetratively recrystallized during the Sveconorwegian Orogeny. Conversely, the preservation of pre-Sveconorwegian titanite suggests that this recrystallization event was either short-lived and/or near the c. 600°C closure temperature of titanite (Tucker et al. 1987; Heaman & Parrish 1991). Although data are limited, the younger titanite age from S. Härene hints that titanite ages may young towards the west. Although attempts to directly date the late interboudin pegmatite have resulted in a complex dataset, it seems likely it crystallized at 949 ± 4 Ma, approximately synchronous with the Sveconorwegian titanite recrystallization event in the eastern ES.

Tectonic implications

Gothian magmatism

Geochronology from the TIB indicates that magmatism occurred intermittently in this belt between c. 1.85 and 1.65 Ga. Larson & Berglund (1992) separated the TIB rocks into three distinct periods of magmatism that they referred to as TIB 1 (1.85–1.77 Ga), TIB 2 (1.7 Ga) and TIB 3 (1.68–1.65 Ga).

The coincident intrusive ages of the ES orthogneisses, Hestra Suite, and later TIB rocks is consistent with previous interpretations that the Gothian magmatic rocks of the ES represent a western extension of TIB 2 and TIB 3 granitoid rocks across the PZ (e.g. Lind & Gorbatchev 1980; Larson et al. 1990; Wahlgren et al. 1994). Comparable intrusive ages from orthogneisses in ES north of Lake Vänern and in the Western Gneiss Region of Norway at 1.69–1.65 Ga (Lind et al. 1994; Tucker et al. 1990), indicate that this period of intrusion was important all along the southwestern part of the Fennoscandian Shield.

The U–Pb age of 1649 + 33/ –19 Ma for a Kongsbergian plutonic rock south-central Norway (Ragnhildstveit et al. 1994) represents the only published zircon data, to our knowledge, on an intrusive rock of this age from southern Norway outside the Western Gneiss Region. In southwest Sweden, there are very few indications of 1.70–1.65 Ga plutonism in the terranes west of the ES (e.g. Hansen et al. 1989). A 1.66 Ma age for the Horred volcanic suite (J. Connelly, unpublished data) indicates that volcanic activity in the Åmål–Horred Belt was approximately synchronous with magmatism in the ES.

Gothian metamorphism, deformation and accretion

The 1.70–1.61 Ga age range for D_1–D_2 indicates that these events were part of the regionally recognized Gothian Orogeny. Post-Gothian recrystallization precludes quantifying PT conditions during the Gothian D_1–D_2. However, the presence of feldspar-rich M_1 leucosomes suggests that approximately upper-amphibolite grade prevailed during at least part of the D_1–D_2 period.

East of the ES (in the TIB), there is no evidence for penetrative deformation related to D_1–D_2, indicating that the early Gothian deformation front approximately coincides with the PZ in the region south of Lake Vänern. This implies that crust at the present erosional level east of the PZ was stable during the Gothian Orogeny. However, the possibility that D_3, which predates 1457 ± 7 Ma and is most probably Gothian, is correlative with the regionally occurring WNW–ESE trending deformation in southeast Sweden, cannot be discounted. If the 1.70 Ga granitoid rocks on either side of the PZ formed as a single magmatic belt, the Gothian deformation front must be intracontinental (Larson et al. 1990).

To the west of the ES in southern Sweden, the earliest Gothian deformation in the Östfold–Marstrand Belt (D_1 of Park et al. 1987) occurred after the 1.76 Ga depositional age but before the intrusion of the cross-cutting 1.59 Ga Göteborg Batholith (Connelly & Åhäll, this volume). Although this permits the D_1 deformation in this terrane to be time-equivalent with the earliest deformation in the ES, the lack of known comparably-aged deformation in the intervening Åmål–Horred Belt, precludes a definite genetic link.

The latest Gothian deformation in the Östfold–Marstrand and Åmål–Horred belts (D_2 according to Åhäll et al. 1990; Åhäll et al. 1995 and Samuelsson & Åhäll 1985) occurred between 1.55 and 1.59 Ga (Connelly & Åhäll, this volume). Since D_2 in the ES must predate the 1.61 Ga aplitic dyke and probably also the 1.66 granite of Larson et al. (1990), it is possible that D_3 in the ES correlates with D2 to the west of the MZ. Constraining the timing of D_3 deformation in the ES is crucial for such a correlation.

Further west of the Östfold–Marstrand Belt, the c. 1.65 Ga metagranite (Ragnhildstveit et al. 1994) from south-central Norway was reported to crosscut older migmatized rocks implying the existence of early Gothian and/or Svecofennian deformation. Correlation with deformations in the ES will not be possible until the age of deformations in Norway is better constrained. Further north in Norway, in the Western Gneiss Region, Tucker et al. (1990) reported several emplacement ages for orthogneisses in the interval 1686 to 1657 Ma. These were interpreted to have been migmatized prior to c. 950 Ma.

Post-Gothian to pre-Sveconorwegian anorogenic events

The 1470 ± 25 Ma thermal event recorded by the upper intercept of the titanite mixing line overlaps with the 1457 ± 7 Ma intrusion of aplitic dykes in the ES that cut across D_3 folds. This magmatism and metamorphism is presently indistinguishable from the c. 1.58–1.45 Ga intrusive rocks found in southern Sweden (Johansson et al. 1993). A composite dyke swarm in the western TIB, approximately 25 km east of the Vistbergen locality, has an age of c. 1.41 Ga (Lundqvist et al. 1992). In addition, Åberg (1988) noted that the K–Ar system of Svecofennian rocks from southeast Sweden closed around 1.4 Ga. Given the uncertainty of both the 1470 Ma (re)crystallization recorded by the titanites in the ES and the intrusive ages, regional magmatism and heating may have been synchronous in southern Sweden between c. 1.50–1.38 Ga (Åhäll et al. this volume). The lack of deformation fabrics attributable to this thermal event would suggest that these events were anorogenic.

Sveconorwegian events

A long and complex Sveconorwegian development has been reported from the SSD (e.g. Larson et al. 1986, 1990; Park et al. 1987, 1991; Starmer 1993; Wahlgren et al. 1994). The Sveconorwegian metamorphic ages reported here represent only part of the thermal history, the last closure of titanite.

Although this thermal record is not easily correlated with specific deformation phases of the Sveconorwegian Orogeny (i.e. D_4–D_5), it is informative to compare the titanite ages from the ES, with those from terranes to the west and east.

The isotopic data of this study indicate that the south-central ES experienced widespread thermal recrystallization and possibly localized leucosome development between 950–930 Ma. Johansson (1993) reports a concordant titanite U–Pb age at c. 1040 Ma in the area south of Göteborg (Fig. 2), whereas titanite from western Värmland, have yielded almost concordant ages at c. 1000 Ma (Hansen et al. 1989). Both these localities lie in the western parts of the Åmål–Horred Belt. Very few titanite ages are reported from the Östfold–Marstrand Belt, but Connelly & Åhäll (this volume) report a titanite age of 1560 ± 10 Ma for a sample of the Göteborg Batholith, an age only slightly younger than the emplacement age for that rock. These U–Pb ages for titanite imply that, at least locally, terranes west of the MZ have not been pervasively subjected to temperatures that affected the U–Pb system in titanite as late as 950 Ma. A variety of other isotopic systems also indicate an early Sveconorwegian metamorphism west of the MZ at 1040–1100 Ma (e.g. Rb–Sr at c. 1087 Ma, Daly et al. 1983; Sm–Nd 1100 Ma, Kullerud & Machado 1991, K–Ar in hornblende at 1040–1080 Ma, Jarl 1992). The preservation of concordant 1.4 Ga titanite in the western TIB (Lundqvist et al. 1992) indicates that the PZ apparently defines the approximate eastern limit of Sveconorwegian thermal effects in this region.

Contractional deformation during the Sveconorwegian Orogeny caused thrusting and crustal stacking within the SSD (e.g. Park et al. 1987; Starmer 1993). However, contrasting Sveconorwegian metamorphic ages and structural grain in terranes on either side of the MZ (e.g. north–south orientated in the western belts and east–west orientated in the ES) indicate that these terranes do not record an identical structural history. The earliest major Sveconorwegian deformation and metamorphism was likely a response to thrusting and thickening of previously stabilized crust. Increased temperatures associated with this initial phase resulted in the initiation of new mineral growth and the opening of various isotopic systems, including the U–Pb system. The initiation of crustal thickening in the SSD during the Sveconorwegian Orogeny is not well constrained, but must have occurred later than the c. 1145 Ma gabbroic intrusion dated by Dahlgren et al. (1990) in the Telemark province.

The contrast in metamorphic ages on either side of the MZ indicates that adjacent terranes experienced different P–T–t paths. These results are compatible with eastward thrusting of a western crustal section over the ES after the initial phase of crustal thickening (along a Sveconorwegian orogenic axis?) in the Östfold–Marstrand Belt and Kongsberg–Bamble Segments. In this model, the Östfold–Marstrand and Åmål–Horred belts were exhumed at the time when temperatures in the lower deck (ES) became elevated. This process accounts for the attainment of closure temperatures in titanite in areas west of the ES at 1040–1000 Ma while the rocks in ES recrystallized and/or lost Pb as late as c. 930 Ma. The loading of the ES from the west would have resulted in a greater thickness and higher temperatures in the western part of the segment resulting in the highest synorogenic topography there. The isostatic re-equilibration of this geometry would require that the western part of the ES cooled later than the eastern part. This is consistent with the younger titanite age found in the west. Johansson & Johansson (1993) also report a 920–930 Ma age from the southernmost MZ, further supporting younger cooling ages for parts of the western ES. Jarl (1992) reported the closure of the K–Ar system in the ES north of Lake Vänern at 915 Ma in hornblende, which suggests that the ES cooled rather slowly, not reaching the c. 525° closure temperature of hornblende (McDougall & Harrison 1988) until 915 Ma. The late, static recrystallization of large areas of the ES probably occurred during this post-thrusting rebound phase.

Although age constraints are presently insufficient to fully constrain the post-thrusting dynamics, the presence of significantly older titanite ages in the area west of the ES may also require late extension along the MZ thus placing cooler, higher crustal levels of the Åmål–Horred Belt and Östfold–Marstrand Belt at the same present level of exposure as the ES. The preservation of higher pre-Sveconorwegian crustal levels in the western terranes through extension along the MZ would also account for the less penetrative Sveconorwegian metamorphism in these terranes (Connelly & Åhäll, this volume).

The correlation of deformation events recognized in the field with thermal events recorded by the U–Pb and K–Ar systematics require further refinement. Presently it is only possible to say that D_4 to D_6 must be younger than 1224 ± 8 Ma (Berglund & Connelly 1994) such that they are interpreted to be Sveconorwegian in age. However, if D_6 fabrics in the MZ were formed during overthrusting and exhumation of the Åmål–Horred Belt over the ES, it must predate the youngest titanite age associated with the unroofing of the ES (932 ± 4 Ma). Ongoing structural work indicates that thrust-related deformation fabrics in the central part of MZ are distinguishable from extensional-dominated deformation fabrics that overprint the southern part of this zone.

The Sm–Nd mineral isochron ages of c. 900 Ma for granulite metamorphism in the ES (Johansson et al. 1991) are enigmatic in light of the titanite ages between 956 ± 4 Ma and 932 ± 4 Ma.

Conclusions

The protolith ages of the ES orthogneisses and Hestra Suite are comparable to TIB 2 and TIB 3 magmatic phases in the adjacent TIB to the east. These ages are consistent with previous models that suggest these rocks formed in a single, long-lived magmatic arc on the present-day western margin of the Fennoscandian craton. The earliest phase of deformation and metamorphism in the ES occurred between 1.70 Ga and 1.61 Ga during the Gothian Orogeny, apparently coeval with at least the later stages of magmatism. Post-Gothian, to pre-Sveconorwegian magmatism may have culminated in this part of the ES by 1470 Ma coincident with elevated temperatures that resulted in titanite (re)crystallization. The distribution of titanite ages is consistent with crustal thickening in the early Sveconorwegian Orogeny followed by the eastward thrusting of the western terranes over the ES. The exhumation of the Åmål–Horred Belt by this process and subsequent extensional collapse would explain the younger titanite ages across the Mylonite Zone.

This project was funded by grants from the Geological Survey of Sweden and the Swedish Natural Research Council to SÅL. Discussions with Greg Dunning regarding the analytical component of this work were very helpful.

Appendix 1: sample localities

Sample number	E–W coordinate*	N–S coordinate	comment
1	132400	644440	S. Härene; palaeosome, granodioritic
2	135670	640610	Hestra Suite granite
3	137600	641440	Vråna; aplitic dyke
4	135780	641320	Vistbergen; granitic palaeosome
5	135780	641320	Vistbergen; 1st leucosome, qz–fsp
6	135780	641320	Vistbergen; aplitic dyke
7	135780	641320	Vistbergen; 2nd leucosome, qz–fsp

*coordinates given in the Swedish national grid system

Appendix 2: U–Pb analytical techniques

The U–Pb data were produced at Memorial University of Newfoundland, St John's, Canada. Rock samples were crushed and minerals were separated under clean conditions using a rock crusher, disc pulverizer, Wilfley panning table, sieves, heavy liquids and Frantz magnetic separator. Mineral fractions were selected for analysis on the basis of microscopic optical properties to ensure that only the highest quality grains were analysed. All fractions analysed have been strongly abraded and subsequently optically re-evaluated. They were washed in distilled 4N nitric acid, water and finally acetone before being loaded into TEFLON capsules with a mixed ^{205}Pb/^{235}U isotopic tracer and HF and HNO$_3$ acid for dissolution. Chemical separation of U and Pb from zircon using 0.055 ml minicolumns after Krogh (1973) resulted in a total procedural blank between 2–5-picograms. A larger, 0.5 ml reservoir cation exchange column was used for titanite for which the total procedural blanks were about 10 picograms. Pb and U were loaded together with phosphoric acid and silica gel onto an outgassed rhenium ribbon and analysed on a multi-collector MAT 262 thermal ionization mass spectrometer, operating in static and/or peak jumping mode with ^{204}Pb measured an axial secondary electron multiplier (SEM)/ion counter. Small samples were measured in peak jumping mode in the axial SEM/ion counter. Ages were calculated using the decay constants of Jaffey et al. (1971). Isotopic results are presented in Table 1 where 2 σ errors on the isotopic ratios are reported. They were calculated by propagating uncertainties in measurement of isotopic ratios, fractionation and amount of blank with a programme modified after an unpublished error propagation programme written by L. Heaman (Royal Ontario Museum, Toronto, Canada). Linear regressions were calculated using the procedure of Davis (1982); the quality of fit is quoted in percent probability where 10% is considered an acceptable lower limit and corresponds to a mean standard weight of the deviations (MSWD) of 2.

References

ÅBERG, G. 1988. Middle Proterozoic anorogenic magmatism in Sweden and world-wide. *Lithos.* **21**, 279–289.

ÅHÄLL, K.-I. & DALY, J. S. 1989. Age tectonic setting and provenance of Östfold–Marstrand belt supracrustals: westward crustal growth of the Baltic shield at 1760 Ma. *Precambrian Research*, **45**, 45–61.

——, —— & SCHÖBERG, H. 1990. Geochronological constraints on Mid-Proterozoic magmatism in the Östfold–Marstrand belt: implications for crustal evolution, SW Sweden. *In*: GOWER, G. F., RIVERS, T. & RYAN, B. (eds) *Mid-Proterozoic Laurentica–Baltica*. Geological Association of Canada Special Paper **38**, 97–115.

——, PERSSON, P-O. & SKIÖLD, T. 1994. Westward accretion of the Baltic Shield: implications from the 1.6 Ga Åmål–Horred Belt, SW Sweden. *Precambrian Research*, **70**, 235–251.

——, SAMUELSSON, L. & PERSSON, P.-O. 1996. Geochronological and structural setting of the 1.38 Ga Torpa granite; implications for charnockite formation in SW Sweden. *This volume*.

BERGLUND, J. & CONNELLY, J. 1994: The Vårgårda intrusion; tectonic setting, age and regional implication. *Abstract volume, 21: a geologiska vintermötet i Luleå.*

BREWER, T. & CONNELLY, J. 1996. Proterozoic geology in SW Sweden; implications for crustal growth anorogenic evolution in the North Atlantic region. Abstract. The 22nd Nordic Geological Winter meeting, Turka Åbo, Norway.

CONNELLY, J. N. & ÅHÄLL, K.-I. 1996. The mid-Proterozoic cratonization of Laurentia-Baltica; new age constraints from SW Sweden. *This volume*.

DAHLGREN, S., HEAMAN, L. & KROGH, T. 1990. Precise U–Pb zircon and baddeleyite age of the Hesjåbutind gabbro, central Telemark are, southern Norway. *Geonytt*, **17**

DALY, J. S., PARK, R. G. & CLIFF, R.A. 1983. Rb–Sr isotopic equilibrium during Sveconorwegian (=Grenvillian) deformation and metamorphism of the Orust dykes, S.W. Sweden. *Lithos*, **16**, 307–318

DAVIS, D. W. 1982. Optimum linear regression and error estimation applied to U–Pb data. *Canadian Journal of Earth Sciences*, **23**, 2141–2149.

GAÀL, G. & GORBATSCHEV, R. 1987. An outline of the Precambrian Evolution of the Baltic Shield. *Precambrian Research*, **35**, 15–52.

GORBATSCHEV, R., LINDH, A., SOLYOM, Z., LAITAKARI, K. A., LOBACH-ZHUCHENKO, S. B. ET AL. 1987. Mafic dyke swarms of the Baltic Shield. *In*: HALLS, H. C. & FAHRIG, W. F. (eds) *Mafic Dyke Swarms*. Geological Association of Canada, Special Paper **34**, 361–372.

GOWER, C. F. 1990. Mid-Proterozoic evolution of the eastern Grenville Province, Canada. *Geologiska Föreningens i Stockholm Förhandlingar*, **112**, 127–139.

—— & OWEN, V. 1984. Pre-Grenvillian and Grenvillian lithotectonic regions in eastern Labrador – correlations with the Sveconorwegian Orogenic Belt in Sweden. *Canadian Journal of Earth Sciences*, **21**, 678–693.

—— & TUCKER, R. D. 1994. Distribution of pre-1400 Ma crust in the Grenville province: Implications for rifting in Laurentia–Baltica during geon 14. *Geology*, **22**, 82–830.

——, RYAN, A. B. & RIVERS, T. 1990, Mid-Proterozoic Laurentia–Baltica: an overview of its geological evolution and a summary of the contributions made by this volume. *In*: GOWER, C. F., RIVERS, T. & RYAN, A. B. (eds) *Mid-Proterozoic Laurentia–Baltica*. Geological Association of Canada, Special Paper **38**, 1–20.

HANSEN, B. T., PERSSON, P.-O., SÖLLNER, F. & LINDH, A. 1989. The influence of recent lead loss on the interpretation of disturbed U–Pb systems in zircons from metamorphic rocks in southwest Sweden. *Lithos*, **23**, 123–136.

HEAMAN, L. & PARRISH, R. 1991. U–Pb Geochronology of Accessory Minerals. *In*: HEAMAN, L. & LUDDEN, J. N. (eds) *Short Course Handbook on Applications of Radiogenic Isotope Systems to Problems in Geology*. Mineralogical Association of Canada, **19**, 59–102.

HOFFMAN, P. F. 1992. Did the breakup of Laurentia turn Gondwanaland inside-out? *Science*, **252**, 1409–1411.

JAFFEY, A. H., FLYNN, K. F., GLENDENIN, L. E., BENTLEY, W. C.& ESSLING, A. M. 1971. Precision measurements of halflives and specific activities of ^{235}U and ^{238}U. *Physics Review* **C4**, 1889–1906.

JARL, L.-G. 1992. New isotope data from the Protogine Zone and southwestern Sweden. *Geologiska Föreningen i Stockholm Föhandlingar*, **114**, 349–350.

JOHANSSON, Å. 1993. U–Pb datings of titanites from Southwest Sweden. Sveriges *Geologiska Undersökning*. Rapporter och meddelanden nr **76**.

——, MEIER, M., OBERLI, F. & WIKMAN, H. 1993. The early evolution of the Southwest Swedish Gneiss Province: geochronological and isotopic evidence from southernmost Sweden. *Precambrian Research*, **64**, 361–388.

JOHANSON, L., LINDH, A. & MÖLLER, C. 1991. Late Sveconorwegian (Grenvillian) high-pressure granulite facies metamorphism in southwest Sweden. *Journal of Metamorphic Petrology*, **9**, 283–292.

JOHANSSON, L. & JOHANSSON, Å. 1993. U–Pb age of titanite in the Mylonite Zone, southwestern Sweden. *Geologiska Föreningen i Stockholm Föhandlingar*, **115**, 1–7.

KROGH, T. E. 1973. A low-contamination method for hydrothermal decomposition of zircon and extraction of U and Pb for age determinations. *Geochimica et Cosmochimica Acta*, **37**, 485–494.

KULLERUD, L. & MACHADO, N. 1991. End of controversy: U–Pb geochronological evidence for significant Grenvillian activity in the Bamble area, Norway. *Terra Abstract*, **4**, 543.

LARSON, S. Å. 1988. Lithological mapsheet 7D

Ulricehamn SV. *Sveriges Geologiska Undersökning*, Ser. Af 159.

—— & BERGLUND, J. 1992. A chronological subdivision of the Trans-scandinavian Igneous Belt – three magmatic episodes? *Geologiska Föreningen i Stockholm Föhandlingar*, **114**, 459–461.

—— & —— 1994. Lithological mapsheet 7D Ulricehamn SO. *Sveriges Geologiska Undersökning*, Ser. Af 178

——, ——, STIGH, J. & TULLBORG, E.-L. 1990. The Protogine Zone, southwest Sweden: a new model – an old issue. *In*: GOWER, C. F., RIVERS, T. & RYAN, A. B. (eds) *Mid-Proterozoic Laurentia–Baltica*. Geological Association of Canada, Special Paper **38**, 317–333.

——, STIGH, J. & TULLBORG, E.-L. 1986. The deformation history of the eastern part of the southwest Swedish gneiss belt. *Precambrian Research*, **31**, 237–257.

LIND, A. & GORBATCHEV, R. 1980. Chemical variation in a Proterozoic suite of granitoids extending across a mobile belt-craton boundary. *Geologische Rundschau*, **73**, 881–893.

——, SCHÖBERG, H. & ANNERTZ, K. 1994. Disturbed radiometric ages and their bearing on interregional correlations in the SW Baltic Shield. *Lithos*, **31**, 65–79.

LUNDQUIST, I. & SKIÖLD, T. 1992. Preliminary age-dating of the Åmål Formation, SW Sweden. Geologiska. *Föreningen i Stockholm Föhandlingar*, **114**, 461–462.

LUNDQVIST, L. & LARSON, S. A., CONNELLY, J. N. & BREWER, T. S. 1992. Midproterozoic intracontinental extension – evidence from an E-W trending, composite dyke swarm, south central Sweden. *Abstract IGCP 257 Precambrian Dyke Swarms. Petrozavodsk 7–17 September.*

MCDOUGALL, I. & HARRISON, T. M. 1988. *Geochronology and Thermocronology by the $^{40}Ar/^{39}Ar$ Method.* Oxford University Press, Oxford.

——, ——, CRANE, A. & Daly, S. 1987. The structural and kinematic evolution of the Lysekil-Marstrand area, *Östfold–Marstrand Belt, southwestern Sweden.* SGU serie C 816.

PERSSON, P.-O. & WIKSTRÖM, A. 1993. U–Pb dating of the Askersund granite and its marginal augen gneiss. *Geologiska Föreningen i Stockholm Föhandlingar*, **115**, 321–329.

RAGNHILDSTVEIT, J., SIGMOND, E. M. O. & TUCKER, R.D. 1994. Early Proterozoic supracrustal rocks west of the Mandal–Ustaoset fault Zone, Hardangervidda, south Norway. TERRA abstracts to *Terra nova*, **6**.

SAMUELSSON, L. & ÅHÄLL, K.-I. 1985. Proterozoic development of Bohuslän, south-west Sweden. *In*: TOBI, A. C. & TOURET, J. L. R. (eds) *The Deep Proterozoic Crust in the North Atlantic Provinces.* NATO ASI **C 158**, 345–357.

——, LARSON, S.-Å., ÅHÄLL, K.-I., LUNDQVIST, I., BROUSELL, J. & BERGLUND, J. 1988. Beskrivning till provisoriska översiktliga berggrundskartan Borås. *Sveriges Geologiska Undersökning*, Ba **41**, 1–32.

STARMER. I. C. 1991. The Proterozoic evolution of the Bamble Sector shear belt, southern Norway: correlations across southern Scandinavia and the Grenvillian controversy. *Precambrian Research*, **49**, 107–139.

—— 1993. The Sveconorwegian Orogeny in Southern Norway, relative to deep crustal structures and events in the North Atlantic Proterozoic Supercontinent. *Norsk Geologisk Tidskrift*, **73**, 109–132.

TUCKER, R. D., KROGH, T. E. & RÅHEIM, A. 1990. Proterozoic evolution and age-province boundaries in the Central Part of the Western Gneiss Region, Norway: results of U-Pb dating of accessory minerals from Trondheimsfjord to Geiranger. *In*: GOWER, C. F., RIVERS, T. & RYAN, A. B. (eds) *Mid-Proterozoic Laurentia–Baltica*. Geological Association of Canada, Special Paper **38**, 149–173.

——, Råheim, A., Krogh, T.E. & Corfu, F. 1987. Uranium–ead zircon and titanite ages from the northern portion of the Western Gneiss Region, south-central Norway. *Earth and Planetary Science Letters*, **81**, 203–211.

VINNERFORS, A., BERGLUND, J. & LARSON, S. L. 1993. The Hammarö (de-)formation. *Sveriges Geologiska Undersökning*. Rapporter och meddelanden nr **76**.

WAHLGREN, C.-H., CRUDEN, A. R. & STEPHENS, M. B. 1994. Kinematics of a major fan-like structure in the eastern part of the Sveconorwegian orogen, Baltic Shield, south-western Sweden. *Precambrian Research*, **70**, 67–91.

$^{40}Ar/^{39}Ar$ geochronological constraints on the tectonothermal evolution of the Eastern Segment of the Sveconorwegian Orogen, south-central Sweden

LAURENCE M. PAGE[1], MICHAEL B. STEPHENS[2] & CARL-HENRIC WAHLGREN[2]

[1] *Department of Geology, Lund University, Sölvegatan 13, S-223 62 Lund, Sweden*
[2] *Geological Survey of Sweden, Box 670, S-751 28 Uppsala, Sweden*

Abstract: A $^{40}Ar/^{39}Ar$ study to constrain the tectonothermal evolution across the Eastern Segment of the Sveconorwegian Orogen has been initiated in the area north and east of lake Vänern, south-central Sweden. This segment of the orogen is confined by two major deformation zones, the Sveconorwegian Frontal Deformation Zone (SFDZ) in the east and the Mylonite Zone in the west. Previous structural work and the prograde character of the metamorphism within the study area suggest that an older (< c. 1.57 Ga), regional foliation was formed by ductile shear deformation in a compressional tectonic regime. The orientation of this foliation was subsequently modified by later rotation along younger ductile shear zones in the easternmost, frontal part of the orogen (SFDZ). The $^{40}Ar/^{39}Ar$ ages for hornblende suggest that the regional foliation is Sveconorwegian. Furthermore, white mica ages demonstrate that the Sveconorwegian tectonothermal overprint continues at least 40 km east of the traditionally accepted limit situated along the 'Protogine Zone'. These results also provide age constraints for different phases of Sveconorwegian tectonothermal evolution with an older group of ages from 1009–965 Ma and a younger set from 930–905 Ma. The older ages are inferred to constrain a minimum age for crustal thickening during which the regional foliation and metamorphism developed, while the younger are associated with later compressional movement along the SFDZ.

The Sveconorwegian Orogen in the Baltic Shield of southwestern Sweden and southern Norway is the tectonic counterpart to the Grenville Orogen in the Laurentian Shield of southeastern Canada and the USA. In southwestern Sweden, the Sveconorwegian Orogen is divided into the Eastern, Median and Western segments which are separated by major ductile deformation zones (Berthelsen 1980; Larson et al. 1990). The presence of these zones (Fig. 1) has been known for a considerable time but only recent structural studies have provided some constraints on the geometry and kinematics of the deformation along and between them (e.g. Larson et al. 1986, 1990; Andréasson & Rodhe 1990; Park et al. 1991; Stephens et al. 1993, in press; Wahlgren et al. 1994). This structural work has naturally inspired an urgent need for tighter control on the tectonothermal evolution of the various crustal segments in this orogen. In particular, there is a need to resolve (1) the relative importance of Sveconorwegian versus older deformation and metamorphism between the major ductile deformation zones; and (2) the age of inferred Sveconorwegian deformation along the zones.

The present study utilizes the $^{40}Ar/^{39}Ar$ mineral age technique to constrain the tectonothermal history in the Eastern Segment of the Sveconorwegian Orogen, as defined in Berthelsen (1980) and Wahlgren et al. (1994). Apart from K–Ar (Magnusson 1960; Jarl 1992) and Rb–Sr (Oen 1982; Verschure et al. 1987) metamorphic mineral ages, a Rb–Sr mineral isochron age (Welin et al. 1980), isolated U–Pb sphene age determinations (Johansson 1990; Johansson & Johansson 1993; Berglund & Connelly 1994) and Sm–Nd ages in the southernmost area (Johansson et al. 1991; Johansson & Kullerud 1993), there is little information pertinent to the timing of deformation and/or uplift history in this crustal segment. The area north and east of lake Vänern (Fig. 1) is of interest since a structural analysis has demonstrated the occurrence of at least two phases of ductile shear deformation in this part of the Eastern Segment (Wahlgren et al. 1994). The crustal section studied (Fig. 2) is spatially constrained by two major deformation zones which overprint older structures in the intervening segment: (1) the Sveconorwegian Frontal Deformation Zone (SFDZ; Wahlgren et al. 1994) in the east, which has been projected south of lake Vättern to coincide with what has traditionally been referred to as the Protogine Zone; and (2) the Mylonite Zone (MZ; Magnusson 1937; Lindh 1974; Stephens et al.

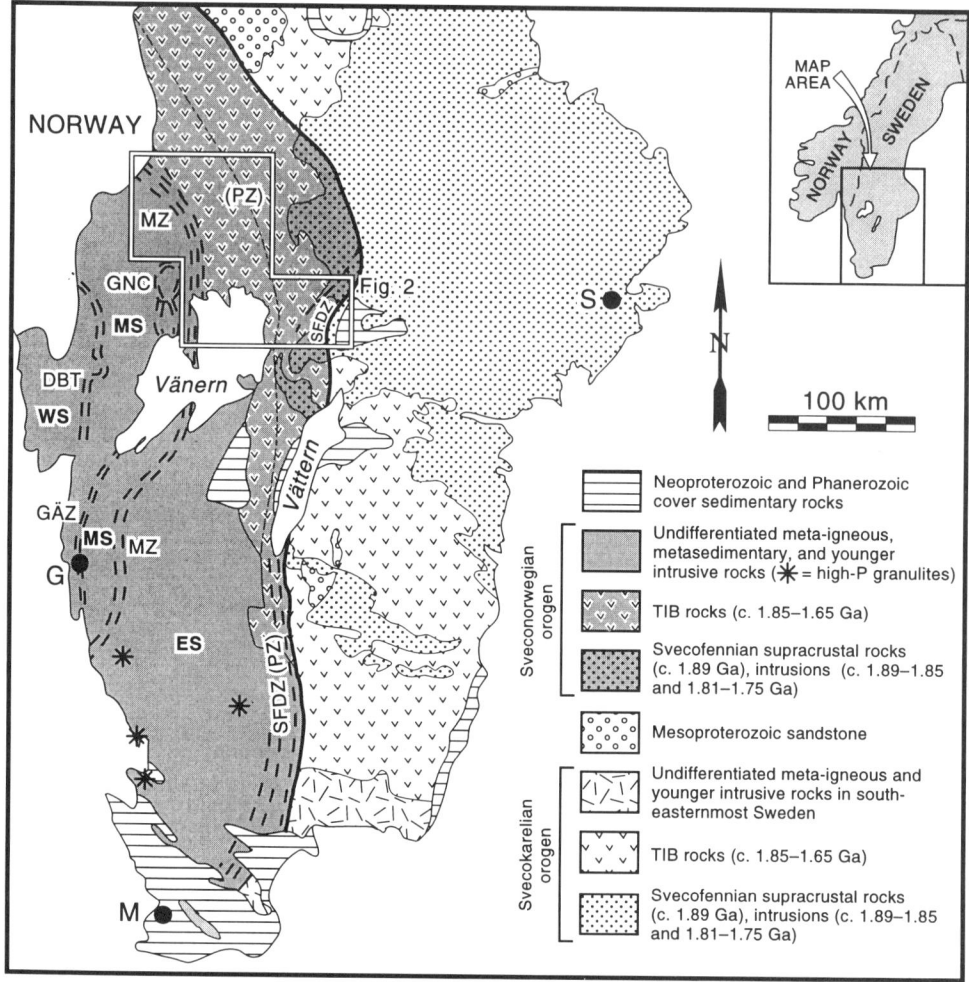

Fig. 1. Major tectonic zones in the Sveconorwegian Orogen of southwestern Sweden. DBT = Dalsland Boundary Thrust, GNC = Glaskogen Nappe Complex, GÄZ = Göta Älv Zone, MZ = Mylonite Zone, SFDZ(PZ) = Sveconorwegian Frontal Deformation Zone (Protogine Zone), ES = Eastern Segment, MS = Medium Segment, WS = Western Segment, G = Göteborg, M = Malmö, and S = Stockholm. Modified after Wahlgren *et al.* (1994).

1993, in press) to the west (Fig. 1). The present work forms part of an ongoing broader ^{40}Ar/^{39}Ar study in the Eastern, Median and Western segments of the Sveconorwegian Orogen, southwestern Sweden.

Geological setting

Lithology

The lithologies in the study area (Fig. 2) are dominated by felsic intrusive rocks which belong to the Transscandinavian Igneous Belt (TIB). With the exception of a few tectonic lenses which show relatively little deformation, these rocks are affected by ductile shear deformation. In the eastern part (Fig. 2), Svecofennian supracrustal rocks and various intrusive rocks which do not belong to the TIB suite are present. These lithologies are affected by regional deformation at *c.* 1.85–1.80 Ga and younger, ductile deformation zones which are separated from each other by tectonic lenses several kilometres across.

The oldest lithologies in the area are the Svecofennian supracrustal rocks (*c.* 1.89 Ga). These are dominated by felsic metavolcanic rocks with local occurrences of calc-silicate, carbonate,

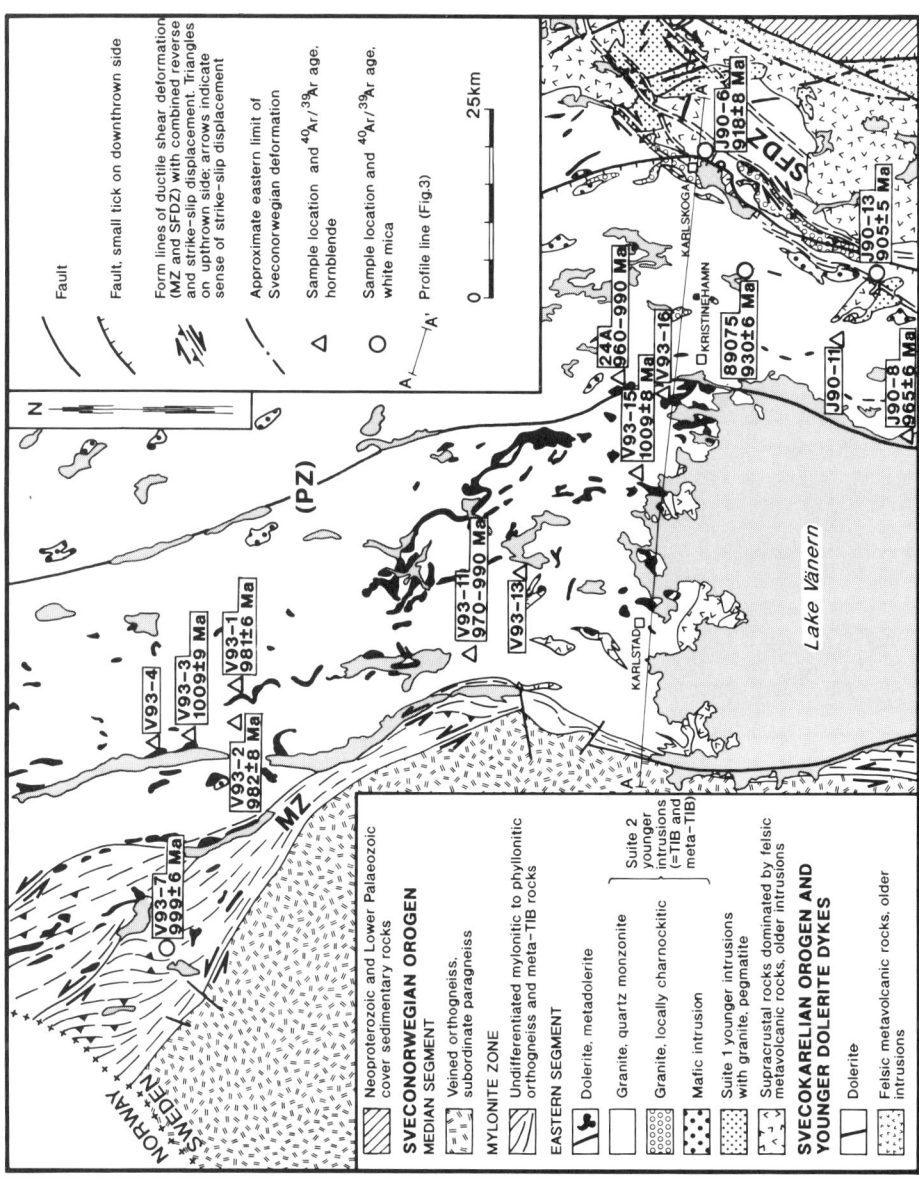

Fig. 2. Geological map of the study area with sample locations, $^{40}Ar/^{39}Ar$ ages and profile line A–A' of Fig. 3. Modified after Wikström (1991), Lundegårdh et al. (1992), Wahlgren (1992, 1993) and Wahlgren et al. (1994).

and semi-pelitic metasedimentary rocks. Calc-alkaline intrusions (c. 1.89–1.85 Ga) range in composition from gabbro to granite and are dominated by granodiorite–tonalite. Two younger suites of intrusions, which are spatially separated from each other, are also present. One of these suites, which occurs in the eastern part of the area (Fig. 2), consists of granites and pegmatites (c. 1.81–1.75 Ga). The second suite comprises the TIB rocks (c. 1.85–1.65 Ga) and dominates the area farther west (Fig. 2).

The oldest TIB lithology is a predominantly equigranular, locally charnockitic granite which, near Karlskoga, is spatially associated with supracrustal rocks metamorphosed under low-pressure, granulite-grade conditions (Andersson et al. 1992). Most of the TIB rocks are finely- to coarsely-porphyritic and vary in composition from quartz monzonite to granite. A subordinate suite of TIB rocks is composed of equigranular granite. Scattered mafic intrusions, varying from gabbro to quartz monzodiorite in composition, as well as subordinate ultramafic rocks occur in the TIB. They commonly display magma-mingling relationships with the felsic TIB rocks.

The youngest rocks in the study area consist of different generations of mafic dykes. These are also variably affected by the ductile shear deformation. Dolerites, which intrude the TIB granitoids in the western part of the area (Fig. 2), have yielded U–Pb zircon, Sm–Nd mineral isochron, and U–Pb baddeleyite ages in the time range c. 1.57–1.47 Ga (Welin et al. 1980; Johansson & Johansson 1990; Welin 1994; Wahlgren et al. in press). However, some of these ages display large errors. The U–Pb zircon age of 1465 ± 11 Ma (Welin 1994) and the U–Pb baddeleyite age of 1568 +30/–8 Ma are considered to provide better estimates of the age of intrusion of these dolerites. In the eastern part of the area, two sets of dolerite dykes are present, trending WNW–ESE and NNW–SSE (Fig. 2). These may correlate with dolerites in the Svecokarelian Orogen even farther east which have yielded Rb–Sr whole-rock/mineral ages of c. 1.56–1.51 and 1.00–0.90 Ga, respectively (Patchett 1978).

Regional structure

Detailed structural mapping of the study area has revealed two episodes of ductile shear deformation which affect the TIB rocks. This deformation preceded movement along brittle faults which are late Neoproterozoic and/or Phanerozoic in age (Wahlgren et al. 1994). Older ductile deformation zones with N–S strike dominate between Karlskoga and Kristinehamn (Fig. 2). These are transected by a younger set in the eastern part of the area which define the SFDZ. The younger zones strike NE–SW, are vertical or dip steeply to the NW, and display west-side-up or reverse dip-slip as well as right-lateral, horizontal components of movement.

The older deformation zones form a fan-like geometry in an E–W cross-section (Fig. 3), with steep westerly to vertical dips in the eastern part around Karlskoga and easterly dips further west around Kristinehamn. The movement sense across

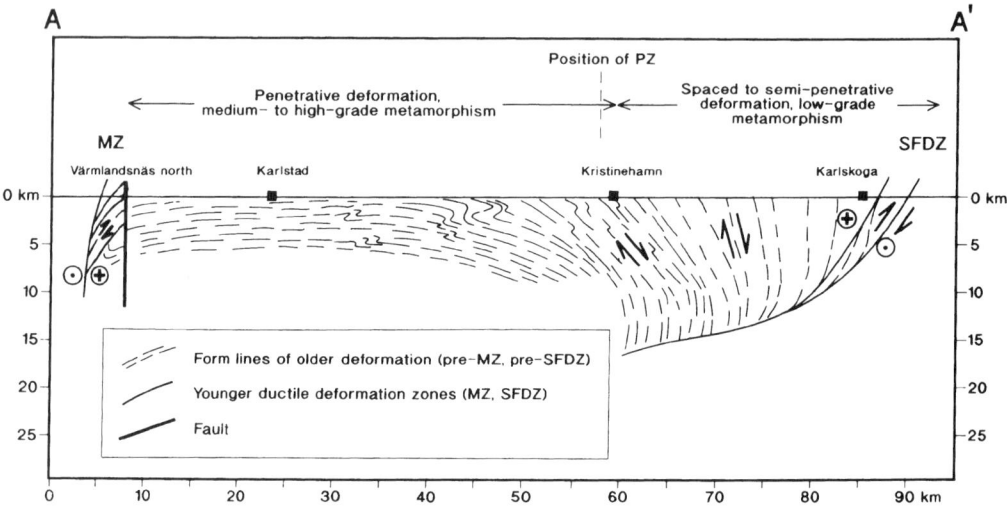

Fig. 3. Schematic structural section across the Eastern Segment of the Sveconorwegian Orogen. Horizontal and vertical scales are identical. Modified after Stephens et al. (in press).

the entire fan-like structure (c. 40 km) is remarkably consistent, with predominantly dip-slip, top-to-the-east sense of shear. In the eastern part of this structure around Karlskoga, the deformation is anastomosing and spaced to semi-penetrative whereas, in the vicinity of Kristinehamn, the shear deformation is penetrative. This variation in the style of deformation corresponds to a westward increase in the grade of syn-deformational metamorphism from greenschist to amphibolite facies (Wahlgren et al. 1994). The broad belt of deformation in the Karlskoga–Kristinehamn area includes the 'Protogine Zone' (Fig. 1). Wahlgren et al. (1994) have demonstrated that the 'Protogine Zone' is not an important regional structure in the study area but merely corresponds to a fault which lies close to the boundary between the semi-penetratively- and penetratively-deformed TIB rocks. West of Kristinehamn towards the MZ, the rocks were metamorphosed under amphibolite-facies conditions and the structures are flat-lying (Fig. 3). Thus, the entire structure in the Eastern Segment, from Karlskoga to the MZ, is strongly asymmetric with the fan-like structure occurring close to the eastern frontal part of the Sveconorwegian Orogen (Fig. 3). Rotation of the older deformation zones into the younger, compressional structures of the SFDZ has been proposed (Wahlgren et al. 1994).

Previous constraints on the timing of deformation and/or uplift

Most constraints for the timing of deformation and/or uplift in the Eastern segment of the Sveconorwegian Orogen occur south of lake Vänern (Fig. 1). Syenites and dolerites affected by ductile shear deformation along the SFDZ in this area have yielded U–Pb zircon ages between 1.22–1.20 Ga (Johansson 1990; Hansen & Lindh 1991; Jarl 1992) and Sm–Nd mineral isochron ages around 1.18 and 0.93 Ga (Johansson & Johansson 1990), respectively. Furthermore, a U–Pb sphene age of 0.94 Ga along the SFDZ was also obtained by Johansson (1990). Further evidence for Sveconorwegian deformation within the Eastern Segment is provided by a U–Pb zircon protolith age of 1224 +8/–7 Ma and a U–Pb sphene age around 0.95 Ga for a strongly- yet non-penetratively-deformed syenite situated c. 20 km east of the MZ (Berglund & Connelly 1994). However, Larson et al. (1986, 1990) have indicated that the penetrative structures in the host banded and veined gneiss are older than c. 1.66 Ga.

North of lake Vänern, the maximum age of the foliation which affects the dolerites and which defines the western part of the fan-like structure is constrained by the c. 1.57–1.47 Ga age of these mafic intrusions (Welin 1994; Wahlgren et al. 1994). Morthorst et al. (1983) argued that these mafic bodies intruded the country rocks after development of a flat-lying regional foliation. However, Wahlgren et al. (in press) suggest that the present structural geometry and the concordant relationships between the dolerites and the country-rocks are due to deformation which postdates the intrusion of these mafic rocks. Thus, deformation prior to c. 1.57 Ga has not been documented with any confidence in the area north of lake Vänern. Although protolith ages of younger rocks affected by the ductile shear deformation in this area are lacking, a Rb–Sr mineral isochron age of c. 0.98 Ga (Welin et al. 1980), Rb–Sr mineral (biotite) ages in the time period c. 0.95–0.90 Ga (Oen 1982; Verschure et al. 1987), and K–Ar mineral (biotite and hornblende) ages in the time-range 0.98–0.90 Ga (Jarl 1992) suggest a Sveconorwegian tectonothermal overprint in and adjacent to the present area of study.

^{40}Ar/^{39}Ar geochronology

Analytical techniques

Samples collected for geochronological analysis were examined in thin section. After rejection of those that were too fine-grained or had unsuitable mineralogy, with alteration or abundant inclusions, 15 were separated for analysis. The samples were crushed and sieved to a 0.25 to 0.15 mm size fraction, and hornblende and white mica were separated using standard magnetic and heavy-liquid techniques. Final purification of the separates was achieved by hand-picking.

30–50 mg of hornblende and 10–20 mg of white mica splits were irradiated along with the flux monitor MMhb-1 (Samson & Alexander 1987) and synthetic salts to permit corrections for interfering nuclear reactions in the McMaster University reactor. Incremental step-heating analyses using a vacuum furnace were performed at the Geochronology Laboratory at the Massachusetts Institute of Technology, USA. A MAP 215–50 mass spectrometer was used for the isotopic analyses. 10–13 steps were run on both micas and hornblendes. Correction factors for interfering isotopes were measured on CaF_2 and K_2SO_4 salts. Ages were calculated using the isotopic ratios and decay constants listed in Steiger & Jäger (1977).

McDougall & Harrison (1988) discuss the generally accepted criteria for determination of a plateau in an age spectrum (cf. Dalrymple & Lanphere 1974; Fleck et al. 1977; Lanphere & Dalrymple 1978). One important criterion involves a minimum of three contiguous steps representing a significant proportion (usually >50%) of the total ^{39}Ar released with concordant ages. Ages are concordant if they do not differ at the 95% confidence level. Isotope peak heights and errors were calculated by least squares linear regression of six cycles to the time at which the mass spectrometer was equilibrated with the inlet section. K/Ca ratios were calculated for each increment from the ^{39}Ar/^{37}Ar ratios.

Results

Eleven hornblende and 4 white mica samples were analysed. Of these, 7 hornblende and all the white mica samples gave interpretable spectra. The samples yield spectra which can generally be placed into three groups:

(1) Samples which yield well-defined plateau ages for over 50% of the gas released include 4 hornblendes and 4 muscovites (V93-3, V93-1, V93-15, J90-8, V93-7, 89075, J90-13, J90-6). One sample (V93-2) gives a plateau age for 39% of the gas released. The hornblendes yield ages of 1009–965 Ma, while 3 white micas associated with the SFDZ yield ages of 930–905 Ma and one associated with the MZ gives an age of 999 Ma. Samples V93-3, V93-2 and V93-1 exhibit spectra yielding older ages in the first few increments which decrease to give good plateau ages. The initial older ages may be due to the presence of an excess Ar component associated with the protolith age of the rock.

(2) Two hornblendes (V93-11 and 24A) do not yield good plateau ages but generally give Sveconorwegian ages for most increments and yield minima consistent with the hornblende samples which produce good plateaus. 24A has a very variable K/Ca ratio possibly indicating different mineral components which may be responsible for the poor plateau.

(3) Four hornblende samples (V93-4, V93-13, V93-16, J90-11) give generally erratic, uninterpretable spectra which yield no conclusive information and which may possibly be due to incomplete recrystallization of the protolith. These samples had a mineralogy similar to those that ran well. However, they generally did not exhibit as much deformation as many of the samples which yielded acceptable results. No useful information was obtained for this suite of samples by using the isotope correlation method as in most of the samples all points cluster too near to the ^{40}Ar/^{39}Ar intercept to yield any statistically meaningful information.

Rock sample descriptions are presented in Appendix 1, sample localities and ages are shown in Fig. 2, release spectra in Figs 4a, b and c, and the heating schedules are presented in Table 1.

Discussion

A Sveconorwegian tectonothermal history

The age of the dolerite dykes in the Kristinehamn area and westward overlaps slightly with the time constraints for the deformation and metamorphism during the Gothian orogeny (Gáal & Gorbatschev 1987) west of the Mylonite Zone (*c*. 1.65–1.56 Ga). However, the age of these dykes is younger and clearly distinct from the >1.66 Ga deformation and metamorphism which has been indicated in the Eastern segment orthogneisses south of lake Vänern (Larson *et al.* 1986, 1990). At *c*. 1.22–1.20 Ga, igneous activity attributed to regional extension, with intrusion of syenites and dolerites, occurred in the Eastern Segment. Igneous activity of this age is common in other North Atlantic shield areas and has been related to the break-up of a Mesoproterozoic supercontinent (Patchett & Bylund 1977; Piper 1983; Gower *et al.* 1990). Wahlgren *et al.* (1994) favoured a compressional model to explain the older (post-1.57 Ga/pre-SFDZ) structures in the present study area but they did not rule out the possibility that these structures could have been related to early- or pre-Sveconorwegian regional extension. Nevertheless, they noted that the TIB rocks and younger mafic intrusions were affected by prograde metamorphism. This argues against deformational processes related to regional extension since extensional environments are typically associated with uplift, cooling and retrograde metamorphism. In summary, deformation and metamorphism in connection with regional compression, either in the last phases of the Gothian Orogeny or during the Sveconorwegian Orogeny, are suggested by the available geological data.

The ^{40}Ar/^{39}Ar ages presented in this study for hornblende strongly suggest a Sveconorwegian age for the foliation development and amphibolite-grade metamorphism in the dolerites. The hornblendes grow in and define the foliation in these intrusions and have been separated from rocks apparently unaffected by deformation related to the retrogressive MZ and SFDZ. It is theoretically possible that the syn-deformational mineral growth occurred during the older Gothian event and did not cool below the ^{40}Ar/^{39}Ar hornblende closure temperature until Sveconorwegian uplift. However, this scenario is considered to be highly unlikely since it would require that the Eastern Segment was stored at depth for over 400 Ma and

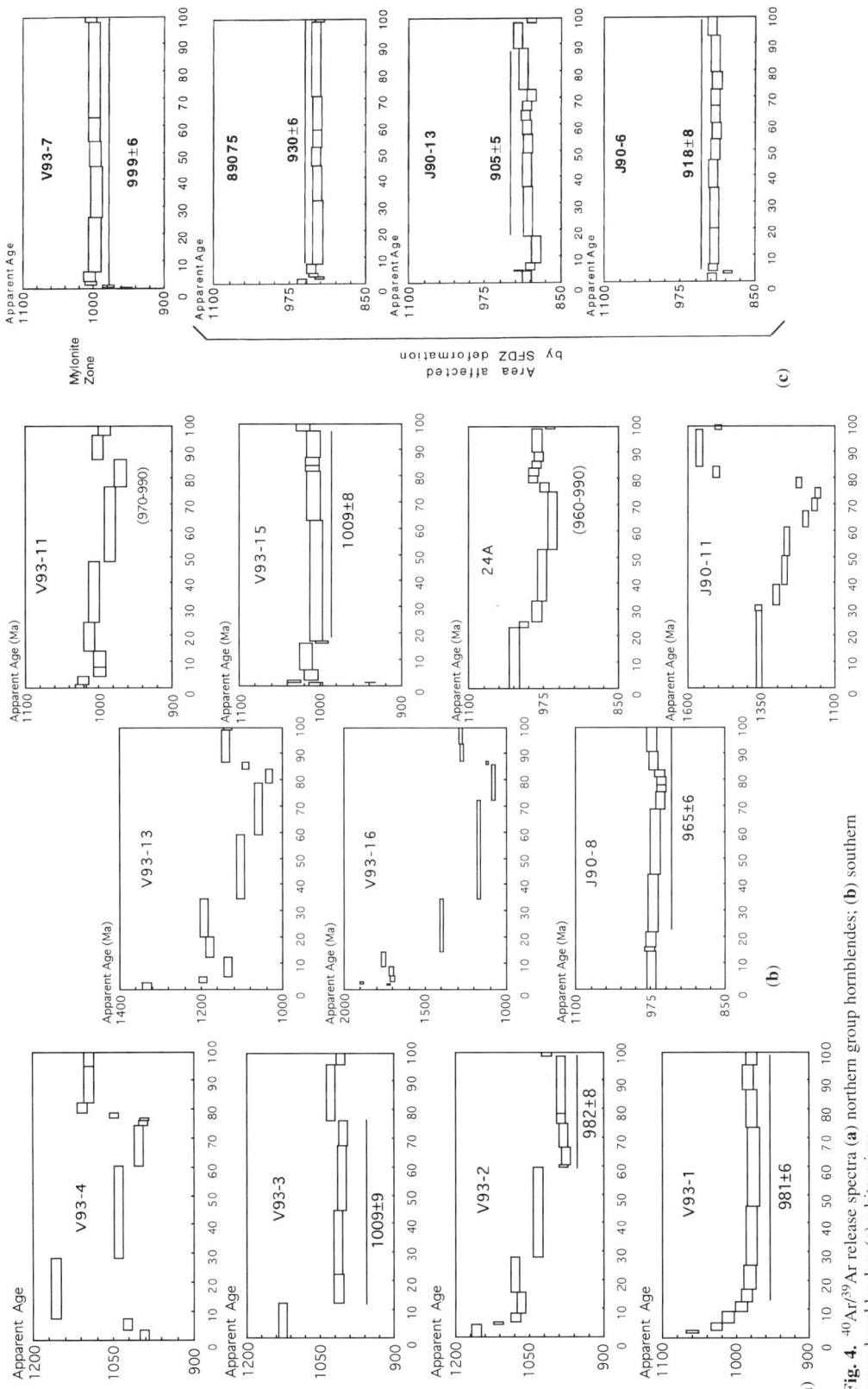

Fig. 4. $^{40}Ar/^{39}Ar$ release spectra (**a**) northern group hornblendes; (**b**) southern group hornblendes; (**c**) white micas.

Table 1 $^{40}Ar/^{39}Ar$ heating schedules and data

Step #	T(K)	$^{39}Ar/^{40}Ar$	$^{36}Ar/^{40}Ar$	$\%^{39}Ar$	$\%^{40}Ar^*$	K/Ca	Age		2s
V93-4 hbl									
J value: 0.00906									
1	1100	0.01197	1.08E-04	3.5	96.8	0.183	992.0	+/−	8.4
2	1210	0.01168	5.65E-05	7.4	98.3	0.098	1022.9	+/−	8.6
3	1250	0.00992	5.46E-05	28.2	98.4	0.073	1156.7	+/−	9.4
4	1270	0.01156	1.75E-05	60.3	99.5	0.073	1039.8	+/−	8.7
5	1285	0.01214	1.32E-05	74.5	99.6	0.078	1002.7	+/−	8.5
6	1295	0.01217	4.85E-05	76.3	98.6	0.090	992.8	+/−	8.4
7	1315	0.01223	3.42E-05	77.2	99.0	0.082	992.5	+/−	8.4
8	1330	0.01148	5.81E-06	78.9	99.9	0.071	1048.0	+/−	8.8
9	1350	0.01066	7.14E-06	82.1	99.8	0.064	1107.9	+/−	9.1
10	1400	0.01075	3.17E-05	94.9	99.1	0.068	1095.2	+/−	9.1
11	1550	0.01071	4.62E-05	100.0	98.7	0.067	1094.2	+/−	9.1
V93-3 hbl									
J value: 0.0103									
1	1240	0.00800	1.10E-03	11.9	67.4	0.103	1126.4	+/−	9.8
2	1255	0.01350	4.90E-05	22.2	98.6	0.083	1012.6	+/−	9.0
3	1270	0.01350	3.41E-05	44.7	99.0	0.084	1013.5	+/−	9.0
4	1285	0.01370	1.70E-05	67.3	99.5	0.085	1005.4	+/−	9.0
5	1300	0.01370	3.79E-05	75.9	98.9	0.086	1004.3	+/−	9.0
6	1400	0.01324	4.10E-05	95.9	98.8	0.077	1028.3	+/−	9.1
7	1550	0.01340	1.01E-04	100.0	97.1	0.074	1007.9	+/−	9.0
V93-2 hbl									
J value: 0.01026									
1	1100	0.01110	9.06E-05	4.1	97.3	0.137	1158.3	+/−	10.0
2	1145	0.01171	8.85E-05	5.0	97.4	0.118	1113.2	+/−	9.7
3	1190	0.01240	4.74E-05	8.2	98.6	0.085	1077.5	+/−	9.5
4	1220	0.01270	4.42E-06	15.5	99.9	0.082	1066.7	+/−	9.4
5	1240	0.01250	2.24E-05	27.9	99.4	0.079	1078.6	+/−	9.5
6	1255	0.01320	2.02E-05	59.3	99.4	0.084	1032.9	+/−	9.2
7	1270	0.01400	4.94E-05	60.3	98.6	0.098	981.2	+/−	8.9
8	1285	0.01420	1.76E-05	66.5	99.5	0.103	975.6	+/−	8.8
9	1300	0.01400	4.22E-05	74.6	98.8	0.098	981.0	+/−	8.9
10	1325	0.01370	1.01E-04	78.4	97.0	0.084	985.2	+/−	8.9
11	1400	0.01400	1.61E-05	98.3	99.6	0.080	987.0	+/−	8.9
12	1550	0.01340	4.52E-05	100.0	98.7	0.077	1014.8	+/−	9.1
V93-1 hbl									
J value: 0.009708									
1	1100	0.00658	9.47E-04	1.4	72.0	0.389	1305.4	+/−	9.6
2	1145	0.01150	1.73E-04	2.3	94.9	0.423	1060.7	+/−	8.3
3	1190	0.01250	5.03E-05	4.9	98.5	0.101	1027.3	+/−	8.1
4	1220	0.01280	3.96E-05	8.8	98.9	0.086	1011.5	+/−	8.0
5	1240	0.01314	2.11E-05	12.3	99.4	0.095	993.0	+/−	7.9
6	1255	0.01320	2.65E-05	16.7	99.2	0.089	985.9	+/−	7.8
7	1270	0.01340	1.81E-05	25.3	99.5	0.088	981.6	+/−	7.8
8	1285	0.01340	1.73E-05	45.6	99.5	0.091	979.1	+/−	7.8
9	1300	0.01350	5.81E-06	73.3	99.8	0.096	976.7	+/−	7.8
10	1325	0.01340	1.26E-05	86.7	99.6	0.098	978.7	+/−	7.8
11	1400	0.01330	3.20E-05	95.4	99.1	0.090	984.1	+/−	7.8
12	1550	0.01320	8.18E-05	100.0	97.6	0.088	978.7	+/−	7.8

Table 1 *Continued*

Step #	T(K)	$^{39}Ar/^{40}Ar$	$^{36}Ar/^{40}Ar$	$\%^{39}Ar$	$\%^{40}Ar*$	K/Ca	Age		2s
V93-11 hbl									
J value: 0.009818									
1	1190	0.01230	1.43E-04	1.1	95.8	0.135	1024.0	+/−	8.0
2	1220	0.01270	5.50E-05	4.0	98.4	0.124	1021.0	+/−	7.9
3	1240	0.01310	3.74E-05	7.8	98.9	0.124	998.1	+/−	7.8
4	1255	0.01320	2.13E-05	13.7	99.4	0.124	998.4	+/−	7.8
5	1270	0.01300	1.66E-05	24.7	99.5	0.124	1012.3	+/−	7.9
6	1285	0.01310	1.48E-05	48.0	99.6	0.126	1005.7	+/−	7.9
7	1300	0.01350	7.69E-06	76.6	99.8	0.130	983.8	+/−	7.7
8	1325	0.01370	1.17E-05	87.2	99.7	0.133	970.0	+/−	7.6
9	1400	0.01320	1.23E-05	96.3	99.7	0.123	1000.7	+/−	7.8
10	1550	0.01319	5.15E-05	100.0	98.5	0.121	991.9	+/−	7.8
V93-1 3 hb									
J value: 0.01026									
1	1100	0.00918	7.20E-05	2.3	97.9	0.192	1332.9	+/−	11.0
2	1145	0.00903	1.10E-04	2.6	96.7	0.184	1337.6	+/−	11.0
3	1190	0.01086	1.75E-05	4.7	99.5	0.124	1195.0	+/−	10.2
4	1220	0.01168	7.13E-06	12.2	99.8	0.111	1135.8	+/−	9.9
5	1240	0.01109	1.22E-05	20.1	99.6	0.110	1178.7	+/−	10.1
6	1255	0.01092	6.54E-06	34.4	99.8	0.114	1193.2	+/−	10.2
7	1270	0.01216	2.58E-06	59.1	99.9	0.117	1102.9	+/−	9.7
8	1285	0.01284	6.64E-07	78.9	100.0	0.119	1059.4	+/−	9.4
9	1300	0.01311	4.24E-05	84.1	98.7	0.122	1033.0	+/−	9.2
10	1325	0.01229	1.24E-05	86.6	99.6	0.117	1091.7	+/−	9.6
11	1400	0.01160	1.09E-05	98.8	99.7	0.109	1140.1	+/−	9.9
12	1550	0.01161	3.40E-05	100.0	99.0	0.109	1133.7	+/−	9.8
V93-15 hbl									
J value: 0.009661									
1	1100	0.01169	3.27E-04	1.1	90.3	0.096	1005.9	+/−	8.0
2	1145	0.01283	3.16E-04	1.2	90.7	0.103	938.9	+/−	7.6
3	1190	0.01242	2.76E-05	1.9	99.2	0.084	1031.9	+/−	8.1
4	1220	0.01279	1.80E-05	5.8	99.5	0.095	1010.8	+/−	8.0
5	1240	0.01274	2.71E-06	15.9	99.9	0.101	1017.2	+/−	8.0
6	1255	0.01240	1.73E-04	16.7	94.9	0.101	998.2	+/−	7.9
7	1270	0.01296	1.03E-06	62.8	100.0	0.111	1004.9	+/−	8.0
8	1285	0.01290	1.27E-08	81.8	100.0	0.110	1008.4	+/−	8.0
9	1300	0.01286	3.82E-06	84.1	99.9	0.106	1009.9	+/−	8.0
10	1325	0.01278	2.77E-05	87.2	99.2	0.100	1009.5	+/−	8.0
11	1400	0.01289	5.07E-06	97.2	99.9	0.103	1008.1	+/−	8.0
12	1550	0.01267	7.96E-06	100.0	99.8	0.102	1020.7	+/−	8.1
V93-16 hbl									
J value: 0.00906									
1	1100	0.00152	3.11E-04	1.6	90.8	0.220	3351.8	+/−	16.8
2	1145	0.00557	4.53E-05	2.0	98.7	0.138	1726.1	+/−	12.3
3	1190	0.00487	1.00E-05	3.0	99.7	0.080	1891.9	+/−	12.9
4	1220	0.00568	6.06E-05	5.0	98.2	0.066	1701.0	+/−	12.2
5	1240	0.00568	3.52E-05	8.4	99.0	0.060	1709.1	+/−	12.2
6	1255	0.00546	1.94E-05	14.0	99.4	0.061	1758.0	+/−	12.4
7	1270	0.00774	6.59E-06	34.4	99.8	0.063	1396.3	+/−	10.7
8	1285	0.00989	7.84E-06	72.0	99.8	0.066	1171.4	+/−	9.5
9	1300	0.01107	1.42E-07	85.6	100.0	0.068	1078.5	+/−	9.0
10	1325	0.01050	1.58E-05	87.1	99.6	0.066	1118.3	+/−	9.2
11	1400	0.00873	2.60E-05	93.7	99.3	0.063	1277.0	+/−	10.1
12	1550	0.00864	3.66E-05	100.0	98.9	0.063	1283.5	+/−	10.1

Table 1 continued overleaf

Table 1 Continued

Step #	T(K)	$^{39}Ar/^{40}Ar$	$^{36}Ar/^{40}Ar$	$\%^{39}Ar$	$\%^{40}Ar^*$	K/Ca	Age		2s

24A hbl
J value: 0.0132

1	1100	0.01716	2.35E-05	22.8	99.3	0.272	1023.8	+/–	8.3
2	1145	0.01750	2.61E-05	25.2	99.2	0.191	1007.9	+/–	8.2
3	1190	0.01787	5.19E-05	33.1	98.5	0.106	986.2	+/–	8.1
4	1220	0.01831	8.95E-06	53.0	99.7	0.064	977.4	+/–	8.0
5	1240	0.01870	1.26E-05	74.7	99.6	0.063	960.9	+/–	7.9
6	1255	0.01818	4.85E-05	78.2	98.6	0.182	973.7	+/–	8.0
7	1270	0.01786	2.26E-05	81.1	99.3	0.216	993.0	+/–	8.1
8	1285	0.01793	1.61E-05	84.0	99.5	0.145	991.6	+/–	8.1
9	1300	0.01798	2.56E-05	86.8	99.2	0.096	987.1	+/–	8.1
10	1325	0.01811	1.94E-05	90.2	99.4	0.064	983.1	+/–	8.0
11	1400	0.01808	1.06E-05	98.7	99.7	0.061	986.4	+/–	8.1
12	1550	0.01789	1.42E-04	100.0	95.8	0.060	964.7	+/–	7.9

J90-8 hbl
J value: 0.0132

1	1100	0.01832	2.22E-05	14.4	99.3	0.172	973.8	+/–	8.3
2	1145	0.01827	2.15E-05	15.9	99.4	0.173	976.2	+/–	8.3
3	1190	0.01835	1.07E-05	21.8	99.7	0.170	975.2	+/–	8.3
4	1220	0.01847	5.00E-06	43.8	99.9	0.172	971.7	+/–	8.3
5	1240	0.01860	2.37E-06	68.6	99.9	0.173	967.1	+/–	8.2
6	1255	0.01876	1.03E-05	75.4	99.7	0.175	958.9	+/–	8.2
7	1270	0.01884	2.57E-06	77.9	99.9	0.173	957.3	+/–	8.2
8	1285	0.01885	2.05E-06	80.9	99.9	0.172	957.0	+/–	8.2
9	1300	0.01875	4.61E-06	83.7	99.9	0.172	960.4	+/–	8.2
10	1325	0.01853	7.31E-07	90.4	100.0	0.172	970.1	+/–	8.2
11	1400	0.01843	7.67E-06	99.6	99.8	0.172	972.5	+/–	8.3
12	1550	0.01870	3.76E-05	100.0	98.9	0.173	955.1	+/–	8.1

J90-11 hbl
J value: 0.0132

1	1100	0.01166	1.59E-05	29.0	99.5	0.153	1361.7	+/–	10.1
2	1145	0.01166	1.04E-05	31.3	99.7	0.123	1362.8	+/–	10.1
3	1190	0.01240	1.92E-05	39.4	99.4	0.084	1302.4	+/–	9.8
4	1220	0.01275	3.16E-05	50.2	99.1	0.079	1273.1	+/–	9.7
5	1240	0.01287	1.95E-05	61.0	99.4	0.078	1268.4	+/–	9.7
6	1255	0.01374	4.66E-05	67.4	98.6	0.145	1202.2	+/–	9.3
7	1270	0.01436	1.27E-05	71.9	99.6	0.198	1172.9	+/–	9.1
8	1285	0.01456	1.41E-05	76.3	99.6	0.115	1160.7	+/–	9.1
9	1300	0.01351	1.15E-05	79.9	99.7	0.082	1226.4	+/–	9.4
10	1325	0.01008	7.22E-06	84.3	99.8	0.049	1507.6	+/–	10.8
11	1400	0.00954	1.57E-05	98.5	99.5	0.050	1562.4	+/–	11.1
12	1550	0.01011	2.77E-05	100.0	99.2	0.060	1499.1	+/–	10.8

V93-7 musc
J value: 0.01033

1	900	0.01450	8.39E-05	0.3	97.5	11.9	953.0	+/–	8.2
2	950	0.01390	1.09E-04	0.9	96.8	25.1	977.8	+/–	8.3
3	1000	0.01380	1.28E-05	2.3	99.6	53.1	1003.3	+/–	8.5
4	1050	0.01380	2.86E-06	6.1	99.9	160.1	1005.3	+/–	8.5
5	1100	0.01390	1.30E-05	26.1	99.6	715.0	997.5	+/–	8.5
6	1150	0.01400	7.85E-06	45.1	99.8	1221.8	995.7	+/–	8.4
7	1200	0.01400	1.35E-06	54.3	100.0	892.7	996.5	+/–	8.5
8	1250	0.01400	9.20E-07	62.9	100.0	221.7	998.1	+/–	8.5
9	1350	0.01400	2.76E-06	97.9	99.9	320.1	998.6	+/–	8.5
10	1550	0.01380	2.10E-05	100.0	99.4	5.2	1003.5	+/–	8.5

Table 1 *Continued*

Step #	T(K)	$^{39}Ar/^{40}Ar$	$^{36}Ar/^{40}Ar$	$\%^{39}Ar$	$\%^{40}Ar*$	K/Ca	Age		2s
89075 musc									
J value: 0.0132									
1	900	0.01790	1.84E-04	1.5	94.6	64.6	955.6	+/-	7.9
2	950	0.01902	1.13E-04	2.3	96.7	89.0	925.9	+/-	7.7
3	1000	0.01910	4.73E-05	3.8	98.6	174.2	936.8	+/-	7.8
4	1050	0.01920	1.62E-05	7.3	99.5	425.8	940.0	+/-	7.8
5	1090	0.01950	1.43E-05	31.1	99.6	6198.9	929.4	+/-	7.7
6	1130	0.01950	8.49E-06	44.5	99.7	2203.7	930.1	+/-	7.7
7	1170	0.01950	1.95E-06	51.5	99.9	944.4	932.2	+/-	7.7
8	1210	0.01950	5.99E-06	58.1	99.8	462.5	930.1	+/-	7.7
9	1250	0.01950	1.05E-05	70.6	99.7	0.0	930.6	+/-	7.7
10	1350	0.01950	1.63E-06	98.7	100.0	9442.7	932.2	+/-	7.7
11	1550	0.01941	4.46E-05	100.0	98.7	8.6	926.5	+/-	7.7
J90-13 musc									
J value: 0.0132									
1	900	0.01910	1.84E-04	4.0	94.6	196.6	905.7	+/-	7.6
2	950	0.01980	2.22E-05	4.7	99.3	73.1	918.4	+/-	7.6
3	1000	0.02020	3.38E-05	7.1	99.0	242.3	899.9	+/-	7.5
4	1050	0.02070	5.24E-06	17.4	99.8	1058.1	890.3	+/-	7.5
5	1075	0.02020	1.40E-05	36.4	99.6	3154.2	903.3	+/-	7.5
6	1100	0.02024	1.65E-06	49.0	100.0	1711.5	905.5	+/-	7.6
7	1125	0.02030	7.58E-06	56.0	99.8	823.5	903.5	+/-	7.5
8	1150	0.02020	1.07E-05	61.0	99.7	551.3	905.6	+/-	7.6
9	1180	0.02011	1.36E-05	65.2	99.6	289.0	907.3	+/-	7.6
10	1210	0.02010	1.44E-05	68.8	99.6	344.5	906.1	+/-	7.6
11	1250	0.02040	1.21E-05	73.2	99.6	487.3	899.1	+/-	7.5
12	1300	0.02000	6.07E-06	88.5	99.8	2241.3	911.8	+/-	7.6
13	1350	0.01980	5.37E-06	98.2	99.8	1210.0	921.5	+/-	7.7
14	1550	0.01980	1.13E-04	100.0	96.7	32.6	898.6	+/-	7.5
J90-6 musc									
J value: 0.0132									
1	900	0.01960	2.44E-05	2.7	99.3	51.0	923.2	+/-	7.7
2	950	0.01950	1.60E-04	3.5	95.3	62.9	897.3	+/-	7.5
3	1000	0.01960	3.11E-05	6.4	99.1	71.3	920.8	+/-	7.6
4	1050	0.01980	8.95E-06	20.0	99.7	97.3	919.2	+/-	7.6
5	1075	0.01980	1.54E-05	35.1	99.5	91.4	919.0	+/-	7.6
6	1100	0.01974	9.87E-06	45.7	99.7	72.4	921.8	+/-	7.7
7	1125	0.01980	2.98E-05	53.7	99.1	68.7	916.8	+/-	7.6
8	1150	0.01980	3.60E-05	59.8	98.9	46.3	914.7	+/-	7.6
9	1200	0.01980	1.39E-05	66.2	99.6	29.2	918.0	+/-	7.6
10	1250	0.01990	1.09E-05	72.5	99.7	22.1	917.0	+/-	7.6
11	1300	0.01990	1.75E-05	79.3	99.5	16.7	914.4	+/-	7.6
12	1350	0.01980	1.24E-05	92.6	99.6	26.5	917.5	+/-	7.6
13	1550	0.01970	1.94E-06	100.0	99.9	9.8	923.4	+/-	7.7

subsequently uplifted without any new, discernible deformation internally within the segment. Further support for regional Sveconorwegian deformation and metamorphism in the Eastern Segment is provided by the U–Pb zircon protolith age of c. 1.22 Ga and the U–Pb sphene age of c. 0.95 Ga for the deformed syenite south of lake Vänern (Berglund & Connelly 1994). The interpretation of the $^{40}Ar/^{39}Ar$ data, with its implications for the major regional significance of Sveconorwegian deformation and metamorphism in the Eastern Segment north of lake Vänern, is in contrast to the tectonic scenario envisaged in the same crustal segment south of the lake, where significant >1.66 Ga deformation has been indicated (Larson et al. 1986, 1990). This difference may reflect the preservation of enclaves which preserve older pre-Sveconorwegian deformation south of the lake.

Eastern limit of Sveconorwegian tectonothermal activity

Following Berthelsen (1980), Wahlgren et al. (1994) argued, on the basis of new structural data, that the eastern boundary of Sveconorwegian deformation and metamorphism extends at least c. 40 km east of the traditionally accepted limit situated along the 'Protogine Zone' in the Kristinehamn area. The 930–905 Ma $^{40}Ar/^{39}Ar$ white mica ages in samples 89075, J90-13 and J90-6 provide independent support for this interpretation and are also in agreement with Rb–Sr mineral (biotite) ages (Oen 1982; Verschure et al. 1987) which occur at a similar tectonostratigraphic level approximately 50 km north of Karlskoga. Thus, there is now considerable evidence that the eastern boundary of Sveconorwegian tectonothermal activity has to be shifted to the east in the Baltic Shield of south-central Sweden (Fig. 1), in particular to the area directly east of Karlskoga in the present area of study (Fig. 2).

Polyphase Sveconorwegian deformation

Early Sveconorwegian tectonothermal event. The hornblende plateau ages range from 1009–965 Ma. These ages constrain the timing of compressional deformation and metamorphism to have begun at or prior to 1009 Ma. Furthermore, there is some tendency for the ages to become younger to the east. The plateau age in the westernmost sample in both the northern and southern groups is older than the plateau ages in the samples to the east. These results provoke two possible interpretations for the early Sveconorwegian tectonothermal event. (1) The difference in hornblende ages may be due to crystallization near to or below the closure temperature of hornblende (500°C, Harrison 1981). Thus, the eastward younging of ages would represent a progression of deformation from west to east, or 2) the entire orogenic wedge was formed at temperatures >500°C and thus ages from the west may represent a higher level in the orogenic wedge which therefore cooled at an older age. As these rocks have been affected by deformation under amphibolite-facies conditions, the second of these possibilities is favoured. Clearly, more data are needed to confirm the validity of the age trend.

One white mica sample (V93-7), which yields an age of 999 ± 6 Ma, is from a meta-igneous rock affected by hydrothermal alteration (Lindh 1974) within an internal duplex of the MZ (Stephens et al. 1993, in press.). This age indicates that these rocks were not affected by tectonothermal processes at temperatures above the closure temperature for argon diffusion (>350°C) after 999 Ma. The age also suggests that the MZ contains at least some rocks affected by an older phase of deformation relative to that along the SFDZ. It is important to emphasize that the majority of the structures associated with the MZ in this area cross-cut and are retrogressive to the amphibolite-grade deformation which contains the hornblendes which have been dated in this study (Stephens et al. 1993, in press). Clearly more $^{40}Ar/^{39}Ar$ dating is required to constrain the tectonothermal history of the MZ; this is the focus of ongoing work by the authors.

Late Sveconorwegian tectonothermal event. The three white mica samples (89075, J90-13, J90-6) which yielded late Sveconorwegian ages come from the area affected by younger structures under greenschist-facies conditions which are associated with the SFDZ. Indeed, the white mica in sample J90-13 defines the younger SFDZ tectonic foliation. These structures are discordant to the foliation in the fan-like structure (Wahlgren et al. 1994) which contains the older hornblende ages further to the west.

Tectonic implications. It is tempting to consider an interpretation of all these ages in which the entire region underwent a single cooling event following Sveconorwegian tectonic imbrication at >1009 Ma, with uplift and cooling below 350°C being complete at c. 905 Ma. However, several important points argue against this interpretation for this portion of the Eastern Segment:

(1) The grade of syn-deformational, peak metamorphism **increases** from greenschist facies in the east around Karlskoga to amphibolite facies west of Kristinehamn. This is illustrated by both the mineral assemblages preserved as well as the deformation mechanism of feldspar megacrysts oriented in the pre-SFDZ foliation (Wahlgren et al. 1994). In eastern areas, these crystals deformed by slip along discrete microfractures while, further west, the feldspar megacrysts recrystallized and only relic outlines of the crystals are preserved.

(2) Previous structural work has demonstrated that there have been two important ductile deformational events in the area between Karlskoga and Kristinehamn (Wahlgren et al. 1994). Older structures with predominantly dip-slip, top-to-the-east sense of shear were rotated into the younger deformation zones belonging to the SFDZ with west-side-up or reverse dip-slip as well as right-lateral, horizontal components of movement.

(3) It is considered unlikely that the large (70–95 Ma) difference in $^{40}Ar/^{39}Ar$ white mica

ages between the three samples along the SFDZ in the east and the single sample along the MZ in the west is a result of a single cooling event.

An alternative interpretation is favoured here which involves polyphase, Sveconorwegian deformation with a possible variation in the timing of the early phase of deformation across the Eastern Segment of the orogen from west to east. In this model, it is suggested that older, compressional deformation and metamorphism related to a crustal thickening event initiated prior to $c.$ 1009 Ma in the Eastern segment. In the vicinity of Kristinehamn, the rocks were not affected by this older deformation at temperatures above the closing temperature for argon diffusion in hornblende ($c.$ 500°C) after $c.$ 965 Ma. This is in contrast to the younger tectonothermal evolution associated with movement along the SFDZ where deformation at temperatures >350°C occurred at or prior to $c.$ 930–905 Ma.

Further evidence for a polyphase Sveconorwegian tectonothermal history has been obtained from $^{40}Ar/^{39}Ar$ geochronology in the southernmost part of the Eastern Segment (Page et al. 1993; in press). In this area, the rocks display a higher grade of metamorphism and represent a deeper crustal section. A bimodal distribution of hornblende ages around 1000–970 Ma and 930–915 Ma has been obtained and has been interpreted to be related to crustal thickening and uplift events, respectively. The regional consistency of ages from the results presented in this paper and by Page et al. (1993, in press), together with new unpublished results along a transect south of lake Vänern, demonstrate the regional tectonic significance of this early ($c.$ pre-1009 Ma) crustal thickening event and associated metamorphism. This regional consistency precludes any suggestion that the older Sveconorwegian ages obtained within the Eastern Segment can be attributed to excess argon. Ongoing work strives to gain better control on the diachroneity at different crustal depths within the Eastern Segment, to examine more white mica samples along the MZ, and to extend the $^{40}Ar/^{39}Ar$ data bank to the Median Segment in order to compare the tectonothermal histories between the Median and Eastern segments.

Conclusions

The results presented in this study are among the first to be obtained using the $^{40}Ar/^{39}Ar$ technique within the Swedish segment of the Baltic Shield and demonstrate the usefulness of this method within this tectonically complex region. Our results suggest:

(1) the ages obtained from hornblendes which define the older, greenschist- to amphibolite-facies regional foliation between the SFDZ and the MZ constrain this deformation to be Sveconorwegian.

(2) The ages obtained from white micas in the area affected by the SFDZ suggest that Sveconorwegian deformation extends at least $c.$ 40 km east of the 'Protogine Zone', thus supporting the conclusions of Berthelsen (1980) and Wahlgren et al. (1994).

(3) The $^{40}Ar/^{39}Ar$ results provide age constraints on a polyphase Sveconorwegian tectonothermal evolution, with an early compressional phase at, or prior to, $c.$ 1009 Ma associated with the main regional deformation and metamorphism, and a later compressional event associated with movement along the SFDZ at, or prior to, $c.$ 930–905 Ma.

We are grateful to the Geological Survey of Sweden for their funding of this research via an internal Research and Development grant to MBS and CHW and an external grant to LMP for field and analytical costs. We also thank Stefan Claesson and David Gee for their comments on an earlier version of this manuscript. The reviewers, Karl-Inge Åhäll and J.-A. Wartho, are gratefully acknowledged for their useful comments and suggestions for improvement of this manuscript.

Appendix 1: sample descriptions and ages

Hornblende: Northern group

V93-4 (Coordinates in national grid system = 665150/135150) – Massive to weakly foliated, generally even-grained metadolerite composed predominantly of hornblende and plagioclase feldspar. Biotite, quartz and opaque minerals occur in minor amounts, and the biotite shows a preferred orientation. Apatite occurs as an accessory mineral. The plagioclase feldspar usually displays a slight sericitization, and the biotite locally shows incipient chloritization.

V93-3 (664750/135310) – Foliated metadolerite in which the main constituents are hornblende, plagioclase and biotite with minor quartz. Accessory minerals include sphene, apatite, epidote and opaques. The plagioclase is often slightly sericitized. The foliation is defined by the alignment of hornblende and biotite. Sphene is also aligned in the foliation. The hornblende and biotite appear to have grown synchronously. This sample yields a plateau age of 1009 ± 9 Ma accounting for 64% of the gas.

V93-2 (664175/135620) – Foliated metadolerite composed primarily of hornblende, plagioclase, biotite and minor quartz. Accessory minerals include apatite, sphene, calcite, opaques and epidote. The foliation is defined by the alignment of hornblende and biotite. Sphene is also aligned in the foliation. A weak crenulation of the foliation is locally visible. The biotite appears to have grown synchronously with the hornblende, but locally appears to have grown after. This sample yields a spectra in which the ages gradually decrease to a plateau accounting for 39% of the gas with an age of 982 ± 8 Ma.

V93-1 (654120/136025) – Massive to weakly foliated metadolerite, composed predominantly of hornblende, plagioclase and biotite, with minor epidote and quartz as well as accessory sphene, apatite and opaques. The hornblende crystals are randomly oriented, usually occuring as aggregates, but some grains are relatively large. The smaller grains of hornblende and biotite occasionally wrap around the larger grains as well as domains of plagioclase and hornblende. The aggregates of smaller grains presumably constitute recrystallized larger grains, related to grain-size reduction. The biotite is younger than and cross-cuts the hornblende. Epidote and sphene form the youngest phases and overgrow both hornblende and biotite. This sample yields a plateau age of 981 ± 6 Ma accounting for 88% of the gas released.

Hornblende: Southern group

V93-11 (660995/136515) – Foliated metadolerite containing hornblende, plagioclase and biotite with minor calcite, epidote, apatite and sphene. The foliation is defined by the alignment of hornblende and biotite which appear to have grown synchronously. This sample does not yield a plateau as the increments do not overlap within uncertainty. However the last four increments, accounting for over 50% of the gas, yield ages within the 970–990 Ma range, which we interpret as the approximate age of the sample.

V93-13 (660250/137620) – Massive to weakly deformed metadolerite in which the main constituents are hornblende and plagioclase feldspar. Biotite and quartz occur in minor amounts. Relatively large sphene grains are common. Accessory minerals include opaques (often as inclusions in sphene), apatite and some epidote. Biotite is partly chloritized and a weak sericitization of the plagioclase is also visible.

V93-15 (658720/138870) – Massive to weakly deformed metadolerite, consisting of hornblende, plagioclase, opaques and minor amounts of biotite with apatite and quartz as accessory minerals. This sample yields a plateau age of 1009 ± 8 Ma accounting for 81% of the gas released.

V93-16 (658460/139800) – Massive to weakly deformed metadolerite composed mainly of hornblende and plagioclase feldspar. The hornblende forms both large grains and aggregates of smaller grains. Porphyroblasts of garnet, which appear to form the youngest mineral phase and overgrow the surrounding minerals, frequently occur. Biotite, quartz and opaque minerals occur in minor amounts. The plagioclase feldspar displays a slight sericitization.

24A (658965/140085) – Massive metadolerite composed of hornblende, plagioclase and biotite with rather frequent opaques and minor amounts of quartz and apatite. Hornblende and biotite appear to have grown synchronously. This sample does not yield a plateau age, but approximately 75% of the gas yields ages between 960–990 Ma, which we consider to represent the approximate age of the sample.

J90-8 (655010/139375) – Strongly deformed and metamorphosed, finely porphyritic, felsic intrusive rock, which contains large syn-deformational quartz-feldspar segregations in which cm-large pargasitic hornblende occurs. This sample yields a plateau age of 965 ± 6 Ma which accounts for 78% of the gas released.

J90-11 (656065/140605) – Weakly foliated metadolerite in the marginal parts of an otherwise isotropic, olivine-bearing dolerite. The metadolerite is mainly composed of hornblende, plagioclase feldspar and biotite. The hornblende is sometimes intergrown with biotite and commonly occurs as aggregates of small grains, but larger grains are also present. Quartz, opaque minrals and chlorite occur in minor amounts. The chlorite forms large grains, comprises the youngest phase in the sample and overgrows biotite and hornblende. Scattered grains of sphene and porphyroblasts of garnet, which predate the chlorite, are also present.

White mica

V93-7 (665080/132515) – Strongly deformed and dynamically recrystallized, mica-rich orthogneiss within the MZ. The phyllonitic orthogneiss is composed of quartz, plagioclase, white mica, chlorite and opaques with minor epidote and garnet. The foliation is defined by the strong alignment of white mica, chlorite and recrystallized quartz ribbons. Plagioclase is typically sericitized. The white mica and chlorite are sometimes intergrown, but are more usually distinct. Pure white mica was isolated from both chlorite and intergrown white mica-chlorite grains during the separation procedure. This sample yields a plateau age of 999 ± 6 Ma accounting for 99% of the gas released.

89075 (657240/141600) – Strongly deformed and dynamically recrystallized, fine-grained, felsic intrusive rock within the area affected by the SFDZ, consisting of quartz, microcline, plagioclase, white mica and biotite with accessory apatite. The plagioclase is slightly sericitized. The foliation is defined by the strong alignment of white mica and biotite, together with the alignment of recrystallized quartz ribbons. The neocrystallized white mica often overgrows the biotite, which is inferred to be mostly reoriented and only partly recrystallized. The sample yields a plateau age of 930 ± 6 Ma for 93% of the gas released.

J90-13 (655550/141405) – Strongly deformed and hydrothermally altered, porphyritic, felsic intrusive rock within the area affected by the SFDZ, consisting of quartz, white mica and biotite with scattered opaques. The foliation is defined by a strong preferred orientation of white mica and biotite. The white mica replaces and overgrows the biotite, and both micas are locally kinked and crenulated. The quartz grains are variably undulose. This sample yields a plateau age of 905 ± 5 Ma which accounts for 71% of the gas released.

J90-6 (657815/143105) – Strongly deformed and dynamically recrystallized, fine-grained, phyllonitic metasupracrustal rock within the area affected by the SFDZ, composed of quartz, plagioclase, white mica and biotite. The white mica is typically fine-grained and fibrous, but larger grains also occur. It is usually intimately intergrown with biotite. Large pre-tectonic garnets now occur as disintegrated porphyroclasts. The foliation is defined by the strong alignment of white mica, biotite and recrystallized quartz ribbons. The micas are locally kinked and crenulated. Together with the quartz ribbons, they bend around large porphyroclasts of plagioclase and garnet as well as aggregates of quartz and feldspar. This sample yields a plateau age of 918 ± 8 Ma representing 96% of the gas released.

References

ANDERSSON, U. B., LARSON, L. & WIKSTRÖM, A. 1992. Charnockites, pyroxene granulites, and garnet-cordierite gneisses at a boundary between Early Svecofennian rocks and Småland-Värmland granitoids, Karlskoga, southern Sweden. *Geologiska Föreningens i Stockholm Förhandlingar*, **114**, 1–15.

ANDRÉASSON, P. G. & RODHE, A. 1990. Geology of the Protogine Zone south of lake Vättern: a re-interpretation. *Geologiska Föreningens i Stockholm Förhandlingar*, **112**, 107–125.

BERGLUND, J. & CONNELLY, J. 1994. The Vårgårda intrusion; Tectonic setting, age and regional implications. *Abstracts 21: a Nordiska Geologiska Vintermötet, Lulea 1994*, 16.

BERTHELSEN, A. 1980. Towards a palinspastic tectonic analysis of the Baltic shield. *In*: COGNE, J. & SLANSKY, M. (eds) *Geology of Europe from the Precambrian to the Post-Hercynian Sedimentary Basins*. International Geologic Congress, Colloquium C6, Paris 1980, 5–21.

DALRYMPLE, G. B. & LANPHERE, M. A. 1974. $^{40}Ar/^{39}Ar$ age spectra of some undisturbed terrestrial samples. *Geochimica et Cosmochimica Acta*, **38**, 715–738.

FLECK, R. J., SUTTER, J. F. & ELLIOT, D. H. 1977. Interpretation of discordant $^{40}Ar/^{39}Ar$ age-spectra of Mesozoic tholeiites from Antarctica. *Geochimica et Cosmochimica Acta*, **41**, 15–32.

GAÁL, G. & GORBATSCHEV, R. 1987. An outline of the Precambrian evolution of the Baltic Shield. *Precambrian Research*, **35**, 15–52.

GOWER, C. F., RYAN, A. B. & RIVERS, T. 1990. Mid-Proterozoic Laurentia–Baltica: an overview of its geological evolution and a summary of the contributions made in this volume. *In*: Gower, C. F., Rivers, T. & Ryan, A. B. (eds) *Mid-Proterozoic Laurentia–Baltica*. Geological Association of Canada Special Paper **38**, 1–20.

HANSEN, B. T. & LINDH, A. 1991. U–Pb zircon age of the Görbjörnarp syenite in Skåne, southern Sweden. *Geologiska Föreningens i Stockholm Förhandlingar*, **113**, 335–337.

HARRISON, T. M. 1981. Diffusion of ^{40}Ar in hornblende. *Contributions to Mineralogy and Petrology*, **78**, 324–331.

JARL, L.-G. 1992. New isotope data from the Protogine Zone and southwestern Sweden (Abstract). *Geologiska Föreningens i Stockholm Förhandlingar*, **114**, 349–350.

JOHANSSON, Å. 1990. Age of the Önnestad Syenite and some gneissic granites along the southern part of the Protogine Zone, southern Sweden. *In*: GOWER, C. F., RIVERS, T. & RYAN, A. B. (eds) *Mid-Proterozoic Laurentia–Baltica*. Geological Association of Canada Special Paper **38**, 131–148.

—— & JOHANSSON, Å. 1990. Isotope geochemistry and age relationships of mafic intrusions along the Protogine Zone, southern Sweden. *Precambrian Research*, **48**, 395–414.

—— & —— 1993. U–Pb age of titanite in the Mylonite Zone, southwestern Sweden. *Geologiska Föreningen i Stockholm Föhandlingar*, **115**, 1–7.

—— & KULLERUD, L. 1993. Late Sveconorwegian metamorphism and deformation in southwestern Sweden. *Precambrian Research*, **64**, 347–360.

——, LINDH, A. & MÖLLER, C. 1991. Late Sveconorwegian (Grenville) high-pressure granulite facies metamorphism in southwest Sweden. *Journal of Metamorphic Geology*, **9**, 283–292.

LANPHERE, M. A. & DALRYMPLE, G. B. 1978. *The use of $^{40}Ar/^{39}Ar$ data in evaluation of disturbed K–Ar systems*. US Geological Survey Open-file Report **78–701**, 241–243.

LARSON, S. Å., BERGLUND, J., STIGH, J. & TULLBORG, E. L. 1990. The Protogine Zone, southwest Sweden: A new model-an old issue. *In*: GOWER, C. F., RIVERS, T. & RYAN, A. B. (eds) *Mid-Proterozoic Laurentia–Baltica*. Geological Association of Canada Special Paper **38**, 317–333.

——, Stigh, J. & Tullborg, E. L. 1986. The deformation history of the eastern part of the Southwest Swedish Gneiss Belt. *Precambrian Research*, **31**, 237–257.

LINDH, A. 1974. The mylonite zone in south-western Sweden (Värmland). A reinterpretation. *Geologiska Föreningens i Stockholm Förhandlingar*, **96**, 205–228.

LUNDEGÅRDH, P. H., LINDH, A. & GORBATSCHEV, R. 1992. Map of the bedrock geology of Värmland County. *Sveriges Geologiska Undersökning*, **Ba 45**.

MAGNUSSON, N. H. 1937. Den centralvärmländska mylonitzonen och dess fortsättning i Norge. *Geologiska Föreningens i Stockholm Förhandlingar*, **59**, 205–228.

—— 1960. Age determinations of Swedish Precambrian rocks. *Geologiska Föreningens i Stockholm Förhandlingar*, **82**, 407–432.

McDOUGALL, I. & HARRISON, T. M. 1988. *Geochronology and Thermochronology by the $^{40}Ar/^{39}Ar$ Method*. Oxford University Press, Oxford.

MORTHORST, J. R., ZECK, H. P. & LUNDEGÅRDH, P. H. 1983. The Proterozoic hyperites in southern

Värmland, western Sweden. *Sveriges Geologiska Undersökning*, **Ba 30**.

OEN, I. S. 1982. Isotopic age determinations in Bergslagen, Sweden: the Filipstad-type granite of Rockesholm, Grythyttan area. *Geologie en Mijnbouw*, **61**, 305–307.

PAGE, L., JOHANSSON, L. & MÖLLER, L. 1993. $^{40}Ar/^{39}Ar$ geochronology across the Mylonite Zone in western Halland and Bohuslän, southern Sweden (abstract). *Sveriges Geologiska Undersökning Rapporter och Meddelanden nr 76*, 14.

——, MÖLLER, L, & JOHANSSON, L. in press. $^{40}Ar/^{39}Ar$ geochronology across the Mylonite Zone and the Southwestern Granulite Province in the Sveconorwegian orogen of S. Sweden. *Precambrian Research*.

PARK, R. G., ÅHÄLL, K. I. & BOLAND, M. P. 1991. The Sveconorwegian shear-zone network of SW Sweden in relation to mid-Proterozoic plate movements. *Precambrian Research*, **49**, 245–260.

PATCHETT, P.J. 1978. Rb/Sr ages of Precambrian dolerites and syenites in southern and central Sweden. *Sveriges Geologiska Undersökning*, **C747**.

—— & BYLUND, G. 1977. Age of Grenville belt magnetization: Rb–Sr and palaeomagnetic evidence from Swedish dolerites. *Earth and Planetary Science Letters*, **35**, 92–104.

PIPER, J. D. A. 1983. Dynamics of continental crust in Proterozoic times. *In*: MEDARIS, L. J., MICKELSON, D. M. & SHANKS, W. C. (eds) *Proterozoic Geology: Selected Papers from an International Proterozoic Symposium*. Geological Society of America Memoir, **161**, 11–34.

SAMSON, S. D. & ALEXANDER, E. C. 1987. Calibration of the interlaboratory $^{40}Ar/^{39}Ar$ dating standard, MMhb-1. *Chemical Geology*, **66**, 27–34.

STEIGER, R. H. & JÄGER, E. 1977. Subcommission on Geochronology; Convention on the use of decay constants in geo- and cosmochronology. *Earth and Planetary Science Letters*, **36**, 259–362.

STEPHENS, M. B. in press. Berggrundskartan (Bedrock map) Karlskoga NO. *Sveriges Geologiska Undersökning* Af 184.

——, WAHLGREN, C-H, WEIJERMARS, R. & CRUDEN, A. R. 1993. Sinistral transpressive deformation along the Mylonite Zone, Sveconorwegian (= Grenvillian) Province, southwestern Sweden. *Terra Abstracts. Abstract supplement No.1 to Terra Nova*, **5**, 321.

——, ——, —— & —— in press. Left-lateral transpressive deformation and its tectonic implications, Sveconorwegian orogen, Baltic Shield, southwestern Sweden. *Precambrian Research*.

VERSCHURE, R. H., OEN, I. S. & ANDRIESSEN, P. A. M. 1987. Isotopic age-determinations in Bergslagen, Sweden: Sveconorwegian Rb–Sr resetting and anomalous radiogenic argon in the Gothian TransScandinavian Småland-Värmland Granitic Belt and bordering parts of the Svecokarelian Bergslagen Region. *Geologie en Mijnbouw*, **66**, 111–120.

WAHLGREN, C.-H. 1992. Berggrundskartan (Bedrock map) Karlskoga NV. *Sveriges Geologiska Undersökning* Af 176.

—— 1993. Berggrundskartan (Bedrock map) Karlskoga SV. *Sveriges Geologiska Undersökning* Af 182.

——, CRUDEN, A. R. & STEPHENS M. B. 1994. Kinematics of a major fan-like structure in the eastern part of the Sveconorwegian orogen, Baltic Shield, south-central Sweden. *Precambrian Research*, **70**, 67–91.

——, HEAMAN, L., KAMO, S. & INGVALD, E. in press. U–Pb baddeleyite dating of dolerite dykes in the eastern part of the Sveconorwegian orogen, south-central Sweden. *Precambrian Research*.

WELIN, E. 1994. Isotopic investigations of Proterozoic igneous rocks in south-western Sweden. *Geologiska Föreningens i Stockholm Förhandlingar*, **116**, 5–86.

——, LUNDEGÅRDH, P. H. & KÄHR, A.-M. 1980. The radiometric age of a Proterozoic hyperite diabase in Värmland county, western Sweden. *Geologiska Föreningens i Stockholm Förhandlingar*, **102**, 49–52.

WIKSTRÖM, A. 1991. Berggrundskartan (Bedrock map) Karlskoga SO. *Sveriges Geologiska Undersökning* Af 183.

Palaeomagnetism and Sm–Nd ages of the Neoproterozoic diabase dykes in Laanila and Kautokeino, northern Fennoscandia

S. MERTANEN[1], L. J. PESONEN[1] & H. HUHMA[2]

[1]*Laboratory for Palaeomagnetism, Geological Survey of Finland, 02150 Espoo, Finland*
[2]*Laboratory for Isotope Geology, Geological Survey of Finland, 02150 Espoo, Finland*

Abstract: The results of palaeomagnetic, rock magnetic and isotope data on the NE–SW trending Neoproterozoic Laanila–Ristijärvi (Finland) and Kautokeino (Norway) dyke swarms, northern Fennoscandia, are reported. Positive baked contact tests in both dyke sets indicate that the characteristic remanence is primary thermoremanence. Rock magnetic and petrological properties show similarity between the two dyke swarms. All the dykes except one (where the dominant carrier is Ti-rich titanomagnetite) have mixed multidomain and single-domain or pseudosingle-domain magnetic carriers of Ti-poor magnetite. From its remanence directions, petrography and rock magnetic properties, the easternmost of the three Laanila dykes represents a deeper crustal level than the other two dykes studied. Palaeomagnetic and aeromagnetic data suggest that the Palaeoproterozoic granulite block containing the easternmost Laanila dyke was uplifted after the dykes were emplaced.

The Sm–Nd isotopic age determinations yield an age of 1042 ± 50 Ma for the Laanila dykes, 1013 ± 32 Ma for the Ristijärvi dykes and 1066 ± 34 Ma for the Kautokeino dykes indicating that the dyke swarms are of similar age within error limits. However, palaeomagnetic pole of the Laanila–Ristijärvi dykes (P_{lat} = 2.1 °S, P_{long} = 212.2 °E, dp = 12.7°, dm = 21.1°, N = 3 dykes) differs significantly from the pole of the Kautokeino dykes (Plat = 35.2 °S, P_{long} = 236.8 °E, $N = 2$ dykes). The isotope and palaeomagnetic ages of the Laanila–Ristijärvi dykes are considered agreeable, but on the basis of younger precisely dated Sveconorwegian palaeopoles, a magnetization age of $c.$ 900 Ma for the Kautokeino dykes is inferred. Three explanations for the discrepancy are suggested: (i) local tectonic movements, (ii) delayed magnetization of the Kautokeino dykes with respect to that of the Laanila–Ristijärvi dykes, (iii) the isotope ages and palaeomagnetic ages are compatible but the Fennoscandian APWP should be revised.

The northern Fennoscandian shield (Fig. 1) is mainly defined by Archaean to Palaeoproterozoic rocks about 3.1–1.7 Ga old. However, igneous activity has taken place since cratonization, as manifested by the roughly 1.0 Ga old dyke swarms, of which the Laanila dyke swarm (Pesonen *et al.* 1986) is a good example. The Laanila dykes were injected along a NE–SW trending fracture zone that partly follows the fracturing of the Palaeoproterozoic (1.95–1.8 Ga) Lapland granulite belt. In the Finnmark area, northern Norway, the Palaeoproterozoic Kautokeino and Karasjok greenstone belts are also cut by NE–SW trending dykes. On aeromagnetic maps both the Laanila and Kautokeino dykes are shown as strong negative anomalies.

One of the topics of the present study is to define the primary remanent magnetization of the Laanila–Ristijärvi and Kautokeino dykes in order to better define the Fennoscandian Apparent Polar Wander Path (APWP) at Sveconorwegian time ($c.$ 1100–850 Ma). Both dyke swarms are ideal for palaeomagnetic studies for three reasons. First, they carry stable remanence directions, which are not affected by later metamorphic events (Vihavainen 1981; Pesonen *et al.* 1986). Second, the dykes are exposed along their contacts, thus allowing baked contact tests to be made in order to establish primary intrusion magnetization. Third, they have been dated isotopically with the Sm–Nd method. Another topic is the correlation of the Laanila and Kautokeino dyke swarms in order to investigate the extent to which Neoproterozoic dyke injection took place in northern Fennoscandia. This is done by using the rock magnetic, petrophysical and petrographic properties of the dykes.

Geological setting

Laanila–Ristijärvi dykes

The Laanila diabase dyke swarm extends for over 100 km from Laanila to Ristijärvi in Finland and into northeastern Norway (Fig. 1). The dykes trend NE–SW in an en echelon-type manner. In the southwest, in Laanila, the dykes cut the Palaeoproterozoic Lapland granulite belt and in the northeast, in Ristijärvi, they cut the Archaean

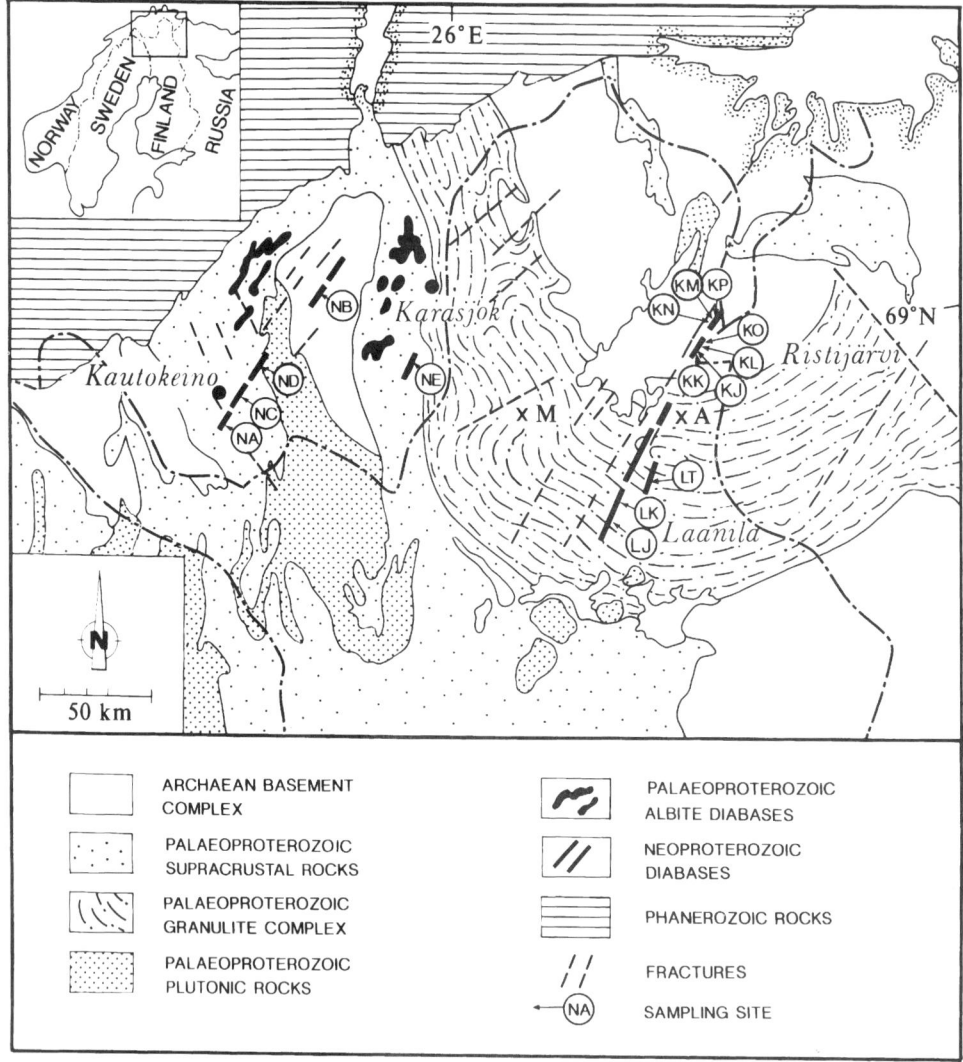

Fig. 1. Generalized geological map of the northern Fennoscandian Shield. A and M denote sampling sites at Akujärvi and Menesjärvi.

basement. The granulite belt is a high-grade metamorphic belt overthrust onto the Karelian basement in the southwest. U–Pb zircon analyses on the meta-igneous rocks of the granulite belt yield ages of 1.95–1.8 Ga, and the Archaean basement surrounding the Lapland granulite belt shows ages of 3.1–2.5 Ga (Meriläinen 1976).

The dyke swarm follows one of the main NE–SW directions of the extensive fracture zone in northern Fennoscandia. The fracture zone has been affected by long lasting repeated magmatic activity, as revealed by the roughly 1.8 Ga old post-orogenic granitoids (e.g. Haapala et al. 1987) and 300 Ma old diabase dykes on the Kola Peninsula (Gorbatschev et al. 1987; Fedotov & Amelin 1992), some 50 km from the area of our study. According to Pihlaja (1987) and Mutanen (pers. comm., 1992), the Laanila dykes represent steep fractures of the crust, extending for hundreds of metres to kilometres into the crust. The range of dyke width is in the Laanila area 1–20 m and in the Ristijärvi area 150–200 m. On aeromagnetic maps, the Laanila–Ristijärvi dykes are seen as strong negative anomalies. In chemical composition, the Laanila–Ristijärvi dykes consist of within-plate, rift-related tholeiitic basalt (Pihlaja 1987).

The diabase in Ristijärvi is an unmetamorphosed, homogeneous, medium-grained rock with ophitic texture. The grain size is fine at the margins and aphanitic right at the contact. The main minerals are plagioclase (andesine–labradorite), clinopyroxene and serpentine, with the latter having completely replaced olivine (see also Pihlaja 1987). The accessories are apatite, titanite, pyrite, pyrrhotite and magnetite. Sericite, chlorite and amphibole occur as alteration minerals and pyrite and magnetite as skeletal grains. In Laanila dyke LJ (Fig. 1), Mutanen (pers. comm., 1979) reports ilmenomagnetite composed of magnetite with ilmenite lamellae. The magnetite is partly altered to leucoxene and in dyke LJ the ilmenomagnetite is hypidiomorphic, not skeletal as in the Ristijärvi dyke. In Laanila dyke LT, the titanomagnetite is hypidiomorphic and skeletal, and has partly altered to leucoxene. Ilmenite occurs as platy separate grains. Dyke LT is better preserved than dyke LJ. However, the alteration is rather irregular from one site to another and is always related to the deuteric alteration associated with the latest phase of magma crystallization.

Kautokeino dykes

Palaeomagnetic investigations were conducted on NE–SW trending basaltic diabase dykes in the Kautokeino area of Finnmark, northern Norway (Fig. 1). Negative aeromagnetic anomalies (Olesen & Sandstad 1993) distinguish these dykes from the older (1815 ± 24 Ma, U–Pb zircon date, Krill *et al.* 1985) albite diabases, which, although similar in orientation (Fig. 1), have positive aeromagnetic anomalies. The Kautokeino dykes cut the Palaeoproterozoic Kautokeino greenstone belt and adjacent Archaean gneisses sharply (Olesen & Solli 1985; Olesen & Sandstad 1993). The Kautokeino greenstone belt forms a 40–80 km wide NNW–SSE trending rift-related synclinorium separated by steeply dipping faults from the Archaean Ráiseatnu Gneiss Complex in the west and the Jergol Gneiss Complex in the east. The area is cut by the NE–SW trending, 230 km long Mierujávri–Sværholt Fault Zone (MSFZ). Since the emplacement of the Kautokeino dykes, the greenstone belt has been tectonically active. The eastern fault of the belt shows sinistral displacement cutting the Kautokeino dykes and causing an offset of 1 km along the fault (Olesen & Sandstad 1993).

The Kautokeino dykes can be followed for about 50 km on low altitude aeromagnetic maps. The length of different dykes ranges from hundreds of metres to tens of kilometres; their width is 20–50 m and their dip varies from 60 to 70° SE. The mineral composition of the Kautokeino diabases corresponds to that of the Laanila diabases. The main minerals are plagioclase, which is is highly altered to sericite and saussurite, and clinopyroxene. Olivine phenocrysts are totally replaced by serpentine and chlorite. The main accessories are biotite, amphibole, magnetite and pyrrhotite. The magnetite occurs partly as skeletal and partly as subhedral or very small grains. The texture is ophitic.

Karasjok dykes

Midtun (1988) reports at least two SW–NE striking diabase dykes cutting the Karasjok greenstone belt some 80 km from the Kautokeino greenstone belt. These dykes, too, show up as strong negative aeromagnetic anomalies. Two pilot samples (provided by O. Olesen) from one of the dykes (NE, Fig. 1) were measured at the Geological Survey of Finland (GSF) and will be reported in this study.

Sampling

Laanila and Ristijärvi dykes

Palaeomagnetic investigations on the Laanila dykes started in 1978 (Vihavainen 1981; Neuvonen, pers. comm., 1986; Pesonen *et al.* 1986), when the first samples were taken from two Laanila dykes and host rocks at sites LJ, LK and LT (see Tables 2 and 4). In 1987, more samples were taken at site LK. Sites LJ and LK (hereafter called dyke LJ) are exposures of the same dyke, from which altogether 30 samples were taken. Five samples were taken from the host granulite at dyke LJ. Dyke LT, where eleven samples were taken, is located about 20 km east of dyke LJ. At site LT, four samples were taken from the host granulite. In 1988, the investigations were extended to Ristijärvi (close to the Russian border), which is a continuation of the Laanila dyke. At Ristijärvi (dyke RL), 26 diabase samples were taken from five sites; 13 samples were taken from host amphibolite at two sites at the contact and one sample was taken from the host Archaean gneiss.

Kautokeino and Karasjok dykes

In the Kautokeino area, diabase samples were taken at two sites (NA and NB, Fig. 1). At site NA, ten samples were taken from the diabase dyke and four samples at varying distances from the host Palaeoproterozoic amphibolite (see Tables 3 and 5). Site NB lies about 30 km northeast from site NA and cuts the Archaean Jergol gneiss. Nine samples were taken from the diabase and four from the gneiss. In addition, O. Olesen (of Geological Survey of Norway) provided three pilot samples from site NB, two from sites NC and ND (Fig. 1) in the Kautokeino area and two from a diabase dyke (NE; Fig. 1) in the Karasjok area.

Sm–Nd Ages

Isotopic dating of the dykes involved Sm–Nd analyses on whole rocks and concentrates of fresh magmatic clinopyroxene and slightly altered plagioclase. The analyses have been performed in three stages: (1) old analyses on the Laanila dyke using non-commercial mass-spectrometer (Huhma 1986); (2) analyses using VG SECTOR 54 with old Faraday buckets (before Aug. 94); and (3) recent analyses after Faraday bucket replacement. We have noticed that analyses with old buckets can be seriously biased, even after standard correction (cf. Thirlwall 1991). Therefore we have not used those data for age determination.

The whole rocks and clinopyroxenes from the Laanila dyke give a Sm–Nd age of 1042 ± 50 Ma with an initial epsilon of $+4.8 \pm 0.5$ (n = 5, MSWD = 1.4). The new data on the Ristijärvi dyke provide an age of 1013 ± 32 Ma $\varepsilon_{Nd} = +4.8$, n = 3). Due to poor analyses on plagioclase (too small sample size) this age is in fact based only on whole rock and pyroxene analyses. All nine analyses listed in Table 1 (and thus including also potentially biased data) give an age of 1005 ± 43 Ma (n = 9, MSWD = 4). The five new analyses on the Kautokeino dyke give an age of 1066 ± 34 Ma

Table 1. *Sm–Nd data on whole rocks and mineral concentrates.*

Sample	Fraction	Sm (ppm)	Nd (ppm)	$^{147}Sm/^{144}Nd$	$^{143}Nd/^{144}Nd \pm 2\sigma m$
Laanila diabase, samples LJ10 (= A540) and LT2:					
[2]LJ10 wr		5.58	17.71	0.1905	0.512835 ± 34
[2]LJ10 cpx	hp	2.43	5.13	0.2870	0.513457 ± 56
[2]LT2 wr		4.49	14.25	0.1905	0.512857 ± 28
[2]LT2 cpx	hp	3.97	9.01	0.2667	0.513363 ± 32
LJ10 cpx#2	hp	2.39	4.90	0.2944	0.513563 ± 10
Ristijärvi diabase, sample A1227 (= KJ2-1B/4821):					
[1]wr		5.47	17.05	0.1941	0.512885 ± 20
[1]wr#2		5.29	16.44	0.1945	0.512898 ± 20
[1]pl	2.62–2.76/nhp	0.18	0.75	0.1414	0.512580 ± 30
[1]pl#2	2.62–2.76/hp	0.17	0.74	0.1423	0.512603 ± 90
[1]cpx	+3.4/m 0.6A/nhp	3.74	7.60	0.2974	0.513600 ± 20
[1]cpx#2	3.4/m 0.6A/hp	3.36	6.79	0.2994	0.513575 ± 20
wr#3		5.31	16.61	0.1931	0.512865 ± 10
pl#3	2.62-2.76	0.21	0.87	0.1486	0.512576 ± 133
cpx#3	+3.4/m 0.6A/nhp	3.90	8.21	0.2877	0.513494 ± 10
Kautokeino diabase A1228 (= NA2-1E/2823):					
[1]wr		5.26	15.30	0.2078	0.513030 ± 20
[1]pl	2.65–2.67/+100/hp	0.94	3.41	0.1663	0.512789 ± 20
[1]cpx	3.3–3.4/m 0.6A/hp	2.83	5.77	0.2970	0.513603 ± 20
wr#2		5.18	15.13	0.2070	0.513023 ± 20
pl#2	2.62–2.75/-160/nhp	0.71	2.46	0.1739	0.512775 ± 11
pl#3	2.62–2.75/-160/nhp	0.45	1.60	0.1693	0.512736 ± 10
cpx#2	3.3–3.4/m 0.6A/hp	3.15	6.81	0.2795	0.513520 ± 16
cpx#3	3.3–3.4/m 0.6A/hp	2.97	6.15	0.2923	0.513596 ± 10

Sample: wr = whole rock, pl = plagioclase, cpx = clinopyroxene, #2 = duplicate.
Fraction: density/m = magnetic/+160 = ∅>160 µm/ hp = handpicked, nhp = not handpicked.
Concentrations were determined using isotope dilution by addition of a mixed ^{149}Sm–^{150}Nd spike without aliquoting (except old analyses on the Laanila dyke/[2]). Based on several duplicated analyses in the laboratory the estimated error in $^{147}Sm/^{144}Nd$ is 0.4%. $^{143}Nd/^{144}Nd$ ratio is normalized to $^{146}Nd/^{144}Nd = 0.7219$. The new set of samples was analysed using VG SECTOR 54 with new Faraday buckets and the average value for La Jolla standard was $^{143}Nd/^{144}Nd = 0.51185 \pm 1$ (SD, n = 20, errors in last significant digits).
[1] = Accuracy problems due to old Faraday buckets in VG SECTOR 54 (analysed before Aug. 94). Based on frequent analyses on La Jolla standard these data should be compatible with La Jolla value $^{143}Nd/^{144}Nd = 0.51185$. The data with new collectors show that this correction was not applicable for all analyses. These data are not shown in figures nor used for age determinations.
[2] = Old analyses using non-commercial mass-spectrometer (for methods see Huhma 1986).

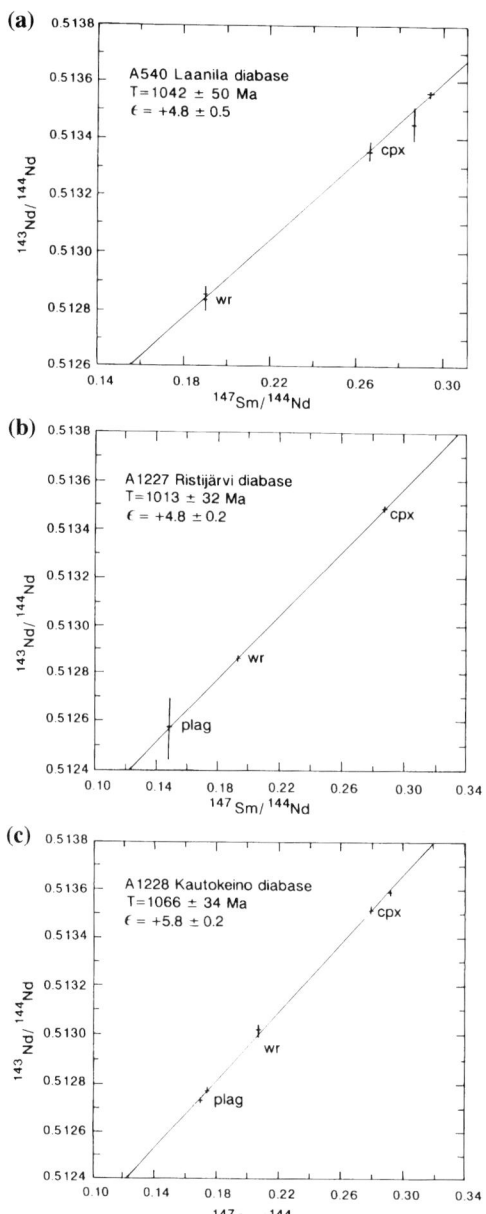

Fig. 2. Sm-Nd diagrams for (a) Laanila dykes, (b) Ristijärvi dykes and (c) Kautokeino dykes.

$\varepsilon_{Nd} = +5.8 \pm 0.2$, MSWD = 2). The old analyses on plagioclase and pyroxene are off the line.

The Sm–Nd ages of the dykes analysed are thus indistinguishable within error. The nearly chondritic Sm/Nd ratios and positive ε_{Nd} values suggest that the mafic magma originated from depleted mantle reservoirs. The ε_{Nd} for the Kautokeino dyke is close to the depleted mantle of DePaolo (1981), whereas ε_{Nd} for the Ristijärvi and Laanila dykes is slightly lower suggesting another source or some crustal contamination.

Palaeomagnetic measurements and analysis

The remanences were measured using a spinner magnetometer built at the Geological Survey of Finland (GSF) (see Pesonen et al. 1983). About half of the specimens from each core were demagnetized with an alternating field (AF) using a Schonstedt or a homebuilt AF demagnetizer. The other half were thermally demagnetized by a Schonstedt thermal demagnetizer. In AF demagnetization, most specimens were demagnetized in 12 steps of 2.5, 5, 10, 20, 30, 40, 50, 60, 70, 80, 90 and 100 mT. Some samples were AF demagnetized up to 110 mT, some only up to 40–60 mT. In thermal demagnetization, temperatures of 200, 300, 350, 400, 500, 520, 540, 560, 570, 580, 600, 620, 640, 660 and 680°C were used. To establish whether mineralogical changes occurred upon heating, bulk susceptibility was measured after demagnetization at temperatures of 200, 300, 400, 500, 570, 600, 620, 640 and 680°C. A joint analysis of stereographic plots, demagnetization decay curves, orthogonal demagnetization diagrams (Zijderveld 1967; Leino 1991) and principal component analysis (Kirschvink 1980) were used to obtain the remanence directions. These were accepted if the angular deviation (MAD) was ≤°6. Mean remanence directions were calculated using Fisher (1953) statistics.

To distinguish the remanence carriers, magnetic hysteresis measurements were conducted on nine specimens (at the Technical Research Centre of Finland) representing characteristic remanence behaviour of the different sites. Another seven specimens were subjected to saturation isothermal remanent magnetization (SIRM) and anhysteretic remanent magnetization (ARM) analysis for conducting Lowrie–Fuller tests. Magnetic minerals were identified measuring thermomagnetic curves (low-field susceptibility versus temperature) on five specimens using a KLY-2 kappabridge connected to a CS-2 furnace (Geofyzika Brno). The measurements were made at a temperature range of 30–700 °C.

Palaeomagnetic results

Laanila–Ristijärvi dykes

According to their palaeomagnetic directions, the sampled Laanila–Ristijärvi dykes represent three

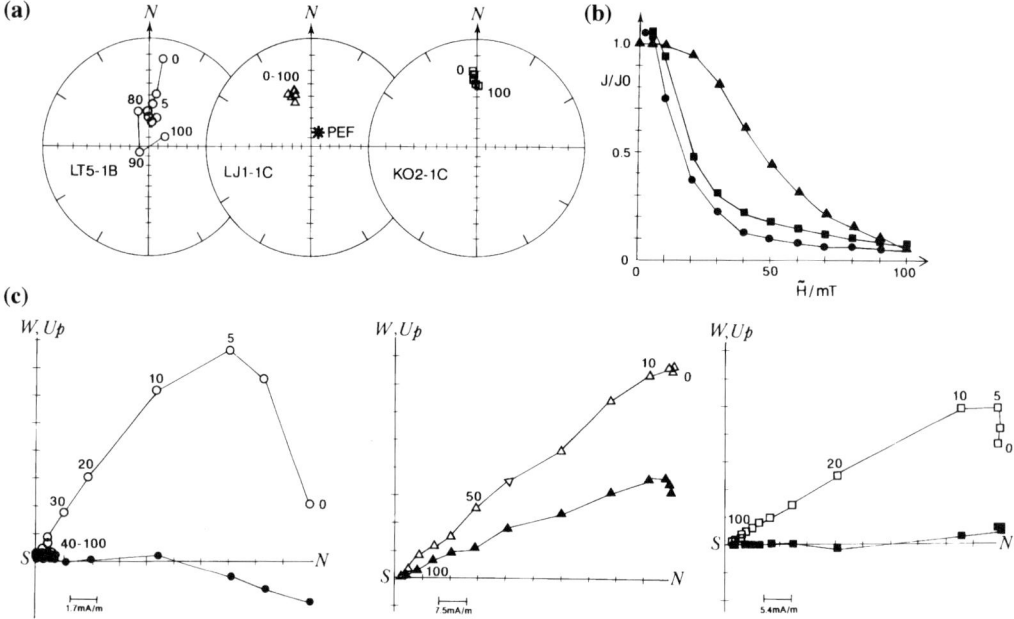

Fig. 3. Examples of AF demagnetization behaviour typical of Laanila–Ristijärvi dykes. (**a**) Stereographic projections of directional data on demagnetization, where PEF shows direction of present Earth's magnetic field; (**b**) relative intensity decay curves; (**c**) orthogonal demagnetization diagrams. Open (closed) symbols denote vertical (horizontal) planes. Numbers by each demagnetization step denote peak alternating field (mT).

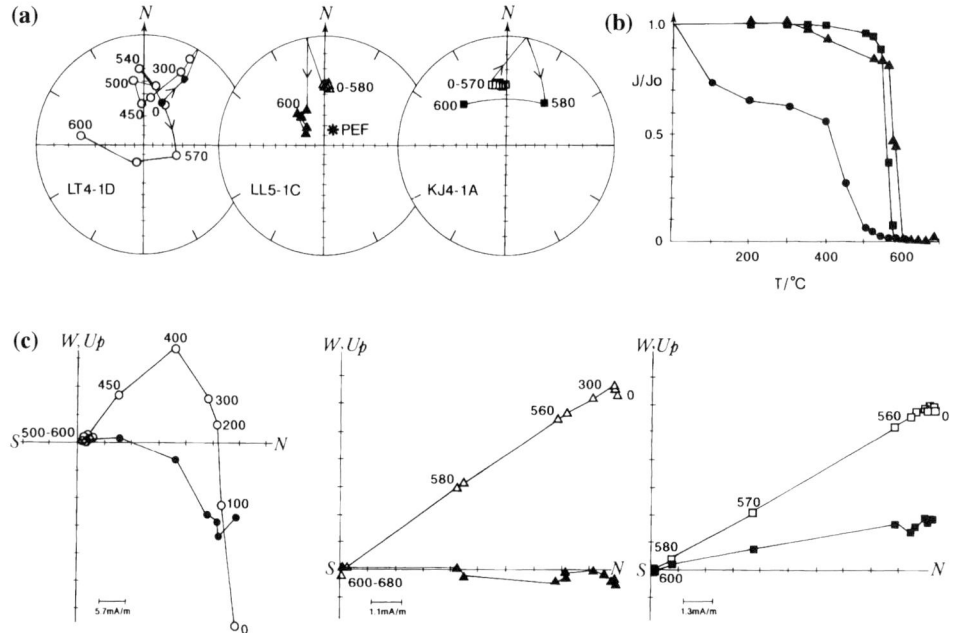

Fig. 4. Examples of thermal demagnetization behaviour typical of Laanila–Ristijärvi dykes. (**a**) Stereographic projections; (**b**) relative intensity decay curve; (**c**) orthogonal demagnetization diagrams. Numbers by each demagnetization step denote temperature (°C). Other symbols as in Fig. 3.

Table 2. Mean palaeomagnetic data on Laanila–Ristijärvi dykes and host rocks

Site/Dyke	B/N/n	D (°)	I (°)	α_{95} (°)	k	P_{lat} (°N)	P_{long} (°E)	dp (°)	dm (°)	(A_{95}) (°)
Diabase										
Mean LT (68.5, 28.15)	1/*9/33	357.3	−52.5	8.7	36.3	−11.6	210.4	8.2	11.9	
				Laanila						
LK	*15/57	350.7	−34.4	3.8	100.9	2.4	216.3	2.5	4.4	
LJ	*13/37	350.3	−34.4	7.7	30.1	2.4	216.7	5.0	8.8	
Mean LJ (68.4, 27.5)	*2/28/84	350.5	−34.4	—	—	2.4	216.5	—	—	
				Ristijärvi						
KJ	*5/17	355.6	−30.3	10.3	56.4	4.7	213.1	6.4	11.4	
KK	*1/3	355.5	−31.1	—	—	4.1	213.2	—	—	
KO	*5/15	1.7	−30.8	2.6	675.6	4.4	207.3	1.6	2.9	
KL	*6/11	4.1	−40.0	5.6	145.0	−1.8	205.2	4.0	6.7	
KN[1]	*4/10	254.8	25.4	14.7	40.3	7.1	317.8	8.5	15.8	
Mean RL (69.0, 28.9)	*4/17/46	359.1	−33.1	6.7	189.9	2.9	209.8	4.3	7.6	
Grand mean (68.7, 28.1)	*3/7/54/163	355.5	−40.0	17.5	50.5	−2.1	212.2	12.7	21.1	(13.8)
				Host rocks						
				Laanila						
Granulite										
LJ8-9, contact	*2/5	7.3	−33.3	—	—	3.2	200.6	—	—	
LJ19, 7 m	*1/4	21.3	−37.9	—	—	−1.0	187.7	—	—	
LT10, contact	*1/1	88.5	−50.6	—	—	—	—	—	—	
				Ristijärvi						
Amphibolite										
KM2-6, contact	*5/10	344.0	−16.6	8.4	84.5	11.7	225.1	4.5	8.6	
Gneiss										
KL11, contact	*1/3	27.3	−10.2	—	—	13.5	180.9	—	—	

NOTE: Numbers below the name of the dyke denote geographical latitude (°N) and longitude (°E) of the sampling site; for the host rocks, distance from the dyke contact is shown; B/N/n is the number of sites/samples/specimens used in statistical calculations; * denotes the statistical level used in mean calculations; [1] not included in mean direction; D and I are mean declination and inclination, respectively; α_{95} is the radius of the circle of 95% confidence; k is Fisher's (1953) concentration parameter; P_{lat} and P_{long} are the palaeolatitude and palaeolongitude of the virtual geomagnetic poles (VGPs) of different sites; the mean pole is the mean of VGPs; dp and dm are the semi-axes of the oval of 95% confidence for each VGP; (A_{95}) is the radius of the circle of 95% confidence of the mean VGP poles. All poles are of normal (N) polarity.

separate dykes: LT and LJ (in the Laanila area) and RL (in the Ristijärvi area). Examples of alternating field (AF) and thermal demagnetizations of the dykes are given in Figs 3 and 4. The dykes carry a coherent northward characteristic remanent magnetization (ChRM) with moderate negative inclination that is most stable and best isolated at sites LJ and LK in Laanila and at site KJ in Ristijärvi (Table 2). In addition, a steep positive component pointing northward was isolated in all specimens at low coercivities and low unblocking temperatures. The direction of this component is close to the present Earth's magnetic field (PEF) and is therefore interpreted as representing recent viscous remanent magnetization (VRM). Most specimens exhibit consistent behaviour upon AF and thermal demagnetization, indicating that both cleaning methods effectively remove the PEF and isolate the ChRM. In some samples, however, AF cleaning was more effective at isolating the ChRM from the PEF component.

Dyke LT has a very large contribution of the present Earth's magnetic field (PEF) component. The inclination of the natural remanent magnetization (NRM) direction before demagnetization is therefore positive. In dykes LJ and RL, the relative amount of PEF is lower and the inclination of NRM is negative at most sites. The mean inclination of the characteristic high coercivity remanence component at site LT is about −50° (Table 2), which is at a high angle (almost opposite) to the PEF direction. In the two other dykes, the inclination is shallower (Table 2) and could be due to incomplete cleaning, the component contaminated by PEF magnetization, which would make the upward characteristic component shallower. However, because both components are sharply isolated in orthogonal plots (Fig. 3) the inclination difference is probably not due to incomplete demagnetization but is real.

Thermal demagnetization decay curves for specimens of dyke LT indicate that the remanence carriers have more widely distributed unblocking temperatures than do those of the two other Laanila dykes (Fig. 4), clearly owing to the larger grain size distribution of the LT specimens. This is consistent with petrophysical and hysteresis properties as will be shown later. The scatter of remanence directions in dyke LT is comparatively high. The two other Laanila dykes show narrow unblocking temperature spectra, the bulk being unblocked between 560° and 600° C indicative of Ti-poor magnetite. The clear change in the slope of the thermal demagnetization curve at about 350° C in some samples may be due to pyrrhotite.

The mean remanence directions at sites LJ and LK in Laanila dyke LJ are almost identical (Table 2) confirming that the sites represent exposures of the same dyke. The remanence direction of Ristijärvi dyke RL agrees with that for dyke LJ of Laanila within the α_{95} confidence level (Fig. 5). In contrast, the inclination of dyke LT is about 20° steeper than that of dykes LJ or RL. The between-dyke scatter may be due to secular variation, but the possibility of tectonic influence or a slight age difference between dykes LT and LJ–RL cannot be ruled out, as will be discussed later. The mean direction and the corresponding palaeomagnetic pole, calculated for a common reference locality, are averaged for the LT, LJ and RL dykes (Table 2, Fig. 5).

Kautokeino dykes

The characteristic remanent magnetization (ChRM) of the Kautokeino dykes after demagnetization has a steep northward negative inclination, typically reached as a stable end point at 100 mT (Fig. 6). All samples also exhibit a pronounced steep downward direction that is removed in low AF fields (5–30 mT). Since it is aligned approximately along the present Earth's field direction (being almost antiparallel to the characteristic remanence direction), it is most likely a viscous remanent magnetization of recent origin. Depending on the magnitude of the PEF component, the intensity decay curves and orthogonal plots behave very

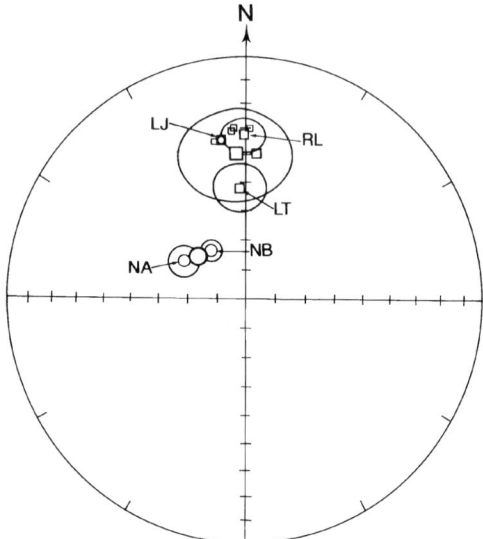

Fig. 5. Mean remanence directions and α_{95} confidence circles for Laanila–Ristijärvi (squares) and Kautokeino (circles) diabase dykes. Smallest symbols denote site means, intermediate symbols dyke means and largest symbols dyke swarm mean directions.

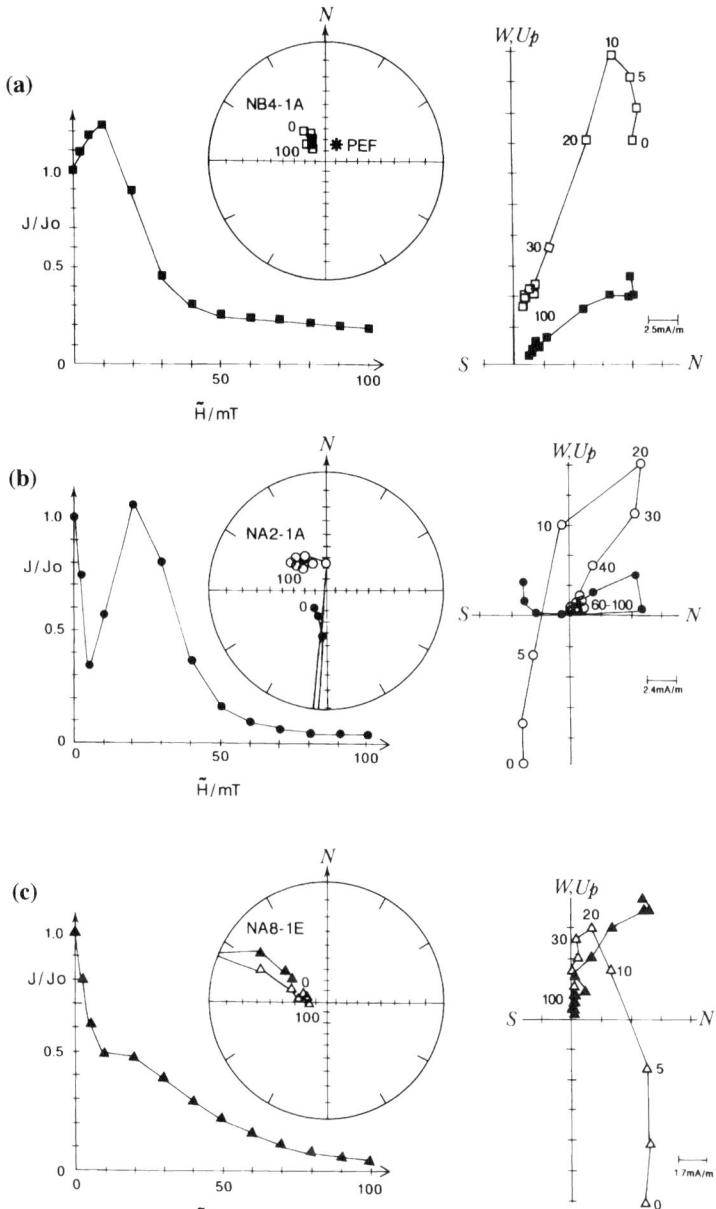

Fig. 6. Examples of AF demagnetization behaviour for the Kautokeino dykes NA and NB. Symbols as in Fig. 3.

differently (Fig. 6a–c) in AF demagnetization. In some of the samples, the magnitude of PEF is so high that the NRM direction (before demagnetization) has a steep to moderate positive inclination. Thermal demagnetization (Fig. 7) does not always isolate the ChRM as effectively as AF demagnetization, and the data are more scattered. Laboratory unblocking temperatures of 520–580° C indicate that titanomagnetite is the remanence carrier.

The mean remanence directions and their statistics are summarized in Table 3. The dyke mean is averaged both for the two dykes (NA and NB) and for all samples. The difference in directions between dykes is small and can be attributed to secular variation.

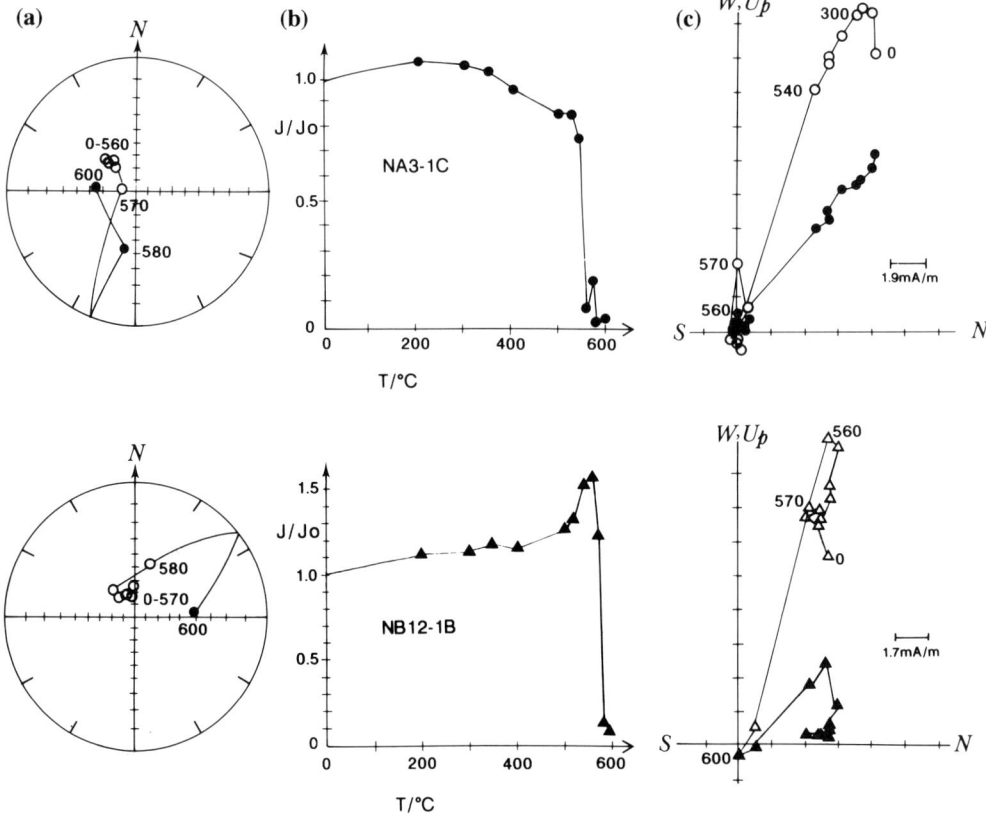

Fig. 7. Examples of thermal demagnetization behaviour typical of Kautokeino dykes. Symbols as in Figs 3 and 4.

Palaeomagnetic data on two pilot samples taken from two other sites of the Kautokeino dyke are also shown in Table 3. They are not included in the statistics because only one sample was measured from both dykes. The remanence directions agree well with those at sites NA and NB. Sample ND was taken from the eastern side of the sinistral fault transecting the diabase dyke (Olesen & Sandstadt 1993), and site NA from the western side of the fault. The remanence directions are identical within the limits of error and, thus, no fault movement can be observed in the palaeomagnetic data. Olesen & Sandstadt (1993) also report NRM results for the Kautokeino dyke, obtaining a direction (D = 336°, I = −56°, α_{95} = 9°, n = 11 samples) consistent with our results, but shallower inclination (12°), presumably due to the large, not erased PEF component (see also pole OS in Fig. 19).

Karasjok dykes

Palaeomagnetic data on two pilot samples from one diabase dyke in the Karasjok area are also given in Table 3 and they are in good agreement with the Kautokeino data. The NRM direction (D = 315°, I = −63°, α_{95} = 7.1°, n = 6 samples) reported by Midtun (1988) for two more diabase dykes in the Karasjok area is also comparable to the mean remanence direction (see also pole MI in Fig. 19) of the Kautokeino dykes. The agreement between the data on the Karasjok dykes, which are located between Kautokeino and Laanila, indicates that dykes with a similar reversed magnetization direction occur over a wide area in northern Fennoscandia and that they are probably contemporary in age and can thus be correlated.

Bulk petrophysics

Petrophysical properties (susceptibility, remanence intensity and density) were studied to correlate different dykes within a dyke swarm (e.g. Laanila) and between dyke swarms (Laanila vs. Kautokeino). Petrophysical properties are informative about the remanence stability, mineralogy and grain size of the rocks. A susceptibility versus

Table 3. *Mean palaeomagnetic data on Kautokeino and Karasjok dykes and host rocks*

Dyke	B/N/n	D (°)	I (°)	α₉₅ (°)	k	P$_{lat}$ (N°)	P$_{long}$ (E°)	dp (°)	dm (°)	(A$_{95}$) (°)
				Kautokeino						
Diabase										
NA	1/*9/29	306.2	−67.4	5.2	97.9	−35.5	242.8	7.2	8.7	
(69.0, 23.3)										
NB	1/*12/19	322.3	−68.8	3.6	150.8	−34.5	230.9	5.1	6.0	
(69.2, 23.8)										
Mean	*2/21/48	314.0	−68.3	—	—	−35.2	236.8	—	—	—
Mean	2/*21/48	315.1	−68.4	3.1	108.6	−35.6	235.7	3.1	5.4	(4.7)
(69.1, 23.6)										
NC[1]	*1/1	328.0	−56.2	—	—	−18.4	230.0	—	—	
(69,0, 23.4)										
ND[1]	*1/1	309.5	−70.3	—	—	−38.83	238.6	—	—	
(69.0, 23.4E)										
				Karasjok						
NE[2]	2/*2	324.0	−65.1	—	—	−29.2	232.4	—	—	
(69.1, 25.2)										
				Host rocks						
				Kautokeino						
Amphibolite										
NA11, contact	*1/2	212.5	−76.1	—	—	−76.1	299.8	—	—	
NA12, 17 m	*1/2	2.2	−53.9	—	—	−13.4	201.4	—	—	
Gneiss										
NB2, contact	*1/1	351.8	−57.6	—	—	−17.6	210.6	—	—	
NB14, 1 m	*1/1	57.7	−31.2	—	—	—	—	—	—	
NB15, 20 m	*1/1	226.5	62.7	—	—	—	—	—	—	

[1] pilot samples from two additional diabase dykes at Kautokeino, [2] pilot samples from one diabase dyke at Karasjok; [1] and [2] are not included in mean direction. For other explanations, see Table 2.

NRM intensity plot is shown in Fig. 8 and a susceptibility versus density plot in Fig. 9. The results are summarized in Tables 4 and 5. The petrophysical properties of the host rocks are discussed in the context of baked contact test.

Laanila–Ristijärvi dykes

There is only little variation in bulk petrophysical properties between different Laanila–Ristijärvi dykes LT, LJ and RL (Fig. 8). On average, the samples carry comparatively high Q values between about 2.5 and 10 (Table 4), indicating that the dykes have stable remanences probably carried by single-domain (SD)/pseudo-single-domain (PSD) magnetite grains.

In dyke LT, the susceptibility is somewhat higher and the NRM intensity and Q values are lower than in dykes LJ or RL. This is obviously due to the large contribution of the VRM component, which reduces the total resultant NRM intensity. The lower NRM and Q values in LT support the observation from demagnetization data that the remanence in LT is carried by larger multidomain (MD) grains, which are therefore less stable and more likely to acquire viscous remanence (e.g. Dunlop 1983) than the grains in other dykes. Unlike dykes LJ and RL, dyke LT does not show up on the aeromagnetic maps because the positive remanence inclination (due to PEF contamination) makes the magnetic anomaly of this site comparable to the anomalies typical of the host granulite. Dykes LJ and RL are seen as strong negative anomalies due to the dominance of the negative inclinations of NRM, with Q values up to 10.4 (Table 4). The very high Q values (> 18) found in three KK samples are probably caused by lightning-induced isothermal remanent magnetization (IRM).

The variation in density between dykes LT, LJ and RL is more pronounced than the variation in NRM (Fig. 9a, Table 4). Dyke LT, characterized by the highest susceptibilities, has the lowest density, which may be due to a higher proportion of titanomagnetite at the expense of pyroxenes. The highest densities occur at Ristijärvi site KL, obviously due

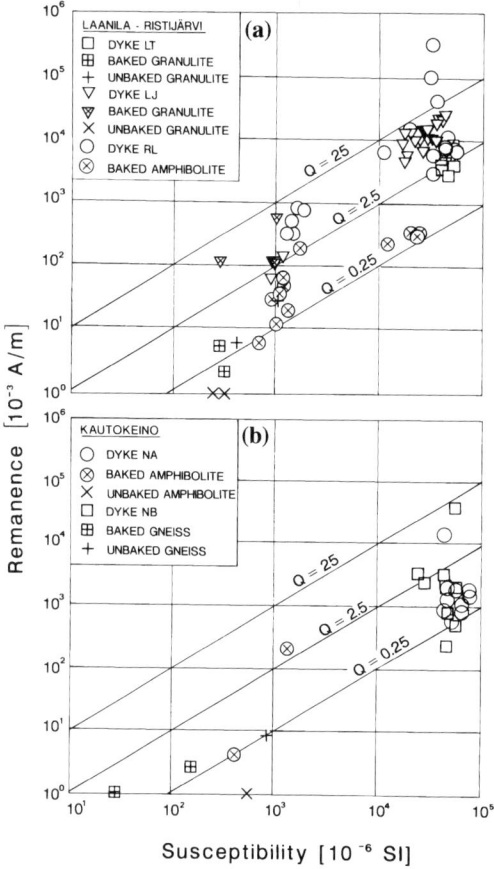

to the high proportion of pyroxenes. Dykes LJ and RL show a clear duality in susceptibility that is not seen in dyke LT, possibly because the magma of dykes LJ and RL may have been more inhomogeneous than that of dyke LT.

Kautokeino dykes

The bulk susceptibility and remanence intensities are similar in dykes NA and NB (Fig. 8b, Table 5). Q values range between 0.25 and 2.5, and are an order of magnitude lower than in the Laanila–Ristijärvi dykes. This is partly explained by the strong VRM component in the PEF direction. In the Kautokeino dykes, the ChRM component is almost opposite the PEF which reduces the total intensity of remanent magnetization. However, in many samples the proportion of the VRM component is so small that the total remanence direction still retains the negative inclination. The uncleaned NRM direction obtained by Olesen & Sandstad (1993) is also close to the direction of ChRM, being steep and negative. Thus, the total NRM intensity, as well as the Q values, of the Kautokeino dykes are slightly lower than those of the Laanila dykes. In the density versus susceptibility plot (Fig. 9b), the densities show only little variation between dykes NA and NB. On average, the densities and susceptibilities accord with those of the Laanila–Ristijärvi dykes; the densities in particular are comparable to those of dyke RL but are slightly higher than those of dykes LT and LJ.

Fig. 8. Remanence–susceptibility diagrams for diabase dykes and host rocks from (**a**) Laanila–Ristijärvi and (**b**) Kautokeino. Koenigsberger's Q ratios are shown as lines.

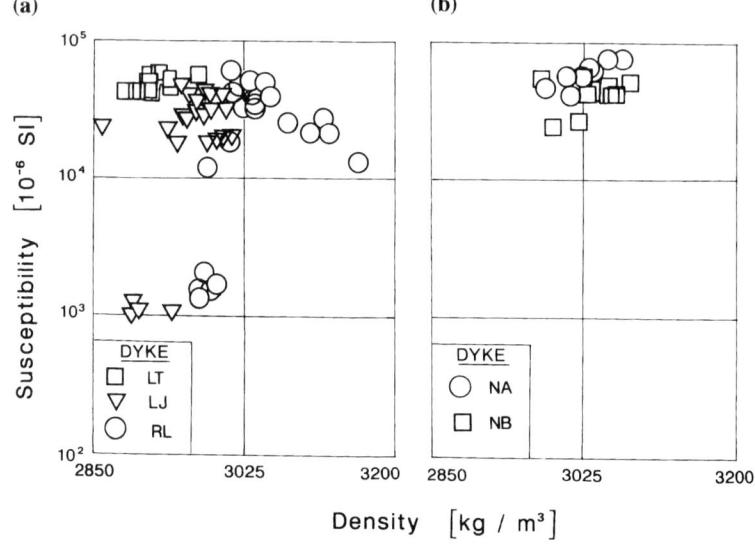

Fig. 9. Susceptibility–density diagrams for diabase dykes from (**a**) Laanila–Ristijärvi and (**b**) Kautokeino.

Table 4. Petrophysical properties of Laanila–Ristijärvi dykes and host rocks

Site/Dyke	B/N/n	D	K	J	Q
		Laanila			
Diabase					
Mean LT	1/*11/14	2920	47571	4705.3	2.5
LK	*16/34	2973	29725	11696.5	10.4
LJ	*14/15	2954	21876	11714.2	9.7
Mean LJ	*2/30/49	2963.5	25800.5	11705.4	10.1
		Ristijärvi			
KJ	*5/18	3017	27084	11076.2	8.4
KK	*4/13	3034	37650	[1]123096.8	[1]18.4
KO	*6/18	3016	41335	6304.9	3.9
KL	*6/11	2983	1621	610.4	9.4
KN	*5/12	3032	45947	7437.1	4.1
Mean RL	*5/26/72	3016	30727	[2]6357.2	[2]6.5
Grand Mean	*3/8/67/135	2966.5	34699.5	7589.3	6.4
		Host rocks			
		Laanila			
Granulite					
LJ7-9, contact	*3/3	2843	787	257.7	8.4
LJ19-20, 7 m	*2/2	2719	288	1.0	0.1
LT10-11, contact	*2/2	2653	315	3.5	0.3
LT13-14	*2/3	2682	798	16.0	0.5
		Ristijärvi			
Amphibolite					
KM1-6, contact	*6/17	3082	1214	59.1	0.8
KP1-7, contact	*7/13	3102	14577	2485.8	0.4
Gneiss					
KL11, contact	*1/4	2663	381	19.8	1.2

NOTE: For host rocks, the distance of the host rock from the dyke is shown; B/N/n is the number of sites/samples/specimens used in the statistical calculations; * denotes the statistical level used in mean calculations, D is density (kg m^{-3}); K is susceptibility (10^{-6} SI); J is remanence intensity (mA m^{-1}); $Q = J/(K \times H)$ where H is the present Earth's magnetic field intensity. [1] not included in mean value, [2] B/N/n = *4/22/59.

Table 5. Petrophysical properties of Kautokeino dykes and host rocks

Site	B/N/n	D	K	J	Q
Diabase					
NA	1/*10/37	3028	55298	2673.4	0.7
NB	1/*10/33	3041	41977	1813.2	1.2
Mean	*2/20/70	3035	48638	2243.3	1.0
		Host rocks			
Amphibolite					
NA11, contact	1/*2	3101	1305	207.4	3.8
NA12, 17 m	1/*4	2919	406	3.8	0.2
NA13-14, 20 m	*2/7	2971	612	0.9	0.1
Gneiss					
NB2, contact	1/*2	2614	140	3.0	0.6
NB14, 1 m	1/*2	2594	28	0.7	1.6
NB15, 20 m	1/*2	2623	826	8.7	0.3

NOTE: For explanations, see Table 3.

Anisotropy of susceptibility

The anisotropy of susceptibility (AMS) was measured on about 30 samples (Table 6). One reason for AMS measurements was to investigate whether the scatter of remanence directions might be partly caused by AMS. The remanence directions may deviate from the ambient geomagnetic field direction owing to either later deformation or magma flow along dyke margins. Palaeomagnetic interpretations of ages and tectonics may therefore be erroneous. However, the AMS does not always indicate that the **remanence** has been deflected, because the remanence carriers may be different from the AMS carriers (e.g. Tarling & Hrouda 1993). There are both SD and MD grains in the Laanila and Kautokeino dykes, and so the larger grains (carrying the VRM component) are mainly responsible for the bulk susceptibilities. On the other hand, the small (SD/PSD) magnetite grains presumably carry the primary characteristic remanence and may not be affected by deflection due AMS (see e.g. Hyodo & Dunlop 1993). The AMS data will therefore mainly indicate the deflection of NRM due to large MD magnetite grains. Here we discuss AMS only.

According to Puranen et al. (1992), AF demagnetization does not cause significant changes in AMS; hence measurements were conducted for both AF demagnetized samples and fresh specimens.

In the Laanila–Ristijärvi and Kautokeino dykes the degree of anisotropy (P, the ratio of maximum and minimum susceptibilities) is typically below 1.05 (Table 6), which is the limiting value above which the remanence may have been deflected (Stephenson et al. 1986). Hence, since the MD grains were not deflected due to AMS, the remanence directions in the Laanila–Ristijärvi and Kautokeino dykes record the ambient geomagnetic field and are suitable for palaeomagnetic and tectonic studies.

The L (lineation)–F (foliation) diagram of AMS data (Fig. 10) shows that both the magnetic foliation and lineation values are below 1.05, indicating that the fabric is due to magma flow and that the dykes have not been affected by later deformation (see e.g. Hrouda & Chlupácôvá 1980; Tarling & Hrouda 1993).

Another use of AMS data is for defining the direction of magma flow. When basic magma moves within the crust, the magnetite grains are aligned with the long axes parallel to the direction

Table 6. *Degree of anisotropy (P) in Laanila–Ristijärvi and Kautokeino dykes*

Laanila–Ristijärvi		Kautokeino	
Specimen	P	Specimen	P
LT1-1D	1.010	NA1-1A*	1.024
LT2-1D*	1.011	NA3-1A*	1.043
LT5-1C	1.014	NA3-1D	1.028
LT7-1C	1.015	NA8-1E*	1.009
LT12-1A*	1.005	NA9-1C	1.057
LJ4-1C	1.017	NA9-1E*	1.029
LJ11-1A	1.026	NB3-1A*	1.013
LJ15-2B	1.006	NB3-1B*	1.014
LL1-1E	1.038	NB5-1A*	1.013
LL4-1D	1.032	NB6-1A*	1.041
LL7-1A	1.019		
LK3-3C	1.019		
LK5-1F	1.031		
LK7-1F	1.039		
KJ4-1E	1.026		
KL14-1A*	1.007		
KL15-2A*	1.008		
KK4-1C*	1.044		
KO2-1B	1.021		
KO5-1A	1.018		
KN1-1A	1.021		
KN4-2B	1.038		

NOTE: P is the ratio between maximum and minimum susceptibility. * = specimen demagnetized with alternating field before anisotropy of susceptibility (AMS) measurement.

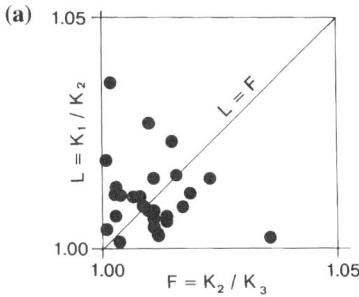 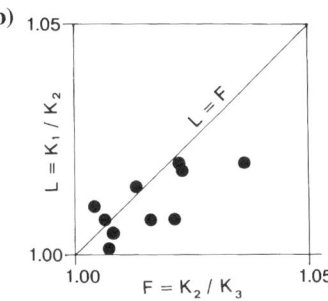

Fig. 10. Magnetic foliation (F) – lineation (L) diagram for (**a**) Laanila–Ristijärvi and (**b**) Kautokeino diabases. F and L values below 1.05 are characteristic of magmatic flow fabrics (see Hrouda & Chlupáčóvá 1980). K_1, K_2, and K_3 are maximum, intermediate and minimum susceptibilities, respectively.

of magma flow and the short axes perpendicular to it. The orientation of the magnetite grains can then be determined by measuring the AMS.

The directions of AMS vary in distinct Laanila–Ristijärvi and Kautokeino dykes (Fig. 11). In most specimens of dyke LT (Fig. 11a), the maximum axes are almost vertical but the minimum axes have shallow inclinations, thus indicating a vertical direction of magma flow. The Kautokeino dykes (Fig. 11b) show similar behaviour as dyke LT. In Laanila dyke LJ (Fig. 11c) and Ristijärvi dyke RL (Fig. 11d), the minimum axes have steep inclinations and the maximum axes are shallow. Thus, horizontal magma flow may be inferred. However, as Tarling & Hrouda (1993) pointed out, weak anisotropies (*P*) of dykes may cause the maximum and intermediate directions of the AMS ellipsoid to interchange direction and therefore the flow direction is not certain. Many workers (e.g. Ernst 1990; Lister & Kerr 1990; Cadman *et al.* 1993) have shown that the flow direction of initially hot magma of dykes is almost vertical because the liquid magma is less dense than the surrounding solidified country rocks (Lister & Kerr 1990). As the magma cools and gets denser, a horizontal or sub-horizontal magma flow direction may result. It is suggested that the vertical AMS directions of Laanila dyke LT and the Kautokeino dykes could reflect steeper crustal depth than Laanila–Ristijärvi dykes LJ and RL, where the horizontal magma flow direction could be due to emplacement at higher, cooler crustal levels. However, the AMS data is limited and quite scattered and therefore the result is inconclusive.

Results of high field measurements

In all palaeomagnetic work it is essential that the remanence is stable and thus reliably records the direction of the past geomagnetic field. The stability of the remanence is heavily dependent on the effective magnetic grain size of the remanence-carrying minerals. The highest stability is maintained by small SD grains and the lowest by large MD grains; PSD size grains have magnetic properties between truly SD and MD particles. In the following, the magnetic grain size was determined by measuring the hysteresis properties (Day *et al.* 1977) and by Lowrie–Fuller tests (Lowrie & Fuller 1971; Dunlop 1983). Acquisition of IRM provides information about magnetic grain size characteristics and aids the identification of the dominant magnetic minerals.

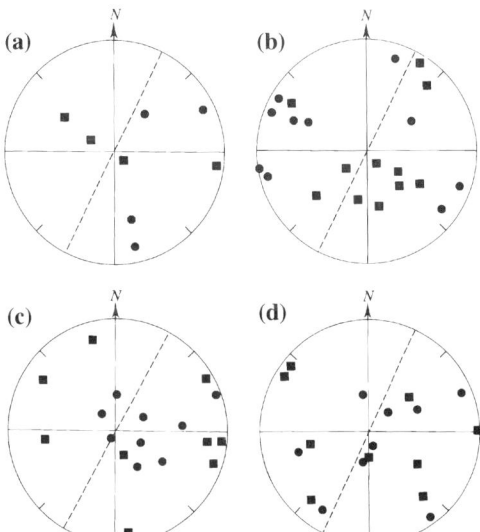

Fig. 11. Directions of minimum (circles) and maximum (squares) susceptibility for specimens of dyke LT (**a**), NA and NB (**b**), LJ (**c**) and RL (**d**), plotted on the lower hemisphere of equal-area stereonets. Average strikes of dykes are given by dashed lines.

The values of saturation remanent magnetization/saturation magnetization (J_{rs}/J_s) and the coercivity of remanence/coercive force (H_{cr}/H_c) of all samples lie in the PSD field (Fig. 12a, Table 7) according to the criteria of Day et al. (1977). In empiricial studies (Parry 1980, 1982), it has been shown that mixtures of SD/PSD and MD grain sizes show linear dispersion depending on the proportions of different particle sizes. In the logarithmic plot (Fig. 12b), the hysteresis ratios of the Laanila and Kautokeino samples follow the line of mixtures of the SD and MD magnetite grains (Parry 1982). AF and thermal demagnetization data, which indicated the presence of a large PEF component superimposed on a stable characteristic component, suggested a mixture of grain sizes of SD/PSD and MD type. In Laanila dyke LT, the large PEF component is consistent with the hysteresis properties, thus suggesting an increased proportion of MD magnetite. The smallest magnetic grain sizes (PSD) are found in LJ, LK and KL samples. Samples from these sites also formed the most coherent group of NRM directions and bulk petrophysical properties (Tables 2 and 4). The hysteresis properties of the samples from the Kautokeino dykes are in agreement with those of the Ristijärvi samples. Ristijärvi sample KL shows complex hysteresis properties, probably due to the dominance of weakly magnetic paramagnetic minerals at the expense of ferrimagnetic minerals.

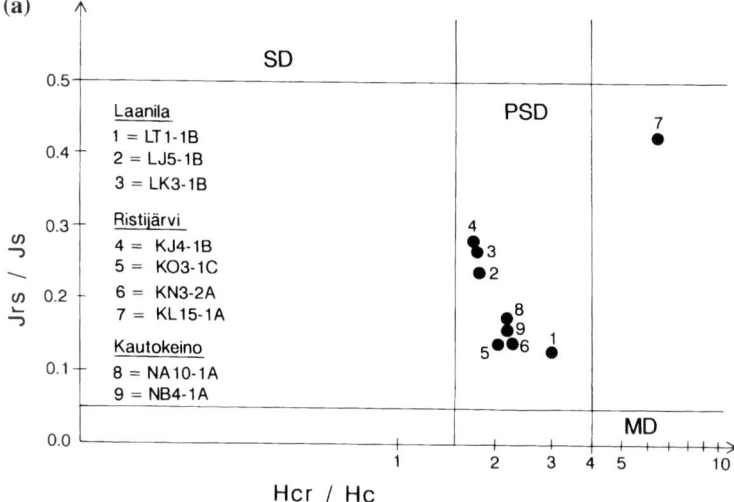

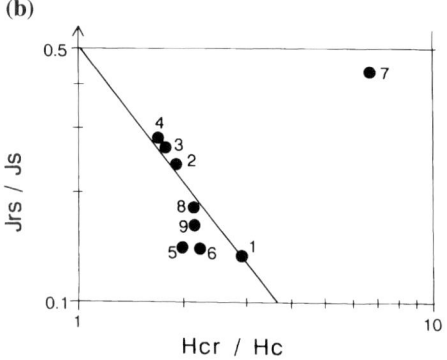

Fig. 12. (a) Half-logarithmic and (b) logarithmic plots of hysteresis parameters for Laanila–Ristijärvi and Kautokeino dykes. J_{rs} = saturation remanence, J_s = saturation magnetization, H_{cr} = remanent coercivity, H_c = coercive force. Single domain (SD), pseudo-single domain (PSD) and multidomain (MD) fields after Day et al. (1977). In (b) the line shows the trend for SD and MD mixture, after Parry (1982).

Table 7. *Hysteresis properties of Laanila–Ristijärvi and Kautokeino dykes*

Specimen	J_s(mT)	J_{rs}(mT)	H_c(T)	H_{cr}(T)	J_{rs}/J_s	H_{cr}/H_c
LT1-1B	2.360	0.309	0.008	0.022	0.131	2.960
LL3-1B	12.466	3.344	0.023	0.040	0.268	1.754
LJ5-1B	2.990	0.711	0.025	0.044	0.240	1.790
KN3-2A	11.338	1.585	0.010	0.023	0.140	2.211
KJ4-1B	9.167	2.546	0.024	0.040	0.278	1.697
KO3-1C	6.185	0.876	0.011	0.022	0.142	2.030
KL15-1A	0.083	0.036	0.017	0.109	0.427	6.550
NA10-1A	8.941	1.454	0.013	0.028	0.163	2.174
NB4-1A	8.659	1.546	0.014	0.030	0.179	2.166

NOTE: J_s = saturation magnetization, J_{rs} = saturation remanence, H_c = coercive force, H_{cr} = coercivity of remanence.

The response of SIRM, ARM and NRM to AF demagnetization are compared in the Lowrie–Fuller test. In this test, if the remanence is carried by SD/PSD grains, the normalized weak field ARM is more resistant to AF demagnetization than the SIRM, but has a similar stability to thermoremanence (TRM); MD grains show the opposite behaviour. The NRM of seven specimens from the Laanila–Ristijärvi and Kautokeino dykes was first AF demagnetized in peak fields up to 100 mT. ARM was then produced in a steady field of 0.05 mT superimposed with an alternating field of 100 mT and then progressively AF demagnetized in the same steps as NRM. The specimens were then given a SIRM in the field of 1.4 T (saturation took place below 0.2 T, see Fig. 14) and then AF demagnetized similarly to NRM and ARM.

The NRM of the LJ and LK samples is harder than ARM, which has higher coercivity than SIRM (Fig. 13a and b). In specimens LT (Laanila), KO (Ristijärvi) and NA and NB (Kautokeino), the proportion of VRM component at low coercivities is so high that their NRM intensity decay curves are not easy to compare with the ARM and SIRM curves. However, as in samples LJ and LK, the ARM curves show higher coercivity than the SIRM curves, which is indicative of SD/PSD magnetic grain sizes.

IRM acquisition curves show saturation of all Laanila and Kautokeino samples at 0.1 or 0.2 T (Fig. 14), characteristic of magnetite as the remanence carrier. The rapid increase in IRM in samples KN, KO of Ristijärvi, NA, NB of Kautokeino and especially LT of Laanila shows that the remanence is carried by relatively coarse MD/PSD grains or by a mixture of large MD grains and small SD/PSD grains. Samples LJ, LK and KL show the more gradual increase in IRM, characteristic of a more SD type of remanence. Sample KL is the hardest one and the remanence can be carried by SD magnetite, small MD magnetite or hematite grains (e.g. Dunlop 1981). Thermal demagnetizations showed that magnetite dominates the remanence; thus small PSD magnetite grains are proposed as the remanence carrier for site KL.

Thermomagnetic analysis

The magnetic minerals of the rocks were identified by thermomagnetic analysis (Curie point analysis) by measuring susceptibility versus temperature. Each Curie temperature shown in the thermomagnetic curve characterizes a ferromagnetic mineral in the rock. The Curie temperature can be determined more accurately using the temperature dependence of low-field susceptibility than saturation magnetization, because the susceptibility change at Curie temperature is much more expressive (e.g. Zapletal 1992). Chemical and mineralogical changes can also be observed using the difference in heating and cooling curves. Thus, the suitability of samples for thermal demagnetization (e.g. Schmidt 1993) and for Thellier palaeointensity runs can also be verified.

Samples LL1-1D from Laanila dyke LJ (Fig. 15b) and NA8-1B from Kautokeino dyke NA (Fig. 15c) show a systematic increase in susceptibility and a pronounced Hopkinson peak in the heating curve just before the Curie temperature, as is characteristic of SD/PSD grains. The heating and cooling curves indicate only one ferrimagnetic mineral with a Curie temperature of about 550° C, indicative of Ti-poor magnetite. In sample NB (Fig. 15d) there is also a decline in susceptibility at about 330 °C due to pyrrhotite, which was also identified in thin sections. The heating and cooling curves are

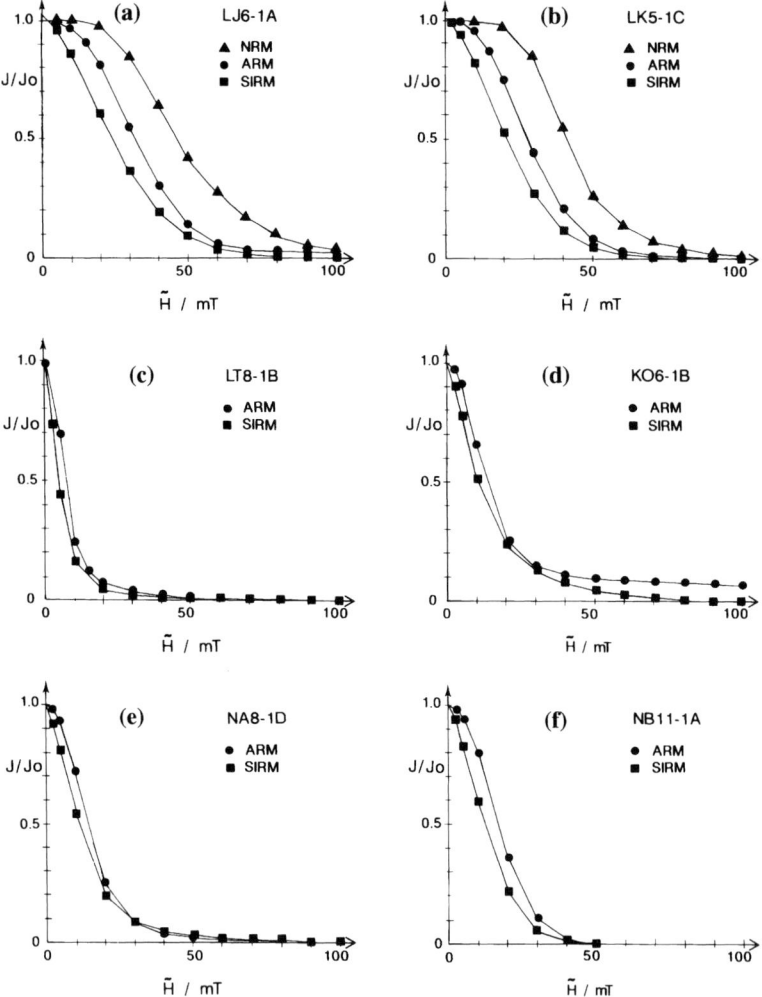

Fig. 13. Lowrie–Fuller test showing NRM, ARM and SIRM curves for specimens of dyke LJ (**a**) and (**b**) and ARM and SIRM curves for specimens of dyke LT (**c**) and RL (**d**) of Laanila–Ristijärvi and NA (**e**) and NB (**f**) of Kautokeino.

irreversible, the cooling curve showing higher susceptibilities. The irreversibility may be due to either changes in internal stresses in the specimen (Zapletal 1992) or the formation of new magnetite (Hrouda 1994) at high temperatures during heating. The difference in the heating and cooling curves is, however, so small that probably no significant chemical changes took place during heating and thus thermal demagnetization of this rock type is suitable.

The thermomagnetic curve for sample LT4-1D (Fig. 15a) is highly irreversible. The Curie temperature determined from the heating curve is about 475°C, indicative of a Ti-rich titano-magnetite, closer in composition to the ulvöspinel end of the magnetite series. In the cooling curve, the susceptibility remains almost constant and is considerably lower than during heating. A slight decrease at about 570 °C may be attributed to the Curie point of Ti-poor magnetite, which formed during heating when titanite or ilmenite were also formed as an alteration product of titanomagnetite.

Petrographic features of LT samples showed that the magnetite phase is almost pure titanomagnetite with no ilmenite lamellae as in dyke LJ. In dyke LT the ilmenite occurs as separate, platy grains. In basic igneous rocks, Ti-rich magnetite is the primary high-temperature mineral (e.g. O'Reilly

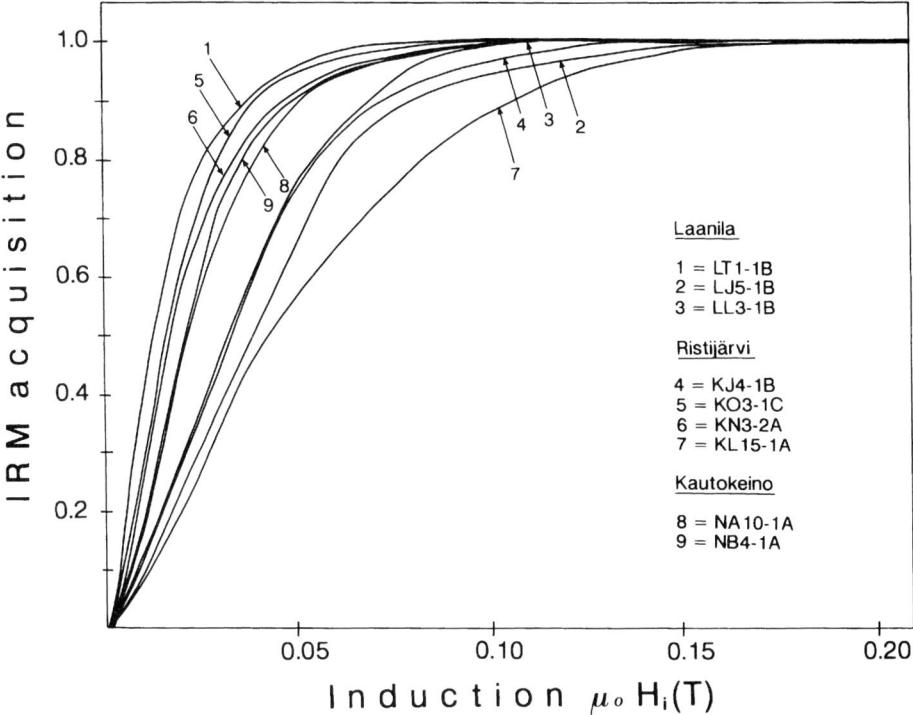

Fig. 14. Normalized IRM acquisition curves for Laanila–Ristijärvi and Kautokeino dykes.

1976). Thus dyke LT represents the initial, most primitive magma of the dyke swarm. The coercive force is known to increase with an increase in the proportion of titanium in titanomagnetite (e.g. Mullins 1977) and therefore susceptibility is also expected to be lower than in pure magnetite. However, the coercive force is lower and the susceptibilities are higher in the LT samples carrying Ti-rich magnetite than in other Laanila samples. The LT samples were dominated by MD magnetite grains (Fig. 12); hence, the grain size dependence of the magnetic properties seems to outweigh the significance of composition. The occurrence of titanomagnetite with no ilmenite lamellae in dyke LT implies that it crystallized in a less oxidizing (O'Reilly 1976) environment than the Laanila–Ristijärvi and Kautokeino dykes.

Results of baked contact tests

The basic requirement for using dyke rocks in palaeomagnetic work is to ensure that the remanence is thermoremanence with a primary ambient field direction acquired during emplacement of the dykes and not during later events. This can be tested by baked contact tests. The baked contact test is based on the idea that when a host rock at the contact of an igneous intrusion is heated above the blocking temperatures of the minerals carrying the remanence, they will acquire a new remanence in the direction of the igneous intrusion. The maximum temperature reached in a host rock close to an igneous intrusion as well as the stability of the baking magnetization decrease with distance from the contact (Everitt & Clegg 1962; McClelland-Brown 1981). As Schmidt (1991) pointed out, palaeomagnetic contact tests demand ideal behaviour of both the dyke and the host rock. The Laanila–Ristijärvi and Kautokeino dykes do behave ideally but, as often happens in Archaean felsic gneiss areas, the stability of the NRM in the host rocks is weak. Therefore, a full baked contact test comprising magnetically stable contact, hybrid and unbaked zones at the sampling sites could not be carried out. However, by compiling known remanence direction data on the host rock from elsewhere in the granulite belt and from the Archaean basement area (e.g. Pesonen et al. 1989) and by combining it with petrophysical information, we can evaluate the primary nature of the remanence in the dykes.

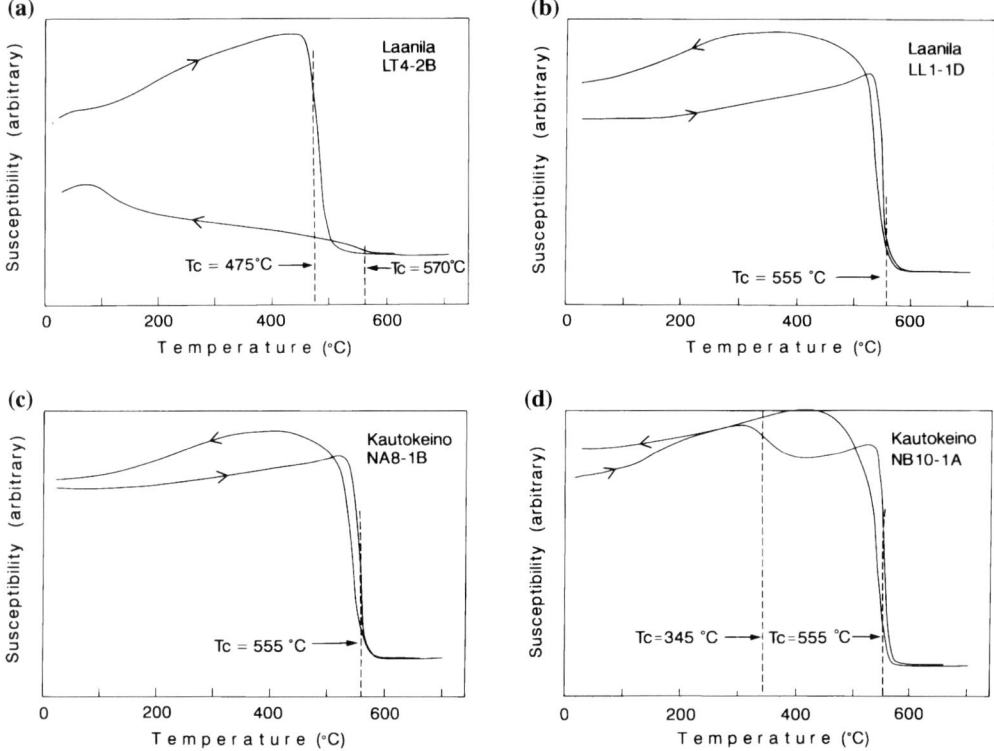

Fig. 15. Thermomagnetic curves (susceptibility/ temperature) for specimens of Laanila–Ristijärvi and Kautokeino dykes. T_c denotes Curie temperature. Specimen LT from dyke LT contains Ti-rich titanomagnetite as main magnetic mineral. All the other dykes contain Ti-poor titanomagnetite.

Laanila–Ristijärvi dykes

Baked contact tests were carried out for all dykes. At site LJ, the remanence direction of the granulite right at the contact (Fig. 16a) is similar to that in the diabase dyke (Table 2) indicating that the remanences are coeval. Further from the contact (Fig. 16b), at the hybrid zone, the baking effect is still seen as negative inclination, although the directions are highly scattered. Probably due to the scattered data and the instability of the host rock, the high temperature component representing the host direction and the low temperature component representing the dyke direction (pTRM) could not be isolated. At site LJ, the baking effect is also seen as increased density, remanence intensity and susceptibility values in the contact, whereas the unbaked sample shows clearly lower values (Table 4, Fig. 8).

At site LT the granulite at the contact exhibits a direction close to that of the dyke, although the magnetization is poorly defined (Fig. 17). Further from the contact no stable directions were obtained. At this site, the density, remanence intensity and susceptibility are slightly lower in the baked sample than in the unbaked sample. The reason for the contrasting baking effects at sites LJ and LT may be the temperature difference between the dyke and host rocks at these sites. At site LJ, the dyke was evidently emplaced in a cooler environment and thermally affected the host rock. At site LT, where the host rock temperature was probably higher, the baking effect of the dyke was relatively low.

At site LT, where the unbaked host granulite was sampled, the remanences were too weak to give reliable results. However, the remanence direction of quartz diorite at Akujärvi (U–Pb zircon age 1925 Ma, age given in Meriläinen 1976; Pesonen & Neuvonen 1981) in the granulite belt, about 15 km from the Laanila dyke, points moderately downwards towards the northwest and thus clearly deviates from the Laanila dyke and baked contact directions. Further west, in the granulite of Menesjärvi (age c. 1850 Ma, Meriläinen 1976; Papunen et al. 1977; Pesonen et al. 1989), about 80 km from Laanila, the remanence direction corresponds to that at Akujärvi.

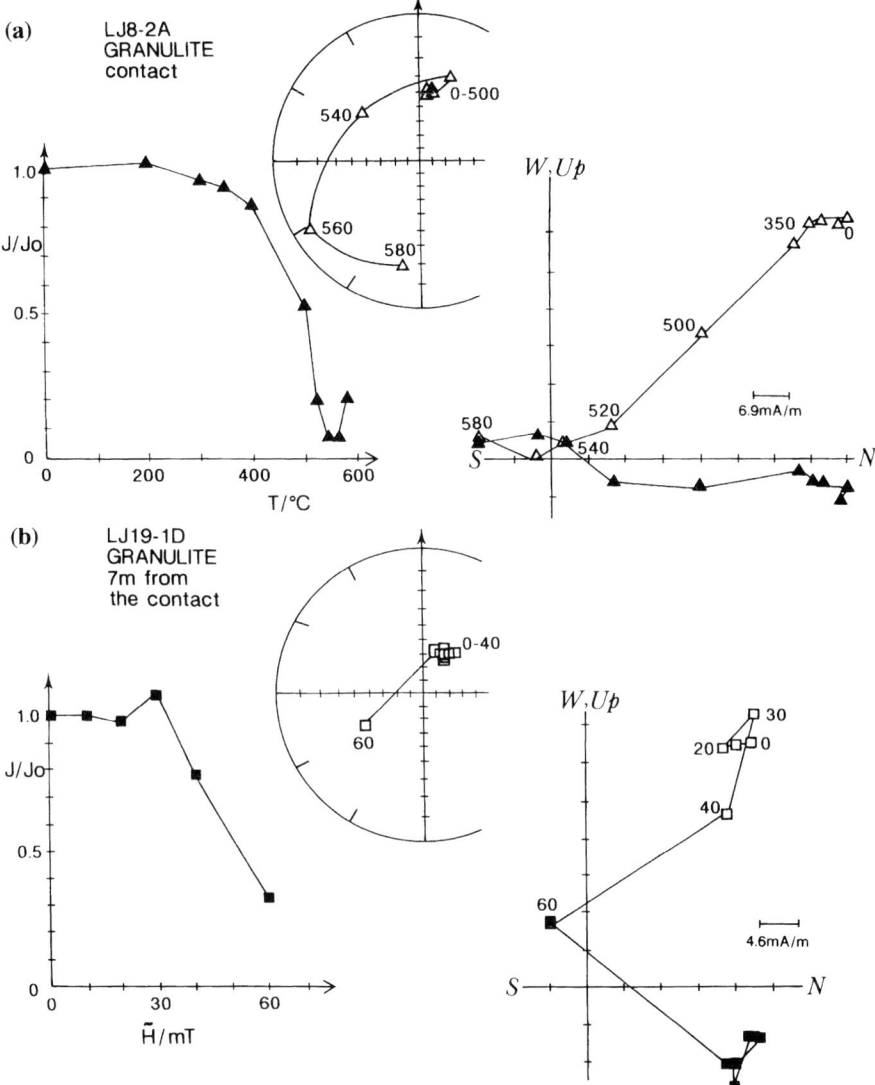

Fig. 16. Baked contact test for dyke LJ (width 20 m) in Laanila. Host granulite right at contact (**a**) carries similar magnetization to that of diabase dyke. Baking effect is still seen c. 7 m from the contact (**b**).

Contact samples were taken at sites KL, KM and KP from Ristijärvi dyke RL. At site KL, the remanence direction of the Archaean gneiss accords with the dyke direction (Table 2) and, hence, at this site, too, the remanences were acquired at the same time. The remanence direction of amphibolite at site KM is close to the mean Ristijärvi dyke direction (Table 2) but the inclination is much shallower and the declination points more to the west. In the amphibolite at site KP, the remanence directions are scattered, and reliable results could not be obtained. At Ristijärvi, samples were not collected far from the dyke and therefore a full baked contact test could not be carried out.

The results of the baked contact tests combined with the results from the unbaked Akujärvi and Menesjärvi rocks indicate that the ChRM is of thermal origin (TRM) in the Laanila–Ristijärvi dykes and baked contacts. Further support for TRM comes from a Thellier palaeointensity determination on one Laanila specimen, which

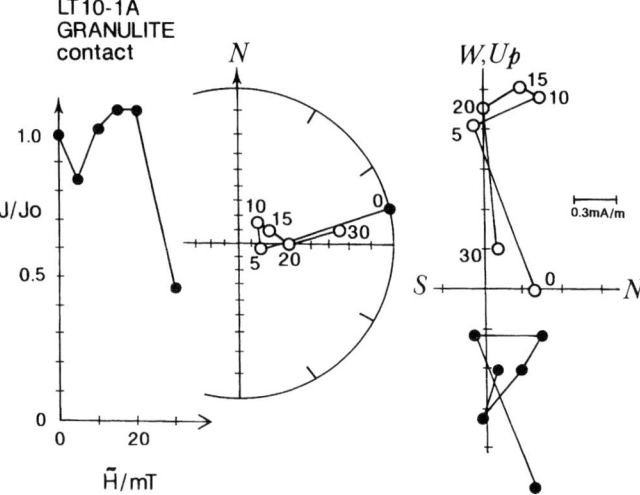

Fig. 17. Baked contact sample from granulite host rock of dyke LT.

yielded a reliable Arai plot with a palaeointensity of 29.6 ± 0.9 μT, consistent with global palaeointensity data about 1.0 Ga ago.

Kautokeino dykes

The Kautokeino dykes have baked the host rocks. This is especially clear in the contact amphibolite of site NA where the remanence direction agrees with the dyke remanence direction (Fig. 18). However, the direction is still close to the dyke direction in a sample taken 17 m from the contact. This can be interpreted in at least two ways. According to baked contact models (e.g. Schwarz 1977), the host rock should acquire partial thermoremanence (pTRM) in an area of about half the dyke width. Therefore, the baking effect of the dyke has extended as far as 17 m from the contact because the width of the dyke is at least 34 m rather than the 22 m consumed initially. Alternatively, a hidden dyke belonging to the same dyke swarm as dyke NA has baked the host rock. The Kautokeino dykes seem to occur in a comparatively narrow zone, but it is possible that the dyke swarm includes a number of dykes that are not exposed and are not seen on the aeromagnetic maps because of the high proportion of VRM aligned in the PEF direction. Stable remanence directions were not obtained from the host amphibolite and therefore, at this site baked contact test is inconclusive. At site NB where the host rock is Archaean gneiss, the remanence direction of the unbaked gneiss differs clearly from the direction of the dyke (Table 3) and thus, baked contact test is positive.

Discussion

Interdyke variation within the Laanila swarm

The easternmost of the Laanila dykes (LT) differs from the westernmost dyke (LJ) on palaeomagnetic, petrographic and bulk petrophysical criteria. Secular variation may explain some of the difference in remanence directions (Fig. 5) but does not explain the other observed differences. The differences in petrography and petrophysics may be due to the fact that dykes LT and LJ represent different crustal levels.

Petrographic and rock magnetic features indicate that the magnetite grains in dyke LT are coarser than those in LJ suggesting that LT crystallized at deeper levels than dyke LJ. Fluid activity is more pronounced at higher crustal levels than at lower crustal levels. This is seen in the Laanila dykes, where LJ (and RL of Ristijärvi) reveals more pronounced hydrothermal alteration (seen as thorough sericitation of plagioclase) than LT. On the basis of AMS data, a vertical magma flow direction for dyke LT and a horizontal direction for dyke LJ can be inferred (although not conclusively). Baked contact tests at dykes LT and LJ show that at the time of dyke injection, the host rock temperature must have been lower than the dyke temperature, that is, below the Curie point of magnetite (*c.* 580 °C). Furthermore, the petrophysical features of the host rocks imply that at site LT the host rock may have been hotter than at site LJ. Low altitude aeromagnetic maps reveal a strikingly clear difference between the magnetic field strengths west and east of dyke LJ. According

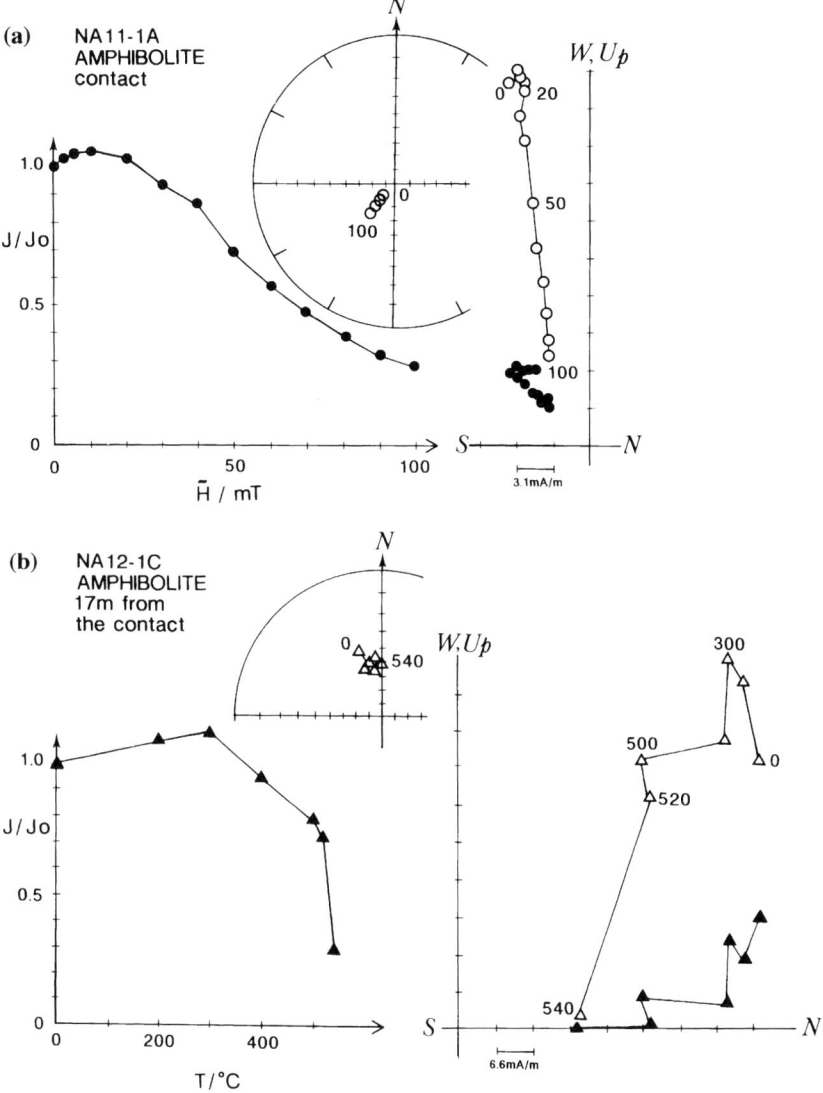

Fig. 18. Baked contact test for dyke NA (width at least 22 m) in Kautokeino. Host amphibolite right at contact (**a**) carries similar magnetization to that of diabase dyke. Similar remanence direction was isolated c. 17 m from contact (**b**).

to magnetic modelling by Airo (1994), this dyke represents a deep fracture between and separating two blocks at different crustal levels.

From the foregoing indications we infer that the remanent magnetization blocking temperature was reached in the LJ dyke prior to have reached in the LT dyke. Since the dykes were emplaced, dyke LT has undergone a greater degree of exhumation than dyke LJ, and thus a deeper crustal level of this intrusion is now exposed. Halls & Palmer (1990) have described a similar case from the 2.45 Ga old Matachewan dykes in Canada, where the N polarity dykes have been relatively uplifted along faults and represent a deeper expression of the swarm than the R polarity dykes. The importance of the finding at Laanila is that the crustal block containing dyke LT must have been uplifted **after** the Laanila dyke was emplaced, that is, after about

1.0 Ga. On the APWP (Fig. 19) the pole LT is slightly younger than poles LJ and RL; this may be accidental but is nonetheless consistent with the above model.

Palaeomagnetic poles

The Laanila–Ristijärvi, Kautokeino and Karasjok dykes trend NE–SW and display a negative anomaly on aeromagnetic maps. Hence, there is a possibility that all the dykes belong to the same diabase dyke swarm with a width of at least 200 km. Sm–Nd ages ranging in 1066–1013 Ma are coeval within the error limits. Bulk petrophysical properties, petrography, thermomagnetic and high field analysis also show a similarity between the Laanila and Kautokeino dykes. Furthermore, preliminary geochemical results show that the Kautokeino and Laanila–Ristijärvi dykes represent same tholeiitic basalts from initial rifting regimes in continental settings (P. Pihlaja, pers. comm., 1995). However, the remanence directions and pole positions of the dyke swarms differ significantly (Figs 5 and 19), suggesting a clearly different magnetization age for the Laanila–Ristijärvi and Kautokeino dykes. From the scatter of virtual geomagnetic poles (VGPs) of three Laanila dykes and from two Kautokeino dykes, it can be considered that the secular variation has been smoothed out for both dyke swarms.

The latest APWP of Fennoscandia (Elming et al. 1993) is based on well-defined palaeopoles. However, due to variety of used dating methods, which in many cases are other than the precise U–Pb method, the ages of the poles are still controversial. Therefore, the APWP cannot be used to give accurate magnetization ages of the poles although their relative ages can be inferred.

The Laanila–Ristijärvi and Kautokeino poles plot among the poles obtained from the Sveconorwegian Southwest Scandinavian Domain in southern Sweden and Norway (see e.g. Poorter 1975; Bylund 1981,1985,1992; Stearn & Piper 1984). Some of the poles derive from Sveconorwegian intrusions, and some record the uplift and cooling of the Sveconorwegian terrain (see e.g. Pesonen et al. 1991). Thus far, the palaeopoles from the Laanila and Kautokeino dykes are the only poles in the Fennoscandian palaeomagnetic database (Pesonen et al. 1991; Elming et al. 1993, 1996) outside and not close to the Sveconorwegian area in Fennoscandia.

Neoproterozoic dyke magmatism in Fennoscandia coincides with the Sweconorwegian–Grenvillian Orogeny 1250–850 Ma ago. The onset of the orogeny occurred in southern–central Fennoscandia and is represented by the Central Scandinavian Dolerite Group in Sweden and Finland with ages of 1270–1200 Ma (Gorbatschev et al. 1987). Dyke complexes of early stages also occur in Lake Ladoga and west of Lake Onega in Russia, in northern Sweden and in southern Norway, but they have not been dated definitely (Gorbatschev et al. 1987).

In the later stages of the Sveconorwegian Orogeny (1050–850 Ma), several dyke swarms intruded in the Fennoscandian shield. Dyke injecton was most extensive in southern Fennoscandia about 1050–850 Ma ago (Rb–Sr and Sm–Nd ages; Patchett & Bylund 1977; Johansson & Johansson 1990). It is represented by the Blekinge–Dalarna dolerite dykes (BDD), which are aligned parallel to the Protogine Zone between the Sveconorwegian and Svecofennian parts of the Fennoscandian shield. Some of the BDD dykes have been dated and their palaeopoles determined within the Protogine Zone, west and east of the zone and also in the Svecofennian host rocks (e.g.

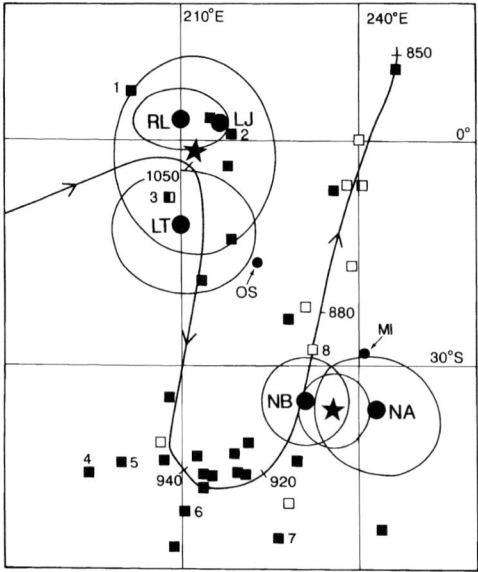

Fig. 19. Apparent Polar Wander Path (APWP) for Fennoscandia (1050–850 Ma) based on data from Elming et al. (1994). Closed (open) symbols denote normal (reversed) polarity; half open symbols denote mixed polarity. Stars represent mean poles for Laanila–Ristijärvi and Kautokeino dykes with their A95 confidence circles. Circles represent dyke mean virtual geomagnetic poles (VGPs) shown with their confidence ovals. Poles OS and MI are calculated from uncleaned NRM data from Kautokeino dykes (Olesen & Sandstad 1993) and Karasjok dykes (Midtun 1988), respectively. Squares denote palaeomagnetic poles from Sveconorwegian rocks of Fennoscandia (Pesonen et al. 1991; Elming et al. 1994). Numbered poles are referred to in text.

Bylund 1981, 1992; Pesonen *et al.* 1991; Bylund & Elming 1992). The Laanila and Kautokeino dyke swarms belong to the same age group as the BDD dykes.

The oldest palaeopoles from the Sveconorwegian Southwest Scandinavian Domain derive from the Bamble intrusions, southern Norway, which comprise basic igneous rocks, amphibolites and hyperites (Poorter 1975; Stearn & Piper 1984). K–Ar dates of the intrusions are in the range 1080–950 Ma. Rb–Sr whole-rock dates give an age of 1139 ± 100 Ma. In the palaeomagnetic grading system (Briden & Duff 1981), three palaeopoles (poles 1–3, Fig. 19) from the Bamble intrusions are regarded well defined but considering the dating methods, the ages of the palaeopoles probably do not record the intrusion ages, but are younger. The poles correspond to the ones from the Laanila–Ristijärvi dykes.

The Sveconorwegian APWP makes a steep loop at 950–900 Ma and turns to lower latitudes from about 920 Ma onwards (Fig. 19). The pole from the Kautokeino dykes is situated on the younger part of this loop. The only poles with precise U–Pb zircon ages are located at the southernmost apex of the loop and they all derive from the Rogaland igneous complex, west of the Protogine Zone in southern Norway. Pole 4 (Fig. 19) derives from the Egersund farsundite and yields U–Pb ages on zircon of 977 ± 36 Ma and 944 ± 84 Ma (Murthy & Deutsch 1975). A new U–Pb age of zircon gives 928 ± 2 Ma for the Egersund anorthosite (Duchesne *et al.* 1993). Pole 5 has been obtained from the Åna-Sira massif yielding a U–Pb age of zircon of 930 ± 4 Ma (Duchesne *et al.* 1993). Pole 6 derives from the Bjerkreim–Sokndal lopolith with a U–Pb age on zircon of 945 ± 40 Ma (Stearn & Piper 1984). Pole 7 represents a mean pole from the Rogaland farsundite with U–Pb age of zircon of 944–977 Ma (Stearn & Piper 1984). A palaeopole at the lower latitudes (pole 8, Fig. 19), near the Kautokeino pole, is from the Egersund olivine dolerite dyke of southern Norway (Storetvedt 1966*a,b*). Palaeomagnetic data suggest a Sveconorwegian age for these dykes (see e.g. Pesonen *et al.* 1991), but a Rb–Sr determination provides an age of 630 Ma (Sundvoll 1987).

Hence, all the palaeopoles at high southern latitudes of the loop yield isotope ages younger than 1.0 Ga except the new pole from Kautokeino dykes which have an exceptionally high Sm–Nd age of 1066 ± 34 Ma. Regarding the markedly different pole position of the Kautokeino dykes compared to the pole of the 1042–1013 Ma old Laanila–Ristijärvi dykes and the above presented precisely dated poles, it is evident that the Kautokeino dykes yield a younger magnetization age than the Laanila–Ristijärvi dykes. On the basis of the dated poles, the age of the Kautokeino pole is *c.* 900 Ma. In order to explain the deviating pole position of the Kautokeino dykes with respect to the Laanila–Ristijärvi dykes, three scenarios are proposed.

First, the Kautokeino palaeopole may record later local tectonic movements which have taken place after the dyke emplacement resulting in new remanence directions. The pole positions are thus fortuitous. Olesen & Sandstad (1993) report large-scale Proterozoic fractures and Phanerozoic postglacial fault zones within the Kautokeino greenstone belt which may be responsible for tilting of the dykes. The Kautokeino dykes dip 60–70° southeast, but tilting to vertical position would steepen the inclination and not make it shallower as required. Furthermore, since the remanence directions from the similarily striking Kautokeino and Karasjok dykes are similar, it is likely that they have been emplaced in their present position. Hence, the present dips of the dykes do not explain the deviations.

Another reason for the younger magnetization age may be due to delayed blocking of the remanence of the Kautokeino dykes. By analogy with the scenario for the Laanila dykes LT and LJ (see before), it is possible that at the present erosion surface the Kautokeino–Karasjok dykes represent a much deeper level than the Laanila–Ristijärvi dykes. At deeper crustal level the magnetization would have been delayed due to prolonged cooling of the dyke at higher ambient temperatures. However, considering the magnetization ages of about 1030 Ma and 900 Ma of the Laanila–Ristijärvi and Kautokeino dykes, respectively (Fig. 19), this model requires blocking of the remanence more than 100 Ma later at deeper parts of the dyke than at upper parts, which is not possible. Moreover, if the difference of crustal levels is due to late vertical uplift of the Kautokeino–Karasjok block, occurrence of steep vertical faults between the Laanila and Karasjok areas as well as differences in the magnetization and gravity anomalies is expected. However, there are no clear indications of such faults and differences.

Third, the isotope ages of the Laanila–Ristijärvi and Kautokeino dykes are consistent with the palaeomagnetic ages but the previous APWP of Fennoscandia (Fig. 19) is wrong and should be revised. This would require a self-closing loop and a much more complex APWP than the previous one.

With the present data, none of the models is favoured. Further isotope age determinations with U–Pb (zircon or baddeleyite) or $^{39}Ar-^{40}Ar$ techniques and palaeomagnetic measurements of both Kautokeino and Laanila–Ristijärvi dykes are required to test the credits of these models.

Conclusions

(1) Baked contact tests of the Laanila–Ristijärvi and Kautokeino dykes show that the characteristic remanence of the dykes is primary TRM acquired during cooling of the magma. TRM is supported by similarity of the ARM with NRM, and enhanced Q values of the baked contact rocks.
(2) The petrophysical, petrographic and magnetic properties of the Kautokeino dykes show similarities with the Laanila dykes and thus both dyke sets may originate from the same magma source. Similar remanences also occur in the dykes of Karasjok, between Laanila and Kautokeino, indicating the vast extent of Neoproterozoic diabase dykes in the northern Fennoscandian shield.
(3) The magnetic and petrographic properties of Laanila dyke LT differ from those of Laanila dyke LJ and Ristijärvi dyke RL. The differences are interpreted as indicating that, at the present erosion surface, dyke LT represents a deeper horizontal level and yields a younger magnetization age than dykes LJ and LK. The granulite block containing dyke LT has been uplifted since the Laanila dykes were emplaced.
(4) Sm–Nd isotope analyses provide ages of 1042 ± 50 Ma, 1013 ± 32 Ma and 1066 ± 34 Ma for the Laanila, Ristijärvi and Kautokeino dykes, respectively. Magnetization ages, suggested from the Fennoscandian APWP show that the Laanila dykes are about 100 Ma older than the Kautokeino dykes. More investigations are needed in order to explain the discrepancy between the palaeomagnetic and isotope ages of the Kautokeino dykes.

Tapani Mutanen and Pekka Pihlaja chose the sampling sites at Laanila and Ristijärvi and carried out some of the petrographic studies. Odleiv Olesen provided sample material from Kautokeino and Karasjok and acted as guide in the field. Kalle Neuvonen provided preliminary data on the Laanila dykes. Meri-Liisa Airo was consulted about the aeromagnetic maps. Matti Leino assisted in many ways in the palaeomagnetic laboratory and Hilkka Kumpunen made the palaeomagnetic measurements. Salme Nässling drafted the figures. The language was corrected by Gillian Häkli. Andrew Cadman and Sten-Åke Elming made valuable comments on the manuscript. To all these people we express our sincere gratitude.

References

AIRO, M.-L. 1994. Late Proterozoic block movement in Laanila, northern Fennoscandia, interpreted from high-resolution aeromagnetic data.

BRIDEN, J. C. & DUFF, B. A. 1981. Pre-Carboniferous paleomagnetism of Europe north of the Alpine orogenic belt. In: MCELHINNY, M. W. & VALENCIO, D. A. (eds) *Paleoreconstruction of the Continents*. American Geophysical Union Geodynamics Series, **2**, 137–150.

BYLUND, G. 1981. Sveconorwegian palaeomagnetism in hyperite dolerites and syenites from Scania, Sweden. *Geologiska Föreningens i Stockholm Förhandlingar*, **103**, 173–182.

—— 1985. Palaeomagnetism of middle Proterozoic basic intrusives in central Sweden and the Fennoscandian apparent polar wander path. *Precambrian Research*, **28**, 283–310.

—— 1992. Palaeomagnetism, mafic dykes and the Protogine Zone, southern Sweden. *Tectonophysics*, **201**, 49–63.

—— & ELMING, S.-Å. 1992. The Dala dolerites, central Sweden, and their palaeomagnetic signature. *Geologiska Föreningens i Stockholm Förhandlingar*, **114**, 143–155.

CADMAN, A. C., PARK, R. G., TARNEY, J. & HALLS, H. C. 1993. Significance of anisotropy of magnetic susceptibility fabrics in Proterozoic mafic dykes, Hopedale Block, Labrador. *Tectonophysics*, **207**, 303–314.

DAY, R., FULLER, M. & SCHMIDT, V. A. 1977. Hysteresis properties of titanomagnetites, grain size and compositional dependence. *Physics of the Earth and Planetary Interiors*, **13**, 260–267.

DEPAOLO, D. 1981. Neodynium isotopes in the Colorado Front Range and crust-mantle evolution in the Proterozoic. *Nature*, **291**, 193–196.

DUCHESNE, J. C., SCHÄRER, U. & WILMART, E. 1993. A 10 Ma period of emplacement for the Rogaland anorthosites, Norway: evidence from U–Pb ages. *Terra abstracts*, **5**, *Abstract supplement No.1 to TERRA nova*, 64.

DUNLOP, D. J. 1981. The rock magnetism of fine particles. *Physics of the Earth and Planetary Interiors*, **26**, 1–26.

——. 1983. Determination of domain structure in igneous rocks by alternating field and other methods. *Earth and Planetary Science Letters*, **63**, 353–367.

ELMING, S.-Å., PESONEN, L. J., LEINO, M. A. H., KHRAMOV, A. N., MIKHAILOVA, N. P., ET AL. 1993. The drift of the Fennoscandian and Ukrainan shields during the Precambrian: a palaeomagnetic analysis. *Tectonophysics*, **223**, 177–198.

——, TORSVIK, T. H., MERTANEN, S., KRASNOVA, A. F., BYLUND, G., ET AL. 1996. *Catalogue of palaeomagnetic directions and poles from Fennoscandia: Archaean to Tertiary*. Second issue.

ERNST, R. 1990. Magma flow directions in two mafic Proterozoic dyke swarms of the Canadian Shield: As estimated using anisotropy of magnetic susceptibility data. In: PARKER, A. J, RICKWOOD, P. C. & TUCKER, D. H. (eds) *Mafic Dykes and Emplacement Mechanisms*. Balkema, Rotterdam, 231–235.

EVERITT, C. W. F & CLEGG, J. A., 1962. A field test of palaeomagnetic stability. *Geophysical Journal Royal Astronomical Society*, **6**, 312–319.

FEDOTOV, ZH. A. & AMELIN, YU. V. 1992. *Dyke magmatism on the Kola Peninsula, as reflecting Proterozoic activity of the Belomorian mobile zone in adjacent stable megablocks*. International Symposium IGCP-project 275 and IGCP-project 257, Petrozavodsk, September 7–17, 1992, Abstracts, 20–22.

FISHER, R. 1953. Dispersion on a sphere. *Proceedings of the Royal Society, London*, **A217**, 293–305.

GORBATSCHEV, R., LINDH, A., SOLYOM, Z., LAITAKARI, I., ARO, K., ET AL. 1987. Mafic dyke swarms of the Baltic shield. *In*: HALLS, H. C. & FAHRIG, W. F. (eds) *Mafic dyke swarms*. Geological Association of Canada, Special Paper, **34**, 361–372.

HAAPALA, I., FRONT, K., RANTALA, E. & VAARMA, M. 1987. Petrology of Nattanen-type granite complexes, northern Finland. *Precambrian Research*, **35**, 225–240.

HALLS, H. C. & PALMER, H. C. 1990. The tectonic relationship of two Early Proterozoic dyke swarms to the Kapuskasing Structural Zone: a palaeomagnetic and petrographic study. *Canadian Journal of Earth Sciences*, **27**, 87–103.

HROUDA, F. 1994. A technique for the measurement of thermal changes of magnetic susceptibility of weakly magnetic rocks by the CS-2 apparatus and KLY-2 Kappabridge. *Geophysical Journal International*, **118**, 604–612.

—— & CHLUPÁCǑVÁ, M. 1980. The magnetic fabric in the Nasavrky massif. *Casopis pro mineralogii a geologii*, **25**, 17–27.

HUHMA, H. 1986. Sm–Nd, U–Pb and Pb–Pb isotopic evidence for the origin of the Early Proterozoic Svecokarelian crust in Finland. *Bulletin of the Geological Survey of Finland*, **337**.

HYODO, H. & DUNLOP, D. J. 1993. Effect of anisotropy on the paleomagnetic contact test for a Grenville dike. *Journal of Geophysical Research*, **98 (B5)**, 7997–8017.

JOHANSSON, L. & JOHANSSON, Å. 1990. Isotope geochemistry and age relationships of mafic intrusions along the Protogine Zone, southern Sweden. *Precambrian Research*, **48**, 395–414.

KIRSCHVINK, J. L.1980. The least-squares line and plane and the analysis of palaeomagnetic data. *Geophysical Journal of the Royal Astronomical Society*, **62**, 699–718.

KRILL, A. G., BERG, S., LINDAHL, I., MEARNS, E. W., OFTEN, M., ET AL. 1985. Rb–Sr, U–Pb and Sm–Nd isotopic dates from the Precambrian rocks of Finnmark. *Geological Survey of Norway, Bulletin*, **403**, 37–54.

LEINO, M. A. H. 1991. *Paleomagneettisten tulosten monikomponenttianalyysi pienimmän neliösumman menetelmällä*. Laboratory for Palaeomagnetism, Department of Geophysics, Geological Survey of Finland, Report Q29.1/91/2 (in Finnish).

LISTER, J. R. & KERR, R. C. 1990. Fluid-mechanical models of dyke propagation and magma transport. *In*: PARKER, A. J, RICKWOOD, P. C. & TUCKER, D. H. (eds) *Mafic Dykes and Emplacement Mechanisms*. Balkema, Rotterdam, 69–80.

LOWRIE, W. & FULLER, M. 1971. On the alternating field demagnetization characteristics of multidomain thermoremanent magnetization in magnetite. *Journal of Geophysical Research*, **76**, 6339–6349.

MCCLELLAND-BROWN, E. 1981. Palaeomagnetic estimates of temperatures reached in contact metamorphism. *Geology*, **9**, 112–116.

MERILÄINEN, K. 1976. The granulite complex and adjacent rocks in Lapland, northern Finland. *Bulletin of the Geological Survey of Finland*, **281**.

MIDTUN, R. D. 1988. Karasjok Greenstone Belt. Regional and geophysical interpretation. *Geological Survey of Norway, Skrifter*, **88**.

MULLINS, C. E. 1977. Magnetic susceptibility of the soils and its significance in soil science – a review. *Journal of Soil Sciences*, **28**, 223–246.

MURTHY, G. S. & DEUTSCH, E. R., 1975. A new Precambrian paleomagnetic pole for northern Europe. *Physics of the Earth and Planetary Interiors*, **11**, 91–96.

OLESEN, O. & SANDSTAD, J. S. 1993. Interpretation of the Proterozoic Kautokeino Greenstone Belt, Finnmark, Norway from combined geophysical and geological data. *Geological Survey of Norway, Bulletin*, **425**, 43–64.

—— & SOLLI, A., 1985. Geophysical and geological interpretation of regional structures within the Precambrian Kautokeino Greenstone Belt, Fibbmark, North Norway. *Geological Survey of Norway, Bulletin*, **403**, 119–129.

O'REILLY, W. 1976. *Rock and Mineral Magnetism*. Blackie, Glasgow.

PAPUNEN, H., IDMAN, H., ILVONEN, E., NEUVONEN, K. J., PIHLAJA, P. & TALVITIE, J. 1977. *Lapin ultramafiiteista*. Geological Survey of Finland, Report of Investigation, **23** (in Finnish).

PARRY, L. G. 1980. Shape-related factors in the magnetization of immobilized magnetite particles. *Physics of the Earth and Planetary Interiors*, **22**, 144–154.

—— 1982. Magnetization of immobilized particle dispersions with two distinct particle sizes. *Physics of the Earth and Planetary Interiors*, **28**, 230–240.

PATCHETT, P. J. & BYLUND, G. 1977. Age of Grenville belt magnetisation: Rb–Sr and palaeomagnetic evidence from Swedish dolerites. *Earth and Planetary Science Letters*, **35**, 92–104.

—— & NEUVONEN, K. J. 1981. Palaeomagnetism of the Baltic shield – implications for Precambrian tectonics. *In*: KRÖNER, A. (ed.) *Precambrian Plate Tectonics*. Elsevier, Amsterdam, 623–648.

——, BYLUND, G., TORSVIK, T. H., ELMING, S.-Å. & MERTANEN, S. 1991. Catalogue of palaeomagnetic directions and poles from Fennoscandia: Archaean to Tertiary. *Tectonophysics*, **195**, 151–207.

——, HUHMA, H. & NEUVONEN, K. 1986. *Palaeomagnetic and Sm–Nd isotopic data of the Late Precambrian Laanila diabase dyke swarm, northeastern Finland*. 17e Nordiska Geologmötet, Helsingfors Universitet 12.-15.5.1986, Abstracts, 149.

——, TORSVIK, T. H., ELMING, S. -Å. & BYLUND, G. 1989. Crustal evolution of Fennoscandia – palaeomagnetic constraints. *Tectonophysics*, **162**, 27–49.

——, LEINO, M. A. H. & LAMMI, A. 1983. Paleomagnetism of the Baltic Shield – a state of art review. *In*: HJELT, S. E. (ed.) *The development of deep geoelectric model of the Baltic Shield, Part 2*. Department of Geophysics, University of Oulu, Report 8, 355–370.

PIHLAJA, P. 1987. The diabase of Laanila. *In*: ARO, K. & LAITAKARI, I. (eds) *Diabases and other mafic dyke rocks in Finland*. Geological Survey of Finland, Report of Investigation, **76**, 189–197 (in Finnish with English abstract).

POORTER, R. P. E. 1975. Palaeomagnetism of Precambrian rocks from southeast Norway and south Sweden. *Physics of the Earth and Planetary Interiors*, **10**, 74–87.

PURANEN, R., PEKKARINEN, L. J. & PESONEN, L. J. 1992. Interpretation of magnetic fabrics in the Early Proterozoic dykes of Keuruu, central Finland. *Physics of the Earth and Planetary Interiors*, **72**, 68–82.

SCHMIDT, P. W., 1991. An attempt to determine uplift of the Sydney Basin, New South Wales, Australia, from the paleomagnetic signatures of dyke contacts from Kiama. *In*: PARKER, A. J, RICKWOOD, P. C. & TUCKER, D. H. (eds) *Mafic Dykes and Emplacement Mechanisms*. Balkema, Rotterdam, 263–271.

—— 1993. Palaeomagnetic cleaning strategies. *Physics of the Earth and Planetary Interiors*, **76**, 169–178.

SCHWARZ, E. J. 1977. Depth of burial from remanent magnetization: the Sudbury irruptive at the time of diabase intrusion (1250 Ma). *Canadian Journal of Earth Sciences*, **14**, 82–88.

STEARN, J. E. F. & PIPER, J. D. A. 1984. Palaeomagnetism of the Sveconorwegian mobile belt of the Fennoscandian shield. *Precambrian Research*. **23**, 201–246.

STEPHENSON A., SADIKUN, S. & POTTER, D. K. 1986. A theoretical and experimental comparison of the anisotropies of magnetic susceptibility and remanence in rocks and minerals. *Geophysical Journal Royal Astronomical Society*, **84**, 185–200.

STORETVEDT, K. M. 1966a. Application of rock magnetism in estimating the age of some Norwegian dikes. *Norges geologiska Tidskrift*, **46**, 193–202.

—— 1966b. *Remanent magnetization of some dolerite intrusions in the Egersund area, southern Norway*. Norske Videnskaps-akademi is Oslo, Geophysical Publication, **26**.

SUNDVOLL, B. 1987. The age of the Egersund dyke-swarm, SW Norway: some tectonic implications. *Terra Cognita*, **7**, 180 (abstract).

TARLING, D. H. & HROUDA, F. 1993. *The Magnetic Anisotropy of Rocks*. Chapman & Hall, London.

THIRLWALL, M. F. 1991. Long-term reproducipility of multicollector Sr and Nd isotope ratio analysis. *Chemical Geology*, **94**, 85–104.

VIHAVAINEN, R. 1981. Paleomagneettinen iänmääritys-menetelmä. Esimerkkinä Laanilan diabaasijuonten magneettinen ikä. *In*: PESONEN, L. J. (ed.) *Petrofysiikan sovellutuksia, jatkokoulutusjulkaisu TKK-V-GEO B9*. Helsinki University of Technology, Economical geology, 1–19 (in Finnish).

ZAPLETAL, K. 1992. Self-reversal of isothermal remanent magnetization in a pyrrhotite (Fe_7S_8) crystal. *Physics of the Earth and Planetary Interiors*, **70**, 302–311.

ZIJDERVELD, J. D. 1967. A.C. demagnetization in rocks: analysis of results. *In*: COLLINSON, D. W. , CREER, K. M. & RUNCORN, S. K. (eds) *Methods in paleomagnetism*. Elsevier, New York, 254–286.

The provenance of pre-Scandian continental flakes within the Caledonide Orogen of south-central Norway

TREVOR F. EMMETT

Division of Chemistry and Geology, Anglia Polytechnic University, East Road, Cambridge CB1 1PT, UK

Abstract: A significant portion of the Caledonian nappe pile of south-central Norway is composed of tectonic flakes of continental crust older than the arkosic sediments of the Upper Precambrian–Lower Cambrian Hedmark and Valdres Groups. Models deriving these flakes from a suture zone S and SE of the Western Gneiss Region have found less favour than those deriving them from a root zone lying off the present-day coast line to the NW. Recent models propose that the largest flake, the Jotun–Valdres Nappe Complex, is a microcontinental suspect terrane structurally bounded above and below by ophiolitic sutures. The pre-Caledonian geological evolution of the Jotun–Valdres Nappe Complex and the Bergen Arcs is shown to be comparable to that of adjacent parts of the autochthonous Fennoscandian Shield, and to a putative root zone along strike from the Lofoten–Vesterålen Province. It is concluded that the Jotun–Valdres Nappe Complex and, by implication, related units in the Bergen Arcs and elsewhere, are not 'suspect' and are portions of the Mid to Late Proterozoic Baltoscandian crust, detached and thrust into position during the *c.* 390 Ma Scandian (= late-Caledonian) orogenic event.

The Caledonian orogenic belt of southern Norway consists of many discrete thrust sheets resting with tectonic contact on the non-Caledonized autochthon of the Telemark block and its attendant Lower Palaeozoic cover (Figs 1 & 2 and Bryhni & Sturt 1985). The nappe pile is disposed in a broad arc around the 'Caledonized' basement of the Western Gneiss Region (=WGR), and is generally preserved in a fault-related downward flexure of the basement known as the Faltungsgraben (Bryhni & Sturt 1985; Hurich & Kristoffersen 1988). The nappe pile contains sheets (tectonic flakes) of high-grade orthogneisses of various types but of a common character that was recognized by Goldschmidt (1916) as *der stamm der Bergen-Jotun-gesteine* (i.e. the Bergen–Jotun kindred, henceforth, for simplicity, the Jotun kindred). Such rocks comprise the most part of the Jotun–Valdres Nappe Complex (= JVNC, Milnes & Koestler 1985), and form the distinctive framework of the Bergen Arcs (Kolderup & Kolderup 1940). Comparable rock types also constitute the higher portions of the Hardanger–Ryfylke Nappe Complex (= HRNC, see Fig. 2 and Bryhni & Sturt 1985). Gneisses ascribed to the Jotun kindred also constitute the Dalsfjord Suite (or Nappe) that rests with tectonic contact upon the WGR (Fig. 2 and Brekke & Solberg 1987). In addition, there are several occurrences of putative Bergen–Jotun kindred gneisses within the WGR complex (Bryhni 1977), but the tectonic significance of these is uncertain and they will not be considered further in this work. Taken together, Jotun kindred rocks occupy a large volume of the Caledonian crustal volume of southern Norway (their outcrop area is about 40% of that of the WGR), and their presence has always demanded explanation in any tectonic model proposed for this portion of the orogen.

The broad regional symmetry of the structural setting of the JVNC (on the NW side thrusts dip to the SE and on the SE side to the NW) has suggested to some authors (e.g. Banham *et al.* 1979; Gorbatschev 1985) that the core of the JVNC was rooted in the Faltungsgraben, a view supported by some geophysical evidence (Smithson *et al.* 1974). In opposition to this view, geological arguments have generally favoured the emplacement of the JVNC as a far-travelled sheet derived from a root zone lying several hundreds of kilometres to the present-day northwest (Fig. 1 and Hossack 1978; Milnes & Koestler 1985). In such a model a suture zone is usually placed at the base of the ophiolite-bearing nappes of the Upper Allochthon of the Trondheim region, which junction overlies the northeastern part of the JVNC (Fig. 2 and Bryhni & Sturt 1985; Hossack & Cooper 1986). In addition, these models regard the gneisses of the WGR as the 'Caledonized' outboard portion of the Baltoscandian craton which is now exposed as a large tectonic window through the Scandian nappe pile. Contrary to this position, geophysical evidence suggests that there must be some relative displacement between the WGR and the Telemark block (Mykkeltveit *et al.* 1980; Hurich & Kristoffersen 1988), and Gorbatschev (1985) has described a tectonic contact between the WGR and presented a cogent and well argued tectonic model for the SW Norwegian Caledonides that indicates the existence of two Lower Palaeozoic ophiolitic sutures, one lying below the JVNC–Dalsfjord Suite and the other above, at the base of the nappes of the Upper Allochthon. Accordingly, these authors propose that the JVNC and the Dalsfjord Suite represent microcontinental fragments that are suspect with respect to autochthonous Baltoscandia, and not simply detached outboard

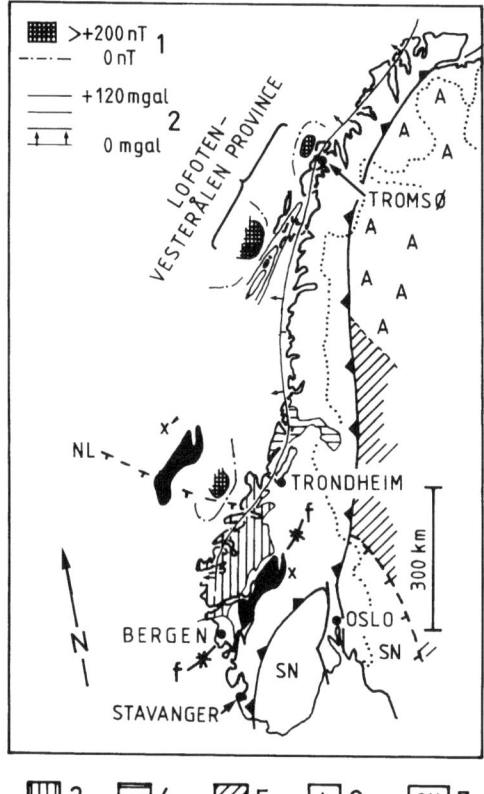

Fig. 1. Regional setting of the Caledonian orogenic belt of western Scandinavia. 1, Total intensity aeromagnetic anomalies (Borg et al. 1985). 2, Bouguer gravity anomalies (Henkel et al. 1985). Intermediate contours are +60 mgal and +100 mgal. 3, Caledonized autochthonous basement of the Western Gneiss Region. 4, Grong–Olden Culmination. 5, Transscandinavian Igneous Belt (Gorbatschev & Bogdanova 1993). 6, Archaean and Svecofennian continental crust. 7, Sveconorwegian continental crust. 8, Caledonian thrust front (barb on orogen side, Hossack & Cooper 1986). 9, Synformal axis of the Faltungsgraben (Bryhni & Sturt 1985). The line labelled NL is the northern limit of Sveconorwegian re-working of Baltoscandia (from Gorbatschev & Bogdanova 1993). The black area marked x is the current position of the main part of the Jotun–Valdres Nappe Complex, and the area marked x' is its calculated pre-Scandian position (Hossack 1978).

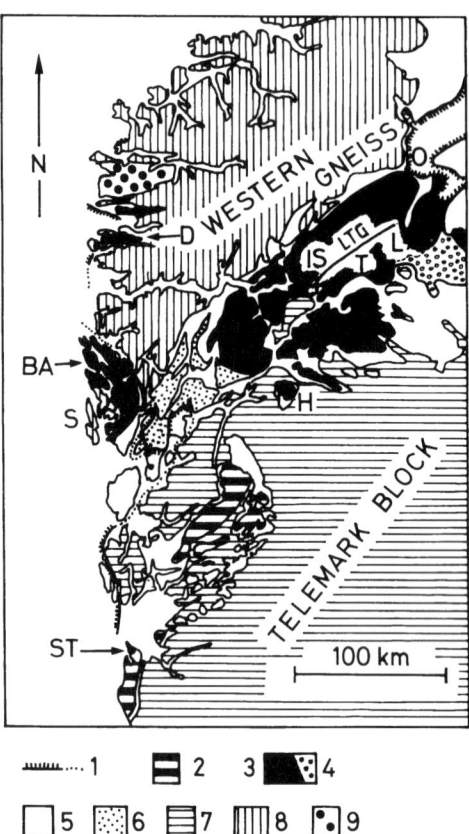

Fig. 2. Geological map of southern Norway, simplified by omission of detail from Bryhni & Sturt (1985). 1, Base of ophiolite-bearing nappes of the Upper Allochthon. Ornament on superincumbent side. 2, Hardanger–Ryfylke Nappe Complex. 3, Mainly orthogneisses of the Bergen–Jotun kindred. 4, Sparagmitic sediments unconformable on Jotun kindred massifs. 5, Mainly sedimentary units of the Lower Allochthon. 6, The Bergsdalen Nappes. 7, Autochthonous basement of the Fennoscandian Shield (sedimentary cover not differentiated). 8, Caledonized autochthon of the Western Gneiss Region. 9, Devonian molasse. BA, Bergen Arcs. D, Dalsfjord Suite (Nappe). H, Hardangerjøkulen. IS, Indre Sogn. L. Leirungsmyran. LTG, Lærdal–Tyin–Gjende Fault. O, Otta. S, Sotra. ST, Stavanger. T, Tyin.

the autochthonous Transscandinavian Granite-Porphyry Belt in the western end of the Grong-Olden Culmination (see Fig. 1). However, these arguments in themselves can neither prove nor disprove the Baltoscandian provenance of the WGR (Hurich & Kristoffersen 1988).

Recently, Andersen & Andresen (1994) have

portions of Baltoscandia. Germane to this argument is the observation that in several localities rocks of the Jotun kindred are unconformably overlain by conglomerates, quartzites, and arkosic sediments ('sparagmites') of the Valdres Group (Bryhni & Sturt 1985; Kumpulainen & Nystuen 1985). Though the provenance of these sediments is not proven, Nystuen & Siedlecka (1988) correlate them, albeit on lithological groups only, with part of the Hedmark Group which itself has correlatives that rest unconformably upon the autochthon (Kumpulainen & Nystuen 1985). Accordingly, these sparagmite basins serve to stitch the Jotun kindred terranes to Baltoscandia during the Late Proterozoic, or, at the latest, Early Cambrian (Nystuen & Siedlecka 1988). Roberts (1988) regards the Hedmark Group as a component part of the 'Baltoscandian miogeocline', so if the Jotun kindred terranes are exotic to Baltoscandia, they must have been accreted to Baltoscandia **before** the onset of deposition of the Hedmark and Valdres Groups and equivalent rocks. The author acknowledges that the correlation and contemporaneity of parts of the Hedmark Group and the Valdres Group is not proven (they occupy distinct depocentres, Kumpulainen & Nystuen 1985), but the possibility is rather summarily dismissed by Andersen & Andresen (1994, pp. 74–75).

All the above tectonic models pay scant, if any, attention to the Jotun kindred rocks themselves. Though there is now a useful volume of basic mapping available, reliable isotopic data for the Bergen Arcs and JVNC are still rather sparse, but enough are available to make reasonable comparisons with surrounding areas. The intention of this paper is to review the geological character and geochronology of the Jotun kindred rocks within the JVNC, the Bergen Arcs, and elsewhere, and then, by comparison with the rest of the Baltic Shield and possible root zones, to demonstrate that this character is essentially Baltoscandian. Localities referred to are identified in Figs 1 & 2, and the geochronological data are summarized in Table 1. Where necessary, published radiometric results have been recalculated using the decay constants recommended by Steiger & Jäger (1977).

Pre-Caledonian evolution of Baltoscandia

Relevant aspects of the geological evolution of the autochthonous Fennoscandian Shield (Baltica) have been reviewed by Åberg (1988), Gaál & Gorbatschev (1987) and Gorbatschev & Bogdanova (1993). Gorbatschev & Bogdanova (1993) described the Svecofennian Orogen (c. 2000–1800 Ma) as being truncated by the Transscandinavian Igneous Belt (c. 1800–1650 Ma), with the period c. 1750–1550 Ma representing the last stage of major crustal growth in the province (the Gothian orogeny of Gaál & Gorbatschev 1987). Park et al. (1991) relate these processes to northward-directed subduction under Laurentia, followed by accretion of the Southwest Scandinavian Microplate to Laurentia at c. 1500 Ma. This crust was re-worked during oblique collision and transform movements during the Svceconorwegian (1200–900 Ma) event (Park et al. 1991; Gorbatschev & Bogdanova 1993). The time period between the Gothian and Sveconorwegian orogenies is characterized by temporally sporadic but widely distributed so-called anorogenic magmatism, with mafic and subordinate granitic magmatism developed, but with most of the igneous activity taking place between 1400 and 1350 Ma (Åberg 1988; Gaál & Gorbatschev 1987); this activity would be coeval with high-grade metamorphism and related crustal anatexis taking place in southwest Sweden, the so-called Hallandian Orogeny. However, the status and significance of this orogeny is not clear (Gaál & Gorbatschev 1987, who refer to it as a 'quasi-orogeny').

The Lofoten–Vesterålen Province (= LVP) of northern Norway (Fig. 1) and the WGR are believed to be portions of the outboard portion of the Fennoscandian Shield exposed in windows through the Caledonian nappe pile. Gorbatschev (1985), though favouring the existence of a Caledonide suture east of the WGR, still makes the point that, for both provinces, the bulk of the geological, geochronological, and geophysical evidence favours this conclusion. The Archaean–Proterozoic crustal evolution of the LVP has been summarized by Griffin et al. (1978). Major crustal growth, including deposition of supracrustal rocks and high-grade metamorphism, took place between 2700 and 1830 Ma, and this was followed by extensive basic and anorthositic–charnockitic magmatism between 1800–1700 Ma. Undated dolerites were intruded prior to a period of crustal anatexis at c. 1400 Ma. According to Gorbatschev & Bogdanova (1993), the LVP lies north of the limit of Sveconorwegian reworking (Fig. 1); this phase (1150–900 Ma) is, in fact, represented in the LVP by emplacement and metamorphism of the Leknes Group schists and retrograde metamorphism of granulite facies rocks. In contrast, the WGR appears to lack the Archaean history of the LVP. Tucker et al. (1987) were able to demonstrate that the continental protolith of the WGR was emplaced, migmatized, and cooled in a very short time period at c. 1657 Ma (Svecofennian events), conclusions confirmed by the review of Kullerud et al. (1986), who also pointed out a concentration of ages between 1250 and 900 Ma (Sveconorwegian events). Initial $^{87}Sr/^{86}Sr$ ratios for

Table 1. *Summary of geochronological data*

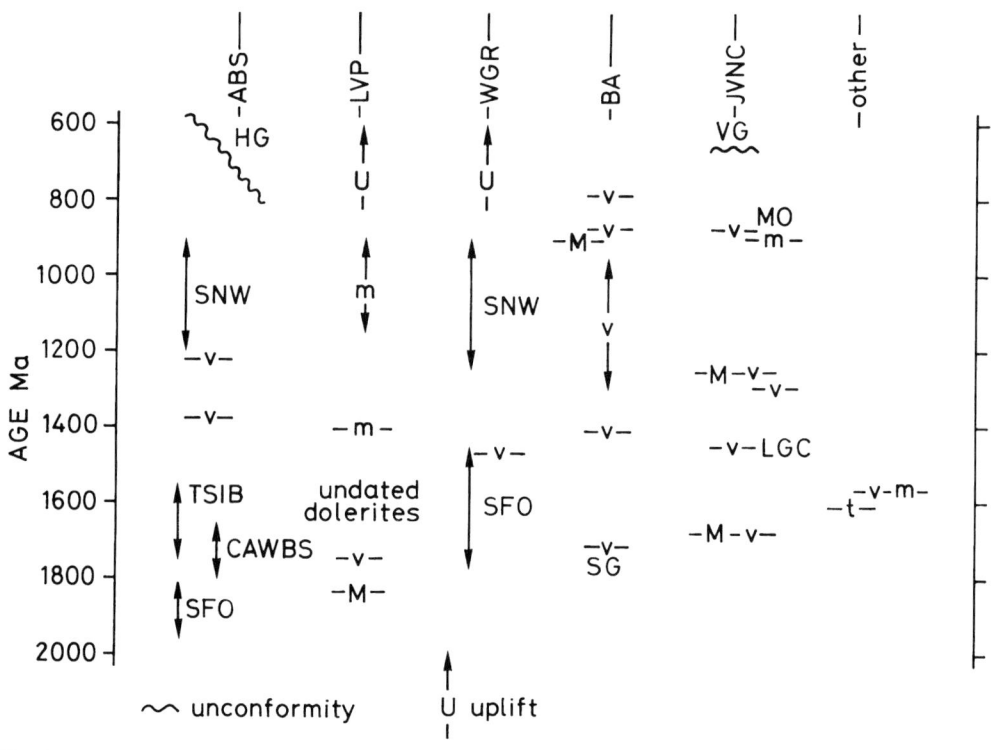

Each column represents one of the terranes discussed in the text. ABS, autochthonous Baltoscandia (Baltic Shield). LVP, Lofoten–Vesterålen Province. WGR, Western Gneiss Region. BA, Bergen Arcs. JVNC, Jotun–Valdres Nappe Complex. Other, Hardanger–Ryfylke Nappe Complex and Hardangerjøkulen. Geological events: m, low grade and/or retrograde metamorphism. M, high grade metamorphism. t, thrusting. v, igneous activity. Other abbreviations: CAWBS, crustal accretion in the western Fennoscandian Shield (Gorbatschev & Bogdanova 1993). HG, Hedmark Group (part of). LGC, Leirungsmyran Gabbroic Complex (Corfu & Emmett 1992). MO, Main Orogeny (Milnes & Koestler 1985). SFO, Svecofennian orogeny. SG, Sotra gneisses (Sturt *et al.* 1975). SNW, Sveconorwegian reworking. TSIB, Transscandinavian Igneous Belt (Gorbatschev & Bogdanova 1993). VG, Valdres Group. All sources are discussed and cited in the main text.

this latter period were higher than for the Svecofennian events, an observation suggestive of the conclusion that the Sveconorwegian events were concerned mainly with the reworking of pre-existing continental crust. The period of time between the Svecofennian and Sveconorwegian events is marked only by sporadic igneous activity (Gorbatschev 1985; Corfu & Emmett 1992).

Pre-Caledonian geological evolution of units containing Jotun kindred gneisses

Bergen Arcs

The relationship between the slices of Jotun kindred gneisses within the Bergen Arcs and the JVNC is rather obscure (Fig. 2). Johns (1978) established that the form of the Arcs is the result of a west-plunging antiform re-folding an earlier N–S trending antiform. The arcs are separated from the JVNC by the large thickness of continental-type gneisses that constitute the Bergsdalen Nappes (Bryhni & Sturt 1985). Qualitatively, a down-the-plunge view of the Faltungsgraben may reveal the Bergen Arcs (and part of the Bergsdalen Nappes and HRNC) as an imbricate zone below the JVNC (see cross-sections in Andresen & Færseth 1982; Bryhni & Sturt 1985; Andersen & Andresen 1994).

The only systematic geochronological examination of the pre-Palaeozoic rocks of the Bergen Arcs is that of Sturt *et al.* (1975), but other useful contributions have been made by Cohen *et al.* (1988) and Burton *et al.* (1995). The oldest rocks appear to be the acidic gneisses of Sotra, which yielded an age of 1738–1713 Ma. This basement complex was reworked during Sveconorwegian events dated at c. 1150 Ma (Burton et al. 1995); the events included the intrusion of mangeritic magmas, and were terminated by intrusion of granitic veins at 890–800 Ma. Both Cohen et al. (1988) and Burton et al. (1995) deduced that high-grade corona formation took place at c. 910 Ma, and this event is correlated this with the 900 Ma reworking recorded in the JVNC by Schärer (1980).

Jotun-Valdres Nappe Complex

Root zone of the JVNC. Hossack (1978) was able to demonstrate that the accumulated strain in the sub-Jotun metasediments indicated a total southeastward displacement of about 300 km. If the nappe is restored to its pre-thrusting position (Hossack 1978 and Fig. 1), it would straddle the southwestward prolongation of major linear gravity and magnetic anomalies that characterize the LVP. These anomalies are distinctly elongated NE–SW and comprise (a) a series of +40 to >+120 mgal Bouguer anomalies and, similarly, (b) a series of >+100 nT magnetic anomalies (Borg et al. 1985; Henkel et al. 1985). These observations suggest to the author that the putative root zone of the JVNC consists of rocks similar to those outcropping in the LVP.

Geological evolution of the JVNC. Radiometric studies of rocks from the JVNC are not great in number, and the only systematic surveys have been those of Schärer (1980) and Koestler (1983). Some useful data were also presented by Corfu (1980) and Koestler (1982). These authors were able to establish the following general framework (localities referred to are located in Fig. 2):

(1) The Jotun kindred gneisses have a magmatic age of c. 1680 Ma. The high-grade metamorphism suffered by these rocks is considered to have occurred either during or very soon after their intrusion.
(2) A period of gabbroic magmatism at c. 1250 Ma. These gabbros show a characteristic development of garnetiferous coronas. Griffin et al. (1985) cited unpublished work suggesting that anorthosites and related gabbroic rocks occurring in the Indre Sogn region of the JVNC had magmatic ages c. 1300 Ma, and underwent a long period of cooling that culminated in corona formation at c. 870 Ma (Nd–Sm methods).
(3) Widespread deformation and partial retrogression at about 900 Ma, possibly associated with intrusion of granodioritic and/or trondhjemitic magmas (Koestler 1982). Rather misleadingly in the author's opinion, events of this stage are referred to as the 'Main Orogeny' by Milnes & Koestler (1985).
(4) Scandian cooling and uplift dated at c. 380–400 Ma by Schärer (1980) and Corfu (1980) involved the development of thrusts and related zones of cataclasis (Koestler 1988; Milnes & Koestler 1985; Milnes et al. 1988).

Corfu & Emmett (1992) reported the results of a U–Pb study on the Leirungsmyran Gabbroic Complex, and these require the established geochronological framework of the JVNC to be slightly modified. The magmatic age of the LGC, 1450 Ma, represents a previously unknown period of activity within the JVNC, and events of this age have not been commonly reported from western Norway (Kullerud et al. 1986). The lack of evidence for any so-called Main Orogeny (c. 900 Ma) overprint is also unexpected; this is difficult to explain except by concluding either that there is no 900 Ma event in this part of the JVNC, or that the effects of the Main Orogeny in this area were less intensive than the subsequent Scandian events (dated at c. 400 Ma by Corfu & Emmett 1992). However, the latter explanation is considered unsatisfactory since, if anything, the Scandian overprint would presumably have been at a lower grade than the Main Orogeny (Schärer 1980; Milnes & Koestler 1985); some evidence of a 900 Ma event should have been preserved in the isotopic systems investigated here, but none was seen. Another curious feature of the results of Corfu & Emmett (1992) is the lack of any c. 1250 Ma effects to correspond with the intrusion of Schärer's coronitic gabbros.

Those portions of the JVNC investigated by Schärer (1980) and Koestler (1983) are both separated from the Leirungsmyran area by major tectonic breaks. The area around Tyin investigated by Schärer lies geologically within a series of thrust sheets that form the footwall of the extensional Lærdal–Tyin–Gjende Fault (Fig. 2 and Battey & MacRitchie 1973). The country rock of the Leirungsmyran Gabbroic Complex is the Svartdalen Gneiss, which forms the highest of these sheets (Corfu & Emmett 1992); the Leirungsmyran area is thus structurally higher than the Tyin region (Battey & MacRitchie 1973). Koestler's area around Indre Sogn is separated from the Leirungsmyran area by the Lærdal–Tyin–Gjende Fault and, possibly, the Utladalen Fault (Battey & MacRitchie 1973, but see also Milnes & Koestler 1985), and the existence of intervening thrusts cannot be excluded. It has to be concluded that the JVNC is a composite body, assembled in at least three phases, pre-1550 Ma, post-900 Ma, and Scandian.

Other data. Felsic gneisses comprising the erosional remnant of the JVNC at Hardangerjøkulen (Fig. 2) have yielded an age of c. 1550 Ma (Priem *et al.* 1968), and the HRNC show evidence for partial assembly prior to c. 1550 Ma (Gabrielsen *et al.* 1979). Anorthositic rocks within the WGR have yielded Rb–Sr ages of 1467 and 1479 Ma, though both isochrons show much scatter due to younger metamorphism, and the error bars are large (Abdel-Monem & Bryhni 1978). There are no age data available for rocks of the Dalsfjord Suite.

Discussion

The foregoing review and the summary provided by Table 1 illustrates that the Jotun kindred gneisses of the JVNC, the Bergen Arcs and, by analogy, the Dalsfjord Suite and portions of the HRNC, have a complex pre-Caledonian history that correlates well both with the indicated root zone, the LVP, and with the southern part of the autochthonous Fennoscandian shield in general. Though the data are not voluminous, it appears that there is nothing about the Jotun kindred rocks themselves that would lead to the conclusion that they are exotic or suspect when compared to the adjacent *in situ* basement. In particular, there is no evidence in this particular region of any significant geological disturbance in the period between the deposition of the Hedmark and Valdres Groups and the emplacement of the tectonic units during the latter stages of the Scandian orogenic event at c. 390 Ma. This observation would seem to preclude the involvement of the JVNC in arc-building events in the Iapetus basin during the Early Ordovician to Early Silurian period as proposed by Andersen & Andresen (1994).

The 'suspect microcontinent' model of Andersen & Andresen (1994) depends intimately upon the existence of sub-JVNC ophiolites. The existence of ophiolites below the JVNC is by no means as certain in the area investigated here as the existence of ophiolites above it. The linkage between the Dalsfjord Suite, which is tectonically underlain by ophiolitic rocks of the Askvoll Group (Brekke & Solberg 1987), and the main portion of the JVNC relies on comparisons between gross lithologies and tectonic position, so the connection is not established. The tectonic contact at the base of the Dalsfjord Suite, the Kvamshesten Fault, is a flat-lying extensional fault (Brekke & Solberg 1987), and so there is no compelling reason to disregard the possibility that the Dalsfjord Suite could represent a tectonic level higher than that represented by the JVNC. The occurrences on Hardangervidda mentioned by Andersen & Andresen (1994) have yet to be properly investigated.

The tectonic position of the Lower Palaeozoic ophiolitic rocks around Vågåmo and Otta (Fig. 2), so often in the past a source of disagreement–see discussions and references in Strand (1972), Bryhni & Sturt (1985), and Andersen & Andresen (1994)–would now, however, appear to be well established (Sturt *et al.* 1991; Bøe *et al.* 1993). The ophiolite-bearing Otta Nappe lies structurally below the JVNC and it contains a Jotun kindred–like basement, the Rudihø Complex, with its own locally autochthonous Valdres Group-like cover, the Heidal Series. This basement-cover 'couplet' (terminology of Bøe *et al.* 1993) is overthrust by the 488–505 Ma Vågåmo Ophiolite, and the tectonically composite terrane so-formed is stitched together by the unconformably overlying conglomerate-bearing Sel Group of Arenig-Llanvirn age (Sturt *et al.* 1991; Bøe *et al.* 1993). These field relationships, together with palaeontological evidence, have been interpreted to indicate the involvement of Jotun kindred-type continental crust in intraoceanic magmatic arcs of Late Cambrian(?) to Ordovician age, but it is the **provenance** of that continental crust which is of issue here. Bøe *et al.* (1993, p. 21) do not rule out the invocation of microcontinents, but they find the suggestion that the basement-cover couplet represents a more 'distal' fragment of the Baltoscandian margin to be 'tempting'. It must also be noted that Hossack & Cooper (1988, p. 300) suggested that ophiolites below the JVNC could be explained by out-of-sequence thrusting, a possibility also entertained by Andersen & Andresen (1994, p. 76). If the Jotun kindred-like fragments of continental crust such as the JVNC and the Rudihø Complex have behaved tectonically as microcontinents during the Lower Palaeozoic, then the author contends that they were probably slivers or flakes of the Baltoscandian margin rifted off the main part of the shield sometime after the deposition of the Hedmark and Valdres groups. The situation of the present-day Lord Howe Rise (continental crust), located between the passive eastern margin of Australia and the convergent margin of the Tonga–New Zealand–Macquarie line (Nur & Ben-Avraham 1982) may be a suitable modern analogue for the Lower Palaeozoic tectonic setting of southern Norway.

Conclusions

(1) The orthogneisses of the JVNC older than the Valdres Group have many similarities with the

LVP, and it is suggested here that they were derived from a root zone contiguous with the LVP.

(2) Elements of the petrological and tectonic evolution of the JVNC can be found all over southern Scandinavia, and the correct provenance for the JVNC and related lithotectonic units is Fennoscandian. Accordingly, there is no evidence from the Jotun kindred rocks themselves that the JVNC is in any way suspect with respect to the autochthonous part of the Fennoscandian Shield.

Fieldwork in Norway was funded by grants from NERC, the Royal Society of London, and Anglia Polytechnic University (and its various predecessors!). The assistance of Dr Brian Sturt (NGU) in the revision of an early draft of this paper is gratefully acknowledged. Correspondence to the author (e-mail: temmett@bridge.anglia.ac.UK).

References

ABDEL-MONEM, A. A. & BRYHNI, I. 1978. A Rb/Sr date from anorthosite-suite rocks of the Gloppen-Eikefjord area, western Norway. *Norsk Geologisk Tidsskrift*, **58**, 229–232.

ANDERSEN, T. B. & ANDRESEN, A. 1994. Stratigraphy, tectonostratigraphy and the accretion of outboard terranes in the Caledonides of Sunnhordland, W. Norway. *Tectonophysics*, **231**, 71–84.

ANDRESEN, A. & FÆRSETH, R. 1982. An evolutionary model for the southwest Norwegian Caledonides. *American Journal of Science*, **282**, 756–782.

ÅBERG, G. 1988. Middle Proterozoic anorogenic magmatism in Sweden and worldwide. *Lithos*, **21**, 279–289.

BANHAM, P. H., GIBBS, A. D. & HOPPER, F. W. M. 1979. Geological evidence in favour of a Jotunheimen Caledonian suture. *Nature*, **277**, 289–291.

BATTEY, M. H. & MACRITCHIE, W. D. 1973. A geological traverse across the pyroxene-granulites of Jotunheimen in the Norwegian Caledonides. *Norsk Geologisk Tidsskrift*, **53**, 237–265.

BORG, K., BERGMARK, T. GEE, D. G. & KUMPULAINEN, R. 1985. Scandinavian Caledonides Magnetic Anomaly Map, 1/2M. *In*: GEE, D. G. & STURT, B. A. (eds) *The Caledonian Orogen–Scandinavia and Related Areas*. J. Wiley & Sons, Chichester.

BREKKE, H. & SOLBERG, P. O. 1987. The geology of Atløy, Sunnfjord, western Norway. *Norges Geologiske Undersøkelse Bulletin*, **410**, 73–94.

BRYHNI, I. 1977. The gneiss region west and northwest of Jotunheimen. *In*: HEIER, K. S. (ed.) *The Norwegian Geotraverse Project*. Norges Geologiske Undersøkelse, Trondheim, 227–246.

—— & STURT, B. A. 1985. Caledonides of southwestern Norway. *In*: GEE, D. G. & STURT, B. A. (eds) *The Caledonian Orogen–Scandinavia and Related Areas*. J. Wiley & Sons, Chichester, 89–107.

BURTON, K. W., KOHN, M. J., COHEN, A. S. & O'NIONS, R. K. 1995. The relative diffusion of Pb, Nd, Sr and O in garnet. *Earth and Planetary Science Letters*, **133**, 199–211.

BØE, R., STURT, B. A. & RAMSAY, D. M. 1993. The conglomerates of the Sel Group, Otta-Vågå area, Central Norway: an example of a terrane-linking succession. *Norges Geologiske Undersøkelse Bulletin*, **425**, 1–24.

COHEN, A. S., O'NIONS, R. K., SIEGENTHALER, R. & GRIFFIN, W. L. 1988. Chronology of the pressure–temperature history recorded by a granulite terrain. *Contributions to Mineralogy and Petrology*, **98**, 303–311.

CORFU, F. 1980. U–Pb and Rb–Sr systematics in a polyorogenic segment of the Precambrian Shield, central southern Norway. *Lithos*, **13**, 305–323.

—— & EMMETT, T. F. 1992. U–Pb age of the Leirungsmyran Gabbroic Complex, Jotun Nappe, southern Norway. *Norsk Geologisk Tidsskrift*, **72**, 369–374.

GAÁL, G. & GORBATSCHEV, R. 1987. An outline of the Precambrian evolution of the Baltic Shield. *Precambrian Research*, **35**, 15–52.

GABRIELSEN, R. H., NATERSTAD, J. & RÅHEIM, A. 1979. A Rb/Sr study of a possible Precambrian thrust zone, Hardanger–Ryfylke Nappe Complex, SW Norway. *Norsk Geologisk Tidsskrift*, **59**, 253–265.

GOLDSCHMIDT, V. M. 1916. Geologische-petrographische studien im hochgebirge de Südlichen Norwegens. IV. Übersicht der eruptivgesteine im Kaledonischen gebirge zwischen Stavanger und Trondheim. *Videnskapsselskapets i Kristiania Skrifter. I. Matematisk-naturvidenskapelig Klasse*, **2**, 1–140.

GORBATSCHEV, R. 1985. Precambrian basement of the Scandinavian Caledonides. *In*: GEE, D. G. & STURT, B. A. (eds) *The Caledonian Orogen – Scandinavia and Related Areas*. J. Wiley & Sons, Chichester, 197–212.

—— & BOGDANOVA, S. 1993. Frontiers in the Baltic Shield. *Precambrian Research*, **64**, 3–21.

GRIFFIN, W. L., MELLINI, M., OBERI, R. & ROSSI, G. 1985. Evolution of coronas in Norwegian anorthosites: re-evaluation based on crystal-chemistry and microstructures. *Contributions to Mineralogy and Petrology*, **91**, 330–339.

——, TAYLOR, P. N., HAKKINEN, J. W., HEIER, K. S., IDEN, I. K. 1978. Archaean and Proterozoic crustal evolution in Lofoten – Vesterålen, north Norway. *Journal of the Geological Society of London*, **135**, 629–647.

HENKEL, H., GEE, D. G. & KUMPULAINEN, R. 1985. Scandinavian Gravity Anomaly Map, 1/2M. *In*: GEE, D. G. & STURT, B. A. (eds) *The Caledonian Orogen – Scandinavia and Related Areas*. J. Wiley & Sons, Chichester.

HOSSACK, J. R. 1978. The correction of stratigraphic sections for tectonic finite strain in the Bygdin area, Norway. *Journal of the Geological Society of London*, **135**, 229–242.

—— & COOPER, M. A. 1986. Collision tectonics in the Scandinavian Caledonides. *In:* COWARD, M. P. & RIES, A. C. (eds) *Collision Tectonics.* Geological Society, London, Special Publication, **19**, 287–304.

HURICH, C. A. & KRISTOFFERSEN, Y. 1988. Deep structure of the Caledonide orogen in southern Norway: new evidence from marine seismic reflection profiling. *Norges Geologiske Undersøkelse Special Publication*, **3**, 96–101.

JOHNS, C. C. 1978. The geology of Sotra and other basement areas west of Bergen. *In:* COOPER, M. A. & GARTON, M. R. (eds) *Tectonic Evolution of the Scandinavian Caledonides.* Proceedings of a conference held at the Department of Geology, City of London Polytechnic, on April 27, 1978, 37–40.

KOESTLER, A. G. 1982. A Precambrian age for the Ofredal granodiorite intrusion, central Jotun Nappe, Sogn, Norway. *Norsk Geologisk Tidsskrift*, **62**, 225–228.

—— 1983. *Zentral Komplex und NW-Randzone der Jotundeck, West-Jotunheimen, Suednorwegen. Struckturgeologie und Geochronologie.* PhD thesis, ETH-Zürich.

—— 1988. Heterogeneous deformation and mylonitisation of a granulite complex, Jotun-Valdres Nappe Complex, central south Norway. *Geological Journal*, **23**, 1–13.

KOLDERUP, C. F. & KOLDERUP, N.-H. 1940. Geology of the Bergen Arc system. *Bergens Museums Skrifter*, **20**, 1–137.

KULLERUD, L., TØRUDBAKKEN, B. & ILEBEKK, S. 1986. A compilation of radiometric age determinations from the Western Gneiss Region, south Norway. *Norges Geologisk Undersøkelse*, **406**, 17–42.

KUMPULAINEN, R. & NYSTUEN, J. P. 1985. Late Proterozoic basin evolution and sedimentation in the westernmost part of Baltoscandia. *In:* GEE, D. G. & STURT, B. A. (eds) *The Caledonian Orogen – Scandinavia and Related Areas.* J. Wiley & Sons, Chichester, 213–232.

MILNES, A. G. & KOESTLER, A. G. 1985. Geological structure of Jotunheimen, southern Norway (Sognefjell – Valdres cross-section). *In:* GEE, D. G. & STURT, B. A. (eds) *The Caledonian Orogen –Scandinavia and Related Areas.* J. Wiley & Sons, Chichester, 457–474.

——, DIETLER, T. N. & KOESTLER, A. G. 1988. The Sognefjord north shore log – a 25 km depth section through Caledonized basement in western Norway. *Norges Geologiske Undersøkelse Special Publication*, **3**, 114–121.

MYKKELTVEIT, S., HUSEBYE, E. S. & OFTEDAHL, C. 1980. Subduction of the Iapetus Ocean crust beneath the Møre Gneiss Region, southern Norway. *Nature*, **288**, 473–475.

NYSTUEN, J. P. & SIEDLECKA, A. 1988. The 'sparagmites' of Norway. *In:* WINCHESTER, J. A. (ed.) *The Later Proterozoic Stratigraphy of the Northern Atlantic Regions.* Blackie, Glasgow, 237–252.

NUR, A. & BEN-AVRAHAM, Z. 1982. Displaced terranes and mountain building. *In:* HSÜ, K. J. (ed.) *Mountain Building Processes.* Academic Press, London, 73–84.

PARK, R. G., ÅHÄLL, K. -I. & BOLAND, M. P. 1991. The Sveconorwegian shear-zone network of SW Sweden in relation to mid-Proterozoic plate movements. *Precambrian Research*, **49**, 245–260.

PRIEM, H. N. A., BOELRIJK, N. A. I. M., VERSCHURE, R. H., HEBEDA, E. H. & VERDURMEN, E. A. TH. 1968. *Second Progress Report on the Isotopic dating Project in Norway.* Nederlandse Organisatie voor Zuiverwetenschappelijk Onderzoek (Z.W.O.) Laboratorium Voor Isotopen-Geologie, Amsterdam.

ROBERTS, D. 1988. The terrane concept and the Scandinavian Caledonides: a synthesis. *Norges Geologiske Undersøkelse Bulletin*, **413**, 93–99.

SCHÄRER, U. 1980. U–Pb and Rb–Sr dating of a polymetamorphic nappe terrain: the Caledonian Jotun Nappe, southern Norway. *Earth and Planetary Science Letters*, **49**, 205–218.

SMITHSON, S. B., RAMBERG, I. B. & GRØNLIE, G. 1974. Gravity interpretation of the Jotun Nappe of the Norwegian Caledonides. *Tectonophysics*, **22**, 205–222.

STEIGER, R. H. & JÄGER, E. 1977. Subcommission on Geochronology: convention on the use of decay constants in geo- and cosmochronology. *Earth and Planetary Science Letters* , **36**, 359–362.

STRAND, T. 1972. The Norwegian Caledonides. *In:* STRAND, T. & KULLING, O. (eds) *Scandinavian Caledonides.* Wiley-Interscience, London, 3–145.

STURT, B. A., RAMSAY, D. M. & NEUMAN, R. B. 1991. The Otta Conglomerate, the Vågåmo Ophiolite – further indications of early Ordovician Orogenesis in the Scandinavian Caledonides. *Norsk Geologisk Tidsskrift*, **71**, 107–115.

——, SKARPENES, O., OHANIAN, A. T. & PRINGLE, I. R. 1975. Reconnaissance Rb/Sr isochron study in the Bergen Arc system and regional implications. *Nature*, **253**, 595–599.

TUCKER, R. D., RÅHEIM, A., KROGH, T. E. & CORFU, F. 1987. Uranium–lead zircon and titanite ages from the northern portion of the Western Gneiss region, south-central Norway. *Earth and Planetary Science Letters*, **81**, 203–211.

Provenance of late Proterozoic Dalradian tillite clasts, Inner Hebrides, Scotland

W. R. FITCHES[1], N. J. G. PEARCE[1], J. A. EVANS[2] & R. J. MUIR[3]

[1] *Institute of Earth Studies, University of Wales, Aberystwyth, Dyfed, Wales, UK*
[2] *NERC Isotope Geosciences Laboratory, Keyworth, Nottingham, UK*
[3] *Isotope Geology Unit, Scottish Universities Research and Reactor Centre, East Kilbride, Glasgow G75 0QU, UK*

Abstract: A geochemical and isotopic study was made of granitoid clasts taken from the c. 660 Ma Dalradian tillites of Islay and the Garvellach Islands. Objectives were to investigate tillite provenance as a means of helping to characterize the ages, compositions and distributions of basement rocks near and beneath the Dalradian Supergroup. Most clasts have closely similar trace element and REE compositions, implying that they were derived from source rocks produced in similar, but not necessarily coeval, geotectonic settings. Most samples have geochemical traits of Within-Plate Granites. Nd isotopic model ages of c. 2.0 Ga were obtained from several clasts. One sample gave a Rb–Sr mica whole-rock age of c. 1.57 Ga, which is probably close to the age of igneous crystallization.

The clasts are petrographically similar to the syenites of the Rhinns Complex, the c. 1.8 Ga alkalic igneous suite exposed locally in the Hebridean region SE of the Great Glen Fault. Most clasts, however, are geochemically distinct from the Volcanic Arc-type Rhinns Complex. The isotopic model ages of the clasts overlap those of the Rhinns Complex, but the c. 1.57 Ga age of one clast is substantially younger. Although most clasts were not eroded from the Rhinns Complex of the type presently exposed, it is possible that they were derived from plutonic rocks closely associated in space and time with the Rhinns Complex. The source plutons and the Rhinns Complex were probably components of the early to mid Proterozoic Svecofennian–Ketilidian–Labradorian–Makkovikian mobile belt which contains voluminous plutonic rocks with diverse ages and compositions.

The basic purpose of this study was to test a hypothesis put forward by Fitches *et al.* (1990) that granitoid clasts in the c. 660 Ma Dalradian tillites of southwest Scotland were derived from local basement rocks now exposed in the Inner Hebrides (Fig.1). If the hypothesis is valid, several fundamental problems concerning the Proterozoic geotectonic evolution of Scotland would be largely resolved. If the hypothesis is not substantiated, those problems persist. The study objectives need to be seen in the context of two main themes of research: the geotectonic setting of the basement rocks of the Inner Hebrides, and the provenance of the tillite clasts.

The crystalline rocks that crop out on the Rhinns of Islay and on Colonsay in the Inner Hebrides were assumed to be part of the late Archaean–early Proterozoic Lewisian Complex which underlies Scotland northwest of the Great Glen Fault. Bentley (1986), Bentley *et al.* (1988) and others, however, pointed out that they are quite different from the basement rocks of northern Scotland. Because they and their late Proterozoic cover of Colonsay Group metasedimentary rocks appeared to have no equivalents in Britain, a West Islay–Colonsay block was identified. That block was interpreted to be an allochthonous terrane which docked against southwest Scotland in the late Proterozoic or early Palaeozoic (Bentley *et al.* 1988). That terrane model was adopted by Rogers *et al.* (1989) who inferred that docking by orthogonal collision was the cause of the Grampian orogeny in the late Proterozoic.

Muir (1990) confirmed that the Islay–Colonsay basement, with its counterparts on the islands of Inishtrahull and Torr Rocks off northwest Ireland, is distinct from any part of the Lewisian Complex. The Hebridean basement, named the Rhinns Complex by Muir *et al.* (1989; 1992*a*, *b*; 1994*a*, *b*), was shown to be a deformed igneous suite composed mainly of syenite with gabbro, minor intrusions of dolerite and diorite, and rare ultramafic bodies. The suite has the petrological and geochemical hallmarks of alkalic assemblages caused by subduction at Andean-type destructive margins. Muir (1990) and Daly *et al.* (1991) used various isotopic methods to show that the syenite was emplaced at c. 1800 Ma, thereby confirming rigorously the preliminary age determinations made by Marcantonio *et al.* (1988). The gabbro,

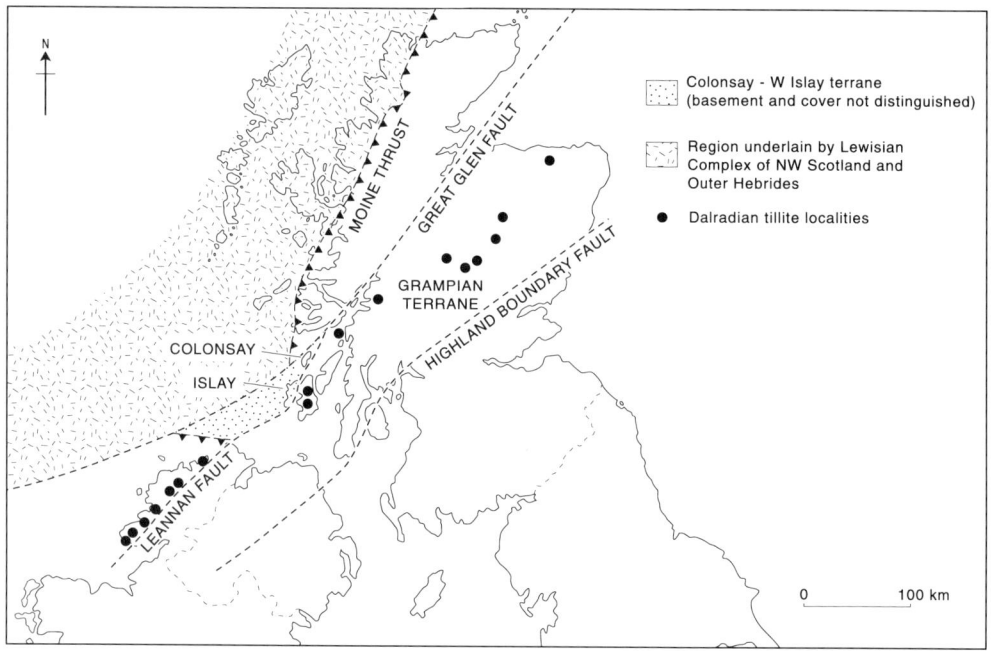

Fig. 1. The Rhinns Complex (basement of Colonsay-W. Islay terrane) and Dalradian tillite outcrops (after Muir 1990). Colonsay and Islay are islands in the Inner Hebrides.

emplaced into already deformed syenite, appears to have been metamorphosed at c. 1.7 Ga (Muir 1990).

Although the Rhinns Complex is distinct from the Lewisian Complex, the contrast does not necessarily imply that the Inner Hebridean block is an allochthonous terrane, as pointed out by Fitches et al. (1990) and by Muir et al. (1992a, in discussion of Dickin & Bowes 1991). It had been inferred by Watson (1975) that the c. 1800 Ma Ketilidian mobile belt of South Greenland might extend under southern parts of the Scottish Highlands, a view supported on isotopic grounds by Marcantonio et al. (1988). This interpretation was developed by Muir et al. (1989, 1992b) who proposed that the Rhinns Complex represents the Scottish link in the 6000 km long, 2.00–1.75 Ga old mobile belt, shown in Fig. 2, which extends from North America (Makkovikian), through South Greenland (Ketilidian) to Scandinavia (Svecofennian). That belt accreted onto the southern flank of the Archaean craton of the North Atlantic region during subduction, magmatic arc construction and allied processes (see, for example, papers in Gower et al. (1990) and in this volume). Accordingly, the Rhinns Complex is neither 'exotic' nor allochthonous: it is situated where expected on most reconstructions of the Proterozoic geotectonic units of the North Atlantic region. On that basis, moreover, the Rhinns Complex or other components of the mobile belt may well underlie much of the southern Highlands region, forming at least part of the otherwise hidden basement to the Dalradian basin. It is possible, for example, that the wedge-shaped mass imaged under the Southwest Highlands on the WINCH seismic profile (Hall et al. 1984) is composed of in situ Rhinns Complex or associated rocks.

A different but linked problem is the provenance of granitoid clasts in the Dalradian tillites. The age of the tillites is uncertain but these deposits are usually assumed, rightly or wrongly, to be coeval with the Varanger glacial deposits of northern Norway. Pringle (1972) obtained a Rb–Sr age of 678 ± 23 Ma (recalculated to c. 655 Ma using modern constants) from shales in the Varanger succession, which he interpreted to be close to the age of deposition.

The clasts in the Dalradian tillites have been considered to be extrabasinal and exotic with respect to the Dalradian basin because they appear to have no sources among exposed British and Irish rock assemblages. The source of the clasts was believed to lie somewhere to the southeast of the Dalradian basin, largely on the basis that no clasts of Lewisianoid material have been found and that there are indications, although insubstantial, of northwesterly inclined palaeoslopes and current flow. Spencer (1971) suggested on these grounds, and the evidence given by a c. 1000 Ma Rb–Sr isotopic age obtained from one clast, that the source was situated in southern parts of the Baltic Shield. That interpretation was suspect, however, because it seemed to require transport of vast numbers of clasts by ice across part of the Iapetus Ocean. Other suggestions were that the clasts were derived from a now-hidden continental block which separated the Dalradian basin from the Iapetus Ocean, or were carried from further afield before the Dalradian basin finally rifted and drifted from the source region as the Iapetus Ocean opened (Anderton 1980, 1982).

These two themes, the geotectonic setting of the Rhinns Complex and the plate tectonic significance of the tillite clasts, were brought together by Fitches *et al.* (1990). They tentatively suggested that the granitoid clasts are not exotic but were locally derived from the Rhinns Complex. If that hypothesis was substantiated, the allochthonous terrane interpretation would be undermined and the problems of deriving clasts from distant and unidentified lands are removed. As important, the clasts might then be treated as samples of the basement which lay adjacent to, and underlies, the Dalradian basin. If so, the clasts provide direct access to components of that basement which otherwise can be characterized only by indirect means, notably geophysics (e.g. Hall *et al.* 1984) and isotopic contamination of Caledonian granites (e.g. Clayburn 1988; Pidgeon & Compston 1992).

The hypothesis that the tillite granitoid clasts were derived from the Rhinns Complex was formulated on grounds of petrographic similarity. The clast types in Islay and the Garvellachs include unfoliated granite and minor granitic gneiss (Spencer 1971), albitized and rapakivi-granite (Spencer 1969, 1975), biotite-alkali granite, granophyre, granite porphyry, gneissose granite, and rare basic igneous rocks (Kilburn *et al.* 1965). In Ireland, the tillites carry vein quartz, gneiss and granite (Howarth *et al.* 1966; Howarth 1971). Sand-sized particles in the tillites, according to Spencer (1971), comprise a wide range of minerals: quartz, chess-board albite, microcline, muscovite, biotite, dolomite, magnetite and pyrite, with small amounts of zircon, rutile, tourmaline and apatite. Rutile-bearing quartz (probably the blue quartz seen in hand specimen) and most of the minerals listed by Spencer (1971) are also common constituents of the lithoclasts.

Similar rock types (porphyritic granite, undeformed and gneissic quartzo-feldspathic rocks) and most of these minerals are also found in the Rhinns Complex. A notable difference is that the dominant component of the complex is quartz-bearing syenite rather than granite but, like the

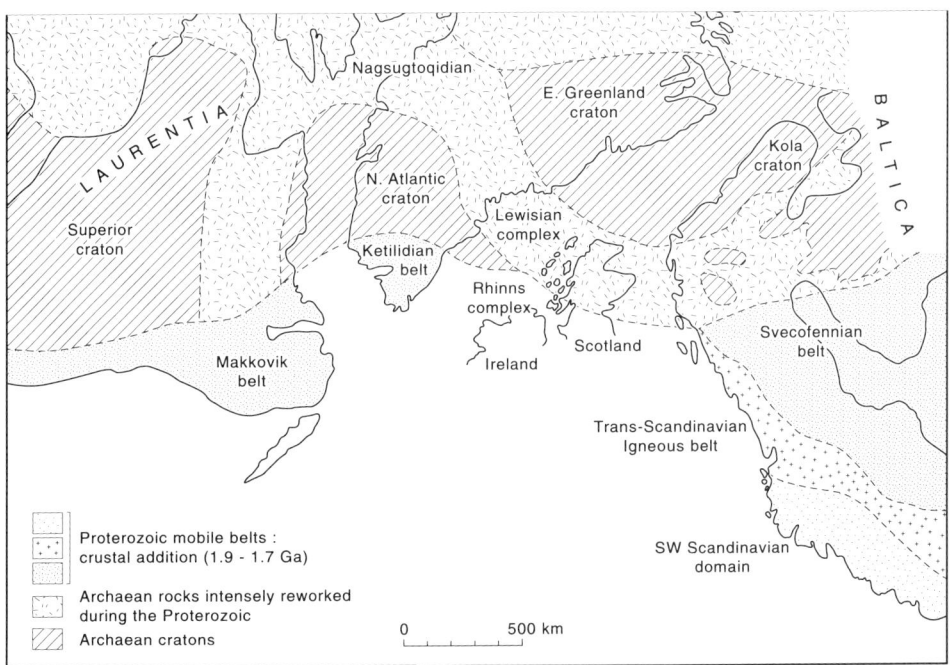

Fig. 2. Reconstruction of the North Atlantic region (after Winchester 1988).

tillite sand-grade material, it is microcline-rich and commonly it contains blue quartz which is also widespread in the tillites. No gabbroic or ultrabasic rocks, two conspicuous but uncommon members of the Rhinns Complex, were found among the tillite clasts in this study, although basic igneous rocks were identified by Kilburn et al. (1965).

Definitive correlation of the tillite detritus with the Rhinns Complex on a simple petrographic basis is not feasible because all of these rocks and minerals are commonplace and there are no unusual types which can be used to make the link. In addition, the tillite clasts are mostly altered to some extent; feldspar to sericite and biotite to chlorite, for example. That alteration was a result of one or more of three processes: hydrothermal activity in the original pluton; weathering before or during transportation of the igneous rocks as tillite clasts; and low-grade regional metamorphism during mild deformation. Most Rhinns Complex rocks are deformed and metamorphosed, with some alteration of feldspars and mafic minerals. Detailed petrographic comparisons of tillite detritus with the rocks of the Rhinns Complex are therefore difficult to make. For these reasons it was considered that geochemical comparability would be a more reliable test, especially using trace elements and rare earth elements (REE) which are usually taken to be stable during subsequent metamorphic and other alteration events (e.g. Winchester & Floyd 1977). Some samples of the Rhinns Complex collected by Muir (1990) were utilized in this study, together with his geochemical and isotopic data.

Full major element analysis of the clast samples has not been carried out because most clasts show signs of mineral alteration, pointing to disturbance of mobile elements. The clasts studied here are nevertheless granites *sensu lato*, containing abundant quartz, alkali feldspars and some mafic minerals.

Petrographic and geochemical data were used to identify clasts which appeared so similar to each other that they are likely to have been derived from a common parent. Two sets of these clast samples, one obtained from outcrops near Port Askaig on Islay (PAT samples in tables and figures), and the other from sites on the Garvellach Islands (GAR samples), were studied isotopically.

Geochemistry

Objectives and method

Sixteen tillite clasts were selected for geochemical analysis on the basis that petrographically they would probably resemble the protoliths of the deformed and metamorphosed Rhinns Complex syenites and granitoids. These clasts were crushed to fine powder at the National Isotope Geochemistry Laboratory (NIGL), and each powder sample was divided for geochemical study at Aberystwyth and isotopic analysis at NIGL.

In the Institute of Earth Studies at Aberystwyth, a glass bead was prepared from each clast powder sample and also from 6 powder samples of syenite collected by Muir from the Rhinns Complex of Islay and Inishtrahull. Geochemical analysis for 18 trace elements and 14 REE was carried out by laser ablation inductively coupled plasma mass spectrometry, using methods devised and under development in the Institute; details of techniques and methods are given by Perkins et al. (1993). The glass bead was ablated by a Nd:YAG Laser focused on the sample surface operating at about 300 mJ for about 5 minutes before analysis. Three separate acquisitions of spectra were made from the bead by scanning the mass spectrometer across the relevant atomic mass range to cover the analyte 400 times for each acquisition. Analyte peak intensities were ratioed to the peak intensity of the internal standard (indium, in this case), and these were compared with calibration lines obtained in the same way from certified reference material. The detection limits of elements in the bead using these methods are approximately 50 ppb; accuracy is typically ± 5–10%

Geochemical results

Trace element and REE data obtained from the samples of tillite clasts and the Rhinns Complex are given in Table 1. The data are used here for two purposes: to compare the compositions of the clasts with those of the Rhinns Complex syenites; and to obtain information on the geotectonic setting of the tillite source rocks and the Rhinns Complex. For these purposes, incompatible and other trace element X–Y plots and REE and trace element 'spidergrams' have been used.

Comparison of incompatible element data, as illustrated by Fig. 3, reveals that the tillite clasts from Port Askaig and the Garvellach Islands are indistinguishable from one another. They are, however, geochemically distinct from the Rhinns Complex syenite samples. That distinction is also shown by the plot of Rb vs. Nb + Y plot (Fig. 4). Figure 4 also demonstrates that the Rhinns Complex syenites fall in the Volcanic Arc Granite field, as defined by Pearce et al. (1984), thereby confirming Muir's (1990) characterization of these rocks, whereas most of the tillite clasts are in the field of Within-Plate Granite. Two of the 16 clast samples are transitional with the Rhinns Complex samples, and three others overlap those of the Rhinns Complex.

Table 1. *Geochemical data obtained from samples of the Rhinns Complex of Islay (RMGC) and Inishtrahull (RMIN), and clasts in the Dalradian Tillite near Port Askaig on Islay (PAT) and the Garvellach Islands (GAR)*

	P₂O₅ wt%	TiO₂ wt%	Rb ppm	Sr ppm	Y ppm	Zr ppm	Nb ppm	Cs ppm	Ba ppm	La ppm	Ce ppm	Pr ppm	Nd ppm	Sm ppm	Eu ppm	Gd ppm	Tb ppm	Dy ppm	Ho ppm	Er ppm	Tm ppm	Yb ppm	Lu ppm	Hf ppm	Ta ppm	Pb ppm	Th ppm	U ppm
RMGC1	0.19	0.82	45.0	657	14.0	400	5.27	0.66	3713	49.6	84.7	10.5	42.5	6.32	4.11	5.86	0.71	2.91	0.60	1.62	0.13	0.99	0.19	7.67	0.44	13.3	0.39	0.67
RMGC2	0.47	0.98	50.9	1005	23.4	473	6.50	0.88	2516	49.4	97.9	12.8	54.5	9.69	2.42	7.98	0.87	5.98	1.02	2.12	0.38	2.17	0.21	9.39	0.36	12.6	0.56	0.97
RMGC4	0.22	0.90	64.7	291	26.7	595	18.8	0.64	1160	61.2	133	17.0	64.3	10.9	1.65	8.44	1.11	6.91	1.20	3.13	0.42	2.44	0.42	11.6	0.97	14.0	2.68	1.26
RMGC10	0.35	0.89	76.1	602	14.7	556	8.09	0.97	2909	33.1	65.0	8.63	35.0	8.27	3.04	5.06	0.63	3.88	0.71	1.64	0.46	2.02	0.29	9.69	0.36	10.2	0.74	0.70
RMGC11	0.10	0.45	42.9	199	7.32	406	2.79	1.21	1491	18.6	29.4	3.89	17.9	3.55	1.72	3.10	0.45	2.01	0.37	1.05	0.29	1.19	0.25	7.11	0.29	9.8	0.58	0.71
RMIN7	0.25	0.93	116	125	21.1	172	11.5	2.25	1036	25.2	46.9	7.00	26.2	5.99	3.19	6.18	0.90	5.03	0.79	2.84	0.33	2.60	0.45	4.41	0.87	18.3	1.3	0.77
PAT1	0.09	0.79	29.3	60.9	41.1	229	26.2	0.69	175	76.4	161	17.5	55.5	9.59	0.71	7.79	1.15	7.78	1.81	4.70	0.72	4.96	0.88	7.20	2.45	11.0	35.5	3.46
PAT2	0.13	0.83	34.2	94.4	38.6	437	21.3	0.79	304	105	202	24.3	85.8	15.8	1.41	9.74	1.68	8.56	1.37	4.72	0.63	4.90	0.61	13.2	1.79	33.4	28.6	3.24
PAT3	0.18	1.60	29.7	60.2	64.6	412	22.4	1.53	104	61.1	125	15.7	57.8	18.1	1.69	13.2	2.57	13.4	3.14	7.50	0.96	6.56	1.56	11.0	2.02	15.3	15.3	2.75
PAT4	0.20	1.41	61.8	120	33.8	186	18.4	1.98	301	35.6	65.4	8.64	32.2	6.95	1.00	7.85	1.25	5.65	1.63	3.81	0.63	3.59	0.71	6.88	2.77	25.9	14.4	5.66
PAT9	0.13	0.89	38.8	81.4	59.2	292	21.0	0.85	217	92.0	186	22.0	83.2	15.8	1.02	13.8	1.93	14.4	2.80	6.94	0.95	5.78	0.87	9.39	1.42	20.5	20.3	1.98
PAT14	0.39	0.47	34.5	89.0	25.7	8.89	4.12	1.02	135	9.37	11.5	1.88	8.16	5.79	0.51	3.63	0.86	6.96	1.30	2.89	0.73	2.86	0.79	2.94	1.24	11.6	5.33	2.95
PAT19	0.20	0.45	62.7	119	14.0	85.6	16.3	1.56	336	30.0	67.1	7.01	26.7	3.56	0.88	4.84	0.81	3.76	0.83	1.61	0.38	2.79	0.41	4.51	1.44	19.9	32.0	4.53
PAT21	0.22	0.43	46.2	129	23.4	125	14.0	1.26	283	36.9	73.0	7.86	34.0	6.30	0.85	5.41	0.95	4.83	0.99	2.89	0.50	3.39	0.56	4.50	1.02	11.0	18.8	2.68
PAT22	0.69	0.84	47.6	73.3	49.4	243	23.3	1.84	317	86.4	167	18.8	62.7	15.1	2.33	10.1	1.97	10.2	2.08	7.29	0.95	7.35	1.79	10.9	3.23	19.0	52.2	6.44
PAT23	0.29	0.34	23.8	87.6	43.0	142	17.1	1.04	107	75.0	154	16.8	58.8	11.2	0.95	9.77	1.61	9.39	1.85	4.86	0.79	4.48	0.86	5.68	2.16	13.6	57.7	3.92
GAR9	0.87	0.71	12.8	927	68.7	415	29.8	1.25	374	76.2	158	18.5	76.5	15.7	1.75	14.5	2.59	16.1	3.58	8.21	1.04	8.13	1.52	12.2	2.19	17.7	15.2	3.64
GAR10	0.04	0.23	42.1	33.6	46.2	254	28.0	0.84	461	73.7	156	18.1	59.4	12.6	0.70	8.42	1.29	7.74	1.57	4.74	0.78	5.82	0.9	8.80	2.69	15.3	36.6	4.99
GAR11	0.08	0.21	17.4	429	28.5	153	21.8	0.53	336	47.1	96.3	11.0	38.3	8.23	0.96	5.77	0.73	4.78	1.02	2.96	0.60	4.21	0.52	5.34	2.53	23.5	29.1	3.74
GAR14	0.10	0.42	29.7	243	41.3	183	25.0	0.97	569	63.2	130	14.1	51.2	10.1	0.67	7.47	1.32	6.79	1.63	4.55	0.74	5.26	0.72	7.42	2.32	31.3	31.6	3.79
GAR15	0.14	0.22	28.4	26.3	42.0	236	29.0	0.84	633	85.0	168	18.7	63.5	11.9	0.76	7.97	1.23	6.77	1.44	5.47	0.88	5.68	0.73	9.66	3.10	15.1	43.9	4.35
GAR18	0.10	0.12	30.2	13.6	45.2	139	15.8	0.83	845	65.8	136	15.0	57.3	10.9	0.95	8.46	1.52	8.52	1.90	5.48	0.83	6.52	0.69	6.08	2.80	10.2	32.9	2.48

Analyses carried out at Institute of Earth Studies, Aberystwyth, using ICP-MS. See text for method outline.

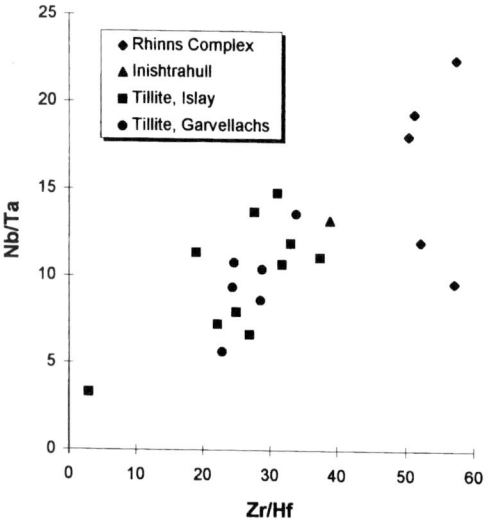

Fig. 3. Incompatible element plot (Nb/Ta vs. Zr/Hf) of data obtained from Rhinns Complex and tillite clasts.

Figure 5 gives chondrite-normalized REE data obtained from tillite clasts and the Rhinns Complex syenites. Again, the Port Askaig and Garvellach Island tillite samples are inseparable but are strongly contrasted with Rhinns Complex syenites. The tillite clasts show slight LREE enrichment, pronounced negative Eu anomalies and flat HREE patterns. These patterns are consistent with their Within-Plate traits indicated by the incompatible

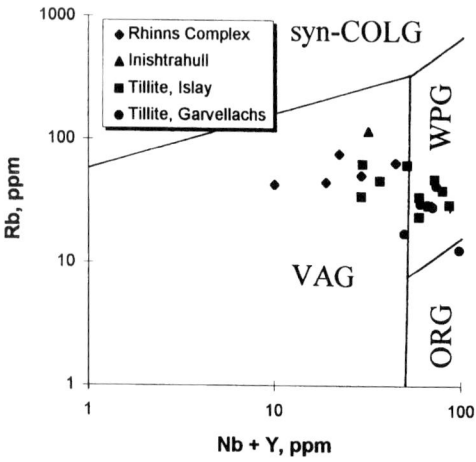

Fig. 4. Tectonic discrimination diagram (Rb vs. Nb + Y) of data obtained from Rhinns Complex and tillite clasts. Based on Pearce et al. (1984). VAG, Volcanic Arc Granite; WPG, Within Plate Granite.

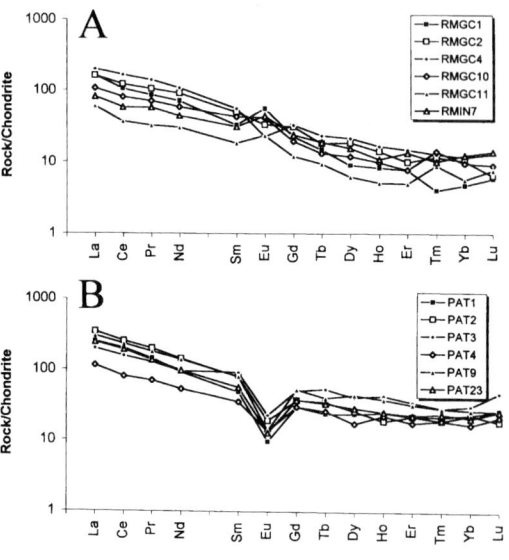

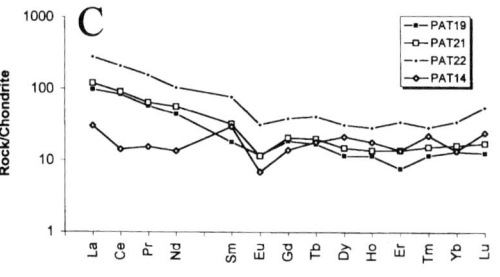

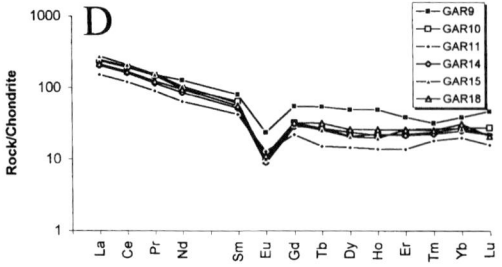

Fig. 5. Chondrite-normalized Rare Earth Element data obtained from Rhinns Complex (**a**) and Tillite Clasts (**b** & **c**); Port Askaig (**d**).

trace element field. In contrast, the Rhinns Complex samples have smooth LREE to HREE paths with no significant Eu anomalies, patterns that conform with Volcanic Arc Granite rocks as indicated by the incompatible trace element data.

On trace element spidergrams (Fig. 6), the Port Askaig and Garvellach Island samples are again indistinguishable from each other but are quite distinct from the Rhinns Complex syenites. The

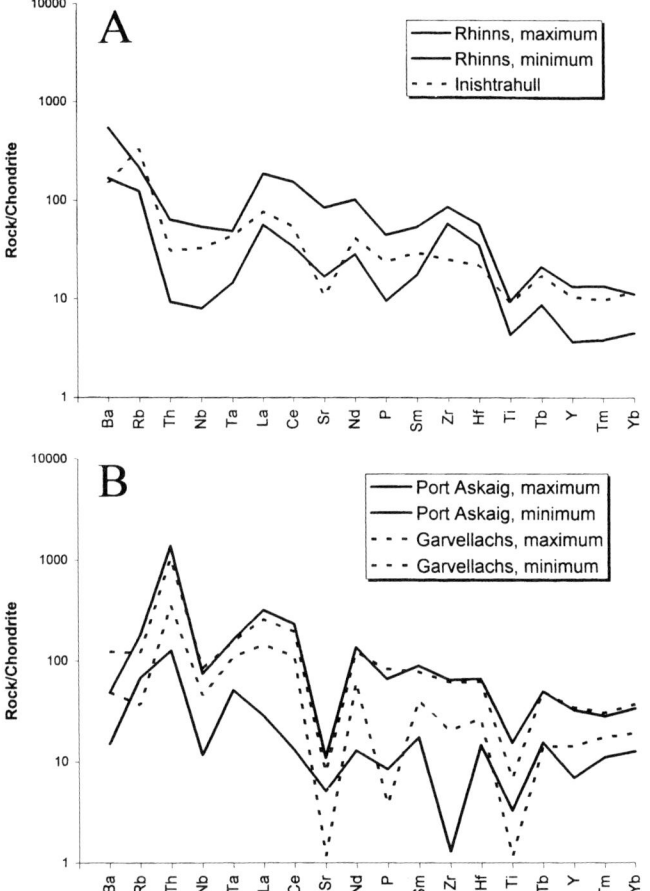

Fig. 6. Chondrite-normalized trace element data obtained from Rhinns Complex (**a**) and tillite clasts (**b**).

Rhinns Complex syenites show the depletion of Th, Nb and Ta which characterizes magmas produced in volcanic arc, supra-subduction zone environments. The tillite clasts, in contrast, have the large positive Th spike and strong depletion in Sr, Zr and Ti of Within-Plate Granites.

Various points arising from these data and observations are discussed in the concluding section.

Isotope geochemistry

Objectives and methods

The basic aim of this study was to constrain the age or ages of the tillite granitoid clasts using Sr and Nd isotope compositions and to compare them with the data acquired from the Rhinns Complex by Muir (1990). Isotopic and other analyses were carried out at NIGL using standard methods which are not described here.

As many of the granitoid clasts contain micas, an initial objective was to obtain Rb–Sr mica whole-rock data to determine the ages of individual clasts. However, the biotite flakes which appear to be fresh in hand-specimen are seen in thin section to be intensely chloritized and unsuitable for isotopic analysis. Muscovite was separated from one sample (GAR 19) for Rb–Sr analysis.

Another objective was to obtain Rb–Sr whole-rock ages, using several clasts and assuming that those clasts had a common igneous source. Six samples from the Port Askaig area and 8 samples from the Garvellach Islands were selected for this purpose on the basis that they form a petrologically and geochemically homogeneous set and thus were perhaps co-magmatic and coeval.

Sm and Nd isotope analysis was carried out on some clast samples to obtain model ages. Those ages might provide a means of comparing samples and placing some constraints on the igneous crystallization ages of the source rocks.

Isotopic results

Analytical results are given in Tables 2 & 3, and Rb–Sr data are presented graphically in Fig. 7.

The Rb–Sr whole-rock data obtained from the tillite clasts display a scatter (MSWD >1000). The scatter may be interpreted in two ways. Firstly, the clasts were derived from several terrains with different ages and isotope compositions. Secondly, the clasts came from a single source but their isotope systems were disturbed to various extents by post-crystallization events such as metamorphism. The second interpretation is preferred here because the clasts, as indicated above, have geochemical compositions consistent with a common igneous source. The alteration of minerals in the clasts is a signature of post-crystallization events, including low-grade metamorphism, during which the Rb–Sr isotope systems were disturbed.

Five samples define a shallow sloping line which gives a minimum age for the clasts of 696 ± 56 Ma (2σ), an initial $^{87}Sr/^{86}Sr$ ratio = 0.722 ± 0.001 (2σ), and MSWD = 60 (Fig.7). The age is consistent with assumptions that the tillite is c. 660 Ma old.

An estimate for the age of igneous crystallization of the clast source rocks is given by GAR 19. This was the only clast sample to contain primary igneous muscovite. In other samples, igneous biotite is present but has been extensively altered to chlorite and magnetite. The muscovite/whole-rock pair from GAR 19 gave an age of 1570 ± 16 Ma (2σ). Mica model ages from this sample, using initial ratios of 0.702 and 0.72, change the age by only ±4 Ma, showing that the age is controlled by the high Rb/Sr ratio of the muscovite and is not greatly affected by the y-axis intercept. This control by the mica composition makes the regression a robust determination of the age of the mica. Muscovite is more resistant to low-grade alteration than the whole-rock samples, in which the feldspar is particularly prone to rapid breakdown. Thus it is inferred that the muscovite in GAR 19 has retained its isotopic integrity even though the host-rock has been extensively disturbed.

Depleted mantle Nd model ages (Table 3) give an average result of 2.0 ± 0.02 Ga (2σ), excluding PAT 14 which has an anomalously high Sm/Nd ratio. This is unlikely to date igneous crystallization but it provides a maximum age for the clasts.

These data and observations are discussed further in the next section.

Interpretations and conclusions

The regional implications of the results of this geochemical and isotopic study are summarized as follows.

The clasts of granitoid rocks collected from various stratigraphic levels of the Dalradian tillites

Table 2. *Rb and Sr isotope data for 9 tillite clasts from the Garvellach Islands (GAR) and 6 clasts from Port Askaig (PAT)*

Sample	Rb (ppm)	Sr (ppm)	Rb/Sr	$^{87}Rb/^{86}Sr$	$^{87}Sr/^{86}Sr$
GAR 9	8.120	69.05	0.118	0.341	0.725941
GAR 10	38.53	80.95	0.476	1.381	0.737024
GAR 11	16.74	118.4	0.141	0.410	0.725973
GAR 13	29.47	102.1	0.289	0.837	0.727465
GAR 14	33.66	64.83	0.519	1.508	0.747174
GAR 15	23.93	60.81	0.393	1.142	0.738534
GAR 18	26.66	32.55	0.819	2.378	0.746178
GAR 19	188.5	54.4	3.466	10.26	0.944708
GAR 19 mica	989.2	22.09	44.78	181.5	4.804455
PAT 2	21.58	78.73	0.274	0.796	0.733725
PAT 4	40.63	105.4	0.385	1.119	0.740351
PAT 10	59.58	57.48	1.037	3.012	0.751135
PAT 14	20.56	66.92	0.307	0.892	0.740431
PAT 21	42.81	120.2	0.356	1.034	0.741820

Rb and Sr concentrations were determined by XRF. The $^{87}Sr/^{86}Sr$ ratio was measured using a Finnigan 262 automated mass spectrometer which gave 0.710262 ± 0.00028 (2σ) for the NBS 987 international Sr standard. Regression calculations were made using the York-Williamson method with errors of 1.0% 2σ on the $^{87}Rb/^{86}Sr$ ratio and 0.02% 2σ on the $^{87}Sr/^{86}Sr$ ratio.

Table 3. Sm and Nd isotope data for 5 tillite clasts

Sample	Sm (ppm)	Nd (ppm)	Sm/Nd	$^{147}Sm/^{144}Nd$	$^{143}Nd/^{144}Nd$	TDM
GAR 9	12.141	64.43	0.1884	0.1139	0.511676	1947
GAR 11	5.013	29.96	0.1673	0.1011	0.511577	1869
GAR 15	7.216	44.79	0.1611	0.0974	0.512076	1222
PAT 2	7.615	47.25	0.1612	0.0974	0.511377	2051
PAT 4	5.251	26.22	0.2003	0.1210	0.511694	2054
PAT 14	1.024	3.63	0.2821	0.1705	0.511975	3127
PAT 21	4.504	25.212	0.1786	0.1080	0.511614	1929
Average depleted mantle excluding PAT 14						1.8
					$1\sigma =$	0.31

Sm and Nd concentrations (measured by isotope dilution using enriched ^{150}Nd and ^{149}Sm) and $^{143}Nd/^{144}Nd$ were determined using a VG 354 automated multicollector mass spectrometer. The data were corrected to $^{146}Nd/^{144}Nd = 0.7219$. La Jolla gave a value of 0.511846 ± 46 2σ ($n = 10$) during the period of analysis. Depleted mantle model ages are calculated using the following values: $^{143}Nd/^{144}Nd_{DM} = 0.513114$; $^{147}Sm/^{144}Nd_{DM} = 0.222$.

in the Port Askaig area of Islay and the Garvellach Islands have closely similar trace element and REE compositions. This implies that they were derived from source rocks which were produced in similar geotectonic settings. It is uncertain, however, if those source rocks were coeval.

The clasts have the geochemical signatures of granitic rocks produced in Within-Plate settings. They bear no geochemical (or petrographic) resemblance to analysed Lewisian gneisses (e.g. fig. 2.8b of Park 1992). As far as we are aware, there are no Precambrian plutonic rocks in Britain or Ireland with the geochemical traits of the tillite clasts analysed here.

The clasts have geochemical compositions which are very different from those of the Rhinns Complex syenites, despite petrographic similarities. Muir's (1990) interpretation that the Rhinns Complex syenites were emplaced in a Volcanic Arc setting is confirmed. By implication the tillite clasts were not derived from the types of rocks that comprise the Rhinns Complex cropping out in the Hebridean region. Nevertheless, it is possible that they were derived from other plutonic rocks closely associated in space and time with the Rhinns Complex as components of a Ketilidian–Svecofennian igneous suite. The c. 1800 Ma plutonic rocks of South Greenland (e.g. the Julianhåb suite; Windley 1993; Chadwick & Garde, this volume) and of Scandinavia (notably the

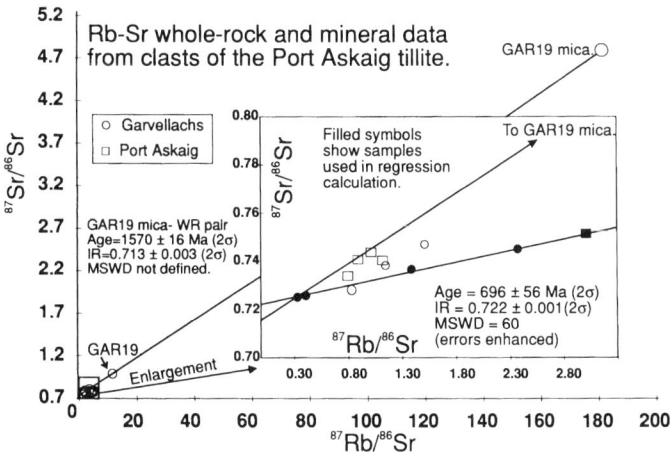

Fig. 7. Rb–Sr whole-rock and mineral data from clasts in the Dalradian tillite.

Transscandinavian Igneous Belt; e.g. Lindh & Persson (1990); Park (1991) contain plutonic rocks of various geotectonic settings, depending on their proximity to inferred subduction zones and their age in relation to subduction processes. Thus it remains feasible that the tillite clasts and the Rhinns Complex represent different components of the same Palaeoproterozoic to early Mesoproterozoic plutonic province.

The Nd model age estimate of the maximum age of the samples is $c.$ 2.0 Ga, consistent with derivation from any of the 2.0–1.65 Ga plutonic suites that comprise large parts of the Svecofennian–Ketilidian–Makkovikian–Labradorian mobile belt. The muscovite age of 1570 ± 16 Ma obtained from GAR 19 is the best estimate of the time of crystallization of this igneous rock. This age can be regarded as the age of crystallization of all of the igneous rock clasts only if it is assumed that their common chemical signature implies that they were derived from the same or consanguinous sources that had the same initial $^{87}Sr/^{86}Sr$ ratio. The age is $c.$ 200 Ma younger than the age of crystallization of the Rhinns Complex syenites as determined by other workers. Nevertheless, the Ketilidian and Svecofennian provinces also contain plutonic rocks of diverse ages, so derivation of a clast from a local, but now hidden, $c.$ 1570 Ma pluton is possible.

Even closer at hand was the Annagh Gneiss Complex which is exposed in northwest Ireland (Menuge & Daly 1994). Parts of that complex have Nd model ages in the range 1.7–2.0 Ga and various younger Proterozoic events down to $c.$ 1000 Ma have been identified. Most of these rocks are high-grade gneisses, very different from the clasts of slightly deformed and low-grade altered igneous rocks that constitute the Dalradian tillites of the Hebrides.

The errorchron age obtained from Rb–Sr whole-rock analysis of 5 clasts suggests that there was isotopic disturbance of their whole-rock systems at 696 ± 56 Ma. That 'event' took place shortly before or during deposition of the tillite which, as indicated in the introduction, is provisionally taken to be $c.$ 660 Ma.

The Rb–Sr and Nd data provide new constraints on the age of the deposition of Dalradian tillite and of the igneous parents of the clasts. It is anticipated that the ages of the igneous parents will be more tightly constrained by the U–Pb analysis of zircons separated from some clasts, now underway at NIGL.

RJM acknowledges a NERC studentship, whilst the other authors were supported by NERC Grant GR/437. This paper has benefitted from the advice of John Winchester and an anonymous referee. Institute of Earth Studies Publication No. 390 and NIGL Publication No.143.

Note added in proof. Peliminary U-Pb analysis of a zircon fraction from one of the tillite boulders (sample PAT 18) collected from near Port Askaig, Islay indicates a Palaeoproterozoic age of formation of this boulder. The zircon fraction, which is only 5% discordant, gives a $^{207}Pb/^{206}Pb$ age of 1870 ± 7 Ma (2σ).

References

ANDERTON, R. 1980. Distinctive pebbles as indicators of Dalradian provenance. *Scottish Journal of Geology,* **16**, 143–152.

—— 1982. Dalradian deposition and the late Precambrian–Cambrian history of the N. Atlantic region: a review of the early evolution of the Iapetus Ocean. *Journal of the Geological Society, London,* **139**, 421–431.

BENTLEY, M. R. 1986. *The Tectonics of Colonsay, Scotland.* PhD thesis, University of Wales, Aberystwyth.

——, MALTMAN, A. J. & FITCHES, W. R. 1988. Colonsay and Islay: a suspect terrane within the Scottish Caledonides. *Geology,* **16**, 26–28.

CLAYBURN, J. A. P. 1988. The crustal evolution of Central Scotland and the nature of the lower crust: Pb, Nd and Sr isotope evidence from Caledonian granites. *Earth and Planetary Science Letters,* **90**, 41–51.

DALY, J. S., MUIR, R. J. & CLIFF, R. A. 1991. A precise U–Pb zircon age for the Inishtrahull syenitic gneiss, County Donegal, Ireland. *Journal of the Geological Society, London,* **148**, 639–642.

DICKIN, A. P. & BOWES, D. R. 1991. Isotopic evidence for the extent of early Proterozoic basement in Scotland and northwest Ireland. *Geological Magazine,* **128**, 385–388.

FITCHES, W. R., MUIR, R. J. & MALTMAN, A. J. 1990. Is the Colonsay–west Islay block of SW Scotland an allochthonous terrane? Evidence from Dalradian tillite clasts. *Journal of the Geological Society, London,* **147**, 417–420.

GOWER, C. F., RIVERS, T. & RYAN, B. (eds). 1990. *Mid-Proterozoic Laurentia–Baltica.* Geological Association of Canada Special Paper **38**.

HALL, J., BREWER, J. A., MATTHEWS, D. H. & WARNER, M. R. 1984. Crustal structure across the Caledonides from "WINCH" seismic reflection profile: influences on the evolution of the Midland Valley of Scotland. *Transactions of the Royal Society of Edinburgh: Earth Sciences,* **75**, 97–109.

HOWARTH, R. J. 1971. The Portaskaig Tillite succession (Dalradian) of Co. Donegal. *Proceedings of the Royal Irish Academy,* **71B**, 1–35.

——, KILBURN, C. & LEAKE, B. E. 1966. The Boulder Bed succesion at Glencolumkille, County Donegal. *Proceedings of the Royal Irish Academy,* **65B**, 117–156.

KILBURN. C., PITCHER, W. S. & SHACKLETON, R. M. 1965.

The stratigraphy and origin of the Portaskaig Boulder Bed Series (Dalradian). *Geological Journal*, **4**, 343–360.

LINDH, A. & PERSSON, P.-O. 1990. Proterozoic granitoid rocks of the Baltic Shield – trends and developments. *In*: GOWER, C. F., RIVERS, T. & RYAN, B. (eds) *Mid-Proterozoic Laurentia–Baltica*. Geological Association of Canada Special Paper **38**, 23–40.

MARCANTONIO, F., DICKIN, A.P., MCNUTT, R. H. & HEAMAN, L. M. 1988. A 1,800- million-year-old Proterozoic gneiss terrane in Islay with implications for the crustal structure and evolution of Britain. *Nature*, **335**, 62–64.

MENUGE, J. F. & DALY, J. S. 1994. The Annagh Gneiss Complex in County Mayo, Ireland *In*: GIBBONS, W. & HARRIS, A. L. (eds) *A Revised Correlation of Precambrian Rocks in the British Isles*. Geological Society, London, Special Report, **22**, 59–62.

MUIR, R. J. 1990. *The Precambrian Basement and Related Rocks of the Southern Inner Hebrides, Scotland*. PhD thesis, University of Wales, Aberystwyth.

——, BENTLEY, M. R., FITCHES, W. R. & MALTMAN, A. J. 1994*a*. Precambrian rocks of the southern Inner Hebrides-Malin Sea region: Colonsay, west Islay, Inishtrahull and Iona. *In*: GIBBONS, W. & HARRIS, A. L. (eds) *A Revised Correlation of Precambrian Rocks in the British Isles*. Geological Society, London, Special Report, **22**, 54–58.

——, FITCHES, W. R. & MALTMAN, A. J. 1989. An Early Proterozoic link between Greenland and Scandinavia in the Inner Hebrides of Scotland. *Terra Abstracts*, **1**, 5.

——, —— & —— 1992*a*. Discussion of isotopic evidence for the extent of early Proterozoic basement in Scotland and northwest Ireland. *Geological Magazine*, **129**, 501–504.

——, —— & —— 1992*b*. The Rhinns Complex: A missing link in the Proterozoic basement of the North Atlantic region. *Geology*, **20**, 1043–1046.

——, —— & —— 1994*b*. The Rhinns Complex: Proterozoic basement on Islay and Colonsay, Inner Hebrides, Scotland, and on Inishtrahull, NW Ireland. *Transactions of the Royal Society, Edinburgh: Earth Sciences*.

PARK, A. F. 1991. Continental growth by accretion: a tectonostratigraphic analysis of the evolution of the western and central Baltic Shield. *Geological Society of America Bulletin*, **103**, 522–537.

PARK, R. G. 1991. The Lewisian Complex. *In*: CRAIG, G. Y. (ed.) *Geology of Scotland*. Geological Society, London, 25–64.

PEARCE, J. A., HARRIS, N. B. W & TINDLE, A. G. 1984. Trace element discrimination diagrams for the tectonic interpretation of granitic rocks. *Journal of Petrology*, **25**, 956–983.

PERKINS, W., PEARCE, N. J. & JEFFREYS, T. 1993. Laser ablation inductively coupled plasma mass spectrometry: a new technique for the analysis of silicates. *Geochimica et Cosmochimica Acta*, **57**, 475–482.

PIDGEON, R. T. & COMPSTON, W. 1992. A SHRIMP ion microprobe study of inherited and magmatic zircons from four Scottish Caledonian granites. *Transactions of the Royal Society of Edinburgh: Earth Sciences*, **83**, 473–483.

PRINGLE, I. R. 1972. Rb-Sr age determinations on shales associated with the Varanger Ice Age. *Geological Magazine*, **109**, 465–472.

ROGERS, G., DEMPSTER, T. J., BLUCK, B. J. & TANNER, P. W. G. 1989. A high precision U–Pb zircon age for the Ben Vuirich granite: implications for the evolution of the Scottish Dalradian Supergroup. *Journal of the Geological Society, London*, **146**, 789–798.

SPENCER, A. M. 1969. Late Pre-Cambrian glaciation in Scotland. *Proceedings of the Geological Society, London*, **1657**, 177–198.

—— 1971. *Late Precambrian glaciation in Scotland*. Memoir of the Geological Society, London, **6**.

—— 1975. Late Precambrian glaciation in the North Atlantic region. *In*: WRIGHT, A.E. & MOSELEY, F. (eds) *Ice Ages: Ancient and Modern*. Seel House Press, 217–240.

WATSON, J. V. 1975. The Lewisian Complex. *In*: HARRIS, A. L. *et al.* (eds) *A Correlation of Precambrian Rocks in the British Isles*. Geological Society, London, Special Report, **6**, 15–29.

WINDLEY, B. F. 1993. Proterozoic anorogenic magmatism and its orogenic connections. *Journal of the Geological Society, London*, **150**, 39–50.

WINCHESTER, J. A. 1988. Later Proterozoic environments and tectonic evolution in the northern Atlantic lands. *In*: WINCHESTER, J. A. (ed.) *Later Proterozoic Stratigraphy of the Northern Atlantic Regions*. Blackie, Glasgow, 253–270.

—— & FLOYD, P. A. 1977. Geochemical discrimination of different magma series and their differentiation products using immobile elements. *Chemical Geology*, **20**, 325–414.

Index

Figures and Tables are in italic.

Abitibi dykes 215, 235, 237
Abloviak shear zone *92*, 93, *95*, 104, 112, 118, 122, 125, 130, 134, 145, 148
  bend in 105, 107, 130, 132, 147
  establishment/development of 132, 146
  structural style 105, 122, 123, 124
Abloviak transect 120–3
  cf. Nachvak and Saglek transects 127–8
accessory minerals 182, 186
accretionary thrust complex/wedge, Tasiuyuk gneiss 129, 131, 132, *133*, 143, 148
Adirondak Highlands (-Morin) terrane 223, 229, 230, 232, 233, 237
Adirondak Lowlands terrane 237
Adirondis, accretion of *210*, 213
Adlavik Brook Fault 170
Adlavik Intrusive Suite 164, 212
Aillik Domain 157, 158–62, 169–70
Alexis River anorthosite 203
Allochthon Boundary Thrust 237
Åmål–Horred Belt (supracrustals/volcanics) 222, 224, 242, 262, *263*, 269, 298, 309, 310, 311
Amazonia 237, 239
AMCG suites 198, 213–14, 215, 235, 292
Ammassalik Mobile Belt 179, 193
Åna–Sira massif 355
Anaktalik Domain 92
Andean-type arcs 7, 8, 18–19, 174
anisotropy of susceptibility (AMS), Laanila/Kautokeino/Karasjok dykes 344–5
Annagh Division 223
Annagh Gneiss Complex 230, 376
Apache Group 235
Apparent Polar Wander Paths 331, 354–5
appinite suite, Julianehåb batholith 181–2, 185
Arc Lake Intrusive Suite 213
arc magmatism 148
  Narsajuaq magmatic arc 13, 18, 19
  Rae margin 93, 108
  Torngat Orogen *109*, *110*, 148
  see also DTG suite; Killinek charnockite suite
arc terranes, Labradorian 174
Arrowhead Lake intrusion 213
Arvidsjaur volcanics 7, 19
Askim Granite 226
Askvoll Group 364
Atikonak River massif 213
attenuation 171, 206
Avalon Composite Terrane 174
Avayalik dykes 96, 98, 118, 120, 144, 147

BABEL surveys *5*, 7, 8
Baby–Howse zone 139, *140*, 147–8
Baffin Orogen 93
Baltic Shield 297
  Archaean–Proterozoic crustal boundary 7–8
  Belmorian Belt 55–66, 69–88
  greenstone belt komatiites 43–50
Baltic–Bothnian shear zone 249–50
Baltica 2, *3*, 18, 219, 222–3, 225, 229, 239, 242

Labradorian–Gothian accretion 224, 226, 241
Mesoproterozoic cratonization of 261–71
  possible movements of 219–20
  separation and rotation of 232, 297
  tectonism and clockwise rotation 230, 231–2
Baltica–Laurentia 275
  collision 235–7, 241
  post-collisional convergence and lateral slip 237–40, 242
  separation 219–20
  tectonic significance, bimodal volcanism 291–2
Baltoscandia, pre-Caledonian evolution 361–2
Baltoscandian craton 359
Bancroft terrane 223, 228, 230
Bandak Group 222, 234, 236, 242, 276, 277, 278
Bell Lake granite 228
Belmorian Belt 55–66
  metamorphism in mafic intrusions 69–88
  U–Pb isotopic studies 59, *60–1*, 62, *63–4*, 64, 65
Belt–Purcell Group 292
Benedict Fault Zone 164, 167, 170, 201
Bergen Arcs 224, 359, 361
  pre-Caledonian evolution of 362–3
Bergsdalen Nappes 362
Bjerkreim–Sokndal lopolith 355
Bläsberget dyke swarm 249
  felsic dyke, U–Pb zircon dating 254–5
  geochemistry 251–4, 258
  mafic/felsic/mixed dykes 250–1, 251–3, 256, 258
Bläsberget granite massif 251
  U–Pb zircon dating 255, *257*
Blekinge–Dalarna dyke swarm 234, 354–5
Blue Ridge terrane (island arc) 235, 237, 239
Border Zone 170, 174, 180, 181, 192
Bothnian Basin 5
Brannigan thrust 124, 147
Breven–Hällefors dyke swarm 224, 226
Brien troctolite–anorthosite 213
Bruce River Group 164
Burwell Domain 91–112, 118, 120, 130, 145–6, 147
  northern segment geology 93–107
  pre-deformational subduction/accretion models 107–10

Caledonides, SW Norwegian, suspect microcontinent model 360, 364
Canadian Shield 237, 239
Cape Harrison Domain 157, 162–4, 170, 173
Cape Harrison Metamorphic Suite (CHMS) 162–3, 166, 170, 173–4
Cape Smith Belt 13, 19
Cape Smith Belt/Narsajuak arc 18
Carthage–Colton Mylonite Zone 237
Central Gneiss Belt 223, 225, 228, 232, 237, 238, 239
Central Granulite Belt 223
  Central Metasedimentary Belt (CMB) 223, 228, 233, 235, 236, 237, 238
Central Scandinavian Dolerites/Dolerite Group 230, 354
Chicoutimi mangerite 214
Chioak zone 139, 140

379

Chukotat Group 13
Chupa alumnious gneisses 55, 57–9, 65
　zircon data 59, *60–1*, 62, *63–4*, 64, 65
clasts, granitoid, in Dalradian tillites 367–76
Coldwell alkaline complex 235
collisional geometries
　Svecofennian 8–9
　Trans-Hudson Orogen 13–14, 15, 17
collisional orogenies 69
　doubly-vergent *see* Torngat Orogen
　world-wide 241–2
collisional tectonism
　in absence of second cratonic block 174
　syn- to post-Grenvillian 214
continental flood basalts 285–6
continental rift zones 287
continental–margin arc 215
　over N-dipping subduction *210*, 212
Coppermine River basalts 230
Cree Lake Zone 15
crust
　continental 69
　　Belmorian 66
　　pre-Caledonian, S Norway 359–65
　　post-Gothian, cratonized 262
　　pre-Labradorian 208, 214
crust formation
　and depletion in granulites 38–9
　Lewisian TTG gneisses 31–5
crustal contamination 48–9, 50, 99, 103
　Vemork Formation 283, 285
crustal shortening 9, 13, 132
crustal stabilization, Southwest Scandinavian Domain 268
crustal thickening 86, 88, 110, 130, 132, 148, 192, 193, 238, 326
　in the SSD 310, 311
　and Trans-Labradorian batholith 209, 212
Cut Throat Island thrust 201

Dal Group supracrustals 222, 234, 238, 242
Dala volcanics 224, 242
Dalen Formation 277
Dalradian tillite clasts, provenance of 367–76
Dalsfjord Suite (Nappe) 359, 360, 364
Dalsland Boundary Thrust–Göta Älv Zone 222
Davy Lake Group 213
De Pas batholith *140*, 141, 148–9
deformation
　Burwell domain 104–5, 107, 112
　Julianehåb batholith 182–5
　Kaipokok Domain 167–9
　Kaipokok Structural Zone/Aillik Domain 169–70
　Lake Melville terrane 202–4
　Rachel zone–Kuujjuaq terrane area 145
　Telemark Supracrustal Suite 276
　Torngat Orogen 147, 148, 149
　Gothian 267–8, 309
　Grenvillian 170, 201, 203, 238
　Inverian 223
　Ketilidian 181
　Laxfordian 9, 223, 225
　Seletsk 69–70
　Sveconorwegian 237, 258, 319, 320, 325, 326
　　amphibolite–facies 98, 107
　　ductile shear 315, 316, 318, 326

mylonite–ultramylonite 147
　strike-slip 76
Degerberg granite 251, 253, 258
　West, U–Pb zircon dating 256, *257*
delamination 2, 9, 132, 219, 222, 225, 230
diatexite 103, 203
diorite, Lewisian 26, *28*, *29*
Disappointment Hill complex 212
Dome Mountain Intrusive Suite 206
Doolough granite 238
Dorset fold belt 93
Double Island magmatic event 201
DTG suite 98, 99, *100*, *101*, 104, 108, 118, 128, 130, 131
　in SECP 143
Duck Island granitoid suite 98
dykes
　amphibolite 79
　aplitic 304–5, 306, 308, 309
　appinite 193
　diabase 332–3
　dolerite 318, 320
　garnetiferous 96
　Julianehåb batholith 181, 182, 185, 192, 193
　mafic
　　Avayalik dykes 96
　　Hawke River terrane 204–5
　　Noodleook complex 102
　　Tolstik intrusion 73, 75–86
　metadolerite 181
　microgranite 206
　Psammite Zone 190
　see also named dykes and dyke swarms

Earl Island granodiorite–diorite domain 204–5
East Greenland aulocogen 230–1, 232, 235, 239, 241
Eastern Churchill Province 157
Eastern Granite–Rhyolite Province (St Francois terrane) 223, 228, 241, 287, 291
ECSOOT seismic line *140*, 144, 145, 148
Egersund anorthosite and Egersund farsundite 355
Egersund olivine dolerite dyke 355
element depletion, TTG gneisses 37–9, 40
Elzevir terrane 223, 230, 232, 236
Elzevirian Orogeny 213–14, 215, 232, 233, 241
exhumation 125
　Åmål–Horred Belt 310, 311
extension 235, 240, 310, 320
　anorogenic 228
　Labrador/Greenland/British Isles 229–30, 232
　lithospheric within-plate 292
Exterior Thrust Belt 198

Falcoz Shear Zone 92, *140*, 145
Faltungsgraben, tectonic flakes in 359
felsic sheets, Lewisian gneisses 26, *28–9*, *29*
Fennoscandian Shield *see* Baltic Shield
Fiace Lake slide 169
Finland, East, komatiites 43, 44, 47, 52
Fjordzone (Oslofjorden) 236, 237, 239, 240, 241
flake tectonics 2, 9, 17, 18, 19
　see also Jotun kindred rocks
Flannan and W mantle reflectors 9–10, 11
Flat-lying Migmatite Zone (Pelite Zone) 172, 180
Flinton Group 235

Flowers River Igneous Suite 213
Folded Migmatite Zone (Psammite Zone) 171–2, 180
foliation
　Eastern Segment, Sveconorwegian Orogen 319
　Katherine River shear zone 107
　Komartorvik shear zone 107
　Tolstik intrusion 76, 78
　axial planar 105
　pinstripe mylonitic 122
foreland, Archaean 179
Four Peaks domain 95, 96, 143–4
　deformation and metamorphism in 104, 105, 146–7
　uplift of 107, 112
Fox River Belt 14
Frontenac terrane 223, 228

Gardar events 223, 229, 232, 235
Gardar Province 179, 223, 235
Gardar Track 242
garnet 125
　Tolstik intrusion and dykes 82, 83–4, 85
garnet fractionation 44
George River shear zone 140, 145
Gilbert Bay pluton 203
Gilbert River belt 202, 204
Gjuve Formation 277
gneiss
　aluminous, Chupa unit 57–9
　augen 164
　'layered' 167
　nebulitic 167–8
　'straightened' 157, 167, 168
gneissosity, migmatitic 104
Göteborg Batholith 264, 269, 298, 310
　calc-alkaline magmatism 267
Gothian Orogeny 222, 261, 308, 320, 361
Gothian–Kongsbergian Orogeny 69
Grady Island intrusion 201
granite 230
　A–type 164
　anatectic 191, 193, 226
　aplitic 73
　Caledonian I–type 166
　high–SiO_2 166
　Ketilidian 181
　Lewisian 29, 29
　post-tectonic 241, 275
　potassic 65, 73, 77–8, 158
　rapakivi 171, 172, 190, 191, 193, 208, 226
　'red granites' 268, 269, 271
　volcanic arc 370
　within-plate 370
Granite Zone (Julianehåb batholith) 171, 174, 180
granite–rhyolite province, US midcontinent, models for 291–2
granodiorite
　Julianehåb batholith 182, 183–4
　Lewisian 26, 28, 29
　Southern Region, formation of 36–7
granulites, TTG gneisses 30
　depletion in 37–9
Grästorp granitoid gneiss 267
Greenland 223, 229, 232, 236, 238, 239
　Labradorian–Gothian accretion 225
greenstone belt komatiites, Baltic Shield 43–52
greenstone belts 65, 66, 331, 333

Grenville Front 141, 145, 157, 163, 199–207, 223, 235, 237, 240
　polyphase shear zone 241
　a through-crust detachment 237
Grenville Province 155, 157, 163, 174
　collision with Baltica 220
　evolution of, E Labrador 197–215
　granitoid gneisses 164
　southwest 225, 230, 239
　　anorogenic potassic-alkaline suites 235
　　arc-backarc association 228
　　deformation and metamorphism 232–3
　　Ottawan Orogeny in 236–7
　　terrane accretion 223
　　tonalite formation 229–30
Grenvillian Ocean 234–5, 292
Grenvillian Orogeny 223, 239, 261, 297, 354
Groswater Bay terrane 198, 209, 214
　structures and metamorphism 201–2

Håland anorthosite-leuconorite 239
Hallandian Orogeny 361
Hällsö diorite 265–7, 268, 270, 271
Häme dykes 224
Hammarö Shear Zone 302
Handy thrust 147
Hardanger–Ryfylke Nappe Complex (HRNC) 224, 359, 364
Harp dykes 213, 229
Harp Lake Complex 228
Hart Jaune terrane 212
Hästfjorden granites 232
'Hastings sequence' 230
Hautavaara greenstone belt komatiites 43, 47, 50, 51
Hawke River terrane 198, 202, 204, 204–6, 209
Hazel Formation 237
Hazel Orogeny 239, 241
Hearne craton 17
　accretion of La Ronge–Lynn Lake arcs 15, 18
Hedmark Group 361
Heidal Series 364
Hestra Suite 298, 299, 302, 304, 308
Hizovaara greenstone belt komatiites 43, 48–9
homogenization, in Tasiuyuk orthogneisses 102–3
hornblende, Eastern Segment, Sweden 320, 321–5, 326, 327–8
hornblende fractionation 35
Horred volcanic suite 309
Huron Supergroup 107
Hutton Anorthositic suite 95, 98–9, 105, 107, 112, 144
hydrothermal alteration 185, 326, 352
hydrothermal post-volcanic fluid flow 49
'hyperite' gabbroids 219, 232
hysteresis properties, Laanila and Kautokeino dykes 346, 347

Iapetus Ocean, opening of 10
Iggavik dykes 170
Iggiuk migmatization 167–8, 169, 172
igneous activity
　Baltic Shield 69
　Belmorian Belt 70, 71
imbrication
　Sveconorwegian 326
　Tasiuyuk gneiss complex 127, 129, 132

indentation tectonics
  oblique collision/indentation, model for 147–50
  Southeastern Churchill Province 137–50
Independence Fjord Group 228
Inter-Sveconorwegian Extensional Period 232, 234
Interior Magmatic Belt 198
intra-arc basins 192, 193
Ireland 223, 230, 236, 238, 239
Island Harbour Bay Plutonic Suite (IHBPS) 158, 166, 167, 172, 174

Jåstad Formation 228
Jatulian Platform Sequence 5
Jergol Gneiss Complex 333
Joe Pond Formation 157
Jotnian graben 230
Jotun kindred rocks 359, 361, 362–4
Jotun–Valdres Nappe Complex 359, 360
  geological evolution of 363–4
Julianehåb batholith 181–5, 192, 375
  as basement to Psammite Zone 188–9, 191
  as root zone of magmatic arc 191
  schistosity and linear fabric 182–4, 191, 193

Kainu Schist Belt 8
Kaipokok Bay Structural Zone 157, *158*, 159, 166, *168*, 169–70, 173
Kaipokok Domain 157–8, *158*, 159
  chronology and character of plutonism 172–3
  reworking and migmatization in 167–9
Kalevian Group 5
Kalevian–Outokumpu Collage 7, 8–9
Kamennoozero greenstone belt 43, 51
Kanairiktok Shear Zone *158*, 167, 169, 173
Karasjok dykes 333, 333–4, 340
Karasjok greenstone belt 331
Karelia, komatiites 43, 44, 47, 50, 51
Karelia Province 5, 65, 71, 86
Karlshamn granite intrusion 226
Katherine River shear zone *95*, 107, 130
Kattsund–Koster dykes 226, 232
Kautokeino dykes 331, 333, 333–4, 334–5, 355
  baked contact tests 349, 352
  bulk petrophysical properties 342, *343*
  palaeomagnetism 338–40, 354
Kautokeino greenstone belt 331, 333
Ketilidian Mobile Belt *156*, 157, 208, 368
  correlation with Makkovik Province 155, 170–2
Ketilidian Orogen 179–94
Ketilidian terrane 17, 223
Keweenawan Track 242
Khetolambina strata 55
Kikkertavak dykes 157, 167, 169
Killarney Igneous Complex 223, 228
Killinek batholith 120
Killinek charnockite suite *95*, 98, 99, 101, 102, 104, 108, 112, 118, 128, 131
  intrusion of 129–30
  in SECP 143
  structures in 105
Kinsarvik Formation 224
Kiruna porphyries 7, 19
Kobbermine Bugt Shear Zone 171, 173, 181, 184
Kola Peninsula komatiites 43, 44, 47, 50, 52
Kola Peninsula Province 86
Kolvitsa folding 70

Komaktorvik shear zone *92*, *95*, 101, 104, 112, 118, 120, 127, 134, 143
  amphibolite and granulite facies blocks 124, 125
  deformation 147
  gneisses redeformed 144
  as a potential suture zone 93
  reworked Nain gneisses in 123, 124, 125
  sinistral motion 147, 149
  structures and metamorphism 105, 107
  uplift over Ramah Group 124
  uplift and transcurrent motion on 131
komatiites, Baltic Shield 43
  formed in response to hot mantle plumes 51–2
  major and trace element geochemistry 44–5, *46–7*
  petrogenesis 48–9
  REE distribution patterns 45, 47–8
  Sm–Nd isotopic systematics 49–51
  spinifex-structured 43
Kongsberg–Bamble Sector 222, 232, 235–6, 310
  as a tectonic wedge 224, 225–6
Kostomuksha greenstone belt komatiites 43, 47, 50, 52
Krummedal sequence 231, 232, 238, 241
Kuhmo–Suommusalmi greenstone belt 43
Kuujjuaq batholith, magmatic arc origin 141
Kuujjuaq terrane *140*, 141, 145, 148
Kvamshesten Fault 364
Kyfanan Lake layered mafic intrusion 207

Laanila–Ristijärvi dykes 331–3, 333, 334, 355
  baked contact tests 349, 350–2
  bulk petrophysical properties 341–2, *343*
  interdyke variation 352–4
  palaeomagnetism 335–8, 354
Labrador 230, 232, 238, 239, 240
  magmatism 228, 229
  southeast, granitoid activity 213–14
Labrador Trough/Geosyncline 139
Labradorian orogenic cycle 208–12
Labradorian Orogeny 198
Labradorian–Gothian Orogeny 226
Lac Allard pyroxene monzonite 213
Lac Joseph terrane 213
Lac Lomier complex *140*, 142, 143, 145
Lac Long Igneous Suite 213
Lac Olmstead thrust 145
Lac Tudor shear zone *140*, 145
Lærdal–Tyin–Gjende Fault 363
Lake Harbour Group *92*, 98, 108, 118, 123, 126–7, 132, 141, 148
Lake Kiki thrust 124
Lake Melville terrane 198, 209, 214
  complex structural geology 203–4
  granitic vein/megacrystic granitoid ages 202–3
Lake Melville–Mealy Mountains boundary 206
Lake Melville–Mealy Mountains–Hawke River boundary 214
Lapland Granulite Belt 71, 331, 332
Lapland–Kola Mobile Belt 69, 71, 86, 87, 107
Laporte terrane 139, *140*
Laurentia 2, *3*, 5, 219, 226, 228, 241, 242
  accretion of SW Scandinavian Microplate 361
  deformation in 239
  E-dipping reflectors on former margin 10
  eastern, rifting of passive margin 212–13
  Palaeoproterozoic orogenic belts 18
  sinistral slip/transpression 241

Trans-Hudson Orogen 10, 12–17
Laurentia–Baltica *see* Baltica–Laurentia
Laxford Front 26, 33
layered intrusions 25–6, 31, 39, 201, 203, 207
Leirungsmyran Gabbroic Complex 363
Leknes Group 361
Letitia Lake Group 213, 229
Lewisian Complex 367, 368
Lewisian TTG gneisses 25–33
   REE geochemistry 27–30
   regional geochemical differences 39
   U loss 37–8
Limestone Lake slide 169
Llano Orogeny 235, 241
Llano Province 223, 237
Llano uplift, Texas 230
Lödingen granite 228
Lofoten–Vesterålen Province 361, 365
Logan Sills 237
Lokhi folding 70
Long Island quartz monzonite 162
Long Range dykes 206
Lower Aillik Group 157, *158*, 159–60, 168, 169, 173
Lower Aillik–Upper Aillik contact, ductile deformation 169–70
Lyagkomina 57
   zircons 59, 62, 64–5

Mackenzie dyke swarm 230
magma flow, and AMS data 344–5, 352
magma ponding 291, 292
magmas
   Bläsberget dyke swarm and granite massif 258
   Lewisian TTG 39–40
   syn-collisional 110
magmatic arcs 364
   emplaced on Nain Province margin 108–10, 112
magmatism
   Baltica 222
   Lake Melville terrane 214
   Östfold–Marstrand and Åmål–Horred belts *263*, 263–4, 268
   Gothian 308–9
   alkaline 235
   AMCG and mafic 213–14, 215
   anorogenic 93, 223, 226, 228, 275, 361
   anothositic–charnockitic 361
   bimodal 73, 224, 225
   calc-alkaline 66, 232, 267, 268
   gabbroic 363
   granitic 203
   mafic 213
   mafic–felsic 209, 212, 213, 268–9
   in pull-apart basins 139
   tonalitic 33
   see also arc magmatism
magnetite (magnetic) grains, Laanila/Kautokeino/Karasjok dykes 341, 344, 345–6
Makkovik Front *see* Kanairiktok Shear Zone
Makkovik Orogen 17
Makkovik Province *140*, 155–74, 208, 225
   correlation with Ketilidian Mobile Belt 170–2
   outstanding problems 172–4
   plutonic suites, geochemistry and affinities of 164–6
   structural and metamorphic development 167–70

Mandel–Ustaoset Shear Zone 222, 231–2, 235, 237, 241
mantle evolution, models for 31–2, *32*
mantle/crustal reflectors 1–2, *3*, 5, 8, 9–10, *11*
Mars Hill terrane 235
Matorssuaq shear zone 183, 191
Mazatzal Orogen 223
Mealy dykes 207, 213, 228
Mealy Mountains Intrusive Suite 206, 225
Mealy Mountains terrane 198, 206–7, 208
Melezes–Schefferville zone 139, *140*
melt intrusion, dilation–controlled 78
metamorphism
   Abloviak, Nachvak and Saglek transects 125–7
   amphibolite facies 57, 236
   granulite-facies 30, 31, 32, 33, 62, 105, 234, 240
   greenschist-facies overprint 167
   Grenvillian 223, 236, 238
   high-grade 212, 363
      Burwell domain 110
      Tostik Peninsula 71, 79–88
   regional, Pelite Zone 191
   SECP 145–7
   Sveconorwegian 310
   syn-deformational 319, 326
metatonalites, mesocratic 101–2
Michael gabbro 201, 202, 213, 228
microcline, Julianehåb granodiorite 184
Mid-Continental Rift 237
Midsommerso–Zig Zag event 232
Mierujávri–Sværholt Fault Zone 333
migmatization 57, 144, 157, 188, 189, 193, 308
   Kaipokok Domain 167–8
   lit-par-lit 187
Mistastin shear zone *140*, 145
Mistibini–Raude domain *140*, 141, 148
mobile belt-foreland relationship 71
Monkey Hill Granite 170
Moonbase Shear Zone *92*, *140*, 145
Moran Lake Group 157, *158*, *159*, 167, 173
Morgedal Formation 277
Moss–Filtvet granite 229
Mount Benedict Intrusive Suite 164, 212
Mugford Bay Shear Zone 92
Mugford Group *92*, 144
Muskox intrusion 230
Mylonite Zone (MZ) 222, 240, 261, 267, 298, 310, 311, 315, 326
mylonite zones 93, 160, 170
   Grenville Front 201
   Kaipokok Domain 168–9
   and ultramylonite zones 183, 184
mylonitization 147, 205, 209

Nachvak Brook thrust 124
Nachvak Fiord thrust 143, 144, 145, 147
Nachvak transect *121*, 123
   cf. Abloviak and Saglek transects 127–8
Nagssugtoqidian mobile belt 17, 137, 179, 193
Nagssugtoqidian–Lewisian belt 224
Nagssugtoqidian Orogen 93
Nain craton 17, 118, 120, 123, 131, 137, 138, 147
   and Killinek/DTG suites 143
   reworking in 127
   within Torngat Foreland zone 144
Nain dolerite dykes 230
Nain Plutonic Suite 213, 229

Nain Province 91, *92*, 107, 147, 157
    magmatic arc scenario 108–10
    metamorphic zones *146*, 147
Nain Province gneiss complex *95*, 96, 104
Nain Province margin 108
    flexural burial of 105, 112
Nain–Rae oblique collision 93, 108, 110, 112, 118, 132, 146, 148
Nakit slide 169
Napatok dykes 118, 123, 144
    sinistral rotation of 124
Narsajuaq magmatic arc 13, 18, 19
Nd isotopic studies
    igneous suites, Makkovik Province 166
    Lewisian TTG gneisses 31–2
Neveisik magmatic event 202, 204
New Quebec Orogen 139–41, 144–5, 149
Newfoundland 238, 239, 241
Noodleook gneiss complex *95*, 104, 118, 120
    lithology 98–102
    structures in 105
North America 223
    Labradorian–Gothian accretion 225
North Arm thrust 124
North Atlantic Craton 137, 166, 173
North Atlantic supercontinent 219–42, *369*
    Baltica 222–3, 224, 225
    British Isles and Greenland 223, 225
    North America 223, 225
North Karelia, komatiites 43, 44, 52
North Pole Intrusive Suite 269
North River Domain *92*, *95*, 104
    structures in 104, 105
Norway, South 228, 230, 235–6, 237, 239, 240
    anorogenic magmatism 275–93
    clockwise rotation 219
    Inter-Sveconorwegian Extensional Period 232, 234
    pre-Caledonian continental crust 359–65
Nutak dykes 213

ophiolites 2, 5, 12, 13, 359, 364
orogenic belts
    Caledonian 359
    N Atlantic region *138*
    Palaeoproterozoic 2, 18
    Phanerozoic 5
orogenic collapse, Torngat Orogen 134
orogenic events, Late Archaean 69
orthogneiss 96, 157, 159, 181
    Eastern Segment, SSD 298–9, 302, 305–7
    fault-bounded slices, Tasiuyuk gneiss 102–3, 108
    Grenville Province 201, 202
    high-strain, Julianehåb batholith 185, 191
    Rae Province 96–8
Orust dykes 232
Osler Group 237
Oslo Region Rocks 222
Östfold–Marstrand Belt 262, *263*, 263–71, 298, 309, 310
Östfold–Marstrand boundary zone 236
Östfold–Marstrand segment 222, 224, 226
Otta Nappe 364
Ottawan Orogeny 233, 235, 236–7, 241
overthrusting
    by Mealy Mountains terrane 208
    of Lapland Granulite Belt 71

Pahrump Group, dolerite sills in 237
palaeomagnetic poles 354–5
Palaj–Lamba greenstone belt 43, 50, 51
Pan-African Orogeny 242
Paradise Arm pluton 205–6
Paradise metasedimentary gneiss belt 205
paragneiss, Tasiuyuk gneiss complex 102, 108
Parent Group 13
partial melting 192, 193
    by DTG magmas 130
    and the Lewisian TTG suite 34–5, 39–40
    of pre-existing continental crust 291
passive continental margins 212, 212–13
passive margin sequences 108, 147, 193
Payne River dykes 139
Pb isotopic studies, Lewisian TTG gneisses 32–3
Pelite Zone 191
Penokean Orogeny 291
Pinware terrane 198, 207–8, 214, 228
Pinwarian orogenesis 212
plagioclase, in schistose granodiorite 183–4
plate convergence, oblique, Ketilidian Orogen 179–94
polarity, of Labradorian subduction 209, 214–15
Post Hill slide 169
Povungnituk Group 13
Protogine Group 234
Protogine Zone (PZ) 222, 224, 225, 232, 234, 236, 237, 240, 261, 298, 308, 315, 326, 354
Psammite Zone 185–90, 191
    basement to 188–9, 191
    structure of 189–90, 193
pseudotachylite 107, 124, 128, 131, 132

Rachel thrust 145
Rachel zone 139, *140*, 141, 145
Rae Province 91, *92*, 112, 137, 138, 147, 148
    collisional interfaces with Superior and Nain cratons 145
    crustal structure and tectonic evolution 145
    thrust onto Kuujjuaq terrane 141
Rae Province gneiss complex, lithology 96–8
Rae Province margin 108, 110, 112
Ráiseatnu Gneiss Complex 333
Ramah Group *92*, 108, 118, 123, 125, 144, 146, 147
    in fold–thrust belt 124, 132, 149
Ranger Bight slide 169
'Rebolian' event 65
recrystallization
    post-Gothian 309
    Sveconorwegian 307, 308, 310
Red Island magmatic event 204, 209
Red Wine Intrusive Suite 213, 229
Reindeer Zone 15, 17, 18
remanent magnetism, Laanila/Kautokeino/Karasjok dykes 331, 335–56
Rhinns Complex 9–10, *11*, 223, 367–8
    and Dalradian tillite clasts 368–76
Rinkian Mobile Belt 179, 193
Rivière Pentecôte anorthosite 213
Rjukan Group 222, 224, 226, 242, 275, 276, 292
    age and metamorphic history 277–8
    correlation, US midcontinent granite–rhyolite province 287–91
    tectonic setting of volcanism 285–7
Rockall Bank 223, 225

Rogaland anorthosite province 222, 239, 240–1
Rogaland farsundite 355
Rogaland igneous complex 355
Rogaland–Vest Agder sector 222, 224, 225
Romain River anorthosite 214
Rönnäng tonalite 264, 265, *266*, 269, 271
   Stora Le–Marstrand xenoliths in 267
Röseskär Dyke *266*, 267, 268, 271
Rudihoø Complex 364

Sâdloq shear zone 183, 184, 191, 193
Saglek transect *121*, 123–5
   cf. Abloviak and Saglek transects 127–8
St François terrane 223, 228, 241, 287, 291
Sand Hill Big Pond granodiorite 205
Scourie dykes 31, 34
Scourie More levered body 39
Seal Lake Group 213, 229
SECP *see* Southeastern Churchill Province
seismic reflection profiling 2, 5, 7–10, *11*, 15, *16*, 17
Sel Group 364
Seletsk deformation/folding 69–70, 71, 86, 87
Seljord Group 222, 228, 276, 277, 292
Seward Subgroup 139
Shabogamo gabbro 213, 228
Sharbot Lake terrane 223
shear/shearing 73, 169
   dextral 145
   ductile 71, 88, 105, 315, 316, 318, 326
   layer-parallel 167
   sinistral 130, 137, 145, 184–5, 191, 235–40, 236, 237, 241
   strike-slip 70
   Tolstik intrusion 79, 88
   transcurrent 124, 125, 130
Sibley Group 225
Siknäs dyke 251
sill swarms, Baby–Howse zone 139, 147–8
Skellefte mineralized zone 7
Sm–Nd analysis
   Belmorian rocks 59, 62, *64*, *65*, 66
   Laanila and Kautokeino dykes 334–5
Sorte Nunatak 188
   supracrustal sequence 186–7
Sortis Group *159*, 170
South Greenland craton, accretion of Julianehåb batholith 191
Southeast Rae craton 118, 127
   reworked gneisses 122–3, 124–5, 126–7, 128, 130
   structures in 123
Southeastern Churchill Province 139–44
   crustal structure and tectonic evolution 144–7, 148–50
Southwest Scandinavian Domain (SSD) 261–3
   constraints, Sveconorwegian orogenesis 269
   Eastern Segment 298–302, 308–11
   geological evolution, implications of the Östfold–Marstrand Belt 267–9
   palaeopoles 354, 355
   regional geology 297–8
   Sveconorwegian events 309–11
   tectonics and trans-Atlantic correlation 269
   U–Pb geochronology, Östfold–Marstrand Belt 264–7
sparagmites 361
Spartan Group 13
Spavinaw terrane 223, 228
Stenkyrka granite, U–Pb dating 265, *266*, 267, 271

Stora Le–Marstrand Formation 222, 224, 225, 226
Strawberry Intrusive Suite 208, 209
Sudbury dykes 232
Suomussalmi greenstone belt komatiites 43, 45, 47, 50
   LREE enriched 48–9
Superior Craton 12, 17, 137, 138, 147
Superior Province 13, 147
Superior (Thompson Belt)–Reindeer Zone suture 17
Susan River quartz diorite 209
suture zones
   Karelia Province–Svecofennides 7, 7–8, 9
   Nain–Southeast Rae cratons 128
   potential, Makkovik Province, location of 173
   Superior craton–allochthonous terranes 12–17
   SW Norwegian Caledonides 359, 360, 361
Svartdalen Gneiss 363
Svecofennian craton 224, 225, 242
Svecofennian Domain 5–6
Svecofennian events 87, 361
Svecofennian Orogen 5–9, 361
Svecofennian Orogeny 62, 64, 69, 291
Svecofennian supracrustal rocks 316, 318
Svecofennian–Archaean boundary zone 5
Svecofennian–Ketilidian–Makkovikian–Labradorian mobile belt 376
Svecokarelian event, secondary REE mobility 50–1
Sveconorwegian belt 219
Sveconorwegian events 361–2
Sveconorwegian Frontal Deformation Zone (SFDZ) 315, 326, 327
Sveconorwegian Orogen 241
   Eastern Segment evolution, S–C Sweden 315–29
Sveconorwegian Orogeny 219, 222, 261, 262, 297, 298, 309–11, 354
Sveconorwegian Province 261
Sweden 226, 230, 232, 234–5, 237–8
   Sveconorwegian metamorphism 236

Tasiuyuk Domain *92*
Tasiuyuk gneiss complex 91, 93, *95*, 118, 124, 127, 128, 132, 145, 148
   charnockitic rocks 103–4
   fault-bound panels, Archaean gneiss 102–3, 108, 120, 143
   lithology 102–4
   origin of 108, 128–9, 132
   in SECP *140*, 142–3
   structures in 104, 105, 120–2, 124, 127
   thrust over Rae crust 146
Telemark Block 359
Telemark region, central, structural units 275–7
Telemark Sector 222, 224, 226, 228, 242
Telemark Supergroup 222
Telemark Supracrustal Suite 276, 277–8
terranes 2, *4*
   arc-related, Grenville Province 209
   Labradorian–Gothian 220–2
   Makkovian, Penokean and Labradorian 223
   Svecofennian 222
   term clarified 197
Thelon Orogen 148
thermomagnetic analysis, Laanila/Kautokeino dykes 347–9
Thompson Belt 15, 17
thrust planes, Karelian–Belmorian boundary 65

thrust wedge
  doubly-vergent 93
  see also accretionary thrust complex/wedge
thrust/nappe stacking 8, 232, 237, 238, 310, 359
Tipasjarvi komatiites 50
titanate 306, 307, 308, 310
Tolstik intrusion
  gabbro outlier 75, *77*
  mafic dykes 73, 75–88
  relation between different melts 73, *75*
  structure and composition 71–3
tonalite 59, 99
  formation, SW Grenville Province 229–30
  Lewisian 26, 27, *28*, *29*, 29
  Rönnäng tonalite 264, 265, *266*, 267, 269, 271
  Southern Region 35–6, 37
  see also DTG suite
Torngat Front 124
Torngat Orogen 138, 141–4, 174
  Abloviak, Nachvak and Saglek transects compared 127–8
  Burwell Domain 91–112
  crustal structure and tectonic evolution 145–7
  doubly vergent collisional orogen 117–34, 142, 145–7, 148
  lithotectonic complexes 93, *95*, 96, 118–20
  metamorphic zonation *146*, 147
  structural evolution of N part 110–12
Tornquist Zone 2
Trans-Hudson Orogen 10, 12–17
  Andean-type arc 18–19
  Hudson Bay Segment 14
  Ungava Segment 12–14, 18, 19
  Western segment 15–17
Trans-Labrador Batholith 163, 164, 201, 223, 225, 269
  accretion of arc-related terranes 209
Transscandinavian Granite–Porphyry Belt 360
Transscandinavian Igneous Belt (TIB) 171, 222, 224, 225, 261, 267, 283, 297, 308, 316, 361, 376
triple junction, Burwell Domain at 93
trondhjemite 26, 28, 29, 36–7
TTG gneisses
  Belmorian Belt 55, 59, 64–5
  Tolstik Peninsula 76
TTG gneisses, Lewisian 25–40
  an amphibolite source for 35, 39
  comparative trace element chemistry 27–33
  genesis involving older continental crust 34
  major element geochemistry 25–6
  Southern region, formation of 35–7
Tuddal Formation 276, 277–8, 285, *288–9*, 290, 292
  major element and trace element analyses 278–81, *283*, *284*, *285*
  petrogenesis of volcanic rocks 283, 284–5

Tupaya Bay, Lake Kovdozero 49, 57, 62

Ukrainian Shield 222
Ullensvang Group 224, 225, 228
underplating, basaltic 191, 192, 193
Ungava Orogen, tectonic history of 13–14
Ungava Peninsula 10, 12
Upper Aillik Group 160–2, 171, 173
Upper Allochthon, Trondheim region 359, 360
Upper North River body/pluton 203, 213
Upper Paradise River pluton 206–7
Utladalen Fault 363

Vågåmo Ophiolite 364
Vågårda granite 302, 308
Vaggeryd syenite *299*, 302
Valdres Group 361
Vallen Group *159*, 170
Vallhamn trondhjemite 269, 271
Varanger glacial deposits 368
Varberg charnockite 226
Vargfors Group 7
Vårmland Hyperite Suite 226
Vatnås granite 229
Vemork Formation 276, 278, 281–5, *286*, *287*, 292
Vihanti–Pyhäsalmi mineralized zone 5, 7
Vistenberg (SSD), samples from 305–7
Vodlozero gneisses 51
Vråna aplitic dyke, age of 304–5, 308

Wakeham Supergroup 212, 213, 230, 232, 235
Warren Creek Formation 157
Wathaman–Chipewyan Batholith 15, 18–19
West Islay–Colonsay block 367
Western Gneiss Region 222–3, 224, 228, 232, 237, 239, 267, 309, 359–60, 361
Western Granite–Rhyolite Province (Spavinaw terrane) 223, 228
White Bear Arm complex 205, 206, 225
white mica, E Segment, Sweden 320, *321–5*, 326, 327–9
Wibork Complex 224
Wilson Lake terrane 201

xenoliths, Julianehåb batholith 182

zircons
  Chupa aluminous gneisses 58–9, *60–1*, 62, *63–4*, 64
  inheritance component 265, 305
  metamorphic, Rae and Hearne provinces 148
  Tasiuyuk gneiss 143
  TTG gneisses 33, 38
  U–Pb dating, Baltic–Bothnian shear zone 254–6